COMBINATORICS, GEOMETRY AND PROBABILITY

COMBINATORICS, GEOMETRY AND PROBABILITY

A tribute to Paul Erdős

Edited by

BÉLA BOLLOBÁS
ANDREW THOMASON

CAMBRIDGE
UNIVERSITY PRESS

PUBLISHED BY THE PRESS SYNDICATE OF THE UNIVERSITY OF CAMBRIDGE
The Pitt Building, Trumpington Street, Cambridge, United Kingdom

CAMBRIDGE UNIVERSITY PRESS
The Edinburgh Building, Cambridge CB2 2RU, UK
40 West 20th Street, New York NY 10011–4211, USA
477 Williamstown Road, Port Melbourne, VIC 3207, Australia
Ruiz de Alarcón 13, 28014 Madrid, Spain
Dock House, The Waterfront, Cape Town 8001, South Africa

http://www.cambridge.org

First published 1997
First paperback edition 2004

Typeset in 10/13pt Monotype Times

A catalogue record for this book is available from the British Library

ISBN 0 521 58472 8 hardback
ISBN 0 521 60766 3 paperback

Contents

Preface

On Friday, 26 March 1993, Paul Erdős celebrated his 80th birthday. To honour him on this occasion, a conference was held in Trinity College, Cambridge, under the auspices of the Department of Pure Mathematics and Mathematical Statistics of the University of Cambridge. Many of the world's best combinatorialists came to pay tribute to Erdős, the universally acknowledged leader of their field.

The conference was generously supported both by the London Mathematical Society and by the Heilbronn Fund of Trinity College. As at former Cambridge Conferences in honour of Paul Erdős, the day-to-day running of this conference was in the able hands of Gabriella Bollobás, with the untiring assistance of Tristan Denley, Ted Dobson, Tom Gamblin, Chris Jagger, Imre Leader, Alex Scott and Alan Stacey. The conference would not have taken place without their dedicated work.

On the eve of Erdős' birthday, a sumptuous feast was held in his honour in the Hall of Trinity College. The words wherein he was toasted are reproduced in the following pages. This volume of research papers was presented to Paul Erdős by its authors as their own toast, gladly offered with their gratitude, respect and warmest wishes.

Sadly, before this book reached its printed form, Paul Erdős died. Whereas it was conceived in joy it appears now tinged with sorrow. We feel his loss tremendously. But it is not appropriate that grief should overshadow this volume. Erdős lived to do mathematics and he died doing mathematics. So this work remains a tribute to the Erdős we fondly remember — the living Erdős — the mathematician.

B.B.
A.G.T.

Preface

Farewell to Paul Erdős

(26/3/1913 – 20/9/1996)

(Paul Erdős died in Warsaw on 20th September 1996. A memorial service was held for him on 18th October 1996 in the Kerepesi Cemetery in Budapest, the traditional resting place of eminent Hungarians. A great number of his friends gathered to mark his passing. Among them were colleagues and former students representing mathematics from many countries and four continents. The orations were given by Ákos Császár, Paul Révész, Gyula Katona, Ron Graham, András Hajnal, George Szekeres, and by Béla Bollobás, whose tribute is reproduced below.)

Paul Erdős was one of the most brilliant and probably the most remarkable of mathematicians of this century. Not only was his output prodigious, with fundamental papers in many branches of mathematics, including number theory, geometry, probability theory, approximation theory, set theory and combinatorics, and not only did he have many more coauthors than anybody else in the history of mathematics, but he was also a personal friend of more mathematicians than anybody else. The vast body of problems he has left behind will influence mathematics for many years to come.

Many of us are lucky to have known him and to have benefited from his incisive mind, fertile imagination and desire to help. But hardly any one of us knew him in his prime, from the mid-thirties to the early sixties. He was hardly twenty when he took the mathematical world by storm, so that the great Issai Schur of Berlin dubbed him *der Zauberer von Budapest*.

Throughout his life, he lived modestly, despising material possessions and coveting no honours, and was always somewhat outside the mathematical establishment. Nevertheless, he was showered with honours. Among others, he was an Honorary Member of the London Mathematical Society and an Honorary Fellow of the Royal Society. These illustrious institutions have sent wreaths to express their grief at his loss. But I am here mainly to represent Paul's many friends, colleagues and, above all, his students.

Thinking of him, David's psalm springs to mind: *"surely goodness and mercy shall follow me all the days of my life."* For decades, he was the window to the West for the Hungarian mathematicians, and has helped more mathematicians all over the world than anybody else. He was especially kind to young people. I was just over fourteen when he called me to him and so changed the course of my life. There is no doubt that I became a combinatorialist only because of him, and I owe him a tremendous debt of gratitude for all his kindness and inspiration. Many people owe their careers to him.

As David in his psalm, he could also have said: *"though I walk in the valley of the shadow of death, I will fear no evil."* Sadly, he was always in the shadow of death. When he was born, his two sisters died; when he was a year-and-a-half his father was taken prisoner of war and spent six years in Siberia; when his father died of a heart attack, he could not come to Hungary to comfort his mother; most of his relatives perished in the

Holocaust; in the fifties even America abandoned him and he was saved only by Israel; finally, the loss of his mother was a terrible blow to him, from which he never really recovered. But whatever happened, he always had a passionate desire to be *free*: he could not tolerate constraint of any kind, he was never willing to compromise.

Perhaps there were only two happy periods in his adult life: from 1934 to 1939 when he was in Manchester and Princeton, and from 1964 to 1971, when he travelled around the world with his beloved mother. I was lucky enough to have known him in this second happy period.

The death of Paul Erdős marks the end of an era. No conference will be the same without the p.g.o.m., the *poor great old man*, as he called himself, no mathematical discussion will be as much fun as it was with him. Our beloved *Pali Bácsi* has left us all orphans.

This exceptional man did think about what will happen after him. Endre Ady, the famous Hungarian poet, wrote: *"Let him be cursed who takes my place!"* Paul's wish was rather different, reflecting his character: *"Let him be blessed who takes my place!"*

Now, when we have to say our final goodbye to Paul Erdős, we all know that there is no chance of that. His death is a tremendous loss to us all, and this sense of loss will stay with us for ever. But we should console ourselves that he has had a marvellous life, in which he has produced an exceptional amount of outstanding mathematics, and we are privileged to have known him.

Kerepesi Cemetery, Budapest, 18/10/1996

Béla Bollobás

Toast to Paul Erdős

(The following is the toast of the Banquet for the 80th Birthday of Paul Erdős, held in Trinity College, Cambridge, on 25 March 1993, the eve of the birthday. The banquet was attended by many of Erdős' other friends, including Lady Jeffreys, Mrs Davenport and Peter Rado, in addition to the conference participants. Trinity College was represented by Sir Andrew Huxley, OM, former president of the Royal Society and former Master of the College, who presided at the feast. Cambridge mathematics was represented by the present and former Sadlerian Professors, John Coates, FRS, and J.W.S. Cassels, FRS.)

Professor Erdős, Sir Andrew, Ladies and Gentlemen,

Mathematics is rich in unusual characters, as everyone here at this dinner will know. Nevertheless, most of us would agree that there is none whose achievement and lifestyle are more extraordinary than those of the man we are celebrating tonight, on the eve of his birthday, following a Hungarian custom. For over 60 years, his fertile mind has maintained a staggering output in many branches of mathematics: he has made notable contributions and broken fresh ground in set theory, number theory, probability theory, classical analysis, geometry, approximation theory and combinatorics. Most of us are particularly aware of his contributions to the last of these subjects: he has done more than anyone else to establish combinatorics; many branches of the subject find their origin in his ideas; the stimulus of his striking theorems and inspiring problems is one that we have all felt, and for which we owe him an incalculable debt of gratitude. It is also true that, as well as being so remarkably gifted intellectually, he has the most admirable and attractive personal qualities. He is generous to a fault, gentle, unassuming, always eager to fight for the downtrodden. Many a young student has been delighted to discover that this famous man is so easily approachable and so interested in their work. He has always made it his business to nurture young talent, possibly his greatest find being Pósa.

What anybody, who has ever heard of this unique man, knows is that he is unceasingly on the move. It is hardly an exaggeration that he has not slept in the same bed for more than a week in over 50 years. As a constant globe-trotter, he is the living link between mathematicians across the world, carrying with him news of theorems, conjectures and problems.

Paul Erdős was a precocious child: at the age of three he was good at arithmetic to the point of discovering for himself negative numbers. Much of Paul's education was done in private; altogether he spent less than four years in schools. At the age of 17, he proceeded to university, where he soon became the focus of a wonderfully talented group of mathematicians.

At the age of 21, he completed his degree, and as was the custom, he looked to spend a year abroad. In the world of 1934, the country that most attracted him was Britain. As an undergraduate, he had corresponded with Louis Mordell, the great American number

xiii

theorist, who by that time had left St John's College, Cambridge to work in Manchester. Mordell offered Erdős a Fellowship in his department, and the offer was gladly accepted. On 1 October 1934, Erdős arrived in London, from where he took the train to Cambridge. At the station he was met by two outstanding young mathematicians who for many years to come were to be his closest friends, Harold Davenport and Richard Rado. Sadly, Harold Davenport and Richard Rado are no longer with us, but it is indeed a pleasure to see Anne Davenport and Peter Rado at this banquet tonight. In fact, it is due to Erdős's friendship with Davenport that my own connection with Trinity came about.

At that time Erdős stayed in Cambridge only for a couple of days, but long enough to meet Hardy and Littlewood, the leading English mathematicians. He then travelled on to Manchester, to Mordell, who became his mentor and friend. In the 1930s Mordell gathered a remarkable group of mathematicians to Manchester: in addition to Erdős, and later Davenport, the group included Mahler, Heilbronn, du Val and Chao Ko. It is extremely fitting that this conference has been supported by Heilbronn's generous bequest to the mathematicians of this college. On looking down on us, Heilbronn must be smiling that we are celebrating his great friend tonight.

Another prominent member of the Manchester group was the eminent fluid dynamicist Miss Swirles, who befriended Paul soon after his arrival. It is a great pleasure that Miss Swirles, by now Lady Jeffreys, can share in this happy celebration tonight.

Paul stayed in Manchester for four years, first as the Bishop Harvey Goodwin Fellow, and then as a Royal Society Fellow. During that time he made frequent visits to Cambridge and other centres of mathematics. In 1938 Paul left England for the States to take up a Fellowship at Princeton. It was to be ten more years before Paul returned to Hungary, and he would never again stay there for more than a few months at a time.

After a year or two at the Institute, the travelling began in earnest, and the now familiar pattern was soon set. In a short space of time, he visited Philadelphia, Purdue, Stanford, Syracuse and Johns Hopkins, and many other universities for even shorter periods.

Since then Paul has been travelling from university to university, from country to country, bringing news, inventing problems, writing joint papers, stimulating the minds of mathematicians everywhere, and generally being the Erdős we know and love so well. By now he has over 300 coauthors, and it has often been said that if a train journey is long enough, he will write a joint paper with the conductor. His 1300 research papers place him in a league of his own among research mathematicians.

It has been said that the world wants geniuses but it wants them to behave just like other people. Paul found this out when one apocryphal, but not too far-fetched, night in Chicago he was out walking by himself. Suddenly a police car appeared and the officers began to question Paul. "So what are you doing out here, all by yourself?" "I am thinking" came the reply. "What do you mean you are thinking? What are you thinking about?" "I am proving a theorem." "You'd better come with us back to the station, Sir."

Back at the station, the officer in charge said "Now, what's all this about your theorem? Tell me about it." "It doesn't matter anymore" grumbled Paul testily, "I've found a counterexample."

In fact, this incident is atypical for, as we know, Paul is remarkably successful in proving theorems. A striking example is quoted by Mark Kac.

"As a mathematician Erdős is what in other fields is called a 'natural'. If a problem can be stated in terms he can understand, though it may belong to a field with which he is not familiar, he is as likely as, or even more likely than, the experts to find a solution. An example of this is his solution of a problem in dimension theory, a part of topology of which in 1939 he knew absolutely nothing. The late Witold Hurewicz and a younger colleague, Henry Wallman, were writing a book on dimension theory which later became an acknowledged classic. They were interested in the unsolved problem of the dimension of the set of rational points in Hilbert space. What all this means is unimportant except that the problem seemed very difficult and that the 'natural' conjectures were that the answer is either zero or infinity. Erdős overheard several mathematicians discussing the problem in the common room of the old Fine Hall at Princeton. "What is the problem?" asked Erdős. Somewhat impatiently he was told what the problem was. "What is dimension?" he asked, betraying complete ignorance of the subject matter. To pacify him, he was given the definition of dimension. In a little more than an hour he came with the answer, which, to everyone's immense surprise, turned out to be '1'!"

In addition to being successful in his own personal research, one of Paul's greatest gifts to mathematics has been his ability to stimulate the creativity of others through his fascinating and penetrating conjectures. His offer of monetary rewards for solutions is legendary. The winner of the largest reward to date is Szemerédi, for finding long arithmetic progressions in sets of positive density. It is a pleasure to see him here tonight. The biggest sum on offer is $10000, for proving that the gap between two consecutive primes is rather large infinitely often. Although Schönhage, Rankin, Maier and Tenenbaum have proved exciting results in this direction, they haven't yet managed to claim the prize. Paul is also offering $3000 for finding long arithmetic progressions in sequences of natural numbers whose reciprocals diverge, and so, in particular, among the primes. A group of Swedish computers has just discovered an arithmetic progression of 22 primes but I doubt that any payment will be forthcoming from Paul.

Paul worked with most of the leading Hungarian mathematicians, especially the number theorist Paul Turán and the probabilist Alfréd Rényi, who were his great friends. Turán's wife, Vera Sós, has also been a close friend and collaborator for many years, and it is fitting that she too should be celebrating tonight.

My own friendship with Paul is also of many years standing. We met when I was 14, and I was tremendously impressed by his willingness to talk to me about his fascinating problems. To me he seemed to be from a different planet, a flamboyant man with an air of the exotic, with his expensive foreign suits and ready cash, brought from the unattainable free Western world. Now I know better; I think it was Paul who inspired the saying: "The man who leaves footprints on the sands of time never wears expensive shoes."

In those days, I also got to know Paul's mother, Annus néni, a charming lady who adored Paul, and was, in turn, adored by her son. She kept his reprints in immaculate order, and sent copies to those who requested them. A year or two later they got to know my family, and were frequent visitors to our house whenever Paul was in Hungary.

In 1964, at the age of 84, Annus néni began to travel with Paul. Their first trip was to Israel; soon Western Europe followed, including England a year later. In 1968, when she was 88, Annus néni accompanied Paul to Hawaii and Australia. When asked whether she liked to travel, she used to reply: "You know I don't travel because I like it, but to be with my son." It was moving too see their affectionate care for each other, catching up with those lost years, when they couldn't see each other. Annus néni greatly enjoyed her role as Queen Mother of mathematics, meeting and entertaining all the people coming to see Paul; her cocoa cake with coffee cream was especially delicious.

Erdős's own tastes in food are well known to be frugal, and he doesn't care for wine, which he calls poison. It has been suggested that the College should on this occasion produce a meal of bread and water. Unfortunately when I checked with the Kitchens, they could not find the recipe, so we had to use the second choice menu.

Paul Erdős has always kept up his close links with Trinity and Cambridge. Some years ago he was a Visiting Fellow Commoner of Trinity College, and in 1991 Cambridge awarded him an Honorary Doctorate – the first citizen of Hungary to receive this honour. At the ceremony it was charming to see the great actor Sir Alec Guinness taking it upon himself to shepherd Paul through the long ritual.

Since his youth, Paul Erdős has had catholic interests: in particular, he has maintained an enthusiasm for history and medicine. It is always fascinating to engage him in discussion of his favourite historical events. Nevertheless, Paul is the quintessential mathematician: he breathes, eats, drinks, and sleeps mathematics, if he sleeps at all. It could have been Erdős, whom Littlewood had in mind when he wrote:

"There is much to be said for being a mathematician. To begin with, he has to be completely honest in his work, not from any superior morality, but because he cannot get away with a fake. It has been cruelly said of arts dons, especially in Oxford, that they believe there is a polemical answer to everything; nothing is really true, and in controversy the object is to prove your opponent a fool. We escape all this. Further, the arts man is always on duty as a great mind; if he drops a brick, as we say in England, it reverberates down the years. After an honest day's work a mathematician goes off duty. Mathematics is very hard work, and dons tend to be above average in health and vigour. Below a certain threshold a man cracks up; but above it, hard mental work makes for health and vigour (also – on much historical evidence throughout the ages – for longevity)."

If hard mental work be the secret of longevity then Paul Erdős will live forever and continue to enrich us all with the brightness of his intellect and the warmth of his heart. In the meantime, we honour him on his 80th birthday.

Ladies and Gentlemen, please rise and toast Paul Erdős.

                                                                              B.B.

List of Contributors

Ron Aharoni
Department of Mathematics, Technion, Haifa 32000, ISRAEL

Rudolf Ahlswede
Universität Bielefeld, Fakultät für Mathematik, Postfach 100131, 33501 Bielefeld, GERMANY

Martin Aigner
Freie Universität Berlin, Fachbereich Mathematik, WE2, Arnimallee 3, 1000 Berlin 33, GERMANY

Noga Alon
Department of Mathematics, Raymond and Beverly Sackler Faculty of Exact Sciences, Tel Aviv University, Tel Aviv, ISRAEL

A. D. Barbour
Institut für Angewandte Mathematik, Universität Zürich, Winterthurerstrasse 190, CH-8057, Zürich, SWITZERLAND

József Beck
Department of Mathematics, Rutgers University, Busch Campus, Hill Center, New Brunswick, NJ 08903, USA

Sergej L. Bezrukov
Fachbereich Mathematik, Freie Universität Berlin, Arnimallee 2-6, D-14195 Berlin, GERMANY

Norman L. Biggs
London School of Economics, Houghton St, London WC2A 2AE, UK

Béla Bollobás
Department of Pure Mathematics and Mathematical Statistics, University of Cambridge, 16 Mill Lane, Cambridge, CB2 1SB, UK and Louisiana State University, Baton Rouge, LA 70803 USA

Ning Cai
Universität Bielefeld, Fakultät für Mathematik, Postfach 100131, 33501 Bielefeld, GERMANY

Peter J. Cameron
School of Mathematical Sciences, Queen Mary and Westfield College, Mile End Road, London, E1 4NS, UK

G. Chen
North Dakota State University, Fargo, ND 58105, USA

Colin Cooper
School of Mathematical Sciences, University of North London, London, UK

Walter A. Deuber
Universität Bielefeld, Fakultät für Mathematik, Postfach 100131, 33501 Bielefeld 1, GERMANY

Michel Deza
CNRS-LIENS, Ecole Normale Supérieure, Paris, FRANCE

Reinhard Diestel
Faculty of Mathematics (SFB 343), Bielefeld University, 4-4800, Bielefeld, GERMANY

J. K. Dugdale
Department of Mathematics, West Virginia University, PO Box 6310, Morgantown, WV 26506-6310, USA

Paul Erdös†
late, Mathematical Institute of the Hungarian Academy of Sciences, Budapest V, HUNGARY

Péter L. Erdös
Centrum voor Wiskunde en Informatica, PO Box 4079, 1009 AB Amsterdam, The NETHERLANDS

R. J. Faudree
Department of Mathematical Science, Memphis State University, Memphis, TN 38152, USA

Hubert de Fraysseix
CNRS, EHESS, 54 Boulevard Raspail, 75006, Paris, FRANCE

Alan Frieze
Department of Mathematics, Carnegie-Mellon University, Pittsburgh, PA 15213, USA

Zoltán Füredi
Department of Mathematics, Massachusetts Institute of Technology, Cambridge, MA 02139, USA

Mario Gionfriddo
Dipartimento di Matematica, Città Universitaria, Viale A, Doria 6, 95125 Catania, ITALY

Michel X. Goemans
Department of Mathematics, Massachusetts Institute of Technology, Cambridge, MA 02139, USA

Viatcheslav Grishukhin
Central Economic and Mathematical Institute of Russian Academy of Sciences (CEMI RAN), Moscow, RUSSIA

Roland Häggkvist
Department of Mathematics, University of Umeå, S-901 87 Umeå, SWEDEN

A. Hajnal
Mathematical Institute of the Hungarian Academy of Sciences, Budapest V, HUNGARY

R. Halin
Mathematisches Seminar der Universität Hamburg, Bundesstraße 55, D-20146, Hamburg, GERMANY

P. L. Hammer
RUTCOR, Rutgers University, New Brunswick, NJ 08903, USA

A. J. W. Hilton
Department of Mathematics, University of Reading, Whiteknights, PO Box 220, Reading
RG6 2AX, UK

Neil Hindman
Department of Mathematics, Howard University, Washington, DC 20059, USA

Christoph Hundack
Institut für Diskrete Mathematik, Universität Bonn, Nassestr. 2, 53113 Bonn, GERMANY

Svante Janson
Department of Mathematics, Uppsala University, PO Box 480, S-751 06, Uppsala,
SWEDEN

Anders Johannson
Department of Mathematics, University of Umeå, S-901 87 Umeå, SWEDEN

William M. Kantor
Department of Mathematics, University of Oregon, Eugene, OR 97403, USA

A. K. Kelmans
RUTCOR, Rutgers University, New Brunswick, NJ 08903, USA

Daniel J. Kleitman
Department of Mathematics, Massachusetts Institute of Technology, Cambridge, MA
02139, USA

R. Klimmek
c/o M. Aigner, Freie Universität Berlin, Fachbereich Mathematik, WE2, Arnimallee 3,
1000 Berlin 33, GERMANY

Y. Kohayakawa
Instituto de Matemática e Estatística, Universidade de São Paulo, Caixa Postal 20570,
01452-990 São Paulo, SP, Brazil

Péter Komjáth
Dept. Comp. Sci. Eötvös University, Budapest, Múzeum krt 6-8, 1088, HUNGARY

János Komlós
Department of Mathematics, Rutgers University, New Brunswick, NJ 08903, USA

Imre Leader
Department of Pure Mathematics and Mathematical Statistics, University of Cambridge,
16 Mill Lane, Cambridge, CB2 1SB, UK

Nathan Linial
Institute of Computer Science, Hebrew University, Jerusalem, ISRAEL

Tomasz Łuczak
Adam Mickiewicz University, Poznań, POLAND

W. Mader
Institut für Mathematik, Universität Hanover, 30167 Hanover, Weifengarten 1, GERMANY

Endre Makai
Mathematical Institute of the Hungarian Academy of Sciences, Budapest V, HUNGARY

A. R. D. Mathias
Peterhouse College, Cambridge, UK

Colin McDiarmid
Department of Statistics, University of Oxford, Oxford, UK

Patrice Ossona de Mendez
CNRS, EHESS, 54 Boulevard Raspail, 75006, Paris, FRANCE

Salvatore Milici
Dipartimento di Matematica, Città Universitaria, Viale A, Doria 6, 95125 Catania, ITALY

Bojan Mohar
Department of Mathematics, University of Ljubljana, Jadranska 19, 61111 Ljubljana, SLOVENIA

Michael Molloy
Department of Mathematics, Carnegie-Mellon University, Pittsburgh, PA 15213, USA

Jaroslav Nešetřil
Department of Applied Mathematics, Charles University, Malostranské nám. 25, 118 00 Praha 1, CZECH REPUBLIC

Edward T. Ordman
Memphis State University, Memphis, TN 38152, USA

János Pach
Department of Computer Science, City University, New York, USA and the Mathematical Institute of the Hungarian Academy of Sciences, Budapest V, HUNGARY

Hans Jürgen Prömel
Institut für Diskrete Mathematik, Universität Bonn, Nassestr. 2, 53113 Bonn, GERMANY

László Pyber
Mathematical Institute of the Hungarian Academy of Sciences, Budapest V, HUNGARY

Pierre Rosenstiehl
CNRS, EHESS, 54 Boulevard Raspail, 75006, Paris, FRANCE

C. C. Rousseau
Department of Mathematical Science, Memphis State University, Memphis, TN 38152, USA

R. H. Schelp
Department of Mathematical Science, Memphis State University, Memphis, TN 38152, USA

Ákos Seress
The Ohio State University, Colombus, OH 43210, USA

M. Simonovits
Mathematical Institute of the Hungarian Academy of Sciences, Budapest V, HUNGARY

V. T. Sós
Mathematical Institute of the Hungarian Academy of Sciences, Budapest V, HUNGARY

Angelika Steger
Institut für Diskrete Mathematik, Universität Bonn, Nassestr. 2, 53113 Bonn, GERMANY

László A. Székely
University of New Mexico, Albuquerque, NM 87131, USA

Endre Szemerédi
Mathematical Institute of the Hungarian Academy of Sciences, Budapest V, HUNGARY

Simon Tavaré
Department of Mathematics, University of Southern California, Los Angeles, CA 90089-113, USA

H. N. V. Temperley
Thorney House, Thorney, Langport, Somerset, UK

Prasad Tetali
AT & T Bell Labs, Murray Hill, NJ 07974, USA

Andrew Thomason
DPMMS, 16, Mill Lane, Cambridge, CB2 1SB, UK

Wolfgang Thumser
Universität Bielefeld, Fakultät für Mathematik, Postfach 100131, 33501 Bielefeld 1, GERMANY

Eberhard Triesch
Forschungsinsitut für Diskrete Mathematik, Nassestraße 2, 5300 Bonn 1, GERMANY

Zsolt Tuza
Computer and Automation Institute, Hungarian Academy of Sciences, H-1111 Budapest, Kende u. 13-17, HUNGARY

Pavel Valtr
Department of Applied Mathematics, Charles University, Malostranské nám. 25, 118 00 Praha 1, CZECH REPUBLIC and Graduiertenkolleg 'Algorithmische Diskrete Mathematik', Fachbereich Mathematik, Freie Universität Berlin, Takustrasse 9, 14195 Berlin, GERMANY

D. J. A. Welsh
Mathematical Institute and Merton College, University of Oxford, Oxford, UK

Herbert S. Wilf
University of Pennsylvania, Philadelphia, PA 19104-6395, USA

Raphael Yuster
Department of Mathematics, Raymond and Beverly Sackler Faculty of Exact Sciences, Tel Aviv University, Tel Aviv, ISRAEL

Yechezkel Zalcstein
Division of Computer and Computation Research, National Science Foundation, Washington, DC 20550, USA

Some Unsolved Problems

PAUL ERDŐS[†]

During my long life I have written many papers on my favourite unsolved problems (see, for example, Baker *et al.* [2]). In the collection below, all the problems are either new ones, or they are problems about which there have been recent developments.

Number theory

1. As usual, let us write $2 = p_1 < p_2 < \cdots$ for the sequence of consecutive primes. I proved in 1934 that there is a constant $c > 0$ such that for infinitely many n we have

$$p_{n+1} - p_n > \frac{c \log n \log \log n}{(\log \log \log n)^2}.$$

Rankin [35] proved that for some $c > 0$ and infinitely many n the following inequality holds:

$$p_{n+1} - p_n > \frac{c \log n \log \log n \log \log \log \log n}{(\log \log \log n)^2}. \tag{1}$$

I offered (perhaps somewhat rashly) $10 000 for a proof that (1) holds for every c. The original value of c was improved by Schönhage [38] and later by Rankin [36]. Rankin's result was recently improved by Maier and Pomerance [30].

2. Let $a_1 < a_2 < \cdots$ be an infinite sequence of integers. Denote by $f(n)$ the number of solutions of $n = a_i + a_j$. Assume that $f(n) > 0$ for all $n > n_0$, i.e. $(a_n)_{n=1}^\infty$ is an *asymptotic basis* of order 2. Turán and I conjectured that then

$$\varlimsup_{n \to \infty} f(n) = \infty \tag{2}$$

and probably $\varlimsup f(n)/\log n > 0$. I offer $500 for a proof of (2). Perhaps (2) and $\varlimsup f(n)/\log n > 0$ already follow if we only assume $a_n < cn^2$ for all n.

Let $a_1 < a_2 < \cdots$ and $b_1 < b_2 < \cdots$ be two sequences of integers such that $a_n/b_n \to 1$ and let $g(n)$ be the number of solutions of $a_i + b_j = n$. Sárközy and I conjecture that if

$g(n) > 0$ for all $n > n_0$ then $\overline{\lim} \, g(n) = \infty$. The condition $a_n/b_n \to 1$ can not entirely be omitted but $1 - \epsilon < a_n/b_n < 1 + \epsilon$ (ϵ small) may suffice.

3. I proved that there is an asymptotic basis of order 2 for which

$$c_1 \log n < f(n) < c_2 \log n$$

(see Halberstam and Roth [26]). I conjecture that

$$\frac{f(n)}{\log n} \to C, \quad (0 < C < \infty),$$

is not possible and I offer \$500 for a proof or disproof of this conjecture. Sárközy and I proved that

$$\frac{|f(n) - \log n|}{\sqrt{\log n}} \to 0$$

cannot hold.

4. Is it true that

$$\sum \frac{1}{n! - 1}$$

is irrational? I conjectured that

$$\sum \frac{1}{2^n - 3}$$

is irrational. This assertion and its generalizations have been proved by Peter Borwein [6]. Denote by $\omega(n)$ the number of distinct prime factors of n. Is it true that

$$\sum \frac{\omega(n)}{2^n}$$

is irrational?

5. Is it true that if $n \not\equiv 0 \pmod 4$ then there is a squarefree natural number θ such that $n = 2^k + \theta$? I could only prove that almost all integers $n \not\equiv 0 \pmod 4$ can be written in the form $2^k + \theta$.

Combinatorics

6. Let $m = m(n)$ be the smallest integer for which there are n-element sets $A_1, \ldots, A_m$ such that $A_i \cap A_j \neq \emptyset$ for all $1 \leqslant i < j \leqslant m$, and such that every set S with at most $n-1$ elements is disjoint from some A_i. (Note that the lines of finite geometry have this property.) I conjectured with Lovász that $m(n)/n \to \infty$, but it is not even known whether $m(n) > 3n$ if n sufficiently large. In the other direction, we could prove only that $m(n) < n^{\frac{3}{2}+\epsilon}$, but Jeff Kahn [27] very recently proved $m(n) < cn$.

Perhaps more is true: for every $C > 0$ there is an $\epsilon > 0$ such that if n is sufficiently large and $m \leqslant Cn$ then for every n-element set $A_1, \ldots, A_m$ with $A_i \cap A_j \neq \emptyset$ there is a set S with $|S| < n(1 - \epsilon)$ which meets all A_i.

7. Is it true that in a finite geometry there is always a blocking set which meets every line in at most c points where c is an absolute constant?

More generally: Let $|\mathscr{S}| = n$, $A_1, \ldots, A_m$ be a family of subsets of $\mathscr{S}$, $|A_i| > c\sqrt{n}$, $c < 1$, $|A_i \cap A_j| \leqslant 1$. Is it then true that there is a set B for which $B \cap A_i \neq \emptyset$ but $|B \cap A_i| < c\prime$ for all i? In other words, is there a set B which meets all the A_i's but none in many points?

8. Here is a problem of Jean Larson and myself [19]. Is it true that there is an absolute constant c so that for every n and $|\mathscr{S}| = n$ there is a family of subsets $A_1, \ldots, A_m$ of $\mathscr{S}$, $|A_i| > n^{1/2} - c$, $|A_i \cap A_j| \leqslant 1$ and every $x, y \in \mathscr{S}$ is contained in some A_i?

Shrikhande and Singhi [39] have proved that every pairwise balanced design on n points in which each block is of size $\geqslant n^{\frac{1}{2}} - c$ can be embedded in a projective plane of order $n + i$ for some $i \leqslant c + 2$ if n is sufficiently large. This implies that if the projective plane conjecture (that the order of every projective plane is a prime power) is true then the Erdős–Larson conjecture is false. But the problem remains for which functions $h(n)$ will the condition $|A_i| > n^{\frac{1}{2}} - h(n)$ make the conjecture true?

Graph theory

9. I offer \$500 for a proof or disproof of the following conjecture of Faber, Lovász and myself. Let $G_1, \ldots, G_n$ be complete graphs (each on n vertices), no two of which have an edge in common. Is it then true that $\chi(\bigcup_{i=1}^{n} G_i) \leqslant n$?

Jeff Kahn [27] recently proved that the chromatic number is $n + o(n)$.

10. Is it true that every triangle-free graph on $5n$ vertices can be made bipartite by the omission of at most $5n^2$ edges? Is it true that every triangle-free graph on $5n$ vertices can contain at most n^5 pentagons? Ervin Győri [25] proved this with $1.03n^5$.

Győri now proved n^5 for $n > n_0$. One could ask more generally: Assume that the number of vertices is $(2r + 1)n$ and that the smallest odd cycle has size $2r + 1$. Is it then true that the number of cycles of size $2r + 1$ is at most n^{2r+1}?

11. Let H be a graph and let G^n be a graph on n vertices which does not contain H as an induced subgraph. Hajnal and I [13] asked whether there is an absolute constant $c = c(H)$ such that G^n contains either a complete graph or an independent set on n^c vertices? If H is C_4 then $\frac{1}{3} \leqslant c < \frac{1}{2}$.

12. Let Q^n be the graph of the n-dimensional cube $\{0, 1\}^n$. I offered \$100 for a proof or disproof of the conjecture that for every $\epsilon > 0$ there is an n_0 such that, for $n > n_0$, every subgraph of Q^n with at least $(\frac{1}{2} + \epsilon)e(Q^n)$ edges contains C_4. It is easy to find subgraphs with more than $\frac{1}{2}e(Q^n)$ edges and no C_4; Guan (see Chung [9]) has constructed an example with $(1 + o(1))(n + 3)2^{n-2}$ edges. Chung has given an upper bound of $(\alpha + o(1))n2^{n-1}$, where $\alpha \approx 0.623$.

I also conjectured that every subgraph of Q^n with $\epsilon e(Q^n)$ edges contains a C_6, for n sufficiently large. Chung [9] and Brouwer, Dejter and Thomassen [7] disproved this by constructing an edge-partition of Q^n into four subgraphs containing no C_6.

13. Suppose that G is a graph of order n with the property that every set of p vertices spans at least q edges. We let $H(n; p, q)$ be the largest integer such that G necessarily contains a clique of that order. In the case where $q = 1$ this corresponds to the standard

finite Ramsey problem: the condition is precisely that G contains no independent set of size p.

Faudree, Rousseau, Schelp and I investigated the behaviour of $H(n; p, q)$ as a function of n. We set

$$c(p, q) = \lim_{n \to \infty} \left(\frac{\log H(n; p, q)}{\log n} \right).$$

Standard bounds on Ramsey numbers (see, for example, Bollobás [5]) tell us that

$$1/(p-1) \leqslant c(p, 1) \leqslant 2/(p+1).$$

We conjecture that with p fixed, $c(p, q)$ is a strictly increasing function of q for $1 \leqslant q \leqslant \binom{p-1}{2} + 1$. It is easy to see that if $q = \binom{p-1}{2} + 1$ then $c(p, q) = 1$, which is as large as possible. For in this case, the complement of G cannot contain any connected subgraphs of size p, so all components of the complement have size less than p. Hence the complement contains at least $n/(p-1)$ independent vertices so G contains a clique of size at least $n/(p-1)$. On the other hand, we have shown that $H(n; p, \binom{p-1}{2}) \leqslant cn^{\frac{1}{2}}$, so $c(p, \binom{p-1}{2}) \leqslant 1/2$.

14. For $\epsilon > 0$, Rödl [37] constructed graphs with chromatic number $\aleph_0$ such that every subgraph of order n can be made bipartite by omitting ϵn edges, for every n; another construction was given by Lovász. Now let $f(n) \to \infty$ as slowly as we please. Is there a graph of chromatic number $\aleph_0$ such that every subgraph of n vertices can be made bipartite by omitting $f(n)$ edges?

Perhaps for every $\epsilon > 0$, there is a graph with chromatic number $\aleph_1$ for which every subgraph of order n can be made bipartite by omitting ϵn edges, but this seems unlikely and I would guess that there is a subgraph of size n which cannot be made bipartite by omitting $nh(n)$ edges, where $h(n) \to \infty$. But perhaps $h(n)$ does not have to tend to infinity fast. See also the paper with Hajnal and Szemerédi [17].

Hajnal, Shelah and I [16] proved that if G has chromatic number $\aleph_1$ then for some n_0 it contains a cycle of length n for every $n > n_0$. Now if $F(n)$ tends to infinity sufficiently fast, then is it true that every graph of chromatic number $\aleph_1$ has a subgraph on at most $F(n)$ vertices with chromatic number n, for all n sufficiently large?

Geometry

15. Let $x_1, \ldots, x_n$ be n distinct points in the plane, and let $s_1 \geqslant s_2 \geqslant \cdots \geqslant s_k$ be the multiplicities of the distances they determine, so

$$\sum_{i=1}^{k} s_i = \binom{n}{2}.$$

I conjectured [12] that

$$\sum_{i=1}^{k} s_i^2 < cn^3 (\log n)^\alpha \tag{3}$$

for some $\alpha > 0$. The lattice points show that we must have $\alpha \geqslant 1$.

In forthcoming papers Fishburn and I conjecture that if $x_1, \ldots, x_n$ form a convex set then (3) can be improved to

$$\sum_{i=1}^{k} s_i^2 < cn^3, \tag{4}$$

and that $\sum s_i^2$ is maximal for the regular n-gon, for $n \geqslant 8$.

A weaker inequality than (3) would follow easily from the following conjecture. Let $A(x_1, \ldots, x_n)$ be the number of pairs x_i, x_j whose distance is 1, and let $f(n)$ be the maximum $A(x_1, \ldots, x_n)$ over all sets of n distinct points in the plane. I conjecture that

$$f(n) < n^{1+c/\log\log n}. \tag{5}$$

The best bound found to date is due to Spencer, Szemerédi and Trotter [40], who proved $f(n) < cn^{\frac{4}{3}}$. It would follow from (5) that $\sum s_i^2 < cn^{3+c/\log\log n}$.

Is it true that the number of incongruent sets of n points with $f(n)$ unit distances exceeds one for $n > 3$ and tends to infinity with n?

Leo Moser and I conjectured that if $x_1, \ldots, x_n$ is a convex n-gon then

$$A(x_1, \ldots, x_n) < cn. \tag{6}$$

Füredi [22] proved that $A(x_1, \ldots, x_n) < cn \log n$; this gives an upper bound of $cn^3 \log n$ in (4). The inequality (4) would follow from (6).

16. Let $x_1, \ldots, x_n$ be n distinct points in the plane. Denote by $F_k(n)$ the maximum number of distinct lines passing through at least k of our points and by $f_k(n)$ the maximum number of lines passing through exactly k of our points. Clearly $f_k(n) \leqslant F_k(n)$. Determine or estimate $f_k(n)$ and $F_k(n)$ as well as possible. Trivially $f_2(n) = F_2(n) = \binom{n}{2}$. The problem with $k = 3$ is the Orchard problem, and really goes back to Sylvester. Burr, Grünbaum and Sloane [8] proved that

$$f_3(n) = \frac{n^2}{6} - O(n) \quad \text{and} \quad F_3(n) = \frac{n^2}{6} - O(n).$$

Determine $\lim_{n\to\infty} F_k(n)/n^2$ and $\lim_{n\to\infty} f_k(n)/n^2$, if they exist. The upper bound $F_k(n) \leqslant \binom{n}{2}/\binom{k}{2}$ follows from an obvious counting argument; a lower bound can be obtained by considering a rectangle of k by n/k lattice points. Are the limits attained by the lattice points?

17. Let $f(n)$ denote the minimum number of distinct distances among a set $\mathscr{C} = (x_i)_1^n$ of points in the plane. In 1946 [12] I proved that

$$\sqrt{n - \frac{3}{4}} - \frac{1}{2} \leqslant f(n) \leqslant cn/\sqrt{\log n},$$

and conjectured that the upper bound gave the true order of $f(n)$. So far, the best lower bound is $n^{\frac{4}{5}}(\log n)^{-c}$, due to Chung, Szemerédi and Trotter [10].

The question also arises whether, in general, a particular point of the configuration is associated with a large number of distances. I conjecture that in any configuration there is some point with at least $cn/\sqrt{\log n}$ distinct distances to other points. In fact this may be true for all but a few of the points.

Altman [1] showed that if $\mathscr{C}$ is convex then there are at least $n/2$ distinct distances between the points; I conjecture that there is some point associated with at least $n/2$ distinct distances. Szemerédi conjectured there are at least $n/2$ distinct distances among the points of $\mathscr{C}$ provided only that $\mathscr{C}$ has no three points collinear, but could only prove this with a bound of $n/3$.

18. Consider two configurations $\mathscr{C} = (x_i)_1^n$, $\mathscr{C}' = (y_i)_1^n$, and define $F(2n)$ to be the minimum over all $\mathscr{C}$ and $\mathscr{C}'$ of the number of distinct distances $\| x_i - y_i \|$. Is F identically 1 in four dimensions (in this case $f(2n) > n^\epsilon$)? How does f/F behave at dimension 2 or 3? Do we have $f/F \to \infty$ or is the ratio bounded? Possibly it is unbounded in $\mathbb{R}^3$ but bounded in $\mathbb{R}^2$.

19. Let $\mathscr{C}$ be a set of points in the plane such that distinct distances between the points always differ by at least 1. I conjecture that the diameter of $\mathscr{C}$ is at least $n - 1$ provided n is large enough. Note that if $\mathscr{C} = ((i, 0))_{i=1}^n$ we obtain equality. However for $n < 10$ some configurations have diameter less than $n - 1$.

The best result in this direction so far is due to Kanold [29], who proved that $diam \, \mathscr{C} \geqslant 0.366 n^{\frac{3}{4}}$.

20. Let $\mathscr{C}$ be a set of n points in Euclidean space among which all distances differ by at least 1. A conjecture independent of dimension is that $diam \, \mathscr{C} \geqslant (1 + o(1))n^2$. Clearly $diam \, \mathscr{C}$ is always at least $\binom{n}{2}$.

The conjecture is settled only for $\mathscr{C} \subset \mathbb{R}$ (not even for $\mathbb{R}^2$). To prove it for $\mathbb{R}$, let $\mathscr{C} = (x_1)_1^n$ with $0 = x_1 < x_2 < \ldots < x_n$. Further, let $y_{k,i} = x_{i+k} - x_i$ and let $Y_k = \sum_{i=1}^{n-k} y_{k,i} = x_n + x_{n-1} + \ldots + x_{n-k+1} < k x_n$. Because the $y_{k,i}$ are all distinct (even over k), we have

$$x_n = Y_1 \geqslant \sum_{i=1}^{n-1} i = \binom{n}{2}$$

$$3x_n > Y_1 + Y_2 \geqslant \binom{2n-2}{2}$$

and for $k \leqslant n/2$,

$$\binom{k+1}{2} x_n > Y_1 + \ldots + Y_K \geqslant \binom{(n-1) + \ldots + (n-k) + 1}{2}$$

Now let $k = \lceil \sqrt{n} \rceil$. Roughly, we get

$$\frac{n}{2} x_n > \left(\frac{\sqrt{n}(n - \sqrt{n})}{2} \right) \approx \frac{n}{2} (n - \sqrt{n})^2$$

so $x_n \geqslant n^2(1 + o(1))$ as desired.

Analysis

21. We let $I = [-1, 1]$ and supose $f : I \to \mathbb{R}$ is a continuous function which we wish to approximate by a polynomial. Suppose we are given, for each $1 \leqslant n < \infty$,

$$-1 \leqslant x_1^{(n)} < x_2^{(n)} < \ldots < x_n^{(n)} \leqslant 1,$$

so we have a triangular matrix $X = (x_i^{(n)})$. Let $l_i^{(n)}$ be the unique polynomial of degree

$n - 1$ satisfying

$$l_i^{(n)}(x_i^{(n)}) = 1 \quad \text{and} \quad l_i^{(n)}(x_j^{(n)}) = 0 \quad \text{if} \quad j \neq i$$

so

$$l_i^{(n)}(x) = \prod_{j \neq i}(x - x_j) \Big/ \prod_{j \neq i}(x_i - x_j).$$

Then we denote by $\mathscr{L}_n(f, X)$, or simply $\mathscr{L}_n(f)$, the polynomial given by

$$\mathscr{L}_n(f)(x) = \sum_{i=1}^{n} f(x_i^{(n)}) l_i^{(n)}(x),$$

so this is the unique polynomial of degree $n - 1$ agreeing with f on $x_1^{(n)} \ldots x_n^{(n)}$.

It is known that for certain choices of X, if f is of bounded variation then $\mathscr{L}_n(f)(x) \to f(x)$ uniformly. However, for more general continuous f the behaviour is not so good and, as we now describe, a number of authors have examined how bad this behaviour can be.

With a fixed choice of X, we can regard $\mathscr{L}_n$ as a linear map from $C(I)$ to itself. Let us write down its norm. Let

$$\lambda_n(x) = \sum_{i=1}^{n} |l_i^{(n)}(x)|.$$

Then we easily see that

$$\max_{\|f\|=1} \|\mathscr{L}_n(f)(x)\| = \lambda_n(x),$$

so if we let

$$\lambda_n = \max_{-1 \leqslant x \leqslant 1} \lambda_n(x),$$

then $\|\mathscr{L}_n\| = \lambda_n$.

Faber [21] proved that for any choice of X, $\overline{\lim}_{n\to\infty} \lambda_n = \infty$. It therefore follows from the Principle of Uniform Boundedness that there exists an f with $\overline{\lim}_{n\to\infty} \|\mathscr{L}_n(f)\| = \infty$.

This result was strengthened by Bernstein [4] who showed that for any X, there exist $f \in C[-1, 1]$ and $x_0 \in [-1, 1]$ such that

$$\overline{\lim_{n\to\infty}} \|\mathscr{L}_n(f)(x_0)\| = \infty, \quad \text{i.e.} \quad \overline{\lim_{n\to\infty}} \lambda_n(x_0) = \infty.$$

In several papers (Bernstein [3], Grünwald ([23], [24]), Marcinkiewicz [31] and Privalov ([32], [33])) it was shown that for particular choices of X, this kind of bad behaviour can occur almost everywhere and, in certain cases, everywhere. In 1980 Vértesi and I [20] showed that given any X, there exists an f with

$$\overline{\lim_{n\to\infty}} \|\mathscr{L}_n(f)(x)\| = \infty \quad \text{for almost all } x.$$

Certainly this result cannot be extended from almost all x to *all* x. For example, if x_0 appears in all but finitely many rows of X – i.e. is equal to some $x_i^{(n)}$ for all $n \geqslant n_0$ – then we have $\mathscr{L}_n(f)(x_0) = f(x_0)$ for $n \geqslant n_0$. Does there, however, exist an X, such that for every f, there is some point x_0 where divergence would be possible, i.e. where

$$\overline{\lim_{n\to\infty}} \lambda(x_0) = \infty \quad \text{yet} \quad \mathscr{L}_n(f)(x_0) \to f(x_0)?$$

22. Let $f(z) = z^n + \ldots$ be a monic polynomial of degree n.

Is it true that the length of $\{z \in \mathbb{C} : |f(z)| = 1\}$ is maximal in the case when $f(z) = z^n - 1$?

This problem was posed, along with many others, in my paper with Herzog and Piranian [18].

23. Let $|z_n| = 1 (1 \leqslant n < \infty)$. Put

$$f_n(z) = \prod_{k=1}^{n} (z - z_k)$$

and

$$M_n = \max_{|z|=1} |f_n(z)|.$$

Is it true that $\overline{\lim} M_n = \infty$? This conjecture was settled by Wagner: he proved that there is a $c > 0$ such that $M_n > (\log n)^c$ holds for infinitely many values of n. I further conjectured that $M_n > n^c$ for some $c > 0$ and infinitely many n and, in fact, for every n we have

$$\sum_{k=1}^{n} M_k > n^{1+c}. \tag{7}$$

Inequality (7), if true, may very well be difficult, so I offer \$100 for a solution.

24. Let $x_1, x_2, \ldots$ be a sequence of real numbers tending to 0. We call $(y_n)_{n=1}^{\infty}$ *similar* to $(x_n)_{n=1}^{\infty}$ if $y_n = ax_n + b$ for some $a, b \in \mathbb{R}$ and all n. Is it true that there is a set $E \subseteq \mathbb{R}$ of positive measure which contains no subsequence $(y_n)_{n=1}^{\infty}$ similar to $(x_n)_{n=1}^{\infty}$?

Komjáth proved that if $x_n \to 0$ slowly ($x_n > c/n$) then there is a set of positive measure which contains no subsequence similar to $(x_n)_{n=1}^{\infty}$.

Set theory

25. I have not included our many problems on set theory with Hajnal since undecidability raises its ugly head everywhere and many of our problems have been proved or disproved or shown to be undecidable (this happened most often). However, I think that the following simple problem is still open. Let α be a cardinal or ordinal number or an order type. Assume $\alpha \to (\alpha, 3)^2$. Is it then true that, for every finite n, $\alpha \to (\alpha, n)^2$ also holds? Here $\alpha \to (\alpha, n)^2$ is the well-known arrow symbol of Rado and myself: if G is a graph whose vertices form a set of type α then either G contains a complete graph K_n or an independent set of type α. See Erdős, Hajnal and Milner [15] and Erdős, Hajnal, Máté and Rado [14].

Group theory

26. Let G be a group. Assume that it has at most n elements which do not commute pairwise. Denote by $h(n)$ the smallest integer for which any such G can be covered by $h(n)$ Abelian subgroups. Determine or estimate $h(n)$ as well as possible. Pyber [34] proved that

$$(1 + c_1)^n < h(n) < (1 + C_2)^n,$$

for some positive constants c_1 and c_2. The lower bound was already known to Isaacs.

References

[1] Altman, (1963), On a problem of P. Erdős, *Amer. Math. Monthly* **70** 148–157.

[2] Baker, A., Bollobás, B. and Hajnal, A. eds. (1990), *A Tribute to Paul Erdős*, Cambridge University Press, xv+478pp.

[3] Bernstein, S. (1918), Quelques remarques sur l'interpolation, *Math. Ann.* **79** 1–12.

[4] Bernstein, S. (1931), Sur la limitation des valeurs d'un polynome, *Bull. Acad. Sci. de l'URSS* **8** 1025–1050.

[5] Bollobás, B. (1985), *Random Graphs*, Academic Press, xiv+447pp.

[6] Borwein, P. B. (1991), On the irrationality of $\sum(1/(q^n + r))$, *J. Number Theory* **37** 253–259.

[7] Brouwer, A. E., Dejter, I. J. and Thomassen C. (1993), Highly symmetric subgraphs of hypercubes (*preprint*).

[8] Burr, S. A., Grünbaum, B. and Sloane, N. J. A. (1974), The orchard problem, *Geom. Dedicata* **2** 397–424.

[9] Chung, F. R. K. (1992), Subgraphs of a hypercube containing no small even cycles, *J. Graph Theory* **16** 273–286.

[10] Chung, F. R. K., Szemerédi, E. and Trotter, W. (1992), The number of different distances determined by a set of points in the Euclidean plane, *Discrete and Computational Geometry* **7** 1–11.

[11] Erdős, P. (1935), On the difference of consecutive primes, *Quart. J. Math. Oxford* **6** 124–128.

[12] Erdős, P. (1946), On sets of distances of n points, *Amer. Math. Monthly* **53** 248–250.

[13] Erdős, P. and Hajnal, A. (1989), Ramsey-type theorems, *Discrete Applied Math.* **25** 37–52.

[14] Erdős, P., Hajnal, A., Máté, A. and Rado, R. (1984), *Combinatorial Set Theory: Partition Relations for Cardinals*, North-Holland Publishing Company, *Studies in Logic and the Foundations of Mathematics*, Vol. 106.

[15] Erdős, P., Hajnal, A. and Milner, E. C. (1966), On the complete subgraphs of graphs defined by systems of sets, *Acta Math. Acad. Sci. Hungaricae* **17** 159–229.

[16] Erdős, P., Hajnal, A. and Shelah, S. (1974), On some general properties of chromatic numbers, in *Topics in Topology* (Proc. Colloq. Keszthely, 1977) *Colloq. Math. Soc. J. Bolyai* **8** 243–255.

[17] Erdős, P., Hajnal, A. and Szemerédi, E. (1982), On almost bipartite large chromatic graphs, *Annals of Discrete Math.* **12**, *Theory and Practice of Combinatorics, Articles in Honor of A. Kotzig* (A. Rosa, G. Sabidussi and J. Turgeon, eds.), North-Holland, 117–123.

[18] Erdős, P., Herzog, F. and Piranian, G. (1958), Metric properties of polynomials, *Journal d'Analyse Mathématique* **6** 125–148.

[19] Erdős, P. and Larson, J. (1982), On pairwise balanced block designs with the sizes of blocks as uniform as possible, *Annals of Discrete Mathematics* **15** 129–134.

[20] Erdős, P. and Vértesi, P. (1980), On the almost everywhere divergence of Lagrange Interpolatory Polynomials for large arbitrary systems of nodes, *Acta Math. Acad. Sci. Hungaricae* **36** 71–89.

[21] Faber, G. (1914), Über die interpolatorische Darstellung stetiger Funktionen, *Jahresber. der Deutschen Math. Ver* **23** 190–210.

[22] Füredi, Z. (1990), The maximum number of unit distances in a convex n-gon, *J. Comb. Theory (Ser. A)* **55** 316–320.

[23] Grünwald, G. (1935), Über die Divergenzersheinungen der Lagrangeschen Interpolationpolynome, *Acta. Sci. Math. Szeged* **7** 207–221.

[24] Grünwald, G. (1936), Über die Divergenzersheinungen der Lagrangeschen Interpolationpolynome stetiger Funktionen, *Annals of Math.* **37** 908–918.

[25] Győri, E. (1989), On the number of C_5's in a triangle-free graph, *Combinatorica* **9** 101–102.

[26] Halberstam, H. and Roth, K. F. (1983), *Sequences*, Springer-Verlag, $xiii$+290pp.

[27] Kahn, J. (1992), Coloring nearly-disjoint hypergraphs with $n + o(n)$ colors, *J. Combinatorial Theory (Ser. A)* **59** 31–39.

[28] Kahn, J. (1993), On a problem of Erdős and Lovász. II $n(r) = o(r)$, *J. Amer. Math. Soc.* **14**.

[29] Kanold, (1981), Über Punktmengen im k-dimensionalen euklidischen Raum, *Abh. Braunschweig. wiss. Ges.* **32** 55–65.

[30] Maier, H. and Pomerance, C. (1990), Unusually large gaps between consecutive primes, *Trans. Amer. Math. Soc.* **322** 201–238.

[31] Marcinkiewicz, J. (1937), Sur la divergence des polynomes d'interpolation, *Acta Sci. Math. Szeged* **8** 131–135.

[32] Privalov, A. A. (1976), Divergence of Lagrange interpolation based on the Jacobi abscissas on sets of positive measure, *Sibirsk. Mat. Z.* **18** 837–859 (in Russian).

[33] Privalov, A. A. (1978), Approximation of functions by interpolation polynomials, in *Fourier analysis and approximation theory* I–II, North-Holland, Amsterdam 659–671.

[34] Pyber, L. (1987), The number of pairwise non-commuting elements and the index of the centre in a finite group, *J. London Math. Soc.* **35** 287–295.

[35] Rankin, R. A. (1938), The difference between consecutive prime numbers, *J. London Math. Soc.* **13**, 242–247.

[36] Rankin, R. A. (1962), The difference between consecutive prime numbers. V, *Proc. Edinburgh Math. Soc.* **13** 331–332.

[37] Rödl, V. (1982), Nearly bipartite graphs with large chromatic number, *Combinatorica* **2** 377–387.

[38] Schönhage, A. (1963), Eine Bemerkung zur Konstruktion grosser Primzahllücken, *Arch. Math.* **14** 29–30.

[39] Shrikhande, S. S. and Singhi, N. M. (1985), On a problem of Erdős and Larson, *Combinatorica* **5** 351–358.

[40] Spencer, J., Szemerédi, E. and Trotter, W. (1984), Unit distances in the Euclidean plane, *Graph Theory and Combinatorics*, Academic Press, London 293–303.

Menger's Theorem for a Countable Source Set

RON AHARONI[†] and REINHARD DIESTEL[‡]

[†]Department of Mathematics, Technion, Haifa 32000, Israel
[‡]Mathematical Institute, Oxford University, Oxford OX1 3LB, England

Paul Erdős has conjectured that Menger's theorem extends to infinite graphs in the following way: whenever A, B are two sets of vertices in an infinite graph, there exist a set of disjoint $A–B$ paths and an $A–B$ separator in this graph such that the separator consists of a choice of precisely one vertex from each of the paths. We prove this conjecture for graphs that contain a set of disjoint paths to B from all but countably many vertices of A. In particular, the conjecture is true when A is countable.

1. Introduction

If there is any conjecture in infinite graph theory whose fame has clearly transcended the boundaries of the field, it is the following infinite version of Menger's theorem, conjectured by Erdős:

Conjecture 1.1. *(Erdős) Whenever A, B are two sets of vertices in a graph G, there exist a set of disjoint $A–B$ paths and an $A–B$ separator in G such that the separator consists of a choice of precisely one vertex from each of the paths.*

Here, G may be either directed or undirected and either finite or infinite, and 'disjoint' means 'vertex disjoint'. If G is finite, the statement is clearly a reformulation of Menger's theorem. A set of $A–B$ paths together with an $A–B$ separator as above will be called an *orthogonal paths/separator pair*.

We remark that the naïve infinite analogue to Menger's theorem, which merely compares cardinalities, is considerably weaker and easy to prove. Indeed, consider any inclusion-maximal set $\mathcal{P}$ of disjoint $A–B$ paths. If $\mathcal{P}$ can be chosen infinite, $\bigcup \mathcal{P}$, which is trivially an $A–B$ separator, still has size only $|\mathcal{P}|$. If not, choose $\mathcal{P}$ of maximal (finite) cardinality, and there is a simple reduction to the finite Menger theorem [5]. This was in fact first observed by Erdős, and seems to have inspired his above conjecture as the 'true' generalization of Menger's theorem.

Although Erdős's conjecture has been proved for countable graphs [2], a full proof still appears to be out of reach. However, no other conjecture in infinite graph theory has inspired as interesting a variety of partial or related results as this has; see [4] for a survey and list of references.

The main aim of this paper is to prove a lemma, which, in addition to implying (with [2]) the results stated in the abstract, might play a role in an overall proof of the conjecture by induction on the size of G. Briefly, the lemma implies that if the conjecture is true for all graphs of size κ, where κ is any infinite cardinal, then it is true also for arbitrary graphs, provided the source set A is no larger than κ. (In particular, we see that the conjecture holds for any graph if A is countable.) Now if $|A| = |G| = \lambda$ and the conjecture holds for all graphs of size $< \lambda$, the lemma enables us to apply the induction hypothesis to G with A replaced by its smaller subsets A'; we may then try to combine the orthogonal paths/separator pairs obtained between these A' and B to one between A and B. We must point out, however, that such a proof of Erdős's conjecture will be by no means straightforward, and it is not the only possible approach.

2. Definitions and statement of the main result

All the graphs we consider will be directed; undirected versions of our results can be recovered in the usual way by replacing each undirected edge with two directed edges pointing in opposite directions. An edge from a vertex x to a vertex y will be denoted by xy. When G is a graph, $\overleftarrow{G}$ denotes the graph obtained from G by reversing all its edges.

Paths, likewise, will be directed, and we usually refer to them by their vertex sequence. If $P = x \ldots y$ is a path and v, w are vertices on P in this order, vPw denotes the subpath of P from v to w. Similarly, we write Pv and vP for initial and final segments of P, $P\mathring{v}$ for $Pv - v$, $\mathring{v}P$ for $vP - v$, and so on. If $Q = y \ldots z$ is another path, and $P \cap Q = \{y\}$, then PyQ denotes the path obtained by concatenating P and Q.

Let X, Y be sets of vertices in a graph. An X–Y path is a path from X to Y whose inner vertices are neither in X nor in Y. If x is a vertex, a set of $\{x\}$–Y paths that are disjoint except in x is an x–Y fan; the fan is *onto* if every vertex in Y is hit. Similarly, a set of X–y paths that are disjoint except in y is an X–y fan.

A *warp* is a set of disjoint paths. When $\mathcal{W}$ is a warp, we write $V[\mathcal{W}]$ for the set of vertices of the paths in $\mathcal{W}$, and $E[\mathcal{W}]$ for the set of their edges. Similarly, we write $in[\mathcal{W}]$ for the set of initial vertices of the paths in $\mathcal{W}$, and $ter[\mathcal{W}]$ for the set of their terminal vertices. For a vertex $x \in V[\mathcal{W}]$, we denote the path in $\mathcal{W}$ containing x by $Q_{\mathcal{W}}(x)$, or briefly $Q(x)$. For $x \notin V[\mathcal{W}]$, we put $Q_{\mathcal{W}}(x) := \{x\}$. A warp consisting of A–B paths is an A–B warp. By $\overleftarrow{\mathcal{W}}$ we denote the warp in $\overleftarrow{G}$ consisting of the reversed paths from $\mathcal{W}$.[†]

[†] Clearly, $\overleftarrow{\overleftarrow{\mathcal{W}}} = \mathcal{W}$. We shall use this fact as an excuse to denote warps in $\overleftarrow{G}$, if they are introduced afresh rather than being obtained from a warp in G, by $\overleftarrow{\mathcal{W}}$ etc. straight away; their reversals in G will then be denoted by $\mathcal{W}$. The idea here is to avoid the counter-intuitive practice of having a warp $\mathcal{W}$ in $\overleftarrow{G}$ and a resulting warp $\overleftarrow{\mathcal{W}}$ in G. This convention, if not its explanation, should help the reader avoid any warps in his or her intuition when such things are discussed briefly in Section 5.

Let $G = (V, E)$ be a graph and $A, B \subseteq V$. Any such triple $\Gamma = (G, A, B)$ will be called a web. The web $(\overleftarrow{G}, B, A)$ is denoted by $\overleftarrow{\Gamma}$. An A–B warp $\mathcal{W}$ with $in[\mathcal{W}] = A$ is a *linkage* in Γ, and Γ is *linkable* if it contains a linkage.

A set $S \subseteq V$ *separates* A from B in G if every path in G from A to B meets S. Note that A and B may intersect, in which case clearly $A \cap B \subseteq S$.

A warp $\mathcal{W}$ in G is called a *wave* in Γ if $V[\mathcal{W}] \cap A = in[\mathcal{W}]$ and $ter[\mathcal{W}]$ separates A from B in G. The wave $\{(a) \mid a \in A\}$ is called the *trivial wave*. If $\mathcal{W}$ is a wave in Γ, then $\Gamma/\mathcal{W}$ denotes the web

$$\left(G - (A \backslash in[\mathcal{W}]) - (V[\mathcal{W}] \backslash ter[\mathcal{W}]), \; ter[\mathcal{W}], \; B \right).$$

In every web $\Gamma = (G, A, B)$ there is a wave $\mathcal{W}$ such that $\Gamma/\mathcal{W}$ has no non-trivial wave. (This is not difficult to see. If $\mathcal{W}_0$ is a wave in Γ and $\mathcal{W}_1$ is a wave in $\Gamma/\mathcal{W}_0$, then $\mathcal{W}_1$ defines a wave in Γ in a natural way: just extend its paths back to A along the paths of $\mathcal{W}_0$. This wave in Γ is 'bigger' than $\mathcal{W}_0$, and chains of waves in Γ with respect to this order tend to an obvious limit wave $\mathcal{W}$, which consists of the paths that are eventually in every wave of the chain. If the chain was maximal, then $\Gamma/\mathcal{W}$ has no non-trivial wave. See [2] for details.)

A wave $\mathcal{W}$ in Γ is a *hindrance* if $A \backslash in[\mathcal{W}] \neq \emptyset$; if Γ contains a hindrance, it is called *hindered*. Note that every hindrance is a non-trivial wave. The following was observed in [2]:

Erdős's conjecture is equivalent to the assertion that every unhindered web is linkable.

We are now in a position to state the main result proved in this paper. (For the reasons explained earlier, and because it is of a technical nature, we call it a lemma, not a theorem.)

Lemma 2.1. *Let $\Gamma = (G, A, B)$ be a web and $\mathcal{J}$ an A–B warp in G (possibly empty). If $|A \backslash in[\mathcal{J}]| > |B \backslash ter[\mathcal{J}]|$, then Γ is hindered.*

Lemma 2.1 will be proved in Sections 3 and 4. Our aim will be to turn the given warp $\mathcal{J}$, step by step, into a hindrance. This will require some alternating path techniques; the definitions and lemmas needed are given in Section 3. Section 4 is devoted to the main body of the proof of Lemma 2.1. In Section 5 we look at the implications of the lemma for Erdős's conjecture.

3. Aternating paths

Let $\Gamma = (G, A, B)$ be a web, and let $\mathcal{J}$ be an A–B warp in G. A finite sequence $P = x_0 e_0 x_1 e_1 \ldots e_{n-1} x_n$ of not necessarily distinct vertices x_i and distinct (directed) edges e_i of G will be called an *alternating path* (with respect to $\mathcal{J}$) if the following three conditions are satisfied:

(i) for every $i < n$, either $e_i = x_i x_{i+1} \in E(G) \backslash E[\mathcal{J}]$ or $e_i = x_{i+1} x_i \in E[\mathcal{J}]$;

(ii) if $x_i = x_j$ for $i \neq j$, then $x_i \in V[\mathcal{J}]$;

(iii) for every i, $0 \leq i < n$, if $x_i \in V[\mathcal{J}]$, then $\{e_{i-1}, e_i\} \cap E[\mathcal{J}] \neq \emptyset$.

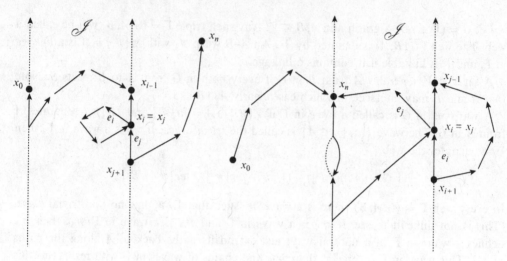

Figure 1 Two alternating paths with respect to $\mathcal{J}$

All the alternating paths we consider in this section will be alternating paths in G with respect to $\mathcal{J}$. Note that, by (iii) above, an alternating path starting at a vertex of $\mathcal{J}$ has its first edge in $\mathcal{J}$. As the edges of an alternating path are pairwise distinct, it can visit any given vertex at most twice, and this happens in essentially only two ways: if $x_i = x_j$ for some $i < j < n$, then $x_i \in V[\mathcal{J}]$ by (ii), so by (iii)

either $e_{i-1}, e_j \in E[\mathcal{J}]$ and $e_i, e_{j-1} \notin E[\mathcal{J}]$ (Figure 1 left)

or $\quad e_i, e_{j-1} \in E[\mathcal{J}]$ and $e_{i-1}, e_j \notin E[\mathcal{J}]$ (Figure 1 right).

Note that initial segments of alternating paths are again alternating paths, but final segments need not be. Finally, an ordinary path which avoids $\mathcal{J}$ or meets it only in its last vertex is trivially an alternating path.

There are analogous alternating versions of the notions of X–Y path, X–y fan and so on.

Lemma 3.1. *If $a \in A \setminus in[\mathcal{J}]$ and $b \in B \setminus ter[\mathcal{J}]$, and if $P = a \ldots b$ is an alternating path with respect to $\mathcal{J}$, then G contains an A–B warp $\mathcal{J}'$ such that $in[\mathcal{J}'] = in[\mathcal{J}] \cup \{a\}$ and $ter[\mathcal{J}'] = ter[\mathcal{J}] \cup \{b\}$ and $\{Q \in \mathcal{J} \mid P \cap Q = \emptyset\} \subseteq \mathcal{J}'$.*

Proof. Consider the graph on $V[\mathcal{J}] \cup V(P)$ whose edge set is the symmetric difference of $E[\mathcal{J}]$ and $E(P)$. The (undirected) components of this graph are all finite. Considering their vertex degrees, we see that they are either A–B paths or cycles avoiding $A \cup B$ (possibly trivial). The assertion follows. □

Lemma 3.2. *Let $P_1 = x_0 e_0 \ldots e_{n-1} x_n$ and $P_2 = y_0 f_0 \ldots f_{m-1} y_m$ be alternating paths. If $x_n = y_0$, then there exists an alternating path P_3 from x_0 to y_m such that $V(P_3) \subseteq V(P_1) \cup V(P_2)$ and $E(P_3) \subseteq E(P_1) \cup E(P_2)$.*

Proof. Let $i \leq n$ be minimal such that there exists a $j \leq m$ with the following two properties:

(i) $x_i = y_j$;
(ii) if $x_i \in V[\mathscr{J}]$, then either $e_{i-1} \in E[\mathscr{J}]$ or $f_j \in E[\mathscr{J}]$.

(Note that such an i exists, because $x_n = y_0$ and P_2 is an alternating path. Moreover, j is easily seen to be unique.) Then $x_0 e_0 \ldots e_{i-1} x_i f_j \ldots f_{m-1} y_m$ is an alternating path as desired. $\square$

4. Proof of the main lemma

We now prove Lemma 2.1. As in the lemma, let $\Gamma = (G, A, B)$ be a web, and let $\mathscr{J}$ be an A–B warp in Γ. Let us write

$$A_1 := in[\mathscr{J}] \quad \text{and} \quad A_2 := A \backslash A_1$$

and

$$B_1 := ter[\mathscr{J}] \quad \text{and} \quad B_2 := B \backslash B_1,$$

and put $\kappa := |B_2|$. We assume that $|A_2| > \kappa$, and construct a hindrance $\mathscr{W}$ in Γ. Again, all the alternating paths considered in this section will be alternating paths in G with respect to $\mathscr{J}$, unless otherwise stated.

Let us quickly dispose of the case when κ is finite. Assume that κ is minimal such that the lemma fails. By Lemma 3.1 and the minimality of κ, there is no alternating path from A_2 to B_2. For each path $Q \in \mathscr{J}$, let $x(Q)$ denote the last vertex of Q that lies on some alternating path starting in A_2; if no such vertex exists, let $x(Q)$ be the initial vertex of Q. We claim that

$$\mathscr{W} := \{Qx \mid Q \in \mathscr{J} \text{ and } x = x(Q)\}$$

is a wave in G; since $|A_2| > \kappa \geq 0$ and hence $in[\mathscr{W}] = in[\mathscr{J}] \subsetneqq A$, this wave $\mathscr{W}$ will be a hindrance and the lemma will be proved.

To show that $\mathscr{W}$ is a wave, we have to prove that $ter[\mathscr{W}]$ separates A from B. So let P be any A–B path. Since P is not an alternating path from A_2 to B_2, it meets $V[\mathscr{J}]$ and hence $V[\mathscr{W}]$; let y be its last vertex in $V[\mathscr{W}]$, and write $Q := Q_{\mathscr{J}}(y)$ and $x := x(Q)$. Suppose P avoids $ter[\mathscr{W}]$. Then $x \neq y$, so there exists an alternating path R from A_2 to y that ends with an edge of $\mathscr{W}$. (Indeed, by definition of $\mathscr{W}$, there is an alternating A_2–x path R'; if x' is the first vertex of R' on $\mathring{y}Qx$, then $R'x'$ followed by yQx' in reverse order is an alternating path from A_2 to y.) By definition of $\mathscr{W}$, R avoids $V[\mathscr{J}] \backslash V[\mathscr{W}]$ and is thus an alternating path with respect to $\mathscr{W}$. Let z be the first vertex of R on yP. Then either $z = y$ or $z \notin V[\mathscr{W}]$, so RzP is again an alternating path with respect to $\mathscr{W}$. By definition of $\mathscr{W}$, RzP avoids $V[\mathscr{J}] \backslash V[\mathscr{W}]$. Therefore zP avoids $V[\mathscr{J}] \backslash \{y\} \supseteq B_1$. (Recall that $y \notin B$, because $y \neq x$ and hence $y \in Q\mathring{x}$.) Thus RzP is an alternating path from A_2 to B_2, a contradiction.

We shall now assume that κ is infinite. To motivate our proof, let us consider the (much easier) case of $\mathscr{J} = \emptyset$. (This is an important special case, and we recommend that the reader remain aware of it throughout the proof of Lemma 2.1.) Assume Erdős's conjecture is true, and let S be an A–B separator as in the conjecture. Then $|S| \leq |B| = \kappa$. Let us think of the vertices in A as being 'to the left' of S, and of those in B as 'to its right'. Which other vertices of G will be to the left of S? Surely those that cannot be separated from A by $\leq \kappa$ vertices, *i.e.* that are joined to A by a fan of size $> \kappa$. We shall call these vertices 'popular'. If a popular vertex is in S, it is the starting vertex of a path to B that contains no other popular vertices; let us call such a path 'lonely', and its starting vertex 'special'. The special vertices, *i.e.* the vertices that are popular and from which we can get to B without hitting any other popular vertex, are in a sense 'rightmost' among the popular vertices, even when they are not in S. As we shall see, they turn out to be 'close enough' to S that they themselves form the set of endvertices of a hindrance in Γ, which is constructable without reference to S.

For the general case, we follow a similar approach, except that now all the relevant paths and fans will be alternating. Let us call a vertex $x \in V(G) \backslash A_1$ *popular* if either $x \in A_2$ or there exists an alternating A_2–x fan of order $> \kappa$. An alternating path P ending in B_2 and with no inner vertex in A_2 is called a *lonely path* if all its vertices are unpopular, *except* possibly its starting vertex and any vertices $x \in V[\mathscr{J}]$ such that, if e is the edge following x on P, then $e \notin E[\mathscr{J}]$. (In the latter case, the edge preceding x on P must be the edge of $\mathscr{J}$ starting in x.) Note that a final segment $x_i e_i x_{i+1} \ldots$ of a lonely path is again lonely if and only if x_i satisfies condition (iii) in the definition of an alternating path, *i.e.* if and only if $e_i \in E[\mathscr{J}]$ when $x_i \in V[\mathscr{J}]$.

Our first lemma is merely a technical argument that will be used twice later and has been extracted for economy. The first time we will use it is in the proof of Lemma 4.2 below, and for motivation the reader may prefer to read Lemma 4.2 and its proof first and then return to Lemma 4.1.

Lemma 4.1. *Let α be a cardinal, $\mathscr{L} = \{L_\beta \mid \beta < \alpha\}$ a family of lonely paths, and $\mathscr{M} = \{M_\beta \mid \beta < \alpha\}$ a family of pairwise disjoint alternating paths starting in A_2. Assume that, for each $\beta < \alpha$, the last vertex of M_β is the starting vertex of L_β. Then $\alpha \leq \kappa$.*

Proof. Suppose $\alpha > \kappa$. For each $\beta < \alpha$, let P_β be an alternating path from the starting vertex of M_β to the final vertex of L_β as provided by Lemma 3.2. Construct an undirected forest $H = \bigcup_{\beta < \alpha} H_\beta$ from these paths, as follows. Let H_0 be the undirected graph underlying P_0. Now let $\beta < \alpha$ be given, and assume that H_γ has been defined for every $\gamma < \beta$. Let z_β denote the first vertex of P_β that is in $B_2 \cup V(H_\beta^-)$, where $H_\beta^- := \bigcup_{\gamma < \beta} H_\gamma$, and let H_β be the union of H_β^- with the undirected graph underlying $P_\beta z_\beta$. (If z_β occurs twice on P_β, we take $P_\beta z_\beta$ to stop at the first occurrence of z_β.) Since $|B_2| \leq \kappa$ and every path P_β ends in B_2, H has at most κ components. One of these components must have size $> \kappa$, so it contains a vertex z of degree $> \kappa$. Then z lies on $> \kappa$ of the paths P_β, so $z = z_\beta$ for every β in some set $\Delta \subseteq \alpha$ of size $> \kappa$.

Let

$$F := \{P_\beta z \mid \beta \in \Delta\}.$$

Note that the paths in F are pairwise disjoint except for z, so F is an alternating A_2–z fan. Hence, z is popular. As the paths M_β are pairwise disjoint, we have $z \notin M_\beta$ for all but at most one $\beta \in \Delta$; let us delete this one β from Δ if it exists. Now for all $\beta \in \Delta$, we have that $z \in L_\beta$ and z is not the starting vertex of L_β (since this is on M_β).

Now consider any $\beta \in \Delta$. Since L_β is a lonely path and z is popular but not the starting vertex of L_β, we have $z \in V[\mathscr{J}]$, and if e denotes the edge following z on L_β, then $e \notin E[\mathscr{J}]$. (Note that e exists, because $z \in V[\mathscr{J}]$ but L_β ends in B_2.) Since L_β is alternating, this means that the edge f preceding z on L_β must be the edge of $\mathscr{J}$ starting at z (and such an edge exists). Since $z \notin M_\beta$, the edge of P_β preceding z is precisely this edge f.

As β was chosen arbitrarily, this is true for every $\beta \in \Delta$ and thus contradicts the fact that for these β the paths $P_\beta \overset{\circ}{z}$ are disjoint. $\qquad \square$

If the starting vertex of a lonely path is popular, this vertex is called *special*; the set of all special vertices outside $V[\mathscr{J}]$ is denoted by S. Special vertices will be our prime candidates for the terminal vertices of the hindrance we are seeking to construct. Since the corresponding paths of the hindrance will have to be constructed from the fans connecting A_2 to these terminal vertices (making them popular), it is important that there are fewer special vertices to be connected in this way than there are connecting paths available from those fans.

Lemma 4.2. *There are at most κ special vertices.*

Proof. Suppose that $\{s_\beta \mid \beta < \kappa^+\}$ is a set of distinct special vertices, where κ^+ is the successor cardinal of κ. For each β, let L_β be a lonely path starting at s_β. Using the popularity of the s_β, we may inductively choose a family $\{M_\beta \mid \beta < \kappa^+\}$ of pairwise disjoint alternating paths M_β from A_2 to s_β. This contradicts Lemma 4.1. $\qquad \square$

Let E denote the set of all those edges in G that lie on some lonely path, and let K be the graph

$$K := \bigcup \mathscr{J} - E.$$

Let $\mathscr{P}$ be the set of all those (undirected) components of K that contain a special vertex or a vertex from A. (Thus, $\mathscr{P}$ is a set of pairwise disjoint subpaths of paths in $\mathscr{J}$.) Let

$$T := \{x \in V(G) \mid x \text{ is the last vertex on some } P \in \mathscr{P}\}.$$

We shall define our desired hindrance $\mathscr{W}$ in such a way that

$$ter[\mathscr{W}] = S \cup T.$$

Lemma 4.3. *If $P = x \dots y$ is a non-trivial component of K and $x \notin A$, then x is special (and hence $P \in \mathscr{P}$ and $y \in T$).*

Proof. Let r be the predecessor and s the successor of x on $Q(x)$. Then $rx \in E$, so there exists a lonely path starting at x with the edge rx. But preceding this path with s does not yield another lonely path (since $xs \in E(P)$, and hence $xs \notin E$). Therefore x must be popular (see the definition of lonely paths), and hence special. $\qquad \square$

To construct $\mathscr{W}$, let us start from $\mathscr{P}$. Let $\mathscr{W}_0$ be the set of all paths $P \in \mathscr{P}$ that start in A. (These paths may be entire paths from $\mathscr{J}$, and they may be trivial.) Our aim is to complete $\mathscr{W}_0$ to our desired wave $\mathscr{W}$ by paths of the form $a\ldots xPy$, where $a \in A_2$ and $P = x\ldots y$ is a path as in Lemma 4.3, together with paths $a\ldots s$, where again $a \in A_2$ and either $s \in S$ or s is a special vertex in $V[\mathscr{J}]$ making up a singleton component of K. It will not be possible to construct $\mathscr{W}$ in exactly this way, because the required paths may interfere with the paths in $\mathscr{W}_0$. However, such interference will be limited by Lemmas 4.1 and 4.2, and can therefore be overcome by the alternating path tools developed in Section 3.

Let

$$S' = \{s_\zeta \mid \zeta < v \leq \kappa\}$$

be a well-ordering of those special vertices that are either in S or are the initial vertex of some (possibly trivial) path $P \in \mathscr{P}$ (cf. Lemma 4.2). For each $\zeta < v$ in turn, we shall choose an alternating path P_ζ from A_2 to s_ζ, with the following properties:

(i) $P_\zeta \cap P_\xi = \emptyset$ for all $\xi < \zeta$;
(ii) $P_\zeta \cap Q_{\mathscr{J}}(s) \subseteq \{s_\zeta\}$ for all $s \in S'$;
(iii) if $Q \in \mathscr{J}$ and $\xi < \zeta$ are such that $P_\xi \cap Q \neq \emptyset$, then $P_\zeta \cap Q \subseteq \{s_\zeta\}$;
(iv) $E(P_\zeta) \cap E = \emptyset$.

Let $\zeta < v$ be given, and assume that paths P_ξ for all $\xi < \zeta$ have been chosen in accordance with (i)–(iv). By (ii), none of these paths contains $(s =) s_\zeta$. Since s_ζ is popular, there is an alternating A_2–s_ζ fan F of size $> \kappa$. Clearly, at most κ of the paths in F meet any of the paths P_ξ ($\xi < \zeta$) or $Q(s)$ for $s \in S'$, except, for the latter, in s_ζ. Similarly, at most κ of the paths in F meet (in a vertex $\neq s_\zeta$) any path $Q \in \mathscr{J}$ that is hit by some P_ξ with $\xi < \zeta$. By Lemma 4.1, at most κ paths of F have an edge in E. (It is straightforward to check that the first edge in E on any path in F starts a lonely path.) We may thus choose P_ζ from the paths in F according to (i)–(iv).

Lemma 4.4. *For every $\zeta < v$, we have $E(P_\zeta) \cap E[\mathscr{J}] \subseteq E[\mathscr{W}_0]$. Thus, P_ζ is in fact an alternating path with respect to $\mathscr{W}_0$.*

Proof. If $e \in E(P_\zeta) \cap E[\mathscr{J}]$, then, by (iv) above, there is a component P of K containing e. By (ii), the initial vertex of P is not in S', and is therefore not special. By Lemma 4.3, therefore, the starting vertex of P must be in A, and so $P \in \mathscr{W}_0$.   □

Applying Lemma 3.1 v times with the paths P_ζ, we now turn $\mathscr{W}_0$ into a warp from A onto $ter[\mathscr{W}_0] \cup S'$ with at most κ initial points in A_2. (Here we use that fact that, by (iii) above, no two of the alternating paths P_ζ use the same path in $\mathscr{W}_0$ to alternate on.) By (ii) above, the paths in $\mathscr{P}$ that start at the vertices in $S'\backslash S$ extend this warp to a warp $\mathscr{W}$. By Lemma 4.3, $ter[\mathscr{W}] = S \cup T$ as desired.

To prove that $\mathscr{W}$ is a wave in Γ, it remains to show that the set $S \cup T$ separates A from B; note that then $\mathscr{W}$ is also a hindrance, since $|in[\mathscr{W}] \cap A_2| = |S'| \leq \kappa$ by construction.

In order to prove that $S \cup T$ separates A from B, consider any A–B path $P = a \dots b$ in G. Suppose P avoids $S \cup T$.

Lemma 4.5. *Either $b \in B_2$, or b is the final vertex of an edge in $E \cap E[\mathcal{J}]$. In either case, b is not special but starts a lonely path.*

Proof. Suppose first that $b \in B_2$. Then b is not special, because $b \notin S$. Moreover, $\{b\}$ is a trivial lonely path.

Suppose now that $b \in B_1$, and let $P' = x \dots b$ be the component of K containing b. As $b \notin T$, we have $P' \notin \mathcal{P}$, so $x \notin A$ and neither x nor b is special. By Lemma 4.3, therefore, P' is trivial, i.e. $b = x \notin A$. The edge e of $\mathcal{J}$ that ends in b is therefore in E, and hence lies on a lonely path. The final segment of this lonely path that starts at b (with e as its first edge) is again a lonely path, because $e \in E[\mathcal{J}]$. $\qquad\square$

Lemma 4.6. *The vertex a does not lie on a lonely path.*

Proof. If $a \in A_2$, then a is popular by definition, so being on – and hence starting – a lonely path would imply $a \in S$. If $a \in A_1$ and a lies on a lonely path, then this path uses the edge of $\mathcal{J}$ starting at a. Then $\{a\}$ is a component of K, and hence a trivial path in $\mathcal{W}_0$ and in $\mathcal{W}$, giving $a \in T$. $\qquad\square$

Let x be the last vertex of P that is not on any lonely path, and let y be the vertex following x on P. Let L be a lonely path containing y. Then

(4.7) $x \notin L$, and xyL is not a lonely path.

Since $y \notin T$, y can only be in $V[\mathcal{J}]$ if the edge of $\mathcal{J}$ ending in y is in E. (Recall that L must use an edge of $\mathcal{J}$ incident with y, and apply Lemma 4.3.) We may therefore make the following assumption:

(4.8)   If $y \in V[\mathcal{J}]$, then L starts at y (with the edge of $\mathcal{J}$ that ends in y).

Lemma 4.9. *The vertex y is popular.*

Proof. If y is not popular, then xyL can fail to be a lonely path only if it fails to be an alternating path. By (4.7) and (4.8), this can happen only if $x \in V[\mathcal{J}]$ and xyL fails to start with an edge of $\mathcal{J}$. But $x \notin B$, so x has a successor q on $Q(x)$. By (4.7) and (4.8), we have $q \neq y$. Now $qxyL$ is a lonely path (possibly containing q twice) that contradicts the choice of x. $\qquad\square$

Let z be the last popular vertex on P. Then $z \neq b$, because b is unpopular by Lemma 4.5. As $z \in yP$ by Lemma 4.9, the choice of x and definition of y imply that z lies on some lonely path. But then $z \in V[\mathcal{J}]$, say $z \in Q \in \mathcal{J}$: otherwise the final segment of this lonely path that starts at z would again be lonely, and the popularity of z would mean that $z \in S$. Let q be the vertex following z on Q and let t be the vertex following z on P.

Lemma 4.10. $zq \notin E$.

Proof. Let p be the vertex preceding z on Q. (This exists, since $z \neq a$.) If $zq \in E$, then $pz \notin E$: otherwise z would be not only popular but special, giving $\{z\} \in \mathscr{P}$ and $z \in T$. But if $pz \notin E$, then $zq \in E$ implies by Lemma 4.3 that $z \in T$, a contradiction.   □

Since $t \in yP$, there is a lonely path M containing t. As with y in (4.8), we may assume the following:

(4.11)   If $t \in V[\mathscr{J}]$, then M starts at t (with the edge of $\mathscr{J}$ that ends in t).

By Lemma 4.10, zq is not an edge of M. By (4.11), this means that $t \neq q$; in particular, zq and zt are distinct edges. Moreover, zt is not an edge of M, since then M would have to use its starting edge again. Therefore, $qztM$ is an alternating path. Since t is not popular (by the choice of z), this means that $qztM$ is even a lonely path. (Note that zt is a 'real' edge, not the reverse of a $\mathscr{J}$-edge, so the popularity of z does not prevent this path from being lonely.) This, however, contradicts Lemma 4.10, completing the proof of Lemma 2.1.

5. Consequences

In this section we apply Lemma 2.1 to deduce some concrete partial results towards Erdős's conjecture. First, we need another lemma.

Lemma 5.1. *Let κ be an infinite cardinal. If Erdős's conjecture holds for all graphs of order $\leq \kappa$, it holds for all webs $\Gamma = (G, A, B)$ such that $|A|, |B| \leq \kappa$.*

Proof. Let $\Gamma = (G, A, B)$ be a web with $|A|, |B| \leq \kappa$, and assume the conjecture holds for every graph of order $\leq \kappa$. Let G' be obtained from G by adding all edges xy such that G contains a set of $> \kappa$ independent x–y paths (i.e. paths that are disjoint except in x and y). To prove the conjecture for Γ, it suffices to find an orthogonal paths/separator pair $(\mathscr{P}, S)$ for $\Gamma' := (G', A, B)$. Indeed, then S is clearly also an A–B separator in G. As for the paths in $\mathscr{P}$, their foreign edges can be replaced inductively by paths in G whose interiors avoid each other and all the paths in $\mathscr{P}$ (since $|\mathscr{P}| \leq \kappa$), giving an A–B warp in G. We thus obtain an orthogonal pair for Γ.

Let G'' be the union of all minimal A–B paths in G'. (A path $P = a \ldots b$ is *minimal* if G' contains no a–b path Q with $V(Q) \subsetneqq V(P)$.) It is now sufficient to find an orthogonal paths/separator pair for $\Gamma'' := (G'', A, B)$, which will clearly also be an orthogonal pair for Γ'. It thus suffices to show that $|G''| \leq \kappa$.

Suppose $|G''| > \kappa$, and consider a set $X \subseteq V(G'') \setminus (A \cup B)$ of size $> \kappa$, say $X = \{x_\beta \mid \beta < \alpha\}$. (Recall that $|A|, |B| \leq \kappa$ by assumption.) For each $\beta < \alpha$, use the definition of G'' to find a minimal A–B path P_β in G' containing x_β. For all $\beta < \alpha$, inductively define P'_β as the maximal final segment of $P_\beta x_\beta$ that meets $\bigcup_{\gamma < \beta} P'_\gamma$ at most in its starting vertex s_β. Since $|A| \leq \kappa$, there is a vertex $s \in G''$ such that $s = s_\beta$ for every β in some set $\Delta \subseteq \alpha$ of size $> \kappa$. Then $F_1 := \bigcup_{\beta \in \Delta} sP_\beta x_\beta$ is a fan from s onto $Y := \{x_\beta | \beta \in \Delta\}$.

Similarly, $\bigcup_{\beta \in \Delta} x_\beta P_\beta$ contains a fan F_2 from some set $Z \subseteq Y$ of size $> \kappa$ to a vertex t. Clearly, F_2 may be chosen so that no two of its paths meet a common path of F_1. It is then easy to combine F_1 and F_2 into a set of $> \kappa$ independent s–t paths in G''. Thus st is an edge of G', by definition of G'. But s and t are non-consecutive vertices on some common path P_β (take any β such that $x_\beta \in Z$), which contradicts the minimality of P_β.   □

Combining Lemma 2.1 and Lemma 5.1, we can now easily prove the following.

Theorem 5.2. *Let κ be an infinite cardinal. If Erdős's conjecture holds for all graphs of order $\leq \kappa$, it holds for all webs $\Gamma = (G, A, B)$ such that $|A| \leq \kappa$.*

Proof. Let $\Gamma = (G, A, B)$ be a web with $|A| \leq \kappa$. Let $\overleftarrow{W}$ be a wave in $\overleftarrow{\Gamma}$ such that $\overleftarrow{\Gamma}' := \overleftarrow{\Gamma}/\overleftarrow{W}$ has no non-trivial wave. Let $B' := \mathrm{ter}[\overleftarrow{W}]$, and let $H \subseteq G$ be such that $\overleftarrow{\Gamma}' = (\overleftarrow{H}, B', A)$. (In other words, take the underlying graph of $\overleftarrow{\Gamma}'$, reverse its edges, and call the resulting graph H.) Since $\overleftarrow{\Gamma}'$ is unhindered, we have $|B'| \leq |A| \leq \kappa$ by Lemma 2.1. Now if the conjecture holds for all graphs of size $\leq \kappa$, then by Lemma 5.1 it holds for $\overleftarrow{\Gamma}'$, and there is a warp $\overleftarrow{\mathscr{J}}$ in $\overleftarrow{\Gamma}'$ together with a B'–A separator S in $\overleftarrow{H}$ consisting of a choice of one vertex from each path in $\overleftarrow{\mathscr{J}}$. But S is also a B–A separator in $\overleftarrow{G}$ (because $\mathrm{ter}[\overleftarrow{W}]$ is one), and hence an A–B separator in G. Thus S, together with $\mathscr{J}$ followed by a suitable subset of W, is an orthogonal paths/separator pair for Γ.   □

Corollary 5.3. *Erdős's conjecture is valid for all webs $\Gamma = (G, A, B)$ in which A is countable.*

Proof. By Theorem 5.2 and the fact that the conjecture holds for countable graphs [2]. □

Unsurprisingly, Corollary 5.3 on its own does not need the full strength of Lemma 2.1. In fact, with hindsight, it is not too difficult to deduce the corollary directly from the main result of [3].

We conclude this section with an application of Lemma 2.1 to webs that come with a partial linkage.

Theorem 5.4. *Let $\Gamma = (G, A, B)$ be a web, and assume that G contains an A–B warp $\mathscr{J}$ such that $A \setminus \mathrm{in}[\mathscr{J}]$ is countable. Then Erdős's conjecture holds for Γ.*

Proof. As in the proof of Theorem 5.2, we let $\overleftarrow{W}$ be a wave in $\overleftarrow{\Gamma}$ such that $\overleftarrow{\Gamma}' := \overleftarrow{\Gamma}/\overleftarrow{W}$ has no non-trivial wave. Let $B' := \mathrm{ter}[\overleftarrow{W}]$, and let $H \subseteq G$ be such that $\overleftarrow{\Gamma}' = (\overleftarrow{H}, B', A)$. Then $\overleftarrow{\Gamma}'$ is unhindered, and the final segments in $\overleftarrow{H}$ of the paths in $\overleftarrow{\mathscr{J}}$ form a B'–A warp $\overleftarrow{\mathscr{J}}'$ in $\overleftarrow{H}$. By Lemma 2.1, we have

$$|B' \setminus \mathrm{in}[\overleftarrow{\mathscr{J}}']| \leq |A \setminus \mathrm{ter}[\overleftarrow{\mathscr{J}}']| = |A \setminus \mathrm{in}[\mathscr{J}]| \leq \aleph_0.$$

But such unhindered 'countable-like' webs as $\overleftarrow{\Gamma}'$ are linkable [3]. Let $\mathscr{L}$ be a B'–A linkage in $\overleftarrow{H}$. The concatenations of the paths in $\mathscr{L}$ with their unique extensions in W then form an A–B warp in Γ, and B' is an A–B separator in G consisting of a choice of one vertex from each path in this warp. $\square$

References

[1] Aharoni, R. (1984) König's duality theorem for infinite bipartite graphs. *J. London Math. Soc.* **29** 1–12.
[2] Aharoni, R. (1987) Menger's theorem for countable graphs. *J. Combin. Theory B* **43** 303–313.
[3] Aharoni, R. (1990) Linkability in countable-like webs. In: Hahn, G. *et al.* (eds.) *Cycles and Rays*, NATO ASI Ser. C, Kluwer Academic Publishers, Dordrecht.
[4] Aharoni, R. (1992) Infinite matching theory. In: Diestel, R. (ed.) *Directions in Infinite Graph Theory and Combinatorics*, Topics in Discrete Mathematics **3**, North-Holland.
[5] König, D. (1936) *Theorie der endlichen und unendlichen Graphen*, Akademische Verlagsgesellschaft, Leipzig (reprinted: Chelsea, New York 1950).

On Extremal Set Partitions in Cartesian Product Spaces

RUDOLF AHLSWEDE and NING CAI

Universität Bielefeld, Fakultät für Mathematik, Postfach 100131, 33501 Bielefeld, Germany

The partition number of a product hypergraph is introduced as the minimal size of a partition of its vertex set into sets that are edges. This number is shown to be multiplicative if all factors are graphs with all loops included.

1. Introduction

Consider $(\mathscr{V}, \mathscr{E})$, where $\mathscr{V}$ is a finite set and $\mathscr{E}$ is a system of subsets of $\mathscr{V}$. For the cartesian products $\mathscr{V}^n = \prod_1^n \mathscr{V}$ and $\mathscr{E}^n = \prod_1^n \mathscr{E}$, let $\pi(n)$ denote the minimal size of a partition of $\mathscr{V}^n$ into sets that are elements of $\mathscr{E}^n$ if a partition exists at all, otherwise $\pi(n)$ is not defined. This is obviously exactly the case if it is so for $n = 1$.

Whereas the packing number $p(n)$, that is the maximal size of a system of disjoint sets from $\mathscr{E}^n$, and the covering number $c(n)$, that is the minimal number of sets from $\mathscr{E}^n$ to cover $\mathscr{V}^n$, have been studied in the literature, this seems to be not the case for the partition number $\pi(n)$.

Obviously, $c(n) \leq \pi(n) \leq p(n)$, if $c(n)$ and $\pi(n)$ are well defined. The quantity $\lim_{n \to \infty} \frac{1}{n} \log p(n)$ is Shannon's zero error capacity [11]. Although it is known only for very few cases (see [7]), a nice formula exists for $\lim_{n \to \infty} (1/n) \log c(n)$ (see [1, 10]).

The difficulties in analyzing $\pi(n)$ are similar to those for $p(n)$. For the case of graphs with edge set $\mathscr{E}$ including all loops, we prove that $\pi(n) = \pi(1)^n$ (Theorem 3). This result is derived from the corresponding result for complete graphs (Theorem 2) with the help of Gallai's Lemma in matching theory [6]. More general results concern products of hypergraphs with non-identical factors. Another interesting quantity is $\mu(n)$, the maximal size of a partition of $\mathscr{V}^n$ into sets that are elements of $\mathscr{E}^n$ (again only hypergraphs $(\mathscr{V}, \mathscr{E})$ with a partition are considered). We also call μ the maximal partition number. It behaves more like the packing number (see example 5). Clearly, $\pi(n) \leq \mu(n) \leq p(n)$. It seems to us that an understanding of these partition problems would be a significant contribution to an understanding of the basic, and seemingly simple, notion of Cartesian products. Another partition problem was formulated in [12]. Among the contributions to this problem, we refer the reader to [5], [9], and [12].

2. Products of complete graphs: first results

For a complete graph $\mathscr{C} = \{\mathscr{V}, \mathscr{E}\}$, let $\mathscr{E}^* = \mathscr{E} \cup \{\{v\} : v \in \mathscr{V}\}$ and define the hypergraph $\mathscr{C}^n = \{\mathscr{V}^n, \mathscr{E}^n\}$, where $\mathscr{V}^n = \prod_1^n \mathscr{V}$ and $\mathscr{E}^n = \prod_1^n \mathscr{E}^*$.

We study the partition number $\pi(n)$, first for $\mathscr{C}^n$, and in later sections extend our results to hypergraphs, which are products of arbitrary graphs including all loops.

First we introduce the map $\sigma : \mathscr{E}^n \to \{0, 1\}^n$, where

$$s^n = \sigma(E^n) = (\log |E_1|, \ldots, \log |E_n|). \tag{2.1}$$

As weight of E^n ($w(E^n)$ for short), we choose the Hamming weight $w_H(s^n) = \sum_{t=1}^n s_t$. Notice that the cardinality $|E^n|$ equals $2^{w(E^n)}$.

Instead of partitions, we consider more generally a packing $\mathscr{P}$ of $\mathscr{C}^n$. We set

$$\mathscr{P}_i = \{E^n \in \mathscr{P} : w(E^n) = i\}, P_i = |\mathscr{P}_i|, \tag{2.2}$$

and call $\{P_i\}_{i=0}^n$ the weight distribution of $\mathscr{P}$.

We associate with $\mathscr{P}$ the set of shadows $\mathscr{Q} \subset \mathscr{L}^n$ defined by

$$\mathscr{Q} = \{E^n \in \mathscr{E}^n : E^n \subset F^n \text{ for some } F^n \in \mathscr{P}\}, \tag{2.3}$$

and its level sets

$$\mathscr{Q}_i = \{E^n \in \mathscr{Q} : w(E^n) = i\}, 0 \le i \le n. \tag{2.4}$$

It is convenient to write $Q_i = |\mathscr{Q}_i|$. $\{Q_i\}_{i=0}^n$ is the weight distribution of $\mathscr{Q} = \text{shad}(\mathscr{P})$.

First we establish some simple connections between these weight distributions.

Lemma 1. *For a packing $\mathscr{P}$ of $\mathscr{C}^n$*

$$\sum_{i=k}^n 2^{i-k} \binom{i}{k} P_i = Q_k. \tag{2.5}$$

Proof. Consider any edge E^n with weight $w(E^n) = i \ge k$. There are exactly $2^{i-k} \binom{i}{k}$ edges contained in E^n with weight k. Therefore we have always

$$\sum_{i=k}^n 2^{i-k} \binom{i}{k} P_i \ge Q_k. \tag{2.6}$$

$\square$

Lemma 2. *For a packing $\mathscr{P}$ of $\mathscr{C}^n$*

$$|\mathscr{P}| = \sum_{i=0}^n P_i = \sum_{k=0}^n (-1)^k Q_k. \tag{2.7}$$

Proof. An edge $E^n \in \mathscr{P}_i$ contributes to $\sum_{k=0}^n (-1)^k Q_k$ the amount

$$\sum_{k=0}^i (-1)^k 2^{1-k} \binom{i}{k} = (2-1)^i = 1. \qquad \square$$

Lemma 3. *For a packing* $\mathscr{P}$ *of* $\mathscr{C}^n$

$$P_0 = \sum_{k=0}^n (-1)^k 2^k Q_k \qquad (2.8)$$

and if in addition $\mathscr{P}$ *is a partition and* $S = |\mathscr{V}|$ *is odd,*

$$\sum_{k=0}^n (-1)^k 2^k Q_k - 1 \geq 0. \qquad (2.9)$$

Proof. An edge $E^n \in \mathscr{P}_i$ contributes to $\sum_{k=0}^n (-1)^k 2^k Q_k$ the amount

$$\sum_{k=0}^i (-1)^k 2^k 2^{i-k} \binom{i}{k} = 2^i (1-1)^i,$$

which equals 1, if $i = 0$, and 0, otherwise.

Therefore (2.8) holds.

Furthermore, if S is odd, then so is S^n and there must be an edge in the partition of odd size, that is, $P_0 \geq 1$ or, equivalently, by (2.8), (2.9) must hold. $\qquad \square$

Remark 1. The last two Lemmas can be derived more systematically from Lemma 1 by Möbius Inversion. Here this machinery can be avoided, but we need it for the more abstract setting of [4].

3. Products of complete graphs: the main results

We shall now exploit Lemma 3 by applying it to classes of subhypergraphs, which we now define. For any $I \subset \{1, 2, \ldots, n\}$ and any specification $(v_j)_{j \in I^c}$, where $v_j \in \mathscr{V}_j$, we set

$$\mathscr{C}^n\big(I, (v_j)_{j \in I^c}\big) = \left(\prod_{i=1}^n \mathscr{U}_i, \prod_{i=1}^n \mathscr{F}_i \right) = (\mathscr{U}^n, \mathscr{F}^n), \qquad (3.1)$$

where

$$\mathscr{U}_i = \begin{cases} \mathscr{V}_i \\ \{v_i\} \end{cases} \quad \text{and} \quad \mathscr{F}_i = \begin{cases} \mathscr{E}_i & \text{for } i \in I \\ \{v_i\} & \text{for } i \in I^c. \end{cases} \qquad (3.2)$$

Clearly, for a partition $\mathscr{P}$ of $\mathscr{C}^n$ and $\mathscr{Q} = \text{shad}\mathscr{P}$, the set $\mathscr{Q}\big(I, (v_j)_{j \in I^c}\big) = \mathscr{Q} \cap \mathscr{F}^n$ is a downset, and the map

$$\psi : \mathscr{F}^n \to \prod_{i \in I} \mathscr{E}_i, \ \psi \left(\prod_{i=1}^n E_i \right) = \prod_{i \in I} E_i \qquad (3.3)$$

is a bijection.

Write $\tilde{\mathcal{Q}} = \psi(\mathcal{Q} \cap \mathcal{F}^n)$ and let $\tilde{\mathcal{Q}}_i$ count the members of $\tilde{\mathcal{Q}}$ of weight i. Since $\tilde{\mathcal{Q}}$ is a downset in $\prod_{i \in I} \mathcal{E}_i$ and its maximal elements form a partition of $\prod_{i \in I} \mathcal{V}_i$, we know that $\tilde{\mathcal{Q}}_0 = S^m$. This fact and Lemma 3 yield

$$S^m + \sum_{k=1}^{m} (-1)^k 2^k \tilde{\mathcal{Q}}_k - 1 \geq 0. \tag{3.4}$$

This is the key to the proof of the following important result.

Theorem 1. *For a partition* $\mathcal{P}$ *of* $\mathcal{C}^n = (\mathcal{V}^n, \mathcal{E}^n)$ *with* $\mathcal{V}^n = \prod_{i=1}^{n} \mathcal{V}_i$, $|\mathcal{V}_i| = S$ *for* $i = 1, 2, \ldots, n$, *the weight distribution* $(Q_k)_{k=0}^{n}$ *of* $Q = \mathrm{shad}\mathcal{P}$ *satisfies, for* $1 \leq m \leq n$,

$$\binom{n}{m} S^n + \sum_{k=1}^{m} (-1)^k \binom{n-k}{m-k} 2^k Q_k - \binom{n}{m} S^{n-m} \geq 0. \tag{3.5}$$

Proof. The map ψ preserves inclusions and weights. The total number of pairs $(I, (v_j)_{j \in I^c})$ with $|I| = m$ equals $\binom{n}{m} S^{n-m}$. Moreover, each $E^n \in \mathcal{Q}$ with $w(E^n) = k$ is contained in exactly $\binom{n-k}{m-k}$ sets of the form $\mathcal{Q}(I, (v_j)_{j \in I^c})$ and thus for the sets of weight k

$$\binom{n-k}{m-k} Q_k = \sum_{(I, (v_j)_{j \in I^c}), |I| = m} \left| \mathcal{Q}_k (I, (v_j)_{j \in I^c}) \right|. \tag{3.6}$$

We have one equation of the form (3.4) for each pair $(I, (v_j)_{j \in I_c})$. Summation of their left-hand sides gives, therefore,

$$\binom{n}{m} S^{n-m} \cdot S^m + \sum_{k=1}^{m} (-1)^k 2^k \binom{n-k}{m-k} Q_k - \binom{n}{m} S^{n-m} \geq 0$$

and hence (3.5). $\qquad\qquad\qquad\qquad\qquad\qquad\qquad\qquad\qquad\qquad\qquad\qquad\qquad\qquad$ $\square$

Now comes the harvest.

Theorem 2. *For a partition* $\mathcal{P}$ *of* $\mathcal{C}^n$

$$|\mathcal{P}| \geq \left\lceil \frac{S}{2} \right\rceil^n.$$

Proof. Since $|E^n| \leq 2^n$, obviously $|\mathcal{P}| \geq S^n/2^n$, and for $S = 2\alpha$ even, the result obviously holds. Now let $S = 2\alpha + 1$.

Summing the left-hand side expressions in (3.5) for $m = 1, 2, \ldots, n$ results in

$$\sum_{m=1}^{n} \binom{n}{m} S^n + \sum_{m=1}^{n} \sum_{k=1}^{m} (-1)^k \binom{n-k}{m-k} 2^k Q_k - \sum_{m=1}^{n} \binom{n}{m} S^{n-m} \geq 0,$$

or in

$$(2^n - 1)S^n + \sum_{k=1}^{n}(-1)^k 2^k Q_k \sum_{m=k}^{n}\binom{n-k}{m-k} - [(S+1)^n - S^n] \geq 0.$$

This is equivalent to

$$2^n \cdot \left[S^n + \sum_{k=1}^{n}(-1)^k Q_k\right] - (S+1)^n \geq 0.$$

As $Q_0 = S^n$, we conclude, with Lemma 2,

$$P \geq (S+1)^n \cdot 2^{-n} = \left[\frac{S}{2}\right]^n, \quad \text{if } S \text{ is odd}. \qquad \square$$

4. Non-identical factors: a generalization

We now consider hypergraphs $\mathscr{C}'^n$ with vertex sets $\mathscr{V}^n = \prod_{t=1}^{n}\mathscr{V}_t$ and edge sets $\mathscr{E}^n = \prod_{t=1}^{n}\mathscr{E}_t$, where the $\mathscr{V}_t$'s are finite sets of not necessarily equal cardinalities S_t. The factors $\mathscr{E}_t$ are such that $(\mathscr{V}_t, \mathscr{E}_t)$ is a complete graph with all loops included. We shall write, with positive integers α_t,

$$|\mathscr{V}_t| = 2\alpha_t + \varepsilon_t, \ \varepsilon_t \in \{0, 1\}. \tag{4.1}$$

Inspection shows that the sizes of factors do not affect the proofs of Lemmas 1 and 2. Also (2.8) in Lemma 2 holds and since $P_0 \geq 1$, if $\varepsilon_t = 1$ for $t = 1, 2, \ldots, n$, we can generalize (2.9) to

$$\sum_{k=0}^{n}(-1)^k 2^k Q_k - \prod_{k=1}^{n}\varepsilon_k \geq 0. \tag{4.2}$$

Theorem 1 in Section 3 generalizes to

Theorem 1'. *For a partition $\mathscr{P}$ of $\mathscr{C}'^n$*

$$\binom{n}{m}\prod_{i=1}^{n}S_i + \sum_{k=1}^{m}(-1)^k\binom{n-k}{m-k}2^k Q_k - \sum_{I:|I|=m}\prod_{i\in I}\varepsilon_i \prod_{j\in I^c}S_j \geq 0. \tag{4.3}$$

Proof. (Sketch) In the proof of Theorem 1, replace S^m by $\prod_{i\in I}S_i$ and inequality (3.4) by

$$\prod_{i\in I}S_i + \sum_{k=1}^{n}(-1)^k 2^k \tilde{Q}_k - \prod_{i\in I}\varepsilon_i \geq 0. \tag{4.4}$$

$$\square$$

Theorem 2'. *For a partition $\mathscr{P}$ of $\mathscr{C}'^n$*

$$|\mathscr{P}| \geq \prod_{i=1}^{n}\left[\frac{S_i}{2}\right]. \tag{4.5}$$

Proof. Summing the expressions on the left-hand side in (4.3) for $m = 1, 2, \ldots, n$ results in

$$
\begin{aligned}
0 &\leq \sum_{m=1}^{n} \binom{n}{m} \prod_{i=1}^{n} S_i + \sum_{m=1}^{n} \sum_{k=1}^{m} \binom{n-k}{m-k} (-1)^k 2^k Q_k - \sum_{m=1}^{n} \sum_{I : |I|=m} \prod_{i \in I} \varepsilon_i \prod_{j \in I^c} S_j \\
&= (2^n - 1) \prod_{i=1}^{n} S_i + \sum_{k=1}^{n} (-1)^k 2^k Q_k \sum_{m=k}^{n} \binom{n-k}{m-k} - \sum_{\phi \neq I} \prod_{i \in I} \varepsilon_i \prod_{j \in I^c} S_i \\
&= 2^n \left[\prod_{i=1}^{n} S_i + \sum_{k=1}^{n} (-1)^k Q_k \right] - \sum_{I} \prod_{i \in I} \varepsilon_i \prod_{j \in I^c} S_j
\end{aligned}
$$

or

$$
|\mathcal{P}| \geq 2^{-n} \sum_{I} \prod_{i \in I} \varepsilon_i \prod_{j \in I^c} S_j. \tag{4.6}
$$

We evaluate the expression on the right-hand side by introducing $J = \{ \ell : 1 \leq \ell \leq n, \; \varepsilon_\ell = 1 \}$ and $I^* = J \setminus I$. Then

$$
\sum_{I} \prod_{i \in I} \varepsilon_i \prod_{j \in I^c} S_j = \sum_{I \subset J} \prod_{j \in I^*} S_j \cdot \prod_{j \in J^c} S_j
$$

$$
= \prod_{j \in J} (S_j + 1) \cdot \prod_{j \in J^c} S_j = \prod_{j=1}^{n} (S_j + \varepsilon_j) \quad \text{and (4.5) follows.} \qquad \square
$$

Corollary 1. *The partition number $\pi(\mathcal{C}'^n)$ equals $\prod_{j=1}^{n} \left\lceil \frac{S_j}{2} \right\rceil$.*

Proof. The partition number of $(\mathcal{V}_j, \mathcal{E}_j)$ is $\left\lceil \frac{S_j}{2} \right\rceil$. Take a product of optimal partitions for the factors. This construction gives the lower bound in Theorem 2'. $\qquad \square$

5. Products of general graphs

We assume now that the factors $\mathcal{G}_t = (\mathcal{V}_t, \mathcal{E}_t)$ $(t = 1, 2, \ldots, n)$ are arbitrary finite graphs with all loops included.

Obviously, we have for the partition number

$$
\pi(\mathcal{G}_t) = |\mathcal{V}_t| - v(\mathcal{G}_t), \tag{5.1}
$$

where $v(\mathcal{G}_t)$ is the matching number of $\mathcal{G}_t$.

Theorem 3. *For the hypergraph product $\mathcal{H}^n = \mathcal{G}_1 \times \ldots \times \mathcal{G}_n$*

$$
\pi(\mathcal{H}^n) = \prod_{t=1}^{n} \pi(\mathcal{G}_t). \tag{5.2}
$$

Here only the inequality

$$\pi(\mathcal{H}^n) \geq \prod_{t=1}^{n} \pi(\mathcal{G}_t) \tag{5.3}$$

is non-trivial. We make use of a well-known result from matching theory.

Gallai's Lemma. *([6] or [8] page 89) If a graph $\mathcal{G} = (\mathcal{V}, \mathcal{E})$ is connected and, for all $v \in \mathcal{V}$, $v(\mathcal{G} - v) = v(\mathcal{G})$, then $\mathcal{G}$ is factor-critical, that is, for all $v \in \mathcal{V}$, $\mathcal{G} - v$ has a perfect matching.*

Proof of 5.3. For every $t \in \{1, 2, \ldots, n\}$ we modify $\mathcal{G}_t$ as follows: remove any vertex $v \in \mathcal{V}_t$ with $v(\mathcal{G}_t - v) < v(\mathcal{G}_t)$ and reiterate this until a graph $\mathcal{G}_t^*$ with $v(\mathcal{G}_t^* - v) = v(\mathcal{G}_t^*)$ for all $v \in \mathcal{V}_t^*$ is obtained.

Notice that (5.1) ensures that

$$\pi(\mathcal{G}_t) = \pi(\mathcal{G}_t^*). \tag{5.4}$$

Denote the set of connected components of $\mathcal{G}_t^*$ by $\{\mathcal{G}_t^{*(j)}\}_{j \in J_t}$. Clearly,

$$\pi(\mathcal{G}_t^*) = \sum_{j \in J_t} \pi(\mathcal{G}_t^{*(j)}). \tag{5.5}$$

Moreover, by Gallai's Lemma each component $\mathcal{G}_t^{*(j)}$ has a vertex set $\mathcal{V}_t^{*(j)}$ of odd size and

$$v(\mathcal{G}_t^{*(j)}) = (|\mathcal{V}_t^{*(j)}| - 1)2^{-1} \triangleq \alpha_t^j, \quad \text{say.}$$

Thus,

$$\pi(\mathcal{G}_t^*) = \sum_j (\alpha_t^j + 1). \tag{5.6}$$

Now, for $\mathcal{H}^{*n} = \prod_1^n \mathcal{G}_t^*$ we have

$$\pi(\mathcal{H}^n) \geq \pi(\mathcal{H}^{*n}), \tag{5.7}$$

because the modifications described above transform a partition of $\mathcal{H}^n$ into a partition of $\mathcal{H}^{*n}$ with no more parts.

Finally, by Theorem 2′, we have for the product $\mathcal{C}^n$ of complete graphs with vertex sets $\mathcal{V}_t^{*(j)}$ that

$$\pi(\mathcal{G}_1^{*(j_1)} \times \ldots \times \mathcal{G}_n^{*(j_n)}) \geq \pi(\mathcal{C}^n) = (\alpha_1^{j_1} + 1) \ldots (\alpha_n^{j_n} + 1). \tag{5.8}$$

Therefore,

$$\pi(\mathcal{H}^{*n}) = \sum_{j_1 \in J_1, \ldots, j_n \in J_n} \pi(\mathcal{G}_1^{*(j_1)} \times \ldots \times \mathcal{G}_n^{*(j_n)})$$

$$\geq \sum_{(j_1, \ldots, j_n)} (\alpha_1^{j_1} + 1) \ldots (\alpha_n^{j_n} + 1)$$

$$= \prod_{t=1}^{n} \sum_{j \in J_t} (\alpha_t^j + 1)$$

$$= \prod_{t=1}^{n} \pi(\mathscr{G}_t^*) = \prod_{t=1}^{n} \pi(\mathscr{G}_t).$$

This and (5.7) imply (5.3).   □

6. Examples for deviation from multiplicative behaviour

First we give two examples of product hypergraphs $\mathscr{H} \times \mathscr{H}'$ for which the partition number π is not multiplicative in the factors. They are due to K.-U. Koschnick.
Example 1.

$$\mathscr{V}_1 = \{0, 1, 2, \ldots, 6\}, \mathscr{E}_1 = \{E \subseteq \mathscr{V}_1 : |E| \in \{1, 4\}\}.$$

Clearly, $\pi(\mathscr{H}_1) = 4$ and the partition

$$\{\{i\} \times \{0, 1, 2, 3\} : i = 0, 1, 2\} \quad \cup \quad \{\{i\} \times \{3, 4, 5, 6\} : i = 4, 5, 6\}$$
$$\cup \quad \{\{0, 1, 2, 3\} \times \{j\} : j = 4, 5, 6\}$$
$$\cup \quad \{\{3, 4, 5, 6\} \times \{j\} : j = \{0, 1, 2\}\}$$
$$\cup \quad \{\{3\} \times \{3\}\}$$

has 13 members. Therefore

$$\pi(\mathscr{H}_1 \times \mathscr{H}_1) \leq 13 < \pi(\mathscr{H}_1)\pi(\mathscr{H}_1) = 16. \tag{6.1}$$

While this example seems to be the smallest possible for identical factors, one can do better with non-identical factors:

$$\mathscr{H}_1 \times \mathscr{H}'_1, \text{ where } \mathscr{V}'_1 = \{0, 1, 2, 3, 4\} \text{ and } \mathscr{E}'_1 = \{E \subset \mathscr{V}'_1 : |E| \in \{1, 3\}\}.$$

Here, by a similar construction, $\pi(\mathscr{H}_1 \times \mathscr{H}'_1) \leq 11$, whereas $\pi(\mathscr{H}_1) \cdot \pi(\mathscr{H}'_1) = 4 \cdot 3 = 12$.
Example 2. Since π is multiplicative for graphs, one may wonder whether it is multiplicative if one factor is a graph.

Consider $G = (\mathscr{V}, \mathscr{E})$ with $\mathscr{V} = \{0, 1, \ldots, 4\}$ and $\mathscr{E} = \{\{i, i+1 \bmod 5\} : i = 0, 1, \ldots, 4\} \cup \{i : 0 \leq i \leq 4\}$, that is, the pentagon with all loops.

Define $\mathscr{H}' = (\mathscr{V}', \mathscr{E}')$ with $\mathscr{V}' = \{1, 2, \ldots, 14\}$ and $\mathscr{E}' = \{E \subset \mathscr{V}' : |E| \in \{1, 9\}\}$.

Notice that $\pi(G) = 3$, $\pi(\mathscr{H}') = 7$, and that the following construction ensures $\pi(G \times \mathscr{H}') \leq 20 < 21 = \pi(G) \cdot \pi(\mathscr{H}')$:

$$\{\{i\} \times \{j + k \bmod 14 : 0 \leq k \leq 8\} : (i, j) \in \{(0, 0), (1, 3), (2, 6), (3, 9), (4, 12)\}\}$$
$$\cup \quad \{\{1, 2\} \times \{j\} : j = 0, 1, 2\} \cup \{\{2, 3\} \times \{j\} : j = 2, 3, 5\}$$
$$\cup \quad \{\{3, 4\} \times \{j\} : j = 6, 7, 8\}$$
$$\cup \quad \{\{4, 0\} \times \{j\} : j = 9, 10, 11\}$$
$$\cup \quad \{\{0, 1\} \times \{j\} : j = 12, 13, 14\}$$

is a set of $5 + 5 \cdot 3 = 20$ edges partitioning $\mathscr{V} \times \mathscr{V}'$.

To help orient the reader, we add three examples, which demonstrate that the covering number c, the packing number p and the maximal partition number μ are not multiplicative in the factors either.

Example 3. $\mathcal{V}_3 = \{0, 1, 2\}$, $\mathcal{E}_3 = \{E \subseteq \mathcal{V} : |E| = 2\}$

We have

$$3 = c(\mathcal{H}_3 \times \mathcal{H}_3) \neq c(\mathcal{H}_3) \cdot c(\mathcal{H}_3) = 4, \tag{6.2}$$

because $\mathcal{C}\{\{0, 1\} \times \{0, 1\}, \{0, 2\} \times \{0, 2\}, \{1, 2\} \times \{1, 2\}\}$ covers $\mathcal{V}_3 \times \mathcal{V}_3$ and there is no covering with 2 edges.

This is the smallest example in terms of the number of vertices.

Remark 2. Quite generally, even in the case of non-identical factors $\mathcal{H}_t = (\mathcal{V}_t, \mathcal{E}_t)$, $t \in \mathbb{N}$, with $\max_t |\mathcal{E}_t| < \infty$, the asymptotic behaviour of $c(n)$ is known [1]:

$$\lim_{n \to \infty} \frac{1}{n} \left(\log c(n) - \sum_{t=1}^{n} \log \left(\max_{q \in \text{Prob}(\mathcal{E}_t)} \min_{r \in \mathcal{E}_t} \sum_{E \in \mathcal{E}_t} 1_E(v) q_E \right)^{-1} \right) = 0,$$

where $\text{Prob}(\mathcal{E}_t)$ is the set of all probability distributions on $\mathcal{E}$, q_E is the probability of E under q and 1_E is the indicator function of the set E.

Example 4. $\mathcal{V}_4 = \{0, 1, 2, 3, 4\}$, $\mathcal{E}_4 = \{\{x, x + 1 \bmod 5\} : x \in \mathcal{V}_4\}$.

Here we have

$$5 = p(\mathcal{H}_4 \times \mathcal{H}_4) \neq p(\mathcal{H}_4) p(\mathcal{H}_4) = 4. \tag{6.3}$$

It was shown in [11] that this is the smallest example in the previous sense. Notice that it is bigger than the previous one.

Example 5. To avoid heavy notation, we will write $\mathcal{H}_5 = (\mathcal{V}_5, \mathcal{E}_5)$ without an index as $\mathcal{H} = (\mathcal{V}, \mathcal{E})$. It is made up of the 5 vertex sets

$$\mathcal{W}_i = \{x_{ij} : j = 1, 2, \ldots, m\}, 3 \leq m(i = 0, 1, 2, \ldots, 4),$$

the 6 edge sets

$$\mathcal{G}_i = \{(x_{ij}, x_{i+1 \bmod 5, j}) : j = 1, 2, \ldots, m\} (i = 0, 1, 2, \ldots, 4),$$

and $\{\mathcal{W}_0, \ldots, \mathcal{W}_4\}$. Thus

$$\mathcal{V} = \bigcup_{i=0}^{4} \mathcal{W}_i, \mathcal{E} = \{\mathcal{W}_0, \ldots, \mathcal{W}_4\} \cup \left(\bigcup_{i=0}^{4} \mathcal{G}_i \right).$$

A look at the pentagon with vertex set $\{x_{01}, x_{11}, x_{21}, x_{31}, x_{41}\}$ shows that a partition of $\mathcal{H}$ must contain at least one of the edges $\mathcal{W}_i$ as a member. On the other hand, the vertices $\mathcal{V} \setminus \mathcal{W}_i$ have a maximal partition of size $2m$. Therefore we have shown that $\mu(\mathcal{H}) = 2m + 1$. We shall next consider $\mu(\mathcal{H} \times \mathcal{H})$. For this we introduce the superedges

$$\mathcal{G}_i^* = \mathcal{W}_i \cup \mathcal{W}_{i+1 \bmod 5} (i = 0, 1, \ldots, 4)$$

in $\mathcal{H}$, and the superedges $\mathcal{G}_i^* \times \mathcal{G}_{i'}^* (i, i' = 0, 1, \ldots, 4)$ in $\mathcal{H} \times \mathcal{H}$. Whereas $\mathcal{G}_i^*$ can be partitioned into m edges, they can be partitioned into m^2 edges.

First we divide $\mathcal{V} \times \mathcal{V}$ into 25 parts $\{\mathcal{W}_i \times \mathcal{W}_{i'} : i, i' = 0, 1, \ldots, 4\}$. Then we pack 5 superedges (as in Shannon's construction) into $\mathcal{V} \times \mathcal{V}$. They cover 20 parts, and the remaining 5 parts are packed with 5 edges of type $\mathcal{W}_i \times \mathcal{W}_{i'}$. Finally, we partition the 5 superedges into the edges of $\mathcal{H} \times \mathcal{H}$. Thus we obtain a desired partition with $5 + 5m^2$ edges. Notice that $\mu(\mathcal{H} \times \mathcal{H}) \geq 5 + 5m^2 > (2m + 1)^2 = \mu(\mathcal{H})^2$ for $m \geq 3$. The smallest example in this class has 15 vertices.

Remark 3. The construction was based on the pentagon. Its vertices were replaced by sets of vertices $\mathcal{W}_i$ with a numbering. The vertices with the same number in the $\mathcal{W}_i$'s form a pentagon. Thus we obtained $m = |\mathcal{W}_i|$ many pentagons. Then we added the $\mathcal{W}_i'$ as further edges. Finally we used the superedges to mimic the original small edges. We can make this construction starting with any hypergraph $\mathcal{H} = (\mathcal{V}, \mathcal{E})$. If it has the property $p(\mathcal{H})^2 < p(\mathcal{H} \times \mathcal{H})$, then for m large enough our construction gives an associated hypergraph for which μ is not multiplicative.

7. Acknowledgement

The authors are very much indebted to Klaus-Uwe Koschnick for constructing beautiful examples.

References

[1] Ahlswede, R. (preprint) On set coverings in Cartesian product spaces, Manuscript 1971. Reprinted in *SFB 343 Diskrete Strukturen in der Mathematik*, Preprint 92–034.

[2] Ahlswede, R. (1979) Coloring hypergraphs: A new approach to multi-user source coding, Pt I. *Journ. of Combinatorics, Information and System Sciences* **4** (1) 76–115.

[3] Ahlswede, R. (1980) Coloring hypergraphs: A new approach to multi-user source coding, Pt II. *Journ. of Combinatorics, Information and System Sciences* **5** (3) 220–268.

[4] Ahlswede, R. and Cai, N. (preprint) On POS partition and hypergraph products. *SFB 343 Diskrete Strukturen in der Mathematik*, Preprint 93–008.

[5] Ahlswede, R., Cai, N. and Zhang, Z. (1989) A general 4-words inequality with consequences for 2-way communication complexity. *Advances in Applied Mathematics* **10** 75–94.

[6] Gallai, T. (1963) Neuer Beweis eines Tutte'schen Satzes. *Magyar Tud. Akad. Mat. Kutato Int. Közl.* **8** 135–139.

[7] Lovasz, L. (1979) On the Shannon capacity of a graph. *IEEE Trans. Inform. Theory* **IT-25** 1–7.

[8] Lovász, L. and Plummer, M. D. (1986) *Matching Theory*, North-Holland Mathematics studies 121, North-Holland.

[9] Mehlhorn, K. and Schmidt, E. M. (1982) Las Vegas is better than determinism in VLSI and distributed computing. *Proceedings 14th ACM STOC* 330–337.

[10] Posner, E. C. and Mc Eliece, R. J. (1971) Hide and seek, data storage and entropy. *Annals of Math. Statistics* **42** 1706–1716.

[11] Shannon, C. E. (1956) The zero-error capacity of a noisy channel. *IEEE Trans. Inform. Theory* **IT-2** 8–19.

[12] Yao, A. (1979) Some complexity questions related to distributive computing. *Proceedings 11th ACM STOC* 209–213.

Matchings in Lattice Graphs and Hamming Graphs

M. AIGNER and R. KLIMMEK

II. Mathematisches Institut, Freie Universität Berlin, Arnimallee 3, D-14195 Berlin
e-mail: aigner@math.fu-berlin.de

In this paper we solve the following problem on the lattice graph $L(m_1,\ldots,m_n)$ and the Hamming graph $H(m_1,\ldots,m_n)$, generalizing a result of Felzenbaum–Holzman–Kleitman on the n-dimensional cube (all $m_i = 2$): Characterize the vectors $(s_1,\ldots,s_n)$ such that there exists a maximum matching in L respectively H with exactly s_i edges in the i-th direction.

1. Introduction

One of the classical enumeration results concerns the number of maximum matchings (1-factors) in the $2n \times 2n$-lattice graph (see e.g. Lovász [2] or Montroll [3]). The answer for higher dimensional lattices is unknown; in fact, no good estimates are known. In this paper we consider and solve a closely related problem concerning the *structure* of maximum matchings.

Consider the simplest case, that of the hypercube Q_n. The vertices are all n-tuples $(u_1,\ldots,u_n)$ with $u_i \in \{0,1\}$. An edge e joins two vertices if they differ in exactly one coordinate. If this coordinate is in the i-th dimension, then we call e an i-edge. Let M be a maximum matching. We associate to M its *type* $(s_1,\ldots,s_n)$, where s_i denotes the number of i-edges in M. Which sequences are types? This question was answered by Felzenbaum-Holzman-Kleitman [1]. They proved that a sequence $(s_1,\ldots,s_n)$ is the type of a maximum matching in Q_n, $n \geqslant 2$, if and only if

(i) $\sum_{i=1}^n s_i = 2^{n-1}$

(ii)   all s_i are even.

The purpose of this paper is to generalize this result to arbitrary lattice graphs and Hamming graphs.

The vertex set $V(L)$ of the *lattice graph* $L(m_1,\ldots,m_n)$ is the set of all n-tuples $(u_1,\ldots,u_n)$ with $0 \leqslant u_i \leqslant m_i - 1$. (Thus $|V(L)| = \prod_{i=1}^n m_i$.) Two vertices $(u_1,\ldots,u_n)$ and $(v_1,\ldots,v_n)$ are joined by an edge if and only if they differ in exactly one coordinate, say i, and $|u_i - v_i| = 1$. The *Hamming graph* $H(m_1,\ldots,m_n)$ has the same vertex-set as $L(m_1,\ldots,m_n)$, and two vertices are joined if they differ in exactly one coordinate.

Thus, keeping all but coordinate i fixed, $L(m_1, \ldots, m_n)$ induces along dimension i the path P_{m_i}, and $H(m_1, \ldots, m_n)$ induces the complete graph K_{m_i}. Clearly, $L(m_1, \ldots, m_n)$ is a subgraph of $H(m_1, \ldots, m_n)$. Again, we speak of the type $(s_1, \ldots, s_n)$ of a maximum matching in $L(m_1, \ldots, m_n)$ or $H(m_1, \ldots, m_n)$.

2. The results

For the rest of the paper we shall assume that $m_1, \ldots, m_n \geqslant 2$ and $n \geqslant 2$. In the next section we shall see that the hypercube result can be generalized in a straightforward way to the case when all m_i are even. But when some of the m_i's are odd, then the situation is different. First we note that in the case when all m_i are odd, a maximum matching M leaves one vertex uncovered; hence in general, we have $|M| = \lfloor \frac{1}{2} \prod_{i=1}^{n} m_i \rfloor$.

Lemma 1. *Suppose there is a maximum matching M of type $(s_1, \ldots, s_n)$ in $L(m_1, \ldots, m_n)$ or $H(m_1, \ldots, m_n)$. Let $A \subseteq \{1, \ldots, n\}$ and suppose m_i is odd for all $i \notin A$. Then*

$$\sum_{i \in A} s_i \geqslant \lfloor \tfrac{1}{2} \prod_{i \in A} m_i \rfloor. \tag{1}$$

Proof. The assertion is trivially true for $A = \emptyset$ since $\sum_\emptyset = 0$ and $\prod_\emptyset = 1$. Let $\emptyset \neq A \subseteq \{1, \ldots, n\}$, and consider $L(m_1, \ldots, m_n)$. We decompose $L(m_1, \ldots, m_n)$ along the dimensions in A into $\prod_{i \in A} m_i$ partial lattices $L' = L(m_i : i \notin A)$. Every such L' has by the hypothesis an odd number of vertices. Hence in every L' there exists at least one vertex which is not covered by an M-edge within L'. The total number of these vertices is therefore at least $\prod_{i \in A} m_i$, and (1) follows. The same proof holds verbatim for $H(m_1, \ldots, m_n)$. □

Definition. A sequence $(s_1, \ldots, s_n)$ is called $(m_1, \ldots, m_n)$-*admissible* if it satisfies (1) and $\sum_{i=1}^{n} s_i = \lfloor \frac{1}{2} \prod_{i=1}^{n} m_i \rfloor$.

When all m_i are even, then (1) is vacuously satisfied. Our main results state that apart from a parity condition (like (ii) for hypercubes) admissibility is all we need to characterize the types.

Theorem 1. *A sequence $(s_1, \ldots, s_n)$ is the type of a maximum matching in $L(m_1, \ldots, m_n)$ if and only if*

(A) $(s_1, \ldots, s_n)$ is $(m_1, \ldots, m_n)$-admissible, and

(B)   if not all m_k are odd, then

$$s_i \equiv \lfloor \frac{m_i}{2} \rfloor \prod_{j \neq i} m_j \pmod 2 \quad (i = 1, \ldots, n).$$

Theorem 2. *A sequence $(s_1, \ldots, s_n)$ is the type of a maximum matching in $H(m_1, \ldots, m_n)$ if and only if*

(C) $(s_1, \ldots s_n)$ is $(m_1, \ldots, m_n)$-admissible.

(D)   if $m_i = 2$, then $s_i \equiv \prod_{j \neq i} m_j \pmod 2$, and

(E)   if $\prod_{j \neq i} m_j$ is even, then $s_i \neq 1$.

Remark 1. Let us prove the necessity of the conditions in Theorem 1. We have seen (A) in Lemma 1. Suppose $\prod_{j \neq i} m_j$ is even. Decompose $L(m_1, \ldots, m_n)$ along dimension i into the $(n-1)$-dimensional lattices $T_0, \ldots, T_{m_i-1}$, where the i-edges connect T_j and T_{j+1} ($j = 0, \ldots, m_i - 2$). Since $|V(T_k)| = \prod_{j \neq i} m_j$ is even, there is an even number of matching edges between T_0 and T_1, hence an even number between T_1 and T_2, and so on, and we conclude $s_i \equiv 0 \pmod 2$. If $\prod_{j \neq i} m_j$ is odd, then we may assume m_i even, and we conclude by an analogous argument $s_i \equiv \frac{m_i}{2} \pmod 2$, i.e. condition (B).

Remark 2. Let us prove the necessity of the conditions in Theorem 2. Again, (C) holds by Lemma 1. If $m_i = 2$, then decomposing $H(m_1, \ldots, m_n)$ along dimension i yields $s_i \equiv \prod_{j \neq i} m_j \pmod 2$. Condition (E) is obvious.

Remark 3. If all m_i are odd, then only conditions (A) and (C) apply, and the characterizations are in this case the same.

Definition. A sequence $(s_1, \ldots, s_n)$ satisfying the conditions of Theorem 1 or of Theorem 2 is called an *L-sequence*, respectively, an *H-sequence* for $(m_1, \ldots, m_n)$.

The hard part of the proof is then to show that a sequence is the type of a maximum matching in $L(m_1, \ldots, m_n)$ or of $H(m_1, \ldots, m_n)$ provided it is an *L*-sequence or, respectively, an *M*-sequence. Note that every *L*-sequence is an *H*-sequence, as it must be since $L(m_1, \ldots, m_n)$ is a subgraph of $H(m_1, \ldots, m_n)$.

3. All m_i are even

We consider first the easy case when all m_i are even. Condition (B) in Theorem 1 and (D) in Theorem 2 reduce to s_i even, and condition (E) in Theorem 2 to $s_i \neq 1$, for all i.

Proposition 1. *Suppose all m_i are even. The sequence $(s_1, \ldots, s_n)$ is the type of a maximum matching in $L(m_1, \ldots, m_n)$ if and only if*

(A') $\sum_{i=1}^n s_i = \frac{1}{2} \prod_{i=1}^n m_i$,
(B') *all s_i are even.*

Proof. We have already seen the necessity. We can decompose the vertex-set of $L(m_1, \ldots, m_n)$ into $m = \frac{m_1}{2} \cdot \ldots \cdot \frac{m_n}{2}$ hypercubes $Q_n^{(j)}$ in an obvious way. Since all s_i are even, we can clearly split $(s_1, \ldots, s_n)$ into m sequences $(t_1^{(j)}, \ldots, t_n^{(j)})$ with $t_i^{(j)}$ even for all i and j, and $\sum_j t_i^{(j)} = s_i$ $(i = 1, \ldots, n)$, $\sum_i t_i^{(j)} = 2^{n-1}$ $(j = 1, \ldots, m)$. By the result of Felzenbaum, Holzman and Kleitman, $(t_1^{(j)}, \ldots, t_n^{(j)})$ can be realized by a matching M_j within $Q_n^{(j)}$ for all j, and we obtain the desired matching in $L(m_1, \ldots, m_n)$ by putting the m matchings M_j together. $\qquad \square$

Proposition 2. *Suppose all m_i are even. A sequence $(s_i, \ldots, s_n)$ is the type of a maximum matching in $H(m_1, \ldots, m_n)$ if and only if*

(C') $\sum_{i=1}^n s_i = \frac{1}{2} \prod_{i=1}^n m_i$,

(D') *if $m_i = 2$, then s_i is even,*

(E') $s_i \neq 1$ *for all i.*

Proof. Again it suffices to prove the sufficiency. We use induction on the number k of indices $i \in \{1, \ldots, n\}$ with $m_i \geqslant 4$. The case $k = 0$ is that of the hypercube, so let $k \geqslant 1$. We may suppose that $m_n \geqslant 4$.

Case 1. s_n is even. We split $(s_1, \ldots, s_n)$ into two sequences $(t_1, \ldots, t_n)$ and $(r_1, \ldots, r_n)$ as follows. Set $t_i = 3$ for s_i odd (i.e. $s_i \geqslant 3$) and $t_i = 0$ for s_i even. Now raise each t_i by even numbers such that $t_i \leqslant s_i$ ($i \leqslant n - 1$), $t_n \leqslant 2$ and $\sum_{i=1}^{n} t_i = \frac{1}{2}(\prod_{i=1}^{n-1} m_i \cdot 2)$. This is possible since by (D') and s_n even

$$3|\{i : s_i \text{ odd}\}| \leqslant 3(k - 1) \leqslant 4^{k-1} \leqslant \prod_{i=1}^{n-1} m_i.$$

The sequence (t_i) satisfies (C'), (D'), (E') for $(m_1, \ldots, m_{n-1}, 2)$ and hence, by the induction hypothesis, is the type of a maximum matching in $H(m_1, \ldots, m_{n-1}, 2)$. The sequence (r_i) with $r_i = s_i - t_i$ satisfies $\sum_{i=1}^{n} r_i = \frac{1}{2} \left(\prod_{i=1}^{n-1} m_i \right) (m_n - 2)$ with all r_i even, and hence is the type of a maximum matching in $L(m_1, \ldots, m_{n-1}, m_n - 2)$ and thus in $H(m_1, \ldots, m_{n-1}, m_n - 2)$ by Proposition 1. Now put the matchings together.

Case 2. s_n is odd. Since $\sum_{i=1}^{n} s_i \equiv 0 \pmod{2}$, there must be an index $j < n$ with s_j odd, so $s_j \geqslant 3$ by (E') and $m_j \geqslant 4$ (by (D')). Consider now the sequence $(s_i') = (s_1, \ldots, s_j + 1, \ldots, s_n - 1)$ and split it as in case 1 into sequences (t_i), (r_i). Since $s_n - 1 \geqslant 2$, $s_j + 1 \geqslant 4$ and both are even, we may assume $t_j > 0$ and $r_j > 0$, $r_n > 0$. As in case 1, (t_i) is the type of a maximum matching M_t in $H_t = H(m_1, \ldots, m_{n-1}, 2)$ and (r_i) is the type of a maximum matching M_r in $L_r = L(m_1, \ldots, m_{n-1}, m_n - 2)$. By Proposition 1, M_r can be obtained from hypercube matchings. Because of $m_j \geqslant 4$ there are two neighboring hypercubes Q^1, Q^2 along dimension j, and since $r_j > 0$ and $r_n > 0$ we may assume that Q^1 contains n-edges in M_r and Q^2 j-edges in M_r.

Choose $u \in V(Q^1)$ and $x \in V(Q^2)$ such that the edge ux is in L_r. By the edge-transitivity of hypercubes we may assume that the M_r-edge uv covering u in Q^1 is an n-edge, and similarly that the M_r-edge xy in Q^2 is a j-edge. The matching M_t in H_t contains a j-edge wz since $t_j > 0$, where by the edge-transitivity in H_t and $m_n \geqslant 4$ we may assume that w and z differ from u and, respectively, x just in the n-th coordinate (see Figure 1).

Now replace the matching edges uv, xy, wz by the edges wv, uy, zx of $H(m_1, \ldots, m_n)$ as in Figure 1. This raises the number of the n-edges in the matching by 1 and decreases the number of the j-edges by 1, whence we obtain our desired matching of type $(s_1, \ldots, s_n)$. $\square$

4. Admissible sequences

The main difficulty in the proof of Theorems 1 and 2 is to check admissibility. By definition, this requires checking (1) for subsets $A \subseteq \{1, \ldots, n\}$. The following result shows how this check can be restricted to a linear number of subsets.

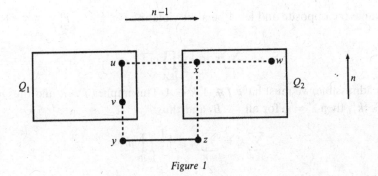

Figure 1

Lemma 2. *Let* $m_1, \ldots, m_r$ *be even,* $m_{r+1}, \ldots, m_n$ *odd. A sequence* $(s_1, \ldots, s_n)$ *with* $\sum_{i=1}^{n} s_i = \lfloor \frac{1}{2} \prod_{i=1}^{n} m_i \rfloor$ *and* $s_{r+1} \leqslant \ldots \leqslant s_n$ *is* $(m_1, \ldots, m_n)$*-admissible if and only if*

$$\sum_{i=1}^{k} s_i \geqslant \lfloor \frac{1}{2} \prod_{i=1}^{k} m_i \rfloor \quad \text{for all} \quad k = r, \ldots, n-1. \tag{2}$$

Proof. The necessity was established in Lemma 1. Suppose (2) is satisfied, and (s_i) is not admissible. Then there exists A with $\{1, \ldots, r\} \subseteq A \subseteq \{1, \ldots, n-1\}$ such that

$$\sum_{i \in A} s_i < \lfloor \frac{1}{2} \prod_{i \in A} m_i \rfloor. \tag{3}$$

Among all such sets A choose one of maximal size. By (2), there are indices j, k with $k \in A$, $j \in \{r+1, \ldots, n\} \setminus A$ and $j < k$. From the maximality of $|A|$ we infer

$$\sum_{i \in A} s_i + s_j \geqslant \lfloor \frac{m_j}{2} \prod_{i \in A} m_i \rfloor,$$

and thus by (3) and $m_j \geqslant 3$ odd

$$s_j > \lfloor \frac{m_j}{2} \prod_{i \in A} m_i \rfloor - \lfloor \frac{1}{2} \prod_{i \in A} m_i \rfloor = \frac{m_j - 1}{2} \prod_{i \in A} m_i \geqslant \prod_{i \in A} m_i.$$

This, however, implies

$$\sum_{i \in A} s_i \geqslant s_k \geqslant s_j > \prod_{i \in A} m_i,$$

a contradiction to (3). $\square$

To check the parity conditions (D) and (E) we will make frequent use of the following lemmas.

Lemma 3. *Let* $m_1, \ldots, m_r$ *be even,* $m_{r+1}, \ldots, m_n$ *odd, and suppose* $(s_1, \ldots, s_n)$ *is* $(m_1, \ldots, m_n)$*-admissible with* $s_{r+1} \leqslant \ldots \leqslant s_n$. *Let* $d \in \{1, 2\}$, *and* $j < k \leqslant n$. *Then* $(s_i') = (s_1, \ldots, s_j + d, \ldots, s_k - d, \ldots, s_n)$ *is* $(m_1, \ldots, m_n)$*-admissible whenever* $s_k \geqslant d$, $r \geqslant 1$ *or* $s_k > d$, $r = 0$.

Proof. Assume the opposite and let A be a set violating (1), $\{1,\ldots,r\} \subseteq A \subseteq \{1,\ldots,n-1\}$, i.e.

$$\sum_{i \in A} s_i' < \lfloor \tfrac{1}{2} \prod_{i \in A} m_i \rfloor. \tag{4}$$

Since (s_i) is admissible, we must have $j \notin A$, $k \in A$. This implies $j > r$, and thus $m_j, m_k \geqslant 3$. Set $B = A \setminus \{k\}$, then $s_i' = s_i$ for all $i \in B$, and thus

$$\sum_{i \in B} s_i' = \sum_{i \in B} s_i \geqslant \lfloor \tfrac{1}{2} \prod_{i \in B} m_i \rfloor. \tag{5}$$

From (4) and (5) we infer

$$
\begin{aligned}
s_k = s_k' + d &= \sum_{i \in A} s_i' - \sum_{i \in B} s_i' + d \\
&< \lfloor \tfrac{1}{2} \prod_{i \in A} m_i \rfloor - \lfloor \tfrac{1}{2} \prod_{i \in B} m_i \rfloor + d \\
&= \frac{m_k - 1}{2} \prod_{i \in B} m_i + d.
\end{aligned}
$$

Since $s_j \leqslant s_k$, this implies by (4) and (5)

$$
\begin{aligned}
\sum_{i \in A} s_i + s_j &\leqslant \sum_{i \in A} s_i + s_k = \sum_{i \in A} s_i' + s_k + d \\
&< \lfloor \tfrac{1}{2} \prod_{i \in A} m_i \rfloor + \frac{m_k - 1}{2} \prod_{i \in B} m_i + 2d \\
&\leqslant \frac{2m_k - 1}{2} \prod_{i \in B} m_i + 2d
\end{aligned}
$$

hence

$$\sum_{i \in A} s_i + s_j < \left(m_k - \frac{1}{2} + \frac{2d}{2^{|B|}} \right) \prod_{i \in B} m_i. \tag{6}$$

If $|B| \geqslant 1$, then (6) yields with $m_j \geqslant 3$, $m_k \geqslant 3$

$$\sum_{i \in A} s_i + s_j < \left(m_k + \frac{3}{2} \right) \prod_{i \in B} m_i \leqslant \frac{3m_k}{2} \prod_{i \in B} m_i \leqslant \frac{m_j}{2} \prod_{i \in A} m_i, \tag{7}$$

and thus $\sum_{i \in A} s_i + s_j < \lfloor \frac{m_j}{2} \prod_{i \in A} m_i \rfloor$ since the difference in (7) is at least 2. So in this case we have a contradiction to the admissibility of (s_i).

Hence suppose $B = \emptyset$, i.e. $A = \{k\}$ and $r = 0$. Note that our assertion is therefore proved for $r \geqslant 1$. Inequality (4) now reads by our assumption $s_k > d$

$$0 < s_k - d \leqslant \frac{m_k - 1}{2} - 1,$$

which implies $m_k \geqslant 4$ and thus $m_k \geqslant 5$. It follows that

$$2s_k \leqslant m_k + 2d - 3.$$

On the other hand, from the admissibility of (s_i) we infer

$$\frac{1}{2}(m_j m_k - 1) \leqslant s_j + s_k \leqslant 2s_k \leqslant m_k + 2d - 3,$$

hence

$$(m_j - 2)m_k \leqslant 4d - 5 \leqslant 3,$$

in contradiction to $m_j \geqslant 3$, $m_k \geqslant 5$. □

Lemma 4. *Let* $m_1, \ldots, m_r$ *be even,* $m_{r+1}, \ldots, m_n$ *odd with* $r \geqslant 1$, *and suppose* $(s_1, \ldots, s_n)$ *is an* $(m_1, \ldots, m_n)$-*admissible sequence with* $s_{r+1} \leqslant \ldots \leqslant s_n$. *Let* $d \in \{1, 2\}$, *and assume*

$$\sum_{i=1}^{r} s_i \geqslant \frac{1}{2} \prod_{i=1}^{r} m_i + d. \tag{8}$$

Then $(s_i') = (s_1, \ldots, s_j - d, \ldots, s_r, s_{r+1} + d, \ldots, s_n)$ *is* $(m_1, \ldots, m_n)$-*admissible for every* $j \leqslant r$ *with* $s_j \geqslant d$.

Proof. If $s_{r+1} + d \leqslant s_{r+2}$, then $s_{r+1}' \leqslant \ldots \leqslant s_n'$ is preserved, and we are done by (8) and Lemma 2. Suppose (s_i') is not admissible. Then there exists $k \geqslant 2$ with

$$s_{r+1} \leqslant s_{r+2} \leqslant \ldots \leqslant s_{r+k} < s_{r+1} + d,$$

for which

$$\sum_{i=1}^{r} s_i + \sum_{i=r+2}^{r+k} s_i - d < \frac{1}{2} \prod_{i=1}^{r} m_i \prod_{i=r+2}^{r+k} m_i =: \alpha. \tag{9}$$

This implies

$$s_{r+1} = \sum_{i=1}^{r+k} s_i - \left(\sum_{i=1}^{r} s_i + \sum_{i=r+2}^{r+k} s_i \right) > \frac{1}{2} \prod_{i=1}^{r+k} m_i - \alpha - d$$

$$= \alpha \left(m_{r+1} - 1 - \frac{d}{\alpha} \right) \geqslant \alpha,$$

since $m_{r+1} \geqslant 3$, $\alpha \geqslant 3$, $d \leqslant 2$. But this implies

$$\sum_{i=r+2}^{r+k} s_i \geqslant (k-1)s_{r+1} \geqslant (k-1)\alpha \geqslant \alpha,$$

in contradiction to (9), since $\sum_{i=1}^{r} s_i \geqslant d$. □

5. The decomposition lemma

To prove Theorems 1 and 2 we apply induction on the number k of odd m_i's. The base case, that of $k = 0$, is supplied by Propositions 1 and 2. The major part of the induction step is provided by the following decomposition lemma.

Decomposition lemma. *Suppose $m_1, \ldots, m_r$ are even amd $m_{r+1}, \ldots, m_n$ are odd for some r, $0 \leqslant r < n$. Let $(s_1, \ldots, s_n)$ be an H-sequence for $(m_1, \ldots, m_n)$ with $s_n \geqslant \max(s_i : r+1 \leqslant i \leqslant n) - 2$. Furthermore,*

$$s_n \equiv \lfloor \frac{m_n}{2} \rfloor \prod_{i=1}^{n-1} m_i \pmod{2}. \tag{10}$$

Then we can split $(s_1, \ldots, s_n)$ into two sequences $(a_1, \ldots, a_n)$, $(b_1, \ldots, b_{n-1})$ such that

(I) $a_i + b_i = s_i$ *for* $1 \leqslant i \leqslant n-1$, *and* $a_n = s_n$,
(II) (a_i) *is an L-sequence for* $(m_1, \ldots, m_{n-1}, m_n - 1)$,
(III) $(b_1, \ldots b_{n-1})$ *is an H-sequence for* $(m_1, \ldots, m_{n-1})$, *and*
(IV) $b_{r+1} \leqslant \ldots \leqslant b_{n-1}$.

Remark 4. Note that the derived sequences $(m_1, \ldots, m_n - 1)$ and $(m_1, \ldots, m_{n-1})$ have one odd length less than $(m_1, \ldots, m_n)$.

Remark 5. As a corollary, the decomposition lemma also holds if we replace an H-sequence by an L-sequence in the hypothesis and the conclusion (III). Indeed, if all m_i are odd, then H-sequences and L-sequences are the same. If one of the m_i's is even (i.e. $r \geqslant 1$) and (s_i) is an L-sequence, then the a_i's of (II) are even by condition (B) of an L-sequence. Since m_n is odd, we conclude by condition (B) for (s_i)

$$b_i = s_i - a_i \equiv s_i \equiv \lfloor \frac{m_i}{2} \rfloor \prod_{j \neq i} m_j \equiv \lfloor \frac{m_i}{2} \rfloor \prod_{\substack{j=1 \\ j \neq i}}^{n-1} m_j \pmod{2}.$$

Proof of the decomposition lemma. We are given $m_1, \ldots, m_r$ even, $m_{r+1}, \ldots, m_n$ odd with $0 \leqslant r < n$, and an H-sequence $(s_1, \ldots, s_n)$ for $(m_1, \ldots, m_n)$ with $s_n \geqslant \max(s_{r+1}, \ldots, s_n) - 2$. We may assume that $s_{r+1} \leqslant \ldots \leqslant s_{n-1}$. The proof proceeds by induction on $\sum_{i=1}^{r} s_i$.

Claim 1. *The conclusion holds if*

$$\text{either} \quad r = 0 \quad (\textit{i.e. all } m_i \textit{ are odd})$$
$$\text{or} \quad r \geqslant 1 \quad \text{and} \quad \sum_{i=1}^{r} s_i = \tfrac{1}{2} \prod_{i=1}^{r} m_i.$$

Note that in the latter case $a_i = 0$, $b_i = s_i$ for $1 \leqslant i \leqslant r$ must hold because of $\sum_{i=1}^{r} b_i \geqslant \tfrac{1}{2} \prod_{i=1}^{r} m_i$.

Proof of Claim 1. We construct the H-sequence $(b_1, \ldots, b_{n-1})$ as follows. First we set up a preliminary H-sequence $(\beta_1, \ldots, \beta_{n-1})$ for $(m_1, \ldots, m_{n-1})$ which does not necessarily satisfy $\beta_i \leqslant s_i$ for all i, and then modify (β_i) suitably.

The β_i's are defined in the following way:

1a. If $r = 0$, then β_1 is minimal with $\beta_1 \equiv s_1 \pmod{2}$, $\beta_1 \geqslant \lfloor \frac{m_1}{2} \rfloor$.
1b. If $r \geqslant 1$, then $\beta_i = s_i$ for $i = 1, \ldots, r$.
2.   Suppose $\beta_1, \ldots, \beta_k$ are already defined, then β_{k+1} is minimal with $\beta_{k+1} \equiv s_{k+1} \pmod{2}$ and

$$\sum_{i=1}^{k+1} \beta_i \geqslant \lfloor \frac{1}{2} \prod_{i=1}^{k+1} m_i \rfloor. \tag{11}$$

It follows that

$$\beta_k = \lfloor \frac{1}{2} \prod_{i=1}^{k} m_i \rfloor - \sum_{i=1}^{k-1} \beta_i + c_k, \quad c_k \in \{0,1\}$$

for $k = r+1, \dots, n-1$, and thus

$$\sum_{i=1}^{k} \beta_i = \lfloor \frac{1}{2} \prod_{i=1}^{k} m_i \rfloor + c_k, \quad c_k \in \{0,1\}, \quad k = r+1, \dots, n-1.$$

$$\sum_{i=1}^{r} \beta_i = \sum_{i=1}^{r} s_i = \frac{1}{2} \prod_{i=1}^{r} m_i \quad \text{if } r \geq 1. \tag{12}$$

Note that because of $\beta_k \equiv s_k$ for all k, (12) implies $\sum_{i=1}^{k} \beta_i \leq \sum_{i=1}^{k} s_i$ for $k = r+1, \dots, n-1$.

Using (12), we have for $k > r$

$$\beta_k = \sum_{i=1}^{k} \beta_i - \sum_{i=1}^{k-1} \beta_i = \frac{m_k - 1}{2} \prod_{i=1}^{k-1} m_i + \delta_k, \quad \delta_k \in \{-1, 0, 1\}, \tag{13}$$

where $\delta_{r+1} \geq 0$ since $\sum_{i=1}^{r} \beta_i = \sum_{i=1}^{r} s_i = \lfloor \frac{1}{2} \prod_{i=1}^{r} m_i \rfloor$.

We want to show next that $(\beta_1, \dots, \beta_{n-1})$ is an H-sequence for $(m_1, \dots, m_{n-1})$ with $\beta_{r+1} \leq \dots \leq \beta_{n-1}$. Condition (D) holds because of $\beta_i \equiv s_i$ and $m_n \equiv 1 \pmod 2$. (E) is true by definition 1b and (13). By (12) and Lemma 2, it remains to show that

$$\sum_{i=1}^{n-1} \beta_{n-1} = \lfloor \frac{1}{2} \prod_{i=1}^{n-1} m_i \rfloor, \quad \text{i.e. } c_{n-1} = 0, \tag{14}$$

$$\beta_{r+1} \leq \dots \leq \beta_{n-1}. \tag{15}$$

To see (14), we note by (10)

$$\sum_{i=1}^{n-1} \beta_i \equiv \sum_{i=1}^{n-1} s_i = \sum_{i=1}^{n} s_i - s_n \equiv \lfloor \frac{1}{2} \prod_{i=1}^{n} m_i \rfloor - \lfloor \frac{m_n}{2} \rfloor \prod_{i=1}^{n-1} m_i$$

$$\equiv \lfloor \frac{1}{2} \prod_{i=1}^{n-1} m_i \rfloor \pmod 2,$$

from which $c_{n-1} = 0$ follows by (12).

To prove (15) we have for $k \geq r+2$ according to (13)

$$\beta_k - \beta_{k-1} = \frac{m_k - 1}{2} \prod_{i=1}^{k-1} m_i - \frac{m_{k-1} - 1}{2} \prod_{i=1}^{k-2} m_i + \delta_k - \delta_{k-1}$$

$$\geq \frac{1}{2} \left(\prod_{i=1}^{k-2} m_i \right) (m_k m_{k-1} - 2m_{k-1} + 1) - 2$$

$$\geq \frac{m_{k-1} + 1}{2} - 2 \geq 0$$

since $m_{k-1}, m_k \geq 3$.

From the H-sequence $(\beta_1, \ldots, \beta_{n-1})$ we now construct the H-sequence $(b_1, \ldots, b_{n-1})$ satisfying (III), (IV) and $b_i \leqslant s_i$ for all i. First we note $\beta_i \leqslant s_i$ for $i = 1, \ldots, r+1$. This is clear for $i \leqslant r$ by definition 1b, and for $i = r+1$ we have $\beta_{r+1} \leqslant s_{r+1}$ if $r = 0$ by definition 1a. For $r \geqslant 1$ we infer from the assumption on $\sum_{i=1}^{r} s_i$ and (13)

$$s_{r+1} \geqslant \frac{1}{2} \prod_{i=1}^{r+1} m_i - \sum_{i=1}^{r} s_i = \frac{m_{r+1} - 1}{2} \prod_{i=1}^{r} m_i \geqslant \beta_{r+1} - 1,$$

and hence $s_{r+1} \geqslant \beta_{r+1}$ because of $s_{r+1} \equiv \beta_{r+1} \pmod 2$.

The following algorithm generates the sequence $(b_1, \ldots, b_{n-1})$:

1. Set $b_i = \beta_i$ for $i = 1, \ldots, r+1$, and $b_i = 0$ for $i = r+2, \ldots, n-1$.
2. For $i = r+2, \ldots, n-1$ distribute β_i step by step on $b_i, b_{i-1}, \ldots, b_{r+1}$, by raising the current b_j ($i \geqslant j \geqslant r+1$) by the maximal possible increment. The i-th distribution step is described as follows:

   initialization $j \leftarrow i$, $\beta_i, s_i, b_i = 0$
   as long as $\beta_i > 0$ repeat
   $c = \min(\beta_i, s_j - b_j)$
   $\beta_i \leftarrow \beta_i - c$, $b_j \leftarrow b_j + c$
   $j \leftarrow j - 1$.

Since $\sum_{i=1}^{k} \beta_i \leqslant \sum_{i=1}^{k} s_i$ for all k, the β_i's are fully distributed, and the loop for the i-th step stops for $j = r+1$ since $b_r = \beta_r = s_r$. Furthermore $b_i \leqslant s_i$ for all i.

$$b_i \equiv s_i \pmod 2 \quad \text{for all } i. \tag{16}$$

This is trivially true for $i \leqslant r+1$. Consider what happens in the i-th step distributing β_i. At the start we have $b_i = 0$. If $\beta_i \leqslant s_i$ then $\beta_i \leftarrow 0$, $b_i \leftarrow \beta_i$ and thus $b_i = \beta_i \equiv s_i \pmod 2$. If $s_i < \beta_i$, then $\beta_i \leftarrow \beta_i - s_i \equiv 0 \pmod 2$ and $b_i \leftarrow s_i$. Any b_j ($j < i$) is thus raised by an even number and hence keeps the same parity as s_j. Note that after the i-th step, $b_i = \min(s_i, \beta_i)$.

$$\text{We have } b_{r+1} \leqslant \ldots \leqslant b_{n-1}. \tag{17}$$

Assume inductively $b_{r+1} \leqslant \ldots \leqslant b_{i-1}$ before the i-th step, $i \geqslant r+2$. Then at the start of the i-th step either $b_i = \beta_i$ or $b_i = s_i$. In the first cast we infer by (15)

$$b_i = \beta_i \geqslant \beta_{i-1} \geqslant \min(s_{i-1}, \beta_{i-1}) = b_{i-1} \geqslant \ldots \geqslant b_{r+1}.$$

In the second case, we have at the start

$$b_i = s_i \geqslant s_{i-1} \geqslant b_{i-1} \geqslant \ldots \geqslant b_{r+1}.$$

Now assume inductively that (17) holds before we reach b_j ($j < i$). If b_j is raised, then by the set-up of the algorithm $b_{j+1} = s_{j+1}$ and thus

$$s_i = b_i \geqslant \ldots \geqslant b_{j+1} = s_{j+1} \geqslant s_j \geqslant b_j \text{ (after)} \geqslant b_j \text{ (pre)} \geqslant \ldots \geqslant b_{r+1}.$$

$$(b_1, \ldots, b_{n-1}) \text{ is } (m_1, \ldots, m_{n-1})\text{-admissible}. \tag{18}$$

This follows immediately from $\sum_{j=1}^{r+1} b_j = \sum_{j=1}^{r+1} \beta_j$, and $\sum_{j=1}^{i} b_j = \sum_{j=1}^{i} \beta_j$ after step i and (12), (14), (17), and Lemma 2.

The conditions (D), (E) for $(b_1, \ldots, b_{n-1})$ are easy. (D) is clear since $b_i \equiv s_i \pmod 2$ and $m_n \equiv 1 \pmod 2$. As for (E) we have $r \geqslant 1$. If $i \leqslant r$, then $b_i = s_i$ and there is nothing to prove. For $i \geqslant r+1$ we see by (17) and (13)

$$b_i \geqslant b_{r+1} \geqslant \beta_{r+1} \geqslant \frac{m_{r+1}-1}{2} \prod_{i=1}^{r} m_i \geqslant 2.$$

Our proof that $(b_1, \ldots, b_{n-1})$ satisfies (III) and (IV) is thus complete.

Now we set $a_i = s_i - b_i$ for $i \leqslant n-1$, $a_n = s_n$, and show that (a_i) is an L-sequence for $(m_1, \ldots, m_{n-1}, m_n - 1)$.

First we note that all $a_i (i \leqslant n-1)$ are even because of (16), and $a_n = s_n \equiv \frac{m_n-1}{2} \prod_{i=1}^{n-1} m_i \pmod 2$ by the assumption (10). So it remains to show that (a_i) is $(m_1, \ldots, m_{n-1}, m_n - 1)$-admissible.

The condition

$$\sum_{i=1}^{n} a_i = \sum_{i=1}^{n} s_i - \sum_{i=1}^{n-1} b_i = \frac{m_n-1}{2} \prod_{i=1}^{n-1} m_i \tag{19}$$

is clear from the construction of the b_i's. Since we do not know the size-ranking of the a_i's, we must verify (1). Note that $a_i = 0$ for $i \leqslant r$. Since $m_n - 1$ is even, the subsets under consideration contain $1, \ldots, r, n$. Suppose there exists $A \subseteq \{r+1, \ldots, n-1\}$ with

$$s_n + \sum_{i \in A} a_i < \frac{m_n-1}{2} \prod_{i=1}^{r} m_i \prod_{i \in A} m_i. \tag{20}$$

Then $A \subsetneqq \{r+1, \ldots, n-1\}$ because of (19).

Set $\alpha = \prod_{i=1}^{r} m_i$ with $\alpha = 1$ if $r = 0$. We infer

$$s_n + \sum_{i \in A} s_i < \frac{\alpha(m_n-1)}{2} \prod_{i \in A} m_i + \sum_{i=r+1}^{n-1} b_i,$$

and, with

$$\sum_{i=r+1}^{n-1} b_i = \sum_{i=1}^{n-1} b_i - \sum_{i=1}^{r} b_i = \frac{\alpha}{2} \left(\prod_{i=r+1}^{n-1} m_i - 1 \right)$$

$$s_n + \sum_{i \in A} s_i < \frac{\alpha}{2} \left((m_n-1) \prod_{i \in A} m_i + \prod_{i=r+1}^{n-1} m_i - 1 \right). \tag{21}$$

With $B = \{r+1, \ldots, n-1\} \setminus A \neq \emptyset$ and $\sum_{i=r+1}^{n} s_i = \frac{\alpha}{2} \left(\prod_{i=r+1}^{n} m_i - 1 \right)$, this yields

$$\sum_{i \in B} s_i = \sum_{i=r+1}^{n} s_i - \left(s_n + \sum_{i \in A} s_i \right)$$

$$> \frac{\alpha}{2} \left(\prod_{i=r+1}^{n} m_i - 1 - (m_n-1) \prod_{i \in A} m_i - \prod_{i=r+1}^{n-1} m_i + 1 \right)$$

$$= \frac{\alpha}{2} \left(\prod_{i \in A} m_i \right) \left(m_n \prod_{i \in B} m_i - (m_n-1) - \prod_{i \in B} m_i \right)$$

$$= \frac{\alpha}{2} (m_n-1) \left(\prod_{i \in A} m_i \right) \left(\prod_{i \in B} m_i - 1 \right).$$

Since $s_n \geqslant \max(s_i : i \geqslant r+1) - 2$, this latter inequality implies

$$s_n \geqslant \frac{1}{|B|} \sum_{i \in B} s_i - 2 > \frac{\alpha(m_n - 1)}{2} \prod_{i \in A} m_i \cdot \frac{1}{|B|} \left(\prod_{i \in B} m_i - 1 \right) - 2$$

$$\geqslant \frac{\alpha(m_n - 1)}{2} \prod_{i \in A} m_i \cdot \frac{1}{|B|} (3^{|B|} - 1) - 2$$

$$\geqslant \frac{\alpha(m_n - 1)}{2} \prod_{i \in A} m_i \cdot 2 - 2,$$

in contradiction to (20). $\qquad \square$

Our claim 1 is thus proved. In particular, the decomposition lemma holds for $r = 0$. We therefore assume from now on $r \geqslant 1$.

Claim 2. *The conclusion holds if*

$$\sum_{i=1}^{r} s_i = \frac{1}{2} \prod_{i=1}^{r} m_i + 1. \tag{22}$$

Proof of Claim 2. Let $(s_1, \ldots, s_n)$ be a sequence which satisfies the hypotheses of the decomposition lemma, where again $s_{r+1} \leqslant \ldots \leqslant s_{n-1}$. Note that s_n is even by (10), and thus $r + 1 < n$ by (22).

$$\text{There exists } j \leqslant r \text{ with } m_j \geqslant 4 \text{ and } s_j \geqslant 3. \tag{23}$$

Indeed, if $r = 1$, then $s_1 = \frac{m_1}{2} + 1$ by (22). If $m_1 = 2$, then $s_1 = 2$. On the other hand, condition (D) for H-sequences implies $s_1 \equiv \prod_{i=2}^{n} m_i \equiv 1 \pmod 2$, a contradiction to $s_1 = 2$. Thus $m_1 \geqslant 4$, $s_1 \geqslant 3$. In case $r \geqslant 2$, we infer $\sum_{i=1}^{r} s_i \equiv 1 \pmod 2$ from (22). Hence there exists $j \leqslant r$ with s_j odd. Condition (E) now implies $s_j \geqslant 3$, and (D) implies $m_j \geqslant 4$.

Consider the sequence $(s_i') = (s_1, \ldots, s_j - 1, \ldots, s_r, s_{r+1} + 1, \ldots, s_n)$. Invoking Lemma 4, we see that (s_i') is $(m_1, \ldots, m_n)$-admissible whenever $s_{r+1} \leqslant s_n$. In the case $s_n < s_{r+1} \leqslant s_n + 2$, the admissibility follows by a simple calculation using (22). Condition (D) is satisfied because of $m_j \geqslant 4$, and (E) is satisfied since $s_j - 1 \geqslant 2$ and $s_{r+1} \geqslant \frac{1}{2} \left(\prod_{i=1}^{r} m_i \right) (m_{r+1} - 1) - 1 \geqslant 1$. Hence, (s_i') is an H-sequence for $(m_1, \ldots, m_n)$ with

$$\sum_{i=1}^{r} s_i' = \frac{1}{2} \prod_{i=1}^{r} m_i.$$

To apply Claim 1 we have to check that $s_n' \geqslant s_i' - 2$ $(r + 1 \leqslant i \leqslant n - 1)$. Suppose, on the contrary, $s_n < s_{r+1} - 1$. Then $s_{r+1} = \ldots = s_{n-1} = s_n + 2 \equiv 0 \pmod 2$, and thus

$$\sum_{i=1}^{r} s_i \equiv \sum_{i=1}^{n} s_i = \frac{1}{2} \prod_{i=1}^{n} m_i \equiv \frac{1}{2} \prod_{i=1}^{r} m_i \pmod 2,$$

in contradiction to (22).

By our Claim 1, there exist appropriate sequences (a_i'), (b_i'). Now consider the sequence

$$(b_i) = (b_1', \ldots, b_j' + 1, \ldots, b_r', b_{r+1}' - 1, \ldots, b_{n-1}').$$

Since $b_i' = s_i'$ for $i \leqslant r$, we have $b_{r+1}' \geqslant \frac{1}{2} \left(\prod_{i=1}^{r} m_i \right) (m_{r+1} - 1) \geqslant 4$ by (23), and thus

$b_{r+1} \geqslant 3$. The admissibility follows now from Lemma 3, and conditions (D) and (E) hold because $b'_i = s'_i$ ($i \leqslant r$) and thus $b_j = b'_j + 1 = s'_j + 1 = s_j \geqslant 3$, $b_{r+1} \geqslant 3$. Note $b_{r+1} \leqslant \ldots \leqslant b_{n-1}$ since $b'_{r+1} \leqslant \ldots \leqslant b'_{n-1}$ holds by Claim 1.

The sequences (a'_i) and (b_i) thus form a desired decomposition, and Claim 2 is proved. $\qquad\square$

Claim 3. *Assume inductively that the conclusion of the decomposition lemma holds for all sequences (s'_i) with $\sum_{i=1}^{r} s'_i \leqslant \gamma$, $\gamma \geqslant \frac{1}{2} \prod_{i=1}^{r} m_i + 1$. Then it holds for a sequence (s_i) with $\sum_{i=1}^{r} s_i = \gamma + 1$.*

Proof of Claim 3. Let (s_i) be a sequence satisfying the hypotheses of the decomposition lemma, and $s_{r+1} \leqslant \ldots \leqslant s_{n-1}$, $\sum_{i=1}^{r} s_i = \gamma + 1$. If $r = 1$, then $s_1 = \gamma + 1 \geqslant \frac{m_1}{2} + 2 \geqslant 3$. For $r \geqslant 2$ we have $s_i \neq 1$ ($i \leqslant r$) by condition (E). Suppose $s_i \in \{0, 3\}$ for all $i \leqslant r$. Since $s_i = 3$ implies $m_i \geqslant 4$ ($i \leqslant r$) by (D), we conclude for $t = |\{i \leqslant r : s_i = 3\}|$

$$3t = \sum_{i=1}^{r} s_i = \gamma + 1 \geqslant \frac{1}{2} \prod_{i=1}^{r} m_i + 2 \geqslant \frac{4^t}{2} + 2,$$

which is impossible for $t \geqslant 1$.

Hence, if $r \geqslant 2$, there exists some $j \leqslant r$ with $s_j = 2$ or $s_j \geqslant 4$. With this j respectively $j = 1$ if $r = 1$, we construct a new sequence (s'_i) as follows.

If $s_{r+1} \leqslant s_n$ (case A), then we set

$$(s'_i) = (s_1, \ldots, s_j - 1, \ldots, s_r, s_{r+1} + 2, \ldots, s_n).$$

By Lemma 4, (s'_i) is $(m_1, \ldots, m_n)$-admissible. If $s_{r+1} > s_n$ (case B), then $s_{r+1}, \ldots, s_{n-1} \in \{s_n + 1, s_n + 2\}$, and we consider instead

$$(s'_i) = (s_1, \ldots, s_j - 1, \ldots, s_r, s_{r+1}, \ldots, s_n + 2).$$

The admissibility is easily shown, and in both cases condition (D) is satisfied since the parities are unchanged, and (E) is satisfied by the choice of s_j. Note, finally, that $s'_n \geqslant \max(s'_i : r + 1 \leqslant i \leqslant n - 1) - 2$ holds in either case. By induction, in either case there exist sequences (a'_i), (b'_i) satisfying the conclusions of the decomposition lemma.

$$\text{Let } t = r + 1 (\text{case A) or } t = n(\text{case B}), \text{ then } a'_t \notin \{0, 1, 3\} \text{ or } b'_t \notin \{0, 1, 3\}. \qquad (24)$$

To see this, note that all a'_i are even, since $r \geqslant 1$ and $m_n - 1$ even (condition (B) of L-sequences). However, if $a'_t = 0$, then $b'_t = s'_t = s_t + 2$. Since $0 \leqslant s_t \neq 1$ (condition (E)), $b'_t \notin \{0, 1, 3\}$ follows.

Case 1. $a'_t \notin \{0, 1, 3\}$. Consider the sequence $(a_i) = (a'_1, \ldots, a'_j + 2, \ldots, a'_r, \ldots, a'_t - 2, \ldots, a'_n)$. Rearranging the a'_i's ($i > r$) in increasing fashion, we use Lemma 3, and conclude that (a_i) is $(m_1, \ldots, m_{n-1}, m_n - 1)$-admissible. Since the parities of a_i and a'_i are the same, (a_i) is an L-sequence, and (a_i), (b'_i) give a desired decomposition of (s_i).

Case 2. $a'_t = 0$, i.e. $b'_t \notin \{0, 1, 3\}$. In this case $t = r + 1$ (Case A), and we use the new seqence $(b_i) = (b'_1, \ldots, b'_j + 2, \ldots, b'_r, b'_{r+1} - 2, \ldots, b'_{n-1})$ and conclude from Lemma 3 that (b_i) is an H-sequence for $(m_1, \ldots, m_{n-1})$. Finally, note $b_{r+1} \leqslant \ldots \leqslant b_{n-1}$ since $b'_{r+1} \leqslant \ldots \leqslant b'_{n-1}$

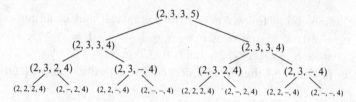

Figure 2

by induction. The sequences (a_i'), (b_i) thus form a desired decomposition of (s_i), and the proof of the decomposition lemma is complete.   □

6. Proof of the theorems

Proof of Theorem 1. Let $(s_1, \ldots, s_n)$ be an L-sequence for $(m_1, \ldots, m_n)$. As before, $m_1, \ldots, m_r$ are even, $m_{r+1}, \ldots, m_n$ odd, and w.l.o.g. $s_{r+1} \leqslant \ldots \leqslant s_n$. We proceed by induction on the number $k = n - r$ of odd m_i's. Proposition 1 settles the case $k = 0$. Hence suppose $k \geqslant 1$. Assume first $r \geqslant 1$. Then $s_n \equiv \lfloor \frac{m_n}{2} \rfloor \prod_{i=1}^{n-1} m_i$ by condition (B), and we can apply the decomposition lemma (see Remark 5). The two L-graphs $L_a = L(m_1, \ldots, m_{n-1}, m_n - 1)$ and $L_b = L(m_1, \ldots, m_{n-1})$ have $k - 1$ odd lengths m_j, and we have $r \geqslant 1$ in both cases. Hence the derived sequences (a_i) resp. (b_i) are by induction the types of a maximum matching in L_a resp. L_b, and can thus be put together to yield a desired matching of type $(s_1, \ldots, s_n)$.

We can conveniently represent the inductive process for $r \geqslant 1$ by means of a *decomposition tree*. Consider as an example $(m_1, m_2, m_3, m_4) = (2, 3, 3, 5)$. The tree is depicted in Figure 2, where the (a_i)-sequences are in the left subtree and the (b_i)-sequences in the right subtree.

Remark 6. The subsequences $(m_1, \ldots, m_n)$ have the same number of odd lengths on each level. The leaves correspond to the subsequences with all m_i even. Note also that the subsets of missing lengths (−) at the leaves correspond precisely to all 2^k subsets of indices i with m_i odd in the starting length sequence $(m_1, \ldots, m_n)$, $k = n - r$.

It remains to consider $r = 0$. In this case any maximum matching leaves one vertex uncovered. To use induction we have to extend the hypothesis.

Claim. *Let all m_i be odd, and suppose $(s_1, \ldots, s_n)$ is an L-sequence for $(m_1, \ldots, m_n)$. Then there exists a maximum matching M of type (s_i) such that the uncovered vertex u satisfies*

$$u_i = \begin{cases} 0 & \text{if } s_i \equiv \frac{m_i - 1}{2} \pmod 2 \\ 1 & \text{otherwise.} \end{cases} \tag{25}$$

Proof of the claim. We use induction on n. For $n = 1$, $L(m_1)$ is a path, $s_1 = \frac{m_1 - 1}{2}$, and we choose as uncovered vertex $u = 0$. Let $n \geqslant 2$, and assume w.l.o.g. $s_1 \leqslant \ldots \leqslant s_n$.

If $s_n \equiv \frac{m_n - 1}{2} \pmod 2$, then (10) is satisfied, and we may apply the decomposition lemma to obtain the sequences (a_i) and (b_i). Now we use induction on n, and infer the existence of appropriate matchings of type (a_i) resp. (b_i). Let $L_a = L(m_1, \ldots, m_{n-1}, m_n - 1)$ with the (a_i)-matching have vertex-set $V_a = \{(v_1, \ldots, v_n) : v_n > 0\}$, and let $L_b = L(m_1, \ldots, m_{n-1})$ with the (b_i)-matching have vertex-set $V_b = \{(v_1, \ldots, v_n) : v_n = 0\}$. The uncovered vertex

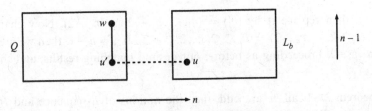

Figure 3

$(u_1, \ldots, u_n) \in V_b$ satisfies $u_n = 0$, and since $s_i \equiv b_i \pmod 2$ for $i \leqslant n-1$, (25) is satisfied. Thus we may put the matchings together.

Suppose now $s_n \not\equiv \frac{m_n-1}{2} \pmod 2$ and consider the sequence $(s_i') = (s_1, \ldots, s_{n-1}+1, s_n-1)$. By Lemma 3 and $s_n \geqslant 1$, (s_i') is admissible and hence an L-sequence for $(m_1, \ldots, m_n)$ with $s_n' \geqslant s_i' - 2$ $(1 \leqslant i \leqslant n-1)$. We can therefore apply the decomposition lemma. Let (a_i), (b_i) be the sequences constructed from (s_i') according to the decomposition lemma, with V_a and V_b the vertex-sets of L_a and L_b as before. Since L_a has an even length $m_n - 1$, we can decompose the matching M_a of type (a_i) in L_a eventually into hypercube matchings as explained above. At least one of these hypercubes, say Q, has full dimension n (see Remark 6). Suppose the induced type of Q is $(c_1, \ldots, c_n)$, where we may assume w.l.o.g. that the vertex-set of Q is $V_Q = \{(v_1, \ldots, v_n) : v_i \in \{0,1\}$ for $i \leqslant n-1$, and $v_n \in \{1,2\}\}$.

Case 1. $c_{n-1} \geqslant 2$, i.e. there are M_a-edges in Q along dimension $n-1$. Let $u = (u_1, \ldots, u_{n-1}, 0)$ be the uncovered vertex in L_b. Note that because of $b_i \equiv s_i \pmod 2$, (25) is satisfied for L_b. Let $u' = (u_1, \ldots, u_{n-1}, 1)$ be the neighbor of u in V_a, thus $u' \in V_Q$. By edge-transitivity we may choose an $(n-1)$-edge in M_a as $u'w$ with $w = (u_1, \ldots, u_{n-2}, 1-u_{n-1}, 1)$. Replace now the M_a-edge $u'w$ by the edge $u'u$ (see Figure 3). Since $u'u$ is an n-edge, our desired matching of type $(s_1, \ldots, s_n)$ satisfying (25) results, since the uncovered vertex w satisfies $w_{n-1} = 1 - u_{n-1}$, $w_n = 1$.

Case 2. $c_{n-1} = 0$, and $c_j \geqslant 2$ for some $j < n-1$, hence $n \geqslant 3$. Since $b_{n-1} = \max(b_i : 1 \leqslant i \leqslant n-1)$, we infer

$$b_{n-1} \geqslant \frac{1}{2(n-1)} \prod_{i=1}^{n-1} m_i \geqslant \frac{3^{n-1}}{2(n-1)} > 2$$

for $n \geqslant 3$. Hence $(b_j') = (b_1, \ldots, b_j + 2, b_{j+1}, \ldots, b_{n-1} - 2)$ is admissible and thus an L-sequence for $(m_1, \ldots, m_{n-1})$ by Lemma 3. By induction there exists a matching $M_{b'}$ of type (b_i') with uncovered vertex $u = (u_1, \ldots, u_{n-1}, 0)$. Now change the type of the hypercube matching in Q to $(c_i') = (c_1, \ldots, c_j - 2, c_{j+1}, \ldots, c_{n-1} + 2, c_n)$ and make an exchange of an $(n-1)$-edge and an n-edge as in case 1. The resulting matching of $L(m_1, \ldots, m_n)$ has type $(s_1, \ldots, s_n)$ and satisfies (25).

Case 3. All $c_j = 0$ $(j \leqslant n-1)$, i.e. $(c_1, \ldots, c_n) = (0, \ldots, 0, 2^{n-1})$. In this case there exists $j < n$ with $a_j \geqslant 2$, since otherwise $a_n = s_n = \frac{m_n-1}{2} \prod_{i=1}^{n-1} m_i$, in contradiction to the assumption $s_n \not\equiv \frac{m_n-1}{2} \pmod 2$. We conclude that there are M_a-edges in L_a along dimension j. Let Q' be a hypercube with matching edges along j. Note that Q' contains dimension n since $m_n - 1$ is even (see Figure 2). Suppose $(q_1, \ldots q_j, \ldots, q_n)$ is the type of the matching in Q', $q_j \geqslant 2$, then replace it by $(q_i') = (q_1, \ldots, q_j - 2, \ldots, q_n + 2)$. Now change (c_i) to $(c_i') = (c_1, \ldots, c_j + 2, \ldots, c_n - 2) = (0, \ldots, 2, \ldots, 2^{n-1} - 2)$. If $j = n - 1$, then we are in case 1, otherwise in case 2. Proceeding as before, our desired matching results, and the theorem is proved. $\qquad\qquad\qquad\qquad\qquad\qquad\qquad\qquad\qquad\qquad\qquad\qquad\qquad\qquad\quad$ $\square$

Proof of Theorem 2. If all m_i are odd, then the notions of L-sequence and H-sequence are equivalent (see Remark 3), and we may apply Theorem 1. Hence we assume $r \geqslant 1$. As before, the proof proceeds by induction on the number $k = n - r$ of odd m_i's. The induction start is provided by Proposition 2. Let $k \geqslant 1$, and $(s_1, \ldots, s_n)$ be an H-sequence for $(m_1, \ldots, m_n)$ with $s_{r+1} \leqslant \ldots \leqslant s_n$. If $s_n \equiv \frac{m_n - 1}{2} \prod_{i=1}^{n-1} m_i \pmod 2$, i.e., s_n even, then (10) is satisfied, and the theorem follows inductively by an application of the decomposition lemma.

So it remains to prove the induction step for s_n odd, where $s_n \geqslant 3$ by condition (E).

$$\text{There exists } j \leqslant n - 1 \text{ with } s_j \geqslant 3, \, m_j \geqslant 3. \tag{26}$$

This is certainly true for $j \in \{r + 1, \ldots, n - 1\}$ by conditions (E) and (D) whenever s_j is odd, since $r \geqslant 1$. Hence suppose $r = n - 1$ or $r < n - 1$ and all s_i ($r + 1 \leqslant i \leqslant n - 1$) are even. If $r \geqslant 2$, then $\sum_{i=1}^{n} s_i = \frac{1}{2} \prod_{i=1}^{n} m_i \equiv 0 \pmod 2$. Hence there is another odd s_j ($j \leqslant r$) beside s_n. Conditions (D) and (E) now imply $m_j \geqslant 4$, $s_j \geqslant 3$. If $r = 1$ and $\frac{m_1}{2} \equiv 0 \pmod 2$, i.e. $m_1 \geqslant 4$, then s_1 is odd with $s_1 \geqslant \frac{m_1}{2} \geqslant 2$, i.e. $s_1 \geqslant 3$. Finally, if $r = 1$ and $\frac{m_1}{2} \equiv 1 \pmod 2$ then s_1 is even, $m_1 \geqslant 6$ (by (D)), and thus $s_1 \geqslant \frac{m_1}{2} \geqslant 3$.

Now choose j as in (26) where we take $j = n - 1$ whenever $s_{n-1} \geqslant 3$. The sequence

$$(s_i') = (s_1, \ldots, s_j + 1, \ldots, s_n - 1)$$

is $(m_1, \ldots m_n)$-admissible by Lemma 3, and thus H-sequence since (D) and (E) are trivially satisfied. Since $s_n' \geqslant s_i' - 2$ ($r + 1 \leqslant i \leqslant n - 1$) we may again apply the decomposition lemma obtaining sequences (a_i) resp. (b_i).

Case 1. m_j odd, i.e. $j = n - 1$. First we note that we may assume $b_{n-1} > 0$. Suppose $b_{n-1} = 0$. If $r \geqslant 2$, then $b_\ell = 2$ or $b_\ell \geqslant 4$ for some $\ell \leqslant r$, since otherwise $b_i \in \{0, 3\}$ for $i \leqslant r$ (condition (E)) which implies for $t = |\{i \leqslant r : b_i = 3\}|$ by condition (D), $\sum_{i=1}^{r} b_i = 3t \geqslant \frac{4^t}{2}$. Hence $t = 1$ and we obtain $\sum_{i=1}^{r} b_i = 3 \geqslant \frac{4 \cdot 2}{2} = 4$ by $r \geqslant 2$, a contradiction. If $r = 1$, then $b_1 = b_1 + b_{n-1} \geqslant \frac{m_1 m_{n-1}}{2} \geqslant 3$, and we set $\ell = 1$. Consider the new sequence

$$(b_i') = (b_1, \ldots, b_\ell - 2, \ldots, b_{n-1} + 2).$$

We have

$$\sum_{i=1}^{r} b_i = \sum_{i=1}^{r} b_i + b_{n-1} \geqslant \frac{1}{2} \prod_{i=1}^{r} m_i \cdot m_{n-1} \geqslant \frac{3}{2} \prod_{i=1}^{r} m_i = \frac{1}{2} \prod_{i=1}^{r} m_i + \prod_{i=1}^{r} m_i \geqslant \frac{1}{2} \prod_{i=1}^{r} m_i + 2.$$

Since $b_{n-1} = 0 = \min(b_i : i > r)$, we may apply Lemma 4, infering that (b_i') is an H-sequence for $(m_1, \ldots, m_{n-1})$. Furthermore,

$$(a_i') = (a_1, \ldots, a_\ell + 2, \ldots, a_{n-1} - 2, a_n)$$

is L-sequence for $(m_1, \ldots, m_n - 1)$ because of $a_{n-1} = s_{n-1}' = s_{n-1} + 1 \geqslant 4$ and Lemma 3. Hence (a_i'), (b_i') form a decomposition of (s_i') with $b_{n-1}' > 0$.

Thus we may assume that (a_i), (b_i) is a decomposition of (s_i') with $b_{n-1} > 0$. Next we show that we may also assume $a_{n-1} > 0$. Suppose $a_{n-1} = 0$. Then $b_{n-1} = s_{n-1}' \geqslant 4$. Furthermore, there must be an index $\ell < n - 1$ with $a_\ell > 0$, since otherwise we would obtain $a_n = s_n' = s_n - 1 = \frac{m_n - 1}{2} \prod_{i=1}^{n-1} m_i$, and hence

$$\sum_{i=1}^{n-1} s_i = \frac{1}{2} \prod_{i=1}^{n} m_i - s_n = \frac{1}{2} \prod_{i=1}^{n-1} m_i - 1,$$

a contradiction to the admissibility of (s_i). Now consider the sequences

$$(b_i') = (b_1, \ldots, b_\ell + 2, \ldots, b_{n-1} - 2)$$
$$(a_i') = (a_1, \ldots, a_\ell - 2, \ldots, a_{n-1} + 2, a_n).$$

The sequence (b_i') is an H-sequence for $(m_1, \ldots, m_{n-1})$ by Lemma 3 and $b_{n-1} \geqslant 4$, whereas (a_i') is an L-sequence for $(m_1, \ldots, m_n - 1)$ by Lemma 4 (if $\ell \leqslant r$) resp. Lemma 3 (if $\ell > r$), since $a_\ell > 0$ even, i.e. $a_\ell \geqslant 2$. Furthermore, $b_{n-1}' > 0, a_{n-1}' > 0$.

Hence we may assume that (a_i), (b_i) is a decomposition of (s_i') with $a_{n-1}, b_{n-1} > 0$. Now we proceed in a similar fashion as in the proof of Theorem 1. Let $H_a = H(m_1, \ldots, m_n - 1)$ be the Hamming graph with lattice-matching M_a of type (a_i), and $H_b = H(m_1, \ldots, m_{n-1})$ the Hamming graph with Hamming matching M_b of type (b_i). As explained above, M_a consists of even length lattice matchings which contain all even length dimensions $1, \ldots, r, n$ and a subset of the other dimensions $\{r + 1, \ldots, n - 1\}$. Each of these subsets appears exactly once. Hence in this decomposition there are lattices $L(m_1, \ldots, m_r, m_n - 1)$ and $L(m_1, \ldots, m_r, m_{n-1}, m_n - 1)$, and they lie w.l.o.g. next to each other in direction $n - 1$. In particular, there is an $(r + 1)$-dimensional hypercube Q_1 with dimensions $1, 2, \ldots, r, n$, and an $(r + 2)$-dimensional hypercube Q_2 with dimensions $1, \ldots, r, n - 1, n$ which are neighbors in direction $n - 1$. For the vertex-sets $V(Q_1)$, $V(Q_2)$ we may assume w.l.o.g.

$$V(Q_1) = \{(v_1, \ldots, v_n) : v_1, \ldots, v_r, v_n \in \{0, 1\}, v_{r+1} = \ldots = v_{n-1} = 0\}$$
$$V(Q_2) = \{(v_1, \ldots, v_n) : v_1, \ldots, v_r, v_n \in \{0, 1\}, v_{n-1} \in \{1, 2\}, v_i = 0 \text{ otherwise}\}.$$

Let $(c_1, \ldots, c_r, c_n)$ be the type of the matching in Q_1 and $(q_1, \ldots, q_r, q_{n-1}, q_n)$ the type of the matching in Q_2. If $c_n = 0$, then we choose $k \leqslant r$ with $c_k > 0$, and another hypercube Q_3 which contains M_a-edges in direction n (recall $a_n = s_n' = s_n - 1 \geqslant 2$). Now delete two k-matching edges in Q_1 and add two n-matching edges, and do the opposite in Q_3. If $q_{n-1} = 0$, then we may apply an analogous exchange since $a_{n-1} > 0$ by assumption.

Let ux be an $(n - 1)$-edge connecting Q_1 and Q_2 and choose the matchings of type (c_i) resp. (q_i) so that u is incident to an n-matching edge uv in Q_1, and x is incident to an $(n - 1)$-matching edge xw in Q_2 (possible by edge-transitivity). Putting the Hamming graph H_a and H_b together we may choose w.l.o.g. an $(n - 1)$-matching edge yz in M_b such that y is adjacent to u and v in $H(m_1, \ldots, m_n)$ and z is adjacent to x, both along dimension n. We thus arrive at the situation shown in Figure 4.

M. Aigner and R. Klimmek

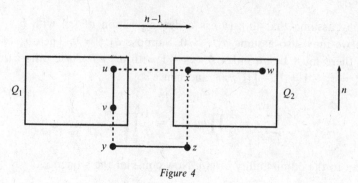

Figure 4

Replacing the matching edges uv, xw, yz by the non-matching edges uw, vy, xz, we decrease the $(n-1)$-edges by 1, increase the n-edges by 1, and thus arrive at a desired matching of type (s_i).

Case 2. m_j even, i.e. $j \leqslant r$. By an analogous argument as in case 1 we may assume $a_j > 0$, $b_j > 0$. Consider again the Hamming graphs H_a and H_b with matchings M_a and M_b. The lattice graph $L(m_1, \ldots, m_n - 1)$ contains in the usual way a sublattice $L(m_1, \ldots, m_r, m_n - 1)$ and therefore, because of $m_j \geqslant 4$, two $(r+1)$-dimensional hypercubes Q_1, Q_2 which are neighbors along dimension j. Let $(c_1, \ldots, c_r, c_n), (q_1, \ldots, q_r, q_n)$ be the types of Q_1 resp. Q_2. As in case 1 we may assume $c_n > 0$, $q_j > 0$ because of $s'_n > 0$, $a_j > 0$. By the assumption $b_j > 0$ we can exchange matching edges and non-matching edges as in case 1 with j in place of $n - 1$. We thus obtain a desired matching in $H(m_1, \ldots, m_n)$ of type (s_i), and the proof is complete. □

References

[1] Felzenbaum, A., Holzman, R. and Kleitman, D. J. (1993) Packing lines in a hypercube, *Discrete Math.* **117** 107–112.
[2] Lovász, L. (1977) *Combinatorial Problems and Exercises.* North-Holland.
[3] Montroll, E. Lattice Statistics. In: *Applied Combinatorial Mathematics* (Beckenbach, E. F. ed.), Wiley.

Reconstructing a Graph from its Neighborhood Lists

MARTIN AIGNER[†] and EBERHARD TRIESCH[‡]

[†]Freie Universität Berlin, Fachbereich Mathematik, WE 2, Arnimallee 3, 1000 Berlin 33, Germany

[‡]Forschungsinsitut für Diskrete Mathematik, Nassestraße 2, 5300 Bonn 1, Germany

Associate to a finite labeled graph $G(V, E)$ its multiset of neighborhoods $\mathcal{N}(G) = \{N(v) : v \in V\}$. We discuss the question of when a list $\mathcal{N}$ is realizable by a graph, and to what extent G is determined by $\mathcal{N}(G)$. The main results are: the decision problem $\mathcal{N} \overset{?}{=} \mathcal{N}(G)$ is NP-complete; for bipartite graphs the decision problem is polynomially equivalent to Graph Isomorphism; forests G are determined up to isomorphism by $\mathcal{N}(G)$; and if G is connected bipartite and $\mathcal{N}(H) = \mathcal{N}(G)$, then H is completely described.

1. Introduction

Consider a finite labeled undirected simple graph $G(V, E)$. Let us associate to G a finite list $P(G)$ of parameters. $P(G)$ may consist of the chromatic number, of the degrees, of the list of cliques, of one-vertex-deleted subgraphs, or whatever we like. To any given list of parameters there arise two natural problems:

(1) *Realizability.* Given P, when is P graphic, *i.e.* when does there exist a graph G with $P = P(G)$?

(2) *Uniqueness.* If $P(G) = P(H)$ for two graphs G and H on the same vertex-set V, what does this tell us about G and H? In particular, when is $G \cong H$ or $G = H$?

The problem when P consists of the degree sequence has been well studied. The famous Erdős–Gallai Theorem [1] gives a (polynomial) characterization of graphic sequences, thereby answering question (1). As far as uniqueness is concerned, more or less anything can happen. In particular, G and H need not be in the least isomorphic. For a recent result in this direction, see [2].

We treat a closely related question, raised by V. Sós at the conference [3]. Associate to every $u \in V$ its neighborhood $N(u) = \{v \in V : uv \in E\}$, and denote by $\mathcal{N}(G) = \{N(u) : u \in V\}$ the *neighborhood list* of G. (The mathematical object is, strictly speaking, a multiset: the sets $N(u)$ are counted with multiplicities and their order does not matter.) In fact, Sós considered the stars $S(u) = N(u) \cup \{u\}$, but the two problems are easily seen to be essentially the same. In fact, a list $\mathcal{N}$ is the neighborhood list of G if and only if $\{V \setminus N : N \in \mathcal{N}\}$ is the system of stars of $\bar{G}$. At the same conference, L. Babai remarked that the realizability problem for $\mathcal{N}$ is at least as hard as the graph isomorphism problem. We will show in Section 2 that it is, in fact, NP-complete. In Section 3 we address the uniqueness question. It may well happen that non-isomorphic graphs G and H on the same labeled vertex-set have the same lists $\mathcal{N}$, but for some classes of graphs we can assert uniqueness up to isomorphism, *e.g.* for complete k-partite graphs and for forests. Furthermore, the question whether for a given connected bipartite graph G there exists a non-isomorphic graph H with the same list $\mathcal{N}$ is also shown to be NP-complete.

2. Realizability of neighborhood lists

Let $\mathcal{N}(G)$ be the neighborhood list of a graph G with vertex-set $V = \{1, \ldots, n\}$. Hence $\mathcal{N}(G)$ consists of n subsets of V (with possible multiplicities). The following characterization of graphic lists is immediate.

Lemma 1. *Let $\mathcal{N}$ be a list of n subsets of $\{1, \ldots, n\}$. $\mathcal{N}$ is graphic if and only if there is a numbering $N_1, \ldots, N_n$ such that for all i and j*

(1) $i \in N_j \iff j \in N_i$,

(2) $i \notin N_i$.

If we write $\mathcal{N}$ as an $n \times n$ incidence matrix Γ with the rows corresponding to the sets, and the columns to the vertices, then Lemma 1 says the following: Γ is graphic iff there exists an $n \times n$ permutation matrix R such that

(1) $R\Gamma$ is symmetric,

(2) all diagonal elements of $R\Gamma$ are zero.

We consider the decision problem NL (neighborhood lists) in this matrix form.

Input: An $n \times n$ matrix Γ with $\{0, 1\}$-entries.

Question: Is Γ graphic?

Theorem 2. NL *is NP-complete.*

Proof. It is clear that NL is in the class NP. To show completeness, we provide a transformation from the problem ORDER 2 FIXED-POINT-FREE AUTOMORPHISM (O2FPFA), which was shown to be NP-complete by A. Lubiw in [4]. This latter problem can be stated as follows:

Input: A graph G.

Question: Is there an involution in the group Aut (G) without fixed points?

Lubiw gives a transformation from 3-SAT. Her construction provides graphs in which every vertex has degree at most $|V| - 2$, hence we can restrict the input of O2FPFA to such graphs without losing NP-completeness.

Now suppose B is the vertex-edge incidence matrix of such a graph, $B \in \mathrm{Mat}_{n \times m}$, and denote by Perm_n the set of $n \times n$ permutation matrices of order n. Note that

$$\mathrm{Aut}\,(G) \cong \{P \in \mathrm{Perm}_n : \exists Q \in \mathrm{Perm}_m \text{ with } PBQ = B\},$$

and further

(1′) P involution $\iff P^T = P$,

(2′) P fixed-point free $\iff$ all diagonal elements are zero.

We denote the $n \times n$ identity matrix by I_n, and the $r \times s$ matrix with all entries equal to t by $(t)_{r \times s}$.

Now consider the $(n + m + 1) \times (n + m + 1)$ matrix

$$\Gamma = \begin{array}{|c|c|c|} \hline I_n & B & (1)_{n \times 1} \\ \hline B^T & (0)_{m \times m} & (0)_{m \times 1} \\ \hline (1)_{1 \times n} & (0)_{1 \times m} & (0)_{1 \times 1} \\ \hline \end{array}$$

Suppose $R \in \mathrm{Perm}_{n+m+1}$ satisfies (1) and (2). Since no row of B contains more than $n - 3$ ones, the last row of Γ is the only row with n ones. Hence, if $R\Gamma = (R\Gamma)^T$ holds, R must fix the last row. But then R permutes the rows $1, \ldots, n$, resp. $n + 1, \ldots, n + m$, among themselves, and we can write

$$R = \begin{array}{|c|c|c|} \hline P & 0 & \\ \hline 0 & Q & 0 \\ \hline 0 & 0 & 1 \\ \hline \end{array}$$

with $P \in \mathrm{Perm}_n$ and $Q \in \mathrm{Perm}_m$.

The condition $R\Gamma = (R\Gamma)^T$ is thus equivalent to the equations

$$P = P^T, \quad PB = BQ^T,$$

i.e. P is an involution with $PBQ = B$. Note that $Q^{-1} = Q^T$ holds for a permutation matrix. Furthermore, $R\Gamma$ has zero-diagonal if and only if P has no fixed points. Hence the proof is complete. $\qquad \square$

There is a natural variant of NL for bipartite graphs. Consider a bipartite graph G on defining vertex-sets U and V, $|U| = m$, $|V| = n$. The neighborhood list (multiset)

$\mathcal{N}_U(G) = \{N(u) : u \in U\}$ consists of m subsets of V, and the list $\mathcal{N}_V(G) = \{N(v) : v \in V\}$ of n subsets of U.

The BIPARTITE NEIGHBORHOOD LISTS (BNL) problem reads as follows:

Input: Lists $\mathcal{N}_U$ of m subsets of V and $\mathcal{N}_V$ of n subsets of U.

Question: When is the pair $(\mathcal{N}_U, \mathcal{N}_V)$ graphic?

BNL might be easier than NL, as the following result suggests.

Theorem 3. BNL *is polynomially equivalent to* GRAPH ISOMORPHISM.

The GRAPH ISOMORPHISM problem is perhaps the most famous decision problem in the class NP, which is neither known to be polynomially solvable nor to be NP-complete. A recent account of the state of the problem is given in [5].

For the proof of Theorem 3, it is convenient to use the following:

Lemma 4. GRAPH ISOMORPHISM (GI) *and* HYPERGRAPH ISOMORPHISM (HI) *are polynomially equivalent.*

Proof. Since each graph is a hypergraph, it suffices to show that HI $\propto$ GI. So choose two hypergraphs $\mathcal{H} = (V, E)$ and $\mathcal{H}' = (V', E')$ and assume, without loss of generality, that $V = V', |V| = n$, and $E = \{e_1, \ldots, e_m\}$, $E' = \{e_1', \ldots, e_m'\}$. Choose sets of new elements $X = \{x_1, \ldots, x_m\}$, $Y = \{y, y_1, \ldots, y_{n+1}\}$, and construct the graph $G = G(\mathcal{H})$ (resp. $G' = G(\mathcal{H}')$) on $V \cup X \cup Y$ as follows: connect x_i to all points in e_i (resp. e_i') and to y $(1 \le i \le m)$; and y to all y_j $(1 \le j \le n + 1)$. No other edges are added.

We claim that G and G' are isomorphic if and only if the corresponding hypergraphs are isomorphic. The 'if'-direction being clear, suppose that G and G' are isomorphic with isomorphism ϕ. By looking at the degrees, we see that ϕ fixes y, and that each orbit of ϕ is contained in one of the sets $V, X, \{y\}$ and $Y \setminus \{y\}$. From this it follows readily that $\mathcal{H}$ and $\mathcal{H}'$ are isomorphic. $\square$

Proof of Theorem 3. In view of the Lemma, it suffices to prove the polynomial equivalence of BNL and HI. By identifying neighborhood systems as well as hypergraphs with their incidence matrices, we see that an input to BNL consists of an $(n \times m)$-matrix Γ and an $(m \times n)$-matrix Υ. The question is do there exist $P \in \text{Perm}_n$ and $Q \in \text{Perm}_m$ such that $(P\Gamma)^T = Q\Upsilon$? But $(P\Gamma)^T = Q\Upsilon$ if and only if $\Gamma^T = Q\Upsilon P$ if and only if the hypergraphs with incidence matrices Υ and Γ^T are isomorphic. The result follows. $\square$

Another interesting decision problem related to neighborhood lists is MATRIX SYMMETRY (MS):

Input: An $n \times n$ matrix A with $\{0, 1\}$-entries.

Question: Does there exist $P \in \text{Perm}_n$ such that $PA = (PA)^T$ holds?

It is easy to see that MS is at least as hard as GRAPH ISOMORPHISM by considering a matrix of the type

0	B_1	1
B_2^T	0	0
1	0	

where B_1 and B_2 are the incidence matrices of graphs (each with maximum degree $\leq |V| - 3$). MS is equivalent to the variant when arbitrary entries are allowed, but we do not know whether MS is NP-complete.

3. Uniqueness of neighborhood lists

Consider two graphs G and H on the same labeled vertex-set V, $|V| = n$. We call G and H *hypomorphic*, denoted by $G \approx H$, if $\mathcal{N}(G) = \mathcal{N}(H)$. Thus $G \approx H$ iff there exists a bijection $\varphi : V \longrightarrow V$ with $N_G(u) = N_H(\varphi u)$ for all $u \in V$. If uv is an edge of G, then for brevity we write $uv \in G$.

Proposition 5.

(i) If $G \approx H$ by means of φ, then

    (A) $u, \varphi u \notin G$

    (B) $uv \in G \Longleftrightarrow \varphi u, \varphi^{-1} v \in G$.

(ii) If $\varphi : V \longrightarrow V$ satisfies (A) and (B), then $G \approx H$, where $E(H) = \{\varphi u, v : uv \in G\}$

Proof.

(i) By the definition of φ, we have $uv \in G \Longleftrightarrow \varphi u, v \in H$. Since $\varphi u \notin N_H(\varphi u) = N_G(u)$, we infer $u, \varphi u \notin G$. To prove (B), we have

$$uv \in G \Longleftrightarrow \varphi u, v \in H \Longleftrightarrow \varphi u \in N_H(v) = N_G(\varphi^{-1} v) \Longleftrightarrow \varphi u, \varphi^{-1} v \in G.$$

(ii) We set up the list $\{N(\varphi u) : u \in V\}$ with $N(\varphi u) = N_G(u)$ for all $u \in V$, and show that it defines a graph H, *i.e.*

$$\varphi u \notin N(\varphi u)$$
$$\varphi u \in N(\varphi v) \Longleftrightarrow \varphi v \in N(\varphi u).$$

We have $\varphi u \notin N(\varphi u)$, since $N(\varphi u) = N_G(u)$ and $\varphi u \notin N_G(u)$ by (A). To prove the second claim, we infer

$$\varphi u \in N(\varphi v) \overset{}{\Longleftrightarrow} \varphi u \in N_G(v) \Longleftrightarrow \varphi u, v \in G \overset{(B)}{\Longleftrightarrow} u, \varphi v \in G$$
$$\Longleftrightarrow \varphi v \in N_G(u) \Longleftrightarrow \varphi v \in N(\varphi u).$$

This completes the proof.   □

Proposition 5 tells us that the possible graphs H, hypomorphic to G, are completely described by mappings $\varphi : V \longrightarrow V$ satisfying conditions (A) and (B) *within G*. Let us call these mappings φ *admissible*. We will write H_φ for the hypomorphic graph induced by φ, and we note

(*) $$uv \in G \Longleftrightarrow \varphi u, v \in H_\varphi \Longleftrightarrow u, \varphi v \in H_\varphi.$$

Lemma 6. *Let φ be an admissible mapping of the graph G, and $O_1, \ldots, O_m$ its orbits on V. Then for all i, j*

(i) $u, v \in O_i \Longrightarrow uv \notin G$,

(ii) Let $|O_i| = k$, $|O_j| = \ell$, $u \in O_i$, $v \in O_j$, $uv \in G$. Then the subgraph of G induced by $O_i \cup O_j$ is a disjoint sum of complete bipartite graphs $K_{r,s}$, where $r = k/\gcd(k,\ell)$, $s = \ell/\gcd(k,\ell)$.

Proof.

(i) Let $v = \varphi^j u$. If $j \equiv 0 \pmod 2$, then by repeated application of (B) it follows that $u, \varphi^j u \in G \Longleftrightarrow \varphi^{j/2} u, \varphi^{j/2} u \in G$, which is impossible. If $j \equiv 1 \pmod 2$, then by the same argument $u, \varphi^j u \in G \Longleftrightarrow \varphi^{(j-1)/2} u, \varphi^{(j+1)/2} u \in G \Longleftrightarrow \varphi^{(j-1)/2} u, \varphi(\varphi^{(j-1)/2} u) \in G$, contradicting (A).

(ii) Easily seen by iterating condition (B).   □

Corollary 7. *Let φ be an admissible mapping of G. If all the induced (bipartite) subgraphs $G(O_i, O_j)$ on the orbits O_i, O_j are either empty or complete bipartite, then $G = H_\varphi$.*

Corollary 8. *If G is a complete k-partite graph, then $H \approx G$ implies $H = G$.*

Before going on, let us look at some examples. The smallest graph G that admits a hypomorphic graph $H \neq G$ is the 5-path. In the following figure, G is shown with full lines and H with dashed lines. The corresponding mapping is $\varphi = \left(\begin{smallmatrix} 1 & 2 & 3 & 4 & 5 \\ 5 & 4 & 3 & 2 & 1 \end{smallmatrix}\right)$. Of course, $H \cong G$.

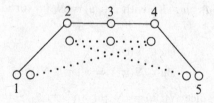

The smallest graph G that admits a non-isomorphic hypomorphic graph H is the 6-cycle $(1,2,3,4,5,6)$, with H being the sum of two disjoint triangles $(1,3,5) + (2,4,6)$. The orbits of G are $\{1,4\}, \{2,5\}, \{3,6\}$.

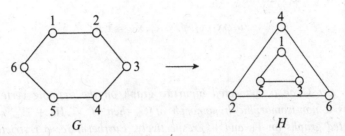

We can immediately generalize this example. Let H_0 be a graph and $H = H_0 + H_0'$ be the disjoint union of two copies of H_0. Denote by u' the vertex in H_0' corresponding to $u \in H_0$. The involution $\varphi : u \longleftrightarrow u'$ is then admissible, and hence $H \approx G_\varphi$. If $H_0 = K_4$, the reader may verify that the corresponding hypomorphic graph G_φ is the 3-cube Q_3, where again $Q_3 \not\cong K_4 + K_4$. Note that the graphs G_φ arising in this way are bipartite.

Proposition 9. *Let φ be an admissible mapping of G with orbits $\mathcal{O} = \{O_1, \ldots, O_m\}$. Denote by $G_{\mathcal{O}}$ the graph with vertex-set $\mathcal{O}$ and $O_i O_j \in G_{\mathcal{O}}$ iff there exist $u \in O_i$, $v \in O_j$ with $uv \in G$. If $G_{\mathcal{O}}$ is bipartite, $G \cong H_\varphi$.*

Proof. Let R and S be the color classes of $G_{\mathcal{O}}$, and define the bijection $\psi : V \longrightarrow V$ by

$$\psi(u) = \begin{cases} u & \text{if} \quad u \in O_i \in R \\ \varphi u & \text{if} \quad u \in O_j \in S. \end{cases}$$

Consider $uv \in G$. By Lemma 6(i), u and v are in different orbits O_i, O_j, hence we may suppose without loss of generality that $O_i \in R$, $O_j \in S$, i.e. $\psi u = u$, $\psi v = \varphi v$. By (*), this yields

$$uv \in G \Longleftrightarrow u, \varphi v \in H_\varphi \Longleftrightarrow \psi u, \psi v \in H_\varphi,$$

implying that ψ is an isomorphism from G onto H_φ. $\qquad\square$

Corollary 10. *If G is a forest, $G \approx H$ implies $G \cong H$.*

Proof. Just observe that any cycle in $G_{\mathcal{O}}$ gives rise to a cycle in G, by iterating condition (B) in Proposition 5. $\qquad\square$

Now let G be an arbitrary connected bipartite graph. In our examples above, we have already seen that G may have non-isomorphic hypomorphs of the form $G \approx H + H$. We now show that these are essentially all the possibilities. First we prove an easy lemma.

Lemma 11. *Let φ be an admissible mapping of G. If u and v are connected in G by a trail of even length, the same holds in H_φ.*

Proof. The trail

$$u = x_0, x_1, x_2, \ldots, x_{2t} = v$$

in G gives rise by (*) to the trail

$$u, \varphi x_1, x_2, \varphi x_3, \ldots, x_{2t} = v$$

in H_φ. □

Theorem 12. *Let G be a connected bipartite graph on the defining vertex-sets V_0, V_0'. If $H = H_\varphi$ is a non-isomorphic hypomorph of G, then $H = H_0 + H_0'$, where H_0 and H_0' are connected graphs on V_0 and V_0', respectively. Furthermore, φ restricted to V_0 is a bijection between V_0 and V_0', and, by identifying $u \equiv \varphi u$ ($u \in V_0$), H_0 is hypomorphic to H_0'. Conversely, if H_0 and H_0' are connected hypomorphic graphs, $H = H_0 + H_0'$ is hypomorphic to a connected bipartite graph.*

Proof. Let φ be an admissible mapping of G. Since any two vertices in V_0 (resp. V_0') are connected in G by a trail of even length, we infer from the Lemma that H_φ is either connected or consists of the two components V_0, V_0'. Let us treat the latter case first.

Let $H_\varphi = H_0 + H_0'$. From $uv \in G \iff \varphi u, v \in H_\varphi = H_0 + H_0'$, we infer that $\varphi u \in V_0'$ for $u \in V_0$, and vice versa. Hence φ restricted to V_0 is a bijection between V_0 and V_0'. Let us identify $u \equiv \varphi u$ ($u \in V_0$), *i.e.* we regard both H_0 and H_0' as graphs on V_0. We claim that $\psi = \varphi^{-2}$ is an admissible mapping of H_0 giving rise to H_0' (on V_0). Indeed, by (*) we have

$$uv \in H_0 \iff u, \varphi^{-1}v \in G \iff \varphi u, \varphi^{-1}v \in H_0' \quad (\text{on } V_0'),$$

which by the identification $u \equiv \varphi u$ means $u, \varphi^{-2}v \in H_0'$ (on V_0), *i.e.* $uv \in H_0 \iff u, \psi v \in H_0'$ (on V_0). Hence ψ is an admissible mapping for H_0 generating H_0'. The proof of the converse statement follows the same lines.

Suppose now that $H = H_\varphi$ is connected with $H \not\cong G$. There must be an edge $uv \in H_\varphi$ with $u \in V_0$, $v \in V_0'$ and $uv \notin G$, since otherwise $H_\varphi = G$. By (*), this means $\varphi^{-1}u, v \in G$, $u, \varphi^{-1}v \in G$, hence $\varphi^{-1}u \in V_0$, $\varphi^{-1}v \in V_0'$ and $\varphi^{-1}u \neq u$, $\varphi^{-1}v \neq v$. Consider the orbits of φ. With $w = \varphi^{-1}u$, $\varphi w = u \in V_0$ and $x = \varphi^{-1}v$, $\varphi x = v \in V_0'$, we know that there is an orbit $O_1 = \{w, u, \ldots\}$ containing two consecutive V_0-vertices, and similarly an orbit O_2 (possibly equal to O_1) containing two consecutive V_0'-vertices. If all orbits are monochromatic, *i.e.* they consist of V_0-vertices or V_0'-vertices only, then $G_\mathbb{O}$ is bipartite, and we are done, by Proposition 9. Hence, assume there is an orbit O_j containing both V_0- and V_0'-vertices. Since G connected implies $G_\mathbb{O}$ is connected, we infer by Proposition 5(B) that *any* orbit contains vertices of both color classes. Consider again the orbit $O_1 = \{w, u = \varphi w, \ldots\}$, $w, u \in V_0$. We now derive a contradiction by showing that w and u are not connected in G.

Let $w = x_1, x_2, x_3, \ldots, x_{2t+1} = u$ be a path in G. Then $x_1, x_3, \ldots, x_{2t+1} \in V_0$ and $x_2, x_4, \ldots, x_{2t} \in V_0'$. From $wx_2 \in G$ and $u = \varphi w$, we conclude $u, \varphi^{-1}x_2 \in G$, $\varphi^{-1}x_2, \varphi x_3 \in G, \ldots, \varphi^{-1}x_{2t}, \varphi u \in G$, hence $\varphi^{-1}x_2 \in V_0'$, $\varphi x_3 \in V_0, \ldots, \varphi u \in V_0$. Since the orbit O_1 also contains V_0'-vertices, we infer $\varphi u \neq w, u$. Now we start the next round. Because of $u, \varphi^{-1}x_2 \in G$, we have $\varphi u, \varphi^{-2}x_2 \in G$, $\varphi^{-2}x_2, \varphi^2 x_3 \in G, \ldots, \varphi^{-2}x_{2t}, \varphi^2 u \in G$ with $\varphi^{-2}x_2 \in V_0'$, $\varphi^2 x_3 \in V_0, \ldots, \varphi^2 u \in V_0$. Hence $\varphi^2 u \neq w, u, \varphi u$. Continuing in this way, we come to the conclusion that the orbit O_1 is infinite, which is absurd. □

Our final theorem is again an NP-completeness result.

Theorem 13. *The decision problem* (NIH) *whether a connected bipartite graph has a non-isomorphic hypomorph is NP-complete.*

For the proof, we need some preparations. Suppose Γ is an $n \times m$ matrix. We call Γ *indecomposable* if there is no partition $A \cup B$ of $\{1, \dots, m\}$ into nonempty subsets such that the column indices of the nonzero entries of each row of Γ are either all contained in A or all in B. We have the following supplement to Theorem 2:

Lemma 14. *The decision problem* NL *is NP-complete even if the input is restricted to indecomposable* $\{0, 1\}$-*matrices.*

Proof. Consider again the matrix Γ constructed in the proof of Theorem 2. Add as the $(n + m + 2)^{\text{nd}}$ row and column, the vectors e and e^T to Γ, where $e = (\underbrace{1, \dots, 1}_{n+m+1}, 0)$, and call the new matrix Γ'. It is easy to see that Γ' is indecomposable, and that the transformation from O2FPFA works with Γ' instead of Γ as well. $\square$

Proof of Theorem 13. We give a transformation from NL with input restricted to inde-composable matrices. Note that a $\{0, 1\}$-matrix Γ is indecomposable if and only if the bipartite graph with adjacency matrix

$$A = A(\Gamma) = \begin{pmatrix} 0 & \Gamma \\ \Gamma^T & 0 \end{pmatrix}$$

is connected.

Claim: An $n \times n$ matrix Γ with $\{0, 1\}$-entries is graphic via $R \in \text{Perm}_n$ if and only if Γ^T is graphic via R^T.

To see this, assume that $\Gamma = (\gamma_{i,j})$ satisfies (1) and (2). $R\Gamma R = \Gamma^T$, hence $\Gamma R = R^T \Gamma^T$, and thus $(R^T \Gamma^T)^T = \Gamma R = R^T \Gamma^T$. If R is the permutation matrix $(\delta_{\sigma(i),j})$ with permutation σ, the element in position (i, i) in $R\Gamma$ and $R^T \Gamma^T$ is $\gamma_{\sigma(i),i}$ and $\gamma_{i,\sigma^{-1}(i)}$, respectively. Hence Γ^T satisfies (1) and (2), and the claim follows in view of $\Gamma^{TT} = \Gamma$.

Now suppose that Γ is an indecomposable input for NL. We claim that the connected bipartite graph G with adjacency matrix $A(\Gamma)$ has a nonisomorphic hypomorph if and only if Γ is graphic. Suppose first that Γ is graphic. Choose $R \in \text{Perm}_n$ satisfying (1) and (2). Then the graph with adjacency matrix

$$\begin{pmatrix} 0 & R^T \\ R & 0 \end{pmatrix} A(\Gamma) = \begin{pmatrix} R^T \Gamma^T & 0 \\ 0 & R\Gamma \end{pmatrix}$$

is a non-isomorphic hypomorph of G (*cf.* the claim above). If, on the other hand, the connected, bipartite graph G has a non-isomorphic hypomorph H, then by Theorem 12 there is a permutation matrix

$$P := \begin{pmatrix} 0 & S \\ R & 0 \end{pmatrix}$$

such that

$$PA(\Gamma) = \begin{pmatrix} S\Gamma^T & 0 \\ 0 & R\Gamma \end{pmatrix}$$

is the adjacency matrix of H. Hence $R\Gamma$ satisfies (1) and (2), and is thus graphic.   □

4. Concluding remarks

Finally, we should like to indicate some possible directions of further research.

First, it would be interesting to know more about the complexity status of the MS problem (see Section 2). Is it NP-complete? Is it ISOMORPHISM-complete?

Our second suggestion is to study the structural differences between hypomorphic graphs $G \approx H$. Since they have the same neighborhood lists, their degree sequences are identical. What about other parameters? Lemma 6 implies that the number of components of H is at most twice the number of components of G. On the other hand, Theorem 12 shows that the coloring numbers can be arbitrarily far apart, and our final example demonstrates that planarity is also not preserved:

Let H_0 be the non-planar graph $K_{3,3}$ with two edges subdivided, and $H = H_0 + H_0'$ as before. Then the bipartite graph G arising from the usual involution $v \longleftrightarrow v'$ is planar.

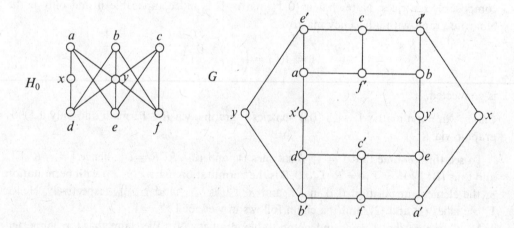

References

[1] Erdős, P. and Gallai, T. (1960) Graphs with prescribed degrees of vertices (Hungarian). *Math. Lapok* **11** 264–274.

[2] Erdős, P., Jacobson, M.S. and Lehel, J. (1991) Graphs realizing the same degree sequences and their respective clique numbers. In: Alavi *et al.* (eds.) *Graph Theory, Combinatorics and Applications*, John Wiley, 439–449.

[3] Hajnal, A. and Sós, V. (1978) Combinatorics. *Coll. Math. Soc. J. Bolyai* **18**, North-Holland.

[4] Lubiw, A. (1981) Some NP-complete problems similar to graph isomorphism. *SIAM J. Computing* **10** 11–21.

[5] Köbler, J., Schöning, U. and Torán, J. (1993) The graph isomorphism problem. *Progress in Theoretical Computer Science*, Birkhäuser.

Threshold Functions for H-factors

NOGA ALON[†] and RAPHAEL YUSTER

Department of Mathematics, Raymond and Beverly Sackler Faculty of Exact Sciences,
Tel Aviv University, Tel Aviv, Israel

Let H be a graph on h vertices, and G be a graph on n vertices. An H-factor of G is a spanning subgraph of G consisting of n/h vertex disjoint copies of H. The *fractional arboricity* of H is $a(H) = \max\{\frac{|E'|}{|V'|-1}\}$, where the maximum is taken over all subgraphs (V', E') of H with $|V'| > 1$. Let $\delta(H)$ denote the minimum degree of a vertex of H. It is shown that if $\delta(H) < a(H)$, then $n^{-1/a(H)}$ is a sharp threshold function for the property that the random graph $G(n, p)$ contains an H-factor. That is, there are two positive constants c and C so that for $p(n) = cn^{-1/a(H)}$, almost surely $G(n, p(n))$ does not have an H-factor, whereas for $p(n) = Cn^{-1/a(H)}$, almost surely $G(n, p(n))$ contains an H-factor (provided h divides n). A special case of this answers a problem of Erdős.

1. Introduction

All graphs considered here are finite, undirected and simple. If G is a graph of order n and H is a graph of order h, we say that G has an H-*factor* if it contains n/h vertex disjoint copies of H. Thus, for example, a K_2-factor is simply a perfect matching.

Let $G = G(n, p)$ denote, as usual, the random graph with n vertices and edge probability p. In the extensive study of the properties of random graphs, (see [5] for a comprehensive survey), many researchers have observed that there are sharp *threshold functions* for various natural graph properties. For a graph property A and for a function $p = p(n)$, we say that $G(n, p)$ satisfies A *almost surely* if the probability that $G(n, p(n))$ satisfies A tends to 1 as n tends to infinity. We say that a function $f(n)$ is a *sharp threshold function* for the property A if there are two positive constants c and C so that $G(n, cf(n))$ almost surely does not satisfy A and $G(n, Cf(n))$ satisfies A almost surely.

Let H be a fixed graph with h vertices. Our concern will be to find the threshold function for the property that $G(n, p)$ contains an H-factor, (assuming, of course, that h divides n). The case $H = K_2$ has been established by Erdős and Rényi in [7]. They showed

† Research supported in part by a United States Israeli BSF grant

that $\log(n)/n$ is a sharp threshold function in this case, and there are many subsequent papers by various authors that supply more detailed information on this problem. In the general case, however, it is much harder to determine the threshold function. Even for the case $H = K_3$ the threshold is not known (*cf.* [3, Appendix B]). In [3, page 243], P. Erdős raised the question of determining the threshold function when $H = H_6$ is the graph on the 6 vertices $a_1, a_2, a_3, b_1, b_2, b_3$ whose 6 edges are a_1b_1, a_2b_2, a_3b_3 and a_1a_2, a_2a_3, a_1a_3. It turns out that in this case $n^{-2/3}$ is a sharp threshold function for the existence of an H-factor. In fact, the graph H_6 is just an element of a large family of graphs H for which we can determine a sharp threshold function for the existence of an H-factor. In order to define this family we need the following definition.

For a simple undirected graph H that contains edges, define the *fractional arboricity* of H as

$$a(H) = \max\left\{\frac{|E'|}{|V'|-1}\right\},$$

where the maximum is taken over all subgraphs (V', E') of H with $|V'| > 1$. Observe that by the well-known theorem of Nash-Williams [13], $\lceil a(H) \rceil$ is just the arboricity of H, *i.e.*, the minimum number of forests whose union covers all edges of H. Denote by $\delta(H)$ the minimum degree of a vertex of H, and let $\mathscr{F}$ be the family of all graphs H for which $a(H) > \delta(H)$ (or, equivalently, the family of all graphs with arboricity bigger than the minimum degree). Our main result is the following

Theorem 1.1. *Let H be a fixed graph in $\mathscr{F}$. Then the following two statements hold:*

1. *There exists a positive constant c such that if $p = cn^{-1/a(H)}$, then almost surely $G(n, p)$ does not contain an H-factor.*

2. *There exists a positive constant C such that if $p = Cn^{-1/a(H)}$, then almost surely $G(n, p)$ contains an H-factor, assuming h divides n.*

Thus Theorem 1.1 asserts that for every $H \in \mathscr{F}$, $n^{-1/a(H)}$ is a sharp threshold function for the property that G contains an H-factor. In particular, the theorem shows that $n^{-2/3}$ is a sharp threshold function in the special case $H = H_6$ mentioned above.

Our method yields the following extension of Theorem 1.1 as well.

Theorem 1.2. *Define the set $\mathscr{G}$ recursively as follows:*

1. *$\mathscr{F} \subset \mathscr{G}$.*

2. *If C_1, C_2 are two of the connected components of some $H' \in \mathscr{G}$ and H is obtained from H' by adding to it less than $a(H')$ edges between C_1 and C_2, then $H \in \mathscr{G}$.*

If $H \in \mathscr{G}$, then $n^{-1/a(H)}$ is a sharp threshold function for the existence of an H-factor.

The proofs are presented in the next section. They rely on the Janson inequalities (*cf.* [6] and [8]), and on a method used by Alon and Füredi in [2].

The last section contains some concluding remarks and open problems.

2. The proofs

Proof of Theorem 1.1. We begin by establishing the first statement in Theorem 1.1, which is not difficult and, in fact, holds even when H is not a member of $\mathscr{F}$.

Lemma 2.1. *Let H be any fixed graph that contains edges, and let $a = a(H)$. There exists a positive constant $c = c(H)$ such that almost surely $G(n, p)$ does not contain an H-factor for $p = cn^{-1/a}$.*

Proof. Let $H' = (V', E')$ be any subgraph of H for which $|E'|/(|V'| - 1) = a$. Denote $|E'| = e'$ and $|V'| = v'$. Let $\{A_i : i \in I\}$ denote the set of all distinct labeled copies of H' in the complete labeled graph on n vertices. Let B_i be the event that $A_i \subset G(n, p)$, and let X_i be the indicator random variable for B_i. Let $X = \sum_{i \in I} X_i$ be the number of distinct copies of H' in $G(n, p)$. It suffices to show that almost surely $X < n/h$. The expectation of X clearly satisfies

$$E[X] \leq \binom{n}{v'} v'! (cn^{-\frac{v'-1}{e'}})^{e'} \leq c^{e'} n,$$

and yet clearly $E[X] = \Omega(n) \to \infty$. Choosing an appropriate constant c, we obtain $E[X] < n/2h$. We next show that $Var[X] = o(E[X]^2)$. This suffices, since by Chebyschev's inequality it implies that almost always $X < n/h$. For two copies A_i and A_j, we say that $i \sim j$ if they share at least one edge. Let $\Delta = \sum_{i \sim j} \Pr[B_i \wedge B_j]$, the sum taken over ordered pairs. Since $Var[X] \leq E[X] + \Delta$ and $E[X] \to \infty$, it remains to show that $\Delta = o(E[X]^2)$. The intersection of any A_i and A_j is a subgraph $H'' = (V'', E'')$ of H' (not necessarily an induced subgraph). We can therefore partition Δ into partial sums Δ'' corresponding to the various possible H''. It suffices to show that for each typical term Δ'', $\Delta'' = o(E[X]^2)$. Denote $|V''| = v''$ and $|E''| = e''$. Then

$$\Delta'' \leq \binom{n}{v'} v'! \binom{n - v'}{v' - v''} (v' - v'')! (cn^{-\frac{v'-1}{e'}})^{(2e'-e'')}$$

$$\leq c^{2e'-e''} n^{2v'-v''-\frac{v'-1}{e'}(2e'-e'')} \leq c^{2e'-e''} n,$$

since $\frac{v''-1}{e''} \geq \frac{v'-1}{e'}$. Hence $\Delta'' = o(E[X]^2)$. $\qquad\square$

Note that by considering the minimal H' for which $|E'|/(|V'| - 1) = a(H)$, we could obtain $\Delta = o(E[X])$ but our estimate suffices.

In order to prove the second part of Theorem 1.1, we need to state the Janson inequalities in our setting. Let Ω be a finite universal set, and let R be a random subset of Ω, where $\Pr[r \in R] = p_r$: these events being mutually independent over $r \in \Omega$. Let $\{A_i : i \in I\}$ be subsets of Ω, with I a finite index set. Let B_i be the event $A_i \subset R$, and let X_i be the indicator random variable for B_i, and $X = \sum_{i \in I} X_i$, the number of $A_i \subset R$. For $i, j \in I$ we write $i \sim j$ if $i \neq j$ and $A_i \cap A_j \neq \emptyset$. We define

$$\Delta = \sum_{i \sim j} \Pr[B_i \wedge B_j],$$

where the sum is taken over ordered pairs. Note that if the B_i were all independent, we would have $\Delta = 0$. The Janson inequalities state that when the B_i are 'mostly' independent, then X is still close to a Poisson distribution with mean $\mu = E[X]$. The first inequality (cf. [6]) applies when Δ is small relative to μ,

Lemma 2.2. *Let B_i, Δ, μ be as above, and assume that $Pr[B_i] \leq \epsilon$ for all i. Then*

$$\Pr[X = 0] \leq \exp\left(-\mu + \frac{1}{1-\epsilon}\frac{\Delta}{2}\right).$$

When $\Delta/2 \geq \mu(1 - \epsilon)$, the bound in Lemma 2.2 is worthless. Even for Δ slightly less, it is improved by the second Janson inequality (cf. [6])

Lemma 2.3. *Under the assumptions of Lemma 2.2 and the further assumption that $\Delta \geq \mu(1 - \epsilon)$,*

$$\Pr[X = 0] \leq \exp\left(-\frac{\mu^2(1 - \epsilon)}{2\Delta}\right).$$

The Janson inequalities play a crucial role in the proof of the next lemma.

Lemma 2.4. *If H is any fixed graph with $h - 1$ vertices and fractional arboricity $a = a(H)$, there exists a constant $C = C(H)$ such that almost surely $G(n, p)$ contains n/h vertex disjoint copies of H, where $p = Cn^{-1/a}$.*

Proof. It suffices to show that almost surely every subset of n/h vertices contains a copy of H. Fix such a subset of vertices, $A \subset \{1, 2, \ldots n\}$. We use a similar notation to that in the proof of Lemma 2.1. That is, X denotes the number of labeled copies of H in A, and Δ'' denotes the sum on all ordered pairs of copies of H whose intersection corresponds to a fixed subgraph H'' of H. However, this time we need to bound $\mu = E[X]$ from below;

$$\mu = E[X] \geq \binom{n/h}{h-1}(Cn^{-1/a})^e \geq (h(h-1))^{1-h}C^e n^{-e/a+h-1},$$

where e is the number of edges of H. Note that $E[X] = \Omega(n)$, since $e/(h - 2) = e/((h - 1) - 1) \leq a$. We now bound Δ'' from above,

$$\Delta'' \leq \binom{n/h}{h-1}(h-1)!\binom{n/h - (h-1)}{h-1-v''}(h-1-v'')!(Cn^{-1/a})^{2e-e''}$$

$$\leq C^{2e-e''}n^{2h-2-v''-(1/a)(2e-e'')}.$$

Claim 1. *If $C > (h(h-1))^{2h-2}2^{h^2+1}$, then*

$$\frac{\mu^2}{2\Delta} > n. \tag{1}$$

Proof. Note that $e'' \geq 1$ in each term Δ''. Hence

$$\frac{\mu^2}{\Delta''} \geq \frac{(h(h-1))^{2-2h}C^{2e}n^{-2e/a+2h-2}}{C^{2e-e''}n^{2h-2-v''-(1/a)(2e-e'')}}$$

$$\geq (h(h-1))^{2-2h}Cn^{v''-e''/a} \geq (h(h-1))^{2-2h}Cn \geq 2^{h^2+1}n.$$

There are less than 2^{h^2} subgraphs of H, so the last inequality (which holds for any Δ'') implies (1). This completes the proof of the claim. □

Returning to the proof of the Lemma with C selected as in the above claim, we proceed as follows. If $\Delta < \mu$, we use Lemma 2.2. Note that in our case we may pick ϵ as an arbitrary small constant, and the lemma implies (since $\mu > 3n$ for our C)

$$\Pr[X = 0] \leq \exp(-\mu/3) \leq \exp(-n).$$

If $\Delta > \mu$, we use Lemma 2.3. Picking $\epsilon = 0.1$ and using the above claim we obtain

$$\Pr[X = 0] \leq \exp(-0.9n).$$

In any case, $\binom{n}{n/h}\Pr[X = 0]$ tends (even exponentially) to zero when n tends to infinity, and this completes the proof of the Lemma. □

Armed with Lemma 2.4 we can now complete the proof of Theorem 1.1. Given $H \in F$, let d be a vertex of minimal degree in H and denote the set of its neighbors by $N(d)$. Set $H' = H \setminus \{d\}$. Note that $a(H') = a(H)$, since $H \in \mathscr{F}$. We apply Lemma 2.4 by first setting $p' = C'n^{-1/a(H)}$, where C' is chosen as in Lemma 2.4. Almost every $G(n, p')$ will have n/h vertex disjoint copies of H'. We now need to match every remaining vertex in G to a copy of H' in such a way that there is an edge between the assigned vertex and each vertex in the set corresponding to $N(d)$ in the matched copy. We use (a modified version of) the method from [2] to do so. We choose the edges of $G(n, p)$ once again (but still keeping the edges of the first selection) with probability $p'' = n^{-1/a(H)}$. Note that this is the same probability space as $G(n, p)$, where $(1-p) = (1-p')(1-p'')$. We define a random bipartite graph with one side being the n/h pairwise disjoint copies of H', and the other side being the remaining vertices of G. There is an edge of the bipartite graph between a copy and a remaining vertex if the vertex can be matched to the copy using only the new randomly chosen edges. The edge probability in this bipartite graph is $n^{-\delta/a}$, where δ is the degree of d. Moreover, crucially, the edges of this bipartite graph are chosen independently, since their existence is determined by considering pairwise disjoint subsets of edges of our random graph. Since $\delta < a$, it follows from the result in [7] that almost always there is a perfect matching. Note also that $p = p' + p'' - p'p''$, and since $p'' < p'$, we may generously set $C = 2C'$. This completes the proof of Theorem 1.1. □

Proof of Theorem 1.2. The fact that the threshold function for the existence of an H-factor is at least $n^{-1/a(H)}$ follows directly from Lemma 2.1. Let $H \in \mathscr{G}$. We must show that there is a constant $C = C(H)$ such that almost always $G(n, Cn^{-1/a(H)})$ contains an H-factor. We prove this by induction on the minimal number of applications of rule 2 in the definition of $\mathscr{G}$ needed to demonstrate the membership of H in $\mathscr{G}$. If no such application is needed, then $H \in \mathcal{F}$ and the result follows from Theorem 1.1. Otherwise, there is an $H' \in \mathcal{G}$ and two connected components C_1 and C_2 of it such that H is obtained from H' by adding a set R of edges between C_1 and C_2, where $r = |R| < a(H')$. It is easy to check that this implies that $a(H) = a(H')$. By the induction hypothesis, there exists a constant C' such that almost surely $G(n, C'n^{-1/a(H)})$ contains an H'-factor. As in the proof of Theorem 1.1, we choose the edges of $G(n, p)$ once again with probability $n^{-1/a(H)}$. We define a random bipartite graph with one side being the n/h pairwise disjoint copies of C_1 and the other side being the n/h pairwise disjoint copies of C_2 (that are also pairwise disjoint with the copies of C_1). There is an edge in the bipartite graph between a vertex corresponding to a copy of C_1 and a vertex corresponding to a copy of C_2 if the edges corresponding to R exist in $G(n, p)$ among the freshly selected edges. The edge probability in this bipartite graph is $n^{-r/a(H)}$ and the choices of distinct edges are mutually independent. Since $r < a(H)$ it follows, as in Theorem 1.1, that $G(n, p)$ almost surely contains an H-factor, where $(1 - p) = (1 - C'n^{-1/a(H)})(1 - n^{-1/a(H)})$. We can now set $C = C' + 1$ and $p = Cn^{-1/a(H)}$ to complete the proof. $\square$

3. Concluding remarks

Somewhat surprising is the fact that there are many regular graphs H that fall into the category of Theorem 1.2. As an example, consider three arbitrary cubic graphs, and subdivide an edge in each of them. Add a new vertex and connect it to the vertices of degree 2 that were introduced by the subdivisions. The resulting graph H is cubic, and satisfies the properties of the graphs in Theorem 1.2, which supplies the appropriate threshold function for the existence of an H-factor.

Our theorems raise a natural algorithmic question. Suppose H is a graph in the family $\mathcal{G}$ defined in Theorem 1.2 and $a(H) = a$. Then, by the theorem, there is a positive constant $C = C(H)$ such that for $p(n) = Cn^{-1/a}$, the random graph $G(n, p)$ contains, almost surely, an H-factor, provided $|V(H)|$ divides n. Can we find such an H-factor efficiently? The proof easily supplies a polynomial time algorithm for every fixed H. Moreover, this algorithm can be parallelized. To see this, observe that in the first step of the proof it suffices to find a *maximal* set of vertex disjoint copies of an appropriate graph H' in our random graph $G(n, p)$, where the maximality is with respect to containment. Such a set can be found in NC (i.e., in polylogarithmic time, using a polynomial number of parallel processors) using any of the known NC-algorithms for the maximal independent set problem (see, e.g., [10], [11], [1]). The rest of the algorithm only has to find perfect matchings in appropriately defined graphs, and this can be done in (randomized) NC by the results of [9] or [12]. Thus, the H-factors whose existence is guaranteed almost surely in Theorems 1.1 and 1.2 can actually be found, almost surely, efficiently (even in parallel).

The methods used in the proofs of the theorems can be used to compute the thresholds for the existence of spanning graphs other than H-factors. For example, let H be the 4 vertex graph consisting of a vertex of degree 1 joined to a triangle. Let Q be the graph obtained from $n/4$ pairwise disjoint copies of H, $H_1, \ldots, H_{\frac{n}{4}}$, where a_i is the vertex of degree 1 in H_i, by adding a cycle of length $n/4$ on the vertices a_i. By Theorem 1.1, $p = Cn^{-2/3}$ is a threshold for the existence of an H-factor. Suppose we now draw edges again with probability of $n^{-2/3}$ (which is much more than needed) in the subgraph of the $n/4$ vertices of degree 1. By the result of Pósa in [14], we will almost always have a Hamilton cycle in this subgraph. Therefore $n^{-2/3}$ is a sharp threshold function for the property that Q is a spanning subgraph of $G(n, p)$. Various similar examples can be given. Here, too, the proof is algorithmic, by applying the result of [4].

The following conjecture seems plausible.

Conjecture 3.1. *Let H be an arbitrary fixed graph with edges. Then the threshold for the property that $G(n, p)$ contains an H-factor, (if h divides n) is $n^{-1/a(H)+o(1)}$.*

We note that Lemma 2.1 shows that the above threshold is at least $n^{-1/a(H)}$. Also, the $o(1)$ term cannot be omitted entirely, because, for example, $\log(n)/n$ is the threshold for a K_2-factor, although $a(K_2) = 1$. Similarly, the threshold for a K_3-factor is at least $\log(n)^{1/3}n^{-2/3}$, since as proved by Spencer [15], this is the threshold for every vertex to lie on a triangle, which is an obvious necessary condition in our case.

Note added in proof. We have recently learned that A. Ruciński, in 'Matching and covering the vertices of a random graph by copies of a given graph' *Discrete Math.* (1992) **105** 185–197, proved, independently (and before us), Theorem 1.1, using similar techniques. He did not prove the more general Theorem 1.2.

References

[1] Alon, N., Babai, L. and Itai, A. (1986) A fast and simple randomized parallel algorithm for the maximal independent set problem. *J. Algorithms* **7** 567–583.

[2] Alon, N. and Füredi, Z. (1992) Spanning subgraphs of random graphs. *Graphs and Combinatorics* **8** 91–94.

[3] Alon, N. and Spencer, J. H. (1991) *The Probabilistic Method*, John Wiley and Sons Inc., New York.

[4] Angluin, D. and Valiant, L. Fast probabilistic algorithms for Hamilton circuits and matchings. *J. Computer Syst. Sci.* **18** 155–193.

[5] Bollobás, B. (1985) *Random Graphs*, Academic Press.

[6] Boppana, R. B. and Spencer, J. H. (1989) A useful elementary correlation inequality. *J. Combinatorial Theory*, Ser. A **50** 305–307.

[7] Erdős, P. and Rényi, A. (1966) On the existence of a factor of degree one of a connected random graph. *Acta Math. Acad. Sci. Hungar.* **17** 359–368.

[8] Janson, S. (1990) Poisson approximation for large deviations. *Random Structures and Algorithms* **1** 221–230.

[9] Karp, R. M., Upfal, E. and Wigderson, A. (1986) Constructing a perfect matching in random NC. *Combinatorica* **6** 35–48.

[10] Karp, R. M. and Wigderson, A. (1985) A fast parallel algorithm for the maximal independent set problem. *J. ACM* **32** 762–773.

[11] Luby, M. (1986) A simple parallel algorithm for the maximal independent set problem. *SIAM J. Computing* **15** 1036–1053.

[12] Mulmuley, K., Vazirani, U. V. and Vazirani, V. V. (1987) Matching is as easy as matrix inversion. *Proc. 19th ACM STOC* 345–354.

[13] Nash-Williams, C. St. J. A. (1964) Decomposition of finite graphs into forests. *J. London Math. Soc.* **39** 12.

[14] Pósa, L. (1976) Hamiltonian circuits in random graphs. *Discrete Math.* **14** 359–364.
[15] Spencer, J. H. (1990) Threshold functions for extension statements. *J. Combinatorial Theory,* Ser. A **53** 286–305.

A Rate for the Erdős–Turán Law*

A. D. BARBOUR† and SIMON TAVARÉ‡

† Institut für Angewandte Mathematik, Universität Zürich, Winterthurerstrasse 190, CH-8057, Zürich, Switzerland

‡ Department of Mathematics, University of Southern California, Los Angeles, CA 90089-1113

The Erdős–Turán law gives a normal approximation for the order of a randomly chosen permutation of n objects. In this paper, we provide a sharp error estimate for the approximation, showing that, if the mean of the approximating normal distribution is slightly adjusted, the error is of order $\log^{-1/2} n$.

1. Introduction

Let σ denote a permutation of n objects, and $O(\sigma)$ its order. Landau [13] proved that $\max_\sigma \log O(\sigma) \sim \{n \log n\}^{1/2}$. In contrast, if σ is a single cycle of length n, $\log O(\sigma) = \log n$, such cycles constituting a fraction $1/n$ of all possible σ's. In view of the wide discrepancy between these extremes, the lovely theorem of Erdős and Turán (1967) comes as something of a surprise: that, for any x,

$$\frac{1}{n!} \# \{\sigma : \log O(\sigma) < \tfrac{1}{2}\log^2 n + x\{\tfrac{1}{3}\log^3 n\}^{1/2}\} \sim \Phi(x),$$

where Φ denotes the standard normal distribution function. In probabilistic terms, their result is expressed as

$$\mathbb{P}[\{\tfrac{1}{3}\log^3 n\}^{-1/2} (\log O(\sigma) - \tfrac{1}{2}\log^2 n) < x] \sim \Phi(x), \tag{1.1}$$

with σ now thought of as a permutation chosen at random, each of the $n!$ possibilities being equally likely. They remark that

'Our proof is a direct one and rather long; but a *first* proof can be as long as it wants to be. It would be however of interest to deduce it from the general principles of probability theory.'

* This work was supported in part by NSF grant DMS90-05833 and in part by Schweizerischer NF Projekt Nr 20-31262.91.

They also entertain hopes of finding a sharp remainder for their approximation.

Shorter probabilistic proofs of (1.1) are given by [5], [6] and [1], the last exploiting the Feller coupling to a record value process. Stein (unpublished) gives another coupling proof, with an error estimate of order $\log^{-1/4} n \{\log \log n\}^{1/2}$, which he describes as 'rather poor'. In fact, [16] sharpens the approach of Erdős and Turán, showing that the first correction to (1.1) is a mean shift of $-\log n \log \log n$, and that the error then remaining is of order at most $O(\log^{-1/2} n \log \log \log n)$. Nicolas also conjectures that the iterated logarithm in the error is superfluous. Our birthday present is to show this, by probabilistic means, not only for the uniform distribution on the set of permutations, but also under any Ewens sampling distribution. Since many combinatorial structures are, in a suitable sense, very closely approximated by one of the Ewens sampling distributions (see [4]), the result carries over easily to many other contexts. A typical example is the l.c.m. of the degrees of the factors of a random polynomial over the finite field with q elements, thus improving upon a theorem of [15].

Consider the probability measure μ_θ on the permutations of n objects determined by

$$\mu_\theta(\sigma) = \frac{\theta^{k(\sigma)}}{\theta_{(n)}}, \tag{1.2}$$

where $k(\sigma)$ is the number of cycles in σ, $\theta > 0$ is a parameter that can be chosen at will, and where rising factorials are denoted by

$$x_{(n)} = x(x+1)\ldots(x+n-1), \quad x_{(0)} = 1.$$

If $\theta = 1$, the uniform distribution is recovered. Under μ_θ, the probability of the set of permutations having a_j cycles of length j, $1 \leq j \leq n$, is given by

$$I\left\{\sum_{j=1}^{n} ja_j = n\right\} \frac{n!}{\theta_{(n)}} \prod_{j=1}^{n} \left(\frac{\theta}{j}\right)^{a_j} \frac{1}{a_j!}, \tag{1.3}$$

as may be verified by multiplying the probability (1.2) by the number of permutations that have the given cycle index.

The joint distribution of cycle counts given by (1.3) is known as the Ewens sampling formula with parameter θ. It was derived by Ewens [8] in the context of population genetics. Ewens [9] provides an account of this theory that is accessible to mathematicians.

Under the Ewens sampling formula, the joint distribution of the cycle counts converges to that of independent Poisson random variables with mean θ/i as $n \to \infty$. Indeed, using the Feller coupling, the cycle counts for all values of n can be linked simultaneously on a common probability space with a single set of independent Poisson random variables with the appropriate means. The following precise statement of this fact comes essentially from [2].

Proposition 1.1. *Let $\{\xi_j, j \geq 1\}$ be a sequence of independent Bernoulli random variables satisfying*

$$\mathbb{P}[\xi_j = 1] = \frac{\theta}{\theta+j-1}. \tag{1.4}$$

Define $(Z_{jm}, j \geqslant 1)$ *by*

$$Z_{jm} = \sum_{i=m+1}^{\infty} \xi_i (1 - \xi_{i+1}) \cdots (1 - \xi_{i+j-1}) \xi_{i+j}, \tag{1.5}$$

and set $Z_j \equiv Z_{j0}$ *and* $Z = (Z_j, j \geqslant 1)$. *Define* $C^{(n)} = (C_j(n), j \geqslant 1)$ *by*

$$C_j(n) = \sum_{i=1}^{n-j} \xi_i (1 - \xi_{i+1}) \cdots (1 - \xi_{i+j-1}) \xi_{i+j} + \xi_{n-j+1}(1 - \xi_{n-j+2}) \cdots (1 - \xi_n)$$

$$= Z_j - Z_{j, \, n-j} + \xi_{n-j+1}(1 - \xi_{n-j+2}) \cdots (1 - \xi_n) \tag{1.6}$$

for $1 \leqslant j \leqslant n$, *setting* $C_j(n) = 0$ *for* $j > n$. *Then* $\mathbb{P}[(C_1(n), \ldots, C_n(n)) = (a_1, \ldots, a_n)]$ *is given by* (1.3), *and the* Z_j *are independent Poisson random variables with* $\mathbb{E}Z_j = \theta/j$. *Furthermore, for* $j \geqslant 1$,

$$Z_j - Z_{jn} - I[J_n + K_n = j+1] \leqslant C_j(n) \leqslant Z_j + I[J_n = j], \tag{1.7}$$

where J_n *and* K_n *are defined by*

$$J_n = \min \{j \geqslant 1 : \xi_{n-j+1} = 1\} \quad and \quad K_n = \min \{j \geqslant 1 : \xi_{n+j} = 1\}. \tag{1.8}$$

With this representation, the order of the random permutation is $O_n(C^{(n)})$, where, for any $a \in \mathbb{N}^{\infty}$,

$$O_n(a) = \text{l.c.m.} \{i : 1 \leqslant i \leqslant n, a_i > 0\} \leqslant P_n(a) = \prod_{i=1}^{n} i^{a_i}.$$

On the other hand, from (1.6), $C_j(n)$ is close to Z_j for each j when n is large, so $\log O_n(C^{(n)})$ might plausibly be well approximated by $\log O_n(Z)$. Now functions involving Z are very much easier to handle than are the same functions of $C^{(n)}$, because the components Z_j of Z are independent and have known distributions. In particular, $\log O_n(Z)$ is close enough for our purposes to $\log P_n(Z) - \theta \log n \log \log n$, and

$$\log P_n(Z) = \sum_{i=1}^{n} Z_i \log i$$

is just a sum of independent random variables. The classical Berry–Esséen theorem [10, p. 544, Theorem 2] can thus be invoked to determine the accuracy of the normal approximation to its distribution.

The above arguments, justified in detail in Section 2, lead to the following result.

Theorem 1.2. *If* $C^{(n)}$ *is distributed according to the Ewens sampling formula* (1.3) *with parameter* θ,

$$\sup_x \left| \mathbb{P}\left[\left\{ \frac{\theta}{3} \log^3 n \right\}^{-1/2} \left(\log O_n(C^{(n)}) - \frac{\theta}{2} \log^2 n + \theta \log n \log \log n \right) \leqslant x \right] - \Phi(x) \right|$$

$$= O(\{\log n\}^{-1/2}).$$

It would not be difficult to give an explicit bound for the constant implied in the error term. Indeed, the leading contributions arise from a Berry–Esséen estimate, for which the necessary quantities are estimated in Proposition 2.4, from inequality (2.1), for which (2.2) and Lemma 2.5 already provide a bound, and from the next mean correction, which requires a more careful asymptotic evaluation following (2.4).

A process variant of Theorem 1.2 can also be formulated. Let W_n be the random element of $D[0, 1]$ defined by

$$W_n(t) = \left\{\frac{\theta}{3}\log^3 n\right\}^{-1/2}\left(\log O_{[n^t]}(C^{(n)}) - \frac{\theta}{2}t^2\log^2 n\right).$$

Theorem 1.3. *It is possible to construct $C^{(n)}$ and a standard Brownian motion W on the same probability space, in such a way that*

$$\mathbb{E}\left\{\sup_{0 \leqslant t \leqslant 1} |W_n(t) - W(t^3)|\right\} = O\left(\frac{\log\log n}{\sqrt{\log n}}\right).$$

2. Proofs

As previously indicated, the proof of Theorem 1.2 consists of showing that $\log O_n(C^{(n)})$ is close enough to $\log O_n(Z)$, which in turn is close enough to $\log P_n(Z) - \theta \log n \log\log n$. The Berry–Esséen theorem then gives the normal approximation for $\log P_n(Z)$.

For vectors a and b, define $|a - b| = \sum_i |a_i - b_i|$. Since $O_n(a) \leqslant O_n(b)n^{|b-a|}$ whenever a and b are vectors with $a \leqslant b$, it follows from (1.7) that

$$\log O_n(Z) - (Y_n + 1)\log n \leqslant \log O_n(C^{(n)}) \leqslant \log O_n(Z) + \log n, \qquad (2.1)$$

where $Y_n = \sum_{j=1}^n Z_{jn}$ is independent of $C^{(n)}$, and

$$\mathbb{E}Y_n = \sum_{j=1}^n \sum_{i > n} \left(\frac{\theta}{\theta + i - 1}\right)\left(\frac{\theta}{\theta + i + j - 1}\right)\prod_{l=i+1}^{i+j-1}\left(\frac{l-1}{\theta + l - 1}\right)$$

$$\leqslant \theta^2 \sum_{j=1}^n \sum_{i > n}\left(\frac{1}{i-1}\right)\left(\frac{1}{i+j-1}\right) \leqslant \theta^2. \qquad (2.2)$$

Inequality (2.1) combined with (2.2) is enough for the closeness of $\log O_n(C^{(n)})$ and $\log O_n(Z)$.

Next, we can compute the difference between $\log O_n(Z)$ and $\log P_n(Z)$ using a formula of [5] and [14]:

$$\log P_n(Z) - \log O_n(Z) = \sum_p{}' \sum_{s \geqslant 1} (D_{n p^s} - 1)^+ \log p, \qquad (2.3)$$

where $\sum'$ and $\sum''$ denote sums over *prime* indices, and

$$D_{nk} = \sum_{j \leqslant n : k|j} Z_j.$$

Considering first its expectation, observe that, since $(d - 1)^+ = d - 1 + I\{d = 0\}$,

$$\mathbb{E}(D_{nk} - 1)^+ = \mathbb{E}D_{nk} - 1 + \mathbb{P}[D_{nk} = 0]$$

$$= \lambda_{nk} - 1 + e^{-\lambda_{nk}} \leqslant (\lambda_{nk} \wedge \tfrac{1}{2}\lambda_{nk}^2), \qquad (2.4)$$

where

$$\lambda_{nk} = \sum_{j \leqslant n : k|j} j^{-1}\theta = \begin{cases} k^{-1}\theta\psi([n/k]+1) & \text{if } k \leqslant n; \\ 0 & \text{if } k > n, \end{cases}$$

and $\psi(r+1) = \sum_{j=1}^{r} j^{-1}$. Hence

$$\mu_n := \mathbb{E}\{\log P_n(Z) - \log O_n(Z)\} = {\sum_{p}}' \sum_{s \geqslant 1} \log p \, \mathbb{E}(D_{np^s}-1)^+$$

$$= {\sum_{p}}' \sum_{s \geqslant 1} \log p(\lambda_{np^s}-1+\exp\{-\lambda_{np^s}\}) = {\sum_{p \leqslant \log n}}' \theta p^{-1}\log p \log n + O(\log n)$$

$$= \theta \log n \log\log n + O(\log n),$$

where the estimates use (2.4), integration by parts, and Theorems 7 and 425 of [11].
For the variability of $\log O_n(Z) - \log P_n(Z)$, we now need two lemmas.

Lemma 2.1. *For $p \neq q$ prime and $s, t \geqslant 1$,*

$$\mathrm{Cov}\,((D_{np^s}-1)^+, (D_{nq^t}-1)^+) \leqslant \frac{\theta(1+\log n)}{p^s q^t}.$$

Proof. Set

$$\lambda_1 = \sum_{\substack{j=1 \\ p^s|j}}^{n} j^{-1}, \quad \lambda_2 = \sum_{\substack{i=1 \\ q^t|i}}^{n} i^{-1} \quad \text{and} \quad \xi = \sum_{\substack{l=1 \\ p^s q^t|l}}^{n} l^{-1} \leqslant \frac{(1+\log n)}{p^s q^t},$$

and write $D_1 = D_{np^s}$ and $D_2 = D_{nq^t}$. Then, in the expansion

$$\mathrm{Cov}\,((D_1-1)^+, (D_2-1)^+) = \mathrm{Cov}\,(D_1, D_2) + \mathrm{Cov}\,(D_1, I[D_2=0])$$
$$+ \mathrm{Cov}\,(I[D_1=0], D_2) + \mathrm{Cov}\,(I[D_1=0], I[D_2=0]),$$

the first contribution is evaluated as

$$\mathrm{Cov}\,(D_1, D_2) = \mathbb{E}\left\{ \sum_{\substack{j=1 \\ p^s|j}}^{n} \sum_{\substack{i=1 \\ q^t|i}}^{n} (Z_j - j^{-1}\theta)(Z_i - i^{-1}\theta) \right\}$$

$$= \sum_{\substack{l=1 \\ p^s q^t|l}}^{n} \mathrm{Var}\, Z_l = \theta \sum_{\substack{l=1 \\ p^s q^t|l}}^{n} l^{-1} = \theta\xi,$$

because of the independence of the Z_j's. For the second contribution, we have

$$\mathrm{Cov}\,(D_1, I[D_2=0]) = \mathbb{P}\left[\bigcap_{\substack{i=1 \\ q^t|i}}^{n} \{Z_i=0\} \right]\{\mathbb{E}(D_1 \mid D_2=0) - \mathbb{E}D_1\} = -\theta\xi e^{-\theta\lambda_2},$$

and similarly for the third, and for the last we have

$$\mathrm{Cov}\,(I[D_1=0], I[D_2=0]) = e^{-\theta(\lambda_1+\lambda_2)}\{e^{\theta\xi}-1\} \leqslant \theta\xi e^{-\theta(\lambda_1+\lambda_2-\xi)}.$$

Hence

$$\mathrm{Cov}\,((D_1-1)^+,(D_2-1)^+) = \theta\xi\{1-e^{-\theta\lambda_1}-e^{-\theta\lambda_2}+e^{-\theta(\lambda_1+\lambda_2-\xi)}\} \leqslant \theta\xi,$$

proving the lemma.

Lemma 2.2.   *For* $1 \leqslant s \leqslant t$,

$$\mathrm{Cov}\,((D_{np^s}-1)^+,(D_{np^t}-1)^+) \leqslant \theta p^{-t}(1+\log n).$$

Proof.   The argument runs as for Lemma 2.1, with λ_1 defined as before, but now with

$$\xi = \lambda_2 = \sum_{\substack{l=1 \\ p^t | l}}^{n} l^{-1} \leqslant p^{-t}(1+\log n).$$

The computations now yield

$$\mathrm{Cov}\,(D_1,D_2) = \theta\xi; \quad \mathrm{Cov}\,(D_1,I[D_2=0]) = -\theta\xi e^{-\theta\lambda_2}; \quad \mathrm{Cov}\,(I[D_1=0],D_2) = -\theta\lambda_2 e^{-\theta\lambda_1}$$

and

$$\mathrm{Cov}\,(I[D_1=0],I[D_2=0]) = e^{-\theta\lambda_1}(1-e^{-\theta\lambda_2}),$$

and thus

$$\mathrm{Cov}\,((D_1-1)^+,(D_2-1)^+) = \theta\xi(1-e^{-\theta\lambda_2})+e^{-\theta\lambda_1}(1-e^{-\theta\lambda_2}-\theta\lambda_2) \leqslant \theta\xi(1-e^{-\theta\lambda_2}) \leqslant \theta\xi.$$

The two lemmas enable us to control the difference between $\log O_n(Z)$ and $\log P_n(Z)$ as follows.

Proposition 2.3.   *For any* $K > 0$,

$$\mathbb{P}[|\log P_n(Z)-\log O_n(Z)-\mu_n| > K\log n] = O\!\left(\frac{(\log\log n)^2}{\log n}\right).$$

Proof.   Write

$$\log P_n(Z)-\log O_n(Z) = \left(\sum_{p \leqslant \log^2 n}' + \sum_{p > \log^2 n}'\right)(D_{np}-1)^+\log p + \sum_{p}'\sum_{s \geqslant 2}(D_{np^s}-1)^+\log p$$
$$= V_1+V_2+V_3,$$

say. Lemmas 2.1 and 2.2 give

$$\mathrm{Var}\,V_1 \leqslant \sum_{p \leqslant \log^2 n}' \frac{\theta(1+\log n)}{p}\log^2 p + \sum_{p \neq q \leqslant \log^2 n}'' \frac{\theta(1+\log n)}{pq}\log p\log q$$
$$= O\,(\log n\,(\log\log n)^2);$$

it follows from (2.4) that

$$\mathbb{E}V_2 \leqslant \frac{\theta^2}{2} \sum_{p > \log^2 n}' p^{-2}\log p(1+\log n)^2 = O(1);$$

and Lemmas 2.1 and 2.2 imply that

$$\operatorname{Var} V_3 \leqslant {\sum_p}' \log^2 p \sum_{s,t \geqslant 2} \frac{\theta(1+\log n)}{p^{s \vee t}} + {\sum_{p \neq q}}'' \log p \log q \sum_{s,t \geqslant 2} \frac{\theta(1+\log n)}{p^s q^t} = O(\log n).$$

Thus, by Chebyshev's inequality,

$$\mathbb{P}[|V_1 - \mathbb{E}V_1| > \tfrac{1}{3}K \log n] = O(\log^{-1} n (\log \log n)^2);$$
$$\mathbb{P}[|V_2 - \mathbb{E}V_2| > \tfrac{1}{3}K \log n] = O(\log^{-1} n),$$

and

$$\mathbb{P}[|V_3 - \mathbb{E}V_3| > \tfrac{1}{3}K \log n] = O(\log^{-1} n),$$

proving the proposition.

We now use the closeness of the quantities $\log O_n(C^{(n)})$, $\log O_n(Z)$ and $\log P_n(Z) - \mu_n$ to prove Theorem 1.2. To do so, we introduce the standardized random variables

$$S_{1n} = \frac{\log P_n(Z) - \dfrac{\theta}{2}\log^2 n}{\sqrt{\dfrac{\theta}{3}\log^3 n}}; \quad S_{2n} = \frac{\log O_n(Z) + \mu_n - \dfrac{\theta}{2}\log^2 n}{\sqrt{\dfrac{\theta}{3}\log^3 n}},$$

and

$$S_{3n} = \frac{\log O_n(C^{(n)}) + \mu_n - \dfrac{\theta}{2}\log^2 n}{\sqrt{\dfrac{\theta}{3}\log^3 n}},$$

whose distributions we shall successively approximate. Since the quantity $\log P_n(Z)$ can be written in the form $\sum_{j=1}^n Z_j \log j$ as a weighted sum of independent Poisson random variables, the normal approximation for S_{1n} follows easily from the Berry–Esséen theorem.

Proposition 2.4. *There exists a constant $c_1 = c_1(\theta)$ such that*

$$\sup_x |\mathbb{P}[S_{1n} \leqslant x] - \Phi(x)| \leqslant c_1 \log^{-1/2} n.$$

Proof. It is enough to note that

$$\sum_{j=1}^n \mathbb{E}(Z_j \log j) = \theta \sum_{j=1}^n j^{-1} \log j = \frac{\theta}{2}(\log^2 n + O(1)),$$

that

$$\sum_{j=1}^n \operatorname{Var}(Z_j \log j) = \theta \sum_{j=1}^n j^{-1} \log^2 j = \frac{\theta}{3}(\log^3 n + O(1))$$

and that

$$\sum_{j=1}^n \mathbb{E}|Z_j - \mathbb{E}Z_j|^3 \log^3 j = O(\log^4 n):$$

indeed, for $j \geqslant \theta$,

$$\mathbb{E}|Z_j - \mathbb{E}Z_j|^3 = \frac{\theta}{j} + \frac{2\theta^3}{j^3} e^{-\theta/j} \leqslant \frac{\theta}{j}[1 + 2e^{-1}],$$

and hence, for $\theta \leqslant 2$,

$$\sum_{j=1}^{n} \mathbb{E}|Z_j - \mathbb{E}Z_j|^3 \log^3 j \leqslant \theta[1 + 2e^{-1}] \sum_{j=1}^{n} j^{-1} \log^3 j = \frac{\theta[1 + 2e^{-1}]}{4}(\log^4 n + O(1)). \qquad (2.5)$$

In order to show that S_{2n} and S_{3n} have almost the same distribution as S_{1n}, because of Proposition 2.3 and (2.1), one further lemma is required.

Lemma 2.5. *Let U and X be random variables with $\sup_x |\mathbb{P}[U \leqslant x] - \Phi(x)| \leqslant \eta$. Then, for any $\epsilon > 0$,*

$$\sup_x |\mathbb{P}[U + X \leqslant x] - \Phi(x)| \leqslant \eta + \frac{\epsilon}{\sqrt{2\pi}} + \mathbb{P}[|X| > \epsilon]. \qquad (2.6)$$

If W and Y are independent random variables with $\mathbb{E}Y < \infty$, and if $|W - U| \leqslant Y$, then

$$\sup_x |\mathbb{P}[W \leqslant x] - \Phi(x)| \leqslant 3\left\{\eta + \frac{4\mathbb{E}Y}{\sqrt{2\pi}}\right\}. \qquad (2.7)$$

Proof. The first part is standard. For the second, let $\delta_y = \mathbb{P}[W \leqslant y] - \Phi(y)$ and set $\Delta = \sup_y |\delta_y|$. Write $\rho = 3\mathbb{E}Y$ and $p = \mathbb{P}[Y > \rho]$, so that $p \leqslant 1/3$. Then, since for any x, $\{U \leqslant x\} \supset \{W + Y \leqslant x\}$, it follows that

$$\mathbb{P}[U \leqslant x] \geqslant \int_{[0, \infty)} \mathbb{P}[W \leqslant x - y] F_Y(dy)$$

$$\geqslant (1 - p)\mathbb{P}[W \leqslant x - \rho] + \int_{(\rho, \infty)} \Phi(x - y) F_Y(dy) - p\Delta,$$

where F_Y denotes the distribution function of Y. Hence, comparing as much as possible to $\Phi(x - \rho)$, it follows that

$$\Phi(x - \rho) + \eta + \frac{\rho}{\sqrt{2\pi}} \geqslant (1 - p)\mathbb{P}[W \leqslant x - \rho] + p\Phi(x - \rho) - \frac{\mathbb{E}\{(Y - \rho)I[Y > \rho]\}}{\sqrt{2\pi}} - p\Delta,$$

implying that

$$(1 - p)\delta_{x-\rho} \leqslant \eta + \frac{4\mathbb{E}Y}{\sqrt{2\pi}} + p\Delta.$$

A similar argument starting from $\{U \leqslant x\} \subset \{W - Y \leqslant x\}$ then gives

$$-(1 - p)\delta_{x+\rho} \leqslant \eta + \frac{4\mathbb{E}Y}{\sqrt{2\pi}} + p\Delta.$$

The choice of x being arbitrary, it thus follows that

$$(1-p)\Delta \leqslant \eta + \frac{4\mathbb{E}Y}{\sqrt{2\pi}} + p\Delta$$

also, and hence that

$$\Delta \leqslant 3\left\{\eta + \frac{4\mathbb{E}Y}{\sqrt{2\pi}}\right\},$$

as claimed.

To complete the proof of Theorem 1.2, apply (2.6) with S_{1n} for U and $S_{2n} - S_{1n}$ for X, taking $\eta = c_1 \log^{-1/2} n$ from Proposition 2.4 and $\epsilon = \log^{-1/2} n$. By Proposition 2.3,

$$\mathbb{P}[|S_{2n} - S_{1n}| > \epsilon] = \mathbb{P}\left[|\log P_n(Z) - \log O_n(Z) - \mu_n| > \epsilon \sqrt{\frac{\theta}{3}\log^3 n}\right] = O\left(\frac{(\log\log n)^2}{\log n}\right),$$

and hence, from (2.6),

$$\sup_x |\mathbb{P}[S_{2n} \leqslant x] - \Phi(x)| \leqslant c_2 \log^{-1/2} n$$

for some $c_2 = c_2(\theta)$. Now we can apply (2.7) with $U = S_{2n}$ and $W = S_{3n}$, since (2.1) implies that $|U - W| \leqslant Y$, with $Y = \{(\theta/3)\log n\}^{-1/2}(Y_n + 1)$, giving

$$\sup_x |\mathbb{P}[S_{3n} \leqslant x] - \Phi(x)| = O\left(\log^{-1/2} n(1 + \mathbb{E}Y_n)\right) = O\left(\log^{-1/2} n\right),$$

in view of (2.2). This is equivalent to Theorem 1.2.

To prove Theorem 1.3, we use essentially the same estimates. First, from (2.1),

$$|\log O_{[n^t]}(C^{(n)}) - \log O_{[n^t]}(Z)| \leqslant (1 + Y_n)\log n$$

for all $0 \leqslant t \leqslant 1$, and then, from (2.3),

$$0 \leqslant \log P_{[n^t]}(Z) - \log O_{[n^t]}(Z)$$

$$= \sideset{}{'}\sum_p \sum_{s \geqslant 1} (D_{[n^t]p^s} - 1)^+ \log p \leqslant \sideset{}{'}\sum_p \sum_{s \geqslant 1} (D_{np^s} - 1)^+ \log p.$$

Hence

$$\mathbb{E}\left\{\sup_{0 \leqslant t \leqslant 1} |\log O_{[n^t]}(C^{(n)}) - \log P_{[n^t]}(Z)|\right\} = O(\log\log n \log n).$$

Now

$$\log P_{[n^t]}(Z) = \sum_{j=1}^{[n^t]} Z_j \log j = \sum_{j=1}^{[n^t]} j^{-1}\theta \log j + \sqrt{\theta \psi(n+1)} \int_0^t s \log n \, dB_n(s),$$

where

$$B_n(t) = \sum_{j=1}^{[n^t]} \left(\frac{Z_j - j^{-1}\theta}{\sqrt{\theta \psi(n+1)}}\right)$$

can be realized as

$$\{\theta\psi(n+1)\}^{-1/2}\{P(\theta\psi([n^t]+1))-\theta\psi([n^t]+1)\}$$

using a Poisson process P with unit rate. Also, since

$$\left|\int_0^t s[dB_n(s)-dB(s)]\right| = \left|t[B_n(t)-B(t)]-\int_0^t \{B_n(s)-B(s)\}\,ds\right| \leqslant 2\sup_{0\leqslant t\leqslant 1}|B_n(t)-B(t)|,$$

the uniform approximation of B_n by a standard Brownian motion B, in the form

$$\mathbb{E}\left\{\sup_{0\leqslant t\leqslant 1}|B_n(t)-B(t)|\right\} = O(\{\log n\}^{-1/2}\log\log n),$$

as carried out using the theorem of Komlós, Major and Tusnády [12] in the case $\theta=1$ in [3], now implies the conclusion of Theorem 1.3: take $W(t^3)=\sqrt{3}\int_0^t s\,dB_n(s)$.

References

[1]   Arratia, R. A. and Tavaré, S. (1992) Limit theorems for combinatorial structures via discrete process approximations. *Rand. Struct. Alg.* **3** 321–345.

[2]   Arratia, R. A., Barbour, A. D. and Tavaré, S. (1992) Poisson process approximations for the Ewens Sampling Formula. *Ann. Appl. Probab.* **2** 519–535.

[3]   Arratia, R. A., Barbour, A. D. and Tavaré, S. (1993) On random polynomials over finite fields. *Math. Proc. Cam. Phil. Soc.* **114** 347–368.

[4]   Arratia, R. A., Barbour, A. D. and Tavaré, S. (1993) Logarithmic combinatorial structures. *Ann. Probab.* (in preparation).

[5]   Best, M. R. (1970) The distribution of some variables on a symmetric group. *Nederl. Akad. Wetensch. Indag. Math. Proc. Ser. A* **73** 385–402.

[6]   Bovey, J. D. (1980) An approximate probability distribution for the order of elements of the symmetric group. *Bull. London Math. Soc.* **12** 41–46.

[7]   Erdös, P. and Turán, P. (1967) On some problems of a statistical group theory. III. *Acta Math. Acad. Sci. Hungar.* **18** 309–320.

[8]   Ewens, W. J. (1972) The sampling theory of selectively neutral alleles. *Theor. Popn. Biol.* **3** 87–112.

[9]   Ewens, W. J. (1990) Population genetics theory – the past and the future. In: Lessard, S. (ed.) *Mathematical and statistical developments of evolutionary theory*, Kluwer Dordrecht, Holland, 177–227.

[10] Feller, W. (1971) *An introduction to probability theory and its applications, Volume II*, 2nd Edition, Wiley, New York.

[11] Hardy, G. H. and Wright, E. M. (1979) *An introduction to the theory of numbers*, 5th Edition, Oxford University Press, Oxford.

[12] Komlós, J., Major, P. and Tusnády, G. (1975) An approximation of partial sums of independent RV's-s, and the sample DF. I. *Z. Wahrscheinlichkeitstheorie verw. Geb.* **32** 111–131.

[13] Landau, E. (1909) Handbuch der Lehre von der Verteilung der Primzahlen. Bd. I.

[14] De Laurentis, J. M. and Pittel, B. (1985) Random permutations and Brownian motion. *Pacific J. Math.* **119**, 287–301.

[15] Nicolas, J.-L. (1984) A Gaussian law on $F_Q[X]$. *Colloquia Math. Soc. János Bolyai* **34** 1127–1162.

[16] Nicolas, J.-L. (1985) Distribution statistique de l'ordre d'un élément du groupe symétrique. *Acta Math. Hungar.* **45** 69–84.

Deterministic Graph Games and a Probabilistic Intuition

JÓZSEF BECK[†]

Department of Mathematics, Rutgers University,
Busch Campus, Hill Center,
New Brunswick, New Jersey 08903 U.S.A.
e-mail: jbeck@aramis.rutgers.edu

There is a close relationship between biased graph games and random graph processes. In this paper, we develop the analogy and give further interesting instances.

1. Introduction

We shall examine the following class of combinatorial games. Two players, Breaker and Maker, with Breaker going first, play on the complete graph K_n of n vertices in such a way that Breaker claims b (≥ 1) previously unselected edges a move, and Maker claims one previously unselected edge a move. Maker wins if he claims all the edges of some graph from a family of prescribed subgraphs of K_n. Otherwise Breaker wins, that is, Breaker simply wants to prevent Maker from doing his job.

As a warm-up consider the following three particular cases. Let $Clique(n; b, 1; r)$ denote the game where Maker wants a complete subgraph of r vertices (from his own edges of course). Denote by $Connect(n; b, 1)$ and $Hamilt(n; b, 1)$ the games where Maker's goal is to select a spanning tree (i.e. a connected subgraph of K_n) and a Hamiltonian cycle of K_n, respectively. Clearly if b is large enough with respect to n, Breaker has a winning strategy; if b is small, Maker has a winning strategy. The following crude heuristic argument, due to Paul Erdős, predicts the asymptotic behaviour of the 'breaking point' with surprising accuracy. The duration of a play allows for approximately $n^2/2(b + 1)$ Maker's edges. In particular, if $b = n/2c \log n$, Maker will have the time to create a graph with $cn \log n$ edges. A *random graph* with n vertices and $cn \log n$ edges is almost certainly connected (and Hamiltonian as well) for $c > 1/2$, and almost certainly disconnected (if fact, it has many isolated points) for $c < 1/2$. Now one suspects that the breaking points for the

[†]Supported by NSF Grant No. DMS-9106631.

games $Connect(n; b, 1)$ and $Hamilt(n; b, 1)$ are around $b = n/\log n$. Similarly, the largest clique in a random graph of n vertices and of parameter $p = 1/2$ has approximately $2\log n/\log 2$ vertices with probability tending to one as n tends to infinity. This suggests that Maker wins the fair game $Clique(n; 1, 1; r)$ if r is around $\log n$. The following results support this probabilistic intuition.

Theorem A. (Erdős–Selfridge [7] and Beck [1]) *Breaker has a winning strategy in the fair game $Clique(n; 1, 1; 2\log n/\log 2)$. On the other hand, given $\epsilon > 0$, if n is sufficiently large, Maker has a winning strategy in $Clique(n; 1, 1; (1 - \varepsilon)\log n/\log 2)$.*

Theorem B. (Erdős–Chvátal [6] and Beck [2],[3])

(i) If $b > (1 + \varepsilon)n/\log n$, Breaker has a winning strategy in $Connect(n; b, 1)$ if n is large enough.

(ii) If $b > (\log 2 - \varepsilon)n/\log n$, Maker has a winning strategy in $Connect(n; b, 1)$ if n is large enough.

(iii) If $b > (\log 2/27 - \varepsilon)n/\log n$, Maker has a winning strategy in $Hamilt(n; b, 1)$ if n is large enough.

The object of this paper is to point out further instances of this exciting analogy between the evolution of random graphs and biased graph games. The following four theorems were motivated by some well-known results in the theory of *random graphs* (see the monograph [5] by Bollobás). For example, Theorem 1 below is the game-theoretic analogue of a result of Bollobás [4] on 'long paths in sparse random graphs'. Needless to say, our proof is essentially different from Bollobás' argument in [4].

The basic idea is that Maker's graph possesses some fundamental properties of random graphs (mostly 'expanding' type properties) provided Maker uses his best possible strategy.

Let us begin with a trivial observation: if $b = 2n$, Breaker can easily prevent Maker even from getting a path of two edges (Breaker blocks the two endpoints of Maker's edge). If $b = \varepsilon n$, $\varepsilon > 0$ constant, then, in view of Theorem B(i), Breaker can force Maker's graph to be disconnected. In fact, Breaker can force at least $(\varepsilon/2)e^{-1/\varepsilon}n$ isolated points in Maker's graph, as shown by the following argument, which is a straightforward adaptation of the Erdős–Chvátal proof of Theorem B(i). Breaker proceeds in two stages. In the first stage, he claims all the edges of some clique K_m^* with $m = \varepsilon n/2$ vertices, such that none of Maker's edges has an endpoint in this K_m^*. In the second stage, he claims all the remaining edges incident with at least $(\varepsilon/2)e^{-1/\varepsilon}n$ vertices of K_m^*, thereby forcing at least $(\varepsilon/2)e^{-1/\varepsilon}n$ isolated points in Maker's graph.

The first stage lasts no more than $m = \varepsilon n/2$ moves, and goes by a simple induction on m. During his first $i - 1$ $(1 \leq i \leq m)$ moves, Breaker has created a clique K_{i-1}^* with $i - 1$ vertices, such that none of Maker's edges has an endpoint in K_{i-1}^*. At this moment there are $i - 1 < \varepsilon n/2$ Maker's edges, hence there are at least two vertices u, v in the complement of $V(K_{i-1}^*)$ that are incident with none of Maker's edges. On his ith move, Breaker claims edge $\{u, v\}$, and all the edges joining u and v to the vertices of K_{i-1}^*, thereby enlarging K_{i-1}^* by two vertices. Then Maker can kill one vertex from this clique K_{i+1}^* by claiming an edge incident with that vertex. Nevertheless, a clique of i vertices still 'survives'.

In the second stage, Breaker has $m = \varepsilon n/2$ *pairwise disjoint* edge-sets: for every $u \in V(K_m^*)$, the edges joining u to all vertices in the complement of $V(K_m^*)$. It is easy to see that Breaker can completely occupy at least $e^{-1/\varepsilon}m$ of these m disjoint edge-sets by the simple rule that he has the same (or almost the same) number of edges from all the 'surviving' edge-sets at any time.

Our first result says that Maker is able to build up a cycle of length at least $(1-e^{-1/200\varepsilon})n$. That is, if Breaker claims εn edges a move, then Maker has an 'almost Hamiltonian cycle' in the sense that the complement is 'exponentially small' (the constant factor of 200 in the exponent is of course very far from the best possible). In this paper we do not make any effort to find the optimal (or even nearly optimal) constants.

Theorem 1. *If $0 < \varepsilon < 1/200$ and Breaker selects εn edges for each move of Maker, then Maker can build a cycle of length at least $(1 - e^{-1/200\varepsilon})n$ on a board K_n.*

Note that if Maker just wants a path P_m of $m = (1 - \text{const}\sqrt{\varepsilon})n$ edges, he can do it in the *shortest way*: in m moves. Indeed, Maker can employ the following simple greedy strategy: keep extending the path by adding that available point (as a new endpoint) that has minimum degree in Breaker's graph. Trivial calculation shows that this greedy procedure does not terminate in $m = (1 - \text{const}\sqrt{\varepsilon})n$ moves.

However, if Maker's object is (say) a *binary tree* BT_m of $m = \text{const} \cdot n$ edges, then this greedy strategy does not seem to work. We formulate it in the following conjecture.

Conjecture. *Let c be an arbitrarily large but fixed positive constant. Consider the game where the board is a complete graph $K_{c \cdot n}$ of $c \cdot n$ points, Breaker and Maker alternately select n and 1 edge(s) a move, respectively, and Maker's goal is a binary tree BT_n of n edges ($n = 2^k$, k integer). If $n > n_0(c)$, Breaker can prevent Maker from getting a copy of BT_n in n moves.*

On the other hand, in *linear time* (i.e. in $\text{const} \cdot n$ moves) Maker can obtain *all trees* with at most n vertices and constant size degrees.

Theorem 2. *Consider the game where the board is K_N with $N = 100dn$. Breaker and Maker alternately select n and 1 previously unselected edges a move, respectively. Maker has a strategy to force his graph to be tree-universal in the sense that it contains every tree with at most n vertices and maximum degree at most d.*

Observe that Theorem 2 is essentially best possible (apart from the constant factor of 100). Indeed, Breaker can prevent Maker from having a degree larger than $2N/b$ on a board K_N: if Maker selects an edge $\{u, v\}$, Breaker occupies $b/2$ edges from u and $b/2$ edges from v.

In general, what can we say about Maker's largest degree, *i.e.*, the largest star Maker can construct? For simplicity we restrict ourselves to the fair case (*i.e.* $b = 1$). This question is apparently due to Erdős (oral communication). Székely [12] proved, by using Lemma 3 in Beck [1], that Breaker can prevent a star of size $(n/2)+\text{const}\sqrt{n \log n}$. In the other direction, we can prove that, for some positive constant, Maker can achieve a star of size $(n/2)+\text{const}\sqrt{n}$. If the complete graph K_n is replaced with the complete *bipartite* graph $K_{n,n}$, we obtain the following interesting 'row-column game'.

Theorem 3. *Consider the game where the board is an $n \times n$ chessboard. Breaker and Maker alternately select a previously unselected cell. Breaker marks his cells blue and Maker marks his cells red. Maker's object is to achieve at least $(n/2) + k$ ($k \geq 1$) red cells in some line (row or column). If $k = \sqrt{n}/32$, Maker has a winning strategy.*

This result is in sharp contrast with the chessboard type alternating two-coloring, where the discrepancy in every line is 0 or 1 depending on the parity of n.

A straightforward modification of the proof of Theorem 3 gives the above-mentioned case where the board is K_n. We leave the details to the reader.

Finally, we study the case where Maker's goal is to build up an *arbitrary* prescribed graph $G_{n,d}$ of n points and maximum degree d. We prove that, for any $b \geq 1$ and $d \geq 1$ there is a $= c(b,d)$ such that Maker's graph contains *all* graphs $G_{n,d}$ of *constant degree* on a board K_N, $N = c \cdot n$, of *linear* size. The quantitative version goes as follows.

Theorem 4. *Consider the game where the board is K_N with $N = 100d^3(3b)^{d+1} \cdot n$, and Breaker and Maker alternately select b (≥ 1) and 1 edge(s) a move. Maker has a strategy to force his graph to be universal, in the sense that it contains all graphs $G_{n,d}$ of n points and maximum degree d.*

Note that the exponential behaviour of $c = c(b,d)$ is necessary: if $b = 1$, $d = n - 1$ and $G_{n,d} = K_n$, we just go back to Theorem A.

2. Proofs of Theorems 1-2 - 'derandomization' of the first moment method

Proof of Theorem 1. We combine the basic idea of Beck [3] with a 'truncation procedure'.

Given a simple and undirected graph G, and an arbitrary subset S of the vertex-set $V(G)$ of G, denote by $\Gamma_G(S)$ the set of vertices in G adjacent to at least one vertex of S. Let $|S|$ denote the number of elements of a set S.

The following lemma is essentially due to Pósa [11] (a weaker version was earlier proved by Komlós–Szemerédi [10]). A trivial corollary of the lemma is that an expander graph has a long path.

Lemma 1. *Let G be a non-empty graph, $v_0 \in V(G)$, and consider a path $P = (v_0, v_1, \ldots, v_m)$ of maximum length that starts from v_0. If $(v_i, v_m) \in G$ ($1 \leq i \leq m - 1$), we say that the path $(v_0, \ldots, v_i, v_m, v_{m-1}, \ldots, v_{i+1})$ arises by a Pósa-deformation from P. Let $end(G, P, v_0)$ denote the set of all endpoints of paths arising by repeated Pósa-deformations from P, keeping the starting point v_0 fixed. Assume that for each vertex-set $S \subset V(G)$ with $|S| \leq k$, $|\Gamma_G(S) \setminus S| \geq 2|S|$. Then $|end(P, G, v_0)| \geq k + 1$.*

In order to use Lemma 1, we need another result.

Lemma 2. *Under the hypothesis of Theorem 1, Maker can guarantee that right after Breaker occupied* $(1/20)\binom{n}{2}$ *edges, Maker's graph G has the following property.*
Let S be a subset of $V(K_n)$ *with* $(1/3)e^{-1/200\varepsilon}n \leq |S| \leq n/4$. *Then*

$$|\Gamma_G(S) \setminus S| \geq 2|S| + e^{-1/200\varepsilon}n.$$

Proof. We apply a general theorem about hypergraph games. Let $\mathcal{H}$ be a hypergraph with vertex set $V(\mathcal{H})$ and edge set $E(\mathcal{H})$, and let $p \geq 1$ and $q \geq 1$ be integers. A $(\mathcal{H}; p, 1; q)$-*game* is a game on $\mathcal{H}$ in which two players, **I** and **II**, select p and 1 previously unselected vertices a move from $V(\mathcal{H})$. The game proceeds until $(1/q)|V(\mathcal{H})|$ vertices have been selected by **I**. Player **II** wins if he occupies at least one vertex from every hyperedge $A \in E(\mathcal{H})$, otherwise **I** wins. In [3] we proved the following result: if

$$\sum_{A \in E(\mathcal{H})} 2^{-|A|/pq} < \frac{1}{2}, \tag{1}$$

then **II** has a winning strategy in the $(\mathcal{H}; p, 1; q)$-game. (The case $p = q = 1$ of this result was proved in Erdős and Selfridge [8]. The proof is based on the method that is now called 'derandomization'.)

In order to apply (1), we introduce some hypergraphs. Let m be an integer satisfying $(1/3)e^{-1/200\varepsilon}n \leq m \leq n/4$, and let $\mathcal{H}(n; m)$ be the set of all complete $m \times (n - 3m - e^{-1/200\varepsilon}n + 1)$ bipartite subgraphs of K_n. The 'vertices' of $\mathcal{H}(n; m)$ are the edges of K_n. Let $\mathcal{H}$ be the union of all these $\mathcal{H}(n; m)$ hypergraphs.

Now to ensure property A, in view of (1) with $p = b = \varepsilon n$ and $q = 20$, it is enough to check the following inequality:

$$\sum_{m=\frac{1}{3}e^{-1/200\varepsilon}n}^{n/4} \binom{n}{m} \binom{n-m}{2m + e^{-1/200\varepsilon}n - 1} 2^{-m(n-3m-e^{-1/200\varepsilon}n+1)/20\varepsilon n} < \frac{1}{2}. \tag{2}$$

Standard calculations show that (2) holds, and Lemma 2 follows from (1) and (2). $\square$

Now we are ready to complete the proof of Theorem 1. We show that if Maker uses the strategy in Lemma 2, and H is Maker's graph at the end, H contains a cycle of $(1 - e^{-1/200\varepsilon})n$ edges.

Let G be Maker's graph right after Breaker occupied $(1/20)\binom{n}{2}$ edges. Assume that there exists a vertex-set $S_1 \subset V(K_n)$ with $|S_1| \leq (1/3)e^{-1/200\varepsilon}n$ such that $|\Gamma_G(S_1) \setminus S_1| < 2|S_1|$. Throwing away the vertices $\Gamma_G(S_1) \cup S_1$ from G, we get a new graph G_1. Again assume that there exists a vertex-set $S_2 \subset V(G_1)$ with $|S_2| \leq (1/3)e^{-1/200\varepsilon}n$ such that $|\Gamma_{G_1}(S_2) \setminus S_2| < 2|S_2|$. Throwing away the vertices $\Gamma_{G_1}(S_2) \cup S_2$ from G_1, we get a new graph G_2, and so on. This truncation procedure terminates (say) in t steps: $G_t = G_{t+1} = \cdots$. That is, for any vertex-set $S \subset V(G_t)$ with $|S| \leq (1/3)e^{-1/200\varepsilon}n$,

$$|\Gamma_{G_t}(S) \setminus S| \geq 2|S|. \tag{3}$$

We claim

$$|V(G_t)| > (1 - e^{-1/200\varepsilon})n. \tag{4}$$

Indeed, otherwise there is an index i ($\leq t$) such that at the ith stage of the truncation, the union $S = S_1 \cup \cdots \cup S_i$ *first* satisfies $(1/3)e^{-1/200\varepsilon}n \leq |S|$, so

$$\frac{1}{3}e^{-1/200\varepsilon}n \leq |S| \leq \frac{2}{3}e^{-1/200\varepsilon}n < n/4$$

and

$$|\Gamma_G(S) \setminus S| < 2|S|,$$

which contradicts property A in Lemma 2.

It follows from (3), (4) and property A that for every set $S \subset V(G_t)$ with $|S| \leq n/4$, we have

$$|\Gamma_{G_t}(S) \setminus S| \geq 2|S|. \tag{5}$$

It immediately follows from (5) that G_t is a *connected* graph. We are going to show that Maker can build up a Hamiltonian cycle on the vertex-set $V(G_t)$.

Let P be a path in G_t of maximum length. Inequality (5) ensures that the truncated graph G_t satisfies the condition of Pósa's lemma with $k = n/4$, so (see Lemma 1) $|end(G_t, P, v_0)| > n/4$, where v_0 is one of the endpoints of P.

Let $end(G_t, P, v_0) = \{x_1, x_2, \ldots, x_k\}$ ($k > n/4$), and denote by $P(x_i)$, $1 \leq i \leq k$ a path arising from P by a sequence of Pósa-deformations, with endpoints v_0 and x_i. By Lemma 1, for every $x_i \in end(G_t, P, v_0)$, we have

$$|end(G_t, P(x_i), x_i)| > n/4. \tag{6}$$

Let

$$close(G_t, P) = \{(x_i, y) : x_i \in end(G_t, P, v_0), \ y \in end(G_t, P(x_i), x_i)\}.$$

By (6) we have

$$|close(G_t, P)| > (n/4)^2/2 = n^2/32.$$

Since at this moment Breaker's graph contains

$$\frac{1}{20}\binom{n}{2}$$

edges, there must exist a previously unselected edge e_1 in $close(G_t, P)$. Let e_1 be Maker's next move. Then Maker's graph $G_t^{(1)} = G_t \cup \{e_1\}$ contains a cycle of length $|P|$. Moreover, $G_t^{(1)} = G_t \cup \{e_1\}$ is *connected*, thus either $|P| = |V(G_t)|$, and we have a Hamiltonian cycle in the truncated vertex-set, or $G_t^{(1)}$ contains a *longer* path (i.e. a path of length $\geq |P| + 1$).

Let P_1 be a path of maximum length in $G_t^{(1)}$. Repeating the argument above, we get that $|close(G_t^{(1)}, P_1)| > n^2/32$. Since at this moment Breaker's graph contains $(1/20)\binom{n}{2} + \varepsilon n < n^2/32$ edges, there must exist a previously unselected edge e_2 in $close(G_t^{(1)}, P_1)$. Let e_2 be Maker's next move. Then Maker's graph $G_t^{(2)} = G_t \cup \{e_1, e_2\}$ contains a cycle of length $|P_1|$. Moreover, $G_t^{(2)} = G_t \cup \{e_1, e_2\}$ is *connected*, thus either $|P_1| = |V(G_t)|$, and we have a Hamiltonian cycle in the truncated vertex-set, or $G_t^{(2)}$ contains a *longer* path (i.e. a path of length $\geq |P_1| + 1$). By repeated application of this procedure, in less than n moves (so the required inequality $(1/20)\binom{n}{2} + n \cdot \varepsilon n < n^2/32$ holds), Maker's graph will certainly contain a Hamiltonian cycle in the truncated vertex-set $V(G_t)$. Theorem 1 follows. $\square$

Proof of Theorem 2. The main difference is that Pósa's lemma is replaced with the following lemma, due to Friedman and Pippenger [9].

Lemma 3. *If H is a non-empty graph such that, for every set $S \subset V(H)$ with $|S| \leq 2n - 2$, we have $|\Gamma_H(S)| \geq (d+1)|S|$, then H contains every tree with n vertices and maximum degree at most d.*

We need the following analogue of Lemma 2.

Lemma 4. *Maker can ensure that at the end of the game his graph satisfies:*
Property B: if $N = 100dn$ and $\subset V(K_N)$ satisfies $2n \leq |S| \leq 4n$, then $|\Gamma_G(S)| \geq (d+1)|S|$, where G is Maker's graph at the end.

Proof. By imitating the proof of Lemma 2, it is enough to check the following inequality (this case is even simpler because $q = 1$):

$$\sum_{m=2n}^{4n} \binom{100dn}{m}\binom{100dn - m}{dm - 1} 2^{-m(100dn - (d+1)m+1)/n} < \frac{1}{2}. \tag{7}$$

Easy calculations show that (7) holds, and Lemma 4 follows. □

We shall now repeat the 'truncation procedure'. We show that if Maker uses the strategy in Lemma 4, then the graph G obtained by Maker at the end contains all trees with n vertices and maximum degree at most d.

Assume that there exists a set $S_1 \subset V(K_N)$ with $|S_1| \leq 2n$ such that $|\Gamma_G(S_1)| < (d+1)|S_1|$. Discarding the set $\Gamma_G(S_1) \cup S_1$ from G, we get a new graph G_1. Again assume that there exists a set $S_2 \subset V(G_1)$ with $|S_2| \leq 2n$ such that $|\Gamma_{G_1}(S_2)| < (d+1)|S_2|$. Discarding the set $\Gamma_{G_1}(S_2) \cup S_2$ from G_1, we get a new graph G_2, and so on. This truncation procedure terminates, say, in t steps, $G_t = G_{t+1} = \cdots$.

We claim that G_t is non-empty. Indeed, otherwise there is an index $i \leq t$ such that at the ith stage of the truncation, the union set $S = S_1 \cup \cdots \cup S_i$ satisfies $2n \leq |S| \leq 4n$ and

$$|\Gamma_G(S)| < (d+1)|S|,$$

which contradicts Property B in Lemma 4.

By definition, the non-empty graph $H = G_t$ has the property that for every vertex-set $S \subset V(H)$ with $|S| \leq 2n$,

$$|\Gamma_H(S)| \geq (d+1)|S|.$$

Thus, by Lemma 3, $H = G_t$ contains *all* trees with n vertices and maximum degree at most d. Since $G \supseteq G_t = H$, Theorem 2 follows. □

3. Proof of Theorem 3 - a 'fake second moment' method

Consider a play according to the rules in Theorem 3. Let $x_1, x_2, \ldots, x_i$ be the blue cells in the chessboard selected by Breaker in his first i moves, and let $y_1, y_2, \ldots, y_{i-1}$ be the red cells selected by Maker in his first $(i-1)$ moves. The question is how to find Maker's optimal ith move y_i. Write

$$X_i = \{x_1, x_2, \ldots, x_i\} \text{ and } Y_{i-1} = \{y_1, y_2, \ldots, y_{i-1}\}.$$

Let A be a line (row or column) of the $n \times n$ chessboard, and introduce the following 'weight':

$$w_i(A) = \left\{ |A \cap Y_{i-1}| - |A \cap X_i| + \frac{\sqrt{n}}{4} \right\}^+,$$

where

$$\{\alpha\}^+ = \begin{cases} \alpha, & \text{if } \alpha > 0; \\ 0, & \text{otherwise.} \end{cases}$$

Let y be an arbitrary unselected cell, and write

$$w_i(y) = w_i(A) + w_i(B),$$

where A and B are the row and the column containing y.

Here is Maker's winning strategy: at his ith move he selects that previously unselected cell y for which the maximum of the 'weights'

$$\max_{y \text{ unselected}} w_i(y)$$

is attained.

The following total sum is a sort of 'variance':

$$T_i = \sum_{2n \text{ lines } A} \left(w_i(A) \right)^2.$$

The idea of the proof is to study the behaviour of T_i as $i = 1, 2, 3, \ldots$, and to show that T_{end} is 'large'.

Remark. The more natural 'symmetric' total sum

$$\sum_{2n \text{ lines } A} \left(|Y_{i-1} \cap A| - |X_i \cap A| \right)^2$$

is of no use because it can be large if in some line Breaker overwhelmingly dominates. This is exactly the reason why we had to introduce the 'shifted and truncated weight' $w_i(A)$.

First we compare T_i and T_{i+1}, that is, we study the effects of the cells y_i and x_{i+1}. We distinguish two cases.

Case 1: the cells y_i and x_{i+1} determine four different lines.
Case 2: the cells y_i and x_{i+1} determine three different lines.

In Case 1, an easy analysis shows that

$$T_{i+1} \geq T_i + 1 \tag{8}$$

except in the 'unlikely situation' when $w_i(y_i) = 0$. Indeed,

$$w_i(y_i) = w_i(A) + w_i(B) \geq w_i(x_{i+1}) = w_i(C) + w_i(D),$$

so

$$T_{i+1} = T_i + 2w_i(y_i) - 2w_i(x_{i+1}) + \{2 \text{ or } 1 \text{ or } 0\} \geq T_i + \{2 \text{ or } 1 \text{ or } 0\},$$

where

$$\{2 \text{ or } 1 \text{ or } 0\} = \begin{cases} 2, & \text{if } w_i(A) > 0, w_i(B) > 0; \\ 1, & \text{if } \max\{w_i(A), w_i(B)\} > 0, \min\{w_i(A), w_i(B)\} = 0; \\ 0, & \text{if } w_i(A) = w_i(B) = 0. \end{cases}$$

Even if the 'unlikely situation' occurs, we have at least equality: $T_{i+1} = T_i$. Because y_i was a cell of maximum weight, for x_{i+1}, and for every other unselected cell x, $w_i(x) = 0$.

Similarly, in Case 2,

$$T_{i+1} \geq T_i + 1 \tag{9}$$

except in the following 'unlikely situation': $w_i(B) = 0$, where A is the line containing both y_i and x_{i+1}, and B is the other line containing y_i. Even if this 'unlikely situation' occurs, we have at least equality: $T_{i+1} = T_i$. Because y_i was a cell of maximum weight, it follows that $w_i(C) = 0$, where C is the other line containing x_{i+1}, and, similarly, for every other unselected cell x in line A, $w_i(D_x) = 0$, where D_x is the other line containing x.

If i is an index for which the 'unlikely situation' in Case 1 occurs, let *unsel(i)* denote the set of all unselected cells after Breaker's ith move. Similarly, if i is an index for which the 'unlikely situation' in Case 2 occurs, let *unsel(i, A)* denote the set of all unselected cells after Breaker's $(i + 1)$st move in line A containing both y_i and x_{i+1}, including y_i and x_{i+1}.

If the 'unlikely situation' occurs in less than $3n^2/10$ moves (*i.e.* in less than 60% of the total time), we are done. Indeed, by (8) and (9),

$$T_{\text{end}} = T_{n^2/2} \geq \frac{n^2}{5}.$$

Since T_{end} is a sum of $2n$ terms, we have

$$\max_{2n \text{ lines } A} \left(w_{n^2/2}(A)\right)^2 \geq \frac{n^2/5}{2n} = \frac{n}{10}.$$

Equivalently, for some line A,

$$w_{n^2/2}(A) = \left\{|A \cap Y_{n^2/2-1}| - |A \cap X_{n^2/2}| + \frac{\sqrt{n}}{4}\right\}^+ \geq \sqrt{n/10},$$

where

$$\{\alpha\}^+ = \begin{cases} \alpha, & \text{if } \alpha > 0; \\ 0, & \text{otherwise.} \end{cases}$$

So

$$|A \cap Y_{n^2/2-1}| - |A \cap X_{n^2/2}| \geq \sqrt{n/10} - \frac{\sqrt{n}}{4} > \frac{\sqrt{n}}{16},$$

and Theorem 3 follows.

If the 'unlikely situation' in Case 1 occurs in more than $n^2/10$ moves (*i.e.* in more than

20% of the time), then let i_0 be the first time when this happens. Clearly

$$|unsel(i_0)| > 2n^2/10 = n^2/5.$$

It follows that there are at least $(n^2/5)/n = n/5$ distinct columns D containing (at least one) element of $unsel(i_0)$ each. So $w_i(D) = 0$ for at least $n/5$ columns D, that is,

$$|D \cap X_i| - |D \cap Y_{i-1}| \geq \frac{\sqrt{n}}{4}$$

for at least $n/5$ columns D. Therefore, after Breaker's i_0th move,

$$\sum_{n \text{ columns } D} \left\{|D \cap X_i| - |D \cap Y_{i-1}|\right\}^+ > \frac{n}{5} \frac{\sqrt{n}}{4}. \tag{10}$$

Since

$$1 + \sum_{n \text{ columns } D} \left\{|D \cap Y_{i-1}| - |D \cap X_i|\right\}^+ = \sum_{n \text{ columns } D} \left\{|D \cap X_i| - |D \cap Y_{i-1}|\right\}^+,$$

by (10),

$$\sum_{n \text{ columns } D} \left\{|D \cap Y_{i-1}| - |D \cap X_i|\right\}^+ \geq \frac{n^{3/2}}{20}.$$

Since the number of terms on the left-hand side is less than $n - n/5 = 4n/5$, after Breaker's i_0th move, we have

$$\max_D \left\{|D \cap Y_{i-1}| - |D \cap X_i|\right\} > \frac{n^{3/2}/20}{4n/5} = \frac{\sqrt{n}}{16}.$$

Obviously Maker can keep this advantage of $\sqrt{n}/16$ for the rest of the game, and again Theorem 3 follows.

Finally, we study the case when the 'unlikely situation' of Case 2 occurs for at least $n^2/5$ moves (*i.e.* for at least 40% of the time). Without loss of generality, we can assume that there are at least $n^2/10$ 'unlikely' indices i when the line A containing both y_i and x_{i+1} is a *row*. We claim that there is an 'unlikely' index i_0 when

$$|unsel(i_0, A)| \geq n/5. \tag{11}$$

Indeed, by choosing y_i and x_{i+1}, in each 'unlikely' move the set $unsel(i, A)$ is decreasing by 2, and because we have n rows, the number of 'unlikely' indices i when $unsel(i, A) < n/5$ is altogether less than $n \cdot \frac{n/5}{2} = n^2/10$.

Now we can complete the proof just as before. We recall that $w_{i_0}(D) = 0$ for those columns D that contain some cell from $unsel(i_0, A)$ (here A is the row containing both y_{i_0} and x_{i_0+1}). So, by (11), $w_{i_0}(D) = 0$ for at least $n/5$ columns D, that is,

$$|D \cap X_i| - |D \cap Y_{i-1}| \geq \frac{\sqrt{n}}{4}$$

for at least $n/5$ columns D. Therefore, after Breaker's i_0th move

$$\sum_{n \text{ columns } D} \left\{|D \cap X_i| - |D \cap Y_{i-1}|\right\}^+ > \frac{n}{5} \frac{\sqrt{n}}{4}. \tag{12}$$

Since

$$1 + \sum_{n \text{ columns } D} \left\{ |D \cap Y_{i-1}| - |D \cap X_i| \right\}^+ = \sum_{n \text{ columns } D} \left\{ |D \cap X_i| - |D \cap Y_{i-1}| \right\}^+,$$

by (12),

$$\sum_{n \text{ columns } D} \left\{ |D \cap Y_{i-1}| - |D \cap X_i| \right\}^+ \geq \frac{n^{3/2}}{20}.$$

Since the number of terms on the left-hand side is less than $n - n/5 = 4n/5$, after Breaker's i_0th move, we have,

$$\max_D \left\{ |D \cap Y_{i-1}| - |D \cap X_i| \right\} > \frac{n^{3/2}/20}{4n/5} = \frac{\sqrt{n}}{16}.$$

Obviously Maker can keep this advantage of $\sqrt{n}/16$ for the rest of the game, and again Theorem 3 follows and the proof is complete. $\qquad \square$

4. Proof of Theorem 4 - a Szemerédi type embedding

The game-theoretical content of the proof is the following lemma.

Lemma 5. *Maker can ensure that at the end of the game his graph satisfies Property C: For any two disjoint subsets U and V of $V(K_N)$ with $u = |U| \geq 100b \log N$ and $v = |V| \geq 100b \log N$, the graph MG constructed by Maker at the end contains more than $u \cdot v/3b$ edges from the $u \cdot v$ edges of the complete bipartite graph $U \times V$.*

Proof. Consider a play according to the rules. After Breaker's ith move, let MG_{i-1} denote the graph of Maker's $i - 1$ edges, and let BG_i be the graph of Breaker's $i \cdot b$ edges. For every $U \times V$ described in property C, let $w_i(U \times V)$ be the 'weight'

$$w_i(U \times V) = (1 + \lambda)^{|BG_i \cap U \times V| - uv(1 - 1/3b)} \cdot (1 - \lambda)^{|MG_i \cap U \times V| - uv/3b},$$

where the parameter λ, $0 < \lambda < 1$, will be specified later. For every unselected edge $e \in K_N$, let

$$w_i(e) = \sum_{U \times V: \, e \in U \times V} w_i(U \times V).$$

Now let Maker's ith move be an unselected edge of *maximum* weight.

Consider the total sum

$$T_i = \sum_{\text{all possible } U \times V \text{ in property C}} w_i(U \times V).$$

We shall verify that the sequence (T_i) is decreasing:

$$T_{i+1} \leq T_i. \tag{13}$$

Let e_i and $f_{i+1,1}, f_{i+1,2}, \ldots, f_{i+1,b}$ denote Maker's ith move and Breaker's $(i + 1)$st move, respectively. That is,

$$\{e_i\} = MG_i \setminus MG_{i-1}, \quad \{f_{i+1,1}, f_{i+1,2}, \ldots, f_{i+1,b}\} = BG_{i+1} \setminus BG_i.$$

Given a product set $U \times V$ in Property C, let $\alpha_i(U \times V)$ be 1 or 0 according to whether $e_i \in U \times V$ or $e_i \notin U \times V$; and let $\beta_{i+1}(U \times V)$ be the number of edges $f_{i+1,j}, 1 \leq j \leq b$ contained in $U \times V$. It is easy to see that

$$T_{i+1} = T_i + \sum_U \sum_V \left((1+\lambda)^{\beta_{i+1}(U \times V)/b} \cdot (1-\lambda)^{\alpha_i(U \times V)} - 1 \right) \cdot w_i(U \times V). \qquad (14)$$

We need the following simple inequality. For arbitrary integers $\alpha = 0, 1$ and $\beta = 0, 1, \ldots, b$,

$$(1+\lambda)^{\beta/b}(1-\lambda)^{\alpha} - 1 \leq \lambda \left(\frac{\beta}{b} - \alpha \right). \qquad (15)$$

Indeed, if $\alpha = 0$, (15) is equivalent to

$$\frac{(1+\lambda)^{\beta/b} - 1}{\lambda} \leq \frac{\beta}{b} \quad (0 \leq \beta \leq b).$$

This inequality is immediate, since the function $y = x^{\delta}$ $(0 \leq \delta \leq 1)$ is concave, so the slope

$$\frac{(1+\lambda)^{\beta/b} - 1}{\lambda}$$

of the chord of $y = x^{\beta/b}$ between $x = 1$ and $x = 1 + \lambda$ is at most β/b, the slope of the tangent line at $x = 1$.

The case of $\alpha = 1$ is even simpler. Indeed, then (15) is equivalent to

$$\left((1+\lambda)^{\beta/b} - 1 \right)(1-\lambda) \leq \lambda \frac{\beta}{b}.$$

This inequality holds, since the case $\alpha = 0$ implies that

$$\left((1+\lambda)^{\beta/b} - 1 \right)(1-\lambda) < (1+\lambda)^{\beta/b} - 1 \leq \lambda \frac{\beta}{b}.$$

This proves (15).

It follows from (14) and (15) that

$$T_{i+1} \leq T_i + \lambda \sum_U \sum_V \left(\frac{1}{b} \beta_{i+1}(U \times V) w_i(U \times V) - \alpha_i(U \times V) w_i(U \times V) \right)$$

$$= T_i + \lambda \left(\frac{1}{b} \sum_{j=1}^{b} w_i(f_{i+1,j}) - w_i(e_{i+1}) \right),$$

and as e_{i+1} was an edge of maximum weight, we obtain (13).

Now assume that during a play Breaker managed to occupy at least $uv(1 - 1/3b)$ edges from some product set $U \times V$ in Property C. Let us say that this happened after his i_0th move. Then clearly $1 \leq T_{i_0}$.

On the other hand, consider $T_{\text{start}} = T_1$, that is, the situation right after Breaker's first move (consisting of b edges). We clearly have

$$T_1 = \sum_{u \geq 100b \log N} \sum_{v \geq 100b \log N} \binom{N}{u} \binom{N-u}{v} (1+\lambda)^{(b-uv(1-1/3b))/b} \cdot (1-\lambda)^{-uv/3b}.$$

Let $\lambda = 1/2$. Since $b \geq 1$, trivial calculations give

$$(1 + \lambda)^{(b - uv(1 - 1/3b))/b} \cdot (1 - \lambda)^{-uv/3b} \geq \left(\frac{9}{8}\right)^{uv/3b}.$$

Using this inequality, one can easily show that $T_1 < 1$.

All in all, $T_1 < 1 \leq T_{i_0}$, which contradicts (13) and so proves Lemma 5. $\qquad\square$

The following purely graph-theoretical result is essentially contained in [7].

Lemma 6. *If a graph H of $N = 100d^3(3b)^{d+1} \cdot n$ vertices satisfies Property C of Lemma 5, then H contains every graph $G_{n,d}$ of n vertices and maximum degree d.*

Proof. We closely follow the argument in [7]. Let $G_{n,d}$ be a graph having n vertices $x_1, x_2, \ldots, x_n$ and maximum degree d. To construct a copy of $G_{n,d}$ in H, we will proceed inductively to choose vertices $y_1, y_2, \ldots, y_n$ from H so that the map $x_i \to y_i$ is an isomorphism.

Let $A_1 \cup A_2 \cup \cdots \cup A_{d+1}$ be an arbitrary partition of the N-element vertex-set of H into disjoint subsets of almost the same size, that is, $|A_i| \approx N/(d + 1)$ $(i = 1, 2, \ldots, d + 1)$. We will choose the points $y_1, y_2, \ldots, y_n$ so that for each $i = 1, 2, \ldots, n$, the following two conditions are satisfied:

(a) If $1 \leq s < t \leq i$ and x_s is adjacent to x_t in $G_{n,d}$, then y_s and y_t come from distinct sets A_j in the partition and y_s is adjacent to y_t in H.

(b) If $i < q \leq n$, $V(q, i) = \{y_t : 1 \leq t \leq i, x_t$ is adjacent to x_q in $G_{n,d}\}$, $v = |V(q, i)|$, $1 \leq p \leq d + 1$ and A_p contains no $y_t \in V(q, i)$, then A_p contains a subset $A_p^* = A_p^*(V(q, i))$ having at least $|A_p|(3b)^{-v}$ points so that every point in A_p^* is adjacent to every $y_t \in V(q, i)$.

At first, condition (b) may seem hopelessly complicated to the reader. (As far as I know it was Szemerédi who first applied conditions like (b) to prove Ramsey type theorems in the early seventies.) However, after some thought it will be clear that this condition is precisely what is needed to ensure that the selection of the vertices $y_1, y_2, \ldots, y_n$ can proceed inductively as claimed.

Here are the details. Suppose that for some nonnegative integer i $(< n)$ the points y_t for $i \leq t \leq i$ have been chosen so that conditions (a) and (b) are satisfied. We show how to make a suitable choice for y_{i+1}. (Note that this definition allows $i = 0$, because the rule for choosing y_1 is the same as for all other values of i.) First choose some j_0 with $1 \leq j_0 \leq d + 1$ so that A_{j_0} does not contain a point from $V(i + 1, i)$ (see (b)), i.e. we choose a set in the partition that does not contain y_t with $1 \leq t \leq i$ for which x_t is adjacent to x_{i+1}. This is possible because x_{i+1} has at most d neighbours. Then let $A_{j_0}^* = A_{j_0}^*\big(V(i + 1, i)\big)$ be the subset of A_{j_0} consisting of those points adjacent to every $y_t \in V(i + 1, i)$. By condition (b) we know that $|A_{j_0}^*| \geq |A_{j_0}|(3b)^{-v}$, where $v = |V(i + 1, i)|$. Since $v \leq d$, we obtain $|A_{j_0}^*| \geq (N/(d + 1))(3b)^{-d} \geq n$. With the choice of any previously unselected point from $A_{j_0}^*$ as y_{i+1}, we would satisfy condition (a). However, some care must be taken to ensure that condition (b) is satisfied. It is clear that we need only be concerned with those values $q > i + 1$ in which x_{i+1} is adjacent to x_q. There are at most d such values. Choose one of these, say q. Then choose $j \in \{1, 2, \ldots, d + 1\}$ with $j \neq j_0$ so that A_j does not contain any $y_t \in V(q, i)$, and let $\mu = |V(q, i + 1)| = 1 + |V(q, i)|$. We already know that A_j contains a subset $A_j^* = A_j^*(V(q, i))$ with at least $|A_j| \cdot (3b)^{-\mu+1}$ points so that every point in A_j^* is adjacent to every point $y_t \in V(q, i)$ (note that $|V(q, i)| = \mu - 1$). We now apply Property C. Note that

$$|A_j^*| \geq |A_j| \cdot (3b)^{-\mu+1} \geq \frac{N}{d+1}(3b)^{-d+1} \geq 100b \log N.$$

It follows from Property C that less than $100b \log N$ points of $A_{j_0}^* = A_{j_0}^*(V(i + 1, i))$ are adjacent to less than $1/3b$ of the points in A_j^*. Fixing q and proceeding through all values of j different from j_0, we would then eliminate at most $d \cdot 100b \log N$ of the points in $A_{j_0}^*$ as candidates for y_{i+1}. If we then range over all possible values for q, we would eliminate at most $d^2 \cdot 100b \log N$ of the points in $A_{j_0}^*$. In addition, we obviously cannot select any of the points in $A_{j_0}^*$ that have been selected previously. This eliminates less than n additional points. Therefore, in order to ensure that the point y_{i+1} can successfully be chosen from $A_{j_0}^*$ (*i.e.* in order to ensure both conditions (a) and (b)), we require that

$$|A_{j_0}^*| \geq \frac{N}{d+1}(3b)^{-d} \geq d^2 \cdot 100b \log N + n.$$

This inequality is satisfied if $N = 100d^3(3b)^{d+1} \cdot n$, and the proof of Lemma 6 is complete.
□

Finally, Theorem 4 immediately follows from Lemmas 5 and 6.   □

References

[1] Beck, J. (1981) Van der Waerden and Ramsey type games. *Combinatorica* **2** 103–116.
[2] Beck, J. (1982) Remarks on positional games - Part I. *Acta Math. Acad. Sci. Hungarica* **40** 65–71.
[3] Beck, J. (1985) Random graphs and positional games on the complete graph. *Annals of Discrete Math.* **28** 7–13.
[4] Bollobás, B. (1982) Long paths in sparse random graphs. *Combinatorica* **2** 223–228.
[5] Bollobás, B. (1985) Random Graphs, Academic Press, London 447ff.
[6] Chvátal, V. and Erdős, P. (1978) Biased positional games. *Annals of Discrete Math.* **2** 221–228.
[7] Chvátal, V., Rödl, V., Szemerédi, E. and Trotter, W. T. (1983) The Ramsey number of a graph with bounded maximum degree. *Journal of Combinatorial Theory* Series B **34** 239–243.
[8] Erdős, P. and Selfridge, J. (1973) On a combinatorial game. *Journal of Combinatorial Theory* Series A **14** 298–301.
[9] Friedman, J. and Pippenger, N. (1987) Expanding graphs contain all small trees. *Combinatorica* **7** 71–76.
[10] Komlós, J. and Szemerédi, E. (1973) Hamilton cycles in random graphs, *Proc. of the Combinatorial Colloquium in Keszthely*, Hungary, 1003–1010.
[11] Pósa, L. (1976) Hamilton circuits in random graphs. *Discrete Math.* **14** 359–64.
[12] Székely, L. A. (1981) On two concepts of discrepancy in a class of combinatorial games. *Colloq. Math. Soc. János Bolyai 37 "Finite and Infinite Sets"* Eger, Hungary. North-Holland, 679–683.

On Oriented Embedding of the Binary Tree into the Hypercube

SERGEJ L. BEZRUKOV

Fachbereich Mathematik, Freie Universität Berlin, Arnimallee 2–6, D-14195 Berlin

We consider the oriented binary tree and the oriented hypercube. The tree edges are oriented from the root to the leaves, while the orientation of the cube edges is induced by the direction from 0 to 1 in the coordinatewise form. The problem is to embed such a tree with l levels into the oriented n-cube as an oriented subgraph, for minimal possible n. A new approach to such problems is presented, which improves the known upper bound $n/l \le 3/2$ given by Havel [1] to $n/l \le 4/3 + o(1)$ as $l \to \infty$.

1. Introduction

Denote by B^n the graph of the n-dimensional unit cube. The vertex set of this graph is just the collection of all binary strings of length n, and two vertices are adjacent if and only if the corresponding sequences differ in one entry only. Let T be a tree. It is easily shown by induction that T is a subgraph of B^n for n sufficiently large. The general question we study here is how to find the minimal such n, which we denote by $\mathbf{dim}(T)$ and call the dimension of T.

Such problems arise in computer science when dealing with multiprocessor systems [6]. The exact answer depends, of course, on the structure of the tree T, rather than on its simple numerical parameters, such as the number of vertices. If one considers trees of bounded vertex degree, which is quite natural for practical applications, one is led to consider the polythomic tree $T^{k,l}$. This is the rooted tree with l levels, where the root has degree k and all the other vertices that are not leaves have degree $k+1$. The dimension of $T^{k,l}$ was studied in [3] (the lower bound) and in [5] (the upper bound), where it is proved that

$$\frac{k \cdot l}{e} \le \mathbf{dim}(T^{k,l}) \le \frac{k \cdot l + k + 2l - 2}{2} \qquad e = 2.718... \ . \tag{1}$$

The lower bound in (1) simply follows from the cardinalities of the sets of vertices at distance at most l from the root, while the upper bound is constructive. Despite several attempts, there have been no improvements of these bounds in the asymptotic sense for arbitrary k, l. For the binary cube it is natural to imagine that the number 2 plays an important role. In accordance with this, let us replace one of the parameters k, l by 2. Then it is known (see [2], [3] respectively) that

$$\mathbf{dim}(T^{2,l}) = l + 2 \quad \text{and}$$
$$\mathbf{dim}(T^{k,2}) = \left\lceil \frac{3k + 1}{2} \right\rceil.$$

It is interesting to notice that, although $T^{2,l}$ has $2^{l+1} - 1 < 2^{l+1}$ vertices, the lower bound $\mathbf{dim}(T^{2,l}) \geq l + 1$, which follows from the cardinalities, is not attainable. Actually, in [4] it is proved that one can even find in B^{l+2} two copies of $T^{2,l}$ joined by an edge connecting their roots.

Therefore, in the simplest cases when one of the parameters k, l equals 2, the problem is completely solved. Let us now consider the oriented version of this problem. We orient the edges of $T^{k,l}$ from the root to the leaves, and the edges of B^n as follows: suppose (v, w) is an edge of B^n such that the sequences v, w differ in the ith entry, where v has 0 and w has 1, then we orient this edge from v to w. Now we look for an oriented subgraph of B^n isomorphic to $T^{k,l}$. In other words, we consider embeddings of $T^{k,l}$ into B^n such that the ith level of $T^{k,l}$ is embedded into the ith level of B^n, for $i = 0, 1, ..., l$. What is the minimal possible n now? We denote this n by $\vec{\mathbf{dim}}(T^{k,l})$.

It is easy to show that the same lower bound (1), following from the inequality

$$\binom{n}{l} \geq k^l, \tag{2}$$

holds. Indeed there is an even better lower bound [1] for $\vec{\mathbf{dim}}(T^{k,l})$, implied by

$$\binom{n}{l} - \binom{n-k}{l} \geq k^l, \tag{3}$$

but it gives no improvement in the asymptotic sense. As it turns out, the upper bound (1) holds for the oriented case as well, as the construction in [5] provides an oriented embedding.

Let us again consider the case when one of the numbers k, l equals 2. If $l = 2$, there is no difference between the oriented and non-oriented cases, as, without loss of generality, one may always assume that the root of $T^{k,2}$ is embedded into the origin of B^n, which forces any embedding to be oriented. It is interesting to note that in this case the lower bound $\vec{\mathbf{dim}}(T^{k,2}) \geq 3k/2$ implied by (3) is asymptotically attained.

The goal of this paper is to study the case $k = 2$. So, we deal with the ordinary binary tree $T^{2,l}$, which we denote T^l for brevity. For this concrete value of k one can get a better lower bound from (2), namely

$$1.2938... \leq \lim_{l \to \infty} \vec{\mathbf{dim}}(T^l)/l. \tag{4}$$

It is easily seen that the trivial upper bound $\vec{\mathbf{dim}}(T^l)/l \leq 2$ equals that given by (1). The best known published upper bound [1] is

$$\lim_{l \to \infty} \vec{\mathbf{dim}}(T^l)/l \leq 3/2. \tag{5}$$

The method of [1] was to find $\vec{\dim}(T^l)$ for $l = 1, ..., 6$, and in particular to prove that T^6 is embeddable into B^9 (here 9/6=3/2). Following this idea, one could try to find a clever embedding of T^{l_0} into B^{n_0} for some l_0, n_0, which would imply the upper bound $\lim_{l\to\infty} \vec{\dim}(T^l)/l \le n_0/l_0$. Here we give a table of $n = n(l) = \vec{\dim}(T^l)$ for small values of l.

l :	1	2	3	4	5	6	7	8	9	10	11
n :	2	4	5	7	8	9	11	12	13	15	16

The entries of this table for $l = 1, ..., 7$ and $l = 10$ are known from [1], while the other three follow from a more detailed analysis, and we give them here without proof. The values for $l = 9$ and $l = 11$ give us an improvement on (5) as $1.444... = 13/9 < 16/11 < 3/2$. We suspect that it is possible to embed T^{12} into B^{17} (at present we are only able to embed T^{12} into B^{18}), in which case we would be able to improve (5) further to $n/l \le 1.416$ for sufficiently large l. But to find an admissible n as l increases is very difficult, and to get a good upper bound in this way is almost hopeless.

Here we present a new approach for obtaining good bounds for the oriented embedding. Our best result is

Theorem 1. $\lim_{l\to\infty} \vec{\dim}(T^l)/l \le 4/3 = 1.333....$

If we consider this result in the light of the old techniques from [1], it becomes apparent that, to prove Theorem 1 using the old approach, one would have to prove that T^{3r} can be embedded into B^{4r} for some $r \ge 13$. To demonstrate this, we computed the function $n(l)$ defined by (3) for $l = 1, ..., 39$ and found that the ratio $n(l)/l$ reaches 4/3 for the first time just when $l = 39$.

Let us mention again that, for $l = 1, ..., 11$, $\vec{\dim}(T^l)$ equals the lower bound given by (3). Moreover, $\vec{\dim}(T^{k,l})$ for $k = 1$ or $l = 1$ is also equal to the lower bound implied by the cardinalities. So as yet, there are no examples where $\vec{\dim}(T^l)$ is not determined by the bound (3).

Conjecture 1. $\vec{\dim}(T^l)$ is determined asymptotically by the inequality (2) as $l \to \infty$.

Conjecture 2. $\vec{\dim}(T^{k,l}) \sim \frac{k \cdot l}{e}$ as $k, l \to \infty$.

2. The new approach

Denote by T_i^l ($i = 0, ..., l$) the ith level of the tree T^l (i.e., the collection of all its vertices at distance i from the root) and by B_i^n ($i = 0, ..., n$) the ith level of B^n (i.e., the collection of all vertices corresponding to sequences with exactly i ones).

We have the trivial upper bound $\vec{\dim}(T^l)/l \le 2$, and thus we may, and shall, assume throughout that T_l^l is embedded above the middle level of B^n. Starting from an embedding of T^l into B^n, let us try to embed T_{l+1}^{l+1} using as few additional dimensions as possible. It is clear that we can always succeed using two additional dimensions. The problem is to try to use just one, as we believe in the following conjecture.

Conjecture 3. $\vec{\dim}(T^{l+1}) > \vec{\dim}(T^l)$ *for all* $l \geq 1$.

It is possible to use only one additional dimension for T^{l+1} if there exists a matching between the image of T_l^l in B_l^n (which we also denote by T_l^l) and B_{l+1}^n. For example T^2 may be embedded into B^4 with the required matching, which implies $\vec{\dim}(T^3) = 5$. Now $\vec{\dim}(T^4) \geq 7$ simply follows from the cardinalities, and our knowledge about $\vec{\dim}(T^3)$ proves $\vec{\dim}(T^4) = 7$ immediately. When can one guarantee the existence of such a matching?

Let $A \subseteq B_k^n$, and x be a given integer. Define an x-*partition* of A to be a partition of A into s parts A_i with $|A_i| \leq x$ $(i = 1, ..., s)$ such that there is a set $M_x(A) = \{a_i : i = 1, ..., s\}$ of distinct vertices of B_{k+1}^n with a_i adjacent to all vertices of A_i $(i = 1, ..., s)$. Call such a set $M_x(A)$ a *covering set* for the x-partition. In particular, if $x = 1$, a covering set for a 1-partition defines a matching between A and B_{k+1}^n.

If there is an embedding of the tree T^l into B^n in such a way that T_l^l has an x-partition, we write $T^l \rightsquigarrow_x B^n$. The arguments above lead us to the following result.

Proposition 1. *If* $T^l \rightsquigarrow_1 B^n$ *then* $T^{l+1} \rightsquigarrow_2 B^{n+1}$.

Proof. Embed T^l into the subcube $x_{n+1} = 0$ in such a way that it has a 1-partition with covering set $A \equiv M_1(T_l^l)$. Set $B = \pi(T_l^l)$ and $C = \pi(A)$, where π is the projection onto the subcube $x_{n+1} = 1$. Now embed T_{l+1}^{l+1} into $A \cup B$ in the obvious way. It is clear that T_{l+1}^{l+1} has a 2-partition with covering set C. $\qquad\qquad\square$

Unfortunately there are examples showing that it is impossible to guarantee that T_l^l has a 1-partition in general, even if $|T_l^l| < |B_{k+1}^n|$. So, the matchings approach does not promise too much, but it is the first step towards more general constructions.

Proposition 2.

(a) If $T^l \rightsquigarrow_2 B^{n_1}$ and $T^k \rightsquigarrow_2 B^{n_2}$, then $T^{l+k} \rightsquigarrow_1 B^{n_1+n_2}$.

(b) If $T^l \rightsquigarrow_1 B^{n_1}$ and $T^k \rightsquigarrow_1 B^{n_2}$, then $T^{l+k+1} \rightsquigarrow_2 B^{n_1+n_2}$.

Proof.

(a) First we build an embedding of T^l into the subcube B_1 of $B^{n_1+n_2}$ based on the first n_1 coordinates, such that T_l^l has a 2-partition. Now for each vertex $v_i \in T_l^l$ we consider the subcube B_2^i based on the last n_2 coordinates, and embed T^k in each such subcube so that T_k^k has a 2-partition. Thus we get an embedding of T^{l+k} into $B^{n_1+n_2}$. Here we mean that the various embeddings of T^k are isomorphic.

To see that T_{l+k}^{l+k} has a 1-partition, we refer to Figure 1. In this picture we represent by a, b and c, d vertices of T^{l+k} in the subcubes B_2^i and B_2^j respectively, such that

(1) these pairs are in the same parts of the second 2-partition, and

(2) the vertex of T^l that is the root of the tree containing a and b is in the same part of the first 2-partition as the vertex corresponding to c and d.

Thus from the embedding of T^k into B_2^i and B_2^j, we deduce that there are vertices

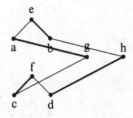

Figure 1

$e \in B_2^i$ and $f \in B_2^j$ that cover the vertices a, b and c, d, respectively. Similarly, there exists at least one vertex g and h that covers a, c and b, d, respectively. More exactly, the edges $(a, g), (b, h), (c, g), (d, h)$ have directions of edges of the subcube B_1, while the edges $(a, e), (b, e)$ are in B_2^i and $(c, f), (d, f)$ are in B_2^j. The required matching between T_{l+k}^{l+k} and $B_{l+k+1}^{n_1+n_2}$ is depicted by the thicker lines.

(b) The proof is similar. $\qquad \square$

Corollary 1. *If* $T^{l_0} \leadsto_2 B^{n_0}$ *then* $\mathbf{dim}(T^l)/l \leq n_0/(l_0 + 1/3) + o(1)$ *as* $l \to \infty$.

Proof. We have $T^{2l_0} \leadsto_1 B^{2n_0}$ by Proposition 2a. Now we apply Proposition 2b with $k = l = 2l_0$ and get $T^{4l_0+1} \leadsto_2 B^{4n_0}$. Therefore each time the cube dimension is multiplied by 4, the height of the tree we can embed increases by a multiple of slightly more than 4. More precisely, if the sequences l_i and n_i are defined by $l_i = 4l_{i-1} + 1$ and $n_i = 4n_{i-1}$ $(i = 1, \ldots)$, then we have $T^{l_i} \leadsto_2 B^{n_i}$ for each i, and $n_i = n_0(l_i + 1/3)/(l_0 + 1/3)$, so $n_i/l_i = n_0/(l_0 + 1/3) + o(1)$ as $i \to \infty$. The result follows. $\qquad \square$

A more detailed analysis of the proof that $\mathbf{dim}(T^6) = 9$ shows that $T^6 \leadsto_2 B^9$, which gives the upper bound $\lim_{l \to \infty} \mathbf{dim}(T^l)/l \leq 9/(6 + 1/3) \approx 1.421$, but some work is still required. Now we present, as the second elementary application of our approach, a simple proof of the bound (5).

Proposition 3. *If* $T^l \leadsto_2 B^n$ *then* $T^{l+2} \leadsto_2 B^{n+3}$.

Proof. First embed the tree T^l into B^n for some n such that T_l^l has a 2-partition with covering set $M_2(T_l^l)$. This is possible by Proposition 1. Now we use this embedding, and its associated 2-partition and covering set, to embed the two extra levels of the binary tree using only three extra dimensions. So, we build the 3-cube growing from each vertex of our n-cube, in particular from each vertex of T_l^l. For each set $\{u_i, v_i\}$ in our 2-partition, let $w_i \in B_{l+1}^n$ be the corresponding vertex in the covering set $M_2(T_l^l)$. This situation is depicted in Figure 2a, where the rectangles represent the 3-cubes growing from vertices u_i, v_i. The corresponding vertices of these 3-cubes are connected, as shown in Figure 2.

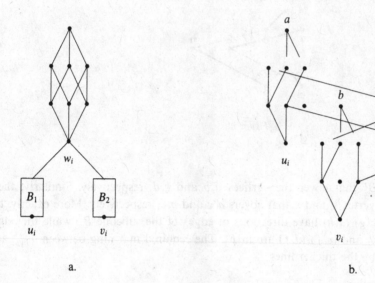

a.   b.

Figure 2

We now embed two copies of T^2 rooted in u_i, v_i into this structure, which will provide an embedding of T^{l+2} into B^{n+3}.

Our embedding scheme is shown in Figure 2b, where we draw the edges of the trees only. Incomplete lines indicate a covering scheme, demonstrating that the embedding has a 2-partition.   □

Now the upper bound (5) follows immediately. We start with an embedding of T^3 into B^5 with a 2-partition, the existence of which was mentioned earlier, and apply Proposition 3. On the ith step of this process, we obtain an embedding of T^{3+2i} into B^{5+3i}, which implies the upper bound $\lim_{l\to\infty} \mathbf{dim}(T^l)/l \le \lim_{i\to\infty}(3i+5)/(2i+3) = 3/2$.

What is important in our approach is that, given an embedding of T^l into B^n, we construct an embedding of $T^{l+\epsilon}$ into $B^{n+\delta}$, even though $\mathbf{dim}(T^\epsilon) > \delta$, which gives the bound $\lim_{l\to\infty} \mathbf{dim}(T^l)/l \le \delta/\epsilon$. We achieve this by using some additional information about the initial embedding, in this case the existence of a 2-partition.

The second important thing is that, in the proof of Proposition 3, it makes no difference which initial embedding we start with. The only thing one needs is a 2-partition, which is in fact easy to guarantee. Indeed, embed T^l into any admissible B^n. Now increase the dimension of the hypercube by 1, adding the subcube $x_{n+1} = 1$. Then each vertex of T^l_i has a neighbour in this subcube, so T^l_i has a 1-partition in B^{n+1}.

Let us finally mention that the proof of the upper bound (5) may be further simplified by using the following result.

Proposition 4. *If $T^l \leadsto_1 B^n$, then $T^{l+2} \leadsto_1 B^{n+3}$.*

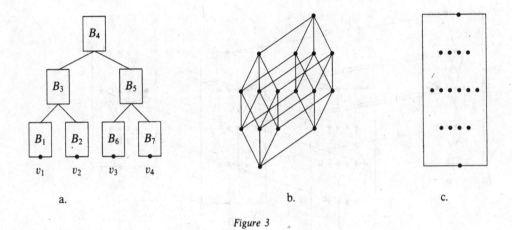

Figure 3

To prove this, one has, in fact, to show that T^2 may be embedded into B^4 so that T_2^2 has a 1-partition, which fact we have already mentioned above.

3. Towards better upper bounds

Our general aim is to get a rational upper bound for $\lim_{l \to \infty} \mathbf{dim}(T^l)/l$, necessarily exceeding 1.29, using constructions involving low-dimensional cubes only. It seems to be impossible to obtain fully satisfactory results using just x-partitions, with the same x before and after the addition of extra levels. So we need some deeper insight.

For $A \subseteq B_l^n$, $t \geq 2$ and a sequence $(x_1, ..., x_t)$ of positive integers, we say that A can be $(x_1, ..., x_t)$-*partitioned* if there are sets $M_0, M_1, ..., M_t$ such that $M_0 = A$, $M_i \subseteq B_{l+i}^n$ for each i, and, for each i, there is an x_i-partition of M_{i-1} with covering set M_i. If there exists an embedding of T^l into B^n such that T_l^l can be $(x_1, ..., x_t)$-partitioned, we write $T^l \rightsquigarrow_{x_1, ..., x_t} B^n$.

Proposition 5. *If* $T^l \rightsquigarrow_{2,2} B^n$ *then* $T^{l+3} \rightsquigarrow_{2,3} B^{n+4}$.

Proof. Now we have to embed four copies of T^3, rooted in vertices $v_1, ..., v_4$, into the structure depicted in Figure 3a, where each box represents the 4-cube. The graph of the 4-cube is as shown in Figure 3b; for convenience we shall normally use the restricted image of it shown in Figure 3c. In this image we show just the vertices of B^4, in the same order from left to right as they are shown in Figure 3b.

The embedding we use here is shown in Figure 4, where, for simplicity, only the subcubes B_1, B_2, B_3 and B_4 are shown, without the edges connecting them. The two copies of T^3 have their roots in vertices v_1, v_2.

Now the top vertices of B_1 and B_2, the vertices of the 3-d level of B_3 and the vertices of the second level of B_4 shown by larger solid circles in Figure 4 form a covering set

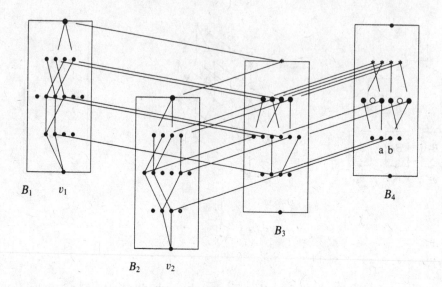

Figure 4

M_1 for a 2-partition of this piece of T_{l+3}^{l+3}, and the vertices labelled by asterisks form a covering set for a 3-partition of M_1. This covering scheme is also presented in Figure 4.

Let us look further at the subcube B_4. In Figure 4 only two vertices a, b of $T^{l+3} \cap B_4$ are depicted. They correspond to the vertices (0010) and (0100) of B_4 respectively (the commas in vectors are omitted). The vertices of B_4 that cover them correspond to (0110) and (1100) respectively. But we also have to embed four vertices coming from the subcube B_5. In order for all these eight vertices to be distinct, we first embed the two other copies of T^3 into B_5, B_6, B_7, and after that use the isometric transformation of these subcubes defined by the permutation $\begin{pmatrix} 1234 \\ 3412 \end{pmatrix}$ of coordinates. This permutation transforms the four mentioned vertices into (1000), (0001), (0011), and (1001) respectively, which guarantees the correct embedding.

For future reference, note that there are two vertices in the second level of B_4, namely (0101) and (1010), that are not in M_1. They are shown as empty circles in Figure 4. The Hamming distance between these two vertices is 4.   □

Using similar techniques one could prove the following properties.

Proposition 6.

(a) If $T^l \leadsto_{2,3} B^n$ then $T^{l+2} \leadsto_{2,2} B^{n+3}$;

(b) If $T^l \leadsto_{2,2,3} B^n$ then $T^{l+3} \leadsto_{2,2,4} B^{n+4}$;

(c) If $T^l \leadsto_{2,2,4} B^n$ then $T^{l+3} \leadsto_{2,3,3} B^{n+4}$;

(d) If $T^l \leadsto_{2,3,3} B^n$ then $T^{l+2} \leadsto_{2,2,3} B^{n+3}$.

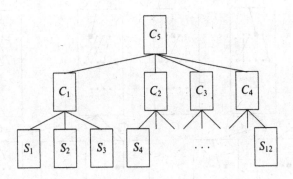

Figure 5

We do not use these properties for the proof of Theorem 1, but believe that they are useful for further research. One could combine them with some others to get new upper bounds. In particular, Propositions 5, 6a and 6b – 6d, respectively, imply the following results.

Corollary 2.

(a) $\lim_{l \to \infty} \overrightarrow{\dim}(T^l)/l \leq 7/5 = 1.4$;

(b) $\lim_{l \to \infty} \overrightarrow{\dim}(T^l)/l \leq 11/8 = 1.375$.

Our main result, Theorem 1, is an immediate consequence of the following result.

Proposition 7. *If* $T^l \rightsquigarrow_{2,2,3,4} B^n$ *then* $T^{l+3} \rightsquigarrow_{2,2,3,4} B^{n+4}$.

Proof. Now we deal with the structure depicted in Figure 5, where $C_1, ..., C_5$ are 4-cubes and $S_1, ..., S_{12}$ are the structures, consisting of seven 4-cubes, depicted in Figure 3a. Each 4-cube C_2, C_3, C_4 is connected with three structures S_i ($i = 4, ..., 12$), as shown in Figure 5 for the cube C_1. We use the image of B^4 shown in Figure 3b, and again reduce it to that shown in Figure 3c.

We start with an embedding of T^l into B^n such that T^l_l can be $(2, 2, 3, 4)$-partitioned, and now embed the three extra levels of our tree by embedding T^3 into each structure S_i, as described in the proof of Proposition 5. The role of the remaining 4-cubes in Figure 5 is to guarantee that T^{l+3}_{l+3} can be $(2, 2, 3, 4)$-partitioned. It was mentioned above that two vertices at distance 4 are free in the subcube B_4 in each structure S_i. Using isometric transformations of the structures S_i, we can establish these free vertices to be just (0011) and (1100) in the structure S_i with $i = 0 \pmod 3$, and the vertices (0101),(1010) and (0110),(1001) in the structures S_i with $i = 1 \pmod 3$ and $i = 2 \pmod 3$, respectively. The free vertices are shown as empty circles in the bottom 4-cubes in Figure 6. These bottom 4-cubes correspond to the subcubes B_4 of the structures S_i (*cf.* Figure 3a).

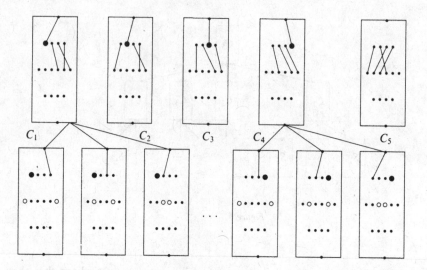

Figure 6

To prove that our embedding of T_{l+3}^{l+3} can be $(2,2,3,4)$-partitioned, we need to construct sets M_1, M_2, M_3 and M_4 such that M_1 is a covering set for a 2-partition of this section of T_{l+3}^{l+3}, M_2 is a covering set for a 2-partition of M_1, M_3 is a covering set for a 3-partition of M_2, and finally M_4 is a covering set for a 4-partition of M_3.

M_1: We use just the same construction for the set M_1 as in the proof of Proposition 5. This set is shown in Figure 4 by the large solid circles.

M_2: We take the top vertices of the subcube B_3 to cover the top vertices of the subcubes B_1, B_2 (*cf.* Figure 4) in each structure S_i, and use all the four vertices in the 3-d level of the subcube B_4 to cover the vertices of the 3-d levels of subcubes B_3, B_5. Now all that remains to be done is to cover the four solid vertices in the second level of the subcube B_4 in each structure S_i (see Figure 4) by the six vertices in the second level of the subcubes C_i ($i = 1,...,4$) (*cf.* Figure 5). The covering scheme is explained in Figure 7a. In this figure we represent the six vertices of C_1 by the top block and the second levels of B_4 in S_1, S_2, S_3 by the three bottom blocks (for other C_i and S_i, the principle is the same). Each vertex of the top block is incident (in B^n) to the three corresponding vertices of the bottom blocks, but we have to choose only two edges to cover all the solid vertices. Now remove the edges shown in Figure 7a. Then the remaining edges between the top and bottom blocks form the required covering.

M_3: As constructed above, M_2 consists of the top vertices of the subcubes B_3, the second levels of the subcubes B_4 (*cf.* Figure 4) and the second levels of the subcubes $C_1,...,C_4$. Now we have to cover all these vertices by the top vertices of the subcubes B_4, the third levels of $C_1,...,C_4$ and the second level of C_5, in such a way that no vertex is matched to more than three from M_2.

Now consider the subcube C_1 and the subcubes B_4 in the structures S_1, S_2, S_3. We

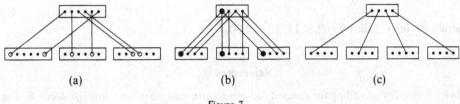

(a)   (b)   (c)

Figure 7

cover the top vertices of the subcubes B_3, B_5 from the top vertex of the subcubes B_4 in each S_i (*cf.* Figure 4) and use the third edge to cover one of the three vertices in the 3-d levels of B_4, as shown in Figure 6 (these three vertices are depicted by small circles). In order to cover the remaining three vertices of B_4's (depicted by large circles in Figure 6), we use the 3-d level of C_1 and the covering scheme as shown in Figure 7b. In this picture the leftmost (large) vertex has degree three, while all the other vertices have degree two. We use the remaining three edges incident to these vertices to cover some three vertices in the second level of C_1, as shown in Figure 6.

Therefore, the vertex of each subcube B_4 represented by the largest circle (see Figure 6) in the structures S_i ($i = 1, 2, 3$) plays a particular role. In other structures, we use a similar principle, and the corresponding vertices of the B_4's are represented in Figure 6 by large circles. Of course, one then has to correct the covering scheme in the subcubes C_2, C_3, C_4, which we do in accordance with Figure 6.

Now consider the 4-cubes $C_1, ..., C_4$ in Figure 6, and notice that the two rightmost vertices in their second levels are already covered from the 3-d levels, and just one of the other four vertices is also covered. In order to cover the remaining three vertices in each subcube $C_1, ..., C_4$, we use the leftmost four vertices of C_5 and the covering graph depicted in Figure 7c. Each vertex of the top block is incident to the corresponding vertex in each bottom block, and removing the depicted edges we get the required covering.

M_4: Finally, we construct M_4 from the top vertices of the subcubes $C_1, ..., C_4$ and the third level of C_5. Indeed, cover the top vertices of the subcubes B_4 in each structure S_j from the top vertex of the corresponding subcube C_i ($i = 1, ..., 4$). Thus we have used three edges for each C_i. The 4th edge is used to cover the large vertices of the 3-d level in each C_i, as shown at the top of Figure 6. The remaining (small) vertices of C_i ($i = 1, ..., 4$) are covered from the 3-d level of C_5 using three edges, with a covering scheme similar to that in Figure 7c. Now each vertex of the third level of C_5 is used to cover some three vertices of M_3, and we use the 4th edge incident to each of them to cover the leftmost four vertices in the second level of C_5 (see Figure 6).

□

We hope that by using similar techniques, it will be possible to operate with larger graphs, and construct an embedding of 10 extra levels in 13 extra dimensions and finally prove the following.

Conjecture 4. $\lim_{l\to\infty} \vec{\dim}(T^l)/l \leq 13/10 = 1.30$.

References

[1] Havel, I. (1982) Embedding the directed dichotomic tree into the *n*-cube. *Rostock. Math. Kolloq.* **21** 39–45.

[2] Havel, I. and Liebl, P. (1972) O vnořeni dichotomického stromu do crychle (Czech, English summary). *Čas. Pěst. Mat.* **97** 201–205.

[3] Havel, I. and Liebl, P. (1973) Embedding the polythomic tree into the *n*-cube. *Čas. Pěst. Mat.* **98** 307–314.

[4] Nebeský, L. (1974) On cubes and dichotomic trees. *Čas. Pěst. Mat.* **99** 164–167.

[5] Ollé, F. (1972) M. Sci. Thesis, Math. Inst., Prague.

[6] Wagner, A. S. (1987) *Embedding trees in the hypercube*, Technical Report 204/87, Dept. of Computer Science, University of Toronto.

Potential Theory on Distance-Regular Graphs

NORMAN L. BIGGS

London School of Economics, Houghton St., London WC2A 2AE

A graph may be regarded as an electrical network in which each edge has unit resistance. We obtain explicit formulae for the effective resistance of the network when a current enters at one vertex and leaves at another in the distance-regular case. A well-known link with random walks motivates a conjecture about the maximum effective resistance. Arguments are given that point to the truth of the conjecture for all known distance-regular graphs.

1. Introduction

We shall be concerned with a graph G regarded as an electrical network in which each edge has resistance 1. A well-known result due to R.M. Foster [6] (see also [3, p.41] and [9]) asserts that if G has n vertices and m edges, the effective resistance between adjacent vertices is $r_1 = (n-1)/m$, provided that all edges are equivalent under the action of the automorphism group. In this paper I shall obtain formulae for r_i, the effective resistance between vertices at distance i, for $i \geq 2$, provided G is distance-transitive (DT). With hindsight, it will be clear that the same formulae hold if we assume only that G is distance-regular (DR). The case $i = 2$ was also studied by Foster [7].

Another well-known fact is that the electrical problem can be regarded as a case of solving Laplace's equation on the graph. As explained in the elegant little book by Doyle and Snell [5], this leads to significant connections with other subjects, in particular the theory of random walks. In that context, the solution to the problem of effective resistances has a simple interpretation in terms of 'hitting times'. Our results can also be applied to questions about the 'cover time', that is, the expected number of steps required to visit all the vertices. In particular, it appears that for all *known* DR graphs (except the cycle graphs), the cover time is $O(n \log n)$.

For the sake of completeness, we gather together here the basic notation and terminology for DT and DR graphs, which will be used in the rest of the paper. The author's book [2], or the standard text of Brouwer, Cohen and Neumaier [4], with its 800 references, should be consulted for details.

We denote the diameter of a connected graph by d, and the distance between vertices v and w by $d(v,w)$. A connected graph G is *distance-transitive* if, for any vertices v, w, x, y satisfying $d(v,w) = d(x,y)$, there is an automorphism of G that takes v to x and w to y. Given integers h, i such that $0 \le h, i \le d$, and vertices v, w, define

$$s_{hi}(v,w) = |\{x \in VG \mid d(x,v) = h \text{ and } d(x,w) = i\}|.$$

In a distance-transitive graph, the numbers $s_{hi}(v,w)$ depend on the distance $d(v,w)$, not on v and w, so we can define the *intersection numbers*

$$s_{hij} = s_{hi}(v,w), \quad \text{where} \quad d(v,w) = j, \quad (h,i,j \in \{0,1,\dots,d\}).$$

Consider the intersection numbers with $h = 1$. For a fixed j, s_{1ij} is the number of vertices x such that x is adjacent to v and $d(x,w) = i$, given that $d(v,w) = j$. The triangle inequality for the distance function implies that $s_{1ij} = 0$ unless $i = j-1, j$, or $j+1$. For the intersection numbers s_{1ij} that are not identically zero, we use the notation

$$c_j = s_{1,j-1,j} \quad a_j = s_{1,j,j}, \quad b_j = s_{1,j+1,j}, \quad (0 \le j \le d),$$

noting that c_0 and b_d are undefined.

The numbers c_j, a_j, b_j have the following simple interpretation. For any vertex v of G let

$$G_i(v) = \{x \in VG \mid d(x,v) = i\}, \quad (0 \le i \le d).$$

It is clear that the sets $\{v\} = G_0(v), G_1(v) \dots, G_d(v)$ form a partition of VG. Given a vertex v and a vertex x in $G_j(v)$, this vertex is adjacent to c_j vertices in $G_{j-1}(v)$, a_j vertices in $G_j(v)$, and b_j vertices in $G_{j+1}(v)$. These numbers are independent of v and x, provided that $d(v,x) = j$.

The *intersection array* of a distance-transitive graph G is

$$\iota(G) = \left\{ \begin{matrix} & c_1 & \dots & c_j & \dots & c_d \\ a_0 & a_1 & \dots & a_j & \dots & a_d \\ b_0 & b_1 & \dots & b_j & \dots & \end{matrix} \right\}.$$

We observe that a distance-transitive graph is vertex-transitive, and consequently regular, of degree k say. Clearly, we have $b_0 = k$ and $a_0 = 0, c_1 = 1$. Further, since each column of the intersection array sums to k, if we are given the first and third rows, we can calculate the middle row. Thus it is convenient to use the alternative notation

$$\iota(\Gamma) = \{k, b_1, \dots, b_{d-1}; 1, c_2, \dots, c_d\}.$$

A *distance-regular graph* is a graph that has the combinatorial regularity implied by distance-transitivity, without (possibly) the prescribed automorphisms. Explicitly, it is a connected graph such that for some positive integers d and k the following holds: there are natural numbers $b_0 = k, b_1, \dots, b_{d-1}, c_1 = 1, c_2, \dots, c_d$, such that for each pair (v, w) of vertices satisfying $d(v,w) = j$ we have

1   the number of vertices in $G_{j-1}(v)$ adjacent to w is c_j $(1 \le j \le d)$;
2   the number of vertices in $G_{j+1}(v)$ adjacent to w is b_j $(0 \le j \le d-1)$.

Clearly, a distance-transitive graph is distance-regular, but the converse is false. The cycle graphs are the only distance-regular graphs with degree $k = 2$; although they are trivial, they are also somewhat anomalous. We exclude them by assuming that $k \geq 3$ always.

In the algebraic theory of DR graphs, it is shown that there is a representation in which the adjacency matrix $\mathbf{A}$ is represented by a tridiagonal matrix $\mathbf{B}$ whose entries are the elements of the intersection array:

$$
\mathbf{B} = \begin{pmatrix}
0 & 1 & 0 & . & . & . & 0 & 0 \\
k & a_1 & c_2 & . & . & . & 0 & 0 \\
0 & b_1 & a_2 & . & . & . & 0 & 0 \\
. & . & . & . & . & . & . & \\
. & . & . & . & . & . & . & \\
. & . & . & . & . & . & . & \\
0 & 0 & 0 & . & . & . & a_{d-1} & c_d \\
0 & 0 & 0 & . & . & . & b_{d-1} & a_d
\end{pmatrix}
$$

Our main result will be formulated in terms of this matrix.

2. Calculation of potentials

In this section we shall obtain formulae for the potentials of the vertices of a DR graph, when a current enters at one vertex and leaves at an adjacent one.

It is convenient to begin with a DT graph. A particular consequence of this assumption is that all pairs of adjacent vertices are equivalent under the action of the automorphism group. Choose an adjacent pair (v, w) once and for all. With respect to this pair we construct the *distance distribution diagram*, or *DDD*. That is, we define

$$
\begin{aligned}
V_i &= \{x \mid d(x, v) = i, d(x, w) = i + 1\}; \\
W_i &= \{x \mid d(x, v) = i + 1, d(x, w) = i\}; \\
Z_i &= \{x \mid d(x, v) = d(x, w) = i\}.
\end{aligned}
$$

In terms of the intersection numbers we have

$$
s_{i\,i+1\,1} = |V_i| = |W_i| = s_{i+1\,i\,1}, \quad |Z_i| = s_{i\,i\,1}.
$$

It should be noted that the numbers of edges joining sets of vertices in the DDD are not completely determined by the intersection array. For example, the number of edges with one end in V_i and one end in W_i is not determined. However, we do have some information.

Lemma 1. *For any vertex x in V_i and any set S of vertices, let $S(x)$ denote the set of edges joining x to vertices in S. Let $|Z_i(x)| = \alpha$, $|W_i(x)| = \beta$, $|Z_{i+1}(x)| = \gamma$. Then we have*

$$
|V_{i-1}(x)| = c_i, \quad |V_{i+1}(x)| = b_{i+1},
$$

$$
\alpha + 2\beta + \gamma = (b_i - b_{i+1}) + (c_{i+1} - c_i).
$$

Proof. For $x \in V_i$, we have $d(v, x) = i$. The number of edges joining x to vertices y such that $d(v, y) = i - 1$ is, by definition, c_i. The construction of the DDD ensures that all vertices adjacent to x and distance $i - 1$ from v are in V_{i-1}, hence the result.

For $x' \in W_i$, we have $d(v, x') = i + 1$. The number of edges joining x' to vertices y' such that $d(v, y') = i + 2$ is, by definition b_{i+1}. The construction of the DDD ensures that all vertices adjacent to x' and distant $i + 2$ from v are in W_{i+1}, hence $|W_{i+1}(x')| = b_{i+1}$. Using the symmetry with respect to v and w, we conclude that $|V_{i+1}(x)| = b_{i+1}$.

For $x \in V_i$ we have $d(x, w) = i + 1$, and the vertices adjacent to x that are at distance i from w are in V_{i-1}, Z_i, and W_i. So we get

$$c_i + \alpha + \beta = c_{i+1}.$$

Similarly, since $d(x, v) = i$, and the vertices adjacent to x and distance $i + 1$ from v are in W_i, Z_{i+1}, and V_{i+1}, we get

$$\beta + \gamma + b_{i+1} = b_i.$$

Adding these two equations gives the result.   □

As we shall now demonstrate, the quantities evaluated in Lemma 1 are sufficient to determine the potential of any vertex when a current J enters at v and leaves at w, given that each edge has unit conductance. Describing the state of the network by means of a potential function automatically ensures that 'Kirchhoff's voltage law' (that the potential drop around any cycle is zero) is satisfied. We assume that all vertices in V_i will be at the same potential, and likewise for W_i and Z_i. The justification for this assumption is that it enables us to solve the equations expressing 'Kirchhoff's current law' (that the net current at any vertex is zero). Since it is known that there is a unique solution to Kirchhoff's equations (see for example [8]), any assumption producing a result must be valid. In the same spirit, we use the symmetry with respect to v and w, and take the potentials on V_i, W_i, Z_i, to be $\phi_i, -\phi_i, 0$, respectively.

Lemma 2. *The potentials ϕ_i satisfy the equations*

$$
\begin{aligned}
c_i \phi_{i-1} - b_i \phi_i &= J - k\phi_0 \quad (1 \le i \le d - 1), \\
c_d \phi_{d-1} &= J - k\phi_0.
\end{aligned}
$$

Proof. We use the standard technique of replacing a set X of vertices that are at the same potential by a single vertex x. Given two such sets X and Y, the edges joining them become parallel edges joining x to y, and can be replaced by a single edge whose conductance is the sum of the conductances. Since we are assuming that each edge has conductance 1, this means that the conductance of the edge xy is equal to the number of edges joining X to Y. In particular, if each vertex in X is joined to the same number λ of vertices in Y, the conductance of xy is $\lambda |X|$.

This technique enables us to regard the DDD as a network equivalent to the given graph. A typical vertex v_i in this network is obtained by identifying all vertices in $V_i (0 \le i \le d - 1)$, and we can define w_i and z_i similarly. The formulae obtained in Lemma 1 can now be interpreted as results about the conductances of edges in the equivalent network. For example, the conductance of the edge joining v_i to v_{i-1} is $c_i|V_i|$, corresponding to the fact that $|V_{i-1}(x)| = c_i$.

Consider first the vertex v in G. The numbers of edges from v to V_1, Z_1, W_0 are b_1, a_1, 1, respectively. Since $|V_0| = 1$, in the equivalent network based on the DDD, these numbers are the conductances of the edges joining v_0 to v_1, z_1, w_0, respectively. Applying Kirchhoff's current law at v_0, we get

$$J = b_1(\phi_0 - \phi_1) + a_1(\phi_0 - 0) + 1(\phi_0 - (-\phi_0)).$$

Since $c_1 + a_1 + b_1 = k$ and $c_1 = 1$, this can be written as

$$c_1\phi_0 - b_1\phi_1 = J - k\phi_0.$$

Similarly, applying Kirchhoff's current law at v_i, we get

$$c_i(\phi_{i-1} - \phi_i) = \alpha(\phi_i - 0) + \beta(\phi_i - (-\phi_i)) + \gamma(\phi_i - 0) + b_{i+1}(\phi_i - \phi_{i+1}),$$

and using the formula for $\alpha + 2\beta + \gamma$, this reduces to

$$c_{i+1}\phi_i - b_{i+1}\phi_{i+1} = c_i\phi_{i-1} - b_i\phi_i.$$

This is valid for $1 \leq i \leq d - 2$, so it follows that the value of $c_i\phi_{i-1} - b_i\phi_i$ is constant for $1 \leq i \leq d - 1$. Comparing with the equation for $c_1\phi_0 - b_1\phi_1$, we see that the constant value is $J - k\phi_0$.

Finally, applying Kirchhoff's current law at v_{d-1}, and using what we have just proved, we get

$$c_d\phi_{d-1} = c_{d-1}\phi_{d-2} - b_{d-1}\phi_{d-1} = J - k\phi_0. \qquad \square$$

When J is given, the d equations obtained in Lemma 2 determine $\phi_0, \phi_1, \ldots, \phi_{d-1}$. The results come out more smoothly if we do a little reorganisation first.

As described in the Introduction, the vertices of G are arranged in disjoint subsets $G_i(v)$ $(0 \leq i \leq d)$, according to distance from a given vertex v. If we write $k_i = |G_i(v)|$, the total number of vertices in G is

$$n = 1 + k_1 + k_2 + \ldots + k_d.$$

Also, by counting the edges joining $G_{i-1}(v)$ to $G_i(v)$ in two different ways, we get a recursion for the sequence (k_i):

$$k_0 = 1, \quad c_i k_i = b_{i-1}k_{i-1} \quad (1 \leq i \leq d).$$

Theorem A. *If G has n vertices and m edges, and the current is $J = 2m$, then the potentials are determined recursively by the equations*

$$\phi_0 = n - 1, \quad b_i\phi_i = c_i\phi_{i-1} - k \quad (1 \leq i \leq d - 1).$$

These equations have the explicit solution

$$\phi_i = \frac{c_1 c_2 \ldots c_i}{b_1 b_2 \ldots b_i}(n - 1 - k_1 - \ldots - k_i) = \frac{k}{b_i k_i}(k_{i+1} + \ldots + k_d).$$

Proof. This is proved by elementary algebra, starting from Lemma 2. □

Examples

(i) The dodecahedron is a distance-transitive graph with $n = 20$ vertices, degree $k = 3$, and intersection array

$$\left\{ \begin{matrix} 1 & 1 & 1 & 2 & 3 \\ 0 & 0 & 1 & 1 & 0 & 0 \\ 3 & 2 & 1 & 1 & 1 \end{matrix} \right\}.$$

Here the potentials are:

$$\phi_0 = 19, \quad \phi_1 = 8, \quad \phi_2 = 5, \quad \phi_3 = 2, \quad \phi_4 = 1.$$

(ii) Similarly, for the cubic DT graph with 102 vertices [4, pp.403-405] and intersection array $\{3, 2, 2, 2, 1, 1, 1; 1, 1, 1, 1, 1, 1, 3\}$, we get

$$\phi_0 = 101, \quad \phi_1 = 49, \quad \phi_2 = 23, \quad \phi_3 = 10, \quad \phi_4 = 7, \quad \phi_5 = 4, \quad \phi_6 = 1.$$

We shall refer to this example later.

(iii) Let *Suz* be Suzuki's simple group of order $2^{13}.3^7.5^2.7.11.13$. As described in [4, pp.410-412], the group *Suz*.2 is the automorphism group of a distance-transitive graph with 22880 vertices and intersection array $\{280, 243, 144, 10; 1, 8, 90, 280\}$. The potentials are:

$$\phi_0 = 22879, \quad \phi_1 = 93, \quad \phi_2 = 29/9, \quad \phi_3 = 1.$$

At this point we can extend our results to distance-regular graphs. Since the equations for the potentials involve only the parameters occuring in the intersection array, it is clear that these equations determine 'potentials' in the DR case. These 'potentials' provide a solution to the network equations and, as has been pointed out, it is known that there is a unique solution. Thus we have the solution to the DR case.

It is intuitively 'obvious' that the potentials ϕ_i form a strictly decreasing sequence. However, we should remember that some things about the flow of electricity are only 'obvious' to those who mistakenly claim to understand what electricity really is (see [5, p. 70]). In fact, the proof that $\phi_i > \phi_{i+1}$ depends upon a property of the intersection array which we have not yet mentioned explicitly.

Lemma 3. *The intersection numbers of a DR graph satisfy*

$$1 = c_1 \leq c_2 \leq \ldots \leq c_d; \quad k = b_0 \geq b_1 \geq \ldots \geq b_{d-1}.$$

Proof. This is a standard result from the early days of DR graph theory [2, p.135]. It also follows from the proof of Lemma 2, specifically the equations

$$c_i + \alpha + \beta = c_{i+1}, \quad \beta + \gamma + b_{i+1} = b_i.$$

□

Theorem B. *The sequence (ϕ_i) of potentials is strictly decreasing.*

Proof. In Theorem A we obtained explicit formulae for ϕ_i, the second of which can be written in the form

$$\phi_i = k\left(\frac{1}{c_{i+1}} + \frac{b_{i+1}}{c_{i+1}c_{i+2}} + \ldots + \frac{b_{i+1}b_{i+2}\ldots b_{d-1}}{c_{i+1}c_{i+2}\ldots c_d}\right).$$

This formula for ϕ_i is the sum of $d - i$ terms. Comparing the jth terms in the formulae for ϕ_i and ϕ_{i+1}, we have

$$\frac{b_{i+1}b_{i+2}\ldots b_{i+j-1}}{c_{i+1}c_{i+2}\ldots c_{i+j}} \geq \frac{b_{i+2}b_{i+3}\ldots b_{i+j}}{c_{i+2}c_{i+3}\ldots c_{i+j+1}},$$

by virtue of the inequalities in Lemma 3. Since ϕ_i contains one term more than ϕ_{i+1}, we have the strict inequality as claimed. $\qquad\square$

A consequence of the formulae obtained above is that, for $0 \leq i \leq d - 1$,

$$\left(\frac{c_{i+1}c_{i+2}\ldots c_d}{k}\right)\phi_i$$

can be written as a sum of $d - i$ monomial terms, each of weight $d - i - 1$. For example, when $d = 4$ we get

$$\begin{aligned}
\left(\frac{c_1c_2c_3c_4}{k}\right)\phi_0 &= c_2c_3c_4 + b_1c_3c_4 + b_1b_2c_4 + b_1b_2b_3, \\
\left(\frac{c_2c_3c_4}{k}\right)\phi_1 &= c_3c_4 + b_2c_4 + b_2b_3, \\
\left(\frac{c_3c_4}{k}\right)\phi_2 &= c_4 + b_3, \\
\left(\frac{c_4}{k}\right)\phi_3 &= 1.
\end{aligned}$$

The rule for forming the expressions for other values of d should be clear. We shall use them in Section 4.

3. The effective resistance vector

We defined the potentials in such a way that the potential difference between v and w is $2\phi_0$. It follows that the equivalent resistance between vertices at distance 1 is $r_1 = 2\phi_0/J$, and Theorem A tells us that this is $(n - 1)/m$. Thus we have verified the formula for r_1 mentioned in the Introduction in the distance-regular case.

We now show that the equations obtained in Theorem A are sufficient to determine the effective resistances r_i ($2 \leq i \leq d$). The trick is to use the 'method of superposition'.

Lemma 4. *If a current J enters G at v, and leaves at a vertex w_i at distance $i + 1$ from v, the potential difference between v and w_i is*

$$2(\phi_0 + \phi_1 + \ldots + \phi_i).$$

Proof. Let us use the term *system* to denote a specific distribution of currents and potentials on a given DR graph. We have already observed that the result is true in the system Σ_0 when a current J enters at v and leaves at w. The system Σ_1 when a current

J enters at v and leaves at $w_1 \in W_1$ is the 'superposition' of Σ_0 and the system Π_1 when J enters at w and leaves at w_1. In Σ_0 the potentials at v, w, w_1 are $\phi_0, -\phi_0, -\phi_1$ respectively, and in Π_1 they are $\phi_1, \phi_0, -\phi_0$ respectively. Thus, in Σ_1 the potentials are $\phi_0 + \phi_1, 0, -(\phi_0 + \phi_1)$, and the required potential difference is $2(\phi_0 + \phi_1)$, as claimed.

More generally, the system Σ_i when a current J enters at v and leaves at w_i is the superposition of Σ_{i-1} and the system Π_i when J enters at w_{i-1} and leaves at w_i. Suppose we make the induction hypothesis that, in Σ_{i-1}, the potentials at v, w_{i-1}, w_i are $\sigma_{i-1}, -\sigma_{i-1}, \phi_0 - \sigma_i$ respectively, where $\sigma_i = \phi_0 + \ldots + \phi_i$. Now Π_i is obtained from Σ_0 by a translation that takes v, w to w_{i-1}, w_i, and a vertex v_i to v, so the corresponding potentials in Π_i are $\phi_i, \phi_0, -\phi_0$. Adding the two sets of potentials we have the result for Σ_i, and the general result follows by induction.   □

Let us denote by $\mathbf{r}$ the row-vector $(r_0, r_1, \ldots, r_d)$ whose ith component r_i is the effective resistance between a pair of vertices at distance i. The results obtained above tell us how to calculate the vector $\mathbf{r}$. For example, for the dodecahedron and the 102-graph (Examples (i) and (ii)) we have

$$\mathbf{r} = \frac{1}{30}(0, 19, 27, 32, 34, 35) \quad \text{and} \quad \mathbf{r} = \frac{1}{153}(0, 101, 150, 173, 183, 190, 194, 195).$$

For the general case we have the following Theorem, in which the matrix $\mathbf{B}$ is the intersection matrix of a DR graph, and $\mathbf{u}, \mathbf{v}$ are the $(d+1)$-dimensional row-vectors

$$\mathbf{u} = (1, 1, 1, \ldots, 1), \quad \mathbf{v} = (n, 0, 0, \ldots, 0).$$

Explicitly $\mathbf{B} - k\mathbf{I}$ is the matrix

$$\begin{pmatrix} -k & 1 & 0 & . & . & . & 0 & 0 \\ k & a_1 - k & c_2 & . & . & . & 0 & 0 \\ 0 & b_1 & a_2 - k & . & . & . & 0 & 0 \\ . & . & . & . & . & . & . & . \\ . & . & . & . & . & . & . & . \\ . & . & . & . & . & . & . & . \\ 0 & 0 & 0 & . & . & . & a_{d-1} - k & c_d \\ 0 & 0 & 0 & . & . & . & b_{d-1} & a_d - k \end{pmatrix}$$

Theorem C. *The effective resistance vector* $\mathbf{r}$ *is the unique solution with* $r_0 = 0$ *of the equation*

$$\mathbf{r}(\mathbf{B} - k\mathbf{I}) = \frac{2}{n}(\mathbf{v} - \mathbf{u}).$$

Proof. We observe that k is a simple eigenvalue of $\mathbf{B}$ with eigenvector $\mathbf{u}$, so it follows that 0 is a simple eigenvalue of $\mathbf{B} - k\mathbf{I}$ with the same eigenvector. Thus the rank of this matrix is $(d+1) - 1 = d$, and the general solution of the equation is of the form $\mathbf{r} = \mathbf{a} + \lambda\mathbf{u}$. It follows that there is a unique solution with $r_0 = 0$.

We can obtain a recursion for r_i as follows. By Lemma 4,

$$r_i = \frac{2}{J}(\phi_0 + \phi_1 + \ldots \phi_{i-1}), \quad (1 \le i \le d),$$

so, taking $r_0 = 0$, we can write

$$\phi_i = \frac{J}{2}(r_{i+1} - r_i), \quad (0 \le i \le d - 1).$$

Substituting in the equations from Theorem A, where $J = 2m$, we get

$$b_i m(r_{i+1} - r_i) = c_i m(r_i - r_{i-1}) + k, \quad (1 \le i \le d - 1).$$

Since $k = 2m/n$ and $a_i = k - b_i - c_i$, it follows that

$$c_i r_{i-1} + (a_i - k)r_i + b_i r_{i+1} = -2/n \quad (1 \le i \le d - 1).$$

These are the components of the given matrix equation, except for the equations derived from the first and last columns of $\mathbf{B} - k\mathbf{I}$, which can be verified separately. □

Using Theorem C, it is possible to obtain a formula for $\mathbf{r}$ in terms of the eigenvalues and eigenvectors of $\mathbf{B}$, but we do not need it here. We can also use the equation $r_{i+1} = r_i + \phi_i/m$ and Theorem A. For example,

$$r_2 = r_1 + \frac{\phi_1}{m} = r_1 + \frac{1}{b_1 m}(n - 1 - k).$$

If the graph has no triangles, $b_1 = k - 1$, and this reduces to

$$r_2 = \frac{2(n-2)}{n(k-1)},$$

a result obtained under slightly different assumptions by Foster [7].

4. Applications to random walks

The analogy between a flow of electric current and a random walk is a reflection of the fact that both can be modelled using Laplace's equation. The random walk process relevant here is the one in which each step consists of a move from a vertex to an adjacent one, with each possible move being equiprobable. In our case the graph is k-regular and so the probability of a given move is $1/k$.

Given two vertices x and v, let q_{xv} denote the expected number of steps required to reach v starting from x, and, in the case of a distance-regular graph, let q_i be the value of q_{xv} when $d(v, x) = i$. In terms of the parameters c_i, a_i, b_i, the probability that the first step from x is a move to a vertex at distance $i - 1, i, i + 1$ from v is, respectively

$$\frac{c_i}{k}, \quad \frac{a_i}{k}, \quad \frac{b_i}{k}.$$

Since the expected number of steps to reach v is one more than the expected number of steps after the first step has been taken, we have the following equation for q_i:

$$q_i = 1 + \frac{1}{k}(c_i q_{i-1} + a_i q_i + b_i q_{i+1}).$$

Multiplying by k/m and rearranging, this gives

$$c_i \left(\frac{q_{i-1}}{m} \right) + (a_i - k) \left(\frac{q_i}{m} \right) + b_i \left(\frac{q_{i+1}}{m} \right) = -\frac{k}{m}.$$

Now $k/m = 2/n$, so we see that q_i/m satisfies the same recursion as the effective resistance. The initial conditions are the same (r_0 and q_0 are both zero), so we conclude that $q_i = mr_i$, for $i = 1, 2, \ldots, d$. Of course, this is a particular case of a relationship that holds for graphs in general [5].

In matrix form, we have the equation

$$\mathbf{q}(\mathbf{B} - k\mathbf{I}) = k(\mathbf{v} - \mathbf{u})$$

for the vector $\mathbf{q} = (0, q_1, \ldots, q_d)$, and in terms of the potentials calculated in Theorem A,

$$q_i = \phi_0 + \phi_1 + \ldots + \phi_{i-1}.$$

A related quantity of some interest in the theory of random walks on graphs is the *cover time* C_G of a graph G. This is the expected number of steps required to visit all the vertices, starting from a given vertex. For our purposes all the vertices are equivalent, so the starting vertex is irrelevant. A survey of work on the cover time has been given by Aldous: in particular, we note the result [1, p.88] that if the maximum of q_{xv} is $O(n)$, the cover time is $O(n \log n)$. In the case of a DR graph of diameter d it is clear that the maximum is $q_d = mr_d$. Thus the problem of the cover time leads us to consider upper bounds for r_d.

The formula obtained in Lemma 4, together with the fact that the sequence (ϕ_i) is strictly decreasing, establishes that r_d is less than dr_1. However, it seems that a much stronger result may hold.

Conjecture 1. *For any DR graph $r_d < Ar_1$, where A is a constant independent of the diameter d.*

It is even possible that $A = 2$. By virtue of the results quoted above, the conjecture would imply that for any DR graph the cover time is $O(n \log n)$, and that the maximum hitting time q_d is not greater than $2(n - 1)$.

The conjecture is partly based on failure to find a counter-example among the large number of graphs and families of graphs listed in [4]. The 'worst' example seems to be the cubic graph with 102 vertices, Example (ii), for which $d = 7$ and $r_7 = (195/101)r_1$. This graph is very exceptional, being one of only three known distance-regular graphs for which the diameter exceeds twice the degree. The partial proof of the conjecture given below points very strongly to the fact that a counter-example would have to be even more exceptional than the 102-graph.

The following is a sketch of how it could be verified that $r_d < 2r_1$ for all *known* DR graphs. This is instructive as far as it goes, but it could by no stretch of the imagination be described as elegant. Even if it were proved that the list of DR graphs in [4] is essentially complete (for $d \geq 6$ would be enough), this proof would certainly not find a place in *The Erdős Book* of ideal proofs.

We begin by proving that $r_d < 2r_1$ for $d \leq 5$. The cases $d = 1$ and $d = 2$ are trivial, and $d = 3$ is easy using the technique described below. In view of the relationship between the effective resistances and the potentials, it is sufficient to prove that

$$\phi_0 > \phi_1 + \phi_2 + \ldots + \phi_{d-1}.$$

We shall use the formulae for potentials obtained in Section 2, and some of the known parameter restrictions for DR graphs.

Lemma 5. *The parameters of a DR graph with $k > 2$ satisfy*

1 $c_i \leq b_j$ *whenever* $i + j \leq d$,
2 $b_1 \geq 2$ *if* $d \geq 3$.

Proof. See [4, p.133 and p.172]. $\qquad\square$

Theorem D. *For any distance-regular graph with diameter 4, $\phi_0 > \phi_1 + \phi_2 + \phi_3$.*

Proof. Using the formulae displayed at the end of Section 2 (and remembering that $c_1 = 1$ always), we have

$$\frac{c_2 c_3 c_4}{k}(\phi_0 - \phi_1 - \phi_2 - \phi_3) = c_2 c_3 c_4 + b_1 c_3 c_4 + b_1 b_2 c_4 + b_1 b_2 b_3$$
$$-c_3 c_4 - b_2 c_4 - b_2 b_3 - c_2 c_4 - c_2 b_3 - c_2 c_3.$$

The terms can be collected as follows:

$$\{(b_1 - 1)b_2 - c_2\}(b_3 + c_4) + c_2 c_3(c_4 - 1) + (b_1 - 1)c_3 c_4.$$

Using both parts of Lemma 5, we have $(b_1 - 1)b_2 - c_2 \geq b_2 - c_2 \geq 0$. Since $c_4 \geq 1$ and $b_1 > 1$, it follows that the entire expression is strictly positive. $\qquad\square$

Theorem E. *For any distance-regular graph with diameter 5, $\phi_0 > \phi_1 + \phi_2 + \phi_3 + \phi_4$.*

Proof. Here we have

$$\left(\frac{c_2 c_3 c_4 c_5}{k}\right)(\phi_0 - \phi_1 - \phi_2 - \phi_3 - \phi_4)$$
$$= c_2 c_3 c_4 c_5 + b_1 c_3 c_4 c_5 + b_1 b_2 c_4 c_5 + b_1 b_2 b_3 c_5 + b_1 b_2 b_3 b_4$$
$$-c_3 c_4 c_5 - b_2 c_4 c_5 - b_2 b_3 c_5 - b_2 b_3 b_4$$
$$-c_2 c_4 c_5 - c_2 b_3 c_5 - c_2 b_3 b_4$$
$$-c_2 c_3 c_5 - c_2 c_3 b_4 - c_2 c_3 c_4.$$

Collecting up the terms on the same lines as in the previous proof we get

$$\{(b_1 - 1)b_2 - c_2\}(c_4 c_5 + b_3 c_5 + b_3 b_4)$$
$$+c_3(c_2 c_4 c_5 + b_1 c_4 c_5 - c_4 c_5 - c_2 c_5 - c_2 b_4 - c_2 c_4).$$

If $b_1 \geq 3$, then $(b_1 - 1)b_2 - c_2 > b_2$. Using this result in the first line, and the monotonicity conditions (Lemma 4), we can find a positive term to dominate each negative one, with one positive term to spare.

If $b_1 = 2$, we have $c_2 \leq b_2 \leq b_1$, so there are three cases: $(b_2, c_2) = (2, 2), (2, 1), (1, 1)$. Using the methods indicated above, we find just two 'intersection arrays' for which the result does not hold:

$$\{k, 2, 1, 1, 1; 1, 1, 1, 1, 1\}, \quad \{k, 2, 2, 2, 2; 1, 2, 2, 2, 2\}.$$

Elementary arguments show that these arrays cannot be realised. In the first case, pick vertices v, x, y such that $d(v, x) = 2, d(v, y) = 3$, and $d(x, y) = 1$. Because $b_2 = c_3 = 1$, x and y can have no common neighbours. This means that $a_1 = 0$, so $k = c_1 + a_1 + b_1 = 3$. But the list of DR graphs with $k = 3$ and $d = 5$ is known to be complete, and it contains no such graph. Similar arguments work for the other array.   □

It is clear that a little more could be squeezed out of these proofs. Probably one could show by similar methods that among all DR graphs with $d \leq 7$ the 102-graph has the maximum value of r_d/r_1. If this were done, the enterprise could be completed as follows.

The known DR graphs with $d \geq 8$ comprise just two 'sporadic' graphs (for which the conjecture can be verified immediately), and a number of infinite families. The families can be arranged into three (not mutually exclusive) classes:

— families with 'classical parameters';
— partition graphs;
— regular near-polygons.

All these families are characterised by the form of their intersection array, and can be dealt with by using the following lemma, or a variant of it.

Lemma 6. *If G is a DR graph with*

$$c_2(d - 2) \leq b_2(b_1 - 1),$$

then $\phi_0 > \phi_1 + \phi_2 + \ldots + \phi_{d-1}$.

Proof. The explicit formulae for ϕ_0, ϕ_1 and ϕ_2 show that

$$\phi_1 = \frac{n - 1 - k}{b_1} < \frac{\phi_0}{b_1}, \quad \phi_2 = \frac{c_2(n - 1 - k - k_2)}{b_1 b_2} < \frac{c_2 \phi_0}{b_1 b_2}.$$

By Theorem B, $\phi_i < \phi_2$ when $i > 2$, so

$$\phi_1 + \phi_2 + \ldots + \phi_{d-1} < \phi_1 + (d - 2)\phi_2$$
$$< \left(\frac{1}{b_1} + \frac{c_2(d - 2)}{b_1 b_2}\right)\phi_0.$$

If $c_2(d - 2) \leq b_2(b_1 - 1)$, the coefficient of ϕ_0 is less than 1, so we have the result.   □

For example, the *Johnson graphs* are the graphs whose vertices are the s-subsets of an r-set, two vertices being adjacent when the subsets have $s - 1$ common members. When $r > 2s$ we have a family of DR graphs with 'classical parameters'

$$b_j = (s - j)(r - s - j), \quad c_j = j^2, \quad (0 \leq j \leq d),$$

where the diameter is $d = s$. It is easy to check that the condition in Lemma 6 is satisfied in this case. Similarly the *doubled odd graphs* form a family of regular near-polygons with diameter $d = 2k - 1$ and $b_2 = b_1 = k - 1$, $c_2 = 1$, so the condition is satisfied here too. Indeed it appears that the condition is satisfied for all the families listed in Chapter 6 of [4].

Of course, *The Erdős Book* would have a proof of the conjecture for all DR graphs with $k \geq 3$, independent of any classification, if such a proof exists.

References

[1] Aldous, D. (1989) An introduction to covering problems for random walks on graphs. *J. Theoretical Probability* **2** 87-120.

[2] Biggs, N. L. (1974) *Algebraic Graph Theory*, Cambridge University Press. (Revised edition to be published in 1993.)

[3] Bollobás, B. (1979) *Graph Theory: An Introductory Course*, Springer-Verlag, Berlin.

[4] Brouwer, A. E., Cohen, A. M. and Neumaier, A. (1989) *Distance-Regular Graphs*, Springer-Verlag, Berlin.

[5] Doyle, P. G. and Snell, J. L. (1984) *Random Walks and Electrical Networks*, Math. Assoc. of America.

[6] Foster, R. M. (1949) The average impedance of an electrical network. In: *Reissner Anniversary Volume – Contributions to Applied Mechanics*, J. W. Edwards, Ann Arbor, Michigan 333-340.

[7] Foster, R. M. (1961) An extension of a network theorem. *IRE Trans. Circuit Theory* **CT-8** 75-76.

[8] Nerode, A. and Shank, H. (1961) An algebraic proof of Kirchhoff's network theorem. *Amer. Math. Monthly* **68** 244-247.

[9] Thomassen, C. (1990) Resistances and currents in infinite electrical networks. *J. Comb. Theory* **B 49** 87-102.

On the Length of the Longest Increasing Subsequence in a Random Permutation

BÉLA BOLLOBÁS[1] and SVANTE JANSON[2]

[1]Department of Pure Mathematics and Mathematical Statistics, University of Cambridge, 16 Mill Lane, Cambridge CB2 1SB, England
email B.Bollobaspmms.cam.ac.uk
[2]Department of Mathematics, Uppsala University, PO Box 480, S-751 06 Uppsala, Sweden
email svante.jansonmath.uu.se

Complementing the results claiming that the maximal length L_n of an increasing subsequence in a random permutation of $\{1, 2, \ldots, n\}$ is highly concentrated, we show that L_n is not concentrated in a short interval: $\sup_l P(l \leqslant L_n \leqslant l' + n^{1/16} \log^{-3/8} n) \to 0$ as $n \to \infty$.

1. Introduction

Ulam [8] proposed the study of L_n, the maximal length of an increasing subsequence of a random permutation of the set $[n] = \{1, 2, \ldots, n\}$. Hammersley [4], Logan and Shepp [7], and Veršik and Kerov [9] proved that $EL_n \sim 2\sqrt{n}$ and

$$L_n/\sqrt{n} \to^P 2 \quad \text{as} \quad n \to \infty. \tag{1.1}$$

Frieze [3] showed that the distribution of L_n is sharply concentrated about its mean; his result was improved by Bollobás and Brightwell [2], who in particular proved that

$$\mathrm{Var}(L_n) = O(n^{1/2}(\log n / \log \log n)^2). \tag{1.2}$$

Somewhat surprisingly, it is not known that the distribution of L_n is not much more concentrated than claimed by (1.2). In fact, it has not previously been ruled out that if $w(n) \to \infty$ then $P(|L_n - EL_n| < w(n)) \to 0$ as $n \to \infty$. Our aim in this paper is to rule out this possibility for a fairly fast-growing function $w(n)$, and to give a lower bound for $\mathrm{Var}(L_n)$, complementing (1.2).

Theorem 1.

$$P(|L_n - EL_n| \leqslant n^{1/16} \log^{-3/8} n) \to 0 \quad \text{as} \quad n \to \infty.$$

More generally, if a_n and b_n are any numbers such that

$$\inf P(a_n \leqslant L_n \leqslant b_n) > 0, \quad \text{then} \quad (b_n - a_n)/n^{1/16} \log^{-3/8} n \to \infty.$$

In particular, for sufficiently large n,

$$\mathrm{Var} L_n \geqslant n^{1/8} \log^{-3/4} n.$$

There is still a wide gap between the upper and lower bounds, and there is no reason to believe that the bounds given here are the best possible. In fact, a boot-strap argument suggests that the range of variation is at least about $n^{1/10}$, see Theorem 2 below, and it is quite possible that the upper bound in (1.2) is sharp up to logarithmic factors, as conjectured in [2].

It is well-known that L_n also can be defined as the height of the random partial order which is in itself defined as follows. Consider the unit square $Q = [0,1]^2$ with the coordinate order. Thus for (x, y), $(x', y') \in Q$, set $(x, y) \leqslant (x', y')$ if and only if $x \leqslant x'$ and $y \leqslant y'$, let $(\xi_i)_{i=1}^{\infty}$ be independent, uniformly distributed random points in Q and consider the induced partial order on the set $(\xi_i)_{i=1}^{n}$.

Let $\mu > 0$ be a constant and let m be the Lebesgue measure in Q. Let us regard a Poisson process with intensity μdm in Q as a random subset of Q. Equivalently, let N be independent of $(\xi_i)_1^{\infty}$, with distribution $P_0(\mu)$, and take the set $\{\xi_i : 1 \leqslant i \leqslant N\}$. Write H_μ for the height of the induced partial order on this set.

The proof of (1.2) by Bollobás and Brightwell in [2] was based on a study of H_n. In particular they proved that

$$\mathrm{P}\left(|H_n - \mathrm{E}H_n| > K_1 \lambda \frac{n^{1/4} \log n}{\log \log n}\right) \leqslant e^{-\lambda^2} \tag{1.3}$$

for some constant K_1, every $n \geqslant 3$ and every λ with $1 \leqslant \lambda \leqslant n^{1/4}/\log \log n$. For larger λ their proof yields

$$\mathrm{P}(|H_n - \mathrm{E}H_n| > K_2 \lambda^2 \log \lambda) \leqslant e^{-\lambda^2}. \tag{1.4}$$

These inequalities hold for non-integer n as well: also if $n \geqslant 3$ and $1 \leqslant \lambda \leqslant n^{1/4}/\log \log n$, then for every $\mu \leqslant n$, we have

$$\mathrm{P}\left(|H_\mu - \mathrm{E}H_\mu| > K_3 \lambda \frac{n^{1/4} \log n}{\log \log n}\right) \leqslant e^{-\lambda^2}. \tag{1.5}$$

It is rather curious that our proof of a lower bound will use these results, together with the following estimate from [2]:

$$0 \leqslant 2n^{1/2} - \mathrm{E}H_n \leqslant K_4 n^{1/4} \log^{3/2} n / \log \log n. \tag{1.6}$$

Remark. It is shown in [2] that (1.3) holds for L_n as well. (The same is true for (1.4) and (1.5).) Similarly, Theorem 1 holds for H_n too; this follows from the proof of Theorem 1 below, with a few simplifications.

The variables L_n and H_n may be defined, more generally, for random subsets of the d-dimensional cube $[0,1]^d$. The results in [2] include this generalization, and it would be interesting to find lower bounds for the variance. Unfortunately, and somewhat surprisingly, the method used here does not work when $d \geqslant 3$. We try to explain this failure at the end of the paper.

2. Proof of Theorem 1

The idea behind the proof is that L_n essentially depends only on the points in a strip of measure $n^{-\alpha}$ for some $\alpha > 0$ ($\alpha = 1/8$ if we ignore logarithmic factors). The number of points in this strip is approximately Poisson distributed with expectation $n^{1-\alpha}$; hence the random variation of this number is of order $n^{(1-\alpha)/2}$ and the relative variation is $n^{-(1-\alpha)/2}$. This ought to correspond to a relative variation in the height of the same order $n^{-(1-\alpha)/2}$, ignoring the further variation due to the random position of the points, which would give a variation of order at least $n^{1/2} \cdot n^{-(1-\alpha)/2} = n^{\alpha/2}$.

We introduce some notation. For a Borel set $S \subset Q$, let

$$N_n(S) = |\{i \leqslant n : \xi_i \in S\}|$$

be the number of our n random points that lie in S, and let $L_n(S)$ be the height of the partial order defined by these $N_n(S)$ points; similarly, let $H_n(S)$ be the height of the partial order defined by the restriction of our Poisson process to S. Finally, let $S_\delta = \{(x,y) \in Q : |x - y| \leqslant \delta\}$ be the strip of width 2δ along the diagonal. We shall deduce our theorem from two lemmas. The first of these claims that the height only depends on the points in S_δ for a fairly small value of δ.

Lemma 1. *If K is sufficiently large, then with $\delta_n = K n^{-1/8} \log^{3/4} n \, (\log \log n)^{-1/2}$ we have*

$$P\left(L_n \neq L_n(S_{\delta_n})\right) \to 0 \quad \text{as} \quad n \to \infty.$$

Proof. We claim that $K = (2K_4)^{1/2}$ will do, where K_4 is the constant in (1.6).

In fact, we shall prove slightly more than claimed, namely that the probability that the set $\{\xi_i : 1 \leqslant i \leqslant n\}$ contains a point $\xi_i \notin S_{\delta_n}$ that belongs to a maximal chain is $o(1)$. Since the probability that a Poisson process Ξ in Q with intensity n has exactly n points is at least $e^{-1} n^{-1/2}$, it suffices to show that the corresponding probability for the Poisson process Ξ is $o(n^{-1/2})$.

Let M be the number of points in $\Xi \setminus S_\delta$ that belong to a maximal chain in Ξ. Then

$$M = \sum_{\xi \in \Xi} f(\xi, \Xi),$$

where

$$f(\xi, \Xi) = I(\xi \notin S_\delta) \cdot I(\xi \text{ belongs to a maximal chain in } \Xi).$$

Hence, using an easily proved formula for Poisson processes (see, e.g; [5, Lemma 2.1], and [6, Lemma 10.1 and Exercise 11.1]),

$$EM = E \sum_{\xi \in \Xi} f(\xi, \Xi) = \int_Q Ef(z, \Xi \cup \{z\}) n \, dm(z)$$

$$= \int_{Q \setminus S_\delta} P(z \text{ belongs to a maximal chain in } \Xi \cup \{z\}) n \, dm(z). \tag{2.1}$$

Fix $z = (x, y) \notin S_\delta$ and let $s = (x + y)/2$, $t = (x - y)/2$, $Q_1 = [0, x] \times [0, y]$ and

$Q_2 = [x, 1] \times [y, 1]$. Then, writing $|R|$ for the area of a set $R \subset Q$, we have

$$|Q_1|^{1/2} + |Q_2|^{1/2} = (s^2 - t^2)^{1/2} + ((1-s)^2 - t^2)^{1/2}$$
$$\leqslant s - \frac{t^2}{2s} + 1 - s - \frac{t^2}{2(1-s)}$$
$$= 1 - \frac{t^2}{2s(1-s)}$$
$$\leqslant 1 - 2t^2$$
$$\leqslant 1 - \frac{1}{2}\delta^2.$$

The random variables $H_n(Q_1)$ and $H_n(Q_2)$ have the same distributions as H_{μ_1} and H_{μ_2}, respectively, with $\mu_i = n|Q_i|$, $i = 1, 2$. Setting $K = (2K_4)^{1/2}$ and $\delta = \delta_n$, inequality (1.6) implies that

$$EH_{\mu_1} + EH_{\mu_2} \leqslant 2\mu_1^{1/2} + 2\mu_2^{1/2} \leqslant 2n^{1/2} - n^{1/2}\delta^2 \leqslant EH_n - 1 - \frac{1}{2}n^{1/2}\delta^2.$$

Hence, by applying (1.5) with $\lambda = (2 \log n)^{1/2}$, we find that

$$P(z \text{ belongs to a maximal chain in } \Xi \cup \{z\}) = P(H_n(Q_1) + H_n(Q_2) + 1 \geqslant H_n)$$
$$\leqslant P(H_{\mu_1} \geqslant EH_{\mu_1} + \frac{1}{6}n^{1/2}\delta^2) + P(H_{\mu_2} \geqslant EH_{\mu_2} + \frac{1}{6}n^{1/2}\delta^2)$$
$$+ P(H_n \leqslant EH_n - \frac{1}{6}n^{1/2}\delta^2)$$
$$\leqslant 3\exp(-2\log n) = 3n^{-2}.$$

Consequently, (2.1) yields $EM \leqslant 3n^{-1}$, and the result follows. □

Our second lemmas states that the height is not too well concentrated.

Lemma 2. *Suppose that* $\delta_n \searrow 0$ *and that* $P(L_n \neq L_n(S_{\delta_n})) \to 0$ *as* $n \to \infty$. *If* (α_n) *is any sequence with* $\alpha_n = o(\delta_n^{-1/2})$ *then*

$$\sup_x P(|L_n - x| \leqslant \alpha_n) \to 0 \quad \text{as} \quad n \to \infty.$$

Proof. It is convenient to use couplings, and we begin by recalling the relevant definitions. A *coupling* of two random variables X and Y (possibly defined on different probability spaces), is a pair of random variables (X', Y') defined on a common probability space such that $X' \stackrel{d}{=} X$ and $Y' \stackrel{d}{=} Y$. The notion of coupling depends only on the distributions of X and Y, so we may as well talk about a coupling of two distributions (which can be formulated as finding a joint distribution with given marginals).

We also define the total variation distance of two random variables X and Y (or, more properly, of their distributions $\mathscr{L}(\mathscr{X})$ and $\mathscr{L}(\mathscr{Y})$) as

$$d_{TV}(X, Y) = \sup_A |P(X \in A) - P(Y \in A)|, \tag{2.2}$$

taking the supremum over all Borel sets A. If (X', Y') is a coupling of X and Y then, clearly, $d_{TV}(X, Y) = d_{TV}(X', Y') \leqslant P(X' \neq Y')$. Conversely, it is easy to construct a coupling of X and Y such that equality holds (such couplings are known as *maximal couplings*). Thus

$$d_{TV}(X, Y) = \min P(X' \neq Y'), \tag{2.3}$$

where the minimum ranges over all couplings of X and Y. Moreover, provided the probability space where X is defined is rich enough, there exists a maximal coupling (X', Y') of X and Y with $X' = X$.

We may assume that $\delta_n < 1$ and $\alpha_n \geqslant \delta_n^{-1/4} \to \infty$. (All limits in the proof are taken as $n \to \infty$.)

Let $r = r(n) = \lceil 6\alpha_n \sqrt{n} \rceil \leqslant 7\alpha_n \sqrt{n}$, and let $\mu = \mu(n) = |S_{\delta_n}|$; thus

$$\delta_n \leqslant \mu \leqslant 2\delta_n.$$

We use the facts that, for any $n, p, \lambda_1, \lambda_2$,

$$d_{TV}\big(\mathrm{Bi}(n, p), \mathrm{Po}(np)\big) \leqslant p$$

and

$$d_{TV}\big(\mathrm{Po}(\lambda_1), \mathrm{Po}(\lambda_2)\big) \leqslant |\lambda_1 - \lambda_2| / \max(\lambda_1, \lambda_2)^{1/2},$$

see e.g. [1, Theorems 2.M and 1.C]. Hence

$$\begin{aligned}
d_{TV}\big(N_n(S_{\delta_n}), N_{n+r}(S_{\delta_n})\big) &= d_{TV}\big(\mathrm{Bi}(n, \mu), \mathrm{Bi}(n + r, \mu)\big) \\
&\leqslant d_{TV}\big(\mathrm{Bi}(n, \mu), \mathrm{Po}(n\mu)\big) + d_{TV}\big(\mathrm{Po}(n\mu), \mathrm{Po}((n + r)\mu)\big) \\
&\quad + d_{TV}\big(\mathrm{Po}((n + r)\mu), \mathrm{Bi}(n + r, \mu)\big) \\
&\leqslant \mu + r\mu/(n\mu)^{1/2} + \mu \\
&= rn^{-1/2}\mu^{1/2} + 2\mu \leqslant 7\sqrt{2}\alpha_n \delta_n^{1/2} + 4\delta_n \leqslant 14\alpha_n \delta_n^{1/2}.
\end{aligned}$$

Choose a maximal coupling (N'_n, N'_{n+r}) of $N_n(S_{\delta_n})$ and $N_{n+r}(S_{\delta_n})$, and let $(\xi'_i)_{i=1}^{\infty}$ be a sequence of independent random points, uniformly distributed in S_{δ_n}; assume also that (ξ'_i) is independent of (N'_n, N'_{n+r}). Let $L'(N)$ be the height of the partial order defined by $\{\xi'_i : i \leqslant N\}$. Then $\big(L'(N'_n), L'(N'_{n+r})\big)$ is a coupling of $L_n(S_{\delta_n})$ and $L_{n+r}(S_{\delta_n})$, and thus

$$\begin{aligned}
d_{TV}\big(L_n(S_{\delta_n}), L_{n+r}(S_{\delta_n})\big) &\leqslant P\big(L'(N'_n) \neq L'(N'_{n+r})\big) \\
&\leqslant P(N'_n \neq N'_{n+r}) = d_{TV}\big(N_n(S_{\delta_n}), N_{n+r}(S_{\delta_n})\big) \\
&\leqslant 14\alpha_n \delta_n^{1/2}.
\end{aligned}$$

Furthermore, using $L_{n+r}(S_{\delta_{n+r}}) \leqslant L_{n+r}(S_{\delta_n}) \leqslant L_{n+r}$, we see that

$$\begin{aligned}
d_{TV}(L_n, L_{n+r}) &\leqslant P(L_n \neq L_n(S_{\delta_n})) + P(L_{n+r} \neq L_{n+r}(S_{\delta_n})) + d_{TV}\big(L_n(S_{\delta_n}), L_{n+r}(S_{\delta_n})\big) \\
&\leqslant P(L_n \neq L_n(S_{\delta_n})) + P(L_{n+r} \neq L_{n+r}(S_{\delta_{n+r}})) + 14\alpha_n \delta_n^{1/2} \to 0.
\end{aligned}$$

Hence a maximal coupling (L'_n, L'_{n+r}) of L_n and L_{n+r} satisfies $P(L'_n \neq L'_{n+r}) \to 0$.

We next define another coupling of L_n and L_{n+r}, now trying to push the variables apart. Observe that necessarily $n\delta_n \to \infty$ since, otherwise, for some $C < \infty$ and arbitrarily large n,

$$EL_n(S_{\delta_n}) \leqslant EN_n(S_{\delta_n}) = n|S_{\delta_n}| \leqslant 2n\delta_n \leqslant 2C,$$

which contradicts $L_n/\sqrt{n} \xrightarrow{\text{P}} 2$ and $\mathrm{P}(L_n \neq L_n(S_{\delta_n})) \to 0$. Hence $r = O(\alpha_n n^{1/2}) = o(\delta_n^{-1/2} n^{1/2}) = o(n)$.

In particular, we may assume that $n > 3r$. Set $Q_1 = [0, \frac{r}{3n}]^2$ and $Q_2 = (\frac{r}{3n}, 1]^2$. Then

$$L_{n+r} \geq L_{n+r}(Q_1) + L_{n+r}(Q_2). \tag{2.4}$$

Moreover, $N_{n+r}(Q_1) \sim \mathrm{Bi}(n+r, (\frac{r}{3n})^2)$ with an expectation of $(n+r)(\frac{r}{3n})^2 > \frac{r^2}{9n} \geq 4\alpha_n^2$; and it follows from Chebyshev's inequality, that

$$\mathrm{P}(N_{n+r}(Q_1) \geq 2\alpha_n^2) \to 1. \tag{2.5}$$

Since the distribution of $L_{n+r}(Q_1)$ conditional on $N_{n+r}(Q_1) = v$ equals the distribution of L_v for any $v \geq 1$, we obtain from (1.1) that

$$\mathrm{P}(L_{n+r}(Q_1) > 2\alpha_n) \to 1. \tag{2.6}$$

Similarly, $n + r - N_{n+r}(Q_2) \sim \mathrm{Bi}(n+r, 1 - (1 - \frac{r}{3n})^2)$ with expectation

$$(n+r)\left(2\frac{r}{3n} - \frac{r^2}{9n^2}\right) = (\tfrac{2}{3} + o(1))r,$$

and thus

$$\mathrm{P}(N_{n+r}(Q_2) \geq n) = \mathrm{P}(n + r - N_{n+r}(Q_2) \leq r) \to 1. \tag{2.7}$$

We define L_n'' to be the height of the partial order defined by the first n of $\xi_1, \xi_2, \dots$ that fall in Q_2; obviously $L_n'' \stackrel{\text{d}}{=} L_n$, so (L_n'', L_{n+r}) is a coupling of L_n and L_{n+r}. Moreover, if $N_{n+r}(Q_2) \geq n$, then $L_{n+r}(Q_2) \geq L_n''$, and thus (2.4), (2.6), (2.7) yield

$$\mathrm{P}(L_{n+r} > L_n'' + 2\alpha_n) \to 1. \tag{2.8}$$

Combining this coupling with a maximal coupling (L_{n+r}', L_n') of L_{n+r} and L_n such that $L_{n+r}' = L_{n+r}$, we obtain a coupling (L_n', L_n'') of L_n with itself, i.e. two random variables L_n' and L_n'' with $L_n' \stackrel{\text{d}}{=} L_n'' \stackrel{\text{d}}{=} L_n$, such that

$$\mathrm{P}(L_n' > L_n'' + 2\alpha_n) \geq \mathrm{P}(L_{n+r} > L_n'' + 2\alpha_n) - \mathrm{P}(L_{n+r} \neq L_n') \to 1.$$

Finally we observe that for any real x,

$$\mathrm{P}(L_n' > L_n'' + 2\alpha_n) \leq \mathrm{P}(L_n' > x + \alpha_n) + \mathrm{P}(L_n'' < x - \alpha_n) = \mathrm{P}(|L_n - x| > \alpha_n)$$

and thus

$$\sup_x \mathrm{P}(|L_n - x| \leq \alpha_n) \leq 1 - \mathrm{P}(L_n' > L_n'' + \alpha_n) \to 0. \qquad \square$$

Theorem 1 follows immediately from the lemmas.

3. Further remarks

Note that the proof of Theorem 1 uses the concentration results in [2], and that stronger concentration results would imply a stronger version of Theorem 1, i.e. less concentration than given above. This leads to the following result, which shows that, at least for some n, the distribution of H_n is not strictly concentrated (with, say, exponentially decreasing tails) with a variation of much less than $n^{-1/10}$. (For simplicity we consider here H_n; presumably the same result is true for L_n.)

Theorem 2. *If $\varepsilon > 0$ is sufficiently small, then there exist infinitely many n such that for some $m \leqslant n$ we have*

$$P(|H_m - EH_m| > \varepsilon n^{1/10}) > n^{-2}.$$

Proof. Assume, to the contrary, and somewhat more generally, that for some γ, $0 < \gamma < 1/2$, and all large n,

$$P(|H_m - EH_m| > n^\gamma) \leqslant n^{-2}, \quad m \leqslant n. \tag{3.1}$$

The argument in the proof of [2, Theorem 9] then yields

$$2n^{1/2} - EH_n = O(n^\gamma) \tag{3.2}$$

and Lemma 1 holds for H_n, by the argument above, with

$$\delta_n = Kn^{\gamma/2 - 1/4}, \tag{3.3}$$

provided K is large enough. Hence Lemma 2 (for H_n) shows that

$$P(|H_n - EH_n| \leqslant \alpha_n) \to 0 \tag{3.4}$$

whenever $\alpha_n = o(\delta_n^{-1/2})$, i.e; when

$$\alpha_n n^{\gamma/4 - 1/8} \to 0. \tag{3.5}$$

If $\gamma < 1/10$, we may take $\alpha_n = n^\gamma$, which then satisfies (3.5), and obtain a contradiction from (3.1) and (3.4). In order to obtain the slightly stronger statement in the theorem, we let $\gamma = 1/10$ and note that if

$$P(|H_n - EH_n| > \varepsilon n^{1/10}) \leqslant n^{-2} < 1/2 \tag{3.6}$$

for every $\varepsilon > 0$ and $n \geqslant n(\varepsilon)$, then there exists a sequence $\varepsilon_n \to 0$ such that

$$P(|H_n - EH_n| > \varepsilon_n n^{1/10}) < 1/2. \tag{3.7}$$

We now choose $\alpha_n = \varepsilon_n n^{1/10}$, which satisfies (3.5), and obtain a contradiction from (3.4) and (3.7). Hence either (3.1) or (3.6), for some $\varepsilon > 0$, fails for infinitely many n, which proves the result. □

Finally, let us see what happens when we try to generalize the results to the random d-dimensional order defined by random points in $Q_d = [0,1]^d$. Lemma 1 holds, with

$$\delta_n = Kn^{-1/4d} \log^{3/4} n \, (\log \log n)^{-1/2}, \tag{3.8}$$

by essentially the same proof; we now define $S_\delta = \{(x_i)^d : |x_i - x_j| \leqslant \delta, \, i < j\}$, and note that $|S_\delta| \asymp \delta^{d-1}$. For Lemma 2, however, we need

$$\alpha_n = o\big(n^{1/d - 1/2} \delta_n^{-(d-1)/2}\big), \tag{3.9}$$

in which case we may take $m = Kn^{1-1/d}\alpha_n$ for some large K. However, (3.8) and (3.9) imply $\alpha_n = o\big(n^{(7-3d)/8d}\big) = o(1)$ for $d \geqslant 3$, so we do not obtain any result at all. (We also need $\alpha_n \geqslant 1$). The method of Theorem 2 yields no result either: we obtain $\delta_n = K n^{\gamma/2-1/2d}$ and by (3.9) we have

$$\alpha_n = o\left(n^{(3-d)/4d-\gamma(d-1)/4}\right), \tag{3.10}$$

which again contradicts $\alpha_n \geqslant 1$ for any $\gamma > 0$ when $d > 3$.

We can explain this failure in terms of the heuristics at the beginning of Section 2. We still have a relative variation of the number of points in the strip S_δ of order $n^{-(1-\alpha)/2}$, for some $\alpha > 0$, but this translates to a variation of the height of order only $n^{1/d-1/2+\alpha/2}$, which does not give any non-trivial result (α is rather small). Of course, this does not preclude the possibility that there is a substantial variation of the height due to the random position of points in the strip.

References

[1] Barbour, A. D., Holst, L. and Janson, S. (1992) *Poisson Approximation*, Oxford Univ. Press, Oxford.

[2] Bollobás, B. and Brightwell, G. (1992) The height of a random partial order: concentration of measure, *The Annals of Applied Probability* **2** 1009–1018.

[3] Frieze, A. (1991) On the length of the longest monotone subsequence in a random permutation, *Ann. Appl. Probab.* **1** 301–305.

[4] Hammersley, J. M. (1972) A few seedlings of research, *Proc. 6th Berkeley Symp. Math. Stat. Prob.* Univ. of California Press, 345–394.

[5] Janson, S. (1986) Random coverings in several dimensions, *Acta Math.* **156** 83–118.

[6] Kallenberg, O. (1983) *Random Measures*, Akademie-Verlag, Berlin.

[7] Logan, B. F. and Shepp, L. A. (1977) A variational problem for Young tableaux *Advances in Mathematics* **26** 206–222.

[8] Ulam, S. M. (1961) Monte Carlo calculations in problems of mathematical physics, *Modern Mathematics for the Engineer*, E. F. Beckenbach Ed., McGraw Hill, N.Y.

[9] Veršik, A. M. and Kerov, S.V. (1977) Asymptotics of the Plancherel measure of the symmetric group and the limiting form of Young tableaux, *Dokl. Akad. Nauk. SSSR* **233** 1024–1028.

On Richardson's Model on the Hypercube

B. BOLLOBÁS[1,2]† and Y. KOHAYAKAWA[3]‡

[1]Department of Pure Mathematics and Mathematical Statistics, University of Cambridge, 16 Mill Lane, Cambridge CB2 1SB, England

[2]Department of Mathematics, Louisiana State University, Baton Rouge, LA 70803, USA

[3]Instituto de Matemática e Estatística, Universidade de São Paulo, Caixa Postal 20570, 01452–990 São Paulo, SP, Brazil

The n-cube Q^n is the graph on the subsets of $\{1,\ldots,n\}$ where two such vertices are adjacent if and only if their symmetric difference is a singleton. Fill and Pemantle [5] started the study of several percolation processes on Q^n and obtained many asymptotic results for $n \to \infty$. As an application of these results, they investigated the contact process with no recoveries in Q^n, also known as Richardson's model for the spread of a disease. They obtained that, in this model, the cover time T_n of Q^n starting from a single infected vertex is bounded in probability: they proved that, with probability tending to 1 as $n \to \infty$, one has $(1/2)\log(2 + \sqrt{5}) + \log 2 + o(1) = 1.414\ldots + o(1) \leqslant T_n \leqslant 4\log(4 + 2\sqrt{3}) + 6 + o(1) = 14.040\ldots + o(1)$. In this note we substantially improve this upper bound by showing that one in fact has $T_n \leqslant 1 + \log 2 + o(1) = 1.693\ldots + o(1)$ in probability.

Introduction

The n-dimensional cube Q^n is the graph whose vertices are the subsets of $[n] = \{1,\ldots,n\}$ where two such vertices are adjacent if and only if their symmetric difference is a singleton. Here we are interested in a stochastic process on Q^n, known as Richardson's model for the spread of a disease. In this model, one vertex of Q^n is at first *infected* and the disease evolves by transmission of the infection according to i.i.d. Poisson processes associated to the edges of Q^n. Every time our Poisson clock associated to an edge xy goes off, if one of x or y is infected at that time, the other vertex becomes infected. The natural questions concerning this model regard the manner in which the disease spreads, and in particular the time it takes for all the vertices of Q^n to become infected. The latter question has been recently studied by Fill and Pemantle [5], who proved that this *covering time* is bounded in probability. Our aim in this note is to give an improved upper bound for this covering time.

† Part of this work was done while this author was visiting the University of São Paulo, supported by FAPESP under grant 92/3169-8.

‡ Research Partially supported by FAPESP under grant 93/0603-1 and by CNPq under grant 300334/93-1.

One may study Richardson's model on Q^n by studying first-passage percolation on this graph where the passage times on the edges are i.i.d. exponential random variables. In fact, we shall study Richardson's model only looking at first-passage percolation. Let $W = (W_e)_{e \in E(Q^n)}$ be a family of independent exponential random variables each with mean 1. If x and y are two vertices of Q^n, we write $d(x, y) = d_{Q^n}(x, y)$ for the distance between x and y in Q^n, and we set

$$d_W(x, y) = \min_P \{\ell(P, W) : P \text{ an } x\text{--}y \text{ path in } Q^n\},$$

where $\ell(P, W) = \sum_{e \in E(P)} W_e$ is the W-length of the path P. The result in [5] concerning Richardson's model on Q^n is that, for a fixed $x_0 \in Q^n$, with probability $1 - o(1)$ as $n \to \infty$, we have $(1/2)\log(2 + \sqrt{5}) + \log 2 + o(1) = 1.414\ldots + o(1) \leqslant \max_{y \in Q^n} d_W(x_0, y) \leqslant 4\log(4 + 2\sqrt{3}) + 6 + o(1) = 14.040\ldots + o(1)$. A corollary of our main result, Theorem 5, is that this upper bound can be improved to $1 + \log 2 + o(1) = 1.693\ldots + o(1)$.

Indeed, we shall show that with probability $1 - o(1)$ as $n \to \infty$, (i) for all fixed $x_0 \in Q^n$ we have that $\max_{y \in Q^n} d_W(x_0, y) \leqslant 1 + \log 2 + o(1)$, and (ii) $\operatorname{diam}(Q^n, W) = \max\{d_W(x, y) : x, y \in Q^n\} \leqslant 1 + 2\log 2 + o(1)$. The methods used in the proof of Theorem 5 are reminiscent of certain techniques first used in [3, 4].

The organisation of this note is as follows. In §1 we present some definitions and preliminary results, and in §2 we prove the main technical lemma (Lemma 4) that is needed in the proof of Theorem 5. This theorem is proved in §3. We close this note with some remarks and further problems.

1. Preliminaries

Throughout this note, we let $W = (W_e)_{e \in E(Q^n)}$ be a family of independent exponential random variables each of mean 1. For this section and the next, we assume that a constant $0 < \varepsilon < 1/6$ has been fixed.

One extremely simple but important idea in our analysis of (Q^n, W) is that we may easily decompose W_e ($e \in E(Q^n)$) into a collection of independent exponential random variables. This decomposition is then used to handle several events independently. Let $X = (X_e)$, $Y = (Y_e)$, and $Z = (Z_e)$ be three families of independent exponential r.v.'s, where the X_e have mean $(1 - 2\varepsilon)^{-1}$, and the Y_e and Z_e have mean ε^{-1}. Writing $a \wedge b$ for the minimum of a and b, if $W'_e = X_e \wedge Y_e \wedge Z_e$ ($e \in E(Q^n)$), clearly W'_e is an exponential r.v. with mean 1. Thus we may study (Q^n, W) by analysing (Q^n, X), (Q^n, Y), and (Q^n, Z).

All the asymptotics below refer to $n \to \infty$. Moreover, we very often tacitly assume that n is large enough in the estimates that follow. As is usual in the theory of random graphs, we use the terms 'almost surely' and 'almost every' to mean 'with probability tending to 1 as $n \to \infty$'. For clarity, we shall denote $\emptyset \in Q^n$ by x_0. The neighbourhood of $x \in Q^n$ is denoted by $\Gamma(x) = \Gamma_{Q^n}(x)$. For graph-theoretical terminology not defined here, we refer the reader to [1].

Our first two lemmas concern (Q^n, X).

Lemma 1. *For all* $x \in Q^n$, *let* $N_x = N_x(\varepsilon, X) \subset \Gamma_{Q^n}(x)$ *be defined by* $N_x = \{y \in \Gamma_{Q^n}(x) : X_{xy} \leqslant \log 2 + 3\varepsilon\}$ *if* $x \neq x_0$, *and* $N_{x_0} = \{y \in \Gamma_{Q^n}(x_0) : X_{x_0 y} \leqslant \varepsilon\}$. *Then there is* *a constant* $c_1 = c_1(\varepsilon) > 0$ *such that with probability* $1 - o(1)$ *we have* $|N_x| \geqslant c_1 n$ *for all* $x \in Q^n$.

Proof. Fix an arbitrary vertex $x \neq x_0$ in Q^n. Then, for a fixed $y \in \Gamma_{Q^n}(x)$,

$$\mathbb{P}(X_{xy} \geqslant \log 2 + 3\varepsilon) = \exp\{-(\log 2 + 3\varepsilon)(1 - 2\varepsilon)\} \leqslant (1 - \alpha)/2,$$

for some $\alpha = \alpha(\varepsilon) > 0$. From standard bounds for the tail of the binomial distribution, we see that for a suitable constant $c_1' > 0$ we almost surely have $|N_x| \geqslant c_1' n$ for all $x \neq x_0$. Similarly, for a suitable constant $c_1'' > 0$, we have that $|N_{x_0}| \geqslant c_1'' n$ almost surely. Thus we may take $c_1 = c_1' \wedge c_1''$. $\square$

We now describe the results from [5] concerning first-passage percolation on Q^n. For all $n \geqslant 1$, let $x_0 = x_0^{(n)} = \emptyset \in Q^n$ and $y_0 = y_0^{(n)} = [n] \in Q^n$. Consider (Q^n, W), and let

$$T_0 = \inf\{T \in \mathbb{R} : d_W(x_0, y_0) \leqslant T \text{ with probability } 1 - o(1)\}. \qquad (1)$$

Equivalently, T_0 is the smallest T ($0 \leqslant T \leqslant \infty$) such that, for any given $\delta > 0$, we have $\limsup_n \mathbb{P}(d_W(x_0, y_0) \geqslant T + \delta) = 0$ Thus T_0 is the first-passage percolation time between two opposite vertices in Q^n. Fill and Pemantle [5] have shown that $T_0 \leqslant 1$, and an argument due to Durrett given in [5] gives that $T_0 \geqslant \log(1 + \sqrt{2}) = 0.881\cdots$. Our next lemma brings the parameter T_0 into our problem. Before we can state this lemma we need some definitions. A *subcube* of Q^n is a subgraph induced in Q^n by a set of the form

$$Q_{S,A} = Q(S, A) = \{x \in Q^n : x \cap S = A\},$$

where $A \subset S \subset [n]$. In the sequel we shall also denote by $Q_{S,A} = Q(S, A)$ the subcube induced by $Q_{S,A}$, since this will not cause any confusion. The *dimension* $\dim(Q_{S,A})$ of $Q_{S,A}$ is $n - |S|$. For an integer $k \geqslant 0$ and a set X, we write $X^{(k)}$ for the set of all k-element subsets of X.

Given x and $y \in Q^n$ we define the subcube $\langle x, y \rangle$ *spanned* by x and y by

$$\langle x, y \rangle = Q((x \triangle y)^c, x \setminus (x \triangle y)) = Q((x \triangle y)^c, y \setminus (x \triangle y)),$$

where $u^c = [n] \setminus u$ and as usual $\triangle$ denotes symmetric difference. If $x \subset y$ then $\langle x, y \rangle$ equals the 'interval' $[x, y]$ in $Q^n = \mathscr{P}([n])$ regarded as a partial order under inclusion, that is $\langle x, y \rangle = [x, y] = \{z \in Q^n : x \subset z \subset y\}$. If $v \in \langle x, y \rangle$, let us call the pair $(\langle x, y \rangle ; v)$ a *rooted cube* with *root* v. For the rest of this note, we let $k_0 = \lceil \log n \rceil$, and set

$$\mathscr{T} = \{(x, y, z) \in Q^n \times Q^n \times Q^n : d(x, y) \geqslant 2k_0 + 1, d(x, z) = 2k_0, z \in \langle x, y \rangle\}.$$

For a triple $(x, y, z) \in \mathscr{T}$ let $\mathscr{F}_{x,y,z}$ be the family

$$\{(\langle v, w \rangle ; v) : v \in \langle x, z \rangle, d(x, v) = k_0, w = v \triangle y \triangle z\}$$

of rooted subcubes of $\langle x, y \rangle$. In visualising $\mathscr{F}_{x,y,z}$, it might be helpful to note that the map $\varphi_{y,z} : u \mapsto u \triangle (y \triangle z)$ gives a natural isomorphism between the subcubes $\langle x, z \rangle$ and $\langle \varphi_{y,z}(x), \varphi_{y,z}(z) \rangle = \langle \varphi_{y,z}(x), y \rangle$. Moreover, note that the rooted subcubes $(\langle v, w \rangle ; v)$ in $\mathscr{F}_{x,y,z}$ are translates of the rooted cube $(\langle x, \varphi_{y,z}(x) \rangle ; x) = (\langle x, x \triangle y \triangle z \rangle ; x)$. In fact, we may clearly regard Q^n as $\langle x, z \rangle \times \langle x, \varphi_{y,z}(x) \rangle$. Let us say that $(\langle v, w \rangle ; v) \in \mathscr{F}_{x,y,z}$ is *inexpensive* if $d_X(v, w)$ is at most $(T_0 + \varepsilon)/(1 - 2\varepsilon)$, and *expensive* otherwise. In the sequel we write $m_0 = \binom{2k_0}{k_0}$. Rather crudely we have $n^{1.38} \leqslant m_0 \leqslant n^{1.39}$.

Lemma 2. *There is a constant $c_2 = c_2(\varepsilon) > 0$ such that the following holds almost surely. For all $(x, y, z) \in \mathcal{T}$ the number of rooted subcubes in $\mathcal{F}_{x,y,z}$ that are inexpensive is at least $c_2 m_0 = c_2 \binom{2k_0}{k_0}$.*

Proof. Let $(x, y, z) \in \mathcal{T}$ be given. Fix v, $w \in Q^\ell$ $(\ell \geqslant 1)$ with $d(v, w) = \ell$, and let $U = (U_e)_{e \in E(Q^\ell)}$ be a family of independent exponential r.v.'s of mean 1. By the definition of T_0, we have that for any given fixed $\delta > 0$ we have $\lim_\ell \mathbb{P}(d_U(v, w) \geqslant T_0 + \delta) = 0$. It follows that there is an integer $\ell_0 = \ell_0(\varepsilon)$ such that $\mathbb{P}(d_U(v, w) \leqslant (T_0 + \varepsilon)/(1 - 2\varepsilon)) \geqslant 1/2$ if $\ell \geqslant \ell_0$. Moreover, there clearly exists a constant $c_2' = c_2'(\varepsilon) > 0$ such that the probability that a rooted cube $(\langle v, w \rangle ; v)$ is inexpensive is at least c_2' if $1 \leqslant \dim \langle v, w \rangle \leqslant \ell_0$. Let us now set $c_2 = (1/2)(c_2' \wedge 1/2)$.

Let $(\langle v_i, w_i \rangle ; v_i)$ $(1 \leqslant i \leqslant m_0)$ be the m_0 rooted cubes in $\mathcal{F}_{x,y,z}$. Note that the $\langle v_i, w_i \rangle$ are pairwise vertex-disjoint and hence the events that $(\langle v_i, w_i \rangle ; v_i)$ $(1 \leqslant i \leqslant m_0)$ should be inexpensive are independent. Thus the number J of inexpensive rooted cubes in $\mathcal{F}_{x,y,z}$ has binomial distribution $\mathrm{Bi}(m_0, p)$ with parameters $m_0 = \binom{2k_0}{k_0}$ and $p \geqslant 2c_2$. It now follows that $\mathbb{P}(J < c_2 m_0) \leqslant \exp\{-pm_0/8\} \leqslant \exp\{-n^{1.3}\} = o(|\mathcal{T}|^{-1})$, as required. $\square$

Our next result is a simple lemma concerning (Q^n, Y). Recall that $k_0 = \lceil \log n \rceil$. If x, $y \in Q^n$ are such that $d(x, y) = 2\ell$, let $L_{x,y}$ denote the *middle layer* $\{z \in \langle x, y \rangle : d(x, z) = d(z, y) = \ell\}$ of $\langle x, y \rangle$.

Lemma 3. *Let $V_0 = V_0(Y)$ be the set of vertices z of Q^n that are incident to at least $(\varepsilon^2/3)n$ edges e with $Y_e \leqslant \varepsilon$. Then almost surely for all $x, y \in Q^n$ with $d(x, y) = 2k_0$ at most n vertices in the middle layer $L_{x,y}$ of $\langle x, y \rangle$ fail to be in V_0.*

Proof. Let $x = x_0 = \emptyset$ and fix $y \in [n]^{(2k_0)}$. For any edge $e \in E(Q^n)$ we have $\mathbb{P}(Y_e \leqslant \varepsilon) = 1 - \exp(-\varepsilon^2) \geqslant \varepsilon^2/2$, and hence for fixed $z \in L_{x,y}$ the probability that z does not belong to V_0 is at most $p_0 = \exp\{-\varepsilon^2 n/36\}$. Thus the probability that we have $|L_{x,y} \setminus V_0| \geqslant n$ is at most

$$\binom{m}{n} \exp\{-\varepsilon^2 n^2/36\} \leqslant \binom{n^{1.4}}{n} \exp\{-\varepsilon^2 n^2/36\} \leqslant \exp\{-\Omega(n^2)\}.$$

Thus the probability that there exist $x, y \in Q^n$ with $d(x, y) = 2k_0$ such that $|L_{x,y} \setminus V_0| \geqslant n$ is at most $4^n \exp\{-\Omega(n^2)\} = o(1)$. $\square$

Let $V_0 = V_0(Y)$ be defined as in Lemma 3. For all $z \in V_0$ we let $N_z' = N_z'(\varepsilon, Y) = \{z' \in \Gamma_{Q^n}(z) : Y_{zz'} \leqslant \varepsilon\}$. Thus $|N_z'| \geqslant (\varepsilon^2/3)n$ for any $z \in V_0$.

2. The main lemma

In this section we study (Q^n, Z), and prove a lemma (Lemma 4) that will be fundamental for the proof of our main result, Theorem 5. We stress that the argument in the proof below is quite crude, and that with a little more patience one may prove a stronger result. However, such a strengthening of this lemma would not, as far as we can see, improve Theorem 5. Before turning to our lemma, recall that $0 < \varepsilon < 1/6$ has been fixed.

Lemma 4. . Set $K = \lfloor (\varepsilon^6 n/650)^{1/7} \rfloor$, and let $m^2 = \omega n^{11/7} \leqslant n^2$, where $\omega = \omega(n) \to \infty$ as $n \to \infty$. Let $x = x_0 = \emptyset$ and fix $y \subset [n]$ with $1 \leqslant k = |y| \leqslant K$. Let $S \subset \Gamma_{Q^n}(x)$ and $T \subset \Gamma_{Q^n}(y)$ with $|S|, |T| \geqslant m$ be given. Then in (Q^n, Z) there is an S–T path P of Z-length $\ell(P, Z) = \sum_{e \in E(P)} Z_e$ at most ε with probability at least $1 - e^{-\omega n/\log \omega}$.

Proof. For all $e \in E(Q^n)$ and $2 \leqslant j \leqslant K + 1$, let $Z_e^{(j)}$ be an exponential r.v. with mean K/ε, and assume that all these variables are independent. Set $Z^{(j)} = (Z_e^{(j)})_{e \in E(Q^n)}$ for $2 \leqslant j \leqslant K + 1$. Clearly Z_e and $\bigwedge_j Z_e^{(j)}$ have the same distribution for all $e \in E(Q^n)$. Let $t_0 = \varepsilon/(2K + k - 2)$. Then for all $2 \leqslant j \leqslant K + 1$ we have

$$p = \mathbb{P}\{Z_e^{(j)} \leqslant t_0\} = 1 - \exp\left\{ \frac{-\varepsilon^2}{K(2K + k - 2)} \right\} \geqslant \frac{\varepsilon^2}{2K(2K + k - 2)} \geqslant \frac{\varepsilon^2}{6K^2}.$$

For all $2 \leqslant j \leqslant K + 1$, let $H_j = H_j(Z^{(j)}) \subset Q^n$ be the spanning subgraph of Q^n whose edge-set is $\{e \in E(Q^n) : Z_e^{(j)} \leqslant t_0\}$. Note that the H_j $(2 \leqslant j \leqslant K + 1)$ are K independent random elements of $\mathscr{G}(Q^n, p)$, the space of random spanning subgraphs of Q^n whose edges are independently present with probability p. Let us set $G_j = H_2 \cup \cdots \cup H_j$ $(2 \leqslant j \leqslant K+1)$. We now claim the following.

Claim. *In G_{K+1} there is an S–T path of length at most $2K + k - 2$ with probability at least $1 - \exp\{-\omega n/\log \omega\}$.*

Note that a path as in the claim above has Z-length at most $t_0(2K + k - 2) = \varepsilon$, and hence to prove our lemma it suffices to prove this claim. To verify this claim we start by setting $S_1 = \{v \in S : v \not\subset y\}$ and $T_1 = \{v \in T : y \subset v\}$. Clearly $|S_1|, |T_1| \geqslant m - k \geqslant m/2$. We shall think of Q^n as $Q^{n-k} \times Q^k$ in the following way. For each $v \subset y^c = [n] \setminus y$, let

$$Q_v = \langle v, v \triangle y \rangle = \langle v, v \cup y \rangle = \{z \subset [n] : z \cap y^c = v\}. \tag{2}$$

Note that Q_v is a cube of dimension $k = |y|$, and that $Q = \langle x_0, y^c \rangle = \{v \in Q^n : v \subset y^c\}$ is a cube of dimension $n - k$. Also, we may naturally regard Q^n as $\langle x_0, y^c \rangle \times \langle x_0, y \rangle$.

Write $n_1 = |y^c| = n - k$ and note that clearly $n_1 = (1 + o(1))n$. Let us set

$$S_2' = \{z \in (y^c)^{(2)} : z \in \Gamma_{Q^n}(S_1) \text{ and } z \cup y \in \Gamma_{Q^n}(T_1)\}.$$

A moment's thought shows that $|S_2'| \geqslant \binom{m/2}{2}$. Thus, setting

$$S_2 = \{z \in (y^c)^{(2)} : z \in \Gamma_{H_2}(S_1) \text{ and } z \cup y \in \Gamma_{H_2}(T_1)\},$$

quite crudely we have $\mathbb{E}(|S_2|) \geqslant \binom{m/2}{2} p^2 \geqslant m^2 p^2/9$. Hence, with probability at least $1 - \exp\{-m^2 p^2/1800\} = 1 - \exp\{-\Omega(\varepsilon^{4/7}\omega n)\}$, we have $|S_2| \geqslant m^2 p^2/10$.

For the rest of the argument, we condition on $|S_2| \geqslant m^2 p^2/10 \geqslant (1/5)(mp/n)^2 \binom{n_1}{2}$. In the sequel, if $U \subset \langle x_0, y^c \rangle$, we write $U \vee y$ for $\{u \cup y : u \in U\}$, namely the translate of U by the map $u \mapsto u \cup y$, which is contained in the cube $\langle y, [n] \rangle$. Let us now fix $3 \leqslant j \leqslant K$

and condition on the existence of $S_2 \subset (y^c)^{(2)}, \ldots, S_{j-1} \subset (y^c)^{(j-1)}$ for which the following conditions hold for all $2 \leqslant i < j$.

(i) $|S_i| \geqslant (4/5)2^{-i} (mp^{i-1}/n)^2 \binom{n_1}{i}$;

(ii) for all $z \in S_i$ there is an S_1-z path of length $i-1$ in $G_i = H_2 \cup \cdots \cup H_i$,

(iii) for all $z \in S_i$ there is a $T_1-(z \cup y)$ path of length $i-1$ in $G_i = H_2 \cup \cdots \cup H_i$.

Now let $S'_j = \Gamma_{Q^n}(S_{j-1}) \cap (y^c)^{(j)} = \{z' \in (y^c)^{(j)} : z \subset z' \text{ for some } z \in S_{j-1}\}$. Then, by the local LYM inequality (see for instance [2], §3), we have that

$$|S'_j| \geqslant |S_{j-1}| \binom{n_1}{j-1}^{-1} \binom{n_1}{j} \geqslant \frac{4}{5}2^{-j+1} \left(\frac{mp^{j-2}}{n}\right)^2 \binom{n_1}{j}.$$

We now set $S_j = \{z \in S'_j : z \in \Gamma_{H_j}(S_{j-1}) \text{ and } z \cup y \in \Gamma_{H_j}(S_{j-1} \vee y)\}$. Then very crudely we have that

$$\mathbb{E}(|S_j|) \geqslant |S'_j|p^2 \geqslant \frac{4}{5}2^{-j+1} \left(\frac{mp^{j-1}}{n}\right)^2 \binom{n_1}{j},$$

and so, with probability at least

$$1 - \exp\left\{-\frac{1}{5}2^{-j}\left(\frac{mp^{j-1}}{n}\right)^2\binom{n_1}{j}\right\},$$

condition (i) above holds for $i = j$, that is $|S_j| \geqslant (4/5)2^{-j}(mp^{j-1}/n)^2 \binom{n_1}{j}$. Note that by the definition of S_j conditions (ii) and (iii) above hold for $i = j$. Let

$$S''_K = \{z \in (y^c)^{(K)} : d_{G_K}(S, z), d_{G_K}(T, z \cup y) \leqslant K - 1\},$$

where d_{G_K} denotes the distance in the graph G_K. Then, using the argument above for $3 \leqslant j \leqslant K$ in turn, we see that, with probability at least

$$P_0 = 1 - \sum_{3 \leqslant j \leqslant K} \exp\left\{-\frac{1}{5}2^{-j}\left(\frac{mp^{j-1}}{n}\right)^2\binom{n_1}{j}\right\},$$

we have

$$|S''_K| \geqslant \frac{4}{5}2^{-K}\left(\frac{mp^{K-1}}{n}\right)^2\binom{n_1}{K}. \tag{3}$$

To estimate P_0, note that if $3 \leqslant j \leqslant K$ then

$$A = 2^{-j}\left(\frac{mp^{j-1}}{n}\right)^2\binom{n_1}{j} \geqslant \frac{m^2}{n^2p^2}\left(\frac{p^2n_1}{2j}\right)^j$$

$$\geqslant \frac{m^2}{n^2p^2}\left\{\left(\frac{p^2n_1}{6}\right)^3 \wedge \left(\frac{p^2n_1}{2K}\right)^K\right\},$$

since the function $f(x) = (B/x)^x$ $(x > 0)$ is log-concave. However,

$$\left(\frac{p^2n_1}{2K}\right)^K \geqslant \left(\frac{1}{73}\frac{\varepsilon^4n}{K^5}\right)^K = \left(\Omega(n^{2/7})\right)^K \geqslant \left(\frac{p^2n_1}{6}\right)^3,$$

and hence $A \geqslant (1/216 + o(1))m^2np^4 = \Omega(\omega n^{10/7})$. Thus, very crudely,

$$P_0 \geqslant 1 - K\exp\{-\Omega(\omega n^{10/7})\} \geqslant 1 - \exp\{-\omega n\}. \tag{4}$$

Let us now condition on (3). To finish the proof of our claim, we shall show that with very high probability there is an $S_K''-(S_K'' \vee y)$ path of length k in H_{K+1}. We shall do this using an extremely crude argument. Consider the family $\mathscr{F} = \{Q_v : v \in S_K''\}$ of k-cubes as defined in (2). Recall that Q_v is a translate of $\langle x_0, y \rangle$ by v, and note that in particular the Q_v $(v \in S_K'')$ are pairwise disjoint. Consider for each $v \in S_K''$ a fixed $v-(v \cup y)$ path $P_v \subset Q_v$ of length k, and note that the probability P_1 that there should not be an $S_K''-(S_K'' \vee y)$ path of length k in H_{K+1} is at most $(1-p^k)^{|S_K''|} \leqslant (1-p^K)^{|S_K''|} \leqslant \exp\{-p^K|S_K''|\}$. However, as (3) holds,

$$p^K|S_K''| \geqslant \frac{4}{5} 2^{-K} p^K \left(\frac{mp^{K-1}}{n}\right)^2 \binom{n_1}{K}$$

$$\geqslant \frac{4}{5} \frac{m^2}{p^2 n^2} \left(\frac{p^3 n}{3K}\right)^K \geqslant \frac{4}{5} \left(\frac{m}{pn}\right)^2 \left(\frac{650}{648}\right)^K,$$

which is much larger than ωn. Thus $P_1 \leqslant e^{-\omega n}$. This bound and inequality (4) finish the proof of the claim, and hence Lemma 4 is proved. $\qquad\square$

3. The main result

Recall that $W = (W_e)$ is a collection of independent exponential r.v.'s each with mean 1. Also, recall that T_0 is the first-passage percolation time between two diametrically opposite vertices in Q^n when the passage times are given by W (cf. (1) in Section 2).

Theorem 5. *The following hold with probability* $1 - o(1)$ *as* $n \to \infty$.

(i) *Writing* $x_0 = \emptyset \in Q^n$, *we have* $\max_{y \in Q^n} d_W(x_0, y) \leqslant T_0 + \log 2 + o(1)$,

(ii) $\operatorname{diam}(Q^n, W) \leqslant T_0 + 2\log 2 + o(1)$.

Proof. (i) Let $0 < \varepsilon < 1/6$ be fixed. We regard W_e as $X_e \wedge Y_e \wedge Z_e$ $(e \in E(Q^n))$, where X_e, Y_e, and Z_e are as in Section 2, i.e. the X_e are exponential with mean $(1 - 2\varepsilon)^{-1}$, the Y_e and Z_e are exponential with mean ε^{-1}, and all these variables are independent. Let us now consider the following events. (We keep the notation introduced in Section 2.)

(a) For all $x \in Q^n$ we have $|N_x| = |N_x(\varepsilon, X)| \geqslant c_1 n$,

(b) for all $(x, y, z) \in \mathscr{T}$, the number of inexpensive rooted subcubes in $\mathscr{F}_{x,y,z}$ is not smaller than $c_2 m_0 = c_2 \binom{2k_0}{k_0}$,

(c) for all $x, y \in Q^n$ with $d(x, y) = 2k_0$, we have $|L_{x,y} \setminus V_0| = |L_{x,y} \setminus V_0(Y)| \leqslant n$.

Note that by Lemmas 1, 2, and 3 conditions (a), (b), and (c) above hold almost surely. For the rest of the argument we condition on (Q^n, X) and (Q^n, Y) satisfying (a), (b), and (c).

Let $S_{x_0} = N_{x_0} = N_{x_0}(\varepsilon, X) = \{y \in \Gamma_{Q^n}(x_0) : X_{x_0 y} \leqslant \varepsilon\}$. For every $x \in Q^n - x_0$, set $S_x = N_x' = N_x'(\varepsilon, Y)$ if $x \in V_0 = V_0(Y)$, and $S_x = N_x = N_x(\varepsilon, X)$ otherwise. For a fixed pair x, $y \in Q^n$ with $d(x, y) \leqslant K = \lfloor (\varepsilon^6 n/650)^{1/7} \rfloor$, if we let $S = S_x$ and $T = S_y$, Lemma 4 tells us that $d_Z(S, T) \leqslant \varepsilon$ with probability at least $1 - \exp\{-n^{5/4}\}$. Thus (d) below holds with probability $1 - o(1)$.

(d) For all $x, y \in Q^n$ with $d(x, y) \leqslant K$ we have

$$d_W(x, y) \leqslant \begin{cases} 3\varepsilon & \text{if } x, y \in V_0 \cup \{x_0\} \\ \log 2 + 5\varepsilon & \text{if } x \in V_0 \cup \{x_0\} \text{ or } y \in V_0 \cup \{x_0\} \\ 2\log 2 + 7\varepsilon & \text{otherwise.} \end{cases}$$

We now condition on (Q^n, W) satisfying (d), and claim that consequently we have $\max\{d_W(x_0, y) : y \in Q^n\} \leqslant T_0 + \log 2 + 3\varepsilon T_0 + 12\varepsilon$. First note that it follows from (d) that $\max\{d_W(x_0, y) : y \in Q^n, d(x_0, y) \leqslant K\}$ is as small as claimed, and hence let $y \in Q^n$ be such that $d(x_0, y) > K \geqslant 2k_0$. Pick $z \in \langle x_0, y \rangle$ with $d(x_0, z) = 2k_0$, and note that then $(x_0, y, z) \in \mathcal{T}$. Since we are conditioning on (b) and (c) above, we may find $v \in L_{x_0, y}$ such that $v \in V_0$, $w = v \triangle y \triangle z \in V_0$, and $(\langle v, w \rangle ; v)$ is an inexpensive rooted cube. But then

$$d_W(x_0, y) \leqslant d_W(x_0, v) + d_W(v, w) + d_W(w, y)$$
$$\leqslant 3\varepsilon + \tfrac{T_0+\varepsilon}{1-2\varepsilon} + \log 2 + 5\varepsilon \leqslant T_0 + \log 2 + 3\varepsilon T_0 + 12\varepsilon.$$

Theorem 5 (i) follows by letting $\varepsilon \to 0$.

(ii) Minor modifications to the above argument gives a proof of Theorem 5 (ii). □

The result above, coupled with the upper bound $T_0 \leqslant 1$ for the first-passage percolation time T_0 established by Fill and Pemantle [5], gives the following corollary.

Corollary 6. *The following hold with probability* $1 - o(1)$ *as* $n \to \infty$.

(i) *Writing* $x_0 = \emptyset \in Q^n$, *we have* $\max_{y \in Q^n} d_W(x_0, y) \leqslant 1 + \log 2 + o(1)$,

(ii) $\mathrm{diam}(Q^n, W) \leqslant 1 + 2\log 2 + o(1)$.   □

4. Concluding remarks and open problems

The best bounds so far for the first-passage percolation time T_0 in Q^n, defined in §1, are the ones given in [5], namely $0.881\ldots = \log(1 + \sqrt{2}) \leqslant T_0 \leqslant 1$. We believe that percolation happens at a sharply defined time.

Conjecture 7. *For all* $\delta > 0$ *we have* $\lim_{n \to \infty} \mathbb{P}\{d_W(x_0, y_0) \leqslant T_0 - \delta\} = 0$.   □

We also believe that the analogous phenomenon happens in Richardson's model. As a starting point, it would be interesting to settle the following problem.

Problem 8. *Do the following two assertions hold with probability* $1 - o(1)$ *as* $n \to \infty$?

(i) *Writing* $x_0 = \emptyset \in Q^n$, *one has* $\max_{y \in Q^n} d_W(x_0, y) = T_0 + \log 2 + o(1)$.

(ii) *One has* $\mathrm{diam}(Q^n, W) = T_0 + 2\log 2 + o(1)$.   □

References

[1] Bollobás, B. (1979) *Graph Theory – An Introductory Course*, Springer–Verlag, New York, *viii*+180pp

[2] Bollobás, B. (1986) *Combinatorics*, Cambridge University Press, Cambridge, *xii*+177pp

[3] Bollobás, B., Kohayakawa, Y. and Łuczak, T. (1996a) On the diameter and radius of random subgraphs of the cube, *Random Structures and Algorithms*, to appear

[4] Bollobás, B., Kohayakawa, Y. and Łuczak, T. (1996b) Connectivity properties of random subgraphs of the cube, *Random Structures and Algorithms*, to appear

[5] Fill, J. A. and Pemantle, R. (1993) Percolation, first-passage percolation, and covering times for Richardson's model on the *n*-cube, *The Annals of Applied Probability* **3** 593–629.

Random Permutations: Some Group-Theoretic Aspects

PETER J. CAMERON† and WILLIAM M. KANTOR‡

†School of Mathematical Sciences, Queen Mary and Westfield College,
Mile End Road, London E1 4NS, U.K.

‡Department of Mathematics, University of Oregon, Eugene, OR 97403, U.S.A.

The study of asymptotics of random permutations was initiated by Erdős and Turán, in a series of papers from 1965 to 1968, and has been much studied since. Recent developments in permutation group theory make it reasonable to ask questions with a more group-theoretic flavour. Two examples considered here are membership in a proper transitive subgroup, and the intersection of a subgroup with a random conjugate. These both arise from other topics (quasigroups, bases for permutation groups, and design constructions).

1. Permutations lying in a transitive subgroup

S_n and A_n denote the symmetric and alternating groups on the set $X = \{1, \ldots, n\}$. A subgroup G of S_n is *transitive* if, for all $i, j \in X$, there exists $g \in G$ with $ig = j$. In a preliminary version of this paper, we asked the following question:

Question 1.1. *Is it true that, for almost all permutations $g \in S_n$, the only transitive subgroups containing g are S_n and (possibly) A_n?*

Here, of course, 'almost all $g \in S_n$ have property P' means 'the proportion of elements of S_n not having property P tends to 0 as $n \to \infty$'.

An affirmative answer to this question was given by Łuczak and Pyber, in [15]. We will discuss the motivation for this question, and speculate on the rate of convergence.

To analyse the question, we make the customary division of transitive subgroups into imprimitive and primitive ones. A subgroup G is *imprimitive* if it leaves invariant some non-trivial partition of X, and *primitive* otherwise. Imprimitive subgroups may be large, but the maximal ones are relatively few in number: just $d(n) - 2$ conjugacy classes, where $d(n)$ is the number of divisors of n. (If the permutation g lies in an imprimitive subgroup, then it lies in a maximal one, which is precisely the stabiliser of a partition of X into s parts of size r, where $rs = n$ and $r, s > 1$.) On the other hand, primitive groups are more mysterious; but it follows from the classification of finite simple groups that

— they are *scarce* (for almost all n, the only primitive groups are S_n and A_n, see [3]);
— they are *small* (order at most $n^{c \log \log n}$ with 'known' exceptions, see [1]).

In addition, many special classes of primitive groups (for example, the doubly transitive groups), have been completely classified.

The number of permutations that lie in some primitive subgroup other than S_n or A_n can be bounded, since such permutations have quite restricted cycle structure (a consequence of minimal degree bounds, see [14] – note that these bounds are a consequence of the classification of finite simple groups – or by more elementary means, as Łuczak and Pyber [15] do). So we will concentrate on imprimitive subgroups, and, in particular, the largest imprimitive subgroups: those preserving a partition of X into two sets of size $n/2$, for n even.

A permutation fixing such a partition must either fix some $(n/2)$-set, or interchange some $(n/2)$-set with its complement. Now a permutation interchanges some $(n/2)$-set with its complement if and only if all its cycles have even length. The number of such permutations is

$$((n-1)!!)^2 = ((n-1)(n-3)\ldots3.1)^2,$$

which is easily seen to be $n!O(1/\sqrt{n})$. (This formula is easily proved using generating function methods. A 'counting' proof is given in [2]. Curiously, it is equal to the number of permutations with all cycles of odd length, see [7, 8, 9, 10]. We are not aware of a 'counting' proof of this coincidence!)

On the other hand, a permutation fixes an $(n/2)$-set if and only if some subfamily of its cycle lengths has sum $n/2$. There seems to be no simple formula for the number of such permutations; but Łuczak and Pyber show that their proportion is at most An^{-c}, where A and c are positive constants. Indeed, more generally, the proportion of permutations fixing some k-set tends to 0 as $k \to \infty$ (as long as $n \ge 2k$).

We turn now to the motivation for this question. A *quasigroup* is a set with a binary multiplication in which left and right division are uniquely defined (equivalently, the multiplication table is a Latin square). In a quasigroup Q, left and right translations are permutations, represented by the rows and columns of the multiplication table of Q. The *multiplication group* $\mathrm{Mlt}(Q)$ of Q is the group generated by these permutations. This group 'controls' the character theory of Q [16]. In particular, if $\mathrm{Mlt}(Q)$ is 2-transitive, then the character theory of Q is trivial. Smith conjectured that this happens most of the time, and this is indeed true.

Theorem 1.2. *For almost all Latin squares A, the group generated by the rows of A is the symmetric or alternating group.*

This is proved in [2], but follows more directly from the affirmative answer to Question 1.1, since the rows of a Latin square obviously generate a transitive permutation group, and the first row of a random Latin square is a random permutation (that is, all permutations occur equally often as first rows of Latin squares).

This suggests several related questions:

1 Is it true that, for almost all Latin squares, the first two rows generate the symmetric or alternating group? (By a theorem of Dixon [4], almost all pairs of permutations generate S_n or A_n; and a positive proportion of these ($1/e$, in the limit) have the property that the second is a 'derangement of the first, and hence occur as the first two rows of a Latin square. But not all derangements occur equally often.) More generally, study further the probability distribution on derangements induced by their frequency of occurrence in Latin squares. What is the ratio of the greatest to the smallest number of completions?

2 Is it true that the multiplication groups of almost all loops are symmetric or alternating? (A *loop* is a quasigroup with identity. Thus we are requiring that the first row and column of the Latin square correspond to the identity permutation, and the deduction of the analogue of Theorem 1.2 from Question 1.1 fails.)

3 What proportion of Latin squares have the property that all the rows are even permutations? (If the limit is zero, the alternating group can be struck out from the conclusion to Theorem 1.2.)

4 Is the proportion of permutations that do lie in a proper transitive subgroup $O(n^{-1/2})$? (By our remarks above, this would be best possible.)

2. Bases and intersections of conjugates

Introducing the next topic requires a fairly long detour. Let G be a permutation group on a set X. A *base* for G is a sequence $(x_1, \ldots, x_r)$ of points of X whose pointwise stabiliser is the identity. It is *irredundant* if no point is fixed by the pointwise stabiliser of its predecessors. Bases are of interest in several fields, including computational group theory.

If G has an irredundant base of size r, then $2^r \le |G| \le n(n-1)\ldots(n-r+1)$, whence $\log_n |G| \le r \le \log_2 |G|$. It is easy to construct examples at or near either side of this inequality. Nevertheless, it is thought that, for many interesting groups, the base size is closer to the lower bound. In particular, certain primitive groups whose order is polynomially bounded should have bases of constant size.

To elucidate this, we look more closely at primitive groups. The *O'Nan–Scott theorem* (see [1]) divides these into several classes. All but one of these classes consist of groups that can be 'reduced' in some way to smaller ones or studied by other means. The one class left over consists of groups G that are *almost simple* (that is, that have a non-abelian simple normal subgroup N such that G is contained in $\text{Aut}(N)$). Using the classification of finite simple groups, it is possible to make some general statements about almost simple primitive groups. For example, the following result holds (see [1, 12]; the latter paper gives $c = 8$).

Theorem 2.1. *There is a constant c with the following property. Let G be an almost simple primitive permutation group of degree n. Then either*

(a) G is known (specifically, G is a symmetric or alternating group S_m or A_m, acting on the
set of k-subsets of $\{1, \ldots, m\}$ or on the set of partitions of $\{1, \ldots, m\}$ into s parts of size
r, or G is a classical group, acting on an orbit of subspaces of its natural module or on
an orbit of pairs of subspaces of complementary dimension); or

(b) $|G| \leq n^c$.

(The methodological point raised by this and similar theorems is that in the study of finite
permutation groups, after the classical divisions into intransitive and transitive groups,
and of transitive groups into primitive and imprimitive groups, one should also divide
primitive groups into 'large' and 'small' groups, the large ones being 'known' in some
sense. This principle applies to both theoretical and computational analysis.)

It is conjectured that *there is a constant c' (perhaps c' = 3) such that, if G is almost
simple and primitive and does not satisfy (a), then almost every c'-tuple of points is a base
for G.*

According to the classification of finite simple groups, the simple normal subgroup N
of G is an alternating group, a group of Lie type, or one of the 26 sporadic groups. In
the first of these three cases, we were able to prove the conjecture (with $c' = 2$).

Theorem 2.2. *Let G be an almost simple group, not occurring under Theorem 2.1(a). If
the simple normal subgroup of G is an alternating group, then almost all pairs of points are
bases.*

We outline the proof.

The first observation is that if G is transitive and H is a point stabiliser, the proportion
of ordered pairs of points that are bases is equal to the proportion of elements $g \in G$ for
which $H \cap H^g = 1$, where H^g is the conjugate $g^{-1}Hg$.

Second, primitivity of G is equivalent to maximality of the subgroup H. Moreover, if
$m \neq 6$, then $\text{Aut}(A_m) = S_m$, so we may assume that $G = S_m$ or A_m. Consider H (the point
stabiliser in the unknown action) acting on $M = \{1, \ldots, m\}$. If H is intransitive, it fixes a
k-subset of M for some k; by maximality, it is the stabiliser of this k-set, and the action
of G is equivalent to that on k-sets. Similarly, if H is transitive but imprimitive, then it is
the stabiliser of a partition, and G acts on partitions of fixed shape. Both of these cases
are included under Theorem 2.1(a). So H is primitive on M. (This is an example of the
'bootstrap principle': note that m is much smaller than n.)

Thus, finally, we need a result about random permutations.

Proposition 2.3. *Let H be a primitive subgroup of S_m, not S_m or A_m. Then, for almost all
permutations $g \in S_m$, we have $H \cap H^g = 1$.*

This is true, and can be shown by a simple counting argument, except in the case of the
largest primitive groups (the automorphism groups of the line graphs of K_r or $K_{r,r}$, with
$m = \binom{r}{2}$ or r^2 respectively), where some special pleading is required. In outline: count
triples (h, k, g) with $h, k \in H$, $h, k \neq 1$, $g \in G$ and $h^g = k$. The number of such triples is not
more than $|H|^2 c$, where c is the largest order of the centraliser of a non-identity element

in H; and it is not less than the number of elements g with $H \cap H^g \neq 1$. Now use the fact that primitive groups are small, and their elements have relatively few fixed points (and so relatively small centralizers).

Remark. There is an analogy between intersections of conjugates and automorphism groups. (For example, if the group G is the automorphism group of a particular structure S, then the intersections of pairs of conjugates of G represent those groups that can be represented in the following way: impose two copies of the structure S on the underlying set, and consider all those permutations which are automorphisms of both structures simultaneously.)

Thus, Proposition 2.3 should be compared with the statement 'almost all graphs have trivial automorphism group' [6]. As the analogue of Frucht's theorem [11], we propose the following conjecture.

Conjecture 2.4. *Let $G_1, G_2, \ldots$ be primitive groups of degrees $n_1, n_2, \ldots$, where $n_i \to \infty$ and $G_i \neq S_{n_i}$ or A_{n_i} for all i. Let X be an abstract group that is embeddable in G_i for infinitely many values of i. Then, for some i, and some permutation $g \in S_{n_i}$, we have $G_i \cap G_i^g = X$.*

This has been proved by Kantor [13] for the family of groups $G_i = \mathrm{P\Gamma L}(i, q)$, $n_i = (q^i - 1)/(q - 1)$, for a fixed prime power q. (In this case, every finite group is embeddable in G_i for all sufficiently large i.) Kantor used this result to show that, for a fixed prime power q, every finite group is the automorphism group of a square $2\text{-}((q^i - 1)/(q - 1), (q^{i-1} - 1)/(q - 1), (q^{i-2} - 1)/(q - 1))$ design for some i.

References

[1] Cameron, P. J. (1981) Finite permutation groups and finite simple groups. *Bull. London Math. Soc.* **13** 1–22.

[2] Cameron, P. J. (1992) Almost all quasigroups have rank 2. *Discrete Math.* **106/107** 111–115.

[3] Cameron, P. J., Neumann, P. M. and Teague, D. N. (1982) On the degrees of primitive permutation groups. *Math. Z.* **180** 141–149.

[4] Dixon, J. D. (1969) The probability of generating the symmetric group. *Math. Z.* **110** 199–205.

[5] Donnelly, P. and Grimmett, G. (to appear) On the asymptotic distribution of large prime factors. *J. London Math. Soc.*

[6] Erdős, P. and Rényi, A. (1963) Asymmetric graphs. *Acta Math. Acad. Sci. Hungar.* **14** 295–315.

[7] Erdős, P. and Turán, P. (1965) On some problems of a statistical group theory, I. *Z. Wahrscheinlichkeitstheorie und verw. Gebeite* **4** 175–186.

[8] Erdős, P. and Turán, P. (1967) On some problems of a statistical group theory, II. *Acta Math. Acad. Sci. Hungar.* **18** 151–163.

[9] Erdős, P. and Turán, P. (1967) On some problems of a statistical group theory, III. *Acta Math. Acad. Sci. Hungar.* **18** 309–320.

[10] Erdős, P. and Turán, P. (1968) On some problems of a statistical group theory, IV. *Acta Math. Acad. Sci. Hungar.* **19** 413–435.

[11] Frucht, R. (1938) Herstellung von Graphen mit vorgegebener abstrakter Gruppe. *Compositio Math.* **6** 239–250.

[12] Kantor, W. M. (1988) Algorithms for Sylow p-subgroups and solvable groups. *Computers in Algebra* (Proc. Conf. Chicago 1985), Dekker, New York 77–90.

[13] Kantor, W. M. (to appear) Automorphisms and isomorphisms of symmetric and affine designs. *J. Algebraic Combinatorics.*

[14] Liebeck, M. W. and Saxl, J. (1991) Minimal degrees of primitive permutation groups, with an application to monodromy groups of covers of Riemann surfaces. *Proc. London Math. Soc.* (2) **63** 266–314.

[15] Łuczak, T. and Pyber, L. (to appear) *Combinatorics, Probability and Computing*.

[16] Smith, J. D. H. (1986) *Representation Theory of Infinite Groups and Finite Quasigroups*, Sém. Math. Sup., Presses Univ. Montréal, Montréal.

Ramsey Problems with Bounded Degree Spread

G. CHEN[†] and R. H. SCHELP[‡]

[†]North Dakota State University, Fargo, ND 58105

[‡]Memphis State University, Memphis, TN 38152

Let k be a positive integer, $k \geq 2$. In this paper we study bipartite graphs G such that, for n sufficiently large, each two-coloring of the edges of the complete graph K_n gives a monochromatic copy of G, with some k of its vertices having the maximum degree of these k vertices minus the minimum degree of these k vertices (in the colored K_n) at most $k - 2$.

1. Introduction

Ramsey's theorem assures a specified local order in the midst of global chaos. Specifically, given graphs G and H, each with no isolates, there exists a number $r(G, H)$ such that every red–blue coloring of K_n with $n \geq r(G, H)$ yields either a red copy of G or a blue copy of H. What are the properties of such monochromatic copies in the global setting of the two-colored complete graphs? In this paper we will investigate the degrees in the two-colored K_n of vertices belonging to such monochromatic copies.

Let $\gamma : E(K_n) \mapsto (R, B)$ denote a two-coloring of the edges of the complete graph of order n using colors red (R) and blue (B), and let $\langle R \rangle$ and $\langle B \rangle$ denote the corresponding monochromatic graphs. For $X \subseteq V(K_n)$, let $\langle X \rangle_R$ and $\langle X \rangle_B$ denote the subgraphs induced by X in $\langle R \rangle$ and $\langle B \rangle$ respectively. Given graphs G and H, each with no isolates, write $X \in R_\gamma(G, H)$ if $|X| = |V(G)|$ and $G \subseteq \langle X \rangle_R$, or $|X| = |V(H)|$ and $H \subseteq \langle X \rangle_B$. By Ramsey's theorem, there exists a number $r(G, H)$ such that $R_\gamma(G, H) \neq \emptyset$ for every $\gamma : E(K_n) \mapsto (R, B)$ whenever $n \geq r(G, H)$. We shall refer to a set in $R_\gamma(G, H)$ as a *Ramsey host*. The following results were obtained in [1] and [2], where the degrees of vertices in Ramsey hosts were investigated.

Theorem 1. (Albertson [1]) *In every two-coloring of the edges of the complete graph of order ≥ 6, there is a monochromatic triangle K_3 for which two vertices have the same degree.*

Theorem 2. (Albertson and Berman [2]) *For all n, there exists a red–blue coloring of the* *edges of K_n that contains no red K_4 and no two vertices of equal degree joined by a blue edge.*

Theorem 2 tells us that Theorem 1 is best possible in the sense that the triangle K_3 cannot be replaced by K_n, $n \geq 4$. Generally, given a graph G, we say that it has the *Ramsey repeated degree property* if for all sufficiently large n every two edge-coloring $\gamma : E(K_n) \mapsto (R, B)$ yields a Ramsey host $X \in R_\gamma(G, G)$ in which there are vertices x, y satisfying $d_\gamma(x) = d_\gamma(y)$. Here d_γ refers to either the degree in $\langle R \rangle$ or in $\langle B \rangle$, since vertices of the same degree in $\langle R \rangle$ have the same degree in $\langle B \rangle$. Recently, Erdős, Chen, Rousseau and Schelp generalized Theorem 1 by proving the following result.

Theorem 3. (Erdős, Chen, Rousseau, and Schelp [3]) *For each $m \geq 1$, the complete bipartite graph $K_{m,m}$ and the odd cycle C_{2m+1} have the Ramsey repeated degree property.*

In the same paper they proved the following result.

Theorem 4. (Erdős, Chen, Rousseau, and Schelp [3]) *In every two-coloring of the edges of the complete graph of order $\geq r(G, H)$, there is a Ramsey host X such that*

$$\max_{x \in X} d(x) - \min_{y \in X} d(y) \leq r(G, H) - 2.$$

Further, the result is best possible in the sense that for every sufficiently large n, there is a two-coloring of the edges of K_n such that for every Ramsey host X the following inequality holds.

$$\max_{x \in X} d(x) - \min_{y \in X} d(y) \geq r(G, H) - 2.$$

Let V_1 and V_2 be two subsets of $V(G)$. We use $E(V_1, V_2)$ to denote the edges with one end vertex in V_1 and the other in V_2. The degree of x in the graph G will be denoted by $d_G(x)$, or simply $d(x)$ if the identity of G is clear from the context. Also the neighborhood of x will be denoted by $N_G(x)$ or $N(x)$ when G is clear. For $Y \subseteq V(G)$ the *degree spread* of Y is defined as

$$\Delta_G(Y) = \max_{y \in Y} d(y) - \min_{y \in Y} d(y).$$

Let k be a positive integer and G be a graph. Let n be a sufficiently large integer and $\gamma : E(K_n) \mapsto (B, R)$ be a two-coloring of the edges of the complete graph K_n. We are interested in a generalization of the Ramsey repeated degree property replacing two vertices by k vertices in the Ramsey hosts for G. Since there are graphs that only have two vertices of the same degree, we do not expect all k of these vertices to have the same degree. Thus our interest is in finding the minimum difference among the degrees of such vertices. To be specific, let H be a graph and $Y \subseteq V(H)$. The degree spread of Y is defined as

$$\Delta(Y) = \max_{y \in Y} d_H(y) - \min_{y \in Y} d_H(y).$$

For the case of a two-colored complete graph with edge coloring γ, the degree spread of Y is the same for $H = \langle B \rangle$ and $H = \langle R \rangle$, and we denote this common value by $\Delta_\gamma(Y)$. In this paper, we will consider

$$\Psi_n^k(G, H) \stackrel{def}{=} \max_\gamma \min_{X \in R_\gamma(G,H)} \min_{|Y|=k} \Delta_\gamma(Y),$$

where the final maximum is taken over all two-colorings $\gamma: E(K_n) \mapsto (R, B)$ and the initial minimum is taken over all the $Y \subseteq X$. Thus $\Psi_n^k(G, H)$ measures the smallest possible degree spread of some k-subset of vertices that must appear as a vertex subset of either a red G or blue H under each two coloring of the edges of K_n.

Let $V = \{v_1, v_2, \cdots v_m\}$, $W = \{w_1, w_2, \cdots w_m\}$, and $E = \{v_i v_j, v_i w_j : 1 \leq i < j \leq m\}$. We call the graph $(V \cup W, E)$ the half-full graph. Note that for every k-vertex set Y in the half-full graph, $\Delta(Y) \geq k - 2$. Thus,

$$\Psi_n^k(G, H) \geq k - 2$$

for all $k \geq 2$.

In this paper we wish to determine the graphs G and H for which $\Psi_n^k(G, H) \leq k - 2$ for all sufficiently large n. When $G = H$, we write $\Psi_n^k(G)$ in place of $\Psi_n^k(G, G)$.

2. Main theorem

Let G be a graph. A vertex subset I of $V(G)$ is called *distance q-independent* if $d(x, y) \geq q+1$ for every pair of vertices x and y in I. Notice that I is distance 1-independent if and only if I is independent. Also if I is distance q-independent with $q \geq 2$, then $N(x) \cap N(y) = \emptyset$ for each $x \neq y$ in I. Let $k \geq 2$ be a positive integer. A bipartite graph $G = (V_1, V_2)$ is called *(q, k)-independent* if for each positive integer $1 \leq m \leq k - 1$, there is a distance q-independent set I of G such that

$$|I \cap V_1| = m \text{ and } |I \cap V_2| = k - m.$$

Notice the following:

— The even cycle C_t with $t \geq 4k$ is $(2, k)$-independent.
— If G is (q, k)-independent and H is a spanning subgraph of G, then H is (q, k)-independent.
— Let G be a connected bipartite graph with diameter $\geq 4k$. Then G is $(2, k)$-independent.

With the above definition we state our main theorem as follows.

Theorem 5. *Let $k \geq 2$ be a positive integer and G be a $(2, k)$-independent bipartite graph. Then there is a positive integer N such that for every positive integer $n \geq N$, $\Psi_n^k(G) \leq k - 2$. Thus each two-coloring of the edges of the complete graph K_n gives a monochromatic copy of G with some k of its vertices of degree spread $\leq k - 2$ in the colored K_n.*

The following results follow directly from the theorem.

Corollary 1. *Let $k \geq 2$ be a positive integer and G be a connected bipartite graph with diameter at least $4k$. Then, for n large, $\Psi_n^k(G) \leq k - 2$.*

Corollary 2. *Let $k \geq 2$ be a positive integer and G be one of the following graphs:*

— *an even cycle C_t with $t \geq 4k$;*
— *a path with more than $4k$ vertices;*

— a tree containing a path of length $\geq 4k$;
— a bipartite graph with at least k components.

Then, for n large, $\Psi_n^k(G) \leq k - 2$.

The following results are needed in the proof of the main theorem. The first lemma generalizes one of the most well-known facts about graphs, namely that in every graph there are two vertices with the same degree.

Lemma 1. (Erdős, Chen, Rousseau, and Schelp [3]) *Let G be a graph of order n and k be a positive integer less than n. Then G contains a set Y of k vertices with degree spread $\Delta(Y)$ at most $k - 2$.*

The following result is an analogue to Ramsey's theorem for bipartite graphs.

Lemma 2. ([4]) *For all positive integers p there exists an N such that for every $n \geq N$ each edge coloring of $K_{n,n}$ with two colors contains a monochromatic $K_{p,p}$. The least such N will be denoted by $r^2(p)$.*

By a well-known argument [5], the following holds.

Lemma 3. *Let ϵ be a positive number and p be a positive integer. There is a positive integer N such that for every graph G of order $n \geq N$, if the vertex subset*

$$C = \{v \ : \ d_G(v) \geq \frac{p^{1/p}}{\epsilon} n^{1-1/p}\}$$

has more than ϵn vertices, then G contains a copy of $K_{p,p}$ with a vertex part in C.

3. Proof of the Main Theorem

Let $G = (V_1, V_2)$ be a $(2, k)$-independent bipartite graph. Let $\max\{|V_1|, |V_2|\} = p$. The 'sufficiently large' nature of n will be assumed throughout the argument, and no attempt will be made to accurately estimate a threshold value of n at which the desired property first appears. Let $\gamma : E(K_n) \mapsto (R, B)$ be a given two-coloring of the edges of K_n. From Lemma 1, we start with k vertices $\{x_1, x_2, \cdots, x_k\}$ for which $|d_\gamma(x_i) - d_\gamma(x_j)| \leq k - 2$. In the remainder of the proof we let $m = n - k$.

Partition the vertex set $V(K_n) - \{x_1, x_2, \cdots, x_k\}$ into 2^k cells, $(A_1, A_2, \cdots, A_k)$, where $A_i \in \{B, R\}$ such that a vertex $v \in (A_1, A_2, \cdots, A_k)$ if the edge vx_1 is colored with the color A_1, vx_2 is colored with the color A_2, ..., vx_k is colored with the color A_k. Two cells $A = (A_1, A_2, ..., A_k)$ and $A^* = (A_1^*, A_2^*, ..., A_k^*)$ are conjugate if $\{A_i\} \cup \{A_i^*\} = \{B, R\}$ for each $i = 1, 2, ..., k$: that is, the cell A^* can be obtained from A by changing the blue colors to red colors and the red colors to blue colors.

First we assume that either $|(R, R, ..., R)|$ or $|(B, B, ..., B)| \geq m/2^k$. Without loss of generality, we assume that $|(R, R, ..., R)| \geq m/2^k$. In fact, in this case we will prove that there is a monochromatic $K_{p,p}$ for which there are k vertices whose degree spread is at most $k - 2$. Let

$$C = \{v \in (R, R, ..., R) \ : \ d_R(v) \geq 2^{k+1} p^{\frac{1}{p}} n^{1-\frac{1}{p}}\}.$$

and $D = V(K_n) - C - \{x_1, x_2, ..., x_k\}$. If $|C| \geq m/2^{k+1}$, then, by Lemma 3, there is a red

$K_{p,p-k}$ in the red graph induced by $V(K_n) - \{x_1, x_2, \ldots, x_k\}$ with the $p-k$ vertex part set in C (since n is large). Combine this with $\{x_1, x_2, \ldots, x_k\}$, giving us a red $K_{p,p}$ with k vertices with degree spread at most $k-2$. Thus $|D| \geq m/2^{k+1}$. Notice that each $v \in D$ has $d_B(v) \geq n - 2^{k+1}p^{\frac{1}{p}}n^{1-\frac{1}{p}} - 1$. Clearly, for n large, D contains k vertices $y_1\, y_2, \ldots, y_k$ of the same blue degree. Further, these k vertices have a large common blue neighborhood and can be easily enlarged to a blue $K_{p,p}$ containing $y_1, y_2, \ldots, y_k$. Thus, we may assume that $|(R, R, \ldots, R)| \leq m/2^k$ and $|(B, B, \ldots, B)| \leq m/2^k$.

Since there are 2^k cells, one of the other cells, say A, must contain at least $m/2^k$ vertices. In this case, we let $r_0 = p$ and $r_1 = r^2(p)$. For every $i \geq 1$, let $r_{i+1} = r^2(r_i)$. We will show that there is a monochromatic copy of G containing the vertex set $\{x_1, x_2, \ldots, x_k\}$ whenever $m \geq 2^{2k}r_{k^2}$. Hence the theorem will hold.

Without loss of generality, assume that

$$A = (\underbrace{R, R, \ldots, R}_{s}, \underbrace{B, B, \ldots, B}_{t}),$$

where s and t are positive integers. For every pair of numbers i and j with $1 \leq i \leq s$ and $1 \leq j \leq t$, $d_R(x_i) \geq |A| \geq m/2^k$ and $d_B(x_{j+s}) \geq |A| \geq m/2^k$. Recall that $|d_R(x_i) - d_R(x_{j+s})| \leq k-2$, and $m = n - k$ is sufficiently large. Thus there exists a cell

$$A_{i,j} = (A_1^{i,j}, A_2^{i,j}, \cdots, A_k^{i,j})$$

with $A_i^{i,j} = B$ and $A_{j+s}^{i,j} = R$ such that

$$|A_{i,j}| \geq |A|/2^{k-2} - (k-2) \geq m/2^{2k}.$$

Since $m/4^k \geq r_{k^2}/4 \geq r_{st}$, there are $A(1,1) \subseteq A$ and $A_{1,1}^* \subseteq A_{1,1}$ such that

$$|A(1,1)| \geq r_{st-1}, \quad |A_{1,1}^*| \geq r_{st-1},$$

and all edges in $E(A(1,1), A_{1,1}^*)$ are colored with the same color.

Since $|A(1,1)| \geq r_{st-1}$ and $|A_{1,2}| \geq r_{st} \geq r_{st-1}$, there are two vertex subsets $A(1,2) \subseteq A(1,1)$ and $A_{1,2}^* \subseteq A_{1,2}$ such that

$$|A(1,2)| \geq r_{st-2}, \quad |A_{1,2}^*| \geq r_{st-2},$$

and all edges in $E(A(1,2), A_{1,2}^*)$ are colored with the same color.

Continuing in the same manner, we can show that there are

$$A(1,1) \supseteq A(1,2) \supseteq \cdots \supseteq A(1,t) \text{ and}$$
$$A_{1,1}^* \subseteq A_{1,1}, \ A_{1,2}^* \subseteq A_{1,2}, \ldots, A_{1,t}^* \subseteq A_{1,t}$$

such that

$$|A(1,i)| \geq r_{st-i} \geq r_{(s-1)t}, \quad |A_{1,i}^*| \geq r_{(s-1)t}$$

for each $i = 1, 2, \ldots, t$, and all edges in $E(A(1,i), A_{1,i}^*)$ are the same color.

For the moment, assume for all $i = 1, 2, \ldots t$ that the edges in $E(A(1,i), A_{1,i}^*)$ are red. Recall that $G = (V_1, V_2)$ is $(2,k)$-independent. Thus there are vertices $y_1, y_2, \ldots, y_s \in V_1$ and $y_{s+1}, y_{s+2}, \cdots, y_k \in V_2$ such that $d(y_i, y_j) \geq 3$ for each $1 \leq i < j \leq k$. Since $r_{(s-1)t} \geq r_0 = p$, we can embed G in the red graph $\langle R \rangle$ such that $V_2 \subseteq A(1,t)$ and

$N(y_{s+1}) \subseteq A_{1,1}^*$, $N(y_{s+2}) \subseteq A_{1,2}^*$, ..., $N(y_{k-1}) \subseteq A_{1,t-1}^*$, $V_1 - \bigcup_{i=1}^{t-1} N(y_{s+i}) \subseteq A_{1,t}^*$. Replacing y_i by the vertex x_i in the embedded graph for $i = 1, 2, ..., k$, we obtain a red graph G with k vertices $x_1, x_2, \cdots x_k$ for which the degree spread is at most $k - 2$. Thus we may assume that there is an i_1 such that all edges in $E(A(1, i_1), A_{1,i_1}^*)$ are blue.

Note that $|A(1, i_1)| \geq r_{(s-1)t}$ and $|A_{2,1}| \geq r_{st} \geq r_{(s-1)t}$. There are $A(2, 1) \subseteq A(1, i_1)$ and $A_{2,1}^* \subseteq A_{2,1}$ such that

$$|A(2, 1)| \geq r_{(s-1)t-1}, \quad |A_{2,1}^*| \geq r_{(s-1)t-1},$$

and all the edges in $E(A(2, 1), A_{2,1}^*)$ are the same color.

Since $|A(2, 1)| \geq r_{(s-1)t-1}$ and $|A_{2,2}| \geq r_{st} \geq r_{(s-1)t-1}$, there are $A(2, 2) \subseteq A(2, 1)$ and $A_{2,2}^* \subseteq A_{2,2}$ such that

$$|A(2, 2)| \geq r_{(s-1)t-2}, \quad |A_{2,2}^*| \geq r_{(s-1)t-2},$$

and the edges in $E(A(2, 2), A_{2,2}^*)$ are the same color.

Continuing in the same manner, there are

$$A(2, 1) \supseteq A(2, 2) \supseteq \cdots \supseteq A(2, t)$$
$$A_{2,1}^* \subseteq A_{2,1}, ..., A_{2,t}^* \subseteq A_{2,t}$$

such that

$$|A(2, i)| \geq r_{(s-1)t-i} \geq r_{(s-2)t}, \quad |A_{2,i}^*| \geq r_{(s-2)t}$$

for each $i = 1, 2, ..., t$, and all edges in $E(A(2, i), A_{2,i}^*)$ are the same color.

Notice that $r_{t(s-2)} \geq r_0 = p$. If for every $i = 1, 2, ..., k$, the edges in $E(A(2, i), A_{2,i}^*)$ are red, in the same manner as argued above, we can show that there is a red copy of G that contains $x_1, x_2, ..., x_k$ whose degree spread at most $k - 2$. Thus we may assume that there is an i_2 such that the edges in $E(A(2, i_2), A_{2,i_2}^*)$ are blue. Notice that $|A(2, i_2)| \geq r_{(s-2)t}$ and $|A_{2,i_2}^*| \geq r_{(s-2)t}$.

Continuing in the same manner, we may assume that there are $2s$ vertex subsets

$$A \supseteq A(1, i_1) \supseteq A(2, i_2) \supseteq \cdots \supseteq A(s, i_s), \text{ and}$$
$$A_{1,i_1}^* \subseteq A_{1,i_1}, A_{2,i_2}^* \subseteq A_{2,i_2}, ..., A_{s,i_s}^* \subseteq A_{s,i_s},$$

such that

$$|A_{1,i_1}^*| \geq |A_{2,i_2}^*| \geq \cdots \geq |A_{s,i_s}^*| \geq r_0 = p,$$

and $|A(s, i_s)| \geq r_0 = p$. Further, for all $j = 1, 2, ..., s$, the edges in $E(A(j, i_j), A_{j,i_j}^*)$ are blue.

Since $G = (V_1, V_2)$ is $(2, k)$-independent, there are vertex sets

$$\{z_1, z_2, ..., z_s\} \subseteq V_1, \text{ and } \{z_{s+1}, z_{s+2}, ..., z_k\} \subseteq V_2$$

such that for each pair of vertices z_i and z_j, $d(z_i, z_j) \geq 3$ whenever $1 \leq i < j \leq k$. Then G can be embedded in the blue graph $\langle B \rangle$ such that $V_1 \subseteq A(s, i_s)$ and

$$N(z_1) \subseteq A_{1,i_1}^*, \; N(z_2) \subseteq A_{2,i_2}^*, \; \ldots \; N(z_{s-1}) \subseteq A_{s-1,i_{s-1}}^*, \; V_2 - \bigcup_{j=1}^{s-1} N(z_j) \subseteq A_{s,i_s}^*.$$

Replacing each z_i by x_i for $i = 1, 2, ... k$, we obtain a blue copy of G containing $x_1, x_2, ... x_k$ whose degree spread at most $k - 2$. This completes our proof.

References

[1] Albertson, M. O. (preprint) *People who know people.*

[2] Albertson, M. O. and Berman, D. M. (preprint) *Ramsey graphs without repeated degrees.*

[3] Erdős, P., Chen, G., Rousseau, C. C. and Schelp, R. H. (1993) Ramsey problems involving degrees in edge-colored complete graphs of vertices belonging to monochromatic subgraphs. *Europ. J. Combinatorics* **14** 183–189.

[4] Graham, R. L., Rothschild, B. R. and Spencer, J. H. (1990) *Ramsey Theory (2nd Edition)*, John Wiley & Sons, New York.

[5] Kővári, T., Sós, V. T. and Túran, P. (1954) On a problem of Zarankiewicz. *Colloq. Math.* **3** 50–57.

Hamilton Cycles in Random Regular Digraphs

COLIN COOPER,[†] ALAN FRIEZE[‡§] and MICHAEL MOLLOY[‡]

[†]School of Mathematical Sciences,
University of North London,
London, U.K.
[‡]Department of Mathematics, Carnegie-Mellon University,
Pittsburgh PA15213, U.S.A.

We prove that almost every r-regular digraph is Hamiltonian for all fixed $r \geq 3$.

1. Introduction

In two recent papers Robinson and Wormald [8, 9] solved one of the major open problems in the theory of random graphs. They proved the following result.

Theorem 1. *For every fixed $r \geq 3$ almost all r-regular graphs are hamiltonian.*

For earlier attempts at this question see Bollobás [2], Fenner and Frieze [5] and Frieze [6], who established the result for $r \geq r_0$.

In [8] ($r=3$) a clever variation on the second moment method was used, and in [9] (for $r \geq 4$) this idea plus a sort of monotonicity argument was used.

In this paper we will study the directed version of the problem. Thus, let $\Omega_{n,r} = \Omega$ denote the set of digraphs with vertex set $[n] = \{1, 2, \ldots, n\}$ such that each vertex has indegree and outdegree r. Let $D_{n,r} = D$ be chosen uniformly at random from $\Omega_{n,r}$.

Theorem 2.

$$\lim_{n \to \infty} Pr(D \text{ is Hamiltonian}) = \begin{cases} 0 & r = 2 \\ 1 & r \geq 3. \end{cases}$$

§ Supported by NSF grant CCR-9024935

The case $r = 2$ follows directly from the fact that the expected number of Hamilton cycles in $D_{n,2}$ tends to zero.

Our method of proof for $r \geq 3$ is quite different from [8, 9] although we will use the idea that for $r \geq 3$, a random r-regular bipartite graph is close, in some probabilistic sense, to a random $(r-1)$-regular bipartite graph plus a random matching.

Our strategy is close to that of Cooper and Frieze [4], who prove that almost every 3-in, 3-out digraph is Hamiltonian.

2. Random digraphs and random bipartite graphs

Given $D_{n,r} = ([n], A)$, we can associate it with a bipartite graph $B = B_{n,r} = \phi(D_{n,r}) = ([n], [n], E)$ in a standard way. Here B contains an edge $\{x, y\}$ iff D contains the directed edge (x, y). The mapping ϕ is a bijection between r-regular digraphs and r-regular bipartite graphs, so B is uniform on the latter space, which we denote by $\Omega_{n,r}^B$.

For $r \geq 3$ we wish to replace $B_{n,r}$ by $B_{n,r-1}$ plus an independently chosen random perfect matching M of $[n]$ to $[n]$. This is equivalent to replacing D by $\Pi_0 \cup \hat{D}$, where Π_0 and $\hat{D}$ are independent and

(i) Π_0 is the digraph of a random permutation,
(ii) $\hat{D} = D_{n,r-1}$.

Of course Π is the union of vertex disjoint cycles. We call such a digraph a *permutation digraph*. Its cycle count is the number of cycles.

The arguments of [9] allow us to make the above replacement. A brief sketch of why this is so would certainly be in order.

Let X_M denote the number of perfect matchings in $B_{n,r}$. Arguments in [9] demonstrate the existence of $\epsilon(b) > 0$ such that for $b > 0$ fixed,

$$\lim_{n \to \infty} Pr(X_M \geq \mathbf{E}(X_M)/b) \geq 1 - \epsilon(b),$$

where $\epsilon(b) \to 0$ as $b \to \infty$.

Now consider a bipartite graph $\mathscr{B} = (\Omega_{n,r-1}^B, \Omega_{n,r}^B, \mathscr{E})$. There is an edge from $G \in \Omega_{n,r-1}^B$ to $G' \in \Omega_{n,r}^B$ iff $G' = G \cup M$, where M is a perfect matching. Now choose (G, G') randomly from $\mathscr{E}$. Let A denote some event defined on $\Omega_{n,r}^B$ and $\hat{A} = \{(G, G') \in \mathscr{E} : G' \in A\}$. Then, since the maximum and minimum degrees of the $\Omega_{n,r-1}^B$ vertices of $\mathscr{B}$ are asymptotically equal to $n! e^{-(r-1)}$ [1],

$$Pr_0(\hat{A}) = (1 + o(1))Pr_1(A),$$

where $o(1)$ refers to $n \to \infty$, Pr_0 refers to the space $\mathscr{E}$ with the uniform measure, and Pr_1 refers to (randomly chosen) $G = B_{n,r-1}$ plus a randomly chosen M, disjoint from $G' = B_{n,r-1}$.

On the other hand, if Pr refers to $B_{n,r}$,

$$Pr_0(\hat{A}) = \sum_{G' \in A} \frac{X_M}{|\mathscr{E}|}$$

$$= \sum_{G' \in A} \frac{X_M}{\mathbf{E}(X_M)|\Omega_{n,r}^B|}$$

$$\geq \quad (Pr(A) - \epsilon(b))/b.$$

Thus

$$Pr(A) \leq \epsilon(b) + (b + o(1))Pr_1(A).$$

Thus, if A is $\{\phi^{-1}(B_{n,r-1} \cup M)$ is non-Hamiltonian$\}$ (M disjoint from $B_{n,r-1}$ here), we can show that $Pr(A) \to 0$ (as $n \to \infty$) by proving that $Pr_1(A) \to 0$ (as $n \to \infty$), since b can be arbitrarily large.

Finally, if Pr_2 refers to $B_{n,r-1}$ plus a randomly chosen M (not necessarily disjoint from $B_{n,r-1}$), then $Pr_2(A) \to 0$ (as $n \to \infty$), implies $Pr_1(A) \to 0$ (as $n \to \infty$) since the probability that M is disjoint from $B_{n,r-1}$ in this case tends to the constant $e^{-(r-1)} > 0$.

We have thus reduced the proof of Theorem 1 to showing that

$$\lim_{n \to \infty} Pr(\Pi_0 \cup \hat{D} \text{ is Hamiltonian}) = 1.$$

In fact we have only to prove the result for $r = 3$ and apply induction. Thus assume $r = 3$ from now on.

We will use a *two phase* method as outlined below.

Phase 0. As Π_0 is a random permutation digraph, it is almost always of cycle count at most $2 \log n$, see, for example [3].

Phase 1. Using $\hat{D}$, we increase the minimum cycle size in the permutation digraph to at least $n_0 = \lceil 100n/\log n \rceil$.

Phase 2. Using $\hat{D}$, we convert the *Phase 1* permutation digraph to a Hamilton cycle.

In what follows inequalities are only claimed to hold for n sufficiently large. The term **whp** is short for *with high probability i.e.* probability $1 - o(1)$ as $n \to \infty$.

3. Phase 1: Removing small cycles

We partition the cycles of the permutation digraph Π_0 into sets SMALL and LARGE, containing cycles C of size $|C| < n_0$ and $|C| \geq n_0$, respectively. We define a Near Permutation Digraph (NPD) to be a digraph obtained from a permutation digraph by removing one edge. Thus an NPD Γ consists of a path $P(\Gamma)$ plus a permutation digraph $PD(\Gamma)$ that covers $[n] \setminus V(P(\Gamma))$.

We now give an informal description of a process that removes a small cycle C from a *current* permutation digraph Π. We start by choosing an (arbitrary) edge (v_0, u_0) of C and delete it to obtain an NPD Γ_0 with $P_0 = P(\Gamma_0) \in \mathscr{P}(u_0, v_0)$, where $\mathscr{P}(x, y)$ denotes the set of paths from x to y in D. The aim of the process is to produce a *large* set S of NPDs such that for each $\Gamma \in S$,

    (i) $P(\Gamma)$ has a least n_0 edges, and

    (ii) the small cycles of $PD(\Gamma)$ are a subset of the small cycles of Π.

We will show that **whp** the endpoints of one of the $P(\Gamma)$s can be joined by an edge to create a permutation digraph with (at least) one less small cycle.

The basic step in an *Out-Phase* of this process is to take an NPD Γ with $P(\Gamma) \in \mathscr{P}(u_0, v)$ and examine the edges of $\hat{D}$ leaving v. Let w be the terminal vertex of such an edge, and assume that Γ contains an edge (x, w). Then $\Gamma' = \Gamma \cup \{(v, w)\} \setminus \{(x, w)\}$ is also an NPD.

Γ' is acceptable if

(i) $P(\Gamma')$ contains at least n_0 edges, and

(ii) any new cycle created (*i.e.* in Γ' and not Γ) also has at least n_0 edges.

If Γ contains no edge (x, w), then $w = u_0$. We accept the edge if $P(\Gamma)$ has at least n_0 edges. This would (prematurely) end an iteration, although it is unlikely to occur.

We do not want to look at very many edges of $\hat{D}$ in this construction and we build a tree T_0 of NPDs in a natural breadth-first fashion, where each non-leaf vertex Γ gives rise to NPD children Γ' as described above. The construction of T_0 ends when we first have $v = \lceil \sqrt{n \log n} \rceil$ leaves. The construction of T_0 constitutes an Out-Phase of our procedure to eliminate small cycles. Having constructed T_0 we need to do a further *In-Phase*, which is similar to a set of Out-Phases.

Then **whp** we close at least one of the paths $P(\Gamma)$ to a cycle of length at least n_0. If $|C| \geq 2$ and this process fails, we try again with a different edge of C in place of (u_0, v_0).

We now increase the formality of our description. We start Phase 1 with a permutation digraph Π_0 and a general iteration of Phase 1 starts with a permutation digraph Π whose small cycles are a subset of those in Π_0. Iterations continue until there are no more small cycles. At the start of an iteration, we choose some small cycle C of Π. There then follows an Out-Phase, in which we construct a tree $T_0 = T_0(\Pi, C)$ of NPDs as follows: the root of T_0 is Γ_0, which is obtained by deleting an edge (v_0, u_0) of C.

We grow T_0 to a depth at most $\lceil 1.5 \log n \rceil$. The set of nodes at depth t is denoted by S_t. Let $\Gamma \in S_t$ and $P = P(\Gamma) \in \mathscr{P}(u_0, v)$. The *potential* children Γ' of Γ, at depth $t + 1$ are defined as follows, with w the terminal vertex of an edge directed from v in $\hat{D}$.

Case 1: w is a vertex of a cycle $C' \in PD(\Gamma)$ with edge $(x, w) \in C'$. Let $\Gamma' = \Gamma \cup \{(v, w)\} \setminus \{(x, w)\}$.

Case 2: w is a vertex of $P(\Gamma)$. Either $w = u_0$, or (x, w) is an edge of P. In the former case, $\Gamma \cup \{(v, w)\}$ is a permutation digraph Π', and in the latter case, we let $\Gamma' = \Gamma \cup \{(v, w)\} \setminus \{(x, w)\}$.

In fact we only admit to S_{t+1} those Γ' that satisfy the following conditions:

C*(i)*, the new cycle formed (Case 2 only), must have at least n_0 vertices, and the path formed must either be empty or have at least n_0 vertices. When the path formed is empty we close the iteration and if necessary start the next with Π'.

Now define W_+, W_- as follows: initially $W_+ = W_- = \emptyset$. A vertex x is added to W_+ whenever we learn any of its out-neighbours in $\hat{D}$, and to W_- whenever we learn any of its in-neighbours. $W = W_+ \cup W_-$. We never allow $|W|$ to exceed $n^{9/10}$.

The only information we learn about $\hat{D}$ is that certain specific arcs are present. The property we need of the random graph $\hat{D}$ is that if $x \notin W_+$ and S is any set of vertices, disjoint from W,

$$Pr(N_+(x) \cap S \neq \emptyset) = \left(1 - \left(1 - \frac{|S|}{n}\right)^2\right)\left(1 + O\left(\frac{1}{n^{1/10}}\right)\right).$$

These approximations are intended to hold conditional on any past history of the algorithm such that $|W| \leq n^{9/10}$. Furthermore, if $x \in W_+$, but only one neighbour y is known then, where $y \notin S$,

$$Pr(N_-(x) \cap S \neq \emptyset | y) = \frac{|S|}{n}\left(1 + O\left(\frac{1}{n^{1/10}}\right)\right).$$

Similar remarks are true for $N_-(x)$. Thus, since W remains small, $N_{\pm}(v)$ are usually (near) random pairs in W.

C(ii) $x \notin W$.

An edge (v, w) satisfying the above conditions is described as *acceptable*. In order to remove any ambiguity, the vertices of S_t are examined in the order of their construction.

Lemma 3. *Let* $C \in SMALL$. *Then*

$$Pr(\exists t < \lceil \log_{3/2} v \rceil \text{ such that } |S_t| \geq v) = 1 - O((\log\log n / \log n)^2).$$

Proof. We assume we stop construction of T_0, in mid-phase if necessary, when $|S_t| = v$, and show inductively that **whp** $(3/2)^t \leq |S_t| \leq 2^t$, for $t \geq 3$. Let t^* denote the value of t when we stop. Thus the overall contribution to $|W|$ from this part of the algorithm is at most $|SMALL| \times 2^{t^*+1} \leq n^{0.86}$.

In general, let X_t be the number of unacceptable edges found when constructing S_{t+1}, $(t = 1, 2, ..., t^*)$. The event of a particular edge (v, w) being unacceptable is stochastically dominated by a Bernouilli trial with probability of success $p < \log\log n/n$. (in general, inequalities are only claimed for sufficiently large n). To see this, observe that there is a probability of at most $201/\log n$ that in Case 2 we create a small cycle or a short path. There is an $O(n^{-1/10})$ probability that $x \in W$. Finally there is the probability that w lies in a small cycle. Now in a random permutation the expected number of vertices in cycles of size at most k is precisely k/n. Thus **whp** Π_0 contains at most $n \log\log n/(2\log n)$ vertices on small cycles, so, given this, the probability that w lies on a small cycle is at most $\log\log n/(2\log n)$.

For $t \leq c$, constant, the probability of 2 or more unacceptable edges in layers $t \leq c$ is $O\left(2^{2c}(\log\log n)^2/(\log n)^2\right)$, and thus $|S_{t+1}| > 2|S_t| - 1 > (3/2)^t$ for $3 \leq t \leq c$ with probability $1 - O((\log\log n/\log n)^2)$.

In order to see this, note that in the case where there is only one acceptable edge at the first iteration, subsequent layers expand by a power of 2, and $|S_1| = 2$ otherwise.

For $t > c, c$ large, the expected number of unacceptable edges at iteration t is at most $\mu = 2p|S_t|$, and thus, by standard bounds on tails of the Binomial distribution,

$$Pr\left(X_t > \lfloor |S_t|/2 \rfloor \Big| |S_t| = s\right) \leq \left(\frac{2e\log\log n}{\log n}\right)^{\lfloor s/2 \rfloor}.$$

This upper bound is easily good enough to complete the proof of the lemma. $\square$

Now, T_0 has leaves Γ_i, for $i = 1, \ldots, v$, each with a path of length at least n_0 (unless we have already successfully made a cycle). We now execute an In-Phase. This involves the construction of trees $T_i, i = 1, 2, \ldots v$. Assume that $P(\Gamma_i) \in \mathscr{P}(u_0, v_i)$. We start with Γ_i and $\mathscr{D}_i$, and build T_i in a similar way to T_0, except that here all paths generated end with v_i. This is done as follows: if a current NPD Γ has $P(\Gamma) \in \mathscr{P}(u, v_i)$, we consider adding an edge $(w, u) \in \hat{D}$ and deleting an edge $(w, x) \in \Gamma$ (as opposed to (x, w) in an Out-Phase). Thus our trees are grown by considering edges directed into the start vertex of each $P(\Gamma)$, rather than directed out of the end vertex. Some technical changes are necessary, however.

We consider the construction of our v trees in two iterations. First of all we grow the trees only enforcing condition C(ii) of success, and thus allow the formation of small cycles. We try to grow them to depth $k = \lceil \log_{3/2} v \rceil$. We also consider the growth of the v trees simultaneously. Let $T_{i,\ell}$ denote the set of start vertices of the paths associated with the nodes at depth ℓ of the ith tree, $i = 1, 2 \ldots, v, \ell = 0, 1, \ldots, k$. Thus $T_{i,0} = \{u_0\}$ for all i. We prove inductively that $T_{i,\ell} = T_{1,\ell}$ for all i, ℓ. In fact, if $T_{i,\ell} = T_{1,\ell}$, the acceptable $\hat{D}$ edges have the same set of initial vertices, and, since all of the deleted edges are Π_0-edges (enforced by C(ii)), we have $T_{i,\ell+1} = T_{1,\ell+1}$.

The probability that we succeed in constructing v trees $T_1, T_2, \ldots T_v$, say, is, by the analysis of Lemma 3, $1 - O((\log\log n / \log n)^2)$. Note that the number of nodes in each tree is at most $2^{k+1} \leq n^{87}$, so the overall contribution to $|W|$ from this part of the algorithm is $O(n^{87} \log n)$.

We now consider the fact that in some of the trees some of the leaves may have been constructed in violation of C(i). We imagine that we prune the trees $T_1, T_2, \ldots T_v$ by disallowing any node that was constructed in violation of C(i). Let a tree be BAD if after pruning it has less than v leaves. Now an individual pruned tree has essentially been constructed in the same manner as the tree T_0 obtained in the Out-Phase. (We have chosen k large enough so that we can obtain v leaves at the slowest growth rate of $3/2$ per node.) Thus

$$Pr(T_1 \text{ is BAD}) = O\left(\left(\frac{\log\log n}{\log n}\right)^2\right)$$

and

$$E(\text{number of BAD trees}) = O\left(v\left(\frac{\log\log n}{\log n}\right)^2\right)$$

and

$$Pr(\exists \geq v/2 \text{ BAD trees}) = O\left(\left(\frac{\log\log n}{\log n}\right)^2\right).$$

Thus

$$Pr(\exists \leq v/2 \text{ GOOD trees after pruning})$$
$$< Pr(\text{failure to construct } T_1, T_2, \ldots T_v) + Pr(\exists \geq v/2 \text{ BAD trees})$$
$$= O\left(\left(\frac{\log\log n}{\log n}\right)^2\right).$$

Thus with probability $1 - O((\log\log n / \log n)^2)$, we end up with $v/2$ sets of v paths, each of length at least $100 n / \log n$, where the ith set of paths have V_i, say, as their set of start vertices, and v_i as a final vertex. At this stage each $v_i \notin W_+$, and each $V_i \cap W_- = \emptyset$. Hence

$$Pr(\text{no } \Pi \text{ edge closes one of these paths}) \leq \left(1 - \frac{2v}{n}\left(1 + O\left(\frac{1}{n^{1/10}}\right)\right)\right)^{v/2}$$
$$= O(n^{-1}).$$

Consequently the probability that we fail to eliminate a particular small cycle is $O((\log \log n / \log n)^2)$ and we have the following.

Lemma 4. *The probability that Phase 1 fails to produce a permutation digraph with minimal cycle length at least n_0 is $o(1)$.*

At this stage we have shown that $\Pi_0 \cup \hat{D}$ almost always contains a permutation digraph Π^* in which the minimum cycle size is at least n_0. We shall refer to Π^* as the *Phase 1* permutation digraph.

4. Phase 2: Patching the Phase 1 permutation digraph to a Hamilton cycle

Let $C_1, C_2, \ldots, C_k$ be the cycles of Π^*, and let $c_i = |C_i \setminus W|$, $c_1 \le c_2 \le \cdots \le c_k$, and $c_1 \ge n_0 - n^{3/4} \ge 99 \log n / n$. If $k = 1$, we can skip this phase, otherwise let $a = n / \log n$. For each C_i, we consider selecting a set of $m_i = 2\lfloor c_i / a \rfloor + 1$ vertices $v \in C_i \setminus W$, and deleting the edge (v, u) in Π^*. Let $m = \sum_{i=1}^k m_i$, and relabel (temporarily) the broken edges as $(v_i, u_i), i \in [m]$ as follows: in cycle C_i, identify the lowest numbered vertex x_i that loses a cycle edge directed out of it. Put $v_1 = x_1$, and then go round C_1 defining $v_2, v_3, \ldots v_{m_1}$ in order. Then let $v_{m_1+1} = x_2$, and so on. We thus have m path sections $P_j \in \mathscr{P}(u_{\phi(j)}, v_j)$ in Π^* for some permutation ϕ. We see that ϕ is an even permutation as all the cycles of ϕ are of odd length.

There is a chance that we can rejoin these path sections of Π^* to make a Hamilton cycle using $\hat{D}$. Suppose we can. This defines a permutation ρ, where $\rho(i) = j$ if P_i is joined to P_j by $(v_i, u_{\phi(j)})$, where $\rho \in H_m$ the set of cyclic permutations on $[m]$. We will use the second moment method to show that a suitable ρ exists **whp**. Unfortunately a technical problem forces a restriction on our choices for ρ.

Given ρ, define $\gamma = \phi\rho$. In our analysis we will restrict our attention to $\rho \in R_\phi = \{\rho \in H_m : \phi\rho = \gamma, \gamma \in H_m\}$. If $\rho \in R_\phi$, we have not only constructed a Hamilton cycle in $\Pi^* \cup \hat{D}$, but also in the *auxiliary digraph* Λ, whose edges are $(i, \gamma(i))$.

Lemma 5. $(m-2)! \le |R_\phi| \le (m-1)!$

Proof. We grow a path $1, \gamma(1), \gamma^2(1), \ldots, \gamma^k(1)$ in Λ, maintaining feasibility in the way we join the path sections of Π^* at the same time.

We note that the edge $(i, \gamma(i))$ of Λ corresponds in $\hat{D}$ to the edge $(v_i, u_{\phi\rho(i)})$. In choosing $\gamma(1)$ we must avoid not only 1, but also $\phi(1)$, since $\gamma(1) = 1$ implies $\rho(1) = 1$. Thus there are $m - 2$ choices for $\gamma(1)$, since $\phi(1) \ne 1$.

In general, having chosen $\gamma(1), \gamma^2(1), \ldots, \gamma^k(1), 1 \le k \le m - 3$, our choice for $\gamma^{k+1}(1)$ is restricted to be different from these choices, and also 1 and ℓ, where u_ℓ is the initial vertex of the path terminating at $v_{\lambda^k(1)}$ made by joining path sections of Π^*. Thus there are either $m - (k+1)$ or $m - (k+2)$ choices for $\gamma^{k+1}(1)$, depending on whether or not $\ell = 1$.

Hence, when $k = m - 3$, there *may* be only one choice for $\gamma^{m-2}(1)$, the vertex h say. After adding this edge, let the remaining isolated vertex of Λ be w. We now need to show that we can complete γ, ρ so that $\gamma, \rho \in H_m$.

Which vertices are missing edges in Λ at this stage? Vertices $1, w$ are missing in-edges, and h, w out-edges. Hence the path sections of Π^* are joined so that either

$$u_1 \to v_h, \quad u_w \to v_w \quad \text{or} \quad u_1 \to v_w, \quad u_w \to v_h.$$

The first case can be (uniquely) feasibly completed in both Λ and D by setting $\gamma(h) = w, \gamma(w) = 1$. Completing the second case to a cycle in Π^* means that

$$\gamma = (1, \gamma(1), \ldots, \gamma^{m-2}(1))(w), \tag{1}$$

and thus $\gamma \notin H_m$. We show that this case cannot arise.

When, $\gamma = \phi\rho$ and ϕ is even, γ and ρ have the same parity. On the other hand, $\rho \in H_m$ has a different parity to γ in (1), which is a contradiction.

Thus there is a (unique) completion of the path in Λ. $\square$

Let H stand for the union of the permutation digraph Π^* and $\hat{D}$. We finish our proof by proving the following.

Lemma 6. *Pr(H does not contain a Hamilton cycle) = $o(1)$.*

Proof. Let X be the number of Hamilton cycles in H resulting from rearranging the path sections generated by ϕ according to those $\rho \in R_\phi$. We will use the inequality

$$Pr(X > 0) \geq \frac{E(X)^2}{E(X^2)}. \tag{2}$$

Here probabilities are now with respect to the $\hat{D}$ choices for edges incident with vertices not in W, and on the choices of the m cut vertices.

Now the definition of the m_i gives

$$\frac{2n}{a} - k \leq m \leq \frac{2n}{a} + k,$$

so

$$(1.99) \log n \leq m \leq (2.01) \log n.$$

Also

$$k \leq m/199, m_i \gtrsim 199 \text{ and } \frac{c_i}{m_i} \geq \frac{a}{2.01} \quad 1 \leq i \leq k.$$

Let Ω denote the set of possible cycle re-arrangements. Then $\omega \in \Omega$ is a *success* if $\hat{D}$ contains the edges needed for the asssociated Hamilton cycle. Thus, where $\epsilon = O(1/n^{1/10})$,

$$
\begin{aligned}
E(X) &= \sum_{\omega \in \Omega} Pr(\omega \text{ is a success}) \\
&= \sum_{\omega \in \Omega} \left(\frac{2}{n}(1 + \epsilon) \right)^m \\
&\geq \left(\frac{2}{n}(1 + \epsilon) \right)^m (m - 2)! \prod_{i=1}^{k} \binom{c_i}{m_i}
\end{aligned}
$$

$$\geq \frac{1-o(1)}{m\sqrt{m}} \left(\frac{2m}{en}\right)^m \prod_{i=1}^k \left(\left(\frac{c_i e}{m_i^{1+(1/2m_i)}}\right)^{m_i} \left(\frac{\exp\{-m_i^2/2c_i\}}{\sqrt{2\pi}}\right)\right)$$

$$\geq \frac{(1-o(1))(2\pi)^{-m/398}}{m\sqrt{m}} \left(\frac{2m}{en}\right)^m \prod_{i=1}^k \left(\frac{c_i e}{(1.02)m_i}\right)^{m_i}$$

$$\geq \frac{(1-o(1))(2\pi)^{-m/398}}{m\sqrt{m}} \left(\frac{2m}{en}\right)^m \left(\frac{ea}{2.01 \times 1.02}\right)^m$$

$$\geq \frac{(1-o(1))(2\pi)^{-m/398}}{m\sqrt{m}} \left(\frac{3.98}{2.0502}\right)^m$$

$$\geq n^{1.3}. \tag{3}$$

Let M, M' be two sets of selected edges that have been deleted in J and whose path sections have been rearranged into Hamilton cycles according to ρ, ρ' respectively. Let N, N' be the corresponding sets of edges that have been added to make the Hamilton cycles. What is the interaction between these two Hamilton cycles?

Let $s = |M \cap M'|$ and $t = |N \cap N'|$. Now $t \leq s$, since if $(v, u) \in N \cap N'$, there must be a unique $(\tilde{v}, u) \in M \cap M'$ that is the unique J-edge into u. We claim that $t = s$ implies $t = s = m$ and $(M, \rho) = (M', \rho')$. (This is why we have restricted our attention to $\rho \in R_\phi$.) Suppose, then, that $t = s$ and $(v_i, u_i) \in M \cap M'$. Now the edge $(v_i, u_{\gamma(i)}) \in N$, and since $t = s$, this edge must also be in N'. But this implies that $(v_{\gamma(i)}, u_{\gamma(i)}) \in M'$, and hence in $M \cap M'$. Repeating the argument, we see that $(v_{\gamma^k(i)}, u_{\gamma^k(i)}) \in M \cap M'$ for all $k \geq 0$. But γ is cyclic so our claim follows.

We adopt the following notation. Let $t = 0$ denote the event that no common edges occur, and (s, t) denote $|M \cap M'| = s$ and $|N \cap N'| = t$. Thus

$$E(X^2) \leq E(X) + (1+\epsilon)^{2m} \sum_\Omega \left(\frac{2}{n}\right)^m \sum_{\substack{\Omega \\ t=0}} \left(\frac{2}{n}\right)^m$$

$$+ (1+\epsilon)^{2m} \sum_\Omega \left(\frac{2}{n}\right)^m \sum_{s=2}^m \sum_{t=1}^{s-1} \sum_{\substack{\Omega \\ (s,t)}} \left(\frac{2}{n}\right)^{m-t}$$

$$= E(X) + E_1 + E_2 \text{ say.} \tag{4}$$

Clearly,

$$E_1 \leq (1+\epsilon)^{2m} E(X)^2. \tag{5}$$

For given ρ, how many ρ' satisfy the condition (s, t)? Previously $|R_\phi| \geq (m-2)!$, and now $|R_\phi(s, t)| \leq (m - t - 1)!$ (consider fixing t edges of Γ').
Thus

$$E_2 \leq (1+\epsilon)^{2m} E(X)^2 \sum_{s=2}^m \sum_{t=1}^{s-1} \binom{s}{t} \left[\sum_{\sigma_1 + \cdots + \sigma_k = s} \prod_{i=1}^k \frac{\binom{m_i}{\sigma_i}\binom{c_i - m_i}{m_i - \sigma_i}}{\binom{c_i}{m_i}}\right] \frac{(m-t-1)!}{(m-2)!} \left(\frac{n}{2}\right)^t.$$

Now,

$$\binom{c_i - m_i}{m_i - \sigma_i} \bigg/ \binom{c_i}{m_i} \leq \binom{c_i}{m_i - \sigma_i} \bigg/ \binom{c_i}{m_i}$$

$$\leq \quad (1+o(1)) \left(\frac{m_i}{c_i}\right)^{\sigma_i} \exp\left\{-\frac{\sigma_i(\sigma_i-1)}{2m_i}\right\}$$

$$\leq \quad (1+o(1)) \left(\frac{2.01}{a}\right)^{\sigma_i} \exp\left\{-\frac{\sigma_i(\sigma_i-1)}{2m_i}\right\},$$

where the $o(1)$ term is $O((\log n)^3/n)$. Also

$$\sum_{i=1}^{k} \frac{\sigma_i^2}{2m_i} \geq \frac{s^2}{2m} \quad \text{for } \sigma_1+\cdots\sigma_k = s,$$

$$\sum_{i=1}^{k} \frac{\sigma_i}{2m_i} \leq \frac{k}{2},$$

and

$$\sum_{\sigma_1+\cdots+\sigma_k=s} \prod_{i=1}^{k} \binom{m_i}{\sigma_i} = \binom{m}{s}.$$

Hence

$$\frac{E_2}{E(X)^2} \leq (1+o(1))e^{k/2} \sum_{s=2}^{m}\sum_{t=1}^{s-1} \binom{s}{t} \exp\left\{-\frac{s^2}{2m}\right\} \left(\frac{2.01}{a}\right)^s \binom{m}{s} \frac{(m-t-1)!}{(m-2)!} \left(\frac{n}{2}\right)^t$$

$$\leq (1+o(1))n^{.01} \sum_{s=2}^{m}\sum_{t=1}^{s-1} \binom{s}{t} \exp\left\{-\frac{s^2}{2m}\right\} \left(\frac{2.01}{a}\right)^s \frac{m^{s-(t-1)}}{(s-1)!} \left(\frac{n}{2}\right)^t$$

$$= (1+o(1))n^{.01} \sum_{s=2}^{m} \left(\frac{2.01}{a}\right)^s \frac{m^s}{s!} \exp\left\{-\frac{s^2}{2m}\right\} m \sum_{t=1}^{s-1} \binom{s}{t} \left(\frac{n}{2m}\right)^t$$

$$\leq (1+o(1))\left(\frac{2m^3}{n^{.99}}\right) \sum_{s=2}^{m} \left(\frac{(2.01)n\exp\{-s/2m\}}{2a}\right)^s \frac{1}{s!} \qquad (6)$$

$$= o(1).$$

To verify that the right-hand side of (7) is $o(1)$, we can split the summation into

$$S_1 = \sum_{s=2}^{\lfloor m/4 \rfloor} \left(\frac{(2.01)n\exp\{-s/2m\}}{2a}\right)^s \frac{1}{s!}$$

and

$$S_2 = \sum_{s=\lfloor m/4 \rfloor+1}^{m} \left(\frac{(2.01)n\exp\{-s/2m\}}{2a}\right)^s \frac{1}{s!}.$$

Ignoring the term $\exp\{-s/2m\}$, we see that

$$S_1 \leq \sum_{s=2}^{\lfloor (.5025)\log n \rfloor} \frac{((1.005)\log n)^s}{s!}$$

$$= o(n^{9/10}),$$

since this latter sum is dominated by its last term.

Finally, using $\exp\{-s/2m\} < e^{-1/8}$ for $s > m/4$, we see that

$$S_2 \leq n^{(1.005)e^{-1/8}} < n^{9/10}.$$

The result follows from (2) to (7). □

Acknowledgement

We thank the referee for his/her comments.

References

[1] Bender, E. A. and Canfield, E. R. (1978) The asymptotic number of labelled graphs with given degree sequences, *Journal of Combinatorial Theory* (A) **24** 296–307.

[2] Bollobás, B. (1983) Almost all regular graphs are Hamiltonian. *European Journal of Combinatorics* **4** 97–106.

[3] Bollobás, B. (1983) *Random graphs*, Academic Press.

[4] Cooper, C. and Frieze, A. M. (to appear) *Hamilton cycles in a class of random digraphs.*

[5] Fenner, T. I. and Frieze, A. M. (1984) Hamiltonian cycles in random regular graphs, *Journal of Combinatorial Theory* B **37** 103–112.

[6] Frieze, A. M. (1988) Finding hamilton cycles in sparse random graphs, *Journal of Combinatorial Theory* B **44** 230–250.

[7] Kolchin, V. F. (1986) *Random mappings*, Optimization Software Inc., New York.

[8] Robinson, R. W. and Wormald, N. C. (1992) Almost all cubic graphs are Hamiltonian, *Random Structures and Algorithms* **3** 117–126.

[9] Robinson, R. W. and Wormald, N. C. (to appear) Almost all regular graphs are Hamiltonian, *Random Structures and Algorithms.*

On Triangle Contact Graphs[†]

HUBERT de FRAYSSEIX, PATRICE OSSONA de MENDEZ
and PIERRE ROSENSTIEHL

CNRS, EHESS, 54 Boulevard Raspail, 75006, Paris, France

It is proved that any plane graph may be represented by a triangle contact system, that is a collection of triangular disks which are disjoint except at contact points, each contact point being a node of exactly one triangle. Representations using contacts of T- or Y-shaped objects follow. Moreover, there is a one-to-one mapping between all the triangular contact representations of a maximal plane graph and all its partitions into three Schnyder trees.

1. Introduction: on graph drawing

An old problem of geometry consists of representing a simple plane graph G by means of a collection of disks in one-to-one correspondence with the vertices of G. These disks may only intersect pairwise in at most one point, the corresponding contacts representing the edges of G. The case of disks with no prescribed shape is solved by merely drawing for each vertex v a closed curve around v and cutting the edges half way. The difficulty arises when the disks have to be of a specified shape. The famous case of circular disks, solved by the Andreev–Thurston circle packing theorem [1], involves questions of numerical analysis: the coordinates of the centers and radii are not rational, and are computed by means of convergent series. This problem is still up to date, and considered in many research works. In the present paper we will consider **triangular** disks. A **contact point** (A, B) is a node of the triangle A and belongs to the side of the triangle B (but is not a node of B). The asymmetry of the pair (A, B) defines an orientation of the corresponding edge. Such an arrangement is called a **triangle contact system** (see Figure 1). It is obvious that any triangle contact system S defines an oriented simple plane graph $G(S)$. Our result is that any simple plane graph may be represented by a triangle contact system, and that these representations for a maximal plane graph are in one-to-one correspondence with the Schnyder partitions (see definition below).

[†] This work was partially supported by the ESPRIT Basic Research Action Nr. 7141 (ALCOM II).

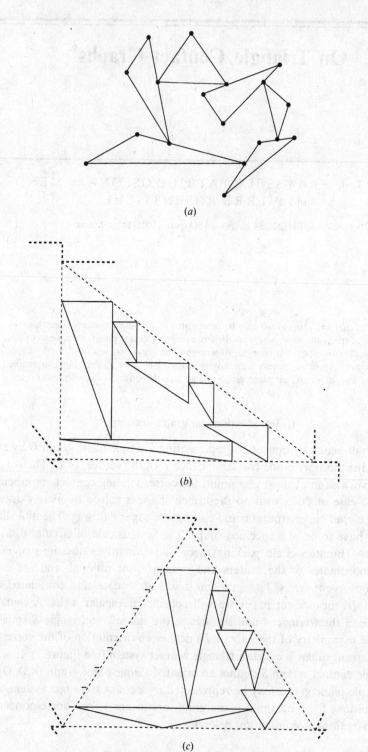

(a)

(b)

(c)

Figure 1   (a) A triangle contact system (b) A common angle representation of the system 1a
(c) An isosceles representation of the system 1a

The main tools we shall make use of are the so-called canonical or shelling order of the vertices of a maximal plane graph G introduced in [6], and a partition of the interior edges of G into three trees, due to Schnyder [9].

2. Triangle contact systems and Schnyder partitions

We consider planar objects defined as closed 2-cells. The instances of planar objects appearing below are closed triangles, segments. We also consider T-shaped objects, or Y-shaped objects as limit cases.

Definition. A **contact system** is a finite family of planar objects such that two objects of the family intersect in at most one point, and that three objects have no common point.

A consideration of the tubular neighborhoods of the objects shows that a contact system S defines a unique simple plane graph $G(S)$. Each bounded face of $G(S)$ corresponds in S to a bounded hole of the representation, the unbounded face of $G(S)$ corresponds in S to the unbounded hole. The system S is **biconnected** if $G(S)$ is biconnected.

Definition. A **triangle contact system** S is a contact system such that every object is a closed triangle, and such that each contact point is a node of exactly one triangle. A **subsystem** of S is a family of triangles of S. A **free node** is a node of a triangle which is not a contact point.

A triangle contact system S is **maximal** if and only if each hole of S is delimited by exactly three triangle sides.

Notice that the graph $G(S)$ defined by a maximal triangle contact system S is a maximal planar graph[†].

Two triangle contact systems S and S' are **isomorphic** if there exists an isomorphism mapping of the sides of the triangles of S into the sides of the triangles of S'.

Definition. A **canonical order**, or **shelling order**, of the vertices of a maximal plane graph G with external face u, v, w is a labelling of the vertices $v_1 = u, v_2 = v, v_3, ..., v_n = w$ meeting the following requirements for every $4 \leq k \leq n$:

— the subgraph $G_{k-1} \subset G$ induced by $v_1, v_2, ..., v_{k-1}$ is 2-connected, and the boundary of its exterior face is a cycle C_{k-1} containing the edge uv;
— the vertex v_k belongs to the exterior face of G_k, and its neighbors in G_{k-1} form a subinterval of the path $C_{k-1} - uv$, with at least two-elements.

It is proved in [6] that such a labelling is always possible and can be computed in linear time by packing the vertices one by one.

[†] The converse is not true, as pointed out by the referee: K_3 may be represented by a non-maximal triangle contact system

Given a biconnected triangle contact system S and a bounded hole H of S, let t_H be the number of triangles adjacent to H and let p_H be the number of free nodes belonging to the boundary of H.

Theorem 2.1. *A biconnected triangle contact system S is isomorphic to a subsystem of a maximal triangle contact system if and only if*

$$t_H - p_H = 3 \tag{1}$$

for any bounded hole H of S.

Actually, one can prove by simple counting that, if (1) holds for any bounded hole, we have on the unbounded hole H_∞:

$$t_{H_\infty} - p_{H_\infty} = -3 \tag{2}$$

In order to prove necessity, we need the following lemma.

Lemma 2.1. *Let G be a maximal planar graph and G' be a biconnected subgraph of G. Then there exists a sequence of k biconnected subgraphs of G: $G_1 = G, \ldots, G_i, \ldots, G_k = G'$ obtained at each step by deleting exactly one vertex.*

Proof. This lemma is a straightforward consequence of the shelling packing order applied to G while starting from G'. $\qquad\square$

This lemma implies that a subsystem S' of a maximal triangle contact system S can be obtained by deleting triangles T_i one by one, keeping the subsystems S_i biconnected.

Proof of theorem 2.1.   The proof of necessity is by induction on the number of removed triangles. The property holds for a maximal triangle contact system, since, for a hole H, $t_H = 3$ and $p_H = 0$. Assume the property holds when the first $i-1$ triangles are removed, and consider the removal of the ith triangle T_i from S_{i-1} and the corresponding hole H. As S_{i-1} is biconnected, each triangle adjacent to T_i in S_{i-1} is adjacent to exactly two holes adjacent to T_i. Therefore, the number d of holes adjacent to T_i equals the number of triangles adjacent to T_i, that is the degree of T_i. Let $H_1, \ldots, H_j, \ldots, H_d$ denote these holes. We have

$$t_H - p_H = \left(\sum_{j=1}^{d} t_{H_j} - 2d\right) - \left(\sum_{j=1}^{d} p_{H_j} + d - 3\right)$$
$$= \sum_{j=1}^{d}(t_{H_j} - p_{H_j}) - 3d + 3,$$

that is $t_H - p_H = 3$. Thus, (1) holds for H.

The condition is sufficient. Let S be a biconnected triangle contact system satisfying condition (1), H a bounded hole of S, and let t_H^i denote the number of triangles adjacent to H and having i free nodes in the boundary of H. We shall prove that condition (1) allows us to fill up H with triangles. There exists a triangle T_0 adjacent to H and without free nodes in H. If the hole H is not yet filled up, T_0 can be chosen in such a way that it is adjacent to a triangle T_1 having at least one free node in H. We consider three cases:

— T_1 has three free nodes in H. Add a triangle T in contact with T_0 and T_1, in the position displayed in Figure 2a; the number t_H^3 decreases.
— T_1 has two free nodes in H. Add a triangle T in contact with T_0 and T_1 in the position displayed in Figure 2b; t_H^2 decreases and t_H^3 remains unchanged.
— T_1 has one free node in H. Add a curve-sided triangle T in contact with T_0 and T_1 in the position displayed in Figure 2c; t_H decreases, while t_H^2 and t_H^3 remain unchanged.

At each step the contact system remains biconnected, and as $t_H^2 + t_H^3 + t_H$ decreases, the iteration stops with $t_H = 3$. Equation (2) allows us to fill up the unbounded hole H_∞ by a similar process. The final system obtained can be redrawn as a maximal triangle system (see Section 3), as required. $\qquad\qquad\qquad\qquad\qquad\qquad\qquad\qquad\qquad\qquad\qquad\qquad$ $\square$

One more definition given in [9]:

Definition. A **Schnyder realizer** of a maximal plane graph is a partition of the interior edges of G in three sets Y_r, Y_g, Y_b of directed edges such that for each interior vertex v

— v has indegree one in each of Y_r, Y_g, Y_b,
— the counterclockwise order of the edges incident on v is: entering in T_r, leaving in T_b, entering in T_g, leaving in T_r, entering in T_b, leaving in T_g.

The first condition of the definition implies that Y_r, Y_g and Y_b are three trees oriented from their roots. Schnyder proved in [9] that any maximal plane graph has a realizer.

Any Schnyder realizer may be extended into a **Schnyder partition** by assigning the edges of the exterior face to the three trees in such a way that all the edges of the graph are partitioned into three trees.

In the following, O_r, O_g and O_b will denote the partial orders on $V(G)$ induced by the three oriented trees of a Schnyder realizer, and $\overline{O_r}, \overline{O_g}$ and $\overline{O_b}$ will denote the reversed partial orders.

Now we relate triangle contact systems and Schnyder partitions.

Theorem 2.2. *A maximal triangle contact system S defines a Schnyder partition of $G(S)$.*

For the proof of this theorem we need a lemma. Given a maximal triangle contact system S, let v be a non-free node of a triangle T, belonging to the side of a triangle T', and let f be the mapping that associates with v the node of T' opposite to v (see Figure 3).

Lemma 2.2. *The mapping f is acyclic.*

Proof. We prove a stronger result: there exists no elementary cycle of nodes $N_1, \ldots, N_k$ such that $N_{i+1} = f(N_i)$ and such that N_1 and N_k belong to a common triangle. Such a cycle of k nodes, if it exists, defines a cycle of k triangles. The deletion of the triangles of

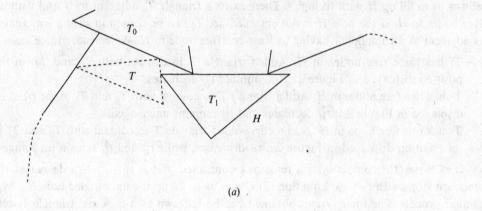

(a)

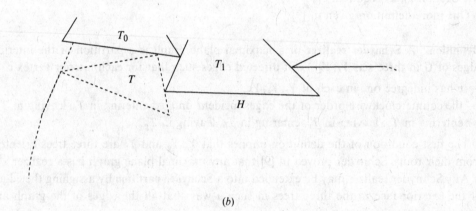

(b)

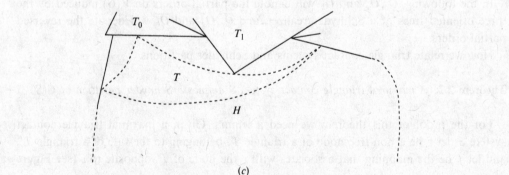

(c)

Figure 2   How to fill up the hole H

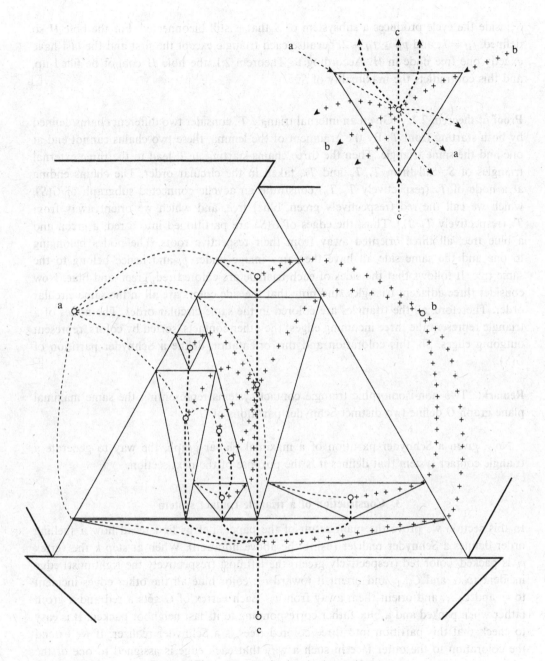

Figure 3 The contact graph edges in three colours

S inside the cycle produces a subsystem of S that is still biconnected. For the hole H so defined, $t_H = k$, and $p_H \geq t_H - 2$, because each triangle except the first and the last have exactly one free node in H. According to Theorem 2.1, the hole H cannot be filled up, and this contradicts the maximality of S.   □

Proof of theorem 2.2.   Given an internal triangle T, consider two different chains defined by both starting from T. By the argument of the lemma, these two chains cannot end at one and the same triangle. Then the three chains starting at T lead to the three external triangles of S; call them T_r, T_g and T_b, taken in the circular order. The chains ending at a node of T_r (respectively T_g, T_b) constitute an acyclic connected subgraph of $G(S)$, which we call the red (respectively green, blue) tree, and which we orient away from T_r (respectively T_g, T_b). Thus, the edges of $G(S)$ are partitioned into a red, a green and a blue tree, all three oriented away from their respective roots. The nodes belonging to one and the same side all have the same image under f, and hence belong to the same tree. It follows that the sides of each triangle are colored red, green and blue. Now consider three adjacent triangles, and note that the side colors are all in the same circular order. Therefore, all the triangles are colored in the same circular order. The nodes of a triangle represent the three incoming edges; the other contacts, sorted by colors, represent outgoing edges. So, this coloration and this orientation define a Schnyder partition of $G(S)$.   □

Remark.   Two non-isomorphic triangle contact systems representing the same maximal plane graph G define two distinct Schnyder partitions of G.

Now, given a Schnyder partition of a maximal planar graph, the way to generate a triangle contact system that defines it is the purpose of the next section.

3. Construction of a triangle contact system

In this section we prove the main result of the paper. Let us first recall how a shelling order defines a Schnyder realizer (using the above notation). When at step k the vertex v_k is packed, color red (respectively green) the leftmost (respectively the rightmost) edge incident to v_k and C_{k-1}, and orient it toward v_k; color blue all the other edges incident to v_k and C_{k-1}, and orient them away from v_k. Each vertex of G gets a red and a green father when packed and a blue father corresponding to its last neighbor packed. It is easy to check that this partition into three colored trees is a Schnyder realizer. If we extend the coloration to the outer face in such a way that each edge is assigned to one of the two trees to which it is incident, we get a Schnyder partition.

We shall first prove a representation result for maximal plane graphs. The general case (Theorem 3.2) will follow immediately.

Theorem 3.1.   *Any maximal plane graph G has a triangle contact representation.*

Proof.   Consider a shelling order and its corresponding Schnyder realizer. In the following,

T_k will denote the triangle representing the vertex v_k and $\phi_r(k)$ (respectively $\phi_g(k), \phi_b(k)$) the shelling label of the red (respectively green and blue) father of v_k in the shelling order.

We start with a maximal triangle contact system formed by the triangles T_1, T_2 and T_n, these triangles having their bases parallel to the x-axis, at ordinates $1, 2$ and n, respectively.

We construct iteratively the representations of the graphs G_k $(2 < k < n)$. Each triangle T_k $(2 < k < n)$ gets its blue base parallel to the x-axis at ordinate k and its opposite node at ordinate $\phi_b(k)$.

At each step, the triangles bounding the representation of G_k will correspond, in the same circular order, to the vertices of C_k, and the intersection of the unbounded hole with the half-plane $y \geq k$ is y-convex (the intersectection with any vertical line is connected).

Assume that the representation of G_{k-1} has been completed according to the previous constraints. By definition of shelling orders, the neighbors $v_{k_1}, \ldots, v_{k_p}$ of v_k in G_{k-1} form an interval of C_{k-1}. As v_k is the blue father of the vertices $v_{k_i}(1 < i < p)$, the free nodes of the corresponding triangles have ordinate k. The vertices v_{k_1} and v_{k_p} are, respectively, the red and green fathers of v_k. As noticed above, their blue fathers are packed after v_k, and have a label greater than k. Let x_r (respectively x_g) denote the abscissa of the point of the right (respectively left) side of triangle T_{k_1} (respectively T_{k_p}) at ordinate k. According to the y-convexity of the intersection of the unbounded hole and the half plane $y \geq k - 1$, the region defined by $y > k$ and $x_r \leq x \leq x_g$ is empty. The triangle T_k is placed (crossing free) with coordinates $(x_r, k), (x_b, k), (\alpha_k x_r + (1 - \alpha_k)x_b, \phi_b(k))$ with $\alpha_k \in [0, 1]$. The two conditions on the representation are obviously preserved for G_k.

The representation of G_n is obtained by adding the already defined triangle T_n to the representation of G_{n-1}. $\qquad\square$

Remark. We may require the triangles to be isosceles (or right-angle) by a proper choice of T_1, T_2 and T_n and the assignment of the value $\frac{1}{2}$ (respectively 0) to all the α_k coefficients. It is easy to check that there are graphs that cannot be represented by a contact system of equilateral triangles.

Theorem 3.2. *Any plane graph G has a triangle contact representation.*

Proof. The graph G can be augmented into a maximal plane graph G' by adding vertices and edges incident to the added vertices. Then a representation of G follows from a representation of G' by deleting the triangles corresponding to the added vertices. $\qquad\square$

Proposition 3.1. *A triangle contact representation of a plane graph G can be computed in $O(n^{2+\epsilon})$ time, with any given $\epsilon > 0$.*

Proof. The graph G can be augmented in linear time into a maximal plane graph G' by adding at most 2 vertices per face. The graph G' and a shelling order of its vertices can be computed in linear time. By choosing a right-angle or isosceles triangle representation, each coordinate needs linear precision. As shown by D. Knuth, each intersection computation may then be achieved in $O(n^{1+\epsilon})$ time. $\qquad\square$

In a representation, the ratio between the largest and the smallest triangle may be exponential, and for some graphs this is unavoidable. In the following, we shall describe another type of contact representation on an $n \times n$ grid.

4. Other representations

From the triangle contact representation of a maximal plane graph G, say the isosceles one, we now deduce another representation.

We construct a T-contact system (contact system of T-shaped objects): for each isosceles triangle T draw the perpendicular height corresponding to its horizontal base and, if T is neither T_1, T_2 nor T_n, extend the base of T on both sides, until a contact with the perpendicular height of its red and green fathers is reached (see Figure 4a).

This representation does not require a triangle contact representation to be achieved. Actually, any x-coordinates of the vertical segments compatible with the shelling order and any y-coordinates of the horizontal segments compatible with $O_r \cap \overline{O_g}$ lead to a T-contact representation (see Figure 4b). As a linear extension of a partial order may be computed in linear time, we have the following theorem.

Theorem 4.1. *Any plane graph may be represented by a T-contact system on a $n \times n$ grid in linear time.* □

From an isosceles triangle representation of a maximal plane graph G, one can deduce a representation of G by a contact system of Y-shaped objects (see Figure 5a). Such a representation can be performed directly by using a procedure similar to the one described for triangles, and we can require that the Y-shaped objects be composed of segments belonging to three fixed directions, as shown in Figure 5b.

From an isosceles triangle representation of a maximal plane graph, one can also derive a tessellation and a rectilinear representation of G (see Figure 6).

5. Final remarks

We have shown that two non-isomorphic triangle contact systems representing the same maximal plane graph G define two distinct Schnyder partitions of G. Moreover, we have shown that a Schnyder realizer obtained by a shelling order allows one to construct a maximal triangle contact system with which the realizer is associated. Actually, any Schnyder realizer may be defined by a shelling order. The following theorem is proved in [3].

Theorem 5.1. *Given a Schnyder realizer (Y_r, Y_g, Y_b) of a maximal plane graph G, a total order of $V(G)$ is a shelling order defining (Y_r, Y_g, Y_b) if, and only if, it is a linear extension of $O_r \cap \overline{O_g} \cap O_b$.*

Because of the different possible constructions of the triangles T_1, T_2 and T_n in the previously described algorithm, we have the following theorem.

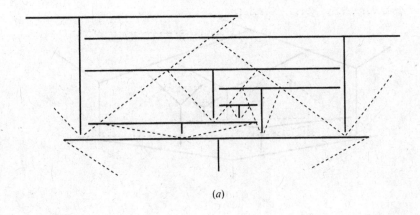

(a)

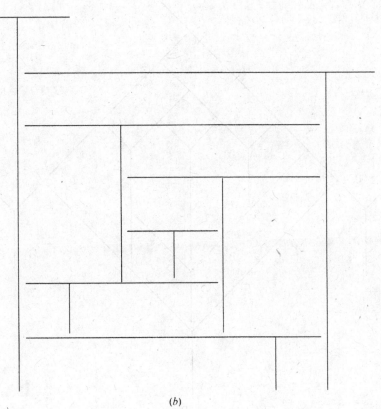

(b)

Figure 4 (*a*) T-contact (*b*) T-contact system on the nxn grid

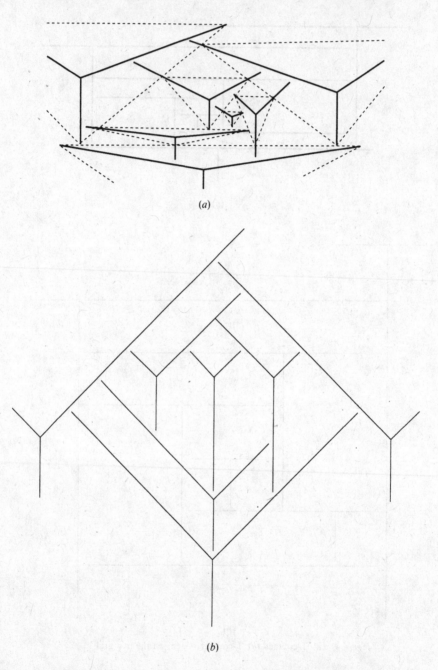

(a)

(b)

Figure 5   (a) Y-contact system representation (b) Y-contact system with 3 fixed directions on the grid

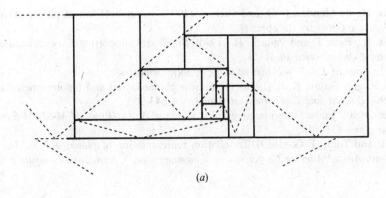

(a)

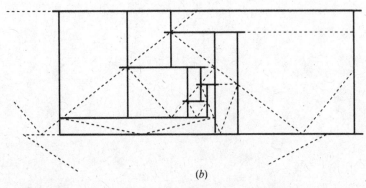

(b)

Figure 6 (a) Tessellation representation (horizontal segments are vertices, vertical segments are faces and rectangles are edges). (b) Rectilinear representation (horizontal segments are vertices, vertical segments are edges)

Theorem 5.2. *The non-isomorphic triangle contact systems representing a maximal plane graph G are in one-to-one correspondence with the Schnyder partitions of G.*

Acknowledgments

We thank the referee and A. Machì for their great attention and many useful suggestions.

References

[1] Andreev, E. M. (1970) On convex polyhedra in Lobacevskii spaces. *Mat. Sb.* **81** 445–478.

[2] Di Battista, G., Eades, P., Tamassia, R. and Tollis, I. G. (1989) Algorithms for drawing planar graphs: an annotated bibliography. Tech. Rep. No. CS-89-09, Brown University, 1989.

[3] de Fraysseix, H. and de Mendez, P. O. (In preparation) *On tree decompositions and angle marking of planar graphs.*

[4] de Fraysseix, H., de Mendez, P. O. and Pach, J. (submitted) *A streamlined depth-first search algorithm revisited.*

[5] de Fraysseix, H., de Mendez, P. O. and Pach, J. (1993) Representation of planar graphs by segments. *Intuitive Geometry* (to appear).

[6] de Fraysseix, H., Pach, J. and Pollack, R. (1990) Small sets supporting Fary embeddings of planar graphs. *Combinatorica* **10** 41–51.

[7] B. Mohar (To appear) *Circle packings of maps in polynomial time.*

[8] Rosenstiehl, P., and Tarjan, R. E. (1986) Rectilinear planar layout and bipolar orientation of planar graphs. *Discrete and Computational Geometry* **1** 343–353.

[9] W. Schnyder (1990) Embedding planar graphs on the grid. In: *Proc. ACM-SIAM Symp. on Discrete Algorithms* 138–148.

[10] Tamassia, R. and Tollis, I. G. (1989) Tessellation representation of planar graphs. In: *Proc. Twenty-Seventh Annual Allerton Conference on Communication, Control, and Computing* 48–57.

A Combinatorial Approach to Complexity Theory via Ordinal Hierarchies

WALTER A. DEUBER and WOLFGANG THUMSER

University of Bielefeld, Faukultät Mathematik, Postfach 10 01 31 33501 Bielefeld 1, Germany

Long regressive sequences in well-quasi-ordered sets contain ascending subsequences of length n. The complexity of the corresponding function $H(n)$ is studied in the Grzegorczyk–Wainer hierarchy. An extension to regressive canonical colourings is indicated.

1. Introduction

For many mathematicians the most noble activity lies in proving theorems. It must have come as a blow for them when Gödel [7] showed that there are unprovable theorems. At the beginning they still could find some consolation in hoping that such culprits might only occur in Peano arithmetics through esoteric diagonalization arguments. Nowadays there is a wealth of the most natural valid theorems that can be stated in the language of finite combinatorics but are not provable within that system.

Mathematicians understand to a certain extent how to find unprovable theorems and how to prove their unprovability within a formal system. In that sense we are relying on the classical work by Gentzen [5], Kreisel [15] and Wainer [31]. Moreover, we shall apply their beautiful ideas to something that seems to be well understood, *viz* to well-quasi-orderings. This is an old concept found in Gordan [6], and Kruskal [16] correctly pointed out that it was 'a frequently discovered concept'. That is why we are not reinventing it and are well aware that any sequence (s_i) of specialists starting with the author must contain an arbitrary long subsequence of experts knowing more than s_0, a fact, which gives a nice theme for this paper. Leeb was one of the first to deal with structural problems of *wqo's*, which are related to this paper [18]. Some beautiful ideas of P. Erdős are valuable for the analysis of such phenomena occurring in all well-quasi-orders. For related combinatorial questions we would also like to draw the reader's attention to the beautiful paper of Nešetřil and Loebl [19].

2. How to use complexity theory

We are interested in first order statements $\forall x \exists y A(x, y)$ in the language of Peano arithmetics where A is primitive recursive. Let $g(x)$ be the smallest y satisfying $A(x, y)$. We are interested in the question of whether g is defined for every x. Let us anticipate the answer, which has been known for a long time: if g grows fast enough, the statement 'g is defined everywhere' is not provable within Peano arithmetics.

In order to specify growth rates in complexity theory, we define a hierarchy of reference functions, There are various hierarchies available and, depending on the combinatorial problems and personal taste, one can make a choice. Here we concentrate on the Wainer–Grzegorcyzk hierarchy, *cf.* [8] and [31].

The first few functions are defined as follows:

$$f_0(n) = n+1$$

$$f_{i+1}(n) = f_i \circ \ldots \circ f_i(n), \text{ where the iteration is } n \text{ fold, and finally}$$

$$f_\omega(n) = f_n(n) \text{ is the Ackermann function defined by diagonalization.}$$

The first few levels are well known: f_0 grows like the identity, f_1 linearly, f_2 exponentially, f_3 is the tower function; f_4 is sometimes called the 'wow'-function [9], ...

Using the Cantor Normal form to define fundamental sequences representing ordinals there is no difficulty extending the hierarchy up to f_{ϵ_0}, for instance

$$f_{\omega+i+1}(n) := f_{\omega+i} \circ \ldots \circ f_{\omega+i}(n) \quad n\text{-times}$$

$$f_{\omega+\omega}(n) := f_{\omega+n}(n) \quad \text{diagonalization}$$

$$f_{\omega^\omega}(n) := f_{\omega^n}(n) \quad \text{diagonalization}$$

$$f_{\epsilon_0}(n) := f_{\omega^{\cdot^{\cdot^\omega}}}(n) \quad \text{diagonalization with the } \omega\text{-tower of height } n.$$

For details, see [31].

One can measure complexity with respect to these reference functions by defining

$$(*) \quad \begin{cases} g > h & \text{iff } \lim_{n \to \infty} \dfrac{h(n)}{g(n)} = 0 \\ \text{and} \quad g \sim f_\alpha & \text{iff } \alpha \text{ is the smallest ordinal with } f_{\alpha+1} > g. \end{cases}$$

One should be aware that this complexity measure is fairly insensitive to small changes but, as we shall see, it will allow rather clean-cut statements on combinatorial complexity.

Theorem (Kreisel). *Let $A(x, y)$ be a primitive recursive formula in the language of Peano arithmetics and $g(x)$ be the smallest witness y for $A(x, y)$. If $g > f_{\epsilon_0}$ or $g \sim f_{\epsilon_0}$ then 'g is defined for all x' is not provable in Peano arithmetic.*

This theorem demonstrates that it might be useful to understand complexity theory with respect to such hierarchies. From the point of view of nonprovability in Peano arithmetic, only certain reference functions such as f_{ϵ_0} are of interest, but we shall see that the other levels of complexity occur in rather natural contexts too. Here we concentrate on surveying some of these results and give examples for combinatorial problems which correspond to various levels.

3. Regressive sequences in wqos

Recall that a *well-quasi-ordering* is a poset $(A, \leqslant)$ that contains no infinite antichains and no infinite strictly descending sequence; thus any infinite sequence of elements of A must contain an infinite weakly ascending subsequence.

Let $(A, \leqslant)$ be a *wqo*-set with an obvious ranking r defined by successively taking minimal elements. Call a sequence $(a_0, a_1, ...)$ **regressive** iff $r(a_i) \leqslant i$ for every $i \in \omega$.

Theorem 3.1. *Let $(A, \leqslant)$ be wqo. Then there exists a function $H_{(A, \leqslant)} : \omega \to \omega$ such that every regressive sequence $(a_0, ..., a_{H(n)})$ contains a weakly ascending subsequence with n terms.*

Harzheim proved this for $(\mathbb{N}, \leqslant)$ [10] and $(\mathbb{N}^d, \leqslant)$ [11]. The general version might be folklore. The following proof should be known to all specialists. It came to the authors mind when teaching on fixed point theorems in compact spaces.

Proof. Consider the space S of regressive ω-sequences over A. Finite sequences should be filled up with minimal elements. Thus with

$$R_i = \{x \in A \mid r(x) \leqslant i\} \text{ one has } S = \Pi_\omega R_i.$$

As a product of the finite sets R_i, the space S is compact in the Tychonoff topology, and as a metric space it is also sequentially compact.

Assuming that the theorem fails, pick a *wqo* set $(A, \leqslant)$ and an $n \in \omega$ such that for every $h \in \omega$ there exists a regressive 'bad' sequence $a(h) = (a_0^{(h)}, ..., a_h^{(h)})$, i.e., an h-term sequence not containing any n term ascending subsequence. Thus the sequence $(a^{(h)})_{h \in \omega}$ has an accumulation point $a \in S$. As A is *wqo*, it follows that a must contain an infinite weakly ascending subsequence, so it contains a weakly ascending subsequence a' with n terms. Of course a' is contained in an initial segment of a, the accumulation point. Thus it is contained in an initial segment of some $a^{(h)}$, yielding the desired contradiction.   □

Of course one could also use König's infinity lemma for a proof. We do not know whether the theorem can be generalized. To start with, finite sets R_i, in order to have a compact S, does not seem to be the most general idea [21].

In this paper we are going to explore the complexity of $H_{(A, \leqslant)}$ for various posets $(A, \leqslant)$. For some of the most natural and commonly occurring wqo's the Wainer–Grzegorczyk hierarchy seems to be quite adequate for neat results.

4. Low complexity levels, product of chains

Harzheim [10] established the following

Theorem 4.1. $H_{(\mathbb{N}, \leqslant)}(n) = 2^{n-1}$. Thus $H_{(\mathbb{N}, \leqslant)} \sim f_1$.

Proof. In order to establish the result, we proceed in the framework of complexity theory and show

i the upper bound $H(n) \leqslant 2^{n-1}$

ii the lower bound $H(n) \geqslant 2^{n-1}$.

For (i) we make use of some beautiful ideas from [4].

Let (a_i) $i = 1, ..., H(n)$ be a regressive sequence of positive integers. So far, $H(n)$ is unknown and we want to show $2^{n-1} \geqslant H(n)$. Define a mapping

$$\ell : \{1, ..., 2^{n-1}\} \to \{1, ..., 2^{n-1}\} \quad \text{by}$$

$$\ell(i) = \begin{cases} \text{length of a weakly ascending sequence} \\ \text{of maximal length with first element in } a_i. \end{cases}$$

Case α. There is an i with $\ell(i) \geqslant n$.

Obviously this shows that an ascending subsequence of length n exists.

Case β. $\ell(i) < n$ for all i.

Thus ℓ may be viewed as a colouring of $\{1, ..., 2^{n-1}\}$ with at most $n-1$ colours.

Definition. A subset X of $\mathbb{N}$ is called **large** iff $|X| > \min X$.

By the pigeon-hole principle, there is a large subset $X = \{i_1, ..., i_{i_1+1}\} \subseteq \{1, ..., 2^{n-1}\}$ that is monochromatic for a certain ℓ. By definition, each of the elements

$$a_{i_1}, ..., a_{i_{i_1+1}} \tag{*}$$

is the starting point for a weakly ascending subsequence of maximal length ℓ. Therefore (*) has to be a strongly descending sequence. (In order to see $a_{i_1} > a_{i_2}$, suppose that, on the contrary $a_{i_1} \leqslant a_{i_2}$. Then a longest sequence starting at a_{i_2} could be extended by a_{i_1} yielding a longest ascending sequence of length $\ell + 1$. The length of (*) is $i_1 + 1$ and its first element has rank $\leqslant i_1$, which gives a contradiction.

It remains to show that 2^{n-1} is such that it allows the application of the pigeon-hole principle. Colour $\{1, ..., a-1\}$ with $n-1$ colours in such a way that no large subset occurs monochromatically. Observe that $f : \{1, ..., 7\} \to \{1, 2, 3\}$, defined by

$$\begin{pmatrix} 1 & 2 & 3 & 4 & 5 & 6 & 7 \\ f_1 & f_2 & f_2 & f_3 & f_3 & f_3 & f_3 \end{pmatrix},$$

is a colouring such that every extension to $8 = 2^3$ would yield either a large set or need a new colour. It is easy to see that any colouring of $\{1, ..., 7\}$ in which the colours do not occur successively either already contains a large set or can be rearranged to the above example, showing that the greedy strategy yields $a = 2^{n-1}$ as an upper bound for $H(n)$.

As for the lower bound (ii), Figure 1 gives an explicit regressive sequence a of length $2^{n-1} - 1$ without weakly ascending subsequence of length $n : a = (0103210)$. $\qquad \square$

Another possibility for establishing upper bounds, which turned out to be useful in more complicated situations, employs a tree argument: a beautiful idea occurring in [1].

$n = 4$:

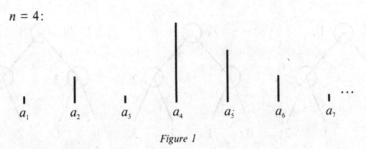

$$a_1 \qquad a_2 \qquad a_3 \qquad a_4 \qquad a_5 \qquad a_6 \qquad a_7$$

Figure 1

Given a regressive sequence a defined on the first few, say, a, integers, we recursively construct a sequence of binary trees $T_1, \ldots, T_j, \ldots, T_a$, in which
— the internal nodes are labelled by $1, \ldots, j$
— the leaves are unlabelled
— the pendant edges (those going into leaves) are labelled by $0, \ldots, j$.
The construction is initiated by Figure 2.

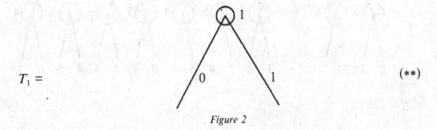

$$T_1 = \qquad\qquad\qquad 0 \qquad\qquad 1 \qquad\qquad\qquad\qquad (**)$$

Figure 2

Given T_j with internal nodes $1, \ldots, j$ and pendant edges labelled $0, \ldots, j$, (*cf.* (**)), define T_{j+1} as follows: as a is regressive, we know that $a(j+1) \leqslant j$.

Thus there is a pendant edge labelled with $a(j+1)$. The corresponding leaf in T_j now becomes an internal node of T_{j+1} labelled $j+1$. Moreover, two new pendant edges are attached to it and labelled by $a(j+1)$, $a(j+1)+1$. The other pendant edges of T_j are kept as such in T_{j+1}, those with labels $< a(j+1)$ going unchanged and for those with labels $> a(j+1)$ the labelling being increased by one. It is immediate from the construction that the function a is always increasing along the paths of internal nodes. If the size of the tree becomes larger than $2^{n-1} - 1$, we cannot help avoiding increasing subsequences of size at least n, which proves the upper bound. The example $a = (0103210)$ of Figure 1 leads to Figure 3.

The complete binary tree of depth $n-1$ may be obtained from the example for the lower bound, which shows that $H(n) \geqslant 2^{n-1}$.

Needless to say, in the simple situation of Harzheim's result, our efforts for proving upper and lower bounds by rather sophisticated looking methods may give an overloaded impression. To us it seems to be the simplest approach (*cf.* [29]), and moreover, it is generalizable to more tricky situations (see Section 6). The general case for products of chains was given by [11].

Theorem 4.2. $H_{\mathbb{N}^d} \sim f_{d-1}$ *for* $d \geqslant 2$.

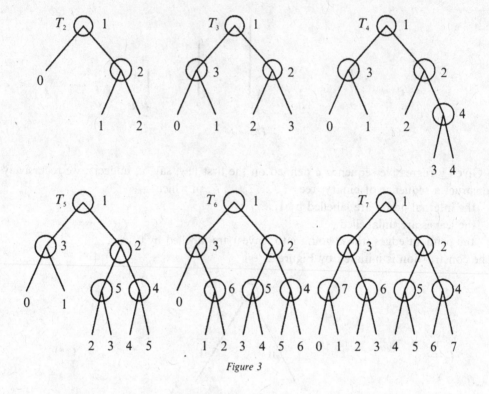

Figure 3

5. The intermediate levels, Higman's theorem for finite alphabets

One of the classical results in wqo theory is Higman's theorem:

If $(X \leqslant)$ is wqo, then $Hig(X, \leqslant)$, the set of finite words over X endowed with embeddability into subwords, is wqo.

We shall indicate the complexity of the corresponding *H*-functions. In doing so, we observed a proof for Higman's theorem, which is as constructive as possible and, astonishingly, avoids minimal bad sequences. The proof makes use of some early observations of [13] on finite sets, but apart from that, should be folklore to the specialists.

The crucial phenomenon is best observed by taking $t = \{1 \leqslant, ..., \leqslant t\}$ to be a finite, linearly ordered alphabet. Let $\vec{a}, \vec{b} \in Hig\, t$. If $\vec{a} \not\leqslant_{Hig} \vec{b}$, then $\vec{b}$ has a certain structure imposed by $\vec{a}$. In order to appreciate the idea, let $t = 10$, $\vec{a} = (2, 10, 7)$ and $\vec{b} = (b_1, ..., b_n)$.

Case 1. $a_1 = 2 \not\leqslant b_i$ for all i (the embedding of $\vec{a}$ fails already in the first place). Then
$\vec{b} \in Hig(a_1 - 1)$.

Case 2. Let i_1 be minimal with $b_{i_1} \geqslant 2$. Thus $\vec{b} = (b_1 ... b_{i_1-1}) * b_{i_1} * (b_{i_1+1} ... b_n)$ with
$(b_1 ... b_{i_1-1}) \in Hig(a_1 - 1)$, $b_{i_1} \geqslant a_1$ and $(b_{i_1+1} ... b_n) \in Hig\, t$ is such that
$(a_2, a_3) \not\leqslant_{Hig} (b_{i_1+1}, ..., b_n)$.

By iteration one obtains the following general result.

Definition. Let X be a poset and $a \in X$. Then $[a)$ is the principal filter $\{z \mid a \leqslant z\}$ generated by a, and $X \backslash [a)$ is the complement of the principal filter $[a)$.

Fact. *Let $(X \leqslant)$ be a poset, $\vec{a} = (a_1 \ldots a_n)$, $\vec{b} = (b_1 \ldots b_m) \in \mathrm{Hig}(X)$ and $\vec{a} \nleqslant_{\mathrm{Hig}} \vec{b}$. Then there exist $\ell < n$, $\vec{b}_0 \in \mathrm{Hig}(X \backslash [a_1]), \ldots, \vec{b}_{\ell+1} \in \mathrm{Hig}(X \backslash [a_{\ell+1}])$ and elements $b_1^* \in [a_1], \ldots, b_\ell^* \in [a_\ell]$ with*

$$\vec{b} = \vec{b}_0 \times b_1^* \times \vec{b}_1 \times b_2^* \ldots \times \vec{b}_{\ell+1}.$$

Basically, the fact says: if $\vec{a} \nleqslant_{\mathrm{Hig}} \vec{b}$, then $\vec{b}$ is contained in a product whose factors are of the form $\mathrm{Hig}(X \backslash [a_i])$ or principal filters. Moreover, the length of all these products has an upper bound depending on the length of $\vec{a}$.

Theorem 5.1. (Higman) *If $(X \leqslant)$ is wqo, then $\mathrm{Hig}(X, \leqslant)$ is wqo.*

Proof. The theorem holds for $X = \varnothing$. So assume that the theorem holds for all complements of principal filters $X \backslash [a]$ of some X. We will show that it holds for X. As such an induction works for wqo's, the theorem follows.

So, let $\vec{a}_0, \vec{a}_1, \vec{a}_2 \ldots$ be a sequence of elements of $\mathrm{Hig} X$. Assume $\vec{a}_0 \nleqslant \vec{a}_i$ for all i (i.e. the greedy approach to show that the sequence is 'good' fails), then each $\vec{a}_i$, $i = 1, 2, \ldots$ has a structure as given by the fact. Thus there exists an ℓ such that for infinitely many i

$$\vec{a}_i \in \prod^{\ell+1} (\text{complement of principal ultrafilter}) \times \prod X.$$

The induction hypothesis and the product lemma [12] imply that there is a weakly increasing subsequence of $\vec{a}_i$s. Thus X is wqo. $\qquad\qquad\square$

The proof looks quite constructive at first sight. A careful analysis reveals that for a general poset X the proof is not at all constructive. Nevertheless, for special Xs it is strong enough that with some additional work [30] one can obtain the following theorem.

Theorem 5.2. *Let $t < \omega$. Then $H_{\mathrm{Hig}(t)} \simeq f_{\omega^{t-1}}$.*

Remark. In the framework of regressive sequences the problem asking for $H_{\mathrm{Hig}(\omega)}$ seems to be ill posed, as the rank function according to our definition ($r(a) = lgth(a)$) does not make sense. it is imaginable that for adequate definitions reasonable results could be obtained for $H_{\mathrm{Hig}(\omega)}$ and beyond. For well-posed but somewhat artificial modifications of this problem [24].

6. The upper levels, the ω-towers

Kanamori–McAloon [14] gave a model theoretic proof for the unprovability of a theorem on regressive colourings of k-element sets. Here we shall analyze the corresponding complexity questions. By doing so we shall explain how 'canonical Ramsey theory', 'large sets' in the sense of Paris and Harrington [22] and 'tree arguments' can be applied in order to obtain sharp complexity results. These are related to the results of [2].

Before generalizing the concept of regressive sequences, we indicate as a combinatorial tool the Erdős–Rado canonization lemma [3]. We need

Definition. Let $n, k \in \omega \cup \{\omega\}$. $\binom{\{1, \ldots, n\}}{k} = \binom{n}{k}$ denotes the set of all k element (respectively infinite) subsets of n. Let $X = \{x_0, \ldots, x_{k-1}\}_<$, $Y = \{y_0, \ldots, y_{k-1}\}_<$ be k-element subsets of M, and I be a subset of $\{0, \ldots, (k-1)\}$. Let $X:I = \{x_i \in X / i \in I\}_<$. Thus we have

$$X:I = Y:I \quad \text{iff} \quad x_i = y_i \quad \text{for all} \quad i \in I.$$

The countable case of the canonical version of Ramsey's theorem can be stated as follows.

Lemma. [3] *Let* $k \in \omega$ *be fixed and* $\Delta : \binom{\omega}{k} \to \omega$ *be a colouring into the natural numbers. Then there exist* $I \subseteq \{0, \ldots, (k-1)\}$ *and an infinite subset* $M \in \binom{\omega}{\omega}$ *of natural numbers such that for all* $X, Y \in \binom{M}{k}$ *the relation*

$$X:I = Y:I \quad \text{holds iff} \quad \Delta(X) = \Delta(Y).$$

Example. In the special case where $k = 2$ the theorem assures the existence of an infinite set such that the restricted colouring is
— constant

$$\Delta(X) = \Delta(Y) \quad \text{for all} \quad X, Y \in M \quad (I = \varnothing),$$

— or injective

$$\Delta(X) = \Delta(Y) \quad \text{iff} \quad X = Y \quad (I = \{0, 1\}),$$

— or depends only on minimum elements

$$\Delta(X) = \Delta(Y) \quad \text{iff} \quad \min(X) = \min(Y) \quad (I = \{0\}),$$

— or depends only on maximum elements

$$\Delta(X) = \Delta(Y) \quad \text{iff} \quad \max(X) = \max(Y) \quad (I = \{1\}).$$

In order to generalize regressive sequences we make use of the following definition

Definition. Fix n and k as above. A colouring $\Delta : \binom{n}{k} \to \omega$ is called **min-regressive** if $\Delta(x) < \min X$ for all $X \in \binom{n}{k}$. For $M \subseteq n$ we call a colouring $\Delta : \binom{M}{k} \to \omega$ **min-homogeneous** if

$$\min X = \min Y \quad \text{implies} \quad \Delta(X) = \Delta(Y) \quad \text{for all} \quad X, Y \in \binom{M}{k}.$$

Note that for $k = 1$ we recover the notion of a regressive sequence.

A classical example is the van der Waerden colouring, which assigns to every arithmetic progression of length k in $\{1, \ldots, n\}$ its first element diminished by one, and which assigns 0 to the other k-tuples.

The following theorem is obvious to all those familiar with canonical Ramsey theory.

Theorem 6.1. *Let $k \in \omega$ be fixed. For every m there exists a smallest $n = H_k(m)$ with the following property:*

Let $\Delta : \binom{n}{k} \to \omega$ be min-regressive. Then there exists an m-element subset M of n such that the restriction $\Delta \upharpoonright \binom{M}{k}$ is min-homogeneous.

Proof. Work with the countable version of the Erdős–Rado canonization lemma. As the colouring Δ is min-regressive the pertinent canonical cases must be min-homogeneous. Finally, apply compactness to obtain the existence of H_k. $\qquad\square$

A natural problem is the analysis of the complexity of H_k. Here we rely on [25], [24] and [29].

Theorem 6.2. *Let $k \geqslant 2$. Then*

$$H_k \sim f_{\omega^{\cdot^{\cdot^{\cdot^{\omega}}}} \text{ tower of height } k-1}.$$

Here we will give an explicit description of the arguments showing $H_2 \sim f_\omega$, the Ackermann function. As in Section 4, the proof consists in giving
(i) a lower bound $\sim f_\omega$
(ii) an upper bound $\sim f_\omega$.
For the lower bound we need the following lemma.

Lemma 6.3. $H_2(Ram(2, m+3, k)) \geqslant f_k(m)$, *where $Ram(2, m+3, k)$ is the ordinary Ramsey number, arrowing $(m+3)^2_k$.*

Proof. Given m, k, let $m^* = Ram(2, m+3, k)$, $n^* = H_2(m^*)$. Observe that for $x < y < \omega$ there exist unique $0 \leqslant k^* < k$ and $1 \leqslant \ell < x$ satisfying

$$f_{k^*}^{(\ell)}(x) := \underbrace{f_{k^*} \circ \ldots \circ f_{k^*}(x)}_{\ell\text{-times}} \leqslant y < \underbrace{f_{k^*} \circ \ldots \circ f_{k^*}(x)}_{\ell+1\text{-times}}.$$

This is well defined as

$$f_{k^*}(x) < f_{k^*+1}(x) = f_{k^*}^{(x)}(x).$$

Now, define a regressive mapping by

$$\Delta(\{x, y\}) = \begin{cases} 0, & \text{if } f_k(x) \leqslant y, \\ \ell & \text{otherwise.} \end{cases}$$

Let $M^* \in \binom{n^*}{m^*}$ be such that Δ is min-homogeneous on M^*. We define a k-colouring $\Delta^* : \binom{M^*}{2} \to k$ by

$$\Delta^*(\{x, y\}) = \begin{cases} 0, & \text{if } f_k(x) \leqslant y, \\ k^* & \text{otherwise.} \end{cases}$$

Let $M \in \binom{M^*}{m+3}$ be such that $\Delta^* \restriction \binom{M}{2}$ is a constant colouring and let $x < y < z$ be the three largest elements of M. Then $m \leqslant x$ and, as the function f_k is increasing, it suffices to show that $f_k(x) \leqslant z$. Assume to the contrary that $f_k(x) > z > y$. Hence $f_k(y) > z$ also, as $f_k(x) \leqslant f_k(y)$. Say, $\Delta(\{x, y\}) = \Delta(\{x, z\}) = \ell$ and $\Delta(\{x, y\}) = \Delta(\{x, z\}) = \Delta(\{x, z\}) = k^*$. Then $f'_{k^*}(x) \leqslant y < z < f'^{+1}_{k^*}(x)$. Apply f_{k^*} to this inequality. Then $z < f'^{+1}_{k^*}(x) \leqslant f_{k^*}(y)$. But this contradicts $f_{k^*}(y) \leqslant z$.

Corollary 6.4. *The function $H_2(m)$ is not primitive recursive.*

Proof. As $Ram(2, m+3, k) \leqslant k^{(m+3) \cdot k}$, it is primitive recursive. But $H_2(Ram(2, m+3, m)) \geqslant f_m(m) = f_\omega(m)$ by the lemma. As the primitive recursive functions are closed under composition, the assertion follows. $\qquad\square$

For the upper bounds we have the following theorem.

Theorem 6.5. $H_2(m) < f_{m-1}(3) < f_\omega(m)$ *for all* $m \geqslant 3$.

Proof. In order to prove this theorem, we consider trees as partially ordered sets, the smallest element being the root. As in Section 4, we use a **tree argument**: For a given regressive mapping $\Delta : \binom{n}{2} \to n$, define a tree $(T_\Delta, \leqslant_T)$ on $\{2, \dots, n-1\}$ by

$$\ell <_T m \text{ iff } \Delta(\{k, \ell\}) = \Delta(\{k, m\}) \text{ for all } k \text{ with } k <_T \ell.$$

For example, the tree depicted in Figure 4 corresponds to regressive mappings $\Delta : \binom{15}{2} \to 15$ such that

$$\Delta(2, 3) = \Delta(2, 4) = \Delta(2, 5) = \Delta(2, 6),$$
$$\Delta(3, 4) = \Delta(3, 5) = \Delta(3, 6),$$
$$\Delta(2, 7) = \Delta(2, 8) = \cdots = \Delta(2, 14),$$
$$\Delta(7, 8) = \cdots = \Delta(7, 14).$$

Nothing is asserted about the remaining pairs.

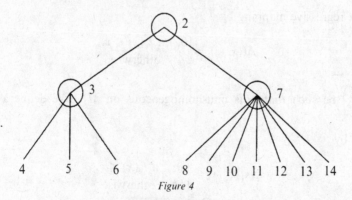

Figure 4

Mills [20] called a tree **small branching** if the successor-degree of each node i is at most i.

The following observation is trivial but useful

Observation 1. *Let* $\Delta:\binom{n}{2} \to n$ *be a regressive mapping and* $(T_\Delta, \leqslant_T)$ *be the associated tree.*

Then:

(i) $k <_T \ell$ *implies that* $k < \ell$

(ii) T_Δ *is small branching*

(iii) *every chain is min-homogeneous.*

For estimating $H_2(m)$ from the above, we ask how large n must at least be such that every small branching tree T_Δ contains a chain of length m. Denote by $M(m)$ the smallest such n.

Figure 4 shows that $M(4) > 14$, and it is easy to see that, in fact, $M(4) = 15$.

Figure 5 indicates that $M(5) > 2^{39}.41 - 2$, and again it is not difficult to see that $M(5) = 2^{39}.41 - 1$. The idea behind Figures 4 and 5 is fairly obvious. To build a large small branching tree without chains of length m, one fills in the branches from left to right by placing smaller numbers as far down on the tree as possible to save vertices higher up for larger numbers. These larger numbers then allow more immediate successors, thus making the tree as big as possible. Such trees are well known in computer science as **balanced preordered trees**.

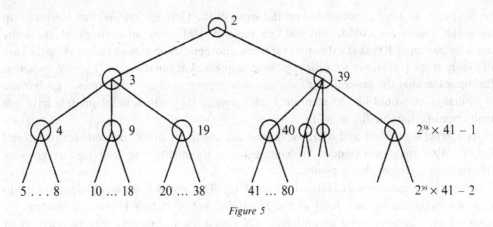

Figure 5

Lemma 5.6. *Let* $n = M(m) - 1$, *and let* T *be a small branching tree defined on* $\{2,...,n\}$ *without any m-element chains. Then* T *is a balanced preordered tree.*

The somewhat technical proof was given in [20] and [29], and a beautifully illustrated version may be found in the highly recommended forthcoming book 'Aspects of Ramsey Theory' [26].

Let $M_m(k)$ be the smallest positive integer n such that every small branching balanced preordered tree on $[k, n]$ contains an m-element chain (thus $M(m) + 1 = M_m(2)$).

Observation 2.

(1) $M_2(k) = k+1$,

(2) $M_{m+1}(k) \leqslant M_m^k(k+1)$, (k-fold iteration).

Proof. (1) is obvious and (2) follows immediately from the construction of small branching trees, cf. Figure 5. $\square$

By boolean combination of definitions, we obtain the following observation.

Observation 3. $M_m(k) \leqslant f_{m-1}(k+1) - 1$ for all $k, m \geqslant 1$.

In order to obtain the upper bound f_ω for H_2 and conclude the proof of Theorem 6.4, it suffices to combine Observations 1–3.

Remark 1. It is possible to extend these arguments and obtain a proof of Theorem 6.2 in general. For details see [29].

Remark 2. Erdős and Mills [2] gave upper bounds for the Paris–Harrington function for colouring pairs with a fixed number of colours; the Ramsey case [27]. The above results cover the canonical min-homogeneous case for pairs and k-tuples in general.

7. Outlook and problems

In this paper we have concentrated on the levels of the Grzegorczyk–Wainer hierarchy up to ϵ_0. Of course we could, and did, go beyond. [28] gives an account of the finite miniaturization of Kruskal's theorem for trees, another classic in wqo theory. For the case of binary trees, [30] shows that for regressive sequences of binary trees $H_{Bin} \sim f_{\epsilon_0}$, whereas [28] indicates that the general case for regressive sequences of arbitrary trees is far beyond f_{Γ_0}. Finally, we would like to mention Leeb's *jungles* [18], which unfortunately have not really been penetrable for us so far.

As a general problem and idea, we suggest searching for other 'natural' combinatorial features that may be extended by compactness arguments and lead to fast growing functions and unprovability results.

Closer to the extension of Harzheim's result (cf. Theorem 4.1, it would be interesting to find orders related to each level of the hierarchy. When stating Higman's theorem, we assumed the alphabet to be an antichain. Of course such alphabets may be partially or totally ordered. How does this order affect the growth of the corresponding H-functions?

References

[1] Erdős, P., Hainal, A., Máté, A. and Rado, R. (1984) Combinatorial set theory: Partition relations for cardinals. *Studies in Logic and the Foundations of Mathematics* **106**. North-Holland.

[2] Erdős, P. and Mills, G. (1981) Some Bounds for the Ramsey–Harrington Numbers. *J. of Comb. Theory*, Ser. A **30**, 53–70.

[3]   Erdős, R. and Rado, R. (1950) A combinatorial Theorem. *Journal of the London Mathematical Society* **25**, 249–255.

[4]   Erdős, P. and Szekeres, G. (1935) A combinatorial problem in geometry. *Composito Math.* **2**, 464–470.

[5]   Gentzen, G. (1936) Die Widerspruchsfreiheit der reinen Zahlentheorie. *Mathematische Annalen* **112**, 493–565.

[6]   Gordan's, P. (1885) Vorlesungen über Invariantentheorie, Hrsg. v. Geo. Kerschensteiner. 1. Bd. Determinanten (XI, 201S.). Teubner, Leibzig.

[7]   Gödel, K. (1931) Uber formal unentscheidbare Sätze der Principia Mathematica und verwandter Systeme, I. *Monatschefte für Mathematik und Physik* **38**, 173–198.

[8]   Grzegorczyk, A. (1933) Some classes of recursive functions. *Rozprawy matematiczne* **4**, Instytut Matematyczny Polskiej Akademie Nauk, Warsaw.

[9]   Graham, R., Rothschild, B. and Spencer, J. (1990) *Ramsey theory*, Wiley, New York.

[10] Harzheim, E. (1967) Eine kombinatorische Frage zahlentheoretischer Art. *Publicationes Mathematicae Debrecen* **14**, 45–51.

[11] Harzheim, E. (1982) Combinatorial theorems on contractive mappings in power sets. *Discrete Math.* **40**, 193–201.

[12] Higman, G. (1952) Ordering by divisibility in abstract algebras. *Proc. London Math. Soc.* **2**, 326–336.

[13] Jullien, P. (1968) Analyse combinatoire – Sur un théorème d'extension dans la théorie des mots. *C.R. Acad. Sci. Paris*, Ser. A **266**, 851–854.

[14] Kanamori, A. and McAloon, K. (1987) On Gödel incompleteness and finite combinatorics. *Annals of Pure and Applied Logic* **33**, 23–41.

[15] Kreisel, G. (1952) On the interpretation of nonfinitistic proofs. *Journal of Symbolic Logic* **17**, II, 43–58.

[16] Kruskal, J. B. (1972) The Theory of Well-Quasi-Ordering: A Frequently Discovered Concept. *Journal of Combinatorial Theory* (A) **13**, 297–305.

[17] Leeb, K. (1973) Vorlesungen über Pascaltheorie. *Arbeitsbericht des Instituts für mathematische Maschinen und Datenverarbeitung*, Friedrich Alexander Universität Erlangen Nürnberg, Bd. 6 Nr. 7.

[18] Leeb, K. Personal communications.

[19] Loebl, M. and Nešsetřil, J. (1991) Unprovable combinatorial statements. In: Keedwell, A. D. (ed.) Surveys in Combinatorics.

[20] Mills, G. (1980) A tree analysis of unprovable combinatorial statements. Model theory of Algebra and Arithmetic. *Springer-Verlag Lecture Notes in Mathematics* **834**, 248–311.

[21] Nešsetřil, J. and Rödl, V. (1990) *Mathematics of Ramsey Theory*, Springer-Verlag, Berlin, Heidelberg.

[22] Paris, J. and Harrington, L. (1977) A mathematical incompleteness in Peano Arithmetic. *Handbook of Mathematical Logic*. In: Barwise, J. (ed.) North-Holland Publishing Company, 1133–1142.

[23] Prömel, H. J., Thumser, W. and Voigt, B. (1989) Fast growing functions based on Ramsey theorems. Forschungsinstitut für Diskrete Mathematik, Bonn (preprint).

[24] Prömel, H. J., Thumser, W. and Voigt, B. (1991) Fast growing functions based on Ramsey theorems. *Discrete Mathematics* **95**, 341–358.

[25] Prömel, H. J. and Voigt, B. (1989) Aspects of Ramsey Theory I: Sets, Report number 87495-OR, Forschungsinstitut für Diskrete Mathematik, Universität Bonn, Germany.

[26] Prömel, H. J. and Voigt, B. (1993) *Aspects of Ramsey Theory*, Springer Verlag, Berlin.

[27] Ramsey, F. P. (1930) On a problem of formal logic. *Proceedings of the London Mathematical Society* **30**, 264–286.

[28] Simpson, S. G. (1987) Unprovable theorems and fast growing functions. In: Simpson, S. G. (ed.) Logic and Combinatorics. *Contemporary Mathematics* **65**, 359–394.

[29] Thumser, W. (1989) On upper Bounds for Kanamori McAloon Function, preprint 89-10, Sonderforschungsbereich 343 "Diskrete Strukturen in der Mathematik", Universität Bielefeld.

[30] Thumser, W. (1992) On the well-order type of certain combinatorial structures, Bielefeld (manuscript, submitted).

[31] Wainer, S. S. (1972) Ordinal recursion and a refinement of the extended Grzegorczyk hierarchy. *Journal of Symbolic Logic* **37** 281–292.

Lattice Points of Cut Cones

MICHEL DEZA[†] and VIATCHESLAV GRISHUKHIN[‡]

†CNRS-LIENS, Ecole Normale Supérieure, Paris

‡Central Economic and Mathematical Institute of Russian Academy of Sciences (CEMI RAN), Moscow.

Let $\mathbb{R}_+(\mathcal{H}_n), \mathbb{Z}(\mathcal{H}_n), \mathbb{Z}_+(\mathcal{H}_n)$ be, respectively, the cone over $\mathbb{R}$, the lattice and the cone over $\mathbb{Z}$, generated by all cuts of the complete graph on n nodes. For $i \geq 0$, let $A_n^i := \{d \in \mathbb{R}_+(\mathcal{H}_n) \cap \mathbb{Z}(\mathcal{H}_n) : d$ has exactly i realizations in $\mathbb{Z}_+(\mathcal{H}_n)\}$. We show that A_n^i is infinite, except for the undecided case $A_6^0 \neq \emptyset$ and empty A_n^i for $i = 0$, $n \leq 5$ and for $i \geq 2$, $n \leq 3$. The set A_n^1 contains $0, 1, \infty$ nonsimplicial points for $n \leq 4$, $n = 5$, $n \geq 6$, respectively. On the other hand, there exists a finite number $t(n)$ such that $t(n)d \in \mathbb{Z}_+(\mathcal{H}_n)$ for any $d \in A_n^0$; we also estimate such scales for classes of points. We construct families of points of A_n^0 and $\mathbb{Z}_+(\mathcal{H}_n)$, especially on a 0-lifting of a simplicial facet, and points $d \in \mathbb{R}_+(\mathcal{H}_n)$ with $d_{i,n} = t$ for $1 \leq i \leq n-1$.

1. Introduction

In this paper we study integral points of cones. Suppose there is a cone C in $\mathbb{R}^n$ that is generated by its extreme rays $e_1, e_2, ..., e_m$, all $e_i \in \mathbb{Z}^n$.

Let d be a linear combination,

$$d = \sum_{1 \leq i \leq m} \lambda_i e_i. \tag{1}$$

We call the expression a $\mathbb{K}$-realization of d if $\lambda_i \in K$, $1 \leq i \leq m$, and K is either of $\mathbb{R}_+$, $\mathbb{Z}, \mathbb{Z}_+$.

If $\lambda_i \geq 0$ for all i, then $d \in C$, and (1) is an R_+-realization of d. If λ_i is an integer for all i, then $d \in L$ where L is a lattice generated by the integral vectors e_i, $1 \leq i \leq m$, and (1) is a $\mathbb{Z}$-realization of d. Obviously $L \subseteq \mathbb{Z}^n$. If $\lambda_i \geq 0$ and is integral for all i, we call the point d an h-point of C. Hence h-points are the points having a Z_+-realization. A

‡ This work was done during the second author's visit to Laboratoire d'Informatique de l'Ecole Normale Supérieure, Paris

point $d \in C \cap L$ is called a quasi-h-point if it is not an h-point. In other words, d is a quasi-h-point if it has $\mathbb{R}_+$- and $\mathbb{Z}$-realizations but no $\mathbb{Z}_+$-realization.

We consider cut cones, *i.e.* those where e_i are cut vectors. Let $\mathcal{K}_n$ be the set of all nonzero cut vectors of a complete graph on n vertices. Then $\mathbb{R}_+(\mathcal{K}_n)$ is the cut cone. The members of the cut cone $\mathbb{R}_+(\mathcal{K}_n)$ are exactly semimetrics, which are isometrically embedded into some l_1-space, *i.e.* into $\mathbb{R}^n$ with the metric $|x - y|_{l_1}$. Between them, the members of *integer* cut cone $\mathbb{Z}_+(\mathcal{K}_n)$ are exactly semimetric subspaces of some hypercube $\{0,1\}_m$ equipped with the Hamming metric. In particular, the graphic metric $d(G)$ belongs to $\mathbb{Z}_+(\mathcal{K}_n), (1/2)\mathbb{Z}_+(\mathcal{K}_n)$ if and only if G is an isometric subgraph of a cube or of a halved cube, respectively. The above equivalences explain the interest of the cut cones, such as $\mathbb{R}_+(\mathcal{K}_n)$ and $\mathbb{Z}_+(\mathcal{K}_n)$. See [12] for a detailed survey of applications of cut polyhedra. As examples, we recall applications for binary addressing in telecomunication networks, the max-cut problem in Combinatorial Optimization, and the feasibility of multicommodity flows. More specifically, the integer cut cone $\mathbb{Z}_+(\mathcal{K}_n)$ provides some tools for Design Theory (see, for example, [9] and Section 8 below) and for the large subject of embedding graphs in hypercubes.

In fact, those problems are related to feasibility problems of the integer program

$$\{A\lambda = d, \ \lambda \in \mathbb{Z}_+^m\}, \tag{2}$$

where A is the $n \times m$ matrix whose columns are the vectors e_i.

In this paper we attack the integer programming aspects of the cut cones, the main general problem of which is to give a criterion of membership in $\mathbb{Z}_+(\mathcal{K}), \mathcal{K} \subseteq \mathcal{K}_n$, for metrics of given class. Examples of possible approaches to it are as follows.

1 Criteria in terms of inequalities and comparisions, as in [3]: $\mathcal{K} = \mathcal{K}_n, (n \leq 5)$; [10], [13]: $\mathcal{K}$ is a simplex, *i.e.* cuts of $\mathcal{K}$ are linearly independent, $\mathcal{K} = Odd\mathcal{K}_n$.
2 Criteria in terms of enumeration, as in [1] for (1,2)-valued d, or in [15] for $d = d(G)$, where G is a distance-regular graph.
3 A polynomial criterion as in [14] for graphic $d = d(G)$ and other of d.

But in this paper we use other concepts (quasi-h-points and scales), which come from the basic concept of the Hilbert base; see Sections 3 and 4, and 8 and 9 below, respectively.

Finally, we also address adjacent problems on cut lattices (characterization and some arithmetic properties), and on the number of representations of a metric in $\mathbb{Z}_+(\mathcal{K}_n)$.

2. Definitions and notation

Set $V_n = \{1, ..., n\}, E_n = \{(i,j) : 1 \leq i < j \leq n\}$, then $K_n = (V_n, E_n)$ denotes the complete graph on n points. Denote by $P_{(i_1, i_2, ..., i_k)} = P_k$ the path in K_n going through the vertices $i_1, i_2, ..., i_k$.

For $S \subseteq V_n, \delta(S) \subseteq E_n$ denote the *cut* defined by S, with $(i,j) \in \delta(S)$ if and only if $|S \cap \{i,j\}| = 1$. Since $\delta(S) = \delta(V_n - S)$, we take S such that $n \notin S$. The incidence vector of the cut $\delta(S)$ is called a *cut vector* and, by abuse of language, is also denoted by $\delta(S)$. Besides, $\delta(S)$ determines a distance function (in fact, a semimetric) $d_{\delta(S)}$ on points of V_n as follows: $d_{\delta(S)}(i,j) = 1$ if $(i,j) \in \delta(S)$, otherwise the distance between i and j is equal to 0. For the sake of simplicity, we set $\delta(\{i, j, k, ...\}) = \delta(i, j, k, ...)$.

We use $\mathcal{K}_n$ to denote the family of all nonzero cuts $\delta(S)$, $S \subseteq V_n$. For any family $\mathcal{K} \subseteq \mathcal{K}_n$, define the cone $C(\mathcal{K}) := \mathbb{R}_+(\mathcal{K})$ as the conic hull of cuts in $\mathcal{K}$. So, by definition, $\mathcal{K}$ is the set of extreme rays of the cone $C(\mathcal{K})$. The cone $C(\mathcal{K})$ lies in the space $\mathbb{R}(\mathcal{K})$ spanned by the set $\mathcal{K}$. We set $C_n := C(\mathcal{K}_n)$.

So, each point $d \in C(\mathcal{K})$ has a representation $d = \sum_{\delta(S) \in \mathcal{K}} \lambda_S \delta(S)$. Since $\lambda_S \geq 0$, the representation is called the $\mathbb{R}_+$-*realization* of d. The number $\sum_{\delta(S) \in \mathcal{K}} \lambda_S$ is called *the size* of the $\mathbb{R}_+$-realization.

The lattice $L(\mathcal{K}) := \mathbb{Z}(\mathcal{K})$ is the set of all integral linear combinations of cuts in $\mathcal{K}$. Let $L_n = L(\mathcal{K}_n)$. The lattice L_n is easily characterized: $d \in L_n$ if and only if d satisfies the following *condition of evenness*

$$d_{ij} + d_{ik} + d_{jk} \equiv 0 \pmod 2, \quad \text{for all } 1 \leq i < j < k \leq n. \tag{3}$$

So, $2\mathbb{Z}^{n(n-1)/2} \subset L_n \subset \mathbb{Z}^{n(n-1)/2}$.

The points of $L(\mathcal{K})$ with nonnegative coefficients, *i.e.*, the points of $\mathbb{Z}_+(\mathcal{K})$, are called *h-points*. We denote the set of h-points of the cone $C(\mathcal{K})$ by $hC(\mathcal{K})$. For $d \in \mathbb{Z}_+(\mathcal{K})$, any decomposition of d as a nonnegative integer sum of cuts is called a $\mathbb{Z}_+$-*realization* of d. An h-point of C_n is (seen as a semimetric) exactly isometrically *embeddable into a hypercube* (or *h-embeddable*) semimetric. This explains the name of an h-point.

For $d \in C_n$, define

$$s(d) := \text{minimum size of } \mathbb{R}_+\text{-realizations of } d,$$

$$z(d) := \text{minimum size of } \mathbb{Z}_+\text{-realizations of } d \text{ if any}.$$

Let $d(G)$ be the shortest path metric of a graph G. We set

$$z_n^t := z(2td(K_n)).$$

For this special case, $G = K_n$, $s(d) = s(2td(K_n))$ is equal to $a_n^t := \frac{tn(n-1)}{\lfloor n/2 \rfloor \lceil n/2 \rceil}$.

A point $d \in C(\mathcal{K})$ is called a *quasi-h-point* of $C(\mathcal{K})$ if d belongs to $L(\mathcal{K})$ but has no $\mathbb{Z}_+$-realization. We set

$$A(\mathcal{K}) := C(\mathcal{K}) \cap L(\mathcal{K}) - \mathbb{Z}_+(\mathcal{K}).$$

Recall (see [18]) that a *Hilbert basis* is a set of vectors $e_1, ..., e_k$ with the property that each vector lying in both the lattice and the cone generated by $e_1, ..., e_k$ is a nonnegative integral combination of these vectors. $A(\mathcal{K}) = \emptyset$ would mean that $\mathcal{K}$ is a Hilbert basis of $C(\mathcal{K})$. Actually, $\mathcal{K}$ would be the minimal Hilbert basis of $C(\mathcal{K})$ if it is a Hilbert basis, since $\mathcal{K}$ is the set of extreme rays of $C(\mathcal{K})$ (see [4]).

Define

$$A^i(\mathcal{K}) := \{d \in C(\mathcal{K}) \cap L(\mathcal{K}) : d \text{ has exactly } i \ \mathbb{Z}_+\text{-realizations}\},$$

$$A_n^i := A^i(\mathcal{K}_n).$$

So, the above defined set $A(\mathcal{K})$ is $A^0(\mathcal{K})$. Define

$$\eta^i(d) := \min\{t \in \mathbb{Z}_+ : td \text{ has } > i \ \mathbb{Z}_+\text{-realizations}\}$$

$$= \min\{t \in \mathbb{Z}_+ : td \notin A^k(\mathcal{K}) \text{ for all } 0 \leq k \leq i\}.$$

A cone $C = \mathbb{R}_+(\mathcal{K})$ is said to be *simplicial* if the set $\mathcal{K}$ is linearly independent; a point $d \in C$ is said to be *simplicial* if d lies on a simplicial face of C, i.e., if d admits a unique $\mathbb{R}_+$-realization.

Let dim $\mathcal{K}$ be the dimension of the space spanned by $\mathcal{K}$. Call $e(\mathcal{K}) := |\mathcal{K}| - \dim \mathcal{K}$, the *excess* of $\mathcal{K}$. Set

$$\mathcal{K}_n^l = \{\delta(S) \in \mathcal{K}_n : |S| = l \text{ or } n - |S| = l\}.$$

For even n we also set

$$Even\mathcal{K}_n = \{\delta(S) \in \mathcal{K}_n : |S|, n - |S| \equiv 0 \pmod 2\},$$

$$Odd\mathcal{K}_n = \{\delta(S) \in \mathcal{K}_n : |S|, n - |S| \equiv 1 \pmod 2\}.$$

For a subset $T \subseteq V_n$ denote

$$Even T \mathcal{K}_n = \{\delta(S) \in \mathcal{K}_n : |S \cap T| \equiv 0 \pmod 2\},$$

$$Odd T \mathcal{K}_n = \{\delta(S) \in \mathcal{K}_n : |S \cap T| \equiv 1 \pmod 2\}.$$

So $Even\mathcal{K}_n = Even T \mathcal{K}_n$, $Odd\mathcal{K}_n = Odd T \mathcal{K}_n$ for $T = V_n$, n even.

Remark that $\mathcal{K}_{2m}^m = \{\delta(S) \in \mathcal{K}_{2m}^m : 1 \notin S\} = \{\delta(S) \in \mathcal{K}_{2m}^m : 1 \in S\}$.

Denote by $\mathcal{K}_n^{i,j}, \mathcal{K}_n^{\neq i}, \mathcal{K}_n^{\neq i}(\text{mod } a)$ the families of $\delta(S) \in \mathcal{K}_n$ with $|S| \in \{i, j, n-i, n-j\}$, $|S| \notin \{i, n-i\}$, $\min\{|S|, n - |S|\} \not\equiv i(\text{mod } a)$, respectively.

We write C_b^a for $C(\mathcal{K}_b^a)$, where a and b are indices or sets of indices.

3. Families of cuts $\mathcal{K}$ with $A(\mathcal{K}) = \emptyset$

Of course $A(\mathcal{K}) = \emptyset$ if $e(\mathcal{K}) = 0$, i.e. if the cone $C(\mathcal{K})$ is simplicial. It is easy to see that $C(\mathcal{K}_n^l)$ is simplicial if and only if either $l = 1$, or $l = 2$, or $(l, n) = (3, 6)$. Also $e(\mathcal{K}_3) = 0$, what is a special case of the formula

$$e(\mathcal{K}_n) = 2^{n-1} - 1 - \binom{n}{2}.$$

Some examples of $\mathcal{K}$ with a positive excess but with $A(\mathcal{K}) = \emptyset$ are:

(a) $\mathcal{K}_4$, $\mathcal{K}_5$ with excess 1 and 5, respectively. The first proof was given in [3]; for details of the proof see [10], where, for any $d \in C_n \cap L_n$, $n = 4, 5$, the explicit $\mathbb{Z}_+$-realization of d is given.

(b) $Odd\mathcal{K}_6$ with the excess 1. For the proof see [10].

(c) (See the case $n = 5$ of Theorem 6.2 below.) The family of cuts (with excess 5) on a facet of $C(\mathcal{K}_6)$ that is a 0-lifting of a simplicial pentagonal facet of $C(\mathcal{K}_5)$.

But $\mathcal{K}_n^{1,2}$ with excess n has $A(\mathcal{K}) \neq \emptyset$ for $n \geq 6$. Below we give some examples of $\mathcal{K}$ with $A(\mathcal{K}) \neq \emptyset$, which are, in a way, close to the above examples of $\mathcal{K}$ with $A(\mathcal{K}) = \emptyset$.

We denote by $Q(b)$ the linear form $\sum_{1 \leq i < j \leq n} b_i b_j x_{ij}$ for $b \in \mathbb{Z}^n$. If $\sum_{i=1}^n b_i = 1$, the inequality $Q(b) \leq 0$ is called a *hypermetric inequality*. We call $d \in \mathbb{R}^{n(n-1)/2}$ a *hypermetric* if it satisfies all the hypermetric inequalities. We denote the hypermetric inequality by $Hyp_n(b)$. It is easy to verify that $\delta(S)$ satisfies all hypermetric inequalities. Moreover, for large classes of parameters b (see [4], [6]) $Hyp_n(b)$ is a facet of $C(\mathcal{H}_n)$. The only known case when a hypermetric face is simplicial is (up to permutation) $Hyp_n(1^2, -1^{n-3}, n-4)$, $n \geq 3$, and (its 'switching' in terms of [6]) $Hyp_n(-1, 1^{n-2}, -(n-4))$. Call the facet $Hyp_n(1^2, -1^{n-3}, n-4)$ the *main n-facet*. Call the facet $Hyp_n(1^2, 0^k, -1^{n-k-3}, n-k-4)$ the *k-fold 0-lifting* of the main $(n-k)$-facet. It is a facet of $C(\mathcal{H}_n)$, because every k-fold 0-lifting of a facet of C_{n-k} is a facet of C_n (see [4]). We call 1-fold 0-lifting simply 0-lifting. We list, up to a permutation, all facets of $C(\mathcal{H}_n)$ for $3 \leq n \leq 6$:

— The unique type of facets of $C(\mathcal{H}_3)$ is the main 3-facet (triangle inequality);
— The unique type of facets of $C(\mathcal{H}_4)$ is the main 4-facet (which is the 0-lifting $Hyp_4(-1, 1^2, 0)$ of a main 3-facet);
— All facets of $C(\mathcal{H}_5)$ are 2-fold 0-liftings of a main 3-facet (i.e. 0-lifting of a main 4-facet), and the main 5-facet $Hyp_5(1^3, -1^2)$, called the *pentagonal* facet;
— All facets of $C(\mathcal{H}_6)$ are: 2-fold 0-liftings of a main 4-facet, 0-lifting of a main 5-facet, the main 6-facet $Hyp_6(2, 1, 1, -1^3)$ and its 'switching' $Hyp_6(-2, -1, 1^4)$.

Lemma 3.1. *If $\mathcal{H}$ is a family of cuts $\delta(S)$, $|S| \leq (n/2)$, lying on a face F of C_n, the family*

$$\mathcal{H}' = \mathcal{H} \cup \{\delta(\{n+1\})\} \cup \{\delta(S \cup \{n+1\}) : \delta(S) \in \mathcal{H}\}$$

is the family of cuts lying on a 0-lifting of the face F. If, for the above $\mathcal{H}$, $C(\mathcal{H})$ is a simplicial facet of C_n, we obtain, for $n \geq 4$,

$$e(\mathcal{H}') = n(n-3)/2.$$

Proof. If $C(\mathcal{H})$ is a simplicial facet of C_n, then dim $\mathcal{H} = |\mathcal{H}| = \binom{n}{2} - 1$. Obviously, $|\mathcal{H}'| = 2|\mathcal{H}| + 1$. Since $\mathcal{H}'$ is a simplicial facet of C_{n+1}, we have, dim $\mathcal{H}' = \binom{n+1}{2} - 1$ also. Hence

$$
\begin{aligned}
e(\mathcal{H}') &= |\mathcal{H}'| - \dim . \mathcal{H}' \\
&= (2|\mathcal{H}| + 1) - \dim \mathcal{H}' \\
&= 2(\binom{n}{2} - 1) + 1 - (\binom{n+1}{2} - 1) \\
&= n(n-3)/2.
\end{aligned}
$$

$\square$

Recall that $A(\mathcal{H}) = \emptyset$ for $\mathcal{H} = \mathcal{H}_5, \mathcal{H}_6^1, \mathcal{H}_6^2, \mathcal{H}_6^3, \mathcal{H}_6^{1,3} = Odd\mathcal{H}_6$, and for the family of any (except triangle) facet of $\mathcal{H}_6$, since $\mathcal{H}_6^i$ is simplicial for $i = 1, 2, 3$, and $\mathcal{H}_5, Odd\mathcal{H}_6$ are examples given at the beginning of this section.

4. Antipodal extension

A fruitful method of obtaining quasi-h-points is the *antipodal extension operation* at the point n. For $d \in \mathbb{R}^{n(n-1)/2}$ we define $ant_\alpha d \in \mathbb{R}^{n(n+1)/2}$ by

$$
\begin{aligned}
(ant_\alpha d)_{ij} &= d_{ij} \text{ for } 1 \leq i < j \leq n, \\
(ant_\alpha d)_{n,n+1} &= \alpha, \\
(ant_\alpha d)_{j,n+1} &= \alpha - d_{jn} \text{ for } 1 \leq j \leq n-1.
\end{aligned}
$$

For $\mathscr{K} \subseteq \mathscr{K}_n$, define

$$ant\mathscr{K} = \{ant_1\delta(S) : \delta(S) \in \mathscr{K}\} \cup \{\delta(n+1)\}.$$

Note that

$$ant_1\delta(S) = \delta(S) \text{ if } n \in S, \text{ and } ant_1\delta(S) \doteq \delta(S \cup \{n+1\}) \text{ if } n \notin S.$$

Hence

$$ant\mathscr{K} = \{\delta(S) : \delta(S) \in \mathscr{K}, n \in S\} \cup \{\delta(S \cup \{n+1\}) : \delta(S) \in \mathscr{K}, n \notin S\}.$$

Observe that if $d \in C(\mathscr{K})$ and $d = \sum_{\delta(S) \in \mathscr{K}} \lambda_S \delta(S)$, then

$$\begin{aligned}
ant_\alpha d &= \sum_{\delta(S) \in \mathscr{K}} \lambda_S ant_\alpha \delta(S) + \alpha(1 - \sum_S \lambda_S)\delta(n+1) \\
&= \sum_{\delta(S) \in \mathscr{K}} \lambda_S ant_1 \delta(S) + (\alpha - \sum_S \lambda_S)\delta(n+1).
\end{aligned} \tag{4}$$

Also, if

$$ant_\alpha d = \sum_{\delta(S) \in \mathscr{K}} \lambda_S ant_1 \delta(S) + \lambda_0 \delta(n+1),$$

then $\alpha = \sum_S \lambda_S + \lambda_0$, and $d = \sum_{\delta(S) \in \mathscr{K}} \lambda_S \delta(S)$ is the projection of $ant_\alpha(d)$ on $\mathbb{R}^{n(n-1)/2}$. So $ant_\alpha d \in \mathbb{R}(ant\mathscr{K})$ if and only if $d \in \mathbb{R}(\mathscr{K})$.

Note that the cone $\mathbb{R}(ant\mathscr{K})$ is the intersection of the triangle facets $Hyp_{n+1}(1^2, -1_j, 0^{n-2})$, where $b_n = b_{n+1} = 1$, $b_j = -1$ and $b_i = 0$ for $i \neq j$, $1 \leq i \leq n-1$.

Proposition 4.1. (Proposition 2.6 of [8])

(i) $ant_\alpha d \in L_{n+1}$ if and only if $d \in L_n$ and $\alpha \in \mathbb{Z}$,

(ii) $ant_\alpha d \in C_{n+1}$ if and only if $d \in C_n$ and $\alpha \geq s(d)$,

(iii) $ant_\alpha d \in hC_{n+1}$ if and only if $d \in hC_n$ and $\alpha \geq z(d)$,

(iv) $ant_\alpha d$ is a simplicial point of C_{n+1} if and only if d is a simplicial point of C_n and $\alpha \geq s(d)$. $\square$

Clearly, $s(ant_\alpha d) = \alpha$ if $ant_\alpha d \in C_{n+1}$ and $z(ant_\alpha d) = \alpha$ if $ant_\alpha d \in hC_{n+1}$. Also, $ant_\alpha d \in A_n^i$ for $i > 0$ if and only if $d \in A_n^i$, $\alpha \in \mathbb{Z}_+$, $\alpha \geq z(d)$.

Proposition 4.1 obviously implies the following important corollary.

Corollary 4.2. Let $d \in hC_n$, and α be an integer such that $s(d) \leq \alpha < z(d)$. Then $ant_\alpha d \in A(ant\mathscr{K}_n) \subset A_{n+1}^0$, i.e. $ant_\alpha d$ is a quasi-h-point in C_{n+1}.

5. Spherical t-extension and gate extension

Let $d \in C_{n+1}$. We write $d = (d^0, d^1)$, where

$$d^0 = \{d_{ij} : 1 \leq i < j \leq n\}, \ d^1 = \{d_{i,n+1} : 1 \leq i \leq n\}.$$

A point $d \in C_{n+1}$ is called the *spherical t-extension*, or simply *t-extension*, of the point $d^0 \in C_n$ if $d = (d^0, d^1)$ and $d^1_{i,n+1} = t$ for all $i \in V_n$. We denote the spherical t-extension of d^0 by $ext_t d^0$.

Let j_n be the n-vector whose components are all equal to 1. Then for the t-extension (d^0, d^1), we have $d^1 = t j_n$.

Proposition 5.1. $ext_t d$ *is a hypermetric if and only if*

(i) d *is a hypermetric,*

(ii) $t \geq (\sum b_i b_j d_{ij})/\Sigma(\Sigma - 1)$

for all integers $b_1, ..., b_n$ with $\Sigma := \sum_1^n b_i > 1$ and g.c.d. $b_i = 1$.

Proof. If $ext_t d$ is hypermetric, then $\sum b_i b_j (ext_t d)_{ij} \leq 0$ for any $b_1, ..., b_n, b_{n+1} \in \mathbb{Z}_+$ with $\sum b_i = 1$, i.e.

$$\sum_{1 \leq i < j \leq n} b_i b_j d_{ij} + \sum_{1 \leq i \leq n} b_i b_{n+1} t \leq 0.$$

Since $b_{n+1} = 1 - \Sigma$, the second term is equal to $-t\Sigma(\Sigma - 1)$. We obtain (i) if $b_{n+1} = 0$ or 1; otherwise $\Sigma(\Sigma - 1) \neq 0$, and we get (ii). $\qquad\square$

Corollary 5.2. $ext_t d$ *is a semimetric if and only if d is a semimetric and $t \geq (1/2) \max_{(ij)} d_{ij}$.*

In fact, apply (ii) above to the case $b_i = b_j = 1$, $b_{n+1} = -1$ and $b_k = 0$ for other b's.

As with Proposition 5.1, one can check that $ant_t d$ is a hypermetric (a semimetric) if and only if d is a hypermetric (a semimetric, respectively) and

$$t \geq (\sum_{1 \leq i < j \leq n} b_i b_j d_{ij})/\Sigma(\Sigma - 1) + \sum_1^n b_i d_{in}/\Sigma$$

for any integers $b_1, ..., b_n$ with $\Sigma := \sum_1^n b_i > 1$ and g.c.d. $b_i = 1$

$(t \geq \frac{1}{2} \max_{1 \leq i < j \leq n-1}(d_{ij} + d_{in} + d_{jn})$, respectively).

There is another operation, similar to antipodal extension operation. We call it the *gate extension operation* at the point n (called the *gate*). For $d \in \mathbb{R}^{n(n-1)/2}$, define $gat_\alpha d \in \mathbb{R}^{n(n-1)/2}$ by

$$(gat_\alpha d)_{ij} = d_{ij} \text{ for } 1 \leq i < j \leq n,$$
$$(gat_\alpha d)_{n,n+1} = \alpha,$$
$$(gat_\alpha d)_{i,n+1} = \alpha + d_{in} \text{ for } 1 \leq i \leq n-1.$$

The following identity shows that $gat_\alpha d$ is, in a sense, a complement of $ant_\alpha d$:

$$ant_\alpha d + gat_{2t - \alpha} d = 2 ext_t d. \tag{5}$$

Recall that we take S in $\delta(S)$ such that $n \notin S$. Hence, for $\mathcal{K} \subseteq \mathcal{K}_n$, we have

$$gat \mathcal{K} = \mathcal{K} \cup \{\delta(n+1)\}.$$

Actually, $ant \mathcal{K}_n = Odd T \mathcal{K}_{n+1}$, $gat \mathcal{K}_n = \{\delta(n+1)\} \cup Even T \mathcal{K}_{n+1}$, for $T = \{n, n+1\}$.

Note that the cone $\mathbb{R}_+(gat\ \mathcal{H})$ is the intersection of the triangle facets Hyp_{n+1} $(1_i, 0^{n-2}, -1, 1_{n+1})$, where $b_i = b_{n+1} = 1$, $b_n = -1$, $b_j = 0$ for $j \neq i$, $1 \leq j \leq n-1$.

It is clear that any $\mathbb{R}_+$-realization of $gat_\alpha d$ (if it belongs to C_{n+1}) has the form $\sum_S \lambda_S \delta(S) + \alpha\delta(n+1)$ where $n+1 \notin S$, and where the above realization is any $\mathbb{R}_+$-realization of d. So, $gat_\alpha d \in L_{n+1}(C_{n+1}, hC_{n+1}, A^i_{n+1}$, respectively) if and only if $d \in L_n(C_n, hC_n, A^i_n$, respectively) and $\alpha \in \mathbb{Z}(\mathbb{R}_+, \mathbb{Z}_+, \mathbb{Z}$, respectively).

Also, $gat_\alpha d$ is a hypermetric (a metric) if and only if $\alpha \in \mathbb{R}_+$ and d is a hypermetric (a metric, respectively).

Hence if $\alpha \in \mathbb{Z}_+$, we have

$$gat_\alpha d \in A^i_{n+1} \iff d \in A^i_n. \tag{6}$$

In particular, $gat_\alpha d$ is a quasi-h-point if and only if d is.

The following facts are obvious.

1 If d_i is the t_i-extension of d^0_i, $i = 1, 2$, then $d_1 + d_2$ is the $(t_1 + t_2)$-extension of $d^0_1 + d^0_2$.

2 If d^0 lies in a facet of the cut cone, the t-extension of d^0 lies in the 0-lifting of the facet.

We call a point $d \in C_n$ even if all distances d_{ij} are even.

Let $d = \sum_S \lambda_S \delta(S)$ be a $\mathbb{Z}_+$-realization of an h-point d. We call the realization *(0,1)-realization* $(2\mathbb{Z}_+$-*realization*) if all λ_S are equal to 0 or 1 (are even, respectively). We have

Fact. *Let d be an h-point. Then $d = d_1 + d_2$, where d_1 has a (0,1)-realization, and d_2 has a $2Z_+$-realization.*

Obviously, if d has a $2\mathbb{Z}_+$-realization, d is even. But if d is even, it can have no $2\mathbb{Z}_+$-realizations.

The following Proposition is an analog of Proposition 4.1.

Proposition 5.3.

(i) $ext_t d \in L_{n+1}$ if and only if $d \in 2Z^{n(n-1)/2}$ and $t \in Z$,

(ii) $ext_t d \in C_{n+1}$ if $d \in C_n$ and $2t \geq s(d)$,

(iii) *suppose that d has $2Z_+$-realizations, and let $z_{even}(d)$ denote their minimal size; then $ext_t d \in hC_{n+1}$ if $d \in hC_n$ and $2t \geq z_{even}(d)$.*

Proof. (i) is implied by the trivial equality $d_{i,n+1} + d_{j,n+1} + d_{ij} = 2t + d_{ij}$, $1 \leq i < j \leq n$.

From (5) we have $ext_t d = (1/2)(ant_\alpha d + gat_{2t-\alpha} d)$. Taking $\alpha = s(d)$ and applying (ii) of Proposition 4.1 we get (ii).

Taking $\alpha = z_{even}(d)$, applying (iii) of Proposition 4.1 and using $ant_{++z_{even}(d)}, gat_{2t-z_{even}(d)} d \in 2\mathbb{Z}_+(\mathcal{H}_{n+1})$, we get (iii). □

Define $ext^m_t d = ext_t(ext^{m-1}_t d)$, where $ext^1_t d = ext_t d$.

Proposition 5.4. *If $2t \geq s(d)$, then $ext^m_t d \in C_{n+m}$ for any $m \in Z_+$, and*

$$\max(s(ext^{m-1}_t d), 2t - \frac{t}{\lceil m/2 \rceil}) \leq s(ext^m_t d) \leq 2t - 2^{-m}(2t - s(d)).$$

Proof. From Proposition 5.3(ii) we get

$$s(ext_t d) \le \frac{1}{2} s(ant_{s(d)} d + gat_{2t-s(d)} d) = t + \frac{1}{2} s(d) \le 2t.$$

By induction on m, we obtain $ext_t^m d \in C_{n+m}$ for all $m \in \mathbb{Z}_+$, and the upper bound for $s(ext_t^m d)$.

The lower bound is implied by the fact that the restriction of $ext_t^m d$ on m extension points is $td(K_m)$. Since $s(td(K_m)) = (1/2)a_m^t$ (see Section 2), we have

$$s(ext_t^m) \ge s(td(K_m)) = \frac{1}{2} \frac{tm(m-1)}{\lfloor m/2 \rfloor \lceil m/2 \rceil} = 2t - \frac{t}{\lceil m/2 \rceil}.$$

$\square$

Remark. So, if $s(d) \le 2t$, then $\lim_{m \to \infty} s(ext_t^m d) = 2t$.

Probably, there exist $m_0 = m_0(t, d)$ such that $s(ext_t^m d) = 2t$ for $m \ge m_0$.

We conjecture that $ext_t^m d \notin C_{n+m}$ for $m > m_1$ if $s(d) > 2t$. For example, if $t = 1$ and $d = d(G)$ ($d(G)$ is the shortest path metric of the graph G), then it can be proved that $m_1 = 2$.

If the conjecture is true,

$$s(d) = 2 \min\{t : ext_t^m d \in C_{n+m} \text{ for all } m \in Z_+\}.$$

Recall, that Proposition 4.1(ii) implies

$$s(d) = \min\{\alpha : ant_\alpha d \in C_{n+1}\}.$$

In terms of $ext_n^m d$ we also have analogs of (i) and (iii) of Proposition 4.1.

Proposition 5.5.

(i) $ext_t^m d \in L_{n+m}$ for all $m \in \mathbb{Z}_+$ if and only if $d \in 2Z^{n(n-1)/2}$ and t is even.

(ii) $ext_t^m d \in hC_{n+m}$ for all $m \in \mathbb{Z}_+$ if and only if t is an even positive integer, and $ext_{t/2}^1 d \in hC_{n+1}$.

Proof. The evenness of t follows from $ext_t^3 d \in L_{n+3}$. So, (i) is implied by Proposition 5.3(i).

Recall the result of [5] that $t \sum_1^n \delta(i)$ is the unique $\mathbb{Z}_+$-realization of $td(K_n)$ for even t and $m \ge (t^2/4) + (t/2) + 3$. Using this fact, we get that any $\mathbb{Z}_+$-realization of $ext_t^m d$ contains $t/2$ cuts $\delta(i)$ for some i if m is large enough. $\square$

6. Quasi-h-points of 0-lifting of the main facet

Consider the main facet

$$F_0(n) = Hyp_n(1^2, -1^{n-3}, n-4) = Hyp_n(b^0),$$

where $b_1^0 = b_2^0 = 1$, $b_i^0 = -1$, $3 \le i \le n-1$, $b_n^0 = n-4$. The cut vectors $\delta(S)$ lying in the facet are defined by equations $b(S) \equiv \sum_{i \in S} b_i = 0$ or 1. We take S not containing n. Then $S \in \mathscr{S}$, where

$$\mathcal{S} = \{\{1\}, \{2\}, \{1i\}, \{2i\}\{12i\} \ (3 \le i \le n-1), \{12ij\} \ (3 \le i < j \le n-1)\}.$$

We set

$$m = |\mathcal{S}| = \frac{n(n-1)}{2} - 1.$$

Every n-facet contains at least m cut vectors. Since the main n-facet contains exactly m cuts, it is simplicial.

The 0-lifting of the main facet is the facet

$$F(n) = Hyp_{n+1}(1^2, -1^{n-3}, n-4, 0).$$

Besides the above cuts $\delta(S), S \in \mathcal{S}$, it contains, according to Lemma 3.1, only the cuts $\delta(S \cup \{n+1\}), S \in \mathcal{S}$, and $\delta(n+1)$.

Note that $A(\mathcal{K}) = \emptyset$ for the main n-facet (as for any simplicial $C(\mathcal{K})$).

Now we consider even points having no $2\mathbb{Z}_+$-realization. The simplest such points are points having a $(0,1)$-realization. We call these points *even $(0,1)$-points*.

Let $d^0 \in F_0(n)$ be an even h-point, and let $\sum_{S \in \mathcal{S}_0} \lambda_S \delta(S)$ be one of its $\mathbb{Z}_+$-realizations. Consider a minimal set of comparisions mod 2 that λ_S's have to satisfy. The comparisions are implied by the conditions $d_{ij} \equiv 0$ for all pairs (ij). Since $d^0 \in L_n$, we have $d_{ij} \equiv d_{ik} + d_{jk}$ (mod 2) for all ordered triples (ijk). Hence independent comparisions are implied by the comparisions $d_{in} \equiv 0$ (mod 2), $1 \le i \le n-1$. The comparisions are as follows. (For the sake of simplicity, we set $\lambda_{\{ij...\}} = \lambda_{ij...}$ and omit the indication (mod 2)).

$$\lambda_{1i} + \lambda_{2i} + \lambda_{12i} + \sum_{3 \le j \le n-1, j \ne i} \lambda_{12ij} \equiv 0, \ 3 \le i \le n-1,$$

$$\lambda_1 + \sum_{3 \le i \le n-1} (\lambda_{1i} + \lambda_{12i}) + \sum_{3 \le i < j \le n-1} \lambda_{12ij} \equiv 0, \tag{7}$$

$$\lambda_2 + \sum_{3 \le i \le n-1} (\lambda_{2i} + \lambda_{12i}) + \sum_{3 \le i < j \le n-1} \lambda_{12ij} \equiv 0.$$

The system of comparisions (7) has $n-1$ equations with $m = n(n-1)/2 - 1$ unknowns. Hence the number of $(0,1)$-solutions distinct from the trivial zero solution is equal to $2^{m-(n-1)} - 1 = 2^{\binom{n-1}{2}-1} - 1$.

This shows that all points of $F_0(3)$ have $2\mathbb{Z}_+$-realizations. The only even $(0,1)$-points of $F_0(4)$ are 2 points $2d(K_3)$ with $d_{13} = 0$ or $d_{23} = 0$, and the point $2d(K_4 - P_{(1,2)})$. There are 31 even $(0,1)$-points in $F_0(5)$.

Since there are exponentially many even $(0,1)$-points in $F_0(n)$, we consider points of the following type and call them *special*.

For these points the coefficients λ_S are

$$\lambda_1 = a_1, \ \lambda_2 = a_2, \ \lambda_{1i} = b_1, \ \lambda_{2i} = b_2, \ \lambda_{12i} = c_1, \ 3 \le i \le n-1,$$

$$\lambda_{12ij} = c_2, \ 3 \le i < j \le n-1.$$

Here $a_i, b_i, c_i, i = 1, 2$, are equal to 0 or 1.

If we set

$$k = n - 3, \ l = \frac{n(n-1)}{2},$$

then for the special points, (7) takes the form

$$b_1 + b_2 + c_1 + (k-1)c_2 \equiv 0,$$

$$a_1 + k(b_1 + c_1) + \frac{k(k-1)}{2}c_2 \equiv 0,$$

$$a_2 + k(b_2 + c_1) + \frac{k(k-1)}{2}c_2 \equiv 0.$$

Since we have 3 equations for 6 variables, we can express 3 variables a_1, a_2, c_1 through the other 3 variables b_1, b_2, c_2.

There are 4 families of the solutions of the system depending on the value of k (mod 4). The solutions are as follows (undefined equivalences are taken by (mod 2)).

$$k \equiv 0 \ (\text{mod } 4), \ a_1 = a_2 = 0, \ c_1 \equiv b_1 + b_2 + c_2,$$

$$k \equiv 1 \ (\text{mod } 4), \ a_1 = b_2, \ a_2 = b_1, c_1 \equiv b_1 + b_2, \ c_2 \text{ arbitrary},$$

$$k \equiv 2 \ (\text{mod } 2), \ a_1 = a_2 = c_2, \ c_1 \equiv b_1 + b_2 + c_2,$$

$$k \equiv 3 \ (\text{mod } 4), \ a_1 \equiv b_2 + c_2, \ a_2 \equiv b_1 + c_2, \ c_1 \equiv b_1 + b_2.$$

In each case we obtain 7 nontrivial special even (0,1)-points.

Turning our attention to the definition of $\mathscr{S}$, for $a = 0, \pm$, we denote by λ^a_{ik}, λ^a_k the k-vectors with the components λ^a_{ij}, $3 \le j \le n-1$, $i = 1, 2$, λ^a_{12j}, $3 \le j \le n-1$, respectively. Similarly, λ^a_l is the l-vector with the components λ^a_{12ij}, $3 \le i < j \le n-1$.

In this notation a special point d^0 has a (0,1)-realization λ^0 such that $\lambda^0_i = a_i$, $\lambda^0_{ik} = b_i j_k$, $i = 1, 2$, $\lambda^0_k = c_1 j_k$ and $\lambda^0_l = c_2 j_l$.

Recall that special points are simplicial. Therefore their size is equal to $\sum_{S \in \mathscr{S}} \lambda_S$. We show below that the t-extension of 2 special points with $(a_1, a_2, b_1, b_2, c_1, c_2) = (1, 1, 0, 0, 0, 1)$ and $(0, 1, 0, 1, 1, 1)$ are quasi-h-points for $n \equiv 2$ (mod 4).

For $n = 6$ the points d^0 are $d(K_6 - P_3)$ and $ant_{10}(ext_4 d(K_4))$. Another example of $d \in A^0_7$ is $ant_6(ext_3 d(K_5)) = d^{5,3}$ in terms of Corollary 6.6 below.

Proposition 6.1. *Let d^0 be one of the 7 special points of the main facet $F_0(n)$. Let t be a positive integer such that $t \ge (1/2)\sum_{S \in \mathscr{S}} \lambda^0_S$. Then the t-extension of d^0 is an h-point if $n \not\equiv 2$ (mod 4), and if $n \equiv 2$ (mod 4), then there is a point d^0 such that its t-extension is a quasi-h-point, namely the point with $(a_1, a_2, b_1, b_2, c_1, c_2) = (1, 1, 0, 0, 0, 1)$.*

Proof. Recall that we can take $\mathscr{S}$ such that $n \notin S$ for all $S \in \mathscr{S}$.

We apply equation (2) to the t-extension d. In this case the matrix A takes the form

$$A = \begin{pmatrix} B & B & 0 \\ D & D & j_n \end{pmatrix}$$

Here the first m columns correspond to sets $S \in \mathscr{S}$, the next m columns correspond to sets $S \cup \{n+1\}$, $S \in \mathscr{S}$, and the last $(2m+1)$th column corresponds to $\{n+1\}$. The size of the matrix B is $\binom{n}{2} \times m$, and $D, \overline{D}$ are $n \times m$ matrices such that $D + \overline{D} = J$, where J is the matrix all of whose elements are equal to 1. Each column of the matrix J is the vector j_n consisting of n 1's. In this notation, we can write J as the direct product $J = j_n \times j_m^T$. Hence for any m-vector a we have $Ja = (j_m, a)j_n$.

The rows of D and $\overline{D}$ are indexed by pairs $(i, n+1)$, $1 \le i \le n$. The S-column of the matrix D is the $(0,1)$-indicator vector of the set S. Since $n \notin S$ for all $S \in \mathscr{S}$, the last row of D consists of 0's only.

We look for solutions of the system (2) for this matrix A such that λ is a nonnegative integral $(2m+1)$-vector. We set

$$\mu_S = \lambda_{S \cup \{n+1\}}, \ S \in \mathscr{S}, \ \gamma = \lambda_{\{n+1\}}.$$

Then the system (2) takes the form

$$B(\lambda + \mu) = d^0,$$

$$D(\lambda - \mu) + (\gamma + (j_m, \mu))j_n = d^1.$$

Now, if we set $\lambda^+ = \lambda + \mu$, $\lambda^- = \lambda - \mu$, $\gamma_1 = \gamma + (j_m, \mu)$, and recall that $d^1 = t j_n$, we obtain the equations

$$B\lambda^+ = d^0,$$

$$D\lambda^- + \gamma_1 j_n = t j_n. \tag{8}$$

Recall that the last row of D is the 0-row. Hence the last equation of the system (8) gives $\gamma_1 = t$, and the equation (8) takes the form

$$D\lambda^- = 0.$$

A solution $(\lambda^+, \lambda^-, \gamma_1)$ is feasible if the vector (λ, μ, γ) is nonnegative. Since

$$\lambda = \frac{1}{2}(\lambda^+ + \lambda^-), \ \mu = \frac{1}{2}(\lambda^+ - \lambda^-), \text{ and } \gamma = t - (j_m, \mu),$$

a solution $(\lambda^+, \lambda^-, \gamma_1)$ is feasible if

$$\lambda^+ \ge 0, \ |\lambda^-| \le \lambda^+, \text{ and } t \ge (j_m, \mu). \tag{9}$$

Since the main facet $F_0(n)$ is simplicial, the system $B\lambda^+ = d^0$ has the full rank m such that $\lambda^+ = \lambda^0$ is the unique solution.

We try to find an integral solution for λ^-. By (9), we have that $|\lambda^-| \le \lambda^0$. This implies that $\lambda_S^- \ne 0$ only for sets S where $\lambda_S^0 \ne 0$. Since λ^0 is a $(0,1)$-vector, an integral λ_S^- takes the value 0 and ± 1 only.

We write the matrix $(D, j_n) \equiv D_n$ explicitly:

$$D_n = \begin{pmatrix} 1 & 0 & j_k^T & 0 & j_k^T & j_l^T & 1 \\ 0 & 1 & 0 & j_k^T & j_k^T & j_l^T & 1 \\ 0 & 0 & I_k & I_k & I_k & G_k & j_k \\ 0 & 0 & 0 & 0 & 0 & 0 & 1 \end{pmatrix}.$$

The first, the second and the last rows of the matrix D_n are indexed by the pairs $(1, n+1), (2, n+1)$ and $(n, n+1)$, respectively. The third row consists of matrices with k rows corresponding to the pairs $(i, n+1)$ with $3 \leq i \leq n-1$. The columns of D_n are indexed by sets $S \in \mathscr{S}_0 \cup \{n+1\}$ in the sequence $\{1\}, \{2\}, \{1i\}, \{2i\}, \{12i\}, 3 \leq i \leq n-1, \{12ij\}, 3 \leq i < j \leq n-1, \{n+1\}$. I_k is the $k \times k$ unit matrix, and G_k is the $k \times l$ incidence matrix of the complete graph K_k. G_k contains exactly two 1's in each column, i.e. $j_k^T G_k = 2j_l^T$. The matrix $D_{n'}$ is an obvious submatrix of D_n, for $n' < n$.

In the above notation, the equation $D\lambda^- = 0$ takes the form

$$\lambda_i^- + j_k^T(\lambda_{ik}^- + \lambda_k^-) + j_l^T \lambda_l^- = 0, \ i = 1, 2,$$

$$\lambda_{1k}^- + \lambda_{2k}^- + \lambda_k^- + G_k \lambda_l^- = 0.$$

Since $j_k^T G_k = 2j_l^T$, the last equality implies that

$$j_k^T(\lambda_{1k}^- + \lambda_{2k}^- + \lambda_k^-) + 2j_l^T \lambda_l^- = 0.$$

Hence the above system implies

$$\lambda_1^- + \lambda_2^- + j_k^T \lambda_k^- = 0.$$

Recall that we look for a $(0, \pm 1)$-solution. Note that if $\lambda_S^+ = 1$ and $\lambda_S^- = 0$, then $\lambda_S = \mu_S = 1/2$ is nonintegral. Hence we shall look for a solution such that $\lambda_S^- = \pm \lambda_S^0$. So, such a solution is nonzero where λ_S^0 is nonzero.

The main part of the above equations is contained in the term $G_k \lambda_l^-$. We can treat the (± 1)-variables $(\lambda^-)_{ij} \equiv \lambda_{12ij}^-$ as labels of edges of the complete graph K_n. Now the problem is reduced to finding such a labelling 'of edges of K_n that the sum of labels of edges incident to a given vertex is equal to a prescribed value, usually equal to 0 or ± 1. The existence of such a solution depends on a possibility of factorization of K_n into circuits and 1-factors.

Corresponding facts can be found in [16, Theorems 9.6 and 9.7].

A tedious inspection shows that a feasible labelling exists for each of the 7 special points if $n \not\equiv 2 \pmod 4$ (i.e. if $k \not\equiv 3 \pmod 4$), and for 5 special points if $n \equiv 2 \pmod 4$. For the other point with $(a_1, a_2, b_1, b_2, c_1, c_2) = (1, 1, 0, 0, 0, 1)$ there is no feasible solution, i.e. there are S such that $\lambda_S^- = 0 \neq \pm \lambda_S^0$.

Now the assertion of the proposition follows. $\qquad\qquad\square$

In the table below, t-extensions of some special points are given explicitly. The last column of the table gives a point of A_{4m-1}^0 for any $m \geq 2$.

$n \pmod 4 \equiv$	3	0	1	2
d_{12}	$n-3$	0	$n-1$	2
$d_{1i}\ (3 \le i \le n-1)$	$\binom{n-4}{2}+1$	$\binom{n-4}{2}$	$\binom{n-4}{2}+2$	$\binom{n-4}{2}+1$
$d_{2i}\ (3 \le i \le n-1)$	$\binom{n-3}{2}$	$\binom{n-4}{2}$	$\binom{n-3}{2}+1$	$\binom{n-4}{2}+1$
$d_{ij}(i \ne j)\ (3 \le i,j \le n-1)$	$2(n-4)$	$2(n-5)$	$2(n-4)$	$2(n-5)$
d_{1n}	$\binom{n-3}{2}$	$\binom{n-3}{2}$	$\binom{n-3}{2}+1$	$\binom{n-3}{2}+1$
d_{2n}	$\binom{n-2}{2}$	$\binom{n-3}{2}$	$\binom{n-2}{2}+1$	$\binom{n-3}{2}+1$
$d_{in}\ (3 \le i \le n-1)$	$n-3$	$n-4$	$n-3$	$n-4$
$d_{in+1}(i \ne n+1)$	$\binom{n-2}{2}/2$	$\binom{n-3}{2}/2$	$(\binom{n-2}{2}+3)/2$	$(\binom{n-3}{2}+3)/2$

Remarks.

(a) For the smallest possible $n \equiv 2\pmod 4$, and $n \ge 6$, (i.e., for $n = 6$) distance d is the 3-extension of $d_6 = 2d(K_6 - P_{(1,6,2)})$, corresponding to the special point $(1,1,0,0,0,1)$. On the other hand, the 3-extension of $2d(K_5 - P_{(1,2,5)})$ by the point 6 is an h-point.
For $n \equiv 0$ and $n \equiv 3 \pmod 4$ this d is an antipodal extension at the point 2, i.e., $d_{in} + d_{2i} = d_{2n}$ for all i.

(b) If we consider λ_i^0 such that $\lambda_{12ij}^0 = 0$ or 1, the problem is reduced to a factorization of the graph whose edges are pairs (ij) such that $\lambda_{12ij}^0 \ne 0$.

(c) In fact, we can take t slightly smaller. By (9), we must have $t \ge (j_m, \mu)$. Let r be the number of $S \in \mathcal{S}_0$ such that $\lambda_S = 1$. Then $(j_m, \mu) \le (1/2)(\sum_{S \in \mathcal{S}_0} \lambda_S^0 - r)$.

Proposition 6.2. *Let $\mathcal{K}$ be the family of cuts lying on the 0-lifting $F(n)$ of the main facet $F_0(n)$. Then $A(\mathcal{K}) = \emptyset$ if and only if $n \le 5$.*

Proof. By Lemma 6.1, $F(6)$ has quasi-h-points, and (6) implies that quasi-h-points exist in all $F(n)$ for $n > 6$. We prove that there is no quasi-h-point on $F(n)$ for $n \le 5$.

We use the above notation and the equations $B(\lambda + \mu) = d^0$, $D_n(\lambda - \mu) + \gamma_1 j_n = d^1$. The first equation has the unique solution $\lambda + \mu = \lambda^0$. Hence $2D_n\lambda - D_n\lambda^0 + \gamma_1 j_n = d^1$, where $\gamma_1 = \gamma + (j_{m_0}, \lambda^0) - (j_{m_0}, \lambda)$. The last row gives $\gamma_1 = d_{n,n+1}$. Hence the ith row of the equation with D_n takes the form

$$(D_n\lambda)_i = \frac{1}{2}((D_n\lambda^0)_i + d_{i,n+1} - d_{n,n+1}).$$

It can be shown that the condition of evenness (3) implies that the right-hand side is an integer for $n \le 5$. Moreover, for $n \le 5$, the matrix D_n is unimodular, i.e., $|\det D'| \le 1$ for each $n \times n$ submatrix D' of D_n. Therefore any solution λ is an integer. This implies that μ and $\gamma = d_{n,n+1} - (j_{m_0}, \mu)$ are integers, too.

So, all points $d \in L_{n+1} \cap F(n)$ have a $\mathbb{Z}_+$-realization (λ, μ, γ) for $n \le 5$. $\square$

We now give some other examples of $\mathbb{Z}_+$-realizations of t-extensions of even h-points.

Using the fact that $\sum_{i \in V_n} \delta(i)$ is the unique $\mathbb{Z}_+$-realization of $2d(K_n)$ for $n \neq 4$, (see [5]), we obtain the following lemma.

Lemma 6.3. *The only $\mathbb{Z}_+$-realizations of $ext_t(2d(K_n))$, $n \geq 5$, $t \in \mathbb{Z}_+$, are*

$$\text{(1)} \qquad \sum_{i \in V_n} \delta(i) + (t-1)\delta(n+1) \text{ for } t \geq 1,$$

$$\text{(1')} \qquad \sum_{i \in V_n} \delta(i, n+1) + (t-n+1)\delta(n+1) \text{ for } t \geq n-1.$$

Proof. Note that $d^0 = 2d(K_n)$ is an even $(0,1)$-point of C_n. The coefficients of its $(0,1)$-realization λ^0 are as follows: $\lambda_S^0 = 1$ if $S = \{i\}$, $1 \leq i \leq n-1$, or $S = V_{n-1}$, and $\lambda_S^0 = 0$ for other S. (Recall that we use S such that $n \notin S$.) Since it is a unique $\mathbb{Z}_+$-realization of d^0, the equation $B\lambda^+ = d^0$ has the unique integral solution $\lambda^+ = \lambda^0$.

The submatrix of D consisting of columns corresponding to S with $\lambda_S^+ \neq 0$, and without the last zero row, has the form $D = (I_{n-1}, j_{n-1})$. Hence the unique (± 1)-solutions of the equation $D\lambda^- = 0$ are as follows:

(1) $\lambda_i^- = 1$, $1 \leq i \leq n-1$, $\lambda_{V_{n-1}}^- = -1$, and
(2) $\lambda_i^- = -1$, $1 \leq i \leq n-1$, $\lambda_{V_{n-1}}^- = 1$.

Since $(j_m, \mu) = 1$ in the first case, and $(j_m, \mu) = n-1$, in the second, we have $\gamma = t-1$, and $\gamma = t-n+1$, respectively. These solutions give the above $\mathbb{Z}_+$-realizations (1) and (1'). $\qquad \square$

If we define $d^{n,t} = ant_{2t} ext_t(2d(K_{n-1}))$, we obtain

$$d_{ij}^{n,t} = 2, \ 1 \leq i < j \leq n-1, \ d_{i,n} = d_{i,n+1} = t, \ 1 \leq i \leq n, \ d_{n,n+1} = 2t.$$

If we apply (4) to (1) and (1') of Lemma 6.3 (where n is interchanged with $n-1$), we obtain (2), and (2) with n and $n+1$ interchanged, of Lemma 6.4 below. Summing these two expressions, we obtain the symmetric expression (3) of that lemma.

Lemma 6.4. *For $d^{n,t}$ the following holds*

$$\text{(2)} \qquad d^{n,t} = \sum_{i \in V_{n-1}} \delta(i, n+1) + (t-1)\delta(n) + (t-n+2)\delta(n+1),$$

$$\text{(3)} \qquad 2d^{n,t} = \sum_{i \in V_{n-1}} (\delta(i, n) + \delta(i, n+1)) + (2t-n+1)(\delta(n) + \delta(n+1)).$$

Lemma 6.5. *For $n \geq 6$, $d^{n,t}$ is h-embeddable if and only if $t \geq n-2$. Moreover, for $t \geq n-2$, the only $\mathbb{Z}_+$-realizations are (2) and its image under the transposition $(n, n+1)$.*

Proof. In fact, if we use Lemma 6.4, the restrictions of an h-embedding of $d^{n,t}$ onto $V_{n+1} - \{n\}$ and V_n has to be of the form (1) and (1') or (1') and (1). $\qquad \square$

The realizations (2) and (3) of Lemma 6.4 imply

Corollary 6.6. $d^{n,t}$ *is a quasi-h-point of* C_n *and* $(antC_n) \cap C_{n+1}^{1,2}$ *having the scale 2 if* $\lceil (n-1)/2 \rceil \leq t \leq n-3$, $n \geq 5$.

In fact, for $n = 7$ we only have to prove that $2d(K_6 - P_{(5,6)})$ is a quasi-h-point of scale 2, and this will be done in Section 7. For $n \geq 8$ we use (2), (3) and Lemma 6.4.

Remark. $d^{n-1,2} = 2d(K_n - P_2)$ and it is a quasi-h-point for $n \geq 6$. Its scale lies in the segment $[\lceil n/4 \rceil, n/2)$. $d^{n-1,2} \in Z(ant\mathcal{H}_{n-1} \cap \mathcal{H}_n^{1,2})$ (see Remark (c) following Lemma 7.1 below) for $n \geq 6$, but $d^{n-1,2} \in R_+(ant\mathcal{H}_{n-1} \cap \mathcal{H}_n^{1,2})$ only for $n = 6$.

The cone $(antC_{n-1}) \cap C_n^{1,2}$ has excess 1. It has $2n - 2$ cuts $\delta(i, n-1), \delta(i, n), \delta(n-1), \delta(n)$, for $i \in V_{n-2}$, its dimension is $2n - 3$, and there is the following unique linear dependency:

$$\sum_{i \in V_{n-2}} \delta(i, n-1) + (n+4)\delta(n) = \sum_{i \in V_{n-2}} \delta(i, n) + (n-4)\delta(n-1).$$

The two sides of this equation differ only by the transposition $(n-1, n)$.

The number of quasi-h-points in $(antC_{n-1}) \cap C_n^{1,2}$ is 0 for $n = 5$ (since it is so for the larger cone C_5) and $\geq n - 2 - \lceil n/2 \rceil = \lfloor n/2 \rfloor - 2$, which is implied by Corollary 6.6. Perhaps, it is exactly 1 for $n = 6, 7$.

7. Cones on 6 points

Consider the following cones generated by cut vectors on 6 points:

$$C_6, \; C_6^1, \; C_6^2 = EvenC_6, \; C_6^3, \; C_6^{1,2}, \; C_6^{1,3} = OddC_6, \; C_6^{2,3}, \; antC_5.$$

Recall (see Section 3) that the facets of C_6 are, up to permutations of V_6, as follows:

(a) 3-fold 0-lifting of the main 3-facet, 3-gonal facet $Hyp_6(1^2, -1, 0^3)$,
(b) 0-lifting of the main 5-facet, 5-gonal facet $Hyp_6(1^3, -1^2, 0)$,
(c) the main 6-facet and its 'switching' (7-gonal simplicial facets) $Hyp_6(2, 1^2, -1^3)$ and $Hyp_6(-2, -1, 1^4)$.

Let

$$d_6 := 2d(K_6 - P_{(5,6)}).$$

Recall that (up to permutations) d_6 is the only known quasi-h-point of C_6.

The following lemma is useful for what follows. It can be checked by inspection. Recall that $V_n = \{1, 2, ..., n\}$.

Lemma 7.1.

(1) *All* $\mathbb{Z}_+$*-realizations of* $2d_6$ *are*

(1a)
$$2d_6 = \sum_{i \in V_4} (\delta(i, 5) + \delta(i, 6)) \in \mathbb{Z}_+(\mathcal{H}_6^2) = \mathbb{Z}_+(Even\mathcal{H}_6),$$

(1b)
$$2d_6 = (\delta(5) + \delta(6)) + \sum_{i \in V_3} (\delta(i, 4, 5) + \delta(i, 4, 6)) \in$$

$$\mathbb{Z}_+(\mathcal{H}_6^{1,3}) = \mathbb{Z}_+(Odd\mathcal{H}_6),$$

(1c) $$2d_6 = \delta(5) + \delta(j,5) + \sum_{i \in V_4 - \{j\}} (\delta(i,j,6) + \delta(i,6)) \text{ for } j \in V_4.$$

(2) *Some representations of $d_6 = 2d(K_6 - P_{(5,6)})$ in L_6 are*

(2a) $$d_6 = \delta(5) + \sum_{i \in V_4} \delta(i,6) - \delta(6) \in L_6^{1,2},$$

(2b) $$d_6 = 2\delta(5) + 2\delta(6) + \sum_{i \in V_4} \delta(i) - \delta(5,6) \in L_6^{1,2},$$

(2c) $$d_6 = \sum_{i \in V_4} \delta(V_4 - \{i\}) - \delta(5,6) - \sum_{i \in V_4}(\delta(i,i+1,6) - \delta(i,i+1)) \in L_6^{2,3}.$$

Here $i+1$ is taken by mod 4.

Remarks.

(a) The projection of 2(a) onto $V_6 - \{1\}$ gives the $\mathbb{Z}_+$-realization $2d(K_5 - P_{(5,6)}) = \delta(5) + \sum_{i=2,3,4} \delta(i,6)$; it and its permutation by the transposition $(5,6)$ are the only $\mathbb{Z}_+$-realizations of the above h-point.

(b) 'Small' pertubations of d_6 do not produce other quasi-h-points. For example, one can check that

$$d_6 + \delta(1,2) = \delta(1) + \delta(2) + \delta(6) + \delta(1,2,5) + \delta(3,5) + \delta(4,5);$$

it and its permutation by the transposition $(5,6)$ are the only $\mathbb{Z}_+$-realizations of this h-point.

(c) Actually, 2(a) is the case $n = 5, \alpha = 4$ of

$$ant_\alpha(2d(K_n)) = \delta(n) + \sum_{i \in V_{n-1}} \delta(i, n+1) - (n - \alpha)\delta(n+1)$$

$$= \sum_{i \in V_{n+1}} \delta(\{i\}) + (\frac{\alpha}{2} - 1)(\delta(\{n\}) + \delta(\{n+1\}) - \delta(\{n, n+1\})).$$

(d) One can check that $L_n^{\neq 1} \subset L_n$ strictly, and $2\mathbb{Z}^{15} \subset L_6^{\neq 1}$ strictly. Note that $L_6^{2,3} = L_6^{\neq 1}$. On the other hand, $L_n^{i,j} = L_n$ if and only if $(i,j) = (1,2)$.

(e) By 1(a) and 1(b) of Lemma 7.1 we have

$$2d_6 \in hC_6^2 \text{ and } 2d_6 \in hC_6^{1,3},$$

but $2d_6 \notin L_6^2 \cup L_6^{1,3} = L(Even \mathcal{K}_6) \cup L(Odd \mathcal{K}_6)$.

We call a subcone of C_n a *cut subcone* if its extreme rays are cuts.

Lemma 7.2. *Let $d \in A(\mathcal{K})$ and let $\mathcal{K}(d)$ be the set of cuts of a minimal cut subcone of C_n containing d. Then*

(i) *$d \in A(\mathcal{K}')$ for any $\mathcal{K}'$ such that $\mathcal{K}(d) \subseteq \mathcal{K}' \subseteq \mathcal{K}$,*

(ii) *$e(\mathcal{K}') = 1$ implies $\mathcal{K}' = \mathcal{K}(d)$.*

Proof. In fact, $d \notin \mathbb{Z}_+(\mathcal{K}(d))$ implies $d \notin \mathbb{Z}_+(\mathcal{K}')$, and $d \in \mathbb{Z}(\mathcal{K}(d)) \cap C(\mathcal{K}(d))$ implies $d \in \mathbb{Z}(\mathcal{K}') \cap C(\mathcal{K}')$, and (i) follows. If $e(\mathcal{K}') = 1$, any proper cut subcone of $C(\mathcal{K})$ is simplicial and has no quasi-h-points. $\qquad\square$

Now we remark that the cone $C_6^{1,2} \cap \mathit{ant}\, C_5$ has excess 1, since it has dimension 9 and contains 10 cuts $\delta(5), \delta(6), \delta(i,5), \delta(i,6),\ 1 \le i \le 4$, with the unique linear dependency

$$\sum_{i \in V_n}(\delta(i,5) - \delta(i,6)) = 2(\delta(5) - \delta(6)).$$

Proposition 7.3. $d_6 = 2d(K_6 - P_2) \in A(\mathcal{K}_6)$ and it is a quasi-h-point of the following proper subcones of C_6: $C_6^{1,2}$, $C_6^{2,3}$, ant C_5, the triangle facet $Hyp(1^2, -1, 0^3)$ and $C_6^{1,2} \cap \mathit{ant}\, C_5$ (which is a minimal cut subcone of C_6 containing d).

Proof. The point d_6, is the antipodal extension $\mathit{ant}_4(d_5)$ of the point $d_5 := 2d(K_5)$. The minimum size of $\mathbb{Z}_+$-realizations of d_5 is equal to $z(d_5) = z_5^1 = 5$, since its only $\mathbb{Z}_+$-realization is the following decomposition $2d(K_5) = \sum_{i=1}^5 \delta(i)$.

The minimum size of $\mathbb{R}_+$-realizations of d_5 is $s(d_5) = a_5^1 = 10/3$, which is given by the $\mathbb{R}_+$-realization $d_5 = (1/3)/\sum_{1 \le i < j \le 5} \delta(ij)$.

Since $10/3 < 4 < 5$, we deduce that $d_6 = 2d(K_6 - P_{\{5,6\}}) \notin \mathbb{Z}_+(C_6)$.

But $d_6 \in C_6 \cap L_6$, from (1) and (2) of Lemma 7.1. So, $d_6 \in A_6^0$. Now, from 1(a) and (2) of the same lemma, we have $d_6 \in C(\mathcal{K}_6^{1,2} \cap \mathit{ant}\, \mathcal{K}_5) \cap L(\mathcal{K}_6^{1,2} \cap \mathit{ant}\, \mathcal{K}_5)$, and so, using (ii) of Lemma 7.2, we get that $\mathcal{K}_6^{1,2} \cap \mathit{ant}\, \mathcal{K}_5$ is a minimal subcone $\mathcal{K}(d)$.

Using (i) of Lemma 7.2, and the fact that $\mathit{ant}\, C_5$ is the intersection of some triangular facets, we get the assertion of Proposition 7.3 for $C_6^{1,2}$, $\mathit{ant}\, C_5$ and the triangle facet. Finaly, 1(a) and 2(c) of Lemma 7.1 imply that $d_6 \in A(\mathcal{K}_6^{2,3})$. $\qquad\square$

Remarks.

(a) On the other hand, the following subcones $C(\mathcal{K})$ of C_6 have $A(\mathcal{K}) = \emptyset$: 5 simplicial cones C_6^i, $i = 1, 2, 3$, both 7-gonal facets, and nonsimplicial cones: C_5, $C_6^{1,3} = OddC_6$, and 5-gonal facet.

(b) Nonsimplicial cones $C_6, C_6^{1,2}, C_6^{2,3}, C_6^{1,3}, C_5, \mathit{ant}\, C_5, Hyp_6(1^2, -1, 0^3),\ Hyp_6(1^3, -1^2, 0)$ have excess 16, 6, 10, 1, 5, 5, 9, 5, respectively. The cones $C_6, C_6^{1,2}, C_6^{2,3}, C^{1,3}, C_5$ have, respectively, 210, 495, 780, 60, 40 facets and the facets are partitioned, respectively, into 4, 5, 8, 1, 2 classes of equivalent facets up to permutations.

8. Scales

In this section we consider the scale $\eta^0(\mathit{ant}_\alpha 2d(K_n))$, which is, by Proposition 4.1(iii), equal to $\min\{t \in \mathbb{Z}_+ : \alpha t \ge z_n^t\}$, especially for two extreme cases $\alpha = 4$ and $\alpha = n - 1$. The number t below is always a positive integer.

Denote by $H(4t)$ a Hadamard matrix of order $4t$, and by $PG(2, t)$ a projective plane of order t.

It is proved in [5] that $t \sum_1^n \delta(\{i\})$ is the unique $\mathbb{Z}_+$-realization of $2td(K_n)$ if $n \ge t^2 + t + 3$, and that for $n = t^2 + t + 2$, $2td(K_n)$ has other $\mathbb{Z}_+$-realizations if and only if there exists a $PG(2,t)$. Below, in $(iv_1) - (iv_3)$ of Theorem 8.1, we reformulate this result in terms of A_n^1, $\eta^1(2d(K_n))$, z_n^t, using the following trivial relations

$$\eta^1(2d(K_n)) \geq t + 1 \Leftrightarrow 2td(K_n) \in A_n^1 \Leftrightarrow z_n^t = nt \Leftrightarrow$$

$$\Leftrightarrow t \sum_1^n \delta(\{i\}) \text{ is the unique } \mathbb{Z}_+\text{-realization of } 2td(K_n).$$

(iii_2) of Theorem 8.1 follows from a result of Ryser (reformulated in terms of z_n^t in [9, Theorem 4.6(1)]) that $z_n^t \geq n - 1$ with equality if and only if $n = 4t$ and there exists an $H(4t)$.

Theorem 8.1.

(i_1) $ant_\beta 2td(K_n) \in C_{n+1}$ if and only if $\beta \geq \frac{tn(n-1)}{\lfloor n/2 \rfloor \lceil n/2 \rceil}$;

(i_2) $ant_\beta 2td(K_n) \in A^0$ if and only if $\frac{tn(n-1)}{\lfloor n/2 \rfloor \lceil n/2 \rceil} \leq \beta < z_n^t$, $\beta \in \mathbb{Z}_+$;

(i_3) $ant_\beta 2td(K_n) \in hC_{n+1}$ if and only if $\beta \geq z_n^t$, $\beta \in \mathbb{Z}_+$;

(i_4) $ant_\alpha 2d(K_n) \in C_{n+1} \cap L_{n+1}$ if and only if $\frac{n(n-1)}{\lfloor n/2 \rfloor \lceil n/2 \rceil} \leq \alpha$, $\alpha \in \mathbb{Z}_+$.

Moreover, if $d = ant_\alpha 2d(K_n) \in C_{n+1} \cap L_{n+1}$, *then*

(ii_1) *either* $n = 3, d \in A_3^1$, *is simplicial,* $d = ant_3 2d(K_4)$ *(so* $\eta^i(d) = 1$ *for* $i \geq 0$*),*

or $d \in A_n^1$, *d is not simplicial,* $\alpha \geq n \geq 4$ *(so* $\eta^0(d) = 1$*), or* $d \in A_n^0$ *(so* $\eta^0(d) \geq 2$*).*

(ii_2) $\eta^0(d) = min\{t : z_n^t \leq \alpha t\}$.

(iii_1) $\eta^0(ant_4 2d(K_n)) = \eta^0(2d(K_{n+1} - P_{(1,2)})) = \eta^0(2d(K_{n \times 2}))$;

(iii_2) $\lceil n/4 \rceil \leq \eta^0(ant_4 2d(K_n)) \leq min\{t \in Z_+ : n \leq 4t \text{ and there exists a } H(4t)\} < n/2$;

(iii_3) *For* $n = 4t, 4t - 1$, *we have* $\eta^0(ant_4 2d(K_n)) = \lceil n/4 \rceil = t$ *if and only if there exists an* $H(4t)$;

(iv_1) $\eta^0(ant_{n-1} 2d(K_n)) = \eta^1(2d(K_n)) \leq min\{n - 3, \eta^1(2d(K_{n+1}))\}$;

(iv_2) $\left\lceil (1/2)(\sqrt{4n-7} - 1) \right\rceil = min\{t \in Z_+ : n \leq t^2 + t + 2\}$

$$\leq \eta^0(ant_{n-1} 2d(K_n))$$

$$\leq min\{t \in Z_+ : n \leq t^2 + t + 2 \text{ and there exists a } PG(2,t)\};$$

(iv_3) *For* $n = t^2 + t + 2$, *we have* $\eta^0(ant_{n-1} 2d(K_n)) = \left\lceil (1/2)(\sqrt{4n-7} - 1) \right\rceil = t$ *if and only if there exists a* $PG(2,t)$.

Remarks.

(a) For $i \geq 0$, we have $\eta^{i+1}(2d(K_4)) = i + 1$, but $\eta^i(ant_3(2d(K_4))) = 1$, since $ant_3(2d(K_4))$ is a simplicial point. For $i \geq 0$ and $n \geq 5$, we have $\eta^{i+1}(2d(K_n)) \leq \eta^i(ant_{n-1}(2d(K_n)))$ with equality for $i = 0$ and for some pair (i, n) with $i \geq 1$. Propositions 5.9–5.11 of [9] imply that

$$\eta^{i+1}(2d(K_5)) = \eta^i(ant_4(2d(K_5))) = 2 \text{ for } i = 0, 1;$$

$$\eta^3(2d(K_5)) = \eta^2(ant_4(2d(K_5))) = \eta^4(2d(K_5)) = 3;$$

$$\eta^5(2d(K_5)) = \eta^4(ant_4(2d(K_5))) = \eta^3(ant_4(2d(K_5))) = 4.$$

(b) Using the well-known fact that $H(4t)$ exists for $t \le 106$, we obtain

$$\eta^0(ant_4(2d(K_n))) = \eta^0(2d(K_{n+1} - P_2)) = \eta^0(2d(K_{n \times 2})) = \lceil n/4 \rceil \text{ for } n \in [4, 424];$$

(c) Using the well-known fact that $PG(2,t)$, $t \le 11$, exists if and only if $t \ne 6, 10$, we get for $a_n = \eta^0(ant_{n-1}(2d(K_n))) = \eta^1(2d(K_n))$, that $6 \le a_n \le 7$ for $33 \le n \le 43$, $10 \le a_n \le 11$ for $93 \le n \le 111$, and $a_n = \lceil (1/2)(\sqrt{4n-7} - 1) \rceil$ for all other $n \in [4, 134]$.

(d) (iii), (iv) of Theorem 8.1 imply that

$$\eta^0(d(K_{2t \times 2})) \ge 2t \text{ with equality if and only if there exists } H(4t),$$

$$\eta^1(d(K_{t^2+t+2})) \ge 2t \text{ with equality if and only if there exists } PG(2,t).$$

Note also that $a_n \le n - 3$ with equality if and only if $n = 4, 5$.

Proof of (iv_1). For $n \ge 4$ we have

$$\left\lceil \frac{1}{2}(\sqrt{4n-7} - 1) \right\rceil \le \eta^1(2d(K_n)) = \eta^0(ant_{n-1}(2d(K_n))) \le n - 3.$$

In fact, we have

$$\eta^1(2d(K_n)) = \min\{t \in Z_+ : z_n^t < nt\},$$

$$\eta^0(ant_N(2d(K_n))) = \min\{t \in Z_+ : z_n^t \le Nt\},$$

since $2td(K_n)$ has the following Z_+-realization $t\sum_1^n \delta(\{i\})$ of maximal size nt, and since $t(ant_N(2d(K_n))) \in hC_{n+1}$ if and only if $2td(K_n)$ admits a Z_+-realization of size at most Nt. Denote

$$p = \eta^1(2td(K_n)), \quad q = \eta^0(ant_{n-1}(2d(K_n))).$$

Then $p \le q$, because $z_n^q \le (n-1)q$ implies $z_n^q \le nq$. Also, $q \le n - 3$, because $2(n-3)d(K_n)$ has the Z_+-realization $\sum_1^{n-1}((n-4)\delta(\{i\}) + \delta(\{i,n\}))$ of size $(n-3)(n-1)$. On the other hand, $p \ge q$, because $z_n^p < np$ implies $z_n^p \le np - (n-3)$, which is proved in [9, Proposition 5.3]. So $z_n^p \le np - q \le np - p$. We have $p \ge \lceil (1/2)(\sqrt{4n-7} - 1) \rceil$, because otherwise $n \ge p^2 + p + 3$, and using [5], $2td(K_n)$ has exactly one Z_+-realization, in contradiction with the definition of p. □

Theorem 8.2. *Let $\eta_n^i = \eta^i(2d(K_n))$. Then*

(i) $\eta_n^0 < \infty$ *for* $d \in L_n \cap C_n$,

(ii) $\eta_n^{i-1}|\eta_n^i$ *for* $i \ge 1$, *and* $\eta_{n-1}^i|\eta_n^i$ *for* $n \ge 5$,

(iii) $\eta^i(ad) = \lceil \eta^i(d)/a \rceil$ *for* $d \in C_n \cup L_n$, $i \ge 0$, $a \in Z_+$.

Proof. (i) Define

$$Y = L_n \cap C_n \cap \left\{\sum \lambda_S \delta(S) : 0 \le \lambda_S \le 1\right\}.$$

Clearly, Y is finite, and one can find $\lambda \in Z_+$ such that λd is an h-point for every $d \in Y$.

Let $d \in L_n \cap C_n$ have an R_+-realization $d = \sum \mu_S \delta(S)$. Clearly the coefficients μ_S are rational numbers. We have $d = d_1 + d_2$, where $d_1 = \sum \lfloor \mu_S \rfloor \delta(S)$, and $d_2 = \sum(\mu_S - \lfloor \mu_S \rfloor)\delta(S)$. By the construction, d_1 is an h-point. Since $d_2 = d - d_1$ and $d \in L_n \cap C_n$, $d_1 \in$

$L_n \cap C_n$, we obtain $d_2 \in Y$. Hence there is λ such that $\lambda d_2 \in hC_n$, and we obtain that $\lambda d = \lambda d_1 + \lambda d_2$ is an h-point, too.

(ii) Obvious.

(iii) Take $\lambda = \eta^i(ad)$, that is $\lambda(ad)$ has at least $i + 1$ $\mathbb{Z}_+$-realizations. Hence $\lambda a \geq \eta^i(d)$ implies $\lambda \geq \lceil \eta^i(d)/a \rceil$, that is, $\eta^i(ad) \geq \lceil \eta^i(d)/a \rceil$.

Now, take $\lambda = \lceil \eta^i(d)/a \rceil$. So, $\lambda - 1 < \eta^i(d)/a \leq \lambda \Rightarrow (\lambda - 1)a < \eta^i(d) \leq \lambda a$. Hence λad has at least $i + 1$ $\mathbb{Z}_+$-realizations, implying that $\lambda \geq \eta^i(ad)$, and so $\lceil \eta^i(d)/a \rceil \geq \eta^i(ad)$. $\square$

Remarks.

(a) $\eta_4^i = \eta^i(2d(K_4)) = i$ for $i \geq 1$; $\eta_n^0 = 1$ if and only if $n = 4, 5$.

(b) For $d \notin L_n$ and $\lambda \in \mathbb{Z}_+$, we have $\lambda d \in L_n$ implies that λ is even (because $(\lambda d_{ij} + \lambda d_{ik} + \lambda d_{jk})/2 = \lambda(d_{ij} + d_{ik} + d_{jk})/2$). Hence, for $d \in \mathbb{Z}^{\binom{n}{2}} - A_n^0$, we have either $d \notin L_n$ (so $\eta^0(d)$ is even), or $\eta^0(d) = 1$ (i.e. $d \in hC_n$). Since $d(G) \notin A_n^0$ for any connected graph G on n vertices (see [14]), we have either $\eta^0(d(G)) = 1$ or $\eta^0(d(G))$ is even. But, for example, $\eta^0(2d(K_{10} - P_2)) = \eta^0(2d(K_{9 \times 2})) = 3$.

It will be interesting to see whether η_n^0 and $\max\{\eta^0(d) : d \in A_n^0\}$ are bounded from above by $const \times n$.

The best-known lower bound for the last number is $\eta^0(d(K_n - P_2))$, which belongs to the interval $[2 \lceil (n-1)/4 \rceil, n-2]$.

It is proved in [19] that for a graphic metric $d = d(G)$, we have

(i) $\eta^0(d) \leq n - 2$ if $d(G) \in C_n$,

(ii) $\eta^0(d) \in \{1, 2\}$, that is, G is an isometric subgraph of a hypercube or a halved cube if $d(G)$ is simplicial.

9. h-points

Recall that any point of $\mathbb{Z}_+(\mathcal{K}_n) = hC_n$ is called an h-point.

A point d is called *k-gonal*, if it satisfies all hypermetric inequalities $Hyp_n(b)$ with $\sum_1^n |b_i| = k$.

The following cases are examples of when the conditions $d \in L_n$ and hypermetricity of d imply that d is an h-point.

(a) [14], [17]: If $d = d(G)$ and G is bipartite, then 5-gonality of d implies that $d \in hC_n$;

(b) [1]: If $\{d_{ij}\} \in \{1, 2\}$, $1 \leq i < j \leq n$, then $d \in L_n$ and 5-gonality of d imply that $d \in hC_n$ (actually, $d = d(K_{1,n-1})$, $d(K_{2,2})$ or $2d(K_n)$ in this case);

(c) [2]: If $n \geq 9$ and $\{d_{ij}\} \in \{1, 2, 3\}$, $1 \leq i < j \leq n$, then $d \in L_n$ and ≤ 11-gonality of d imply that $d \in hC_n$.

So, the cases (a), (b), (c) are among known cases when the problem of testing membership of d in hC_n can be solved by a polynomial time algorithm. The polynomial testing holds for any $d = d(G)$ (see [19]) and for 'generalized bipartite' metrics (see [7] which generalizes the cases (b) and (c) above).

Cases (a), (b) and (c) imply (i), (ii) and (iii), respectively, of

Corollary 9.1. *If* $d \in A_n^0$, *then none of the following hold*

(i) $d = d(G)$ *for a bipartite graph* G,

(ii) $\{d_{ij}\} \in \{1, 2\}$, $1 \le i < j \le n$,

(iii) $\{d_{ij}\} \in \{1, 2, 3\}$, $1 \le i < j \le n$, *if* $n \ge 9$.

A point $d \in \mathbb{Z}_+(\mathcal{H}_n) = hC_n$ is called *rigid* if d admits a unique $\mathbb{Z}_+$-realization. In other words, d is rigid if and only if $d \in A_n^1$. Clearly, if $d \in hC_n$ is simplicial, d is rigid. Rigid nonsimplicial points are more interesting. Hence we define the set

$$\tilde{A}_n^1 := \{d \in A_n^1 : d \text{ is not simplicial}\},$$

and call its points *h-rigid*.

Theorem 9.2.

(i) $A_n^0 = \emptyset$ for $n \le 5$, $2d(K_6 - P_2) \in A_6^0$, $|A_n^0| = \infty$ for $n \ge 7$,

(ii) $\tilde{A}_n^1 = \emptyset$ for $n \le 4$, $\tilde{A}_5^1 = \{2d(K_5)\}$, $|\tilde{A}_n^1| = \infty$ for $n \ge 6$,

(iii) for $i \ge 2$, $A_n^i = \emptyset$ if $n \le 3$, $|A_n^i| = \infty$ if $n \ge 4$.

Proof. (i) and (ii) The first equalities in (i) and (ii) are implied by results in [3]. The inclusion in (i) is implied by [1]. The second equality in (ii) is proved in [13]. We have $|A_n^0| = \infty$ for $n \ge 7$, because $A_6^0 \ne \emptyset$ and $|A_{n+1}^i| = \infty$ whenever $A_n^i \ne \emptyset$ from (6).

We prove the third equality of (ii): $|\tilde{A}_n^1| = \infty$ for $n \ge 6$. The equality is implied by the fact that $ant_\alpha(2d(K_n)) \in \tilde{A}_{n+1}^1$ for any $n \ge 5$, $\alpha \in \mathbb{Z}_+$, $\alpha \ge n$. We prove the inclusion.

Recall that $2td(K_n)$ has the unique $\mathbb{Z}_+$-realization of size tn if $n \ge t^2 + t + 3$. (See [5] or the beginning of Section 8). For $t = 1$ we obtain the equality $z(2d(K_n)) = n$ for $n \ge 5$. Using the fact that $2d(K_n)$ is not simplicial for $n \ge 4$, and (iv) of Proposition 4.1 we obtain the required inclusion.

(iii) Since C_3 is simplicial, $A_3^i = \emptyset$ for $i \ge 2$. Consider now $n = 4$. We show that $A_4^i = \{2(i-1)d(K_4) + d : d \text{ is a simplicial h-point of } C_4\}$. This follows from the fact that the only linear dependency on cuts of C_4 is, up to a multiple, $\delta(1) + \delta(2) + \delta(3) + \delta(4) = \delta(1, 4) + \delta(2, 4) + \delta(3, 4)$.

So, $|A_4^i| = \infty$, because there are an infinity of simplicial points, e.g., $\lambda d(K_{2,2})$ for $\lambda \in \mathbb{Z}_+$. Finally we use (6). □

Some questions.

(a) Is it true that all 10 permutations of $d_6 = 2d(K_6 - P_2)$ are only quasi-h-points of C_6? If yes, these 10 points and 31 nonzero cuts from $\mathcal{H}_6$ form a Hilbert basis of C_6.

(b) Does there exist a ray $\{\lambda d : \lambda \in \mathbb{R}_+\} \subset C_n$ containing an infinite set of quasi-h-points? Recall that we got in Section 6 examples of rays $\{d^0 + td^1 : t \ge 0\}$ containing infinitely many quasi-h-points.

Lemma 9.3. *Let* $d \in A_n^0$, *and let* $d = ant_\alpha d'$ *where* $d' \notin A_{n-1}^0$. *Then* d' *is an h-point and* $z(d') \ge \lceil s(d') \rceil + 1$.

Proof. In fact, $d \in C_n \cap L_n$, so $d' \in C_{n-1} \cap L_{n-1}$. But $d' \notin A_{n-1}^0$, so d' is an h-point of C_{n-1}. Hence by Proposition 4.1(ii), $\alpha \in \mathbb{Z}_+$, $s(d') \le \alpha < z(d')$.

Note that for $n \geq 5$ we have $2d(K_{n \times 2}) \in A_{2n}^0$, $2d(K_{n \times 2}) = ant_4 d'$, where $d' \in A_{2n-1}^0$ and $d' = ant_4 d''$ for $d'' \in A_{2n-2}^0$, and so on.

So, d' is neither a simplicial point nor an antipodal extension (*i.e.* $d' \notin \mathbb{R}_+(ant\mathcal{K}_{n-2})$), nor $d' \in \mathbb{Z}_+(\mathcal{K}_{n-1}^m)$, $m = \lfloor (n-1)/2 \rfloor$, because in each of these 3 cases we have for an h-point d', $z(d') = s(d')$; this also implies that by Proposition 4.1(iv), d itself is not simplicial. $\square$

The following proposition makes plausible the fact that the metric $d_6 = 2d(K_6 - P_2)$ is the unique (up to permutations) quasi-h-point of C_6.

Proposition 9.4. *Let* $d \in A_6^0$, $d = ant_\alpha d'$ *and* $d \neq d_6$. *Then*

(a) *both* d *and* d' *are not simplicial;*
(b) $d' \notin \mathbb{R}_+(ant\mathcal{K}_4)$, $d' \notin \mathbb{Z}_+(\mathcal{K}_5^2)$;
(c) $d' \neq \lambda d(G)$ *for any* $\lambda \in \mathbb{Z}_+$ *and any graph* G *on 5 vertices;*
(d) d' *has at least two* $\mathbb{Z}_+$-*realizations.*

Proof. Since $A_5^0 = \emptyset$ by [3], we can apply Proposition 9.3, and (a) and (b) follow. One can see by inspection, that among all 21 connected graphs on 5 vertices, the only graphs G with nonsimplicial $d(G) \in C_6$ are the following 3 graphs: K_5, $K_5 - P_2$, and $K_4.K_2 = K_4$ with an additional vertex adjacent to a vertex of K_4. For these graphs, $\lambda d(G)$ is an h-point if and only if $\lambda \in 2\mathbb{Z}_+$.

Since $2d(K_5 - P_2) = ant_4(2d(K_4))$, then, according to (b), $d' \neq \lambda d(K_5 - P_2)$.

Since for any $\lambda \in \mathbb{Z}_+$ we have $z(2\lambda d(K_4.K_2)) = 5\lambda = s(2\lambda d(K_4.K_2))$, and (by Proposition 9.3) $s(d') < z(d')$, then $d' \neq \lambda d(K_4.K_2)$.

There remains the case $d' = \lambda d(K_5)$. We have $s(d') = \lambda 5/3$, $z(d') = 5$ for $\lambda = 2$ and $z(d') = s(d')$ for $\lambda \in 2\mathbb{Z}_+$, $\lambda > 2$. (See [9, Proposition 5.11]). So $s(d') \leq \alpha < z(d')$ implies $\lambda = 2$, $\alpha = 4$, *i.e.*, exactly the case $d = ant_4(2d(K_5))$. This proves (c).

Finally, (d) follows from the fact (see [13]) that $2d(K_5)$ is the unique nonsimplicial h-point of C_5 with unique $\mathbb{Z}_+$-realization. $\square$

References

[1] Assouad, P. and Deza, M. (1980) Espaces métriques plongebles dans un hypercube: aspects combinatoires *Annals of Discrete Math.* **8** 197–210.

[2] Avis, D. (1990) On the complexity of isometric embedding in the hypercube: In Algorithms. *Springer-Verlag Lecture Notes in Computer science,* **450** 348–357.

[3] Deza, M. (1960) On the Hamming geometry of unitary cubes. *Doklady Academii Nauk SSSR* **134**, 1037–1040 (in Russian) *Soviet Physics Doklady* (English translation) **5** 940–943.

[4] Deza, M. (1973) Matrices de formes quadratiques non négatives pour des arguments binaires. *C. R. Acad. Sci. Paris* **277** 873–875.

[5] Deza, M. (1973) Une proprieté éxtremal des plans projectifs finis dans une classe de codes equidistants. *Discrete Mathematics* **6** 343–352.

[6] Deza, M. and Laurent, M. (1992) Facets for the cut cone I, II. *Mathematical Programming* **52** 121–161, 162–188.

[7] Deza, M. and Laurent, M. (1991) Isometric hypercube embedding of generalized bipartite metrics, Research report 91706-OR, Institut für Discrete Mathematik, Universität Bonn.

[8]   Deza, M. and Laurent, M. (1992) Extension operations for cuts. *Discrete Mathematics* **106-107** 163–179.

[9]   Deza, M. and Laurent, M (1992) Variety of hypercube embeddings of the equidistant metric and designs. *Journal of Combinatorics, Information and System sciences* (to appear).

[10] Deza, M. and Laurent, M. (1993) The cut cone: simplicial faces and linear dependencies. *Bulletin of the Institute of Math. Academia Sinica* **21** 143–182.

[11] Deza, M., Laurent, M. and Poljak, S. (1992) The cut cone III: on the role of triangle facets. *Graphs and Combinatorics* **8** 125–142.

[12] Deza, M. and Laurent, M. (1992) Applications of cut polyhedra, Research report LIENS 92-18, ENS. *J. of Computational and Applied Math* (to appear).

[13] Deza, M. and Singhi, N. M. (1988) Rigid pentagons in hypercubes. *Graphs and Combinatorics* **4** 31–42.

[14] Djokovic, D.Z. (1973) Distance preserving subgraphs of hypercubes. *Journal of Combinatorial Theory* **B14** 263–267.

[15] Koolen, J. (1990) On metric properties of regular graphs, Master's thesis, Eindhoven University of Technology.

[16] Harary, F. (1969) *Graph Theory*, Addison-Wesley.

[17] Roth, R. L. and Winkler, P. M. (1986) Collapse of the metric hierarchy for bipartite graphs. *European Journal of Combinatorics* **7** 371–375.

[18] Schrijver, A. (1986) *Theory of linear and integer programming*, Wiley.

[19] Shpectorov, S. V. (1993) On scale embeddings of graphs into hypercubes. *European Journal of Combinatorics* **14**.

The Growth of Infinite Graphs: Boundedness and Finite Spreading

REINHARD DIESTEL[†] and IMRE LEADER[‡]

[†]Faculty of Mathematics (SFB 343), Bielefeld University, D-4800 Bielefeld, Germany
[‡]Department of Pure Mathematics, University of Cambridge,
16 Mill Lane, Cambridge, CB2 1SB England

An infinite graph is called bounded if for every labelling of its vertices with natural numbers there exists a sequence of natural numbers which eventually exceeds the labelling along any ray in the graph. Thomassen has conjectured that a countable graph is bounded if and only if its edges can be oriented, possibly both ways, so that every vertex has finite out-degree and every ray has a forward oriented tail. We present a counterexample to this conjecture.

1. The conjecture

For two $\mathbb{N} \to \mathbb{N}$ functions f and g, let us say that f *dominates* g if $f(n) \geq g(n)$ for every n greater than some $n_0 \in \mathbb{N}$.

An infinite graph G is called *bounded* if for every labelling of its vertices with natural numbers, there is an $\mathbb{N} \to \mathbb{N}$ function that dominates every labelling along a ray (one-way infinite path) in G. More precisely, G is bounded if for every labelling $\ell: V(G) \to \mathbb{N}$ there is a function $f: \mathbb{N} \to \mathbb{N}$ such that for every ray $x_0 x_1 \ldots$ in G the function $n \mapsto \ell(x_n)$ is dominated by f. Otherwise G is *unbounded*.

Let us see some examples of bounded and unbounded graphs.

Every locally finite connected graph is bounded. Indeed, given a labelling ℓ, and given any fixed vertex v of G, it is easy to define a function f_v that dominates all the rays starting at v: just take as $f_v(n)$ the largest label of the vertices at distance at most n from v. Now G has only countably many vertices, so there are only countably many functions f_v, say $f^0, f^1, \ldots$. Setting $f(n) = \max_{i \leq n} f^i(n)$, we obtain a function $f: \mathbb{N} \to \mathbb{N}$ that dominates every f_v, and hence dominates every ray in G.

The complete graph on a countably infinite set of vertices, K_ω, is clearly unbounded: just choose any labelling that uses infinitely many distinct labels, and there will be

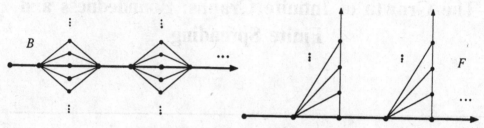

Figure 1 The unbounded graphs B and F

rays whose labellings grow faster than any fixed $\mathbb{N} \to \mathbb{N}$ function. The regular tree of countably infinite degree, T_ω, is another simple example of an unbounded graph: just label its vertices injectively, that is, so that any two labels are different.

Two further examples of unboundedness are found in the graphs B and F shown in Figure 1; again, any injective labelling will show that these graphs are unbounded.

Bounded graphs were first introduced by Halin around 1964, in connection with Rado's well-known paper on *Universal graphs and universal functions* [4]. Halin conjectured that a countable graph is bounded if and only if it has no subgraph isomorphic to a subdivision of any of the three graphs T_ω, B and F. Halin himself proved this for some special cases [2, 3]; the conjecture was recently proved by the authors [1]. (We remark that [1] also contains an uncountable version of this result. In the present paper, however, we are only interested in countable graphs.) An interesting aspect of this 'bounded graph theorem', typical for a characterization by forbidden configurations, is that it provides us with simple 'certificates' for unboundedness: all we need do to convince someone of the unboundedness of a particular countable graph is to exhibit in it one of the three types of forbidden subgraph. For boundedness, by contrast, no such 'certificates' are known.

C. Thomassen has recently proposed the following attractive conjecture, which would have provided not only another elegant characterization of the bounded graphs but also something like a certificate for boundedness:

Conjecture. (Thomassen) *A countable graph is bounded if and only if its edges can be oriented, each in one or both or neither of its two directions, so that every vertex has finite out-degree and every ray has a forward oriented tail.*

(A *tail* of a ray $x_0 x_1 \ldots$ is a subray $x_n x_{n+1} \ldots$, and it is *forward oriented* if every edge $x_m x_{m+1}$ $(m \geq n)$ is oriented from x_m towards x_{m+1} (and possibly, but not necessarily, also from x_{m+1} towards x_m).)

An orientation as above will be called *admissible*. We remark that any admissible orientation can be extended to one in which every edge has at least one direction: since the graph has only countably many vertices, $v_0, v_1, \ldots$ say, local finiteness will be preserved if every unoriented edge $v_i v_j$ with $i < j$ is oriented from v_j to v_i.

Intuitively, an admissible orientation identifies in the graph a locally finite substructure mapping out the preferred directions of rays: eventually, every ray in the graph will follow a ray indicated by the orientation. Much of the attractiveness of Thomassen's conjecture lies in its promise that the boundedness of any bounded graph can be tied to such a definite and simple substructure – one that is obviously itself bounded (by local finiteness), and at the same time accounts for the boundedness of the entire graph.

The 'if' direction of Thomassen's conjecture is clearly true: to prove it, we just imitate the proof that locally finite connected graphs are bounded. More precisely, given an admissible orientation of the graph and any labelling of its vertices, we first find a function f that dominates every forward oriented ray (as in our local finiteness proof); the function g defined by

$$g: n \mapsto f(1) + \cdots + f(2n)$$

then dominates every ray in the graph.

Note also that the conjecture is trivially true for locally finite graphs themselves, as we may simply orient every edge both ways. The provision for 2-way orientations in the definition of admissible is, however, essential: the infinite ladder is an example of a bounded graph whose edges cannot be 1-way oriented in such a way that every ray has a forward oriented tail.

Finally, it is not difficult to prove the conjecture for trees; this was first observed by Thomassen[5].

Unfortunately, Thomassen's conjecture is not true in general: in the next section we shall exhibit a graph that is bounded but allows no admissible orientation of its edges.

2. The counterexample

Let S be the graph constructed as follows (see Fig. 2). For every $n \in \mathbb{N}$, let $R^n = v_0^n v_1^n v_2^n \ldots$ be a ray. Let these rays be pairwise disjoint, except that $v_0^n = v_n^0$ for every n. For every odd n, make the pair (R^n, R^{n+1}) into a ladder by adding the edges $v_i^n v_i^{n+1}$ for all $i > 0$, as rungs. Finally, for every even $n > 0$, add a new vertex x^n and join it to every vertex of R^n except v_0^n.

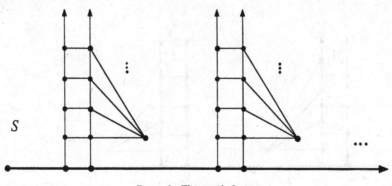

Figure 2 The graph S

Theorem. *The graph S is bounded but allows no admissible orientation of its edges.*

Proof. It is not difficult to see that the edges of S cannot be admissibly oriented. Indeed, as the vertices x^n all have infinite degree, any admissible orientation would leave each x^n incident with an edge $e^n = x^n v_i^n$ (for some i) that is *not* oriented from x^n towards v_i^n. It is then easy to find a ray in S that traverses every such edge e^n from x^n towards v_i^n, that is against its (possible) orientation.

It remains to show that S is bounded. Using the above-mentioned bounded graph theorem, all we need to show is that S contains no subdivision of T_ω, B or F. This is easily done. The cases of T_ω and B are trivial. Now suppose we have embedded a subdivision of F into S. The bottom ray of F will then be mapped to a ray $R \subset S$ that contains infinitely many of the vertices x^n, since these are the only vertices of S that have infinite degree. For each of those n (except possibly the first), the initial segment Rx^n of R separates its tail $x^n R$ from all but finitely many neighbours of x^n in S. As this is not the case for the bottom ray and the vertices of infinite degree in F, we have a contradiction. □

Actually, it is not much more difficult to verify the boundedness of S directly. Let $\ell : V(G) \to \mathbb{N}$ be a labelling of S; we shall define a function $f : \mathbb{N} \to \mathbb{N}$ that dominates every ray in S with respect to ℓ. Let g and f be defined by

$$g : n \mapsto \sum_{i,j \le n} (\ell(x_{2i+2}) + \ell(v_j^i)) \qquad \text{and} \qquad f : n \mapsto g(2n).$$

Note that g is increasing and dominates every R^n. Therefore f dominates every ray that has a tail in

$$\tilde{S} = S - \{x_2, x_4, \ldots\}.$$

Now let R be an arbitrary ray in S. If R has a tail in $\tilde{S}$, then f dominates R. Otherwise, R contains infinitely many x^n. It is easily seen that g dominates any ray that starts at v_0^0 and contains infinitely many x^n. Since R contains a tail of such a ray, it follows that f dominates R.

Of course, the question now arises as to which graphs *can* be admissibly oriented. To give this property a proper name (at last), let us say that a countable graph is *finitely*

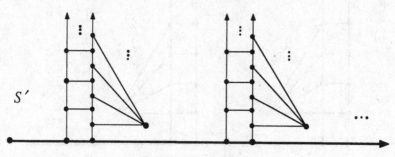

Figure 3   The graph S'

spreading if its edges can be admissibly oriented. Thus finitely spreading graphs are bounded, but not vice versa.

It is natural to ask whether S is essentially the only counterexample to Thomassen's conjecture. More precisely, is it true that every bounded graph that is not finitely spreading contains a subdivision of the graph S' of Fig. 3? (Note that S contains a subdivision of S', but not conversely.) It turns out that this is indeed the case, and a proof will be given elsewhere by the first author.

References

[1] Diestel, R. and Leader, I. (1992) A proof of the bounded graph conjecture. *Invent. Math.* **108** 131–162.

[2] Halin, R. (1989) Some problems and results in infinite graphs. In: Andersen, L. D. *et al.*, (eds.) Graph Theory in Memory of G. A. Dirac. *Annals of Discrete Mathematics* **41**.

[3] Halin, R. (1992) Bounded graphs. In: Diestel, R. (ed.) Directions in infinite graph theory and combinatorics. *Topics in Discrete Mathematics* **3**.

[4] Rado, R. (1964) Universal graphs and universal functions. *Acta Arith.* **9** 331–340.

[5] Thomassen, C. (private communication).

Amalgamated Factorizations of Complete Graphs

J. K. DUGDALE[†] and A. J. W. HILTON[‡]

[†] Department of Mathematics, West Virginia University, PO Box 6310, Morgantown WV 26506-6310, U.S.A.

[‡] Department of Mathematics, University of Reading, Whiteknights, PO Box 220, Reading RG6 2AX. U.K.

We give some sufficient conditions for an (S, U)-outline T-factorization of K_n to be an (S, U)-amalgamated T-factorization of K_n. We then apply these to give various necessary and sufficient conditions for edge coloured graphs G to have recoverable embeddings in T-factorized K_n's.

1. Introduction

In [9] (see also [12]) the second author developed the idea of an outline latin square, and showed that every outline latin square is an amalgamated latin square. In [4] the second author and A. G. Chetwynd described various analogues of this result for symmetric latin squares. Since a latin square can be viewed as a proper edge colouring of $K_{n,n}$ with n colours, it is also very natural to consider similar analogues for edge-coloured K_n's. This has already been done to a limited extent by Andersen and Hilton [1, 2, 3] and later by Rodger and Wantland [21] (who were concentrating on other aspects) but a more rounded and complete account is given here.

The ideas here began with the joint work of L. D. Andersen and the second author on Generalized Latin Squares [1, 2, 3]. The amalgamation idea has been taken further in different directions by various authors. As well as the authors already mentioned, further developments have been due, at least in part, to R. Häggkvist, A. Johanson, C. St J. A. Nash-Williams and J. Wojciechowski (see the references).

Graphs will in general contain loops and multiple edges.

Given a graph G, an *amalgamation* of G is a graph G^* and a surjective map $\phi: V(G) \to V(G^*)$ together with a bijective map $\eta: E(G) \to E(G^*)$ such that if $e \in E(G)$ and $e = vw$ then

$\eta(e) = \phi(v)\,\phi(w)$. Intuitively in an amalgamation of G, various of the vertices are amalgamated, or stuck together, whilst the original set of edges all remain distinct.

An *edge colouring* of a graph G is here simply a function $\psi: E(G) \to \mathscr{C}$, where $\mathscr{C}$ is a set of colours.

If G has an edge colouring ψ, then an *amalgamated edge coloured G* is an amalgamation G^* of G, as above, together with an edge colouring ψ^* of G^*, $\psi^*: E(G^*) \to \mathscr{C}^*$, and a surjection $\zeta: \mathscr{C} \to \mathscr{C}^*$ such that $\zeta(\psi(e)) = \psi^*(\eta(e))$ ($\forall e \in E(G)$). Intuitively, if G is edge coloured, then in an amalgamated edge coloured G, the vertices are amalgamated, as above, and various of the colours on the edges are combined [one could imagine, for example, that the distinction between light blue, medium blue and dark blue edges is forgotten, and that these edges are simply taken together as being the blue edges].

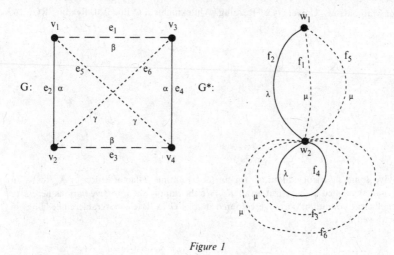

Figure 1

This is illustrated in Figure 1. The amalgamation of G is given by $\phi(v_1) = w_1$, $\phi(v_2) = \phi(v_3) = \phi(v_4) = w_2$ and $\eta(e_i) = f_i$ ($1 \le i \le 6$). The colour set for G is $\mathscr{C} = \{\alpha, \beta, \gamma\}$ and the colour set for G^* is $\mathscr{C}^* = \{\lambda, \mu\}$. The surjection $\zeta: \mathscr{C} \to \mathscr{C}^*$ is given by $\zeta(\alpha) = \lambda$, $\zeta(\beta) = \zeta(\gamma) = \mu$ and it has the required property that

$$\zeta(\psi(e_1)) = \zeta(\beta) = \mu = \psi^*(f_1) = \psi^*(\eta(e_1))$$
$$\zeta(\psi(e_2)) = \zeta(\alpha) = \lambda = \psi^*(f_2) = \psi^*(\eta(e_2))$$
$$\zeta(\psi(e_3)) = \zeta(\beta) = \mu = \psi^*(f_3) = \psi^*(\eta(e_3))$$
$$\zeta(\psi(e_4)) = \zeta(\alpha) = \lambda = \psi^*(f_4) = \psi^*(\eta(e_4))$$
$$\zeta(\psi(e_5)) = \zeta(\gamma) = \mu = \psi^*(f_5) = \psi^*(\eta(e_5))$$
$$\zeta(\psi(e_6)) = \zeta(\gamma) = \mu = \psi^*(f_6) = \psi^*(\eta(e_6)).$$

If each vertex of a graph G is incident with the same number, say t, of edges from a set F of edges, then F is called a *t-factor* of G. If G is regular and if each colour class of an edge colouring of G is a t-factor, then the edge colouring is called a *t-factorization* of G. Similarly if G is edge coloured with the q colours $k_1, \ldots, k_q$, and, for $1 \le i \le q$, the edges of k_i form a t_i-factor, then the edge colouring of G is called a T-factorization of G, where $T = (t_1, \ldots, t_q)$ is a composition of $d(G)$, the degree of G.

A *composition* of a positive integer n is a vector whose components are positive integers that sum to n. Let I denote the composition $(1, 1, \ldots, 1)$ (the appropriate value of n will always be clear from the context).

If $T = (t_1, \ldots, t_q)$ and $U = (r_1, \ldots, r_u)$ are two compositions of the same number n, and, letting $x_0 = 0$, if there is a composition $X = (x_1, \ldots, x_n)$ of q such that

$$t_{x_{j-1}+1} + \ldots + t_{x_j} = r_j \quad (1 \leqslant j \leqslant u),$$

then we call U an *amalgamation* of T.

We are concerned in this paper with T-factorizations of K_n (so that $T = (t_1, \ldots, t_q)$ is a composition of $n-1$). Given a T-factorization of $G = K_n$, an edge coloured amalgamation G^* of G may conveniently be described in the following way. We may suppose that the vertices of K_n are $v_1, \ldots, v_n$ and that the colours used on $E(G)$ are $\kappa_1, \ldots, \kappa_q$ (the colour class κ_i is a t_i-factor). Similarly we may suppose that the vertices of G^* are $w_1, \ldots, w_s$ and that the colours used on $E(G^*)$ are $c_1, \ldots, c_u$. Let $|\phi^{-1}(w_i)| = p_i$ $(1 \leqslant i \leqslant s)$ and $|\zeta^{-1}(c_k)| = x_k$ $(1 \leqslant k \leqslant u)$. Let $t_{x_{j-1}+1} + \ldots + t_{x_j} = r_j$ $(1 \leqslant j \leqslant u)$ so that $U = (r_1, \ldots, r_u)$ is an amalgamation of T. Without loss of generality, we may suppose that $\phi^{-1}(w_j) = \{v_{p_{i-1}+1}, \ldots, v_{p_i}\}$ for $1 \leqslant i \leqslant s$, taking $p_0 = 0$. Let $S = (p_1, \ldots, p_s)$. Clearly much of the essence of the edge coloured G^* is described by S and U, given that the T-factorization of K_n is known. We say that the edge coloured G^* is the (S, U)-amalgamation of the T-factorized K_n.

Some obvious properties of an (S, U)-amalgamation of a T-factorized K_n are given in the following proposition.

Proposition 1. *Let K_n be given a T-factorization. Let $S = (p_1, \ldots, p_s)$ be a composition of n and let $U = (r_1, \ldots, r_u)$ be an amalgamation of $T = (t_1, \ldots, t_q)$. Then the (S, U)-amalgamation of the T-factorized K_n has the properties:*

- *(Pi)*   *colour c_k occurs on $r_k p_i$ edges incident with w_i (counting loops as two edges) $(1 \leqslant k \leqslant u, 1 \leqslant i \leqslant s)$;*
- *(Pii)*   *there are $p_i p_j$ edges joining w_i to w_j $(1 \leqslant i < j \leqslant s)$;*
- *(Piii)*   *there are $\binom{p_i}{2}$ loops incident with w_i $(1 \leqslant i \leqslant s)$.*

Proof. After the T-factorization of K_n is amalgamated to form a U-factorization, colour c_k occurs on r_k edges incident with each vertex v. When the vertices are amalgamated, a vertex w_i is formed by amalgamating p_i of the vertices of the K_n, so the number of edges of colour c_k incident with w_i is $p_i r_k$ (counting a loop as two edges). This proves (Pi). Between the two disjoint sets of p_i vertices and p_j vertices in K_n which are amalgamated to form w_i and w_j, there are $p_i p_j$ edges, so there are $p_i p_j$ edges joining w_i and w_j. This proves (Pii). Between any two of the set of p_i vertices in K_n which are amalgamated to form w_i there are $\binom{p_i}{2}$ edges, and these become loops on w_i. This proves $(Piii)$. $\square$

Suppose now that G^* is an edge coloured graph with s vertices, $w_1, \ldots, w_s$, whose edges are coloured with u colours, $c_1, \ldots, c_u$, and suppose that there are compositions $S = (p_1, \ldots, p_s)$ of n and $U = (r_1, \ldots, r_u)$ of $n-1$, where U is an amalgamation of $T = (t_1, \ldots, t_q)$, such that

G^* satisfies $(P\,i)$, $(P\,ii)$ and $(P\,iii)$ of Proposition 1. Then the edge coloured G^* is called an (S, U)-outline $(t_1, \ldots, t_q)$-factorized K_n. If $t_1 = \ldots = t_q = t$ then this would be abbreviated to an (S, U)-outline t-factorized K_n.

The idea here is that if G^* is an (S, U)-outline $(t_1, \ldots, t_q)$-factorization of K_n, then G^* satisfies all the numerical conditions that we know that an (S, U)-amalgamated $(t_1, \ldots, t_q)$-factorization of K_n would have, but we are not informed as to whether G^* is actually an (S, U)-amalgamated $(t_1, \ldots, t_q)$-factorization or not. It is not hard to construct examples where an outline factorization is not an amalgamated factorization. However for some values of the various parameters, an (S, U)-outline $(t_1, \ldots, t_q)$-factorization of K_n is an (S, U)-amalgamated $(t_1, \ldots, t_q)$-factorization of K_n.

We give various such values of the parameters in the following theorem which is proved in Section 3.

Theorem 2. *Let* $S = (p_1, \ldots, p_s)$ *be a composition of* n, *let* $T = (t_1, \ldots, t_q)$ *be a composition of* $n - 1$, *and let* $U = (r_1, \ldots, r_u)$ *be an amalgamation of* T. *Suppose also that* $X = (x_1, \ldots, x_u)$ *is a composition of* q *such that, for* $k \in \{1, \ldots, u\}$, $r_k = t_{x_{k-1}+1} + \ldots + t_{x_k}$. *If either*

(i) $p_1, \ldots, p_s$ *are even (so that* n *is even), or*

(ii) $u = q$ *(so that* $U = T$*), or*

(iii) *for* $k \in \{1, \ldots, u\}$ *either* $r_k = t_{x_{k-1}+1}(= t_{x_k})$ *or* $t_{x_{k-1}+1}, \ldots, t_{x_k}$ *are even,*

then any (S, U)-*outline* T-*factorization of* K_n *is the* (S, U)-*amalgamation of a* T-*factorized* K_n.

2. Preliminary definitions and results about edge-colourings

We first need to give a number of definitions and results concerning edge colourings of graphs.

Suppose that a multigraph G is edge coloured with colours $\mathscr{C} = \{c_1, \ldots, c_u\}$. For $k \in \{1, \ldots, u\}$ and $v \in V(G)$, let $E_k(v)$ be the set of edges incident with v of colour c_k, and for $v, w \in V(G)$, $v \neq w$, let $E_k(v, w)$ be the set of edges joining v and w of colour c_k; if $v = w$ then $E_k(v, w)$ denotes the set of loops of colour k incident with v. We let $|E_k(v)|$ denote the number of edges of colour c_k incident with v, counting each loop as two edges. The edge colouring of G is called *equitable* if

$$\|E_j(v)| - |E_k(v)\| \leqslant 1 \quad (\forall v \in V(G) \quad \text{and} \quad \forall j, k \in \{1, \ldots, u\}).$$

The edge colouring of G is called *balanced* if in addition to being equitable, the edge colouring has the property that

$$\|E_j(v, w)| - |E_k(v, w)\| \leqslant 1 \quad (\forall v, w \in V(G), \quad \text{and} \quad \forall j, k \in \{1, \ldots, u\}).$$

If G is edge coloured with $\mathscr{C} = \{c_1, \ldots, c_u\}$, then, for $k \in \{1, \ldots, u\}$, let E_k denote the set of edges of colour c_k. An edge colouring is called *equalized* if

$$\|E_j| - |E_k\| < 1 \quad (1 \leqslant j < k \leqslant u).$$

An edge colouring with the property that no vertex has more than one edge of any colour incident with it is called a *proper edge colouring*.

For a graph G, let $\Delta(G)$ and $\delta(G)$ denote the maximum and minimum degree of G respectively.

The first lemma we need is due to de Werra (see [24, 25]; another proof may be found in [3]).

Lemma 3. *Let $k \geqslant 1$ be an integer and let G be a bipartite multigraph (so G has no loops). Then G has a balanced edge colouring with k colours.*

The next lemma is essentially due to Petersen [20]; see also [4].

Lemma 4. *Let G be a multigraph in which loops are permitted, and let k be a positive integer such that, for each $v \in V(G)$, either $(1/k)\,d_G(v)$ is an even integer or $(1/k)(d_G(v)+1)$ is an even integer. Then G has an equitable edge colouring with k colours.*

Here as elsewhere in this paper, a loop counts two towards the degree of a vertex.

A special case of Lemma 4 is the following well known theorem of Petersen [2].

Lemma 5. *Let G be a regular multigraph in which loops are permitted. Let $d(G) = 2k$. Then G can be 2-factorized.*

Lemma 6. *Let $x_1, \ldots, x_l$ be positive even integers. Let H be a graph satisfying*

$$\Delta(H) \leqslant x_1 + \ldots + x_l$$

and

$$|E(G)| \geqslant a(x_1 + \ldots + x_l),$$

where a is a positive integer. Then H has an edge colouring with l colours $\kappa_1, \ldots, \kappa_l$ such that, with H_i denoting the spanning subgraph of H whose edges are the edges of H coloured κ_i,

$$\Delta(H) \leqslant x_i \quad (1 \leqslant i \leqslant l)$$

and

$$|E(H_i)| \geqslant a\,x_i \quad (1 \leqslant i \leqslant l).$$

Proof. The number of vertices of H with odd degree is even, say $2y$. From H form a graph H^+ by adding in y further edges in such a way that the degree of each vertex of H^+ is even. Now form a graph H^{++} by adding sufficiently many loops at each vertex so that H^{++} is regular of degree $x_1 + \ldots + x_l$ [each loop counts two]. By Petersen's theorem (Lemma 5) H^{++} can be 2-factorized; thus we can edge colour H^{++} with colours $\gamma_1 + \ldots + \gamma_p$, where $p = \frac{1}{2}(x_1 + \ldots + x_l)$, in such a way that the spanning subgraph, whose edges are the edges of H^{++} coloured γ_i, is regular of degree two $(1 \leqslant i \leqslant p)$. Let J_i $(1 \leqslant i \leqslant p)$ denote the spanning subgraph of H whose edges are coloured γ_i. Then $\Delta(J_i) \leqslant 2$ $(1 \leqslant i \leqslant p)$.

We can now change the edge colouring of H with $\gamma_1, \ldots, \gamma_p$ so that it is equalized and still has the property that the maximum degree in each colour is at most two. For suppose there are two colours, say γ_1, and γ_2, such that $|E(J_1)| \geqslant |E(J_2)| + 2$. Consider the graph $J_1 \cup J_2$. If this has $2y$ vertices of odd degree, form $(J_1 \cup J_2)^+$ by inserting y edges so that, in $(J_1 \cup J_2)^+$, each vertex has even degree. Then each component of $(J_1 \cup J_2)^+$ is Eulerian. Going round an Eulerian cycle in each component and leaving out the y extra edges, produces trails in $J_1 \cup J_2$ which begin and end with distinct vertices of odd degree, and cycles. Note that if a cycle is regular of degree four it has an even number of edges. We colour each cycle and trail alternately γ_1 and γ_2. If a cycle has an odd number of edges, we ensure that some vertex of degree 2 is the starting and finishing vertex (so the two edges incident with it receive the same colour). Let J_1' and J_2' denote the spanning subgraphs of $J_1 \cup J_2$ coloured γ_1 and γ_2 after this recolouring. Then $\Delta(J_1') \leqslant 2$ and $\Delta(J_2') \leqslant 2$. We may arrange that the cycles and trails with an odd number of edges were paired off (with possibly one over) so that if one has one more γ_1 edge than γ_2 edges, the other has one more γ_2 edge than γ_1 edges. If this is done then $\|E(J_1')\| - |E(J_2')\| \leqslant 1$.

Repeating this as often as necessary, produces an edge colouring in which

$$\|E(J_i'')| - |E(J_j'')\| \leqslant 1 \quad (1 \leqslant i < j \leqslant p),$$

where J_i'' denotes the spanning subgraph of H coloured γ_i eventually obtained $(1 \leqslant i \leqslant p)$. Since $|E(G)| \geqslant a(x_1 + \ldots + x_t) = 2ap$ it follows that

$$|E(J_i'')| \geqslant 2a \quad (1 \leqslant i \leqslant p).$$

Now for $1 \leqslant j \leqslant l$ we form disjoint unions of $\frac{1}{2}x_j$ of these colour classes. Calling these unions $H_1, \ldots, H_l$ we have that $H_1 \cup \ldots \cup H_l = H$,

$$\Delta(H_j) \leqslant x_j \quad (1 \leqslant j \leqslant l)$$

and

$$|E(H_j)| \geqslant ax_j \quad (1 \leqslant j \leqslant l)$$

as required.   □

3. Proof of Theorem 2

In this section we prove several lemmas, and these lead to a proof of Theorem 2.

Lemma 7. Let $S = (p_1, \ldots, p_s)$ be a composition of n, let $T = (t_1, \ldots, t_q)$ be a composition of $n-1$, and let $U = (r_1, \ldots, r_u)$ be an amalgamation of T. Let $X = (x_1, \ldots, x_u)$ be a composition of q such that

$$r_k = t_{x_{k-1}+1} + \ldots + t_{x_k}.$$

If, for each $k \in \{1, \ldots, u\}$, either

$$r_k = t_{x_{k-1}+1}$$

or $t_{x_{k-1}+1}, \ldots, t_{x_k}$ are all even, then any (S, U)-outline T-factorization of K_n is the (S, U)-amalgamation of an (S, T)-outline T-factorization of K_n.

Proof. Consider an (S, U)-outline T-factorization G^* of K_n. Let the vertices of G^* be $w_1, \ldots, w_s$ and the colours used be $c_1, \ldots, c_u$. If $r_k = t_{x_{k-1}+1}$ for all $k \in \{1, \ldots, u\}$ then $T = U$ and there is nothing to prove. If $r_k \neq t_{x_{k-1}+1}$ for some k then $t_{x_{k-1}+1}, \ldots, t_{x_k}$ are, by hypothesis, all even. Let G_k denote the subgraph of G^* induced by the edges coloured c_k. Then, by property

$(P\,i)$, in G_k each vertex w_i has degree $p_i r_k$. We give G^* an equitable edge colouring with $\frac{1}{2}r_k$ colours, $\gamma_1, \dots, \gamma_{\frac{1}{2}r_k}$; such an edge colouring exists by Lemma 4. Then, for each $j \in \{x_{k-1}+1, \dots, x_k\}$, we combine a disjoint set of $\frac{1}{2}t_j$ of $\gamma_1, \dots, \gamma_s$ together to produce colours $\beta_{x_{k-1}+1}, \dots, \beta_{x_k}$. Then there are $t_j p_i$ edges of colour β_j incident with each vertex w_i. Doing this for each colour c_k for which $r_k \neq t_{x_{k-1}+1}$ produces an (S, T)-outline T-factorization of K_n of which G^* is an (S, U)-amalgamation. □

Lemma 8. *Let* $S = (p_1, \dots, p_n)$ *be a composition of* $2n$ *with* $p_1, \dots, p_n$ *all even, let* $T = (t_1, \dots, t_q)$ *be a composition of* $2n-1$, *and let* $U = (r_1, \dots, r_u)$ *be an amalgamation of* T. *Then any* (S, U)-*outline* T-*factorization of* K_{2n} *is the* (S, U)-*amalgamation of an* (S, T)-*outline* T-*factorization of* K_{2n}.

Proof. Consider an (S, U)-outline T-factorization Y of K_{2n}. For each colour c_k where $r_k > 1$, let G_k denote the subgraph of Y induced by the edges coloured c_k. Then, by property $(P\,i)$, vertex w_i has degree $r_k p_i$. We give G_k an equitable edge colouring with r_k colours; such an edge colouring exists by Lemma 4, since $p_1, \dots, p_s$ are all even. Doing this for each colour c_k produces an (S, I)-outline 1-factorization of K_{2n}. By amalgamating colours appropriately we obtain an (S, T)-outline T-factorization of K_n of which the original (S, U)-outline T-factorization of K_{2n} is the (S, U)-amalgamation. □

Lemma 9. *Let* $S = (p_1, \dots, p_s)$ *be a composition of* n *and let* $U = (r_1, \dots, r_u)$ *be a composition of* $n-1$. *Any* (S, U)-*outline* U-*factorization of* K_n *is the* (S, U)-*amalgamation of a* U-*factorization of* K_n.

Proof. Suppose we have an (S, U)-outline U-factorization G^* of K_n. Let the vertices of G^* be $w_1, \dots, w_s$ and the colours be $c_1, \dots, c_u$ and suppose that G^* satisfies properties $(P\,i)$, $(P\,ii)$ and $(P\,iii)$ of Proposition 1. If $S = I$ there is nothing to prove, so we may suppose that $S \neq I$. We may suppose without loss of generality that $p_s \geq 2$.

Our object will be to change the edge coloured graph G^* on the vertices $w_1, \dots, w_s$ to an edge coloured graph G^{**} on vertices $w_1, \dots, w_{s-1}, w_{s1}, w_{s2}$ by 'splitting' the vertex w_s into two vertices w_{s1} and w_{s2}. For $i \in \{1, \dots, s-1\}$ the $p_i p_s$ edges joining w_i to w_s will be redistributed so that p_i of them join w_i to w_{s1} and the remaining $(p_s - 1)p_i$ join w_i to w_{s2}; $(p_s - 1)$ of the $\binom{p_s}{2}$ loops on w_s in G^* will become edges joining w_{s1} to w_{s2}, and the remaining $\binom{p_s}{2} - (p_s - 1) = \binom{p_s - 1}{2}$ loops on w_s in G^* become loops on w_{s2} in G^{**}. The colours on the subgraphs induced by $w_1, \dots, w_{s-1}$ in G^* and in G^{**} are the same. For $i \in \{1, \dots, s-1\}$ and $k \in \{1, \dots, u\}$, the number of edges of colour c_k joining w_i to w_s in G^* equals the number of edges of colour c_k joining w_i to w_{s1} or w_{s2} in G^{**}. Also the number, say $l(k)$, of loops of each colour c_k on w_s in G^* equals the number of edges of colour c_k joining w_{s1} to w_{s2} in G^{**} plus the number of loops of colour c_k on w_{s2} in G^{**}. Finally we arrange that the number of edges of colour c_k incident with w_{s1} is r_k and that the number of edges of colour c_k incident with w_{s2} is $(p_s - 1)r_k$ (counting loops as two edges). [Recall that, by $(P\,i)$, the number of edges of colour c_k incident in G^* with w_s was $p_s r_k$.] This process keeps the number of edges of colour c_k incident with each of $w_1, \dots, w_{s-1}$ the same as it was in G^{**}.

It is easy to check that if this process is carried out successfully, then the resulting edge coloured graph G^{**} is an (S', U)-outline U-factorization, where $S' = (p_1, \ldots, p_{s-1}, 1, p_s - 1)$, and it is easy to see that G^* is an amalgamation of G^{**}. Repetition of this process will eventually produce an (I, U) outline U-factorization of which G^* is the (S, U)-amalgamation.

Of course the construction of G^{**} from G^* is no problem; the aspect we need to concentrate on is the colouring of the edges incident with w_{s1} and w_{s2}. As an aid in seeing how to colour these edges, we construct the following bipartite graph H. We let the vertex sets of H be $\{p'_1, \ldots, p'_{s-1}\}$ and $\{r'_1, \ldots, r'_u\}$, and we join vertex p'_i to vertex r'_k with x edges if there are exactly x edges of colour c_k joining w_i and w_s in G^* ($1 \leqslant k \leqslant u$, $1 \leqslant i \leqslant s-1$). Using Lemma 3 we give H an equitable edge colouring with the p_s colours $\gamma_1, \ldots, \gamma_{p_s}$. [Note that in various analogues we must require at this point that H has further properties. For example in [4] it is required that the analogous graph be balanced; however this is not needed here.]

If there are z edges in H joining p'_i to r'_k coloured γ_1, then we colour z edges of G^{**} joining w_{s1} to w_1 with colour c_k, and corresponding to each of the remaining $r_k p_s - 2l(k) - z$ edges of H incident with r'_k we colour an edge of G^{**} joining w_{s2} to w_i with colour c_k ($1 \leqslant i \leqslant s-1$, $1 \leqslant k \leqslant u$). Since $d_H(p'_i) = p_i p_s$, there are p_i edges coloured γ_1 incident with p'_i, and so the p_i edges in G^{**} joining w_i to w_{s1} each receive a colour, and similarly the $p_i(p_s - 1)$ edges joining w_i to w_{s2} each receive a colour. The number y_k of edges of H incident with r'_k coloured γ_1 satisfies

$$\left\lfloor \frac{1}{p_s} d_H(r'_k) \right\rfloor \leqslant y_k \leqslant \left\lceil \frac{1}{p_s} d_H(r'_k) \right\rceil = \left\lceil \frac{1}{p_s}(r_k p_s - 2l(k)) \right\rceil \leqslant r_k.$$

Therefore the number (y_k) of edges of G^{**} incident with w_{s1} coloured c_k is at most r_k. In order to make the number of edges of G^{**} incident with w_{s1} coloured c_k be exactly r_k, $r_k - y_k$ edges joining w_{s1} to w_{s2} are coloured c_k. The number of edges (excluding loops) incident with w_{s2} coloured r_k is therefore

$$(r_k p_s - 2l(k) - y_k) + (r_k - y_k) = r_k(p_s - 1) - 2(y_k + l(k) - r_k).$$

But

$$y_k + l(k) - r_k \geqslant \left\lfloor \frac{1}{p_s} d_H(r'_k) \right\rfloor + l(k) - r_k$$

$$= \left\lfloor \frac{1}{p_s}(p_s r_k - 2l(k)) \right\rfloor + l(k) - r_k$$

$$= \left\lfloor -\frac{2l(k)}{p_s} \right\rfloor + l(k)$$

$$= l(k) - \left\lceil \frac{2l(k)}{p_s} \right\rceil$$

$$\geqslant 0.$$

Therefore the number of edges of G^{**} incident with w_{s2} coloured c_k (excluding loops) is at most $(p_s - 1) r_k$.

We now colour the loops on w_{s2} in such a way that the c_k-degree of w_{s2} (i.e. the number of edges coloured c_k incident with w_{s2}, counting loops as two edges) is $(p_s - 1) r_k$. Since the number of edges of colour c_k in G^* joining w_s to $\{w_1, \ldots, w_{s-1}\}$ equals the number of edges coloured c_k joining $\{w_{s1}, w_{s2}\}$ to $\{w_1, \ldots, w_{s-1}\}$, and since the c_k-degree of w_s in G^* $(p_s r_k)$ equals the c_k-degree of w_{s1} (r_k) in G^{**} plus the required c_k-degree of w_{s2} $((p_s - 1) r_k)$, the number of loops on w_{s2} we need to colour c_k equals $l(k) - $(the number of edges coloured c_k joining w_{s1} to w_{s2}), i.e. $l(k) - (r_k - y_k) = l(k) + y_k - r_k \geq 0$ (as above). This colouring is clearly possible and exactly uses up all the $\binom{p_s}{2}$ loops on w_{s1}.

Proof of Theorem 2. In case (ii), namely when $u = q$, so that $T = U$, Theorem 2 is simply Lemma 9.

In case (i), when $p_1, \ldots, p_s$ are all even, then, by Lemma 8, any (S, U)-outline T-factorization of K_{2n} is the (S, U)-amalgamation of an (S, T)-outline T-factorization of K_{2n}. By Lemma 9, any (S, T)-outline T-factorization of K_{2n} is the (S, T)-amalgamation of a T-factorization of K_{2n}. Therefore any (S, U)-outline T-factorization is the (S, U)-amalgamation of a T-factorized K_{2n}.

In case (iii), by Lemma 9, any (S, U)-outline T-factorization of K_n is the (S, U)-amalgamation of an (I, U)-outline T-factorization of K_n. By Lemma 7, when either $t_{x_{k-1}+1} = r_k$ or $t_{x_{k-1}+1}, \ldots, t_{x_k}$ are all even, each (I, U)-outline T-factorization of K_n is the (I, U)-amalgamation of a T-factorization of K_n. Therefore, in case (iii), any (S, U)-outline T-factorization of K_n is the (S, U)-amalgamation of a T-factorization of K_n. [In case (iii), we could equally well apply Lemma 7 first and Lemma 9 afterwards.]

4. Embedding an edge coloured K_r

Theorem 2 has a number of interesting applications to do with embedding edge coloured graphs G inside T-factorizations of K_n, where $T = (t_1, \ldots, t_q)$ and the colours used are $c_1, \ldots, c_q$. The embeddings are such that the i-th colour class, consisting of the edges coloured c_i, becomes part of the corresponding t_i-factor of K_n.

The simplest such application is the following result, which generalizes a theorem of Andersen and Hilton [2, Corollary 4.3.4]. The Andersen–Hilton result was rediscovered recently by Rodger and Wantland [21].

Theorem 10. *Let $T = (t_1, \ldots, t_q)$ be a composition of $n - 1$. Let K_r be edge coloured with q colours, $c_1, \ldots, c_q$, and let G_i be the spanning subgraph of K_r whose edges are the edges of K_r coloured c_i $(1 \leq i \leq q)$. Then the edge coloured K_r can be embedded in a T-factorized K_n, with G_i forming part of the corresponding t_i-factor, if and only if*

(i) $|E(G_i)| \geq t_i r - \frac{1}{2} t_i n$ $(1 \leq i \leq q)$,

(ii) $t_i n$ *is even* $(1 \leq i \leq q)$,

and

(iii) $\Delta(G_i) \leq t_i$ $(1 \leq i \leq q)$.

Proof. It follows from Proposition 1 and Theorem 2(ii) that the edge coloured K_r can be embedded in K_n with each G_i in the corresponding t_i-factor if and only if we can construct an (S, T)-outline T-factorization G^* of K_n, with $S = (1, 1, \ldots, 1, n-r)$ being a composition of n, in the following way. Let $w_1, \ldots, w_r$ be the vertices of K_n and let w_{r+1} be a further vertex. Join w_{r+1} to w_i $(1 \leqslant i \leqslant r)$ by $n-r$ edges, and place $\binom{n-r}{2}$ loops on w_{r+1}. For each $i \in \{1, \ldots, r\}$ and $j \in \{1, \ldots, q\}$, colour $t_j - d_{G_j}(w_i)$ edges joining w_i to w_{r+1} with colour c_j. If this is done, then since $\sum_{j=1}^{q} (t_j - d_{G_j}(w_i)) = (n-1) - (r-1) = n-r$, all edges between w_i and w_{r+1} are coloured. This colouring is possible if and only if $t_j \geqslant \Delta(G_j) \geqslant d_{G_j}(w_i)$, which is condition (iii). After this, colour sufficient loops incident with w_{r+1} with colour c_i so that the number of edges of colour c_i incident with w_{r+1} becomes equal to $t_i(n-r)$. [Here a loop counts as two edges.] If this is done then, since $\sum_{j=1}^{q} t_j(n-r) = (n-1)(n-r)$, all loops and edges incident with w_{r+1} are coloured. This is possible if and only if

$$\sum_{i=1}^{r} (t_j - d_{G_j}(w_i)) \leqslant t_j(n-r)$$

and

$$t_j(n-r) - \sum_{i=1}^{r} (t_j - d_{G_j}(w_i)) \equiv 0 \pmod{2} \quad (1 \leqslant j \leqslant q).$$

The first condition here can be rearranged to give

$$(2r-n)t_j \leqslant \sum_{i=1}^{r} d_{G_j}(w_i) = 2|E(G_j)|$$

which is equivalent to (i), and the second condition yields

$$(2r-n)t_j - 2|E(G_j)| \equiv 0 \pmod{2}$$

which is equivalent to (iii). This proves Theorem 10. □

We can use Theorem 10 to show that an edge coloured graph of order r can be embedded in a T-factorized K_n if $n \geqslant 2r$. The argument to show this is essentially the same as that used by Evans [7] to deduce from Ryser's theorem [22] that an incomplete latin square of side r can be placed in a complete latin square of side n if $n \geqslant 2r$.

Theorem 11. *Let $T = (t_1, \ldots, t_q)$ be a composition of $n-1$, and let $t_i n$ be even $(1 \leqslant i \leqslant q)$. Let G be a simple graph with r vertices, where $r \leqslant \frac{1}{2}n$. Let G be edge coloured with colours $c_1, \ldots, c_q$ in such a way that, with G_i denoting the spanning subgraph of G whose edges are the edges of G coloured c_i, $\Delta(G_i) \leqslant t_i$ $(1 \leqslant i \leqslant q)$. Then the edge coloured graph G can be embedded in a T-factorized K_n, with each G_i becoming part of the corresponding t_i-factor.*

Proof. We may first extend the edge colouring of G to an edge colouring of K_r with $c_1, \ldots, c_q$ with the property that, with H_i denoting the spanning subgraph of K_r whose edges are the edges of K_r coloured c_i, $\Delta(H_i) \leqslant t_i$ $(1 \leqslant i \leqslant q)$. We do this by colouring the edges of $\bar{G}$ one by one. If an edge $e = \{u, v\}$ of $\bar{G}$ has yet to be coloured then there are at most $2(r-2)$

coloured edges incident with one or other of u and v, and so there is a colour c_i which is neither used on t_i edges incident with u, nor on t_i edges incident with v. We may therefore colour e with this colour c_i. Proceeding in this way, all edges of K_r are coloured. Since, for $1 \leqslant i \leqslant q$, $t_i r - \frac{1}{2} t_i n \leqslant 0$ as $r \leqslant \frac{1}{2}n$, it follows that all of the conditions (i)–(iii) of Theorem 10 are satisfied, and so the edge coloured K_r can be embedded in a T-factorized K_n. Therefore the edge coloured graph G is embedded in a T-factorized K_n, as required. $\square$

5. Recoverable embeddings of edge coloured graphs

Now consider a simple graph G with vertex set $\{v_1, \ldots, v_r\}$ which is edge coloured with s colours. Suppose that G can be embedded inside a T-factorized K_n, where $T = (t_1, \ldots, t_q)$ and $q \geqslant s$, which has vertex set $\{v_1, \ldots, v_n\}$ in such a way that the edges of G coloured c_i are all within the corresponding t_i-factor, for each $i \in \{1, \ldots, s\}$. We say that the edge-coloured graph G is *recoverable* from the T-factorized K_n if the colours used on the edges of $\bar{G}$ (the complement of G with respect to the vertex set $V(G)$) are all in $\{c_{s+1}, \ldots, c_q\}$; we say that G is *recoverably embedded* in K_n. The word recoverable is used because if all the new vertices (i.e. the vertices $v_{r+1}, \ldots, v_n$) are removed from K_n and the edges with the new colours (i.e. $E(\bar{G})$) are also removed, the original edge coloured graph G is what remains.

Theorem 12. *Let $T = (t_1, \ldots, t_q)$ be a composition of $n-1$, with $t_{s+1}, \ldots, t_q$ all being even. Let G be a simple graph with r vertices, and let G be edge coloured with s colours $c_1, \ldots, c_s$. Let G_i be the spanning subgraph of G whose edges are the edges of G coloured c_i ($1 \leqslant i \leqslant s$). Then the edge coloured graph G can be recoverably embedded in a T-factorized K_n, with G_i forming part of the corresponding t_i-factor ($1 \leqslant i \leqslant s$), if and only if*

(i) $\displaystyle \binom{r}{2} - |E(G)| \geqslant \sum_{i=s+1}^{q} (t_i r - \tfrac{1}{2} t_i n)$,

(ii) $|E(G_i)| \leqslant t_i r - \frac{1}{2} t_i n$ $(1 \leqslant i \leqslant s)$,

(iii) $t_i n$ is even $(1 \leqslant i \leqslant s)$,

(iv) $\Delta(G_i) \leqslant t_i$ $(1 \leqslant i \leqslant s)$,

(v) $r - \delta(G) - 1 \leqslant t_{s+1} + \ldots + t_q$.

We give two proofs of Theorem 12. The first is more elegant, but the second is useful as a model for some later results. For the second proof we need Lemma 6, which is not needed for the first.

First proof of Theorem 12

By Theorem 2 (iii), since $t_{s+1}, \ldots, t_q$ are even, G can be recoverably embedded in a T-factorized K_n in the way described if and only if G can be embedded in an (S, U)-outline T-factorization of K_n, where $S = (1, \ldots, 1, n-r)$ is a composition of n and $U = (t_1, \ldots, t_s, t_{s+1} + \ldots + t_u)$.

Let the vertex set of G be $\{w_1, \ldots, w_r\}$. Let $t^* = t_{s+1} + \ldots + t_q$. Let $\bar{G}$ be the complement of G with respect to $\{w_1, \ldots, w_r\}$. Colour each edge of $\bar{G}$ with colour c^*. For $1 \leqslant i \leqslant r$, join w_i to a further vertex w^* with $n-r$ edges, and introduce $\binom{n-r}{2}$ loops onto w^*.

For $1 \leqslant i \leqslant r$, colour sufficient edges from w^* to w_i with colours c_j $(1 \leqslant j \leqslant s)$ and c^* so that the number of edges of colour c_j incident with w_i is t_j, and the number of edges of colour c^* is t^*. Since $(r-1)+(n-r) = n-1 = t_1+\ldots+t_q = t_1+\ldots+t_s+t^*$, there are exactly the right number of edges joining w_i to w^* for this to be possible, with every edge receiving a colour. So this process is possible if and only if

$$d_{G_j}(w_i) \leqslant t_j \quad (1 \leqslant j \leqslant s, 1 \leqslant i \leqslant r)$$

and

$$d_{\bar{G}}(w_i) \leqslant t^* \quad (1 \leqslant i \leqslant r),$$

or in other words, if and only if condition (iv) is satisfied and

$$r - \delta(G) - 1 = \Delta(\bar{G}) \leqslant t^* = t_{s+1}+\ldots+t_q,$$

which is condition (v), is satisfied.

When this is done, colour the loops on w^* in such a way that the number of edges of colours c_j $(1 \leqslant j \leqslant s)$ and c^* incident with w^* is $t_j(n-r)$ and $t^*(n-r)$ respectively, where a loop counts as two edges. The number of non-loop edges of colours c_j and c^* incident with w^* is $rt_j - 2|E(G_j)|$ and $rt^* - 2|E(\bar{G})|$ respectively, so the number of loops we need to colour is

$$\frac{1}{2}\left\{ \sum_{j=1}^{s} [(n-r)t_j - (rt_j - 2|E(G_j)|)] + [(n-r)t^* - (rt^* - 2|E(\bar{G})|)] \right\}$$

$$= \frac{1}{2}\left\{ (n-2r)\left[\sum_{j=1}^{s} t_j + t^*\right] + 2\left[\sum_{j=1}^{s} |E(G_j)| + |E(\bar{G})|\right] \right\}$$

$$= \frac{1}{2}\left\{ (n-2r)(n-1) + 2\binom{r}{2} \right\}$$

$$= \binom{n-r}{2},$$

so there are exactly the right number of loops on w^* for this to be possible, with each loop receiving a colour. It follows that we may colour the loops in the way described if and only if

(a) there are not already too many edges of some colour incident with w^*, and

(b) the number of edges of each colour joining w^* to $\{w_1, \ldots, w_s\}$ has the right parity.

Condition (a) is, more precisely, that

$$rt_j - 2|E(G_j)| \leqslant t_j(n-r) \quad (1 \leqslant j \leqslant s),$$

and

$$rt^* - 2|E(\bar{G})| \leqslant t^*(n-r),$$

and these are equivalent to conditions (ii) and (i) respectively. Condition (b) is, more precisely, that

$$rt_j - 2|E(G_j)| \equiv t_j(n-r) \pmod 2 \quad (1 \leqslant j \leqslant s)$$

and

$$rt^* - 2|E(\bar{G})| \equiv t^*(n-r) \pmod 2;$$

the second of these is always true, and the first is true if and only if condition (iii) is true.

□

Second proof of Theorem 12

Necessity. Suppose G can be recoverably embedded in K_n. Then $\bar{G}$ is edge coloured with $c_{s+1}, \ldots, c_q$, so G and $\bar{G}$ together constitute a K_r edge coloured with $c_1, \ldots, c_q$, and the embedding can be viewed as being of K_r embedded in a T-factorization of K_n with each G_i forming part of the corresponding t_i-factor of K_n $(1 \leqslant i \leqslant q)$. [Here, for $1 \leqslant i \leqslant q$, G_i is the spanning subgraph of G whose edges are the edges of K_r coloured c_i.] Conditions (ii), (iii) and (iv) now follow from Theorem 10. By Theorem 10 (i)

$$|E(G_i)| \geqslant t_i r - \tfrac{1}{2} t_i n \quad (s+1 \leqslant i \leqslant q),$$

so

$$|E(\bar{G})| \geqslant \sum_{i=s+1}^{q} (t_i r - \tfrac{1}{2} t_i n),$$

from which condition (i) follows. By Theorem 10 (iii),

$$\Delta(G_i) \leqslant t_i \quad (s+1 \leqslant i \leqslant q),$$

so that

$$\Delta(\bar{G}) \leqslant t_{s+1} + \ldots + t_q,$$

which yields condition (v).

Sufficiency. Suppose (i)–(v) hold. Since, by (v), $\Delta(\bar{G}) \leqslant t_{s+1} + \ldots + t_q$, since $t_{s+1}, \ldots, t_q$ are all even, and since, by (i),

$$|E(\bar{G})| \geqslant \sum_{i=s+1}^{q} t_i(r - \tfrac{1}{2}n),$$

it follows from Lemma 6 that we can edge colour $\bar{G}$ with $q-s$ colours $c_{s+1}, \ldots, c_q$ in such a way that, with G_i denoting the spanning subgraph $\bar{G}$ whose edges are the edges of $\bar{G}$ coloured c_i,

$$|E(G_i)| \geqslant t_i(r - \tfrac{1}{2}n) \quad (s+1 \leqslant i \leqslant q)$$

and

$$\Delta(G_i) \leqslant t_i \quad (s+1 \leqslant i \leqslant q).$$

This edge colouring of $\bar{G}$, together with the given edge colouring of G, constitute an edge colouring of K_r which satisfies Theorem 10. Therefore this edge coloured K_r can be embedded in a T-factorized K_n, where each G_i forms part of the corresponding t_i-factor. Clearly the embedding of G that is included by the embedding of K_r is recoverable. $\quad\square$

Of course, Theorem 10 can be considered to be a corollary of Theorem 12 (just put $s = q$). A further pair of corollaries concerns the recoverable embedding of an edge coloured graph G into a 2-factorized K_{2n+1} or a 2-factorized K_{2n}^*, where K_{2n}^* denotes the graph obtained from K_{2n} by removing a 1-factor (K_{2n}^* is sometimes known as the 'cocktail party graph').

The first is straightforward from Theorem 12.

Corollary 13. *Let G be a simple graph of order r which is edge coloured with s colours $c_1, \dots, c_s$ in such a way that, if G is the spanning subgraph of G whose edges are the edges of G coloured c_i then $\Delta(G_i) \le 2$ $(1 \le i \le s)$. Then the edge coloured G can be recoverably embedded in a 2-factorized K_{2q+1} with each G_i forming part of a distinct 2-factor if and only if*

(i) $\displaystyle\binom{r}{2} - |E(G)| \le (q-s)(2r-2q-1)$,

(ii) $|E(G_i)| \ge 2r-2q-1$ $(1 \le i \le s)$,

(iii) $r - \delta(G) - 1 \le 2(q-s)$.

The second corollary is also fairly straightforward if you think of a 2-factorized K_{2n}^* as being a T-factorized K_{2n}, where $T = (1, 2, \dots, 2)$ is a composition of $2n-1$.

Corollary 14. *Let G be a simple graph of order r which is edge coloured with s colours $c_1, \dots, c_s$ in such a way that, if G_i is the spanning subgraph of G whose edges are the edges of G coloured c_i, then $\Delta(G_i) \le 2$ $(1 \le i \le s)$. Then the edge coloured G can be recoverably embedded in a 2-factorized $K_{2(q+1)}^*$, with each G_i forming part of a distinct 2-factor, if and only if $\bar{G}$ has a partial matching F^* such that*

(i) $\displaystyle\binom{r}{2} - |E(G)| - |F^*| \ge (q-s)(2r-2q-2)$,

(ii) $|E(G_i)| \ge 2r-2q-2$ $(1 \le i \le s)$,

(iii) $|F^*| \ge r-q-1$,

(iv) $r - \delta(G \cup F^*) - 1 \le 2(q-s)$.

Proof. If G can be recoverably embedded in $K_{2(q+1)}^*$ and if F represents the missing 1-factor, then for some $F^* \subset F$, with $V(F^*) \subset V(G)$, $G \cup F^*$ is recoverably embedded in a T-factorization of $K_{2(q+1)}^*$, where $T = (1, 2, \dots, 2)$ is a composition of $2q+1$. Then conditions (i)–(iv) follow from Theorem 12.

Conversely if conditions (i)–(iv) are satisfied, then $G \cup F^*$ can be recoverably embedded in a T-factorization of $K_{2(q+1)}$, and so G can be recoverably embedded in a 2-factorization of $K_{2(q+1)}^*$. $\quad\square$

If $t_{s+1}, \ldots, t_q$ in Theorem 12 are not all even, then the analogous result may not be true; it depends on the structure of $\bar{G}$. We can replace (i) and (v) by requirements about the graph $\bar{G}$.

Theorem 15. *Let* $T = (t_1, \ldots, t_q)$ *be a composition of* $n-1$. *Let* G *be a simple graph with* r *vertices and let* $\bar{G}$ *be its complement. Let* G *be edge coloured with* s *colours* $c_1, \ldots, c_s$ *and, for* $1 \leqslant i \leqslant s$, *let* G_i *be the spanning subgraph of* G *whose edges are the edges of* G *coloured* c_i. *Then the edge coloured graph* G *can be recoverably embedded in a* T*-factorized* K_n, *with* G_i *forming part of the corresponding* t_i*-factor* $(1 \leqslant i \leqslant s)$, *if and only if*

(i) $|E(G_i)| \geqslant t_i(r - \frac{1}{2}n)$ $(1 \leqslant i \leqslant s)$,

(ii) $t_i n$ *is even* $(1 \leqslant i \leqslant q)$,

(iii) $\Delta(G_i) \leqslant t_i$ $(1 \leqslant i \leqslant s)$,

(iv) $\bar{G}$ *can be edge coloured with colours* $c_{s+1}, \ldots, c_q$ *so that, with* G_i *denoting the spanning subgraph of* $\bar{G}$ *whose edges are the edges of* $\bar{G}$ *coloured* c_i,

   (a) $|E(G_i)| \geqslant t_i(r - \frac{1}{2}n)$, $(s+1 \leqslant i \leqslant q)$

   (b) $\Delta(G_i) \leqslant t_i$ $(s+1 \leqslant i \leqslant q)$.

Proof. This is easy to see from the second proof of Theorem 12. $\square$

Whether or not $t_{s+1}, \ldots, t_u$ are all even, conditions (i)–(iv) of Theorem 12, together with strict inequality in condition (v), are sufficient for the edge coloured graph G to be recoverably embeddable in a T-factorized K_n.

Theorem 16. *Let* $T = (t_1, \ldots, t_q)$ *be a composition of* $n-1$. *Let* G *be a simple graph with* r *vertices and let* G *be edge coloured with* s *colours* $c_1, \ldots, c_s$. *Let* G_i *be the spanning subgraph of* G *whose edges are the edges of* G *coloured* c_i $(1 \leqslant i \leqslant s)$. *Then:*

I. *If* $r - \delta(G) \leqslant t_{s+1} + \ldots + t_q$, *the edge coloured graph* G *can be recoverably embedded in a* T*-factorized* K_n, *with* G_i *forming part of the corresponding* t_i*-factor, if and only if*

(i) $\dbinom{r}{2} - |E(G)| \geqslant \displaystyle\sum_{i=s+1}^{q} (t_i r - \tfrac{1}{2} n t_i)$,

(ii) $|E(G_i)| \geqslant t_i r - \tfrac{1}{2} n t_i$ $(1 \leqslant i \leqslant s)$,

(iii) $t_i n$ *is even* $(1 \leqslant i \leqslant q)$, *and*

(iv) $\Delta(G_i) \leqslant t_i$ $(1 \leqslant i \leqslant s)$.

II. *If* $r - \delta(G) - 2 \geqslant t_{s+1} + \ldots + t_q$, *there is no such recoverable embedding.*

Proof. First suppose that $r - \delta(G) \leqslant t_{s+1} + \ldots + t_q$ and that conditions (i)–(iv) all hold.

If n is odd then condition (iii) implies that $t_{s+1}, \ldots, t_q$ are all even, and then Theorem 16 is implied by Theorem 12. So we may suppose that n is even.

By Vizing's theorem [23] and the fact that $\Delta(\bar{G}) + 1 = r - \delta(G) \leqslant t_{s+1} + \ldots + t_q$, we can properly edge colour $\bar{G}$ with $p = t_{s+1} + \ldots + t_q$ colours $\gamma_1, \ldots, \gamma_p$. If there is a colour, say γ_1, which appears on at least two more edges than some other colour, say γ_2, then there is an odd length path coloured alternately γ_1 and γ_2 with one more γ_1 edge than γ_2 edges. Now interchange colours on such a path. Repeating this as necessary with all the various colours, we eventually obtain a proper equalized edge colouring with p colours. In view of condition (i) and the fact that n is even, each colour class has at least $r - \frac{1}{2}n$ edges. Now, for $j = s+1, \ldots, u$, form disjoint unions of t_j of these colour classes. Recolouring the edges of the j-th union with colour c_j and letting G_j denote the edges of $\bar{G}$ coloured c_j $(s+1 \leqslant j \leqslant u)$, we have that

$$\Delta(G_j) \leqslant t_j \quad (s+1 \leqslant j \leqslant u),$$

and

$$|E(G_j)| \geqslant t_j(r - \tfrac{1}{2}n) \quad (s+1 \leqslant j \leqslant u).$$

It now follows from Theorem 10 that G can be recoverably embedded in the required way.

If $r - \delta(G) \leqslant t_{s+1} + \ldots + t_q$ and G can be recoverably embedded in the way described, then the argument given in the necessity part of the second proof of Theorem 12 shows that (i)–(iv) all hold.

If $r - \delta(G) - 2 \geqslant t_{s+1} + \ldots + t_q$ then $\Delta(\bar{G}) > t_{j+1} + \ldots + t_q$, so $\bar{G}$ cannot be edge coloured in the required way with $\Delta(G_i) \leqslant t_i$ $(s+1 \leqslant i \leqslant q)$, so there is no recoverable embedding.

We remark that the equalizing argument used in the proof of Theorem 16 goes back to McDiarmid [17] and de Werra [24, 25].

References

[1]   Andersen, L. D. and Hilton, A. J. W. (1980) Generalized latin rectangles I: construction and decomposition. *Discrete Math.* **31** 125–152.

[2]   Andersen, L. D. and Hilton, A. J. W. (1980) Generalized latin rectangles II: embedding. *Discrete Math.* **31** 235–260.

[3]   Andersen, L. D. and Hilton, A. J. W. (1979) Generalized latin rectangles. In: *Graph Theory and Combinatorics, Research Notes in Mathematics*. Pitman, London, 1–17.

[4]   Chetwynd, A. G. and Hilton, A. J. W. (1991) Outline symmetric latin squares. *Discrete Math.* **97** 101–117.

[5]   Cruse, A. B. (1974) On embedding incomplete symmetric latin squares. *J. Combinatorial Theory, Ser. A* **16** 18–22.

[6]   Cruse, A. B. (1974) On extending incomplete latin rectangles. *Proc. 5th Southeastern Conf. on Combinatorics, Graph Theory and Computing*. Florida Atlantic University, Boca Raton, Florida, 333–348.

[7]   Evans, T. (1960) Embedding incomplete latin squares. *Amer. Math. Monthly* **67** 958–961.

[8]   Häggkvist, R. and Johanson, A. (to appear 1994) (1, 2)-factorizations of general Eulerian nearly regular graphs. *Combinatorics, Probability and Computing*.

[9]   Hilton, A. J. W. (1980) The reconstruction of latin squares, with applications to school timetabling and to experimental design. *Math. Programming Study* **13** 68–77.

[10] Hilton, A. J. W. (1981) School timetables. In: Hansen, P. (ed.) *Studies on graphs and Discrete Programming*. North-Holland, Amsterdam, 177–188.

[11] Hilton, A. J. W. (1982) Embedding incomplete latin rectangles. *Ann. Discrete Math.* **13** 121–138.

[12] Hilton, A. J. W. (1987) Outlines of Latin squares. *Ann. Discrete Math.* **34** 225–242.

[13] Hilton, A. J. W. (1984) Hamiltonian decompositions of complete graphs. *J. Combinatorial Theory, Ser. B* **36** 125–134.

[14] Hilton, A. J. W. and Rodger, C. A. (1986) Hamiltonian decompositions of complete regular *s*-partite graphs. *Discrete Math.* **58** 63–78.

[15] Hilton, A. J. W. and Wojciechowski, J. (1993) Weighted quasigroups. Surveys in Combinatorics. *London Mathematical Society Lecture Note Series* **187** 137–171.

[16] Hilton, A. J. W. and Wojciechowski, J. (submitted) Simplex Algebras.

[17] McDiarmid, C. J. H. (1972) The solution of a time-tabling problem. *J. Inst. Math. Applics.* **9** 23–34.

[18] Nash-Williams, C. St J. A. (1986) Detachments of graphs and generalized Euler trials. In: *Proc. 10th British Combinatorics Conf., Surveys in Combinatorics* 137–151.

[19] Nash-Williams, C. St J. A. (1987) Amalgamations of almost regular edge-colourings of simple graphs. *J. Combinatorial Theory, Ser. B* **43** 322–342.

[20] Petersen, J. (1891) Die Theorie der regulären Graphen. *Acta Math.* **15** 193–220.

[21] Rodger, C. A. and Wantland, E. (to appear) Embedding edge-colourings into *m*-edge-connected *k*-factorizations. *Discrete Math.*

[22] Ryser, H. J. (1951) A combinatorial theorem with an application to latin squares. *Proc. Amer. Math. Soc.* **2** 550–552.

[23] Vizing, V. G. (1960) On an estimate of the chromatic class of a *p*-graph (in Russian). *Diskret. Analiz.* **3** 25–30.

[24] de Werra, D. (1971) Balanced schedules. *INFOR* **9** 230–237.

[25] de Werra, D. (1975) A few remarks on chromatic scheduling. In: Roy, B. (ed.) *Combinatorial Programming: Methods and Applications.* Reidel, Dordrecht. 337–342.

Ramsey Size Linear Graphs

PAUL ERDŐS[†], R. J. FAUDREE[‡§], C. C. ROUSSEAU[‡]
and R. H. SCHELP[‡]

[†]Mathematical Institute, Hungarian Academy of Sciences, Budapest V, Hungary

[‡]Department of Mathematical Science, Memphis State University, Tenn. 38152 USA

A graph G is *Ramsey size linear* if there is a constant C such that for any graph H with n edges and no isolated vertices, the Ramsey number $r(G, H) \leq Cn$. It will be shown that any graph G with p vertices and $q \geq 2p - 2$ edges is not Ramsey size linear, and this bound is sharp. Also, if G is connected and $q \leq p + 1$, then G is Ramsey size linear, and this bound is sharp also. Special classes of graphs will be shown to be Ramsey size linear, and bounds on the Ramsey numbers will be determined.

1. Introduction

Only finite graphs without loops or multiple edges will be considered. The general notation will be standard, with specialized notation introduced as needed. For a graph G, the vertex set and edge set will be denoted by $V(G)$ and $E(G)$ respectively, and the *order* of G (the number of vertices in $V(G)$) and the *size* of G (the number of edges in $E(G)$) will be denoted by $p(G)$ and $q(G)$ respectively. For graphs G and H, the *Ramsey number* $r(G, H)$ is the smallest positive integer n such that if the edges of a K_n are colored either red or blue, there will always be a red copy of G or a blue copy of H.

The following Ramsey bound theorem was conjectured by Harary, and proved by Sidorenko in [12].

Theorem 1. *For any graph H_n of size n and without isolated vertices,*

$$r(K_3, H_n) \leq 2n + 1.$$

§ Research partially supported by O.N.R. Grant No. N00014-91-J-1085 and N.S.A. Grant No. MDA 904-90-H-4034

Since $r(K_3, T_{n+1}) = 2n + 1$ (see [2]), and $r(K_3, nK_2) = 2n + 1$ (see [11]), the bound in Theorem 1 cannot be lowered. Thus, for $G = K_3$ the Ramsey number $r(G, H)$ has an upper bound that is linear in the number of edges in H. It is natural to ask for which graphs G is this true. This motivated the following definition.

Definition 1. *A graph G is Ramsey size linear if there is a constant C such that for any graph H_n of size n without isolated vertices,*

$$r(G, H_n) \leq C \cdot n.$$

In Section 2, the maximum number of edges in a graph G that is Ramsey size linear will be determined, as well as the minimum number of edges in a connected graph G that is not Ramsey size linear. In particular, it will be shown that if $q(G) \geq 2p(G) - 2$, then G is not Ramsey size linear. We will also prove that if G is connected and $q(G) \leq p(G) + 1$, then G is Ramsey size linear. Both Ramsey size linear graphs and graphs that are not Ramsey size linear exist in the interval $p(G) + 1 < q(G) < 2p(G) - 2$. In fact, examples will be described to show that for each $p + 2 \leq q \leq 2p - 3$, there are connected graphs G_i with $p(G_i) = p$ and $q(G_i) = q$ for $i = 1, 2$ such that G_1 is Ramsey size linear and G_2 is not Ramsey size linear.

In Section 3, special classes of graphs will be shown to be Ramsey size linear; more specifically graphs G with extremal number $ext(G, n) = O(n^{3/2})$ will be shown to be Ramsey size linear. In Section 4, some upper bounds for the Ramsey numbers $r(G, H_n)$ will be verified for some special graphs G, such as even cycles, where H_n denotes a graph of size n without isolated vertices. Some open questions related to the results of the paper will be discussed in Section 5.

2. Extremal problems

We start this section with a proof of Theorem 2, which will be the basis for showing that a graph of order p and size at least $2p - 2$ is not Ramsey size linear.

Theorem 2. *Let G be a fixed graph with $p(G) = p \geq 3$ and $q(G) = q$. There exists a positive constant C such that for n sufficiently large,*

$$r(G, K_n) > C \left(\frac{n}{\log n} \right)^{(q-1)/(p-2)}.$$

An immediate consequence of Theorem 1 is the following.

Corollary 1. *If $p(G) \geq 3$ and $q(G) \geq 2 \cdot p(G) - 2$, then G is not Ramsey size linear.*

Theorem 2 can be found in [6], but we include it here since it is central to the results of this paper. In this proof, $[N]^k$ will denote the set of all k-element subsets of $\{1, 2, \cdots, N\}$. Any 2-coloring of the edges $[N]^2$ of the complete graph with vertices $[N]$ will be denoted by (R, B), with R as the red graph and B as the blue graph. If $S \subset [N]$, then the red subgraph (blue subgraph) induced by $R(B)$ will be denoted by $\langle S \rangle_R$ ($\langle S \rangle_B$). Central to this proof is the following result of Erdős and Lovász (see [8]). The form used in this paper can be found in [14].

Lemma 1. **(Erdős–Lovász)** *Let* $C_1, C_2, \ldots, C_n$ *be events with probabilities* $P(C_i)$, $i = 1,$ 2, $\ldots, n-1, n$. *Suppose there exist corresponding positive numbers* $x_1, x_2, \ldots, x_n$ *such that* $x_i \cdot P(C_i) < 1$ *and*

$$\log x_i > \sum x_j P(C_j), \quad i = 1, 2, \ldots, n,$$

where the sum is taken over all $j \neq i$ *such that* C_i *and* C_j *are dependent. Then*

$$P(\bigcap \overline{C}_i) > 0.$$

With Lemma 1, we can give the proof of Theorem 2.

Proof of Theorem 2. The proof uses the Lovász-Spencer method (see [14]). For an appropriately large N, we will verify the existence of a two-coloring (R, B) of $[N]^2$ such that $R \not\supseteq G$ and $B \not\supseteq K_n$. Randomly two-color $[N]^2$, each edge being red with independent probability r. For each $S \subset [N]^p$ let A_S denote the event $\langle S \rangle_R \supset G$. Similarly, for each $T \subset [N]^n$, let B_T denote the event $\langle T \rangle_B \supset K_n$.

The fundamental result to be used here is the Erdős–Lovász local lemma (Lemma 1). To implement Lemma 1 in the setting previously described, we make the following simplification.

For each $C_i = A_S$, let $x_i = a$, and for each $C_i = B_T$, let $x_i = b$. For a fixed A_S, let N_{AA} denote the number of $S' \neq S$ such that A_S and $A_{S'}$ are dependent. Similarly, define N_{AB} to be the number of T such that A_S and B_T are dependent. In exactly the same way, define N_{BA} and N_{BB}. Letting A and B denote typical A_S and B_T respectively, note that the desired conclusion follows if there exist positive numbers a and b such that $a \cdot P(A) < 1$, $b \cdot P(B) < 1$,

$$\log a > N_{AA} \cdot a \cdot P(A) + N_{AB} \cdot b \cdot P(B), \tag{1}$$

and

$$\log b > N_{BA} \cdot a \cdot P(A) + N_{BB} \cdot b \cdot P(B). \tag{2}$$

Note that A_S and B_T are dependent only if $|S \cap T| \geq 2$. A similar observation holds for the pairs $(A_S, A_{S'})$ and $(B_T, B_{T'})$.

For the purpose of this calculation, it suffices to use the following bounds:

$$N_{AA} \leq \binom{p}{2}\binom{N-2}{p-2} = O(N^{p-2}),$$

$$N_{AB}, N_{BB} \leq \binom{N}{n} < N^n,$$

$$N_{BA} \leq \binom{n}{2}\binom{N-2}{p-2} = O(n^2 \cdot N^{p-2}),$$

$$P(A) \geq p! r^q, \quad \text{and} \quad P(B) = (1-r)^{\binom{n}{2}}.$$

Let $s = (p-2)/(q-1)$ and set

$$p = C_1 \cdot N^{-s}, \quad n = C_2 \cdot N^s \cdot \log N,$$

$$a = C_3 > 1 \quad \text{and} \quad b = e^{C_4 \cdot N^s \cdot (\log N)^2},$$

where C_1 through C_4 are positive constants. Then $\log a > 0$,

$$N_{AA} \cdot a \cdot P(A) = O(N^{p-2} N^{-sq}) = o(1)$$

and

$$N_{AB} \cdot b \cdot P(B) < e^{((C_2 + C_4 - C_1 C_2^2/2) N^s (\log N)^2)} = o(1)$$

if $C_1 C_2^2/2 > C_2 + C_4$. Similarly, both sides of equation (2) are of order

$$c \cdot N^s \cdot (\log N)^2$$

for an appropriate constant c. The constants C_1 through C_4 may be chosen so that equation (2) holds. Thus, there is a two-coloring of $[N]^2$ with no red G and no blue K_m, where $n = C_2 \cdot N^s \cdot \log N$. Solving for N in terms of n, we get the stated result. This completes the proof of Theorem 2. $\quad\square$

There are graphs of order p and size $q = 2p - 3$ that are Ramsey size linear, as the following result confirms.

Theorem 3. *Let T_{p-1} be any tree on $p-1$ vertices ($p \geq 2$), $G_p = K_1 + T_{p-1}$, and H_n be any graph of size n. Then,*

$$r(G_p, H_n) \leq 2n(p-2) + p(H_n).$$

If H_n has no isolated vertices, then $p(H_n) \leq 2n$. Thus, one immediate consequence of Theorem 3 is the following corollary.

Corollary 2. *If H_n is a graph of size n without isolated vertices, then*

$$r(G_p, H_n) \leq 2(p-1)n.$$

Proof of Theorem 3. The proof will be by induction on n. The result is trivial for $n = 1$, since $r(G_p, H_n) = \max\{p, p(H_1)\}$. Proceed by induction on n.

Let v be a vertex of H_n of smallest degree, and let $H'_n = H_n - v$. Two color the edges of a $K_{2n(p-2)+p(H_n)}$, and assume that there is no red G_p or blue H_n. By the induction assumption, there is a blue copy of H'_n.

Let N be the neighborhood of v in the graph H'_n.

By assumption, each vertex of $K_{2n(p-2)+p(G_n)} - H'_n$ is adjacent in red to at least one vertex of N. On the other hand, no vertex of N can have red degree $r(T_{p-1}, K_{p(H_n)}) = (p-2)(p(H_n) - 1) + 1$ in $K_{2n(p-2)+p(H_n)}$, since this would ensure either a red G_p or a blue H_n. Therefore, using these counts on the number of red edges emanating from N gives the following inequalities.

$$2n(p-2) + 1 \leq |N|(p-2)(p(H_n) - 1) \leq \frac{2n}{p(H_n)}(p-2)(p(H_n) - 1).$$

This gives a contradiction that completes the proof of Theorem 3. □

The next result gives the minimum number of edges in a connected graph that is not Ramsey size linear.

Theorem 4. *If G is a connected graph with $q(G) \leq p(G) + 1$, then G is Ramsey size linear. In addition, there is a graph G with $q(G) = p(G) + 2$ that is not Ramsey size linear.*

Some preliminary results and examples will be needed in the proof of Theorem 4, which we now give.

Lemma 2. *If G_1 and G_2 are Ramsey size linear graphs, then the graph $G_1 \cdot G_2$ obtained by identifying precisely one vertex from each graph is also Ramsey size linear.*

Proof. Let H_n be a graph of size n. We can assume for $i = 1, 2$ that $r(G_i, H_n) \leq c_i n$ for positive integers with $c_1 \leq c_2$. We will show that

$$r(G_1 \cdot G_2, H_n) \leq (c_2 p(G_1) + c_1)n,$$

and so $G_1 \cdot G_2$ is Ramsey linear. Let $m = (c_2 p(G_1) + c_1)n$, and 2-color the edges of a K_m with red and blue. Assume there is no blue copy of H_n. Therefore, using $r(G_1, H_n) \leq c_1 n$, there must be $c_2 n$ vertex disjoint red copies of G_1. If v_1 is the vertex of G_1 that is to be identified with the vertex v_2 of G_2, let S be the set of $c_2 n$ vertices that represent v_1 in each of the $c_2 n$ copies of G_1. Since $r(G_2, H_n) \leq c_2 n$, there is a red copy of G_2 using only vertices in S. In this red copy of G_2, the vertex identified with v_2 is the same as the vertex identified with v_1 in some copy of G_1, and this gives a red copy of $G_1 \cdot G_2$. This completes the proof of Lemma 2. □

An immediate consequence of Lemma 2 is the following.

Corollary 3. *If G is a graph such that each of its blocks is Ramsey size linear, then G is Ramsey size linear.*

Next we describe a family of examples that will be needed to verify the sharpness of the result in Theorem 4.

Example 1. *Let G be any graph that contains K_4 as a subgraph. Then, since $r(K_4, H_n) > C(\frac{n}{\log n})^{5/2}$, any graph G that contains a K_4 is not Ramsey size linear. Thus for any tree T_{p-3} on $p - 3$ vertices, the graph $K_4 \cdot T_{p-3}$ is a connected graph of order p and size $p + 2$ that is not Ramsey size linear. Clearly, any graph G that is a supergraph of $K_4 \cdot T_{p-3}$ is not Ramsey size linear.*

Proof of Theorem 4. If there is a vertex v in G such that $G - v$ has no cycles, then G is Ramsey size linear by Corollary 2. This will always be true if $q(G) \leq p(G)$. If $q(G) = p(G) + 1$, and no such vertex v exists, then G will contain two vertex disjoint cycles. Thus, each block of G will be either an edge or a cycle, both of which are Ramsey size linear. Hence, by Corollary 3, G is Ramsey size linear. Example 1 gives the sharpness of the result. This completes the proof of Theorem 4. □

Let G be a connected graph of order p and size q. To summarize, we know that: if $q \leq p + 1$, then G is Ramsey size linear; if $p + 2 \leq q \leq 2p - 3$, then it could be Ramsey size linear, but may not be; and if $q \geq 2p - 2$, then it is not Ramsey size linear.

3. Special classes of graphs

In this section special classes of graphs will be shown to be Ramsey size linear. We start with a class of graphs defined by their Turán extremal numbers. Recall that $ext(G, n)$, the Turán extremal number, is the maximum number of edges in a graph of order n that does not contain a copy of G. An excellent survey of results in Turán extremal theory can be found in [13], and a more general survey of extremal theory in [1].

Theorem 5. *If $ext(G, n) \leq cn^{3/2}$, then for every graph H_n of size n without isolates,*

$$r(G, H) \leq (32c^2 + 8)n.$$

Proof. Let (R, B) be a two-coloring of the edges of K_N, where $N \geq (32c^2 + 8)n$. Suppose $\langle R \rangle \not\supseteq G$. Then $\langle R \rangle$ has at most $cN^{3/2}$ edges. Sequentially delete vertices of degree at least $2c\sqrt{N}$ in the current red graph until none remain. After M vertices have been deleted, at least $2c\sqrt{N}M$ red edges have been removed from the original two-colord K_N, so at most $N/2$ vertices are deleted before the process terminates.

Now we have a two-colored complete graph with at least $N/2$ vertices in which the red graph has no vertices of degree $2c\sqrt{N}$ or more. We wish to show that there is an embedding of H_n into the blue graph. Embed H_n into the blue graph one vertex at a time, starting with the largest degree vertex of H_n and continuing so the sequence is non-increasing by degree. Suppose that this process terminates. Then some induced subgraph of H_n has been embedded and the process cannot be continued because there is no external vertex that can play the role of the next vertex of H_n in the sequence. We may suppose that H_n has p vertices altogether; since H_n has no isolates, $p \leq 2n$. Suppose that the vertex needed to continue the embedding has degree k in H_n. Thus H_n has $k + 1$ vertices of degree k or more, so $k(k + 1) \leq 2n$ and $k < \sqrt{2n}$. In the two-colored complete graph, there are more than $N/2 - 2n$ vertices external to the subgraph of H_n that is embedded. By assumption, there are k vertices in the embedded subgraph of H_n that in the blue graph have no common neighbor among these external vertices. Thus, at least one of the k vertices has degree $\lceil (N/2 - 2n)/k \rceil$ or more in the red graph, and we have

$$2c\sqrt{N} > \left\lceil \frac{N/2 - 2n}{k} \right\rceil > \frac{N/2 - 2n}{\sqrt{2n}},$$

so

$$\frac{N}{n} - 4c\sqrt{\frac{2N}{n}} < 4.$$

But the left-hand side is an increasing function of N/n for $N/n \geq 8c^2$, so, since $N/n \geq 32c^2 + 8$,

$$\frac{N}{n} - 4c\sqrt{\frac{2N}{n}} \geq 32c^2 + 8 - 32c^2\sqrt{1 + \frac{1}{4c^2}} > 32c^2 + 8 - 32c^2\left(1 + \frac{1}{8c^2}\right) = 4,$$

and a contradiction has been obtained. This completes the proof of Theorem 5. $\square$

The Turán extremal numbers for bipartite graphs have been studied extensively, and there are several graphs of interest that have extremal numbers $O(n^{3/2})$, and thus are Ramsey size linear. In [9] and [13] many families of such examples can be found, and some particular families can be found in [3] and [4]. For example it is known that $K_{3,3} - e$, and $Q_3 - e$ (where Q_3 is the 3-dimensional cube) have Turán extremal numbers equal to $O(n^{3/2})$.

Using Corollary 1, Theorem 4, Corollary 2, and Theorem 5, it can be determined with just one exception if a graph of order at most 5 is Ramsey size linear. All graphs of order at most 4 are Ramsey size linear with the exception of K_4, which is not Ramsey size linear. All graphs of order 5 that do not contain a K_4 or have at least 8 edges can be shown to be Ramsey size linear with the exception of $K_5 - (K_2 \cup K_{1,2})$. It is not known if this graph is Ramsey size linear. Also, it is not known if $K_{3,3}$, a graph with 6 vertices and 9 edges, is Ramsey size linear.

More generally, it would be of interest to know if a graph G is Ramsey size linear if it satisfies the density condition that each subgraph H of order m has size at most $2m - 3$.

The graph K_4 is not Ramsey size linear, but the deletion of any edge leaves the graph B_2, which is Ramsey size linear. Graphs with this property are of interest, and thus we give the following definition.

Definition 2. *A graph G is minimal Ramsey size linear if G is not Ramsey size linear, but if any edge is deleted, then the resulting graph is Ramsey size linear.*

If any of the graphs $K_5 - (K_2 \cup K_{1,2})$, $K_{3,3}$, and the 3-dimensional cube Q_3 are not Ramsey size linear, then they would be minimal, since all of their proper subgraphs are Ramsey size linear.

4. Upper bounds for special graphs

In this section we will consider some special graphs G that we know are Ramsey size linear, and determine an upper bound on the Ramsey number $r(G, H_n)$, where H_n is a graph of size n with no isolated vertices. Of course, Corollary 2 gives upper bounds for the books and fans (where the book $B_k = K_1 + K_{1,k}$ and the fan $F_k = K_1 + P_k$). We have

$$r(B_k, H_n) \leq 2(k + 1)n \quad \text{and} \quad r(F_k, H_n) \leq 2kn.$$

We next look at even cycles C_{2k}, and in particular, C_4.

Theorem 6. *If $k \geq 2$ and H_n is a connected graph of size n, then for n sufficiently large*

$$r(C_{2k}, H_n) \leq n + 22k\sqrt{n}.$$

An immediate consequence of Theorem 6 is the following corollary.

Corollary 4. *If $k \geq 2$ and H_n is a graph of size n without isolated vertices, then for n sufficiently large*

$$r(C_{2k}, H_n) \leq 2n + k - 1.$$

Note that if $H_n = nK_2$, then $r(C_{2k}, H_n) \geq 2n + k - 1$. If a K_{2n+k-2} is colored such that there is a blue K_{2n-1} and the remaining edges are red, then there is no red C_{2k} and no blue nK_2. Thus the bound in Corollary 4 is sharp.

Before we give the proof of Theorem 6, we will prove a technical lemma needed in the proof.

Lemma 3. *If a subgraph F of $K_{2n, \sqrt{n}}$ has $6kn$ edges, then F contains a C_{2k}.*

Proof. Let A and B be the parts of the bipartite graph F, with $|A| = 2n$ and $|B| = \sqrt{n}$. Delete any vertices of A of degree less than $2k$ and then delete any vertices of B of degree less than $2k\sqrt{n}$. Continue to do this until no more vertices can be deleted.

This results in a graph F'. Note that F' is non-empty, since fewer than $(2k\sqrt{n})\sqrt{n}+4kn = 6kn$ edges have been deleted, and this is less than the number of edges in F. Let A' and B' be the corresponding parts of F'. Each vertex in A' has degree at least $2k$, and each vertex in B' has degree at least $2k\sqrt{n}$. Select a vertex b in B', and let N be the neighborhood of b in A'. Let N' be a subset of N with $2k\sqrt{n}$ vertices, and let G be the subgraph of F' induced by $N' \cup (B' - b)$.

Thus G is a graph with at most $4k\sqrt{n}$ vertices and at least $4k^2\sqrt{n}$ edges. Therefore, by a result of Erdős and Gallai [7], G has a path with at least $2k$ vertices. This path (actually $2k - 1$ vertices of this path), along with the vertex b, will give a C_{2k}. This completes the proof of Lemma 3.   □

Proof of Theorem 6. Let F be a complete graph on $n + 22k\sqrt{n}$ vertices whose edges are colored either red or blue. We will assume that there is no red C_{2k} in F, and we will show that there is a blue H_n.

Let L be the vertices of F of red degree at least $7k\sqrt{n}$. If the number of vertices in L is as large as $\sqrt{n}$, then there is a red bipartite graph with $\sqrt{n}$ vertices in one part, $n + 22k\sqrt{n} - \sqrt{n}$ vertices in the other part, and at least $\sqrt{n}6k\sqrt{n} = 6kn$ edges. Then by Lemma 3, there must be a red C_{2k} in this bipartite graph, and thus in F. Note that for n sufficiently large, $22k\sqrt{n} \leq n$, and additional vertices can be added to the large part of the bipartite graph to get $2n$ vertices, and so Lemma 3 applies. Let $F' = F - L$. Thus, we can assume that each vertex of F' has red degree less than $7k\sqrt{n}$.

Order the vertices of H_n in non-increasing degree order, say $h_1, h_2, \ldots, h_p$. For each j, $(1 \leq j \leq p)$, let S_j be the subgraph of H_n induced by the vertices $\{h_1, h_2, \ldots, h_j\}$. Assume that there is an embedding σ of S_r into the blue graph of F', but there is no embedding of S_{r+1}. Let N be the neighborhood of h_{t+1} in S_r, so $N' = \sigma(N)$ is the corresponding set of vertices in F'.

From [5] and an unpublished result of Szemerédi we know that there are constants c and c' such that

$$r(C_4, K_m) \leq \frac{cm^2}{(\log m)^2},$$

and for $k > 2$,

$$r(C_{2k}, K_m) \leq c'm^{(k+1)/k} \leq \frac{cm^2}{(\log m)^2}.$$

Therefore, since there is no red C_{2k} in F, there is a constant c'' such that there is a blue $K_{c''\sqrt{n}\log n}$. This implies that $S_{c''\sqrt{n}\log n}$ can be embedded in the blue subgraph of F', and so $r \geq c''\sqrt{n}\log n$.

We will first consider the case when $r \leq n/2$. Each vertex of $F' - S_t$ is adjacent in red to a vertex of N'. Therefore, there will be at least $(n/2) - \sqrt{n}$ red adjacencies emanating from N'. On the other hand, because of the ordering of the vertices in H_n, each vertex of S_t has degree at least $|N'|$, so $|N'|c''\sqrt{n}\log n \leq 2n$. This implies that $|N'| \leq 2\sqrt{n}/c''\log n$. Since each vertex of F' has red degree at most $7k\sqrt{n}$, there will be at most $(7k\sqrt{n})(\sqrt{n}/c''\log n) = 7kn/c''\log n$ red edges emanating from N'. This implies $(n/2) - \sqrt{n} \leq 7kn/c''\log n$, a contradiction for n sufficiently large. This completes the proof of the case when $r \leq n/2$.

Next, we consider the case when $r > n/2$. First observe that, $|N'|r < 2n$ means $|N'| < 4$. Since there are at least $21k\sqrt{n}$ vertices in $F' - S_r$, there are at least $21k\sqrt{n}$ red edges emanating from N'. This implies there is a vertex of N' of red degree at least $(21k\sqrt{n})/3 = 7k\sqrt{n}$, which gives a contradiction and completes the proof of Theorem 6. $\square$

It should be noted that the bound $n + 22k\sqrt{n}$ could be improved by more careful counting, but this would not give any improvement in Corollary 4, so the additional space and effort is not warranted.

Sharper bounds can be obtained for the case $k = 2$. It is known (see [10]) that $r(C_4, K_{1,n}) \leq n + 1 + \lceil \sqrt{n} \rceil$, with equality for an infinite number of values of n. However, if H_n is connected and not very 'star like', a sharper bound on $r(C_4, H_n)$ can be determined. Using exactly the same techniques and proof structure as in Theorems 5 and 6, the following two theorems can be proved. Due to the similarity to the previous proofs, the details will not be given.

Theorem 7. *Let H_n be a connected graph of size n and order at most $n - 12\sqrt{n}$. If n is sufficiently large,*

$$r(C_4, H_n) \leq n + 2.$$

Theorem 8. *Let H_n be a connected graph of size n and $\Delta(H_n) \leq \delta\sqrt{n}$. If n is sufficiently large, and $\delta < 1/8$,*

$$r(C_4, H_n) \leq n + 2.$$

Using Theorems 7 and 8, the proof techniques of these results, and a much more detailed analysis, the following result can be proved.

Theorem 9. *Let H_n be a connected graph of size n and $\Delta(H_n) \leq \epsilon n$. If ϵ is sufficiently small and n is sufficiently large,*

$$r(C_4, H_n) \leq n + 2.$$

If the edges of a K_{n+1} are colored such that the red subgraph is a star $K_{1,n}$ and the blue subgraph is a complete graph K_n, there is no red C_4 and no blue connected graph of order $n + 1$. Thus, $r(C_4, H_n) > n + 1$ for any tree H_n of size n. Thus, each of the Theorems 7, 8, and 9 is sharp.

5. Questions

There are many questions left unanswered. The following density question may be very difficult, but it is certainly of interest.

Question 1. *If every subgraph S of G satisfies $q(S) \leq 2p(S) - 3$, is G necessarily Ramsey size linear?*

If the answer to the previous question is yes, the minimal graphs $K_{3,3}$, $G_5 = K_5 - (K_{1,2} \cup K_2)$, and Q_3 are Ramsey size linear. Thus, a subquestion of the previous question is the following.

Question 2. *Are the graphs $K_{3,3}$, $G_5 = K_5 - (K_{1,2} \cup K_2)$, and Q_3 Ramsey size linear?*

Trees and complete graphs could play central roles in determining if a graph G is Ramsey size linear. In particular, consider the following question.

Question 3. *If there is a constant c such that for each integer n, $r(G, T_n) \leq cn$ and $r(G, K_n) \leq cn^2$, is G Ramsey size linear?*

In the upper bound on the Ramsey numbers for cycles, only even cycles were considered. Thus, the following questions is of interest.

Question 4. *For which constants c is $r(C_{2k+1}, H_n) \leq c(2k + 1)n$, where H_n is a graph of size n without isolated vertices?*

A much more difficult problem is to extend the result of Sidorenko and of Corollary 4 on triangles to cycles of arbitrary length.

Question 5. *Is $r(C_m, H_n) \leq 2n + \lfloor (m-1)/2 \rfloor$, where $m \geq 3$, and H_n is a graph of size n without isolated vertices?*

Question 6. *Is there an infinite family of minimal Ramsey size linear graphs, or more specifically, is there a minimal Ramsey size linear graph other than K_4?*

References

[1] Bollobás, B. (1978) *Extremal Graph Theory*, Academic Press, London.

[2] Chvátal, V. (1977) Tree-Complete Graph Ramsey Numbers. *J. Graph Theory* **1** 93.

[3] Erdős, P. (1965) On some Extremal Problems in Graph Theory. *Israel J. Math.* **3** 113–116.

[4] Erdős, P. (1965) On an Extremal Problem in Graph Theory. *Colloquium Math.* **13** 251–254.

[5] Erdős, P., Faudree, R. J., Rousseau, C. C. and Schelp, R. H. (1978) On Cycle-Complete Graph Ramsey Numbers. *J. Graph Theory* **2** 53–64.

[6] Erdős, P., Faudree, R. J., Rousseau, C. C. and Schelp, R. H. (1987) A Ramsey Problem of Harary on Graphs with Prescribed Size. *Discrete Math* **67** 227–233.

[7] Erdős, P. and Gallai, T. (1959) On Maximal Paths and Circuits of Graphs. *Acta Math. Acad. Sci. Hungar.* **10** 337–356.

[8] Erdős, P. and Lovász, L. (1973) Problems and Results on 3-Chromatic Hypergraphs and Some Related Questions. *Infinite and Finite Sets* **10**, Colloquia Mathematica Societatis János Bolyai, Keszthely, Hungary 609–628.

[9] Faudree, R. J. (1983) On a Class of Degenerate Extremal Graph Problems. *Combinatorica* **3** 83–93.

[10] Parsons, T. D. (1975) Ramsey Graphs and Block Designs. *Trans. Amer. Math. Soc.* **209** 33–44.

[11] Lorimer, P. (1984) The Ramsey Numbers for Stripes and One Complete Graph. *J. Graph Theory* **8** 177–184.

[12] Sidorenko, A. F. (manuscript) *The Ramsey Number of an N-Edge Graph Versus Triangle is at Most $2N + 1$.*

[13] Simonovits, M. (1983) Extremal Graph Theory. In: Beineke, L. W. and Wilson, R. J. (eds.) *Selected Topics in Graph Theory II*, Academic Press, New York 161–200.

[14] Spencer, J. (1952) Asymptotic Lower Bounds for Ramsey Functions. *Discrete Math.* **20** 69–76.

Turán–Ramsey Theorems and K_p-Independence Numbers

P. ERDŐS, A. HAJNAL[†], M. SIMONOVITS[†], V. T. SÓS[†]

and E. SZEMERÉDI[†]

Mathematical Institute of the Hungarian Academy of Sciences, Budapest.

Let the K_p-independence number $\alpha_p(G)$ of a graph G be the maximum order of an induced subgraph in G that contains no K_p. (So K_2-independence number is just the maximum size of an independent set.) For given integers $r, p, m > 0$ and graphs $L_1, \ldots, L_r$, we define the corresponding Turán–Ramsey function $RT_p(n, L_1, \ldots, L_r, m)$ to be the maximum number of edges in a graph G_n of order n such that $\alpha_p(G_n) \leq m$ and there is an edge-colouring of G with r colours such that the jth colour class contains no copy of L_j, for $j = 1, \ldots, r$.

In this continuation of [11] and [12], we will investigate the problem where, instead of $\alpha(G_n) = o(n)$, we assume (for some fixed $p > 2$) the stronger condition that $\alpha_p(G_n) = o(n)$. The first part of the paper contains multicoloured Turán–Ramsey theorems for graphs G_n of order n with small K_p-independence number $\alpha_p(G_n)$. Some structure theorems are given for the case $\alpha_p(G_n) = o(n)$, showing that there are graphs with fairly simple structure that are within $o(n^2)$ of the extremal size; the structure is described in terms of the edge densities between certain sets of vertices.

The second part of the paper is devoted to the case $r = 1$, i.e., to the problem of determining the asymptotic value of

$$\theta_p(K_q) = \lim_{n \to \infty} \frac{RT_p(n, K_q, o(n))}{\binom{n}{2}},$$

[‡]for $p < q$. Several results are proved, and some other problems and conjectures are stated.

0. Notation

In this paper we will consider graphs without loops and multiple edges. Given a graph G, $e(G)$ will denote the number of edges, $v(G)$ the number of vertices, $\chi(G)$ the chromatic number, and $\alpha(G)$ the maximum cardinality of an independent set in G. More generally, given an integer $p > 1$, $\alpha_p(G)$ denotes the *p-independence number* of G: the maximum cardinality of a set S such that the subgraph of G spanned by S contains no K_p. Given a

† Supported by GRANT 'OTKA 1909'.

‡ This notation, where we put $o(n)$ in place of $f(n)$ is slightly imprecise. It means that any function $f(n) = o(n)$ and will be clarified in Section 2.

graph, the (first) subscript will mostly denote the number of vertices: G_n, S_n, will always denote graphs on n vertices. For given graphs $L_1, \ldots, L_r$, $R(L_1, \ldots, L_r)$ will denote the usual Ramsey number, that is, the minimum t such that for every edge-colouring of K_t in r colours, for some v the vth colour contains an L_v[†]. If we partition n vertices into q classes as equally as possible and join two vertices iff they belong to different classes, we obtain the so-called Turán graph on n vertices and k classes, denoted by $T_{n,k}$. This graph is the (unique) k-chromatic graph on n vertices with the maximum number of edges.

For a set Q, we will use $|Q|$ to denote its cardinality. Given two disjoint vertex sets, X and Y, in a graph G_n, we use $e(X, Y)$ to denote the number of edges in G_n joining X and Y, and $d(X, Y)$ to denote the edge-density between them:

$$d(X, Y) = \frac{e(X, Y)}{|X| \cdot |Y|}.$$

Given a graph G and a set U of vertices of G, we use $G[U]$ to denote the subgraph of G induced (spanned) by U. The number of edges in a subgraph spanned by a set U of vertices of G will be denoted by $e(U)$. We will say that X is *completely joined* to Y if every vertex of X is joined to every vertex of Y.

Given two points x, y in the Euclidean space $\mathbf{E}^h$, we use $\rho(x, y)$ to denote their ordinary distance.

1. Introduction

Ramsey's Theorem [23] and Turán's Extremal Theorem [33, 34] are both among the most well-known theorems of graph theory. Both served as starting points for whole branches of graph theory. (For Ramsey Theory, see the book by R. L. Graham, B. L. Rothschild and J. Spencer [21], and for Extremal Graph Theory, see the book by Bollobás [2], or the survey by Simonovits [29].) In the late 1960's a new theory emerged connecting these fields. Perhaps the first paper in this field is due to V. T. Sós [30], and quite a few results have been found since then.

The 'historical' part of the introduction of this paper is slightly condensed, to avoid too much repetition. For some further information see [12]. Some important references can be found at the end of the paper, see [3, 11, 12, 18, 20, 31].

In [11] P. Erdős, A. Hajnal, V. T. Sós, and E. Szemerédi investigated the following problem:

Suppose that a so-called forbidden graph L and a function $f(n) = o(n)$ are given. Determine

$$RT(n, L, f(n)) = \max \left\{ e(G_n) \ : \ L \nsubseteq G_n \text{ and } \alpha(G_n) < f(n) \right\}.$$

They showed that this number depends (in some sense) primarily on the so-called Arboricity of L (which is a slight modification of the usual arboricity of L). In a continuation [12] of that paper, we started investigating the following problem:

[†] This is the only case when the (first) subscript is not the number of vertices: *i.e.* when we speak of the excluded graphs L_i.

Let G_n be a graph on n vertices the edges of which are coloured by r colours $\chi_1, \ldots, \chi_r$, so that the subgraph of colour χ_v contains no complete subgraph K_{p_v}, $(v = 1, \ldots, r)$. Let a function $f(n)$ be given, (mostly $f(n) = o(n)$) and suppose that $\alpha(G_n) \leq f(n)$. What is the maximum number of edges in G_n under these conditions?

In *this* continuation of [11] and [12] we will investigate the problem where, instead of $\alpha(G_n) = o(n)$, we assume a stronger independence condition: that the maximum cardinality of a K_p-free induced subgraph of G_n is $o(n)$:

$$\alpha_p(G_n) = o(n).$$

The concept of $\alpha_p(G)$ was introduced long ago by A. Hajnal, and also investigated by Erdős and Rogers, see [16]. (A similar 'independence notion' is investigated for random graphs in a paper of Eli Shamir [24], where he generalizes some results on the chromatic number of random graphs.)

The general problem

Assume that $L_1, \ldots, L_r$ are given graphs, and G_n is a graph on n vertices, the edges of which are coloured by r colours $\chi_1, \ldots, \chi_r$, and

$$(*) \quad \begin{cases} \text{for } v = 1, \ldots, r \text{ the subgraph of colour } \chi_v \text{ contains no } L_v \\ \text{and } \alpha_p(G_n) \leq m. \end{cases}$$

What is the maximum of $e(G_n)$ under these conditions?

The maximum will be denoted by $RT_p(n, L_1, \ldots, L_r, m)$. The graphs attaining the maximum in this problem will be called *extremal* graphs for $RT_p(n, L_1, \ldots, L_r, m)$. It may happen that there exist no graphs satisfying our conditions. Then we will say that the maximum is 0.

Of course, for fixed m and large n – by Ramsey's theorem – there are no graphs with the above properties: the maximum is taken over the empty set. However, we are interested mainly in the case $m \to \infty$, $m = f(n) = o(n)$, but $m/n \to 0$ very slowly.

The existence of graphs satisfying $(*)$ is far from being trivial. We will use a theorem of Erdős and Rogers to prove the existence of such graphs for the case of one colour and when the forbidden graph is a complete graph. We will sketch a constructive proof of the Erdős–Rogers theorem in Section 4, and return to this question in a more general setting in the Appendix, where we will characterize the cases when $(*)$ can be satisfied (for 2-connected forbidden graphs). Among others, we will see that $(*)$ can always be satisfied when all the forbidden graphs L_i are complete graphs of more than p vertices and $m = n^{1-c}$ for some small $c > 0$.

Some motivation Our problems are motivated by the classical Turán and Ramsey Theorems [33, 34, 23], and also (indirectly) by some applications of the Turán Theorem to geometry, analysis (in particular, potential theory) [35, 36, 37, 13, 14, 15], and probability theory (see, for example, Katona, [22], or Sidorenko, [25, 26]), (see also [38]).

In [12] we proved (among others), for the problem of $RT_2(n, K_{\ell_1}, \ldots, K_{\ell_r}, o(n))$, the existence of a sequence of asymptotically extremal graph sequences of relatively simple structure[†].

Assume now, that $\alpha(G_n)$ is *much smaller* than cn, for example $\alpha(G_n) \leq \sqrt{n}$. Then we know (since $R(K_3, K_k) < k^2/\log k$) that for every fixed $c > 0$, every set of $> cn$ vertices of G_n will contain not only an edge, but also a K_3. Similarly, if we choose even smaller upper bounds for $\alpha(G_n)$, we can ensure the even stronger conditions that every induced subgraph of G_n of at least cn vertices contains a larger complete graph K_p. This also leads to the problems of the present paper, though apart from Theorem 2.1 we will deal only with the simplest case $f(n) = o(n)$.

Some basic definitions ⁓It is probably hopeless to give an exact description of the maximum in the general problem. Therefore we will try to find an asymptotically extremal sequence of graphs of relatively simple structure. The definitions listed here are needed to make precise what we consider 'relatively simple'.

Notation. For any given function f, let

$$\vartheta_\varepsilon = \vartheta_{\varepsilon,p,f}(L_1, \ldots, L_r) = \limsup_{n \to \infty} \frac{e(G_n)}{\binom{n}{2}} \quad \text{and} \quad \vartheta_{p,f} = \lim_{\varepsilon \to 0} \vartheta_\varepsilon$$

where the limsup is taken for the r-coloured graphs G_n satisfying (*) with $m = \varepsilon f(n)$:

$$\begin{cases} \text{for } v = 1, \ldots, r \text{ the subgraph of colour } \chi_v \text{ contains no } L_v \\ \text{and } \alpha_p(G_n) \leq \varepsilon f(n). \end{cases}$$

(If the limsup is taken over the empty set (of graphs), it is defined to be 0.) Clearly, if $\varepsilon \to 0$, the lim sup above will converge, since it is monotone in ε. One can easily see the following claim.

Claim 1.1.

(a) If $\varepsilon_n \to 0$,

$$\limsup_{n \to \infty} \frac{RT_p(n, L_1, \ldots, L_r, \varepsilon_n f(n))}{\binom{n}{2}} \leq \vartheta_{p,f}(L_1, \ldots, L_r).$$

(b) There exist a sequence $\varepsilon_n^* \to 0$ and an infinite sequence $(S_n : n \in \mathbb{N}_0)$ $(\mathbb{N}_0 \subseteq \mathbb{N})$ of graphs with the property (*) for $m = \varepsilon_n^* f(n)$ where the equality holds in (a).

(c) For every $\varepsilon_n \geq \varepsilon_n^*, \varepsilon_n \to 0$,

$$\lim_{\substack{n \in \mathbb{N}_0 \\ n \to \infty}} \frac{RT_p(n, L_1, \ldots, L_r, \varepsilon_n f(n))}{\binom{n}{2}} = \vartheta_{p,f}(L_1, \ldots, L_r).$$

Proof. Here (a) is trivial from the definition, (c) is trivial from (a) and (b), by monotonicity, and (b) follows by an easy diagonalization.

Indeed, assume that for $k = 1, \ldots, t-1$ we have already fixed S_{n_k}. Now we fix $\varepsilon = \varepsilon_t = 1/t$ and find an S_{n_t} with the following properties: $n_t > n_{t-1}$,

$$e(S_{n_t}) \geq \left(\vartheta_\varepsilon - \frac{1}{t} \right) \binom{n_t}{2},$$

and $\alpha_p(S_{n_t}) \leq (1/t)f(n_t)$. □

[†] The definitions can be found below.

Unfortunately, we cannot prove the corresponding assertions for *all $n \geq n_0$*: we cannot exclude the possibility that

$$\frac{RT_p(n, L_1, \ldots, L_r, \varepsilon_n f(n))}{\binom{n}{2}}$$

jumps up and down as $n \to \infty$.

We will often speak of the problem of determining $RT_p(n, L_1, \ldots, L_r, o(n))$, meaning the determination of $\vartheta_{p,f}(L_1, \ldots, L_r)$, for $f(n) = n$. This slightly imprecise notation will cause no problems. Similarly, if $f(n) = n$, we will often use the notation $\vartheta_p(L_1, \ldots, L_r)$ instead of $\vartheta_{p,f}(L_1, \ldots, L_r)$. Observe that ϑ is monotone: if we replace L_1 by an $L_1^* \supseteq L_i$, then $\vartheta_{p,f}(L_1, \ldots, L_r) \leq \vartheta_{p,f}(L_1^*, \ldots, L_r)$. In particular, $\vartheta_p(K_q)$ is monotone increasing in q.

Definition 1.2. (Asymptotically extremal graphs)   Suppose that the forbidden graphs $L_1, \ldots, L_r$, and the function f are given. An infinite sequence of graphs, (S_n), will be called an *asymptotically extremal sequence* (for $L_1, \ldots, L_r$ and f) if the edges of each S_n can be r-coloured so that the vth colour contains no L_v, ($v = 1, \ldots, r$), $\alpha_p(G_n) = o(f(n))$, and

$$\frac{e(S_n)}{\binom{n}{2}} = \vartheta_{p,f}(L_1, \ldots, L_r) + o(1).$$

In Section 2 we will formulate some theorems asserting that, for any r, there are always asymptotically extremal graph sequences of *fairly simple* structure. To formulate these theorems, we have to introduce the notion of matrix graphs, and matrix graph sequences.

We will say that two disjoint vertex sets X and Y *are joined ε-regularly* in the graph G if for every subset $X^* \subseteq X$ and $Y^* \subseteq Y$ satisfying $|X^*| > \varepsilon|X|$ and $|Y^*| > \varepsilon|Y|$, we have

$$|d(X^*, Y^*) - d(X, Y)| < \varepsilon.$$

In the following $A = (a_{ij})$ will always be a *symmetric* matrix with all $a_{ij} \in [0, 1]$.

Definition 1.3. (*A*-matrix graph sequences) Given a $t \times t$ symmetric matrix $A = (a_{ij})$, a graph sequence (S_n) – defined for infinitely many n but not necessarily defined for every $n > n_0$ – is said to be an *A-matrix graph sequence* if the vertices of S_n can be partitioned into t classes $V_{1,n}, \ldots, V_{t,n}$ so that in S_n

— $e(V_{i,n}) = o(n^2)$, for every $i = 1, \ldots, t$,
— $d(V_{i,n}, V_{j,n}) = a_{ij} + o(1)$ for every $1 \leq i < j \leq t$ and
— the classes $V_{i,n}$ and $V_{j,n}$ are joined ε_n-regularly for every $1 \leq i < j \leq t$ for some $\varepsilon_n \to 0$.

We will associate a quadratic form $\mathbf{u}A\mathbf{u}^T$ to A and maximize it over the simplex $\sum u_i = 1, u_1, \ldots, u_i \geq 0$:

$$g(A) := \max\{\mathbf{u}A\mathbf{u}^T : \sum u_i = 1, u_i \geq 0\}.$$

The quadratic form will be used to measure the number of edges in the corresponding matrix graph sequence. The vectors attaining the maximum will be called *optimum vectors*. (Optimum below will always mean maximum.)

Definition 1.4. (Dense matrices) A matrix A is *dense*, if for any i deleting the ith row and the ith column of the matrix A we get an A' with $g(A') < g(A)$.

One can easily see [4] that if A is dense, it has a *unique* optimum vector and all the coordinates of this optimum vector are positive. The uniqueness implies that the symmetries of the matrix leave the optimum vector invariant: the corresponding coordinates are equal. This means that if a permutation π of $1,\dots,t$ applied to the rows and to the columns of A leaves A invariant, then π applied to the optimum vector also leaves it unchanged. Further, if $g(A') < g(A)$ for some symmetric minor A' of A, there exists an A'' obtained from A by deleting just one row and the corresponding column and satisfying $g(A'') < g(A)$. For a more detailed description of this function $g(A)$ see [4, 7].

Definition 1.5. (Asymptotically optimal A-matrix-graph sequences) Let A be a fixed matrix and $\mathbf{u} = (u_1,\dots,u_t)$ be an optimum vector for A. We will call an A-matrix graph sequence (S_n) *asymptotically optimal* if the classes $V_{i,n}$ can be chosen so that $|V_{i,n}|/n = u_i + o(1)$, for $i = 1,\dots,t$.

Clearly, an optimal matrix graph has

$$\frac{1}{2}g(A)n^2 + o(n^2)$$

edges. If the matrix A has a submatrix A' such that $g(A') = g(A)$, we can always replace the matrix graph sequence corresponding to A by the simpler matrix graph sequence corresponding to A'. This is why we are interested only in *dense* matrices.

2. Main results

We start with the existence of the limit.

Theorem 2.1. *For any* $p_1,\dots,p_r$ *and for* $f(n) = n$, *for any* $\varepsilon_n \to 0$:
(a) Let (S_n) *be an extremal graph sequence for* $RT_p(n, K_{p_1},\dots,K_{p_r}, \varepsilon_n n)$. *Then*

$$\limsup_{n\to\infty} \frac{e(S_n)}{\binom{n}{2}} \le \vartheta_{p,f}(K_{p_1},\dots,K_{p_r}). \tag{1a}$$

(b) There exists an $\varepsilon_n^* \to 0$ *for which on the left-hand side of (1a) the limit exists and*

$$\lim_{n\to\infty} \frac{e(S_n)}{\binom{n}{2}} = \vartheta_{p,f}(K_{p_1},\dots,K_{p_r}). \tag{1b}$$

(c) For every $\varepsilon_n \to 0$ *with* $\varepsilon_n \ge \varepsilon_n^*$ *the same – namely, (1b) – holds.*

Here $f(n) = n$ means that we consider the case $\alpha_p(G_n) = o(n)$. The difference between this theorem and Claim 1.1 is that there we regard *all possible forbidden graphs*, here only complete graphs, and there we assert only the existence of a *sparse subset of integers* along which a limit exists, (*i.e.*, we assert that the limsup can be obtained in some specific way) here we assert that the actual limit exists.

The meaning of Theorems 2.2 and 2.3 below is that in the general case there are asymptotically extremal graph sequences of fairly simple structure, where 'simple' means that the structure depends on n weakly. This is a weak generalization of the Erdős–Stone–Simonovits Theorem (from ordinary extremal graph theory) [17, 19]. The optimal matrix graph sequences – in some sense – generalize the Turán graphs, while the matrix graphs generalize the complete t-partite graphs. (See also [8], and [28]).

Theorem 2.2. *For any $p_1, \ldots, p_r$ let $\varepsilon_n \to 0$ sufficiently slowly (which means that the condition of (c) of Theorem 2.1 is satisfied). Then there exists a dense $\Omega \times \Omega$ matrix A with $\Omega < R(K_{p_1}, \ldots, K_{p_r})$ and an asymptotically extremal sequence (S_n) for $RT_p(n, K_{p_1}, \ldots, K_{p_r}, \varepsilon_n n)$ that is an asymptotically optimal A-matrix graph sequence.*

For general $L_1, \ldots, L_r$ we have the following theorem.

Theorem 2.3. *Let r forbidden graphs $L_1, \ldots, L_r$ be fixed. Let $f(n) \to \infty$ $(f(n) = O(n))$ be an arbitrary function for which for every $c \in (0, 1)$ there exists an $\eta = \eta_{f,c} > 0$ such that*

$$f(cn) > \eta_{f,c} f(n).$$

Then there exist a dense matrix $A = (a_{ij})_{\Omega \times \Omega}$ – for some $\Omega < R(L_1, \ldots, L_r)$, and an asymptotically extremal sequence (S_{n_ℓ}) (for $L_1, \ldots, L_r$ and f, for some subsequence of integers) that is an asymptotically optimal A-matrix graph sequence.

This means that the structure of some asymptotically extremal sequences is simple. The matrix A depends on the function f: for different f's we get different extremal densities. The matrix depends primarily on the sample graphs and on f. However, we are unable to exclude the possibility that A must, even in the simplest case $f = n$, depend on the actual subsequence of integers as well: that there is no common A for all $n > n_0$. The condition $f(cn) > \eta_{f,c} f(n)$ is a 'smoothness' condition, which is satisfied in 'all the reasonable cases'.

Remark 2.4. We are primarily interested in functions of type $f(n) = n^\gamma$. By the quantitative Ramsey Theorem, for every family $L_1, \ldots, L_r$ we can fix a $\Gamma > 0$ so that if $\alpha(G_n) < f(n) = n^\Gamma$, then every r-colouring of G_n contains an L_v of colour v for some $v \leq r$ (since it contains a large clique of colour v): no graphs satisfy (∗).

Remark 2.5. When we assert the existence of a matrix A in Theorems 2.2 and 2.3, we do not know too much about this A. The only thing we know is that it is dense and (therefore, by Lemma 3.3) its off-diagonal entries are all positive.

Unfortunately, most of the non-trivial results for the K_p-free case $(p > 2)$ are related to the special case when all the forbidden subgraphs L_v are complete graphs. So in Sections 4-6 we will assume that the graphs L_i are complete graphs. In Section 4 we will prove some general upper and lower bounds for the case of one colour $(r = 1)$. The following result is a direct generalization of the Erdős–Sós Theorem from [18].

Theorem 2.6.

(a) For any integers $p > 1$ and $q > p$ we have

$$\vartheta_p(K_q) \le \frac{1}{2}\left(1 - \frac{p}{q-1}\right).$$

(b) For every k, for $q = pk + 1$ this is sharp:

$$\vartheta_p(K_q) = \frac{1}{2}\left(1 - \frac{p}{q-1}\right) = \frac{1}{2}\left(1 - \frac{1}{k}\right).$$

To get the lower bound in Theorem 2.6 (*i.e.*, Theorem 2.6(b)) we will use a geometric construction of Erdős and Rogers [16]. Here we formulate their theorem, but the verification is postponed to Section 4.

Erdős–Rogers Theorem. Let $p \ge 2$ be an integer. There are a constant $c = c_p > 0$ and an $n_0(p, c)$, such that for every $n > n_0(p, c)$, there exists a graph Q_n not containing K_{p+1}, but satisfying $\alpha_p(Q_n) \le n^{1-c}$.

Construction 2.7. Let $q = pk + 1$. Take k vertex-disjoint Erdős–Rogers graphs of size $(n/k) + o(n)$ (described in the previous theorem) and join each vertex to all the vertices in the other graphs. (We will sometimes describe this as putting (p, ε)-Erdős–Rogers graphs into each class of a $T_{n,k}$.) Thus we get a graph sequence (S_n) with $\alpha_p(S_n) \le kn^{1-c}$ for some $c > 0$ and $K_{pk+1} \not\subseteq S_n$.

This proves the lower bound in Theorem 2.6. For $q = p + \ell$, $\ell = 2, 3, 4, 5$ we can improve the upper bound of Theorem 2.6, see Theorem 2.11.

Remark 2.8. Now, for $p \ge 2$ fixed, we know the value of every p^{th} $\vartheta_p(K_q)$. Perhaps the other values have a 'pseudo-periodical' behaviour similar to that of $\vartheta_2(K_q)$: the extremal structure is determined by the residue of q mod p. The situation is analogous to that in the Erdős–Hajnal–Sós–Szemerédi [11] Theorem, where the case of odd values of q was much simpler (and also much simpler to prove) than the case of even q's.

In Section 5, we investigate some special cases that seem to be interesting, because they suggest some conjectures for the general case. Perhaps the following conjecture holds.

Conjecture 2.9. The asymptotically extremal graphs for $RT_p(n, K_q, o(n))$ have the following structure: Let $q = pk + \ell$, $(\ell = 1, 2, \ldots, p)$. Then n vertices are partitioned into $k + 1$ classes $V_{0,n}, \ldots, V_{k,n}$. For each pair $\{i, j\} \ne \{0, 1\}$, $V_{i,n}$ is almost completely joined to $V_{j,n}$ in the sense that every $x \in V_{i,n}$ is joined to every $y \in V_{j,n}$ with a possible exception of $o(n^2)$ pairs xy. Further, $d(V_{0,n}, V_{1,n}) = ((\ell - 1)/p) + o(1)$ (as $n \to \infty$), and $V_{0,n}, V_{1,n}$ are joined $o(1)$-regularly. Finally, $e(V_{i,n}) = o(n^2)$, $i = 1, \ldots, k$.

Remark 2.10. For graphs of this kind the optimal sizes of the classes V_i can easily be computed: the optimal class-sizes are as follows. The edges in $G[V_i]$ can be neglected,

$$|V_i| = \frac{1}{2 + (k-1)(2 - \frac{\ell-1}{p})} n + o(n) \text{ for } i = 0, 1$$

and

$$|V_i| = \frac{(2 - \frac{\ell-1}{p})}{2 + (k-1)(2 - \frac{\ell-1}{p})} n + o(n) \text{ for } 2 \le i \le k.$$

From this, $e(S_n)$ can easily be calculated: if S_n is the graph described in the conjecture, it is almost regular, and the degrees in V_2 are $n - |V_2|$. Hence

$$e(S_n) \approx \frac{1}{2}(n - |V_2|) n \approx \left(1 - \frac{(2p - \ell + 1)}{k(2p - \ell + 1) - \ell + 1}\right)\binom{n}{2}.$$

We describe some cases below, where we can prove the upper bound in Conjecture 2.9.

Theorem 2.11. *Let $\ell = 2, 3, 4$ or 5 and $\ell \le p + 1$. If $K_{p+\ell} \nsubseteq G_n$ and $\alpha_p(G_n) = o(n)$, then*

$$e(G_n) \le \frac{\ell - 1}{4p} n^2 + o(n^2).$$

By Theorem 2.6, we know that $\vartheta_p(K_{p+1}) = 0$ and $\vartheta_p(K_{2p+1}) > 0$. Here one of the main problems is:

Problem 2.12. *For fixed p determine the minimum ℓ for which*

$$\vartheta_p(K_{p+\ell}) > 0.$$

In particular, is $\vartheta_p(K_{p+2}) > 0$ or not? If $\vartheta_p(K_{p+\ell}) > 0$, how large is it?

Theorem 2.13. *For any $p \ge 2$, $\vartheta_p(K_{2p}) \ge \frac{1}{8}$.*

It is worth observing that replacing K_{2p} by K_{2p+1} we get by Theorem 2.6(b), for any $p \ge 2$, $\vartheta_p(K_{2p+1}) = 1/4$.

For $p = 2$ Theorem 2.13 is sharp: $\vartheta_2(K_4) \le 1/8$ was proved by Szemerédi [31] and $\vartheta_2(K_4) \ge 1/8$ was settled by Bollobás and Erdős in [3], via a high-dimensional geometric construction. In a slightly different form, Bollobás and Erdős did the following. Fix a high-dimensional sphere S^h and partition it into $n/2$ domains $D_1, \ldots, D_{n/2}$, of equal measure and diameter $(1/2)\mu$, with $\mu = \varepsilon/\sqrt{h}$. Choose a vertex $x_i \in D_i$ and an $y_i \in D_i$ for $i = 1, \ldots, n/2$ and put $X = \{x_1, \ldots, x_{n/2}\}$ and $Y = \{y_1, \ldots, y_{n/2}\}$. Let $X \cup Y$ be the vertex-set of our S_n, and

$$\begin{array}{ll} \text{join an } x \in X \text{ to a } y \in Y & \text{if } \rho(x, y) < \sqrt{2} - \mu; \\ \text{join an } x \in X \text{ to a } x' \in X & \text{if } \rho(x, x') > 2 - \mu; \\ \text{join a } y \in Y \text{ to a } y' \in Y & \text{if } \rho(y, y') > 2 - \mu. \end{array}$$

For $p \geq 3$, our result follows from a generalization of this construction. Theorem 2.13 may also be sharp for $p \geq 3$, but we cannot prove it. Let $p = 3$. Our results show only that

$$0 \leq \vartheta_3(K_5) \leq \frac{1}{12}$$

and

$$\frac{1}{8} \leq \vartheta_3(K_6) \leq \frac{1}{6}.$$

One of the most intriguing problems is to determine the values and some asymptotically extremal graphs for $RT_3(n, K_5, o(n))$ and $RT_3(n, K_6, o(n))$. Unfortunately, this task seems to be too difficult. We do not know the answer to the simplest subproblem if $\vartheta_3(K_5) > 0$.

The last section contains some further open problems.

The basic proof techniques include primarily the application of Szemerédi's Regularity Lemma, [32], a modification of Zykov's symmetrization method, [39] and multigraph extremal-graph results [4, 5, 6, 7] (in the background).

Remark 2.14. It is difficult to find the places in this paper that would distinguish between the conditions '(+) S_n contains no L_i' and '(++) S_n can be coloured in r colours so that the vth colour contains no L_v'. The reason for this is that the limit constants are the ones that are different: we have the existence theorems in the same generality for the more general case (++).

3. Proofs of Theorems 2.1–2.3

The aim of this section is to prove Theorems 2.1–2.3. We will start with the simpler Theorem 2.1, move on to the proof of Theorem 2.3 and then return to the proof of Theorem 2.2.

Proof of Theorem 2.1. Again, as in the 'proof' of Claim 1.1, (a) is trivial, (c) follows from (a) and (b) and the only thing to be proved is that if the forbidden graphs are complete graphs and we have an infinite sequence (S_{m_t}), as described in Claim 1.1(b), then we can extend this sequence to every $n > n_0$.

First fix an $\varepsilon > 0$. Assume that S_{m_t} is an extremal graph for $RT_p(m_t, K_{p_1}, \ldots, K_{p_r}, \varepsilon m_t)$. We may choose this sequence so that

$$e(S_{m_t}) \geq (\vartheta_\varepsilon - \varepsilon) \binom{m_t}{2}.$$

So S_{m_t} has an r-colouring in which the vth colour contains no K_{p_v} and $\alpha_p(S_{m_t}) \leq \varepsilon m_t$ if $t > t_0(\varepsilon)$. Below, we will sometimes abbreviate m_t to m. Let h be an arbitrary integer and put $Z_{mh} = S_m \otimes I_h$, that is, let Z_{mh} be the graph obtained from S_m by replacing each vertex by h independent vertices and joining two new vertices in colour v iff the original vertices have been joined in colour $v^\dagger$. Since the forbidden graphs are complete graphs,

---

† Here I_h means the complementary graph of K_h.

the r-coloured Z_{mh} will contain no K_{p_v} (either) in the vth colour. Further, trivially,

$$\frac{e(Z_{mh})}{(mh)^2} = \frac{e(S_m)}{m^2},$$

and

$$\frac{\alpha_p(Z_{mh})}{mh} = \frac{\alpha_p(S_m)}{m}.$$

(Indeed, each K_p-independent set increases by a factor h, and each K_p-independent set X of Z_{mh} induces a K_p-independent set of S_m of size at least $(1/h)|X|$.)

As described in the proof of Claim 1.1, we may choose a sequence S_{m_t} with $\varepsilon_t = (1/t)$, $\alpha_p(S_{m_t}) \leq \varepsilon_t m_t$, and (for $f = n$ and $\vartheta = \vartheta_{p,f} = \lim \vartheta_\varepsilon$)

$$e(S_{m_t}) \geq \left(\vartheta - \frac{1}{t} \right) \binom{m_t}{2}.$$

Now, for every $n > m_1^2$ choose the largest $m_t \leq \sqrt{n}$. Then choose $h = \lceil n/m_t \rceil$ and delete $n - m_t h$ vertices of $Z_{m_t h}$ to get a graph S_n^*. Clearly, $m_t \to \infty$. Since we have deleted at most $m_t h - n = o(n)$ vertices from $Z_{m_t h}$, we obtain a sequence S_n^* with $\alpha_p(S_n^*) = o(n)$ and

$$e(S_{m_t}) = \vartheta \binom{m_t}{2} + o(m_t^2).$$

$\square$

(As $m_t \to \infty$, we cannot get all the integers in the form hm_t. Therefore we must approximate some n's by $hm_t > n$: to delete $\leq h = o(m_t h)$ vertices from some of the $Z_{m_t h}$'s.)

One of the basic methods we use to handle Turán–Ramsey type problems is the Regularity Lemma [32].

The Regularity Lemma The regularity condition means that the edges behave (in some weak sense) as if they were random. The Regularity Lemma asserts that the vertices of the graph can be partitioned into a bounded number of classes $V_0, \ldots, V_k$ such that almost every pair is ε-regular.

The Regularity Lemma. (See, for example, [32].) *For every $\varepsilon > 0$ and integer κ there exists a $k_0(\varepsilon, \kappa)$ such that every graph G_n, the vertex set $V(G_n)$ can be partitioned into sets $V_0, V_1, \ldots, V_k$ – for some $\kappa < k < k_0(\varepsilon, \kappa)$ – so that $|V_0| < \varepsilon n$, $|V_i| = m$ (is the same) for every $i > 0$, and for all but at most $\varepsilon \binom{k}{2}$ pairs (i, j), for every $X \subseteq V_i$ and $Y \subseteq V_j$ satisfying $|X|, |Y| > \varepsilon m$, we have*

$$|d(X, Y) - d(V_i, V_j)| < \varepsilon.$$

Remark 3.1. The role of V_0 is purely technical: it makes it possible for all the other classes to have *exactly* the same cardinality. Indeed, having a κ and choosing $\kappa' > \kappa, \varepsilon^{-2}$ and applying the Regularity Lemma with this κ, one can distribute the vertices of V_0 evenly among the other classes so that $|V_i| \approx |V_j|$ and the ε-regularity will be preserved with a slightly larger ε. So from now on (for the sake of simplicity) we will assume that $V_0 = \emptyset$. The role of κ is to make the classes V_i sufficiently small, so that the number of edges inside those classes are negligible. The partitions described in the Regularity Lemma, or here, will be called *Regular Partitions* of G_n.

Now we turn to the second tool used in our proof: the application of matrix graph sequences.

Dense matrices, matrix graph sequences

Lemma 3.2. *Let A be a symmetric matrix, τ and π, $\tau \neq \pi$ be given integers, and let $a_{\tau,\pi} = 0$. Then deleting either*

— *the τth row and column, or*
— *the πth row and column*

we get a matrix A' with

$$g(A') = g(A).$$

This implies

Lemma 3.3. *If a symmetric matrix A is dense, then all its off-diagonal entries are positive.*

The lemma is a variant of Zykov's symmetrization [39], and its proof can be found, for example, in [4]. Hence we only sketch its proof here[†].

Proof of Lemma 3.2. (Sketched) Let $\mathbf{u}$ be an optimum vector for A, *i.e.*,

$$g(A) = \max \left\{ \mathbf{u}A\mathbf{u}^T \ : \ u_i \geq 0 \quad (i = 1, \ldots, \ell) \text{ and } \sum u_i = 1 \right\}.$$

We define $\mathbf{u}(h)$ to be the vector where the τth coordinate of the optimum vector $\mathbf{u}$ is decreased by h and the πth is increased by h. Clearly,

$$\varphi(h) = \mathbf{u}(h)A\mathbf{u}(h)^T = (a_{\pi,\pi} + a_{\tau,\tau})h^2 + c_1 h + c_2$$

for some constants c_1, c_2 (because $a_{\tau,\pi} = a_{\pi,\tau} = 0$). For any interval, such functions attain their maximum at some end of the interval (and maybe, inside as well). Hence we may choose either $h = u_\tau$ or $h = -u_\pi$ and still get the same maximum $g(A)$. But now one of the coordinates is 0, therefore the value of $g(A)$ is the same as if we had deleted the τth or π^{th} row and column: $g(A) = g(A')$. $\qquad\square$

Lemma 3.4. *Assume that $f(n)$ satisfies the condition of Theorem 2.3. Then for every sequence $\varepsilon_n \to 0$ we can find a sequence $\beta_n \to 0$ such that*

$$f(\beta_n n) \geq \sqrt{\varepsilon_n} f(n). \tag{2}$$

[†] A. Sidorenko [27] has found a generalization of this lemma, providing a necessary and sufficient condition for being dense.

Proof. Let t be an integer and $\beta = 1/t$. Then $f(\beta n) \geq \eta_{f,\beta} f(n)$. If $n > n_t$ then $\varepsilon_n \leq \eta_{f,\beta}^2$. Thus $f(\beta n) \geq \sqrt{\varepsilon_n} f(n)$. We may assume $n_{t+1} > n_t$. Define $\beta_n = 1/t$ for $n \in [n_t, n_{t+1})$, $t := 1, 2, 3 \ldots$ Then $\beta_n \to 0$ and (2) holds. $\qquad\square$

Proof of Theorem 2.3. For every fixed $\varepsilon > 0$, for some infinite set of integers $\mathbb{N}_\varepsilon$, for every $n \in \mathbb{N}_\varepsilon$, we may fix an S_n satisfying

(i) $\quad\begin{cases} \text{for } v = 1, \ldots, r \text{ the subgraph of colour } \chi_v \text{ contains no } L_v \\ \text{and } \alpha_p(S_n) \leq \varepsilon f(n), \end{cases}$

and

(ii) $\quad \vartheta_\varepsilon = \lim_{n \in \mathbb{N}_\varepsilon} \frac{e(S_n)}{\binom{n}{2}}$.

Apply the Regularity Lemma to this sequence (S_n) with this ε and $\kappa = 1/\varepsilon$ (where κ is the lower bound on the number of classes). Thus we get a $k_0 = k_0(\varepsilon)$ such that the vertices of S_n can be partitioned into the classes $V_{1,n}, \ldots, V_{k,n}$ for some $\kappa < k < k_0$ so that

(iii) $\quad$ all but $\varepsilon \binom{k}{2}$ pairs are ε-regular, $(k = k(n).)$ [†]

Using a diagonalization, we may find an infinite set of integers $\mathbb{N}^*$ and for each $n \in \mathbb{N}^*$ an r-coloured graph S_n, with a Regular Partition $\{V_{1,n}, \ldots, V_{k(n),n}\}$, satisfying

(i*) $\quad\begin{cases} \text{for } v = 1, \ldots, r \text{ the subgraph of colour } \chi_v \text{ contains no } L_v \\ \text{and } \alpha_p(S_n) \leq \varepsilon_n f(n), \end{cases}$

with some $\varepsilon_n \to 0$, and

(ii*) $\quad \vartheta = \lim_{n \in \mathbb{N}^*} \frac{e(S_n)}{\binom{n}{2}}$, and

(iii*) $\quad$ all but $\varepsilon_n \binom{k}{2}$ pairs are ε_n-regular in the corresponding Regular Partition.

Here ε_n usually tends to 0 very slowly, but still it tends to 0! We may assume that $\vartheta > 0$. Next, delete the edges $(x, y) : x \in V_{i,n}, y \in V_{j,n}$ if

(a) either $(V_{i,n}, V_{j,n})$ is nonregular, or

(b) $d(V_{i,n}, V_{j,n}) < 2\varepsilon_n$.

Thus we have deleted by (a) at most $\varepsilon_n \binom{k}{2}(n/k)^2 < (1/2)\varepsilon_n n^2$ edges and by (b) at most $2\varepsilon_n (n/k)^2 \binom{k}{2}$ edges. In this way we have ensured that all the pairs $(V_{i,n}, V_{j,n})$ are ε_n-regular. The number of edges has been changed by at most $(3/2)\varepsilon_n n^2$. Denote the resulting graph by T_n.

There is a matrix $A = A_n$ of $k < k_0(\varepsilon_n)$ rows (and columns), corresponding to this graph T_n (and its ε_n-regular partition), where $a_{ij} = d(V_{i,n}, V_{j,n})$ (this value being the density in T_n). Clearly, if $\mathbf{e}$ is the k-dimensional vector each coordinate of which is n/k, then

[†] $V_0 = \emptyset$ is assumed, by Remark 3.1.

$(1/2)\mathbf{e}A\mathbf{e}^T$ counts the edges between the classes (but it does not count the edges within the classes) and

$$e(T_n) < \frac{1}{2}\mathbf{e}A\mathbf{e}^T + \varepsilon_n n^2 \le \frac{1}{2}g(A)n^2 + \varepsilon_n n^2.$$

Thus

$$\vartheta\binom{n}{2} - \varepsilon_n n^2 \le e(S_n) < \frac{1}{2}g(A)n^2 + 3\varepsilon_n n^2,$$

and therefore

$$g(A) \ge 2\vartheta - 8\varepsilon_n.$$

In the following $m = |V_1|$, $M = |\cup_{i \in I} V_i|$. We will find a subgraph H_M of T_n, equally dense (but possibly much smaller), spanned by the union of some $\Omega = \Omega_n \le R(L_1, \ldots, L_r)$ classes $V_{i,n}$. (This makes the problem bounded in some sense.) For any subset $\{V_{i,n} : i \in I\}$ of $\{V_{i,n}\}$ we have a symmetrical minor (submatrix) A' of A and a corresponding number $g(A')$. We will choose an I for which $g(A') \ge g(A)$ and $|I|$ is the minimum. (Since $g(A') \le g(A)$, we will actually have $g(A') = g(A)$.) By Lemma 3.2, all the densities between these classes are positive in T_n, and therefore are at least 2ε. Further, the resulting matrix A' is dense:

So, if we end up with Ω classes, any two of which are joined by density $> 2\varepsilon_n$, then, by a very standard application of the Regularity Lemma, $T_n \supset K_\Omega{}^\dagger$. (See, for example, [11]) Hence $\Omega < R = R(L_1, \ldots, L_r)$. In other words, we end up with a bounded number of classes (independently of n and ε).

Originally, when $n \to \infty$, we have $\varepsilon_n \to 0$, and the number of classes in the Regular Partition could have tended to ∞ and the entries a_{ij} to 0. Now the situation is nicer, the numbers of rows and columns in the matrices A' are bounded, independently of ε and n. So we can take a convergent subsequence of these matrices, while $n \to \infty$: we may assume that the matrices A'_n converge to a matrix A^*. Still, it can happen that A^* is not dense. In that case we can take a dense submatrix A_0 of A^*. (Otherwise $A_0 = A^*$.)

Now we have a (mostly very sparse) sequence of integers n_t and the corresponding graphs S_{n_t} with their Regular Partitions (described in the Regularity Lemma) and the corresponding matrices A_{n_t} with their dense submatrices A'_{n_t} converging to A^*. We consider only the dense submatrix A_0 of A^*. Let A_0 be an $\Omega \times \Omega$ matrix. It has an optimum vector $\mathbf{u}$ and each coordinate of $\mathbf{u}$ is positive, say at least $\gamma > 0$. So we can fix the corresponding $\Omega \le R(L_1, \ldots, L_r)$ classes, say $V_1, \ldots, V_\Omega$, and the corresponding $u_i m$ vertices in them, thus getting an optimal A_0-matrix graph sequence

$$H_m \subseteq S\left[\bigcup_{i \le \Omega} V_{i,n_t}\right].$$

Since each class of $W_i := V_i \cap V(H_m)$ of H_m has at least γm vertices, the W_i's will be joined to each other $(1/\gamma)\varepsilon_n$-regularly: they will induce an optimal A_0-matrix graph sequence.

We have to prove four things:

---

† Here we need that ε is small in terms of the Ramsey number $R(L_1, \ldots, L_r)$.

(α) The corresponding graphs can be coloured in r colours so that the vth colour contains no L_v;

(β) $\alpha_p(H_m) = o(f(m))$,

(γ) this matrix graph sequence has enough edges to be asymptotically extremal.

(δ) $e(W_i) = o(m^2)$.

(α) This is trivial, since $H_{m_t} \subseteq S_{n_t}$ and the S_n's have this colouring property.

(β) Up to this point we have used one fixed sequence ε_n. Replacing this sequence by another $\varepsilon'_n > \varepsilon_n$ tending to 0, everything above remains valid (with the same regular partition). Given the original sequence ε_n, we fix a sequence β_n as described in Lemma 3.4. For any fixed ε the upper bound k_0 of the Regularity Lemma is a constant. So we may find an $\varepsilon''_n \to 0$ (very slowly) for which, for ε''_n and $\kappa = 1/\varepsilon''_n$ we have $k_0(\kappa, \varepsilon''_n) < 1/\beta_n$. If $\tilde{\varepsilon}_n = \max\{\sqrt{\varepsilon_n}, \beta_n, \varepsilon''_n\}$, then with this $\tilde{\varepsilon}_n \to 0$ we have for every induced subgraph $H_m \subseteq S_n$ of at least n/k vertices

$$\alpha_p(H_m) \leq \sqrt{\varepsilon_n} f(n) \leq f(n/k) \leq f(m).$$

(γ) This follows by a simple computation: we have $g(A'_{n_t}) \geq \vartheta - 8\varepsilon_t$. Hence $g(A_0) \geq \vartheta$. So for an A_0-graph H_m, we would know that $e(H_m) \geq (1/2)g(A_0)m^2$. Now the subgraph of S_{n_t} spanned by the selected classes $V_{i,n_t} : i \in I_{n_t}$ is only a 'nearly'-A_0-graph: the entries in A'_{n_t} tend to the corresponding entries of A_0, but they are not equal. Thus we have only

$$e(H_m) \geq \vartheta \binom{m}{2} - o(m^2).$$

However, this is enough to ensure that (H_m) is an asymptotically extremal graph sequence.

(δ) In principle, some classes of H_m could contain too many edges (in terms of m). Now we exclude this. By the construction, $g(A_0) = \vartheta_{p,f}(L_1, \ldots, L_r) = \vartheta$. Hence, on the one hand, for $W_i = V_{i,n} \cap V(H_m)$,

$$e(H_m) \geq \frac{1}{2}(g(A_0) - o(1))m^2 + \sum e(W_i) = \frac{1}{2}(\vartheta - o(1))m^2 + \sum e(W_i).$$

On the other hand,

$$e(H_m) \leq \frac{1}{2}\vartheta m^2 + o(m^2).$$

Thus $\sum e(W_i) = o(m^2)$. $\qquad\qquad\square$

Remark 3.5. This remark is aimed primarily at those who know the Zykov symmetrization. Here we try to explain something of the background of the above proof. In constructing (finding) the 'good'' subgraph $H_m \subseteq S_n$, we have basically used a modification of Zykov's 'symmetrization' method [39]. The original Zykov type symmetrization means that (instead of deleting vertices) we change the edges incident with some vertices, obtaining a graph with the same number of vertices, but of simpler, more symmetric structure. This method breaks down because the symmetrization may increase the independence number

($\alpha(G_n)$ or $\alpha_p(G_n)$), and that is not allowed here. Further, symmetrization can introduce unwanted subgraphs: it may happen for example that G_n contains no $K_3(10, 10, 10)$ but after several symmetrizations it will. Deleting vertices, we can replace the original method of symmetrization: unless we delete too many of them, $\alpha_p(G_n) = o(f(n))$ will be preserved, and of course, no new subgraphs occur. At the same time, the structure becomes simpler and, in some very vague sense, more symmetric.

Proof of Theorem 2.2. We know that there is a sequence of graphs (described in the proof of Theorem 2.3) that is for some *fixed* matrix A an optimal A-matrix graph sequence. We need to show that for each $n > n_0$ the *same* matrix A can be used. As in the proof of Theorem 2.1, we will blow up some good graphs S_{m_t}.

If we have an infinite sequence (S_{m_t}) and a fixed matrix A such that (S_{m_t}) is an optimal A-matrix graph sequence, and asymptotically extremal for some $\varepsilon_t \to 0$, for $RT_p(m_t, K_{p_1}, \ldots, K_{p_r}, \varepsilon_t m_t)$, then $Z_{m_t h} = S_{m_t} \otimes I_h$ will also be optimal A-matrix graphs.

Hence, fix the matrix A obtained in the proof of Theorem 2.3 for a sequence $\varepsilon_t \to 0$ and some sequence m_t. For every n, take the largest $m_t \le \sqrt{n}$, then put $h = \lceil n/m_t \rceil$ and delete $(hm_t - n)$ vertices of $Z_{m_t h} = S_{m_t} \otimes I_h$. The resulting A-matrix graph sequence (S_n^*) proves Theorem 2.2. $\qquad\square$

4. Quantitative results for one colour

In this section we obtain various estimates for $\vartheta_p(K_q)$.

Proof of Theorem 2.6. In the following, the constants $c_0, c_1, c_2, \ldots$ are positive and independent of n, m. Assume indirectly that there exist a constant $c_0 > 0$ and infinitely many graphs G_n not containing K_q, satisfying $\alpha_p(G_n) = o(n)$ and yet having many edges:

$$e(G_n) > \left(1 - \frac{p}{q-1} + c_0\right)\binom{n}{2}.$$

By a standard argument, for some constants c_1, $c_2 > 0$, there exist subgraphs $H_m \subseteq G_n$ with minimum degree

$$d_{\min}(H_m) > \left(1 - \frac{p}{q-1} + c_1\right)m, \qquad m > c_2 n \text{ and } \alpha_p(H_m) = o(m). \qquad (3)$$

By a 'saturation argument', we may assume that $H_m \supset K_{q-1}$: if not, add edges to it one by one, until it does. Clearly, (3) remains valid. Fix a $K_{q-1} \subseteq H_m$. Now

$$e(K_{q-1}, H_m - K_{q-1}) > (q - p - 1 + c_1)(m - q + 1).$$

Therefore, for some $c_3 > 0$, there exists a set U of $c_3 m$ vertices of $H_m - K_{q-1}$, each joined to the *same* $q - p$ vertices of this fixed K_{q-1}. By the assumption, $\alpha_p(G_n) = o(n)$, if n (and therefore m) is sufficiently large, then there is a $K_p \subset U$. This K_p, together with the fixed $q - p$ vertices of K_{q-1} forms a $K_q \subseteq H_m \subseteq G_n$. This contradiction proves (a). As we have mentioned, Construction 2.7 provides the lower bound, i.e. (b). $\qquad\square$

For $q = p + 1$, Theorem 2.6 reduces to the following claim.

Claim 4.1. *For any $p > 1$,* $\vartheta_p(K_{p+1}) = 0$.

This also has a trivial direct proof.

Proof. (Direct) Suppose that (G_n) is a graph sequence with

$$K_{p+1} \nsubseteq G_n \text{ and } \alpha_p(G_n) = o(n).$$

If x is an arbitrary vertex, then its neighbourhood $N(x)$ contains no K_p. Therefore $d(x) = |N(x)| \leq \alpha_p(G_n) = o(n)$. Hence $e(G_n) \leq n\alpha_p(G_n) = o(n^2)$. $\square$

Now we can return to the proof of Theorem 2.11, which improves Theorem 2.6 in some special cases. We will need the following two lemmas.

Lemma 4.2. *For any integers $p \geq 2$ and $0 \leq \gamma < p$, and constant $c > 0$, there exists a constant $M_{p,c}$ with the following properties. Let $\varepsilon > 0$ be fixed and $\eta \geq M_{p,c}\varepsilon$. Suppose $\alpha_p(H_n) = o(n)$ and $B_e \subseteq H_n$ be a bipartite graph with colour classes V_1 and V_2 that are joined ε-regularly. Let $|V_1| = |V_2| > cn$ and $d(V_1, V_2) \geq (\gamma/p) + \eta$ and $n > n_0(p, c, \eta)$. Then $H_n \supset K_{p+\gamma+1}$.*

Obviously, we are thinking of the case when we apply the Regularity Lemma to a large graph and V_1, V_2 are two classes in the resulting partition connected to each other regularly and with a sufficiently high density.

Proof. For n large enough, all but at most εn vertices of V_1 are joined to at least $((\gamma/p) + (1/2)\eta)|V_2|$ vertices of V_2. Hence V_1 contains a K_p joined with at least $(\gamma + (1/2)p\eta)|V_2|$ edges to $|V_2|$. Thus (for some fixed constant $c_1 > 0$) V_2 contains at least $c_1 n$ vertices joined to the same $\gamma + 1$ vertices of *this* $K_p \subset V_1$. They form a $K_{\gamma+1} \subseteq K_p \subseteq V_1$. The $c_1 n$ vertices in V_2 contain a K_p completely joined to $K_{\gamma+1} \subseteq V_1$: $K_{p+\gamma+1} \subseteq H_n$. $\square$

Lemma 4.3. *For any integers $p, k \geq 2$, and $0 \leq \gamma < p$ and constant $c > 0$ there exists a constant $M_{p,c,k}$ with the following properties. Let $\varepsilon > 0$ be fixed and $\eta \geq M_{p,c,k}\varepsilon$. Let $\alpha_p(H_n) = o(n)$ and $V_1, \ldots, V_k \subseteq V(H_n)$, $V_i \cap V_j = \emptyset$, $|V_i| > cn$. Assume that for every $1 \leq i < j \leq k$ the pairs of classes (V_i, V_j) are ε-regular, and $d(V_i, V_j) > \eta$. If $d(V_1, V_2) \geq (\gamma/p) + \eta$ and $n > n_0(p, c, \eta)$, then $H_n \supset K_{p+\gamma+k-1}$.*

Proof. For $j = k, k - 1, \ldots, 3$ we fix, recursively, a vertex $x_j \in V_j$, so that they form a complete $k - j + 1$-graph and are joined completely to some sets $V_{ij} \subseteq V_i$ $(i < j)$ and $|V_{ij}| > c_j^* n$ for some constant $c_j^* > 0$. For $j = 3$ we get a complete $(k - 2)$-graph joined completely to some sets $V_1^* \subseteq V_1$ and $V_2^* \subseteq V_2$, $|V_1^*|, |V_2^*| > c^* n$, for some constant $c^* > 0$. (We use $\eta \geq M_{p,c,k}\varepsilon$ to ensure that all the sets V_{ij} above are large enough to apply the ε-regularity of the Regularity Lemma iteratively.) Applying Lemma 4.2

to the corresponding bipartite graph $H(V_1^*, V_2^*)$ (with $\eta - k\varepsilon$ instead of η), we get a $K_{p+\gamma+1+k-2} = K_{p+\gamma+k-1} \subseteq H(V_1 \cup V_2 \cup \ldots \cup V_k)$. $\qquad \square$

Proof of Theorem 2.11.

(a) Let $\alpha_p(G_n) = o(n)$ and $K_{p+\ell} \not\subseteq G_n$. Fix an $\varepsilon > 0$ and put $\eta = M_{p,c,k}\varepsilon$. We apply the Regularity Lemma to G_n, with this ε. Thus we get a partition $V_1, \ldots, V_k$ of the vertices into $k \leq k_0(\varepsilon, \kappa)$ sets of size $\approx n/k$ (see Remark 3.1 on V_0).

(b) For any graph G let

$$\Phi(G) = e(G) \Big/ \binom{v(G)}{2}.$$

We apply symmetrization in the sense described in the proof of Theorem 2.3: we find a subset of the classes V_i, say $V_1, \ldots, V_t$ so that the density between any two of them is at least 2η and the density for the obtained $G_M = G[\cup_{i \leq t} V_i]$ is high:

$$\Phi(G_n) < \Phi(G_M) + 2\eta.$$

There is a unique integer γ such that for these t classes the largest density occuring is $\geq (\gamma/p) + \eta$ but $\leq ((\gamma + 1)/p) + \eta$. The density $\Phi(G_M) = e(G_M)/\binom{M}{2}$ can be estimated as follows:

$$\Phi(G_M) \leq \frac{1}{2}\left(1 - \frac{1}{t}\right)\left(\frac{\gamma + 1}{p} + \eta\right).$$

Here $G_M \supseteq K_{p+t+\gamma-1}$ and $G_M \not\supseteq K_{p+\ell}$. Therefore $\gamma \leq \ell - t$, so

$$\Phi(G_M) \leq \frac{1}{2}\left(1 - \frac{1}{t}\right)\left(\frac{\ell - t + 1}{p} + \eta\right). \qquad (4)$$

Put

$$h(t, \ell) = \left(1 - \frac{1}{t}\right)\frac{\ell - t + 1}{2p}.$$

For $t = 2$ we get the conjectured density: $h(2, \ell) = (\ell - 1)/4p$. What we have to prove is that for $\ell = 2, 3, 4$ and 5, $h(t, \ell) \leq h(2, \ell)$:

$$\frac{1}{2}\left(1 - \frac{1}{t}\right)\frac{\ell - t + 1}{p} \leq \frac{1}{4}\frac{\ell - 1}{p},$$

which follows from

$$h(2, \ell) - h(t, \ell) = \frac{(t - 2)(2t - \ell - 1)}{2tp} > 0.$$

$\qquad \square$

Proof of Theorem 2.13 In proving the lower bound on $RT_2(n, K_4, o(n))$, Bollobás and Erdős used a geometric, or more precisely, an 'isoperimetric' theorem. Theorem 2.13 is a generalization of the Bollobás–Erdős result. So it is natural to prove Theorem 2.13 using a generalization of the original Isoperimetric Inequality. This generalization was conjectured by Erdős and proved by Bollobás [1].

We need the following definition.

Definition 4.4. ([1]) For $k \geq 2$ define the *k-diameter* of a set A in a metric space by

$$d_k(A) = \sup_{x_1,\ldots,x_k \in A} \min_{i<j} \rho(x_i, x_j).$$

(In other words, this is the kth 'packing constant' of A.)

A spherical cap is the intersection of an h-dimensional sphere $\mathbf{S}^h$ and a halfspace Π.
Bollobás Theorem. ([1])*Let A be a nonempty subset of the h-dimensional sphere $\mathbf{S}^h$ of outer measure $\mu^{\bullet}(A)^{\dagger}$, and let C be a spherical cap of the same measure. Then $d_k(A) \geq d_k(C)$ for every $k \geq 2$.*

In the following, whenever we speak of 'measure', we will always consider relative measure, which is the measure of the set on the sphere $\mathbf{S}^h$ divided by the measure of the whole sphere.

Denote by $\delta = \delta_p$ the diameter of a p-simplex. ($\delta_2 = 2$, $\delta_3 = \sqrt{3},\ldots$)
Corollary 1 of Bollobás Theorem. *Let the integer p and two small constants ε and $\eta > 0$ be fixed. Then for $h > h_0(p,\varepsilon,\eta)$, if A is a measurable subset of $\mathbf{S}^h$ of relative measure $> \varepsilon$, there exist p points $x_1,\ldots,x_p \in A$ such that all $d(x_i, x_j) > \delta_p - \eta$.*

Proof. Indeed, if A does not contain such a p-tuple, its p-diameter is at most $\delta_p - \eta$. Hence – by the Bollobás theorem – the outer measure of A is at most as large as that of a spherical cap of p-diameter $\delta_p - \eta$. For some constant $c_{p,\eta} > 0$ the ordinary diameter of such a cap is at most $2 - c_{p,\eta}$, independent of the dimension h. Hence the relative measure of such a spherical cap is at most $(Q_{p,\eta})^h$ for some constant $0 < Q_{p,\eta} < 1$ and so the relative measure of A is at most $(Q_{p,\eta})^h \leq \varepsilon$ if $h > h_0(p,\varepsilon,\eta)$, a contradiction. $\square$

Corollary 2 of Bollobás Theorem. (Erdős–Rogers Theorem) *For any integer p, there exists a sequence (S_n) of graphs with $K_{p+1} \nsubseteq S_n$ but $\alpha_p(S_n) = O(n^{1-c})$ for some $c > 0$.*

Proof of the Erdős–Rogers Theorem. Let δ_p be the edge-length of the regular p-simplex in $\mathbf{S}^{p-1} \subseteq \mathbf{R}^{p-1}$:

$$\delta_p := \sqrt{\frac{2p}{p-1}}. \tag{5}$$

$\ddagger$Clearly, $\delta_p \searrow \sqrt{2}$.

For a given $\varepsilon > 0$, we fix a sufficiently high-dimensional sphere $\mathbf{S}^h$ and fix an $n \gg h$. We partition the surface of $\mathbf{S}^h$ into n domains D_i ($i = 1,\ldots,n$) of equal measure and of diameter

$$\leq \frac{\delta_p - \delta_{p+1}}{4}.$$

(This can be done if n is sufficiently large.) Then we choose n vertices $x_i \in D_i$ ($i = 1,\ldots,n$).

$\dagger$ We will only use 'nice sets', but Bollobás formulated his result in this generality. The reader can replace 'outer measure' by 'measure'.

$\ddagger$ (5) is taken from [16], and will be obtained (as a by-product) in the proof of Theorem 2.13.

They will be the vertices of our graph Q_n. We join x_i and x_j if

$$\rho(x_i, x_j) > \frac{\delta_{p+1} + \delta_p}{2} > \delta_{p+1}.$$

Trivially, $K_{p+1} \nsubseteq Q_n$. If we choose εn vertices x_ℓ of Q_n and A is the union of the corresponding D_i's, then the relative measure of A is at least ε, and – by Bollobás Theorem – A contains some $w_1, \ldots, w_p$ with $\rho(w_i, w_j) > \delta_p$, $(1 \le i < j \le p)$. Replacing each $w_i \in D_i$ by the corresponding vertex $x_i \in A_i$, we still have $\rho(x_i, x_j) > (1/2)(\delta_p + \delta_{p+1})$, i.e. we have found a K_p in the subgraph induced by these n^{1-c} vertices: $\alpha_p(Q_n) = \le \varepsilon n$. As $\varepsilon \to 0$, the dimension $h \to \infty$ and $\alpha_p(Q_n) = o(n)$. Using a more careful calculation, we get $\alpha_p(Q_n) = O(n^{1-c})$. □

Proof of Theorem 2.13. We will use a Bollobás–Erdős type construction (see [3]) to get a graph sequence (B_n) to prove Theorem 2.13. Fix a high-dimensional sphere S^h and partition it into $n/2$ domains $D_1, \ldots, D_{n/2}$, of equal measure and diameter $(1/2)\mu$, with $\mu = \varepsilon/\sqrt{h}$. This can always be done if $\varepsilon > 0$ is first fixed, h is then chosen to be sufficiently large, and, finally, $n > n_0(\varepsilon, h)$.

Choose a vertex $x_i \in D_i$ and a $y_i \in D_i$ (for $i = 1, \ldots, n/2$), and put $X = \{x_1, \ldots, x_{n/2}\}$ and $Y = \{y_1, \ldots, y_{n/2}\}$. Let $X \cup Y$ be the vertex-set of our B_n and

> join an $x \in X$ to a $y \in Y$ if $\rho(x, y) < \sqrt{2} - \mu$;
> join an $x \in X$ to a $x' \in X$ if $\rho(x, x') > \delta_p - \mu$;
> join a $y \in Y$ to a $y' \in Y$ if $\rho(y, y') > \delta_p - \mu$.

(a) First we show that $\alpha_p(B_n) = o(n)$. To show this, choose εn vertices of B_n. At least $(1/2)\varepsilon n$ vertices belong to (say) X and the union of the corresponding D_i's has relative measure $\ge (1/2)\varepsilon$. Denote by A the union of the D_i's corresponding to these x_i's. By Bollobás Theorem, if $d_p(A) \le (1/2)(\delta_p + \delta_{p+1})$, then $\mu(A) < \varepsilon$, provided that $h > h_0$. So we may choose $w_1, \ldots, w_p \in A$ such that for each $i \ne j$, $\rho(w_i, w_j) > \delta_p - (1/2)\mu$, and therefore $\rho(x_i, x_j) > \delta_p - \mu$, yielding a K_p in the subgraph of B_n spanned by these εn vertices.

(b) Now we show that the resulting graph B_n contains no K_{2p}. Clearly, if $2p$ vertices form a $K_{2p} \subseteq B_n$, then p of them must be in X and the other p in Y, since – for sufficiently small ε – neither X nor Y contains a K_{p+1}. Suppose that $a_1, \ldots, a_p \in X$ and $b_1, \ldots, b_p \in Y$ form a K_{2p}. In the following, a_i's and b_j's are unit vectors and points of the sphere at the same time. The idea of the proof is as follows. We will show that the existence of such a K_{2p} implies that $(\sum a_i - \sum b_j)^2 < 0$, which is a contradiction. To get this, we will estimate $\sum a_i a_j$, and $\sum b_i b_j$ from above, and $\sum a_i b_j$ from below. Let $d = \delta_p - \mu$ and $t = \sqrt{2} - \mu$. Now, $|a_i| = 1$, $|b_j| = 1$, and $|a_i - a_j| > d$. Therefore

$$2 \sum_{1 \le i < j \le p} a_i a_j = \sum_{1 \le i < j \le p} \left((a_i^2 + a_j^2) - (a_i - a_j)^2 \right) < \binom{p}{2}(2 - d^2).$$

The same holds for the b_i's. Hence

$$2 \sum_{1 \le i < j \le p} (a_i a_j + b_i b_j) < (p^2 - p)(2 - d^2).$$

Let us now turn to the mixed terms. By $|a_i - b_j| < t$, we have

$$2 \sum_{i=1}^{p} \sum_{j=1}^{p} a_i b_j = \sum_{i=1}^{p} \sum_{j=1}^{p} (a_i^2 + b_j^2) - \sum_{i=1}^{p} \sum_{j=1}^{p} (a_i - b_j)^2 > p^2(2 - t^2).$$

This implies that

$$\left(\sum_{i=1}^{p} a_i - \sum_{j=1}^{p} b_j \right)^2 = \sum_i a_i^2 + \sum_j b_j^2 + 2 \sum_{1 \le i < j \le p} (a_i a_j + b_i b_j) - 2 \sum_{i=1}^{p} \sum_{j=1}^{p} a_i b_j$$

$$< 2p + 2(p^2 - p) - (p^2 - p)d^2 - (2p^2 - p^2 t^2)$$

$$= p^2 t^2 - (p^2 - p)d^2 = p^2(\sqrt{2} - \mu)^2 - (p^2 - p)(\delta_p - \mu)^2$$

$$= (2p^2 - (p^2 - p)\delta_p^2) - 2\left(\sqrt{2}p^2 - (p^2 - p)\delta_p\right)\mu + p\mu^2.$$

To avoid clumsy calculations involving δ_p, observe that in all the above formulas we have equality if $\varepsilon = 0$, $\mu = 0$, that is, a_i's are the vertices of a regular p-simplex and b_j's are the vertices of another. Indeed, in this case $\sum a_i = 0$ and $\sum b_j = 0$. Hence $2p^2 - (p^2 - p)\delta_p^2 = 0$, that is,

$$\delta_p = \sqrt{\frac{2p}{p - 1}}.$$

Returning to the $\mu > 0$ case, we get

$$0 \le \left(\sum a_i - \sum b_j \right)^2 < -2\left(\sqrt{2}p^2 - (p^2 - p)\sqrt{\frac{2p}{p - 1}} \right)\mu + p\mu^2$$

$$= -2\sqrt{2}p\left(p - \sqrt{p^2 - p} \right)\mu + p\mu^2 < 0,$$

provided that μ is sufficiently small, is a contradiction. This shows that $B_n \not\supseteq K_{2p}$.

(c) Each vertex has degree $(n/4) + o(n)$, since each a_i is joined to the b_j's on an 'approximate half-sphere' and thus the the surface considered has measure $\ge (1/2) - O(\varepsilon)$ and the number of vertices b_j is proportional to this measure. So

$$\frac{n^2}{8} - O(\varepsilon n^2) \le e(B_n) \le \frac{n^2}{8} + O(\varepsilon n^2).$$

This completes the proof.

$\square$

5. Two special cases

The last problems we discuss here are:

How large are $\vartheta_3(K_8)$ and $\vartheta_3(K_9)$?

Conjecture 2.9 asserts that $\vartheta_3(K_8) = 3/11$ and $\vartheta_3(K_9) = 3/10$. The conjectured extremal structures (described in Conjecture 2.9) in both cases have 3 classes and are as follows. Put $x = (3n/11) + o(n)$ vertices in the classes V_1, V_2 and $y = (5n/11) + o(n)$ vertices into

V_3. Then join V_1 and V_2 with $d(V_1, V_2) = 1/3$, $o(1)$-regularly, and join V_3 completely to the other two classes. The classes V_i contain some edges to ensure $\alpha_3(G_n) = o(n)$. However, the problem is that we are unable to find such graphs.

One reason that we cannot prove Conjecture 2.9 (even for $p = 3$, $q = 8, 9$) is that we are unable to construct bipartite graphs analogous to the Bollobás–Erdős [3], or Erdős–Rogers graph [16], but with density $1/3$ (or $2/3$) instead of $1/2$. Here the 'analogous' means that we fix, for some t, a $t \times t$ matrix $D = (d_{ij})$ of positive elements, and on a high-dimensional sphere S^h, we choose some sets $X_1, \ldots, X_t$, each uniformly distributed on the sphere in some sense, and join two vertices $u \in X_i$, and $v \in X_j$ if their Euclidean distance $\rho(u, v) \approx d_{ij}$, or $\rho(u, v) \geq d_{ij}, \ldots$

So we have only an upper bound on the number of edges.

Theorem 5.1. $\vartheta_3(K_8) \leq \dfrac{3}{11}$.

In the proofs of this and the next theorem we need some case-distinction. In many cases we know that the graph structure considered is dense, and we can easily calculate the edge-densities by solving a small system of linear equations. Here we formulate a lemma, which covers most of the cases we need. (It has a more general form as well.)

Lemma 5.2. *Let* $A = A_{h,k,\lambda,\varphi,\beta}$ *be a symmetric* $(h + k) \times (h + k)$ *matrix satisfying*

$$a_{i,j} = \begin{cases} \lambda & \text{if} \quad 1 \leq i < j \leq h, \\ \varphi & \text{if} \quad h < i < j \leq h + k, \\ \beta & \text{else.} \end{cases}$$

If A *is dense, its optimum vector* **w** *has coordinates*

$$w_i = \frac{\beta k - \varphi(k - 1)}{2\beta hk - \varphi h(k - 1)) - \lambda k(h - 1)} \qquad (i \leq h), \tag{6a}$$

and

$$w_i = \frac{\beta h - \lambda(h - 1)}{2\beta hk - \varphi h(k - 1)) - \lambda k(h - 1)} \qquad (i > h). \tag{6b}$$

The density is

$$g(A) = \frac{\beta^2 hk - \lambda\varphi(h - 1)(k - 1)}{2\beta hk - \varphi h(k - 1)) - \lambda k(h - 1)}. \tag{7}$$

Proof. Assume that H_n is an optimal matrix graph corresponding to A. Let the classes of H_n be $V_1, \ldots, V_{h+k}$. Then $|V_i| \approx w_i n$. When counting the sizes of the classes in an optimal matrix graph, it is enough to take into account that the degrees must be asymptotically equal – provided that the matrix is dense[†] (see, for example [4]). Let the first h coordinates of the optimum vector be x, the others y[‡]. Now the vertices in the first h classes will have

---

[†] For dense matrices this condition is necessary and sufficient.

[‡] Because of the symmetry, the first h class sizes will be asymptotically the same, and the same holds for the other k classes.

degree $(\lambda(h-1)x + \beta ky)n$, while in the last k classes the degrees will be $(\beta hx + \varphi(k-1)y)n$. Furthermore, $hx + ky = 1$. Solving this system of linear equations, we get (6a) and (6b). Now, $g(A)$ is the common degree divided by n, (the edge-density is half of this). This proves (7)[†]. □

Remark 5.3. These formulas become much simpler if, for example, $h = 1$ or $k = 1$. For $k = 1$, φ drops out and we get

$$g(A) = \frac{\beta^2 h}{2h\beta - (h-1)\lambda}. \tag{8}$$

Proof of Theorem 5.1. Let us fix an η as described in Lemma 4.3. Using the argument of the proof of Theorem 2.11, we get some sets $V_1, \ldots, V_t$, and we define γ to be an integer for which the largest density between these classes is between $(\gamma/p)+\eta$ and $((\gamma+1)/p)+\eta$. By (4), applied with $p = 3$, $\ell = 5$, we have

$$\Phi(G_M) \leq \frac{1}{2}\left(1 - \frac{1}{t}\right)\left(\frac{6-t}{3} + \eta\right) \leq \frac{3}{11},$$

if $t > 3$ and η is small enough. Therefore we may assume that $t \leq 3$.

With $t = 2$ the maximum density is $1/4 < 3/11$. So we may suppose that $t \geq 3$, that is, $t = 3$.

(i) If the classes are V_1, V_2, V_3 and $d(V_1, V_2) \leq (1/3)+\eta$, then the density is the maximum if the other two densities are 1, *i.e.* (by (8) applied with $\lambda = 1/3$ and $\beta = 1$, $h = 2$) the maximum is at most $(3/11) + O(\eta)$ and we are home.

(ii) If, for example, $d(V_3, V_1) > (2/3) + \eta$, and $d(V_3, V_2) > (2/3) + \eta$, then we are home: we may choose a K_3 in V_3 and a subset $V_i' \subseteq V_i$ of $c_1 n$ vertices in both other classes, completely joined to this K_3. By Lemma 4.2, we find a K_5 in $V_1' \cup V_2'$, and we are home again.

(iii) In the remaining case there is a class adjacent to the other 2 classes with density $\leq (2/3) + \eta$. We may assume that $d(V_3, V_1) \leq (2/3) + \eta$, and $d(V_3, V_2) \leq (2/3) + \eta$. By (8) (applied with $h = 2$, $\beta = (2/3) + \eta$, $\lambda = 1$) the edge-density is at most $(4/15) + O(\eta) < 3/11$.

□

Theorem 5.4. $\vartheta_3(K_9) \leq \dfrac{3}{10}$.

We know that $\vartheta_3(K_9) \geq 2/7$ because we may fix 3 classes V_1, V_2, V_3 of sizes $2n/7, 2n/7, 3n/7$, join V_3 to $V_1 \cup V_2$ completely and build a graph on $V_1 \cup V_2$ as described in the proof of Theorem 2.13. Put an Erdős–Rogers graph into V_3. The resulting graph contains no K_9, since V_3 contains no K_4 and $G[V_1 \cup V_2]$ contains no K_6.

[†] Of course, the proof can be given entirely in the language of Linear Algebra without mentioning graphs.

Proof of Theorem 5.4. (Sketched.) Again, as above, we have to end up with at least $t \geq 3$ classes after the symmetrization, and if we have $t \geq 5$, then, by (4), the density is smaller than $3/10$. So we may assume that $t \leq 4$.

The case of 3 classes is easy. Now at least one of the 3 densities is at most $(2/3) + 2\eta$, otherwise we have a $K_9 \subseteq G_n$. So the density is at most $(3/10) + O(\eta)$ (by (8), applied with $\lambda = (2/3) + \eta$, $\beta = 1$, $h = 2$), and we are home. Hence we may assume that $t = 4$.

We will distinguish 3 types of connections between V_i and V_j:

— if $d(V_i, V_j) < (1/3) + \eta$, we will call (V_i, V_j) a $(1/3)$-pair;
— if $(1/3) + \eta \leq d(V_i V_j) < (2/3) + \eta$, we will call (V_i, V_j) a $(2/3)$-pair;
— if $d(V_i V_j) > (2/3) + \eta$, we will call (V_i, V_j) a 1-pair.

We may assume that there is at least one 1-pair, otherwise the density could be estimated by

$$\frac{1}{2}\left(1 - \frac{1}{4}\right)\left(\frac{2}{3} + \eta\right) < \frac{3}{10}.$$

How many 1-pairs can we have on 4 classes? If we have two adjacent 1-pairs, (V_a, V_b) and (V_a, V_c), then (V_b, V_c) must be a $(1/3)$-pair: otherwise – by the proof of Theorem 5.1 – we could find $K_8 \subseteq V_a \cup V_b \cup V_c$, extendable into a K_9.

This immediately implies that we may have at most 4 1-pairs. If we have exactly 4 1-pairs, they form a 4-cycle and the remaining 2 densities are $1/3$. Applying Lemma 5.2 with $h = k = 2$, $\lambda = \varphi = 1/3$ and $\beta = 1$ we get that the edge-density is at most $7/24 < 3/10$.

Here, unfortunately, we have to distinguish some cases.

(i)   If $t = 4$ and there are 3 1-pairs meeting in one class, the other 3 pairs form a $(1/3)$-triangle. Applying (8) with $h = 3$, $\lambda = (1/3) + \eta$, $\beta = 1$ we get that the edge-density is at most $9/32 < 3/10$, and we are home again.

(ii) Suppose that we have on 4 classes 3 1-pairs that do not meet. Now they form a path, say $V_1 V_2 V_3 V_4$. The density is the highest when

$$d(V_1, V_3) = d(V_2, V_4) = \frac{1}{3} + \eta$$

and

$$d(V_1, V_4) = \frac{2}{3} + \eta.$$

An easy calculation shows that the optimal weights (for $\eta = 0$) are $1/6$, $1/3$, $1/3$, $1/6$, the density is $5/18 < 3/10$. (Or we can reduce this case to the case when the 1-edges form a C_4.)

(iii) We have settled the case when the number of 1-pairs is 4 or 3. The case of one 1-pair or when we have 2 independent 1-pairs can be majorized by the case when we have 2 independent 1-pairs and all the other pairs are $(2/3)$-pairs. By Lemma 5.2, applied with $k = h = 2$, $\lambda = \varphi = 1$, $\beta = 2/3$ we again get that the edge-density is smaller than $(7/24) + O(\eta) < 3/10$.

(iv) The only remaining case to be settled is when we have 2 adjacent 1-pairs, say (V_1, V_2) and (V_1, V_3). Now we know that we get the maximum density if $d(V_2, V_3) = (1/3) + \eta$

and $d(V_i, V_4) = (2/3) + \eta$. One can easily check (by determining the optimum vector of this structure) that the maximum density is $(11/39) + O(\eta) < 3/10$.

$\square$

6. Open problems

Various open problems are stated in [12] and we have already stated the above Problem 2.12. Here we list some others. The first two of these are the simplest special cases of Conjecture 2.9, where we got stuck.

Problem 6.1. *How large is $\vartheta_3(K_{11})$?*

Problem 6.2. *How large is $\vartheta_3(K_{14})$?*

Conjecture 2.9 states that $\vartheta_3(K_{11}) = 11/32$ and $\vartheta_3(K_{14}) = 8/21$.

Problem 6.3. *Can one always find a matrix A such that one has a graph sequence $(S_n : n > n_0)$ obeying the partition rules of the matrix A and being asymptotically extremal for $RT_p(n, L_1, \ldots, L_r, o(n))$ (and not only for an infinite sequence of integers n_{k_ℓ})?*

The answer to this problem is very probably YES. (If it were not, it would probably mean that the extremal structure sharply depends on some parameters such as, for example, the divisibility properties of n, which are not really graph theoretic properties.)

Problem 6.4. *Is there a finite algorithm to find the limit*

$$\vartheta_p(L) = \lim \frac{RT_p(n, L, o(n))}{\binom{n}{2}}?$$

We have shown in our previous paper that there is a finite algorithm for finding $\vartheta_2(L_1, \ldots, L_r)$ if the sample graphs L_i are complete graphs. A paper of Brown, Erdős and Simonovits [7] shows that for the digraph extremal problems without parallel arcs (which seem to be very near to the Turán-Ramsey problems) there is an algorithmic solution, though far from being trivial. What is the situation in case of $\vartheta_p(L_1, \ldots, L_r)$?

Some hypergraph problems (and results) on Turán-Ramsey problems can be found in [18, 20].

Appendix A. Are there graphs satisfying (∗)?

In the above, the forbidden graphs were complete graphs, here we discuss the general case, where $L_1, \ldots, L_r$ are arbitrary graphs.

We are interested in two strongly connected problems. Given either a family $\mathscr{L}$ or r families of excluded graphs, $\mathscr{L}_1, \ldots, \mathscr{L}_r$ and a graph sequence (G_n) with $\alpha_p(G_n) = o(n)$. Under what conditions on $\mathscr{L}$ or the families $\mathscr{L}_i$ can we assert that there exists a graph sequence (G_n) such that

(i) G_n contains an $L \in \mathscr{L}$ for $n > n_0$; or

(ii) there is an r-colouring of G_n so that for no colour v is there a v-coloured $L \in \mathscr{L}_v$?

The case $p = 2$ is easy. In both problems, if no L is a tree, such graphs exist. On the other hand, if (in each $\mathscr{L}_i$) some L is a tree, those graphs do not exist. Indeed, in [9] Erdős has proved that for every ℓ there exist a $c = c_\ell$ ($0 < c_\ell < 1$) and an n_ℓ such that for every $n > n_\ell$ there exist graphs S_n with girth greater than ℓ and independence number $\alpha(S_n) < O(n^{1-c})$. This implies that if none of the graphs $L \in \mathscr{L}$ is a tree or a forest, and $\ell = \max_{L \in \mathscr{L}} v(L)$, the above graphs S_n will contain no L's and $\alpha(S_n) < O(n^{1-c})$. This answers (i) and (ii) also, since $\alpha(G_n) = o(n)$ implies that for all r-colourings of G_n some colours contain all the trees of at most ℓ vertices for $n > n_\ell$. For $p > 2$ the situation is similar, but somewhat more complicated. First we will solve the problem (i). We start with some definitions.

Definition A.1. A graph T is a *p-forest* if

(a) it is the union of complete graphs of order p, having no common edges and
(b) for every integer $t > 1$, the union of any t of these K_p's has at least $pt - t + 1$ vertices; or
(c) it is a subgraph of a graph described in (a) and (b).

Definition A.2. (Girth)

(1) We will say that the *girth* of a *p*-uniform hypergraph H is at least ℓ if the union of any $t < \ell$ hyperedges has at least $pt - t + 1$ vertices.
(2) We will say that the *p-girth* of a graph G is at least ℓ if every subgraph of G of fewer than ℓ vertices is a *p*-forest.

Clearly, the 2-forests are exactly the ordinary forests and the 2-girth of a graph is the ordinary girth.

Erdős–Hajnal Theorem. ([10, Theorem 13.3]) *For every given p, and ℓ and suitable constants $c_1, c > 0$ (for $n > n_0(p, \ell, c, c_1)$) there exist p-uniform hypergraphs H_n for which*

— *any two hyperedges intersect in at most one vertex (such hypergraphs are sometimes called* linear hypergraphs),
— *any set of $c_1 n^{1-c}$ vertices contains a hyperedge, and*
— *the union of any $t < \ell$ hyperedges has at least $pt - t + 1$ vertices. (In other words, the p-girth of H_n is at least ℓ.)*

The proof used random hypergraphs.

Let us call a graph U_n the *shadow* of a *p*-uniform hypergraph H_n if H_n and U_n have the same vertex-sets, and (x, y) is an edge of U_n iff there is a hyperedge in H_n containing both x and y. We will call the shadow S_n of H_n of [10, Theorem 13.3] the *Erdős–Hajnal Random Graph*.

As for the shadow, one can easily see that if the girth of H_n is at least 4, (which implies also that H_n is a 'linear hypergraph'), then H_n can easily and uniquely be reconstructed from U_n. The following claim is an immediate consequence of Theorem 13.3 of [10].

Claim A.3. *There exist a constant $c = c_{p,\ell} > 0$ and an integer $n_{p,\ell}$ such that for every $n > n_{p,\ell}$ there exist graphs S_n with p-girth greater than ℓ and independence number $\alpha_p(S_n) = O(n^{1-c})$.*

Indeed, the Erdős–Hajnal Random graph (S_n) proves Claim A.3. This implies the following claim.

Claim A.4. *If no $L \in \mathscr{L}$ is a p-forest, then there exist graph sequences (S_n) with $\alpha_p(S_n) = O(n^{1-c})$ (for some $c > 0$) and with $L \nsubseteq S_n$ $(L \in \mathscr{L})$.*

This is sharp:

Claim A.5. *If (S_n) is a graph sequence with the property that $\alpha_p(S_n) = o(n)$ and L is a p-forest, then $L \subseteq S_n$ for $n > n_0$.*

The case of many colours In the following, we will use the notation $R(\mathscr{L}_1, \ldots, \mathscr{L}_r)$ in the obvious way. Clearly, if $\alpha_p(G_n) = o(n)$ and $n > n_0$, then $K_p \subseteq G_n$

Since $\alpha_p(S_n) = o(n)$ implies $K_p \subseteq S_n$, if $p \geq R(\mathscr{L}_1, \ldots, \mathscr{L}_r)$, then any r-colouring of S_n has for some v an $L \in \mathscr{L}_v$ of colour v.

This trivial assertion is sharp for 2-connected excluded graphs.

Theorem A.6. *Assume that the excluded graphs in all the $\mathscr{L}_v$'s $(v = 1, \ldots, r)$ are 2-connected and $p < R(\mathscr{L}_1, \ldots, \mathscr{L}_r)$. Then there exist graph sequences (G_n) with $\alpha_p(G_n) = O(n^{1-c})$ (for some constant $c > 0$) such that the graphs G_n are r-colourable such that no monochromatic copies of any $L \in \mathscr{L}_v$ in the vth colour occurs $(v = 1, \ldots, r)$.*

Proof. Let

$$\ell > \max_{L \in U_v \mathscr{L}_v} v(L).$$

We can take the Erdős–Hajnal Random graph $G_n = S_n$ with p-girth larger than ℓ, and edge-colour each $K_p \subseteq S_n$ in r colours without monochromatic L's, since $p < R(\mathscr{L}_1, \ldots, \mathscr{L}_r)$. If $L \subseteq S_n$ is 2-connected, $L \in \mathscr{L}_v$ is in a uniquely defined $K_p \subseteq S_n$ and therefore cannot be monochromatic, of colour v. $\square$

Some similar results can be formulated for the case when the 2-connectedness of the graphs L_v is dropped. In fact, one can define a p-tree W_K of size $K(p, \ell)$ such that if all the excluded graphs are of order at most ℓ, then (∗) can be satisfied iff W_K can be coloured in v colours without having $L \in \mathscr{L}_v$ in the vth colour. The details are easy and omitted here.

Acknowledgement

We would like to thank the referee for many helpful suggestions.

References

[1] Bollobás, B. (1989) An extension of the isoperimetric inequality on the sphere. *Elemente der Math.* **44** 121–124.

[2] Bollobás, B. (1978) Extremal graph theory, Academic Press, London.

[3] Bollobás, B. and Erdős, P. (1976) On a Ramsey–Turán type problem. *Journal of Combinatorial Theory* B **21** 166–168.

[4] Brown, W. G., Erdős, P. and Simonovits, M. (1973) Extremal problems for directed graphs. *Journal of Combinatorial Theory* B **15** (1) 77–93.

[5] Brown, W. G., Erdős, P. and Simonovits, M. (1978) On multigraph extremal problems. In: Bermond, J. *et al.* (ed.) *Problèmes Combinatoires et Théorie des Graphes*, (Proc. Conf. Orsay 1976), CNRS Paris 63–66.

[6] Brown, W. G., Erdős, P. and Simonovits, M. (1985) Inverse extremal digraph problems. Finite and Infinite Sets, Eger (Hungary) 1981. *Colloq. Math. Soc. J. Bolyai* **37**, Akad. Kiadó, Budapest 119–156.

[7] Brown, W. G., Erdős, P. and Simonovits, M. (1985) Algorithmic Solution of Extremal Digraph Problems. *Transactions of the American Math Soc.* **292/2** 421–449.

[8] Erdős, P. (1968) On some new inequalities concerning extremal properties of graphs. In: Erdős, P. and Katona, G. (ed.) *Theory of Graphs* (Proc. Coll. Tihany, Hungary, 1966), Acad. Press N. Y. 77–81.

[9] Erdős, P. (1961) Graph Theory and Probability, II. *Canad. Journal of Math.* **13** 346–352.

[10] Erdős, P. and Hajnal, A. (1966) On chromatic number of graphs and set-systems. *Acta Math. Acad. Sci. Hung.* **17** 61–99.

[11] Erdős, P., Hajnal, A., Sós, V. T. and Szemeredi, E. (1983) More results on Ramsey–Turán type problems. *Combinatorica* **3** (1) 69–82.

[12] Erdős, P., Hajnal, A., Simonovits, M., Sós, V. T. and Szemerédi, E. (1993) Turán–Ramsey theorems and simple asymptotically extremal structures. *Combinatorica* **13** 31–56.

[13] Erdős, P., Meir, A., Sós, V. T. and Turán, P. (1972) On some applications of graph theory I. *Discrete Math.* **2** (3) 207–228.

[14] Erdős, P., Meir, A., Sós, V. T. and Turán, P. (1971) On some applications of graph theory II. *Studies in Pure Mathematics (presented to R. Rado)*, Academic Press, London 89–99.

[15] Erdős, P., Meir, A., Sós, V. T. and Turán, P. (1972) On some applications of graph theory III. *Canadian Math. Bulletin* **15** 27–32.

[16] Erdős, P. and Rogers, C. A. (1962) The construction of certain graphs. *Canadian Journal of Math* 702–707. (Reprinted in Art of Counting, MIT PRESS.)

[17] Erdős, P. and Simonovits, M. (1966) A limit theorem in graph theory. *Studia Sci. Math. Hungar.* **1** 51–57.

[18] Erdős, P. and Sós, V. T. (1969) Some remarks on Ramsey's and Turán's theorems. In: Erdős, P. *et al* (eds.) Combin. Theory and Appl. Mathem. *Coll. Soc. J. Bolyai* **4**, Balatonfüred 395–404.

[19] Erdős, P. and Stone, A. H. (1946) On the structure of linear graphs. *Bull. Amer. Math. Soc.* **52** 1089–1091.

[20] Frankl, P. and Rödl, V. (1988) Some Ramsey–Turán type results for hypergraphs. *Combinatorica* **8** (4) 323–332.

[21] Graham, R. L., Rothschild, B. L. and Spencer, J. (1980) *Ramsey Theory*, Wiley Interscience, Ser. in Discrete Math.

[22] Katona, G. (1985) Probabilistic inequalities from extremal graph results (a survey). *Annals of Discrete Math.* **28** 159–170.

[23] Ramsey, F. P. (1930) On a problem of formal logic. *Proc. London Math. Soc.*, 2nd Series **30** 264–286.

[24] Shamir, E. (1988) *Generalized stability and chromatic numbers of random graphs* (preprint, under publication).

[25] Sidorenko, A. F. (1980) *Klasszi gipergrafov i verojatnosztynije nyeravensztva*, Dokladi **254**/3,

[26] Sidorenko, A. F. (1983) (Translation) Extremal estimates of probability measures and their combinatorial nature. *Math. USSR - Izv* **20** N3 503–533 MR 84d: 60031. (Original: *Izvest. Acad. Nauk SSSR. ser. matem.* **46** N3 535–568.)

[27] Sidorenko, A. F. (1989) Asymptotic solution for a new class of forbidden *r*-graphs. *Combinatorica* **9** (2) 207–215.

[28] Simonovits, M. (1968) A method for solving extremal problems in graph theory. In: Erdős, P. and Katona, G. (ed.) *Theory of Graphs* (Proc. Coll. Tihany, Hungary, 1966), Acad. Press N. Y. 279–319.

[29] Simonovits, M. (1983) Extremal Graph Theory. In: Beineke and Wilson (ed.) *Selected Topics in Graph Theory*, Academic Press, London, New York, San Francisco 161–200.

[30] Sós, V. T. (1969) On extremal problems in graph theory. *Proc. Calgary International Conf. on Combinatorial Structures and their Application* 407–410.

[31] Szemerédi, E. (1972) On graphs containing no complete subgraphs with 4 vertices (in Hungarian). *Mat. Lapok* **23** 111–116.

[32] Szemerédi, E. (1978) On regular partitions of graphs. In: Bermond, J. *et al.* (ed.) *Problèmes Combinatoires et Théorie des Graphes*, (Proc. Conf. Orsay 1976), CNRS Paris 399–401.

[33] Turán, P. (1941) On an extremal problem in graph theory (in Hungarian). *Matematikai Lapok* **48** 436–452.

[34] Turán, P. (1954) On the theory of graphs. *Colloq. Math.* **3** 19–30.

[35] Turán, P. (1969) Applications of graph theory to geometry and potential theory. In: *Proc. Calgary International Conf. on Combinatorial Structures and their Application* 423–434.

[36] Turán, P. (1972) Constructive theory of functions. *Proc. Internat. Conference in Varna, Bulgaria, 1970*, Izdat. Bolgar Akad. Nauk, Sofia.

[37] Turán, P. (1970) A general inequality of potential theory. *Proc. Naval Research Laboratory*, Washington 137–141.

[38] Turán, P. (1989) *Collected papers of Paul Turán* Vol 1–3, Akadémiai Kiadó, Budapest.

[39] Zykov, A. A. (1949) On some properties of linear complexes. *Mat Sbornik* **24** 163–188. (*Amer. Math. Soc. Translations* **79** (1952)).

Nearly Equal Distances in the Plane

PAUL ERDŐS[†], ENDRE MAKAI[†] and JÁNOS PACH[‡§]

[†] Mathematical Institute of the Hungarian Academy of Sciences

[‡] Department of Computer Science, City College, New York, and Mathematical Institute of the Hungarian Academy of Sciences

For any positive integer k and $\epsilon > 0$, there exist $n_{k,\epsilon}$, $c_{k,\epsilon} > 0$ with the following property. Given any system of $n > n_{k,\epsilon}$ points in the plane with minimal distance at least 1 and any t_1, $t_2, ..., t_k \geq 1$, the number of those pairs of points whose distance is between t_i and $t_i + c_{k,\epsilon} \sqrt{n}$ for some $1 \leq i \leq k$, is at most $(n^2/2)(1 - 1/(k+1) + \epsilon)$. This bound is asymptotically tight.

1. Introduction

Almost fifty years ago the senior author [1] raised the following problem: given n points in the plane, what can be said about the distribution of the $\binom{n}{2}$ distances determined by them? In particular, what is the maximum number of pairs of points that determine the same distance? Although a lot of progress has been made in this area, we are still very far from having satisfactory answers to the above questions (*cf.* [4], [6], [7] for recent surveys).

Two distances are said to be *nearly the same* if they differ by at most 1. If all points of a set are close to each other, all distances determined by them are nearly the same (nearly zero). Therefore, throughout this paper we shall consider only *separated* point sets P, *i.e.*, we shall assume that the minimal distance between two elements of P is at least 1. In [3] we have shown that the maximum number of times that nearly the same distance can occur among n separated points in the plane is $\lfloor n^2/4 \rfloor$, provided that n is sufficiently large. In fact, a straightforward generalization of our argument gives the following.

Theorem 1. *There exists $c_1 > 0$ and n_1 such that, for any set $\{p_1, p_2, ..., p_n\} \subseteq \mathbb{R}^2$ $(n \geq n_1)$ with minimal distance at least 1 and for any real t, the number of pairs $\{p_i, p_j\}$ whose distance $d(p_i, p_j) \in [t, t + c_1 \sqrt{n}]$ is at most $\lfloor n^2/4 \rfloor$. (Evidently, the statement is false with, say, $c_1 = 2$.)*

§ Research supported by NSF grant CCR-91-22103 and OTKA-1907, 4269 and 326 04 13.

The aim of the present note is to establish the following result.

Theorem 2. *Given any positive integer k and $\epsilon > 0$, one can find a function $c(n)$ tending to infinity and an integer n_0 satisfying the following condition: for any set $\{p_1, p_2, ..., p_n\} \subseteq \mathbb{R}^2$ ($n \geq n_0$) with minimal distance at least 1 and for any reals $t_1, t_2, ..., t_k$, the number of pairs $\{p_i, p_j\}$ whose distance*

$$d(p_i, p_j) \in \bigcup_{r=1}^{k} [t_r, t_r + c(n)]$$

is at most $(n^2/2)(1 - 1/(k+1) + \epsilon)$.

To see that this bound is asymptotically tight, let $P = \{(iN, j): 0 \leq i \leq k, 1 \leq j \leq n/(k+1)\}$, where N is a very large constant. Now $|P| \leq n$, and the distance between any two points of P with different x-coordinates is nearly iN for some $1 \leq i \leq k$. Hence, there are at least $(n^2/2)(1 - 1/(k+1) + o(1))$ point pairs such that all distances determined by them belong to the union of the intervals $[iN, iN+1]$, $1 \leq i \leq k$.

Let $K_{k+2}^{(m)}$ denote a $(k+2)$-uniform hypergraph whose vertex set can be partitioned into $k+2$ parts $V(K_{k+2}^{(m)}) = V_1 \cup V_2 \cup ... \cup V_{k+2}$, $|V_i| = m$ ($1 \leq i \leq k+2$), and $K_{k+2}^{(m)}$ consists of all $(k+2)$-tuples containing exactly one point from each V_i. Our proof is based on the following two well-known facts from extremal (hyper)graph theory.

Theorem A. [5, Ch. 10, Ex. 40]. *Any graph with n vertices and $(n^2/2)(1 - 1/(k+1) + \epsilon)$ edges has at least $\epsilon((k+1)!/(k+1)^{k+1})n^{k+2}$ complete subgraphs on $k+2$ vertices.*

Theorem B. [2] *For $n \geq (k+2)m$, any $(k+2)$-uniform hypergraph with n vertices and at least $n^{k+2-(1/m)^{k+1}}$ hyperedges contains a subhypergraph isomorphic to $K_{k+2}^{(m)}$.*

In the final section, we will show that Theorem 2 is valid with $c(n) = c_{k,\epsilon} \sqrt{n}$ for a suitable constant $c_{k,\epsilon} > 0$. Our main tool will be a straightforward generalization of Szemerédi's Regularity Lemma. Given a graph G whose edges are coloured by k colours, and two disjoint subsets $V_1, V_2 \subseteq V(G)$, let $e_r(V_1, V_2)$ denote the number of edges of colour r with one endpoint in V_1 and the other in V_2. The pair $\{V_1, V_2\}$ is called δ-*regular* if

$$\left| \frac{e_r(V_1', V_2')}{|V_1'| \cdot |V_2'|} - \frac{e_r(V_1, V_2)}{|V_1| \cdot |V_2|} \right| < \delta \quad \text{for every} \quad 1 \leq r \leq k,$$

and for every $V_1' \subseteq V_1$, $V_2' \subseteq V_2$ such that $|V_1'| \geq \delta|V_1|$, $|V_2'| \geq \delta|V_2|$. We say that the sizes of V_1 and V_2 are *almost equal* if $||V_1| - |V_2|| \leq 1$.

Theorem C. [8] *Given any $\delta > 0$ and any positive integers k, f, there exist $F = F(\delta, k, f)$ and $n_0 = n_0(\delta, k, f)$ with the property that the vertex set of every graph G with $|V(G)| > n_0$, whose edges are coloured by k colours, can be partitioned into almost equal classes $V_1, V_2, ..., V_g$ such that $f \leq g \leq F$ and all but at most δg^2 pairs $\{V_i, V_j\}$ are δ-regular.*

2. Proof of Theorem 2

The proof is by induction on k. For $k = 1$ the assertion is true (Theorem 1), so we can assume that $k \geqslant 2$, $\epsilon > 0$, and that we have already proved the theorem for $k - 1$ with an appropriate function $c_{k-1,\epsilon}(n) \to \infty$.

Fix a set $P = \{p_1, p_2, \ldots, p_n\} \subseteq \mathbb{R}^2$ with minimal distance at least 1, and suppose that there are reals $t_1, t_2, \ldots, t_k$ such that the number of pairs $\{p_i, p_j\}$ with

$$d(p_i, p_j) \in \bigcup_{r=1}^{k} [t_r, t_r + c(n)]$$

is at least $(n^2/2)(1 - 1/(k+1) + \epsilon)$. We are going to show that one can specify the function $c(n) \leqslant c_{k-1,\epsilon}(n)$ tending to infinity so as to obtain a contradiction if n is sufficiently large.

Lemma 2.1. *If $c(n) = o(\sqrt{n})$, then $\min\limits_{1 \leqslant r \leqslant k} t_r/\sqrt{n} \to \infty$ as n tends to infinity.*

Proof. Assume that, for example, $t_k \leqslant C\sqrt{n}$. For any p_i, the number of points p_j with $d(p_i, p_j) \in [t_k, t_k + c(n)]$ is at most $100\,(t_k + c(n))\,c(n)$. Hence the number of point pairs whose distances belong to $\bigcup_{r=1}^{k-1}[t_r, t_r + c_{k-1,\epsilon}(n)] \supseteq \bigcup_{r=1}^{k-1}[t_r, t_r + c(n)]$ is at least

$$\frac{n^2}{2}\left(1 - \frac{1}{k+1} + \epsilon\right) - 50n(t_k + c(n))\,c(n) > \frac{n^2}{2}\left(1 - \frac{1}{k} + \epsilon\right),$$

provided that n is sufficiently large. This contradicts the induction hypothesis. $\qquad\square$

Lemma 2.2. *Suppose $c(n) = o(\sqrt{n})$. Then one can choose disjoint subsets $P_i \subseteq P$ ($1 \leqslant i \leqslant k+2$) such that $|P_i| > b_{k,\epsilon}(\log n)^{1/(k+1)}$ for a suitable constant $b_{k,\epsilon} > 0$, and the following condition holds: for any $1 \leqslant i \neq j \leqslant k+2$, there exists $1 \leqslant r(i,j) \leqslant k$ such that $r(i,j) = r(j,i)$ and*

$$d(p_i, p_j) \in [t_{r(i,j)}, t_{r(i,j)} + c(n)] \quad \text{for all} \quad p_i \in P_i, p_j \in P_j.$$

Proof. Let G denote the graph with vertex set P, whose two vertices are connected by an edge if and only if their distance belongs to $\bigcup_{r=1}^{k}[t_r, t_r + c(n)]$. By Theorem A (in the Introduction), we know that G contains at least $\epsilon(n/(k+2))^{k+2}$ complete subgraphs K_{k+2} on $k+2$ vertices. Since for a random partition $\{P_1, \ldots, P_{k+2}\}$ of P the number of the above K_{k+2}'s meeting each P_i in one point is at least $d(k)\,\epsilon(n/(k+2))^{k+2}$, where $d(k) > 0$ we can suppose this inequality for a fixed partition $\{P_1, \ldots, P_{k+2}\}$ of P.

Let K_{k+2} be such a subgraph with vertices $p_{s_1}, p_{s_2}, \ldots, p_{s_{k+2}}$ ($p_{s_i} \in P_i$). Then for any $1 \leqslant i \neq j \leqslant k+2$, there exists $1 \leqslant r(i,j) \leqslant k$ such that $d(p_{s_i}, p_{s_j}) \in [t_{r(i,j)}, t_{r(i,j)} + c(n)]$. The symmetric array $(r(i,j))_{1 \leqslant i \neq j \leqslant k+2}$ is said to be the *type* of K_{k+2}. Since the number of different types is at most $k^{\binom{k+2}{2}}$, we can choose at least $d(k)(\epsilon/k^{(k+2)^2})n^{k+2}$ complete subgraphs K_{k+2} having the same type. Applying Theorem B to the $(k+2)$-uniform hypergraph H formed by the vertex sets of these $d(k)(\epsilon/k^{(k+2)^2})n^{k+2}$ complete subgraphs, we can show that H contains a subhypergraph isomorphic to $K_{k+2}^{(m)}$ with $m \geqslant b'_{k,\epsilon}(\log n)^{1/(k+1)}$, for a suitable constant $b'_{k,\epsilon} > 0$. From this the assertion readily follows. $\qquad\square$

In what follows, we shall analyze the relative positions of the sets P_i $(1 \leqslant i \leqslant k+2)$ described in Lemma 2.2. Consider two sets (P_1 and P_2 say), and assume that all distances between them belong to the interval $[t_1, t_1 + c(n)]$. For any $p, p' \in P_1$, all elements of P_2 must lie in the intersection of two annuli centred at p and p'. If $d(p, p') < 2t_1$, $c(n) = o(\sqrt{n})$, then (by Lemma 2.1) the area of this intersection set is at most

$$\frac{20t_1^2 c^2(n)}{d(p, p') \sqrt{4t_1^2 - d^2(p, p')}},$$

and, using the notation $m(n) = b_{k, \epsilon}(\log n)^{1/(k+1)}$, we have

$$m(n) \leqslant |P_2| \leqslant \frac{50t_1^2 c^2(n)}{d(p, p') \sqrt{4t_1^2 - d^2(p, p')}}.$$

Assuming that $c(n) = o(\sqrt{m(n)})$, this immediately implies that $d(p, p')/t_1$ is either close to 0 or close to 2. More exactly,

$$d(p, p') \in \left[1, \frac{50c^2(n)}{m(n)} t_1\right] \cup \left[\left(2 - \frac{50c^2(n)}{m(n)}\right) t_1, 2t_1 + 2c(n)\right]$$

for any $p, p' \in P_1$, provided that n is large enough.

Now pick any point $q \in P_2$. P_1 must be entirely contained in the annulus around q whose inner and outer radii are t_1 and $t_1 + c(n)$, respectively. Thus, if P_1 has two elements with $d(p, p') \geqslant (2 - 50c^2(n)/m(n)) t_1$, all other points of P_1 must lie in the union of the two circles of radius $(50c^2(n)/m(n)) t_1$ centred at p and p'. In any case, there is an at least $m(n)/2$-element subset $P_1' \subseteq P_1$ whose diameter

$$\text{diam } P_1' \leqslant \frac{50c^2(n)}{m(n)} t_1 = o(1) t_1.$$

Repeating this argument $(k+2$ times), we obtain the following.

Lemma 2.3. *Let* $m(n) = b_{k, \epsilon}(\log n)^{1/(k+1)}$, $c(n) = o(\sqrt{m(n)})$. *Then one can choose disjoint subsets* $Q_i \subseteq P$, $|Q_i| \geqslant m(n)/2$ $(1 \leqslant i \leqslant k+2)$ *such that the following conditions are satisfied:*
(i) For any $1 \leqslant i \neq j \leqslant k+2$, *there exists* $1 \leqslant r(i, j) = r(j, i) \leqslant k$ *such that*

$$d(p_i, p_j) \in [t_{r(i, j)}, t_{r(i, j)} + c(n)] \quad \text{for all} \quad p_i \in Q_i, \quad p_j \in Q_j;$$

(ii) For any $1 \leqslant i \leqslant k+2$,

$$\text{diam } Q_i = o(1) \min_{j \neq i} t_{r(i, j)};$$

(iii) There is a line ℓ *such that the angle between* ℓ *and any line* $p_i p_j$ $(p_i \in Q_i, p_j \in Q_j, i \neq j)$ *is* $o(1)$.

Proof. We only have to prove part (iii). Fix two subsets Q_i and Q_j $(i \neq j)$. By (ii),

$$\max(\text{diam } Q_i, \text{diam } Q_j) = o(1) t_{r(i, j)},$$

so the angle between any two lines $p_i p_j$ and $p_i' p_j'$ $(p_i, p_i' \in Q_i; p_j, p_j' \in Q_j; i \neq j)$ is $o(1)$.

Let q_i and q_i' be two elements of Q_i whose distance is maximal. Clearly, for any $j \neq i$,

$$\frac{\sqrt{m(n)}}{10} \leqslant d(q_i, q_i') = \operatorname{diam} Q_i \leqslant o(1) t_{r(i,j)}.$$

It is sufficient to show that for any $p_j \in Q_j$, the lines $q_i q_i'$ and $q_i p_j$ are almost perpendicular. Indeed,

$$|\cos(\angle q_i' q_i p_j)| = \left| \frac{(d(q_i, p_j) - d(q_i', p_j))(d(q_i, p_j) + d(q_i', p_j)) + d^2(q_i, q_i')}{2d(q_i, p_j) d(q_i, q_i')} \right|$$

$$\leqslant \frac{c(n)(2t_{r(i,j)} + 2c(n))}{2t_{r(i,j)}(\sqrt{m(n)}/10)} + \frac{d(q_i, q_i')}{2t_{r(i,j)}} = o(1). \qquad \square$$

We need the following key property of the sets Q_i constructed above.

Lemma 2.4. *Suppose* $c(n) = o(\sqrt{n})$. *Let* $s \geqslant 3$ *be fixed, and suppose that*

$$\operatorname{diam}(Q_1 \cup Q_2 \cup \dots \cup Q_s) = d(p_1, p_2) \quad \text{for some} \quad p_1 \in Q_1, \quad p_2 \in Q_2.$$

Then for any $1 \leqslant i \neq j \leqslant s$, $r(i,j) = r(1,2)$ *if and only if* $\{i, j\} = \{1, 2\}$.

Proof. Suppose, in order to obtain a contradiction, that there are two points $p_i' \in Q_i$, $p_j' \in Q_j$, $2 \leqslant i \neq j \leqslant s$ such that

$$d(p_i', p_j') \in [t_{r(1,2)}, t_{r(1,2)} + c(n)].$$

By Lemma 2.1 and Lemma 2.3 (iii), all points of $Q_2 \cup Q_3 \cup \dots \cup Q_s$ lie in a small sector (of angle $o(1)$) of the annulus around p_1 whose inner and outer radii are $\sqrt{n}$ and $d(p_1, p_2)$, respectively. Obviously, the diameter of this sector is $d(u, v)$, where u (resp. v) is the intersection of one (the other) boundary ray with the inner (outer) circle of the annulus. But then we have

$$d(p_1, p_2) - d(p_i', p_j') \geqslant d(p_1, p_2) - d(u, v) = d(p_1, v) - d(u, v)$$

$$= \frac{2d(p_1, u) d(p_1, v) \cos(\angle up_1 v) - d^2(p_1, u)}{d(p_1, v) + d(u, v)}$$

$$\geqslant d(p_1, u) \cos(\angle up_1 v) - \frac{d^2(p_1, u)}{2d(p_1, v)}$$

$$\geqslant \sqrt{n}(1 - o(1)) - \frac{n}{2t_{r(1,2)}} > \frac{\sqrt{n}}{2} > c(n),$$

which is the desired contradiction. $\qquad \square$

Now we can easily complete the proof of Theorem 2. For the sake of simplicity we assume the intervals are disjoint, but the same arguments work in the general case as well. Assume, without loss of generality, that the diameter of $Q = Q_1 \cup Q_2 \cup \dots \cup Q_{k+2}$ is attained between a point of Q_1 and a point of Q_{j_1}, for some $j_1 > 1$. By Lemma 2.4, no distance determined by the set $Q' = Q_2 \cup Q_3 \cup \ldots \cup Q_{k+2}$ belongs to the interval $[t_{r(1,j_1)}, t_{r(1,j_1)} + c(n)]$. Suppose that the diameter of Q' is attained between a point of Q_2 and a point of $Q_{j_2}, j_2 > 2$. Applying the considerations of the lemma again, we obtain that none of the distances determined by $Q'' = Q_3 \cup Q_4 \cup \ldots \cup Q_{k+2}$ is in $[t_{r(2,j_2)}, t_{r(2,j_2)} + c(n)]$, where $r(2,j_2) \neq r(1,j_1)$. Proceeding like this, we can conclude that no distance determined by $Q_{k+1} \cup Q_{k+2}$ belongs to

$$\bigcup_{i=1}^{k} [t_{r(i,j_i)}, t_{r(i,j_i)} + c(n)],$$

where $\{r(i,j_i) : 1 \leqslant i \leqslant k\} = \{1, 2, \ldots, k\}$. In other words, there exists no integer $r(k+1, k+2)$ satisfying the condition in Lemma 2.3(i). This contradiction completes the proof of Theorem 2 for any function $c(n) = o((\log n)^{1/(2k+2)})$. In fact, our argument also shows that there is a small constant $c_{k,\epsilon} > 0$ such that the theorem is true with $c(n) = c_{k,\epsilon}(\log n)^{1/(2k+2)}$. $\qquad\square$

3. Strengthening of Theorem 2

In this section we are going to modify the above arguments to show that Theorem 2 is valid for any function $c(n) = o(\sqrt{n})$. Notice that in the previous section we have not really used the fact that *all* distances between Q_i and Q_j (in Lemma 2.3) belong to the interval $[t_{r(i,j)}, t_{r(i,j)} + c(n)]$. It is sufficient to require that *many* distances have this property, and there are much larger subsets Q_i $(1 \leqslant i \leqslant k+2)$ satisfying this weaker condition. As a matter of fact, we can assume that $|Q_i| \geqslant m(n) = b_{k,\epsilon}^* n$ for a suitable constant $b_{k,\epsilon}^* > 0$, and follow essentially the same argument as before for any $c(n) = o(\sqrt{m(n)}) = o(\sqrt{n})$.

In the following we shall assume that $k, \epsilon < 1$ and $\delta < (\epsilon/100k)^{k+5}$ are fixed, $c(n) = o(\sqrt{n})$, and n is very large, and again we will argue by contradiction. We want to apply Theorem C (in the Introduction) to the graph G on the vertex set P whose two points p, p' are connected by an edge of colour r whenever

$$d(p,p') \in [t_r, t_r + c(n)], \quad 1 \leqslant r \leqslant k,$$

and r is minimal with this property. Then Lemma 2.3 can be replaced by the following.

Lemma 3.1. *There is a constant* $b = b(k, \epsilon, \delta)$ *such that there exist disjoint subsets* $Q_i \subseteq P$, $|Q_i| > bn$ $(1 \leqslant i \leqslant k+2)$ *satisfying the following conditions.*

(i) *For any* $1 \leqslant i \neq j \leqslant k+2$, *one can find* $1 \leqslant r(i,j) = r(j,i) \leqslant k$ *such that*

$$\frac{e_{r(i,j)}(Q_i, Q_j)}{|Q_i| \cdot |Q_j|} \geqslant \frac{\epsilon}{20k}.$$

(ii) *For any* $1 \leqslant i \leqslant k+2$,

$$\operatorname{diam} Q_i = o(1) \min_{j \neq i} t_{r(i,j)}.$$

(iii) *There is a line* ℓ *such that the angle between* ℓ *and any line* $p_i p_j$ $(p_i \in Q_i, p_j \in Q_j)$ *is* $o(1)$.

Proof. Consider a partition $V(G) = P = V_1 \cup V_2 \cup \ldots \cup V_g$ meeting the requirements of Theorem C with $f = \lceil 10/\epsilon \rceil$. Let G^* denote the graph with vertex set $V(G^*) = \{V_1, V_2, \ldots, V_g\}$, where V_i and V_j are joined by an edge if $\{V_i, V_j\}$ is a δ-regular pair and

$$\frac{e_{r(i,j)}(V_i, V_j)}{|V_i| \cdot |V_j|} \geq \frac{\epsilon}{10k} \tag{1}$$

for some $1 \leq r(i,j) = r(j,i) \leq k$. Clearly,

$$\frac{n^2}{2}\left(1 - \frac{1}{k+1} + \epsilon\right) \leq |E(G)| \leq \left(|E(G^*)| + \delta g^2 + \binom{g}{2}\frac{\epsilon}{10}\right)\left\lceil\frac{n}{g}\right\rceil^2 + g\binom{\lceil n/g \rceil}{2},$$

whence

$$|E(G^*)| \geq \frac{g^2}{2}\left(1 - \frac{1}{k+1} + \tfrac{1}{2}\epsilon\right).$$

By Theorem A (or by Turán's theorem [T]), this implies that G^* has a complete subgraph on $k+2$ vertices, say, $V_1, V_2, \ldots, V_{k+2}$.

Assume, without loss of generality, that $r(1,2) = 1$, $t_1 = \min_{j \neq 1} t_{r(1,j)}$, and let G_r denote the subgraph of G consisting of all edges of colour r. By (1), at least $(\epsilon/10k)|V_1| \cdot |V_2|$ edges of G_1 run between V_1 and V_2. Therefore, we can pick a point $p_2 \in V_2$ connected to all elements of a subset $P_1 \subseteq V_1$, $|P_1| \geq (\epsilon/10k)|V_1|$. Clearly, P_1 lies in an annulus centred at p_2 with inner radius t_1 and outer radius $t_1 + c(n)$. Using the fact that $\{V_1, V_2\}$ is a δ-regular pair, it can be shown by routine calculations that there are $(\epsilon/100k)^4 |P_1|^2$ pairs $\{p_1, p_1'\} \subset P_1$ such that p_1 and p_1' have at least $(\epsilon/100k)^2 |V_2| \geq (1/F(\delta,k,f))(\epsilon/100k)^2 n$ common neighbours in G_1. As in the proof of Lemma 2.3, we can argue that, for any such pair,

$$d(p_1, p_1') = o(1)\, t_1 \quad \text{or} \quad d(p_1, p_1') = (2 - o(1))\, t_1.$$

Hence, we can find a point $p_1 \in P_1$ such that

$$|\{p_1' \in P_1 : d(p_1, p_1') = o(1)\, t_1\}| \geq \left(\frac{\epsilon}{100k}\right)^4 |P_1|,$$

or

$$|\{p_1' \in P_1 : d(p_1, p_1') = (2 - o(1))\, t_1\}| \geq \left(\frac{\epsilon}{100k}\right)^4 |P_1|.$$

Let $Q_1 \subseteq P_1$ denote the larger of these two sets. Then

$$|Q_1| \geq \left(\frac{\epsilon}{100k}\right)^4 |P_1| > \left(\frac{\epsilon}{100k}\right)^5 |V_1| \geq \frac{1}{F(\delta,k,f)}\left(\frac{\epsilon}{100k}\right)^5 n, \tag{2}$$

and diam $Q_1 = o(1)\, t_1$. Repeating the same argument for every V_i ($1 \leq i \leq k+2$), we obtain $Q_i \subseteq V_i$ satisfying conditions (i) and (ii).

To establish (iii), notice that the angle between any two lines $p_i p_j$ and $p_i' p_j'$ ($p_i, p_i' \in Q_i$; $p_j, p_j' \in Q_j$; $i \neq j$) is $o(1)$. Using the fact that $\{V_1, V_j\}$ is δ-regular for all $2 \leq j \leq k+2$, one can recursively pick $p_j \in Q_j$ so that

$$|\{q \in Q_1 : qp_j \in E(G_{r(i,j)}) \quad \text{for all} \quad 2 \leqslant j \leqslant k+2\}|$$

$$\geqslant \left(\frac{\epsilon}{100k}\right)^{k+1} |Q_1| \geqslant \left(\frac{\epsilon}{100k}\right)^{k+6} |V_1|$$

$$\geqslant \frac{1}{F(\delta,k,f)} \left(\frac{\epsilon}{100k}\right)^{k+6} n.$$

Thus, two elements of this set, q_1 and q_1' (say), are relatively far away from each other:

$$\sqrt{\frac{1}{F(\delta,k,f)} \left(\frac{\epsilon}{100k}\right)^{k+6}} \, n/10 \leqslant d(q_1,q_1') \leqslant \operatorname{diam} Q_1 = o(1) \min_{j \neq 1} t_{r(1,j)}.$$

This in turn implies, in the same way as in the proof of Lemma 2.3 (iii), that

$$|\cos(\angle q_1' q_1 p_j)| = o(1) \quad (2 \leqslant j \leqslant k+2),$$

i.e., every line $p_1 p_j$ ($p_1 \in Q_1, p_j \in Q_j, j \neq 1$) is almost perpendicular to the line $q_1 q_1'$. Applying the same argument for $Q_2, Q_3, \ldots$ (instead of Q_1), we obtain (iii). $\qquad\square$

Using Lemma 3.1, (i) and $|Q_i| > (\epsilon/100k)^5 |V_1| \geqslant \delta|V_1|$ we can see (using induction), that there are const $(k,\epsilon,\delta) n^{k+2}$ $(k+2)$-tuples $(q_1, \ldots, q_{k+2})$, with $q_i \in Q_i$, such that

$$d(q_i, q_j) \in [t_{r(i,j)}, t_{r(i,j)} + c(n)].$$

(For details *cf.* [7].) Fix one of them. Then repeating the considerations of Lemma 2.4 for this $(k+2)$-tuple only, we get that the assertion of Lemma 2.4 is valid also now. Then the proof of Theorem 2 can be completed in exactly the same way as in the previous section with any function $c(n) = o(\sqrt{n})$. As a matter of fact, in order to apply our argument, it is sufficient to assume that $c(n) \leqslant c_{k,\epsilon} \sqrt{n}$ for a suitable constant $c_{k,\epsilon} > 0$. $\qquad\square$

References

[1] Erdős, P. (1946) On sets of distances of n points. *Amer. Math. Monthly* **53**, 248–250.

[2] Erdős, P. (1965) On extremal problems for graphs and generalized graphs. *Israel J. Math.* **2**, 183–190.

[3] Erdős, P., Makai, E., Pach, J. and Spencer, J. (1991) Gaps in difference sets and the graph of nearly equal distances. In: Gritzmann, P. and Sturmfels, B. (eds.) *Applied Geometry and Discrete Mathematics, the Victor Klee Festschrift*, DIMACS Series **4**, AMS-ACM, 265–273.

[4] Erdős, P. and Purdy, G. (to appear) Some extremal problems in combinatorial geometry. In: *Handbook of Combinatorics*, Springer-Verlag.

[5] Lovász, L. (1979) *Combinatorial problems and exercises*, Akad. Kiadó, Budapest, North Holland, Amsterdam–New York–Oxford.

[6] Moser, W. and Pach, J. (1993) Recent developments in combinatorial geometry. In: Pach, J. (ed.) *New Trends in Discrete and Computational Geometry*, Springer-Verlag, Berlin 281–302.

[7] Pach, J. and Agarwal, P. K. (to appear) *Combinatorial Geometry*, J. Wiley, New York.

[8] Szemerédi, E. (1978) Regular partitions of graphs. In: *Problèmes Combinatoires et Théorie de Graphes*, Proc. Colloq. Internat. CNRS, Paris 399–401.

[9] Turán, P. (1941) Eine Extremalaufgabe aus der Graphentheorie. *Mat. Fiz. Lapok* **48**, 436–452. (Hungarian, German summary.)

Clique Partitions of Chordal Graphs[†]

PAUL ERDŐS[‡], EDWARD T. ORDMAN[§]
and YECHEZKEL ZALCSTEIN[¶]

[‡] Mathematical Institute, Hungarian Academy of Sciences ,

[§] Memphis State University, Memphis, TN 38152 U.S.A.

[¶] Division of Computer and Computation Research, National Science Foundation,
Washington, D.C. 20550, U.S.A.

To partition the edges of a chordal graph on n vertices into cliques may require as many as $n^2/6$ cliques; there is an example requiring this many, which is also a threshold graph and a split graph. It is unknown whether this many cliques will always suffice. We are able to show that $(1 - c)n^2/4$ cliques will suffice for some $c > 0$.

1. Introduction

We consider undirected graphs without loops or multiple edges. The graph K_n on n vertices for which every pair of distinct vertices induces an edge is called a *complete* graph or a *clique* on n vertices. If G is any graph, we call any complete subgraph of G a *clique* of G (we do not require that it be a maximal complete subgraph). A *clique covering* of G is a set of cliques of G that together contain each edge of G at least once; if each edge is covered exactly once we call it a *clique partition*. The *clique covering number* $cc(G)$ and *clique partition number* $cp(G)$ are the smallest cardinalities of, respectively, a clique covering and a clique partition of G.

The question of calculating these numbers was raised by Orlin [13] in 1977. DeBruijn and Erdős [6] had already proved, in 1948, that partitioning K_n into smaller cliques required at least n cliques. Some more recent studies motivating the current paper include [11, 14, 2, 7, 9].

It is widely known that a graph on n vertices can always be covered or partitioned by no more than $n^2/4$ cliques; the complete bipartite graph actually requires this many.

[†] This work was done at Memphis State University.

[§] Partially supported by U.S. National Science Foundation Grant DCR-8503922.

[¶] Partially supported by U.S. National Science Foundation Grant DCR-8602319 at Memphis State University.

Turán's theorem states that if G has more than $n^2/4$ edges, it must contain a clique K_3; if it has more than $n^2(c-2)/(2c-2)$ edges it must contain a K_c. (For a more precise statement and proof, see *e.g.* [3, Chapter 11].)

A subgraph H of a graph G is an *induced* subgraph if for any pair of vertices a and b of H, ab is an edge of H if and only if it is an edge of G.

Two classes of graphs we shall refer to here are chordal graphs and threshold graphs. A graph is *chordal* (or often triangulated; [10, Chapter 4]) if every cycle of size greater than 3 has a chord (no set of more than 3 vertices induces a cycle). A graph G is *threshold* ([10, Chapter 10; 4; 5; 12]) if there exists a way of labelling each vertex A of G with a nonnegative integer $f(A)$ and there is another nonnegative integer t (the threshold) such that a set of vertices of G induces at least one edge if and only if the sum of their labels exceeds t.

A graph is *split* if its vertices can be partitioned into two sets A and B such that the vertices A form a clique and the vertices B induce no edges. (Two vertices, of which one is in A and one is in B, may or may not induce an edge.)

All threshold graphs are split and all split graphs are chordal. In a sense, most chordal graphs are split [1]. Induced subgraphs of chordal graphs are chordal; similar results hold for split graphs and threshold graphs.

2. Preliminary results on split graphs

A *complete matching* in a graph G is a set of edges such that each vertex of G lies on exactly one edge in the set. It is well known that the $t(2t-1)$ edges of K_{2t} can be edge-partitioned by a set of $2t-1$ matchings, each of t edges. By the *join* of two graphs G and H, we mean the graph made by taking the disjoint union of the two graphs and adding all edges of the form gh, where g is a vertex of G and h is a vertex of H.

By the graph $K_n - K_m$, for $n > m$, we mean a graph made by taking K_n and deleting all the edges induced by some particular m of the vertices. Equivalently, this is the join of K_{n-m} with the complement of K_m (a collection of m isolated vertices).

Lemma 2.1. *Let* $G = K_{4t} - K_{2t}$. *Then* $cp(G) \le t(2t+1)$.

Proof. Think of G as a complete graph $A = K_{2t}$ joined completely to an empty graph C on $2t$ vertices. Partition A into $2t-1$ disjoint matchings; join each matching to a different vertex in C, each matching yielding t triangles. The remaining vertex in C lies on $2t$ single edges to A. Thus we partition G by $t(2t-1)$ triangles and $2t$ single edges, a total of $2(2t+1)$ cliques.   □

In fact, $cp(G) = t(2t+1)$. See, for example, [7].

Lemma 2.2. *In the graph G of the previous lemma, suppose r edges are deleted. Then this new graph has clique partition number not exceeding* $t(2t+1) + r$.

Proof. Start with the same partition as above. Each edge deletion at worst demolishes one triangle, requiring it to be replaced in the partition by two edges. □

3. Preliminary results on chordal graphs

We will rely heavily on the following lemma of Bender, Richmond, and Wormald, which gives a means of constructing an arbitrary chordal graph.

Lemma 3.1. *[1, Lemma 1.] For each chordal graph G and each clique R of G there is a sequence*

$$R = G_r, G_{r+1}, \ldots, G_n = G$$

of graphs such that G_{i+1} is obtained from G_i by adjoining a new vertex to one of its cliques.

Corollary 3.2. *If G is a chordal graph on n vertices with largest clique of size r, then G can be covered by at most $n - r + 1$ cliques.*

It is easy to see that the bound in the corollary cannot be improved; $K_n - K_{n-r+1}$ is an example requiring $n - r + 1$ cliques to cover.

Covering G may require less than $n - r + 1$ cliques. If G consists of two copies of K_t with a single vertex in each identified, G has $2t - 1$ vertices, the largest clique is of size t, this corollary produces a covering by $(2t - 1) - t + 1 = t$ cliques, but obviously there is a covering (and for that matter a partition) by two cliques.

We now utilize this construction with one additional specialization: we begin with a clique of maximum possible size in G. Supposing this clique to be of size r, each subsequently added vertex will add, at the time it is adjoined, at most $r - 1$ edges (or it would form a clique of more than r vertices).

Corollary 3.3. *A chordal graph on n vertices with a largest clique having r vertices has at most $(n-r)(r-1)$ edges outside that clique.*

Theorem 3.4. *Let G be a chordal graph on n vertices and $1/4 > d > 0$. Suppose G has at least dn^2 edges. Then G contains a clique with at least $(1 - \sqrt{1 - 2d})n > dn$ vertices.*

Proof. If the largest clique in G contains cn vertices, then that clique contains $cn(cn-1)/2$ edges and each of the remaining $n - cn$ vertices of G can be added to G adding at most $cn - 1$ edges at each stage. Hence the total number of edges of G is at least dn^2 and at most $cn(cn - 1)/2 + (cn - 1)(n - cn)$, so $dn \le (2c - c^2)(n/2) + (c - 2)/2$ and $dn < (2c - c^2)(n/2)$ since $c \le 1$. Hence $d < (2c - c^2)/2$ and $c > 1 - \sqrt{1 - 2d} > d$ as needed. □

The result of this theorem turns out to be essentially best possible, not only for chordal graphs, but for split graphs and threshold graphs as well.

Example 3.5. Let $0 < c < 1$. Consider the graph $K_n - K_k$ where $k = n - cn + 1$, that is, the base clique has $cn - 1$ vertices and forms a clique on cn vertices with each other vertex. Clearly there are

$$(cn - 1)(cn - 2)/2 + (n - cn + 1)(cn - 1) = (c - c^2/2)n^2 - (1 - c/2)n$$

edges. So a graph can be threshold (hence split and chordal) and have almost $(c - c^2/2)n^2$ edges and no clique on more than cn vertices.

4. Clique partitions of chordal graphs

An arbitrary graph on n vertices may require $n^2/4$ cliques to cover or partition it [8]. We saw above that a chordal graph on n vertices may always be covered by fewer than n cliques. It may, however, still require a large number of cliques to partition it. The examples in [7] with high clique partition numbers are chordal graphs.

Example 4.1. [7] The graph $K_n - K_{2n/3}$ requires $n^2/6 + n/6$ cliques to partition it and $2n/3$ cliques to cover it. Thus for a chordal graph, both $cp(G)$ and $cp(G) - cc(G)$ can be approximately $n^2/6$.

We note that for a different example, the ratio of $cp(G)$ to $cc(G)$ may be larger.

Example 4.2. [7] The graph G_n composed of 3 cliques $K_{n/3}$, with all vertices of the first clique attached by edges to all vertices of the second and third, is a chordal graph (but not a split graph or threshold graph). As n increases, $cp(G_n)/cc(G_n)$ grows at least as fast as cn^2 for some $c > 0$.

We do not know if $cp(G)$ can significantly exceed $n^2/6$ for a chordal graph, or even for a split graph or a threshold graph.

Conjecture 1. *The clique partition number of a chordal graph, split graph, or threshold graph on n vertices cannot exceed $n^2/6$ (except by a term linear in n).*

It is even possible that $K_n - K_{2n/3}$ is literally the best example. (Some very minor adjustments to $n^2/6 + n/6$ may be needed because of round-off error). However, it is unclear how one would go about proving the following:

Conjecture 2. *No chordal, threshold, or split graph on n vertices requires more than $cp(K_n - K_{2n/3})$ cliques to partition it.*

For chordal graphs in general, we are very far from proving that $n^2/6$ cliques will suffice for a partition. In fact, we can improve only slightly on $n^2/4$.

Theorem 4.3. *There is a constant $c > 0$ such that if G is a chordal graph with n vertices, G may be partitioned into no more than $(1 - c)n^2/4$ cliques.*

Proof. As the details are messy, we first give an outline; we follow this by some indication of more precise calculations, which the reader may choose to ignore, and a few numeric indications. As the result is clear for $n < 5$, we assume $n \geq 5$ in the proof. Let the largest clique in G have $(1 + a)n/2$ vertices (a may be negative). Pick such a clique and call it A. Let C denote the subgraph of G induced by those vertices not in A; the set of edges not in A or C will be denoted B.

In case 1, the large clique is larger or smaller than half the vertices by a reasonable amount ($a^2 > c$). By Corollary 3.3, there are so few edges outside A that we can cover them by single edges. In case 2, A has close to half the vertices, and C has a significant number of edges. By Theorem 3.4, C contains a large clique C'; we can cover by A, C', and single edges. In case 3, A has close to half the vertices and C has few edges; in this case the graph must be very similar in form to $K_n - K_{n/2}$ and Lemma 2.2 can be used to construct a partition with 'little more than' $n^2/8$ triangles and edges.

We now give somewhat more precise calculations.

1 If $a^2 > c$, we can cover A with one clique and each edge not in A by a single edge. The number of edges outside A is at most

$$(1 - a)(n/2)((1 + a)n/2 - 1) < (1 - a^2)n^2/4 < (1 - c)n^2/4$$

as desired. Hereafter, we suppose $a^2 \leq c$.

2 If C has very many edges, we can cover A with a clique, the largest clique in C with a clique, and all other edges singly. Suppose C has dn^2 edges. Then, since C is an induced subgraph of G, it is a chordal graph with $v = (1 - a)n/2$ vertices and $dn^2 = (dn^2/((1 - a)n/2)^2)v^2$ edges; so by Theorem 3.4 it contains a clique with at least $(dn^2/((1 - a)n/2)^2)v = 2dn/(1 - a)$ vertices and $(2(dn)^2 - dn(1 - a))/(1 - a)^2$ edges. Covering this clique by itself, A by a clique, and each remaining edge with an edge, we get a number of cliques guaranteed to be less than

$$2 + (1 - a^2)n^2/4 - (1 - a)n/2 - (2(dn)^2 - dn(1 - a))/(1 - a)^2$$

$$= (1 - a^2 - 8d^2/(1 - a)^2)n^2/4 + 2 - (1 - a)n/2 + dn/(1 - a)$$

Now supposing $c < .01$, $|a| < .1$, $n > 4$, and $d < .04$, we see that

$$2/n + d/(1 - a) + a/2 < 1/2,$$

so

$$2 - (1 - a)n/2 + dn/(1 + a) < 0$$

and we need only have

$$1 - a^2 - 8d^2/(1 - a)^2 < 1 - c$$

to finish, which is clearly true if $d^2 > (c - a^2)(1 - a)^2/8$. If that condition is met, we are done. Hereafter, we assume that $d^2 \leq (c - a^2)(1 - a)^2/8$, and hence that $d^2 < c(1 + \sqrt{c})^2/8$. In particular, as c nears 0, so does d.

3 In the remaining case, we will cover the edges in C by single edges, and cover the edges in B and A by triangles and single edges using the technique of Lemma 2.2. Consider the number of edges in B. Since B and C together must have at least $(1 - c)n^2/4$

edges and C has no more than dn^2 edges, we see that B has at least $(1-c-4d)n^2/4$ edges. In the 'complete' graph $H = K_n - K_{(1-a)n/2}$ there are $(1-a^2)n^2/4$ edges in B, so we see that if we can partition H by edges and triangles, we can partition G with only a few extra cliques: dn^2 for the edges in C and an allowance of at most $(1-a^2)(n^2/4) - (1-c-4d)(n^2/4) = (c-a^2+4d)n^2/4$ for the 'missing' edges of B.

We now set out to clique-partition H. We neglect some constant multiples of n to reduce the bulk of the expressions below. As in Lemma 2.1, partition $A = K_{(1+a)n/2}$ into $(1+a)(n/2) - 1$ matchings of $(1+a)(n/4)$ edges each (if $(1+a)n/2$ is odd, there is an extra linear factor in n neglected below). We must consider two subcases, $a \geq 0$ and $a < 0$.

If $a \geq 0$, we join $(1-a)n/2$ of these matchings to distinct points in C to form $(1-a^2)n^2/8$ triangles consuming all the connecting (B) edges of H; this leaves $(2a)(1+a)n^2/8$ edges of A unused and we cover them with single edges. Thus we partition H with $(1-a^2)(n^2/8) + a(1+a)(n^2/4)$ triangles and edges. This means we obtain a clique partition of G using no more cliques than

$$(1-a^2)(n^2/8) + a(1+a)(n^2/4) + dn^2 + (c-a^2+4d)n^2/4$$

$$= (n^2/4)[(1/2)(1-a^2) + a(1+a) + 4d + (c-a^2+4d)].$$

But it is easy to see that as c approaches 0 so that a and d also approach 0, this expression approaches $(n^2/4)[1/2 + 0 + 0 + 0]$, so it can clearly be made less than $(n^2/4)[1-c]$ as required.

If $a < 0$, we are able to join all the $(1+a)(n/2)-1$ matchings in A to distinct points in C. The resulting $(1+a)^2n^2/8$ triangles consume all (except a constant multiple of n) of the edges of A but only $(1+a)^2n^2/4$ edges of B, leaving as many as $(1-a^2)n^2/4-(1+a)^2n^2/4$ to cover with single edges. Thus we partition H into

$$(1+a)^2(n^2/8) + (1-a^2)(n^2/4) - (1+a)^2(n^2/4)$$

cliques (which approaches $(1/2)n^2/4$ as c approaches 0), and the rest of the argument goes exactly as in the prior paragraph.

A somewhat more careful calculation suggests that letting $c = 1/400$ will easily suffice for $n \geq 5$, forcing $|a| < .05$ by case 1 and $d < .02$ by case 2. Unfortunately, linear terms neglected here, such as $(1+a)n/4$, complicate the actual calculation of c badly for low values of n. □

If we require G to be threshold, or split, the situation simplifies somewhat, since C will contain no edges and case (2) becomes unnecessary. Still, this method appears to produce only a marginal improvement in the c in these cases. The first two authors and Guan-Tao Chen have made some further progress in the case that G is a split graph, but are still not close to $n^2/6$; this will be pursued elsewhere.

References

[1] Bender, E. A., Richmond, L. B. and Wormald, N. C. (1985) Almost all chordal graphs split. *J. Austral. Math. Soc.* (A) **38** 214–221.

[2] Caccetta, L., Erdős, P., Ordman, E. and Pullman, N. (1985) The difference between the clique numbers of a graph. *Ars Combinatoria* **19** A 97–106.

[3] Chartrand, G. and Lesniak, L. (1986) *Graphs and Digraphs*, 2nd. Edition, Wadsworth, Belmont, CA.

[4] Chvátal, V. and Hammer, P. (1973) *Set packing and threshold graphs*, Univ. of Waterloo Research Report CORR 73–21.

[5] Chvátal, V. and Hammer, P. (1977) Aggregation of inequalities in integer programming. *Ann. Discrete Math* **1** 145–162.

[6] DeBruijn, N. G. and Erdős, P. (1948) On a combinatorial problem. *Indag. Math.* **10** 421–423.

[7] Erdős, P., Faudree, R. and Ordman, E. (1988) Clique coverings and clique partitions. *Discrete Mathematics* **72** 93–101.

[8] Erdős, P., Goodman, A. W. and Posa, L. (1966) The representation of a graph by set intersections. *Canad. J. Math.* **18** 106–112.

[9] Erdős, P., Gyárfás, A., Ordman, E. T. and Zalcstein, Y. (1989) The size of chordal, interval, and threshold subgraphs. *Combinatorica* **9** (3) 245–253.

[10] Golumbic, M. (1980) *Algorithmic Graph Theory and Perfect Graphs*, Academic Press, New York.

[11] Gregory, D. A. and Pullman, N. J. (1982) On a clique covering problem of Orlin. *Discrete Math.* **41** 97–99.

[12] Henderson, P. and Zalcstein, Y. (1977) A graph theoretic characterization of the PVchunk class of synchronizing primitives. *SIAM J. Comp.* **6** 88–108.

[13] Orlin, J. (1977) Contentment in Graph Theory: covering graphs with cliques. *Indag. Math.* **39** 406–424.

[14] Wallis, W. D. (1982) Asymptotic values of clique partition numbers. *Combinatorica* **2** (1) 99–101.

On Intersecting Chains in Boolean Algebras

PÉTER L. ERDŐS[†], ÁKOS SERESS[‡] and LÁSZLÓ A. SZÉKELY[§]

[†]Centrum voor Wiskunde en Informatica, P.O. Box 4079, 1009 AB Amsterdam, The Netherlands

[‡]The Ohio State University, Columbus, OH 43210

[§]University of New Mexico, Albuquerque, NM 87131

Analogues of the Erdős–Ko–Rado theorem are proved for the Boolean algebra of all subsets of $\{1, \ldots n\}$ and in this algebra truncated by the removal of the empty set and the whole set.

1. Introduction

One of the basic results in extremal set theory is the Erdős–Ko–Rado (EKR) Theorem [5]: if $\mathcal{F}$ is an intersecting family of k-element subsets of $[1, n] = \{1, 2, \ldots, n\}$ (*i.e.* every two members of $\mathcal{F}$ have non-empty intersection) and $n \geq 2k$, then $|\mathcal{F}| \leq \binom{n-1}{k-1}$ and this bound is attained. We can consider k-subsets of $[1, n]$ as length-k chains in the (total) order $1 < 2 < \ldots < n$: using this terminology, the EKR theorem is a result about intersecting k-chains in a special partially ordered set.

Erdős, Faigle, and Kern [3] pointed out that certain results of Deza, Frankl [2, Theorem 5.8], and Frankl and Füredi [7] on intersecting sequences of integers may be interpreted as results on intersecting families of chains in some partially ordered sets.

The purpose of this note is to prove analogues of the EKR theorem in two other partially ordered sets: in the Boolean algebra $\mathcal{B}_n$ of all subsets of $[1, n]$ (with $A \leq B$ if $A \subset B$), and in the truncated Boolean algebra $\mathcal{B}_n^- := \mathcal{B}_n \setminus \{\emptyset, [1, n]\}$. We say that $\mathcal{L} = (L_1, L_2, \ldots, L_k)$ is a k-chain in $\mathcal{B}_n$ if $L_i \in \mathcal{B}_n$ for all $1 \leq i \leq k$ and L_i is a proper subset of L_{i+1} for all $1 \leq i \leq k-1$. A family $\mathcal{F}$ of k-chains in $\mathcal{B}_n$ is *intersecting* if any two elements of $\mathcal{F}$ have non-empty intersection.

k-chains and an intersecting family in $\mathcal{B}_n^-$ are defined analogously. Let $f(n, k)$ and

[‡]Research partially supported by NSF Grant CCR-9201303 and NSA Grant MDA904-92-H-3046.
[§]Research partially supported by ONR Grant N-0014-91-J-1385.

$f^-(n,k)$ denote the maximum size of intersecting families of k-chains in $\mathscr{B}_n$ and $\mathscr{B}_n^-$, respectively.

Obviously, the family $\mathscr{F}(A)$ of all k-chains containing some fixed $A \in \mathscr{B}_n$ is an intersecting family, and the same is true for the family $\mathscr{F}^-(A)$ of all k-chains in $\mathscr{B}_n^-$ containing some fixed $A \in \mathscr{B}_n^-$. Our main result is the following.

Theorem 1.1. *For any k,n, we have*

(i) $f(n,k) = |\mathscr{F}(\emptyset)|$ and

(ii) $f^-(n,k) = |\mathscr{F}^-(\{1\})|$.

Moreover, for $2 \le k \le n+1$, the only extremal families in $\mathscr{B}_n$ are $\mathscr{F}(\emptyset)$ and $\mathscr{F}([1,n])$.

The most well-known proof techniques for the original EKR Theorem are *shifting* and the *kernel method*. (For a brief introduction to these methods, see *e.g.*, the survey papers of Frankl [6] and Füredi [8].) The kernel method usually ensures short and easy proofs, but rarely gives the exact range of the result. Shifting gives exact (but perhaps slightly more complicated) proofs.

The situation is very similar in our case: Z. Füredi (personal communication) showed, using only the kernel method, that for $n \ge 6k \ln k$ Theorem 1.1(i) holds. In our proof of Theorem 1.1, we use an analogue of the shifting method and obtain a result without any restrictions on the parameters.

We remark, however, that to obtain sharp results in the case of t-intersecting families of chains, or the poset obtained by deleting the top m and bottom m levels in $\mathscr{B}_n$ for some $m < n/2$, it seems to be necessary to combine the two methods. Hilton-Milner type generalizations are also possible. Moreover, we have a common generalization of the original EKR theorem and Theorem 1.1. We shall return to these problems in a forthcoming paper.

Let $S(p,q)$ denote the Stirling numbers of the second kind, *i.e.* $S(p,q)$ is the number of partitions of a p-element set into q nonempty parts. It is easy to see that $|\mathscr{F}^-(\{1\})| = k! S(n-1,k)$, since each $\mathscr{L} = (L_1, L_2, ..., L_k) \in \mathscr{F}^-(\{1\})$ corresponds to an ordered partition $(L_2 \setminus L_1, L_3 \setminus L_2, ..., L_k \setminus L_{k-1}, [1,n] \setminus L_k)$ of $[2,n]$. Similarly, $|\mathscr{F}(\emptyset)| = (k-1)! S(n+1,k) = (k-1)! S(n,k-1) + k! S(n,k)$, the two last terms corresponding to the number of k-chains in $\mathscr{F}(\emptyset)$ containing and not containing $[1,n]$, respectively.

In the proofs, we shall often use the well-known recursion

$$S(n,k) = S(n-1,k-1) + kS(n-1,k)$$

(see *e.g.*, [9, Chapter 1]). In particular, $|\mathscr{F}(\emptyset)| = (k-1)! S(n+1,k)$.

2. Shifting

In this section we begin the proof of Theorem 1.1. We reduce the problem to the examination of so-called *compressed* sets of chains and prove that these satisfy a strong intersection property.

Let $\mathscr{F}$ be a family of pairwise intersecting k-chains from $\mathscr{B}_n$ or $\mathscr{B}_n^-$, and let $1 \le i < j \le n$ be integers. The (i,j) *chain-shift* $S_{ij}(\mathscr{F})$ of the family $\mathscr{F}$ is defined as follows.

For every k-chain $\mathscr{L} = (L_1, \ldots L_k) \in \mathscr{F}$, let $S_{ij}(\mathscr{L}) = (L_1', \ldots, L_k')$, where

$$L_l' = \begin{cases} L_l \setminus \{j\} \cup \{i\} & \text{if } j \in L_l \text{ and } i \notin L_l, \\ L_l & \text{otherwise.} \end{cases}$$

We say that L_l' is the *shift* of L_l. Shifting preserves set containment, so $S_{ij}(\mathscr{L})$ is a k-chain. The shifted family $S_{ij}(\mathscr{F})$ is obtained by the following rule: replace every k-chain $\mathscr{L} \in \mathscr{F}$ by $S_{ij}(\mathscr{L})$ if and only if

(1) $S_{ij}(\mathscr{L}) \neq \mathscr{L}$ and
(2) $S_{ij}(\mathscr{L}) \notin \mathscr{F}$.

It is clear from the definition that $|S_{ij}(\mathscr{F})| = |\mathscr{F}|$. Moreover, shifting preserves the intersection property.

Lemma 2.1. *If $\mathscr{F}$ is an intersecting family of k-chains in $\mathscr{B}_n$ or $\mathscr{B}_n^-$, then $S_{ij}(\mathscr{F})$ is also intersecting.*

Proof. Let $\mathscr{L}_1, \mathscr{L}_2 \in S_{ij}(\mathscr{F})$; we have to prove that they contain a common element. We distinguish three cases:

Case 1: $\mathscr{L}_1, \mathscr{L}_2 \in \mathscr{F}$. In this case it is obvious that $\mathscr{L}_1$ and $\mathscr{L}_2$ intersect.

Case 2: $\mathscr{L}_1, \mathscr{L}_2 \notin \mathscr{F}$. In this case, there are $\mathscr{L}_3, \mathscr{L}_4 \in \mathscr{F}$ such that $\mathscr{L}_1 = S_{ij}(\mathscr{L}_3)$ and $\mathscr{L}_2 = S_{ij}(\mathscr{L}_4)$. Let $M \in \mathscr{L}_3 \cap \mathscr{L}_4$. Then the shift of M (which may be M itself) is a common element of $\mathscr{L}_1$ and $\mathscr{L}_2$.

Case 3: $\mathscr{L}_1 \notin \mathscr{F}$ and $\mathscr{L}_2 \in \mathscr{F}$. Then let $\mathscr{L}_3 \in \mathscr{F}$ such that $\mathscr{L}_1 = S_{ij}(\mathscr{L}_3)$. There may be two reasons why $\mathscr{L}_2$ was not replaced. If $\mathscr{L}_2 = S_{ij}(\mathscr{L}_2)$ then let $M \in \mathscr{L}_2 \cap \mathscr{L}_3$. The shift of M is itself (since $\mathscr{L}_2 = S_{ij}(\mathscr{L}_2)$) so $M \in \mathscr{L}_2 \cap S_{ij}(\mathscr{L}_3) = \mathscr{L}_2 \cap \mathscr{L}_1$ as well.

The other reason is that $\mathscr{L}_2 \neq S_{ij}(\mathscr{L}_2)$ but $S_{ij}(\mathscr{L}_2) \in \mathscr{F}$. In this subcase, let $M \in \mathscr{L}_3 \cap S_{ij}(\mathscr{L}_2)$. It is impossible that $j \in M$ and $i \notin M$ since M is the shift of some element of $\mathscr{L}_2$. Also, it is impossible that $i \in M$ and $j \notin M$ because there is some $K \in \mathscr{L}_3$ such that $j \in K$ and $i \notin K$ (because $S_{ij}(\mathscr{L}_3) \neq \mathscr{L}_3$) and one of K, M must contain the other. So M is a set containing either both of i, j or neither of i, j. In either case, from $M \in S_{ij}(\mathscr{L}_2)$ we have $M \in \mathscr{L}_2$ so $M \in \mathscr{L}_1 \cap \mathscr{L}_2$. $\square$

We say that the family $\mathscr{F}$ of intersecting k-chains is *compressed* if $\mathscr{F}$ is invariant for all chain-shift operations S_{ij}, $1 \le i < j \le n$. By Lemma 2.1, for any intersecting family $\mathscr{F}$, repeated applications of chain-shifts result in a compressed family of the same size.

Compressed families satisfy a strong intersection property. We say that $M \in \mathscr{B}_n$ (or $M \in \mathscr{B}_n^-$) is an *initial segment* if $M = [1, m]$ for some $1 \le m \le n$ or $M = \emptyset$.

Lemma 2.2. *Let $\mathscr{F}$ be a compressed family of intersecting k-chains. Then for any $\mathscr{L}_1, \mathscr{L}_2 \in \mathscr{F}$, $\mathscr{L}_1$ and $\mathscr{L}_2$ intersect in an initial segment.*

Proof. Suppose that the lemma is not true and let $\mathscr{L}_1 \in \mathscr{F}$ be a minimal counterexample in the sense that

(i) there exists $\mathscr{L}_2 \in \mathscr{F}$ such that $\mathscr{L}_1 \cap \mathscr{L}_2$ contains no initial segment

(ii) $\sum_{L\in\mathscr{L}_1}\sum_{x\in L} x$ is minimal among all $\mathscr{L}_1$ satisfying (i).

Let $M \in \mathscr{L}_1 \cap \mathscr{L}_2$. Since M is not an initial segment, there exist $1 \le i < j \le n$ such that $i \notin M$ and $j \in M$. Then $S_{ij}(\mathscr{L}_1) \neq \mathscr{L}_1$, so $S_{ij}(\mathscr{L}_1)$ is not a counterexample. Therefore, there exists an initial segment $K \in S_{ij}(\mathscr{L}_1) \cap \mathscr{L}_2$. It is impossible that $j \in K$ and $i \notin K$, since K is an initial segment. Also, it is impossible that $i \in K$ and $j \notin K$, because $K, M \in \mathscr{L}_2$, so one of them must contain the other. So K is a set containing both of i, j or neither of i, j. In either case $K \in \mathscr{L}_1$, which is a contradiction.   □

In the next two sections, we prove Theorem 1.1 for $\mathscr{B}_n$ and $\mathscr{B}_n^-$, respectively. By Lemma 2.1, it is enough to consider compressed families.

3. Chains in $\mathscr{B}_n$

We prove by induction on n that $f(n,k) = (k-1)!S(n+1,k)$. The base case $n = 1$ is trivial. Suppose we are done for $n - 1$ (with all values of k) and let $\mathscr{F}$ be a compressed family of chains in $\mathscr{B}_n$. We distinguish two cases:

Case 1: $\mathscr{F}$ contains a chain $\mathscr{L}$ such that the only initial segment in $\mathscr{L}$ is $[1, n]$. Then, since each chain in $\mathscr{F}$ must intersect $\mathscr{L}$ in an initial segment (see Lemma 2.2), all chains contain $[1, n]$ and we are done.

Case 2: There is no chain in $\mathscr{F}$ such that $[1, n]$ is the only initial segment in the chain. Then delete n from each element of each chain. Each chain is transformed into either a k-chain or a $(k-1)$-chain and so we obtain an intersecting family $\mathscr{C}_{k-1}$ of chains of length $k - 1$ in $\mathscr{B}_{n-1}$ and an intersecting family $\mathscr{C}_k$ of chains of length k in $\mathscr{B}_{n-1}$.

We claim that each $(L_1, ..., L_{k-1}) \in \mathscr{C}_{k-1}$ can be obtained from $\le k - 1$ chains of $\mathscr{F}$. This is true since we have to add the set $L_i \cup \{n\}$ to the chain for some $1 \le i \le k - 1$ and add n to the sets $L_{i+1}, ..., L_{k-1}$. The value of i uniquely determines the chain in $\mathscr{F}$. Furthermore, $i = 0$ is impossible, since then the only initial segment would be $[1, n]$.

We also claim that each $(L_1, ..., L_k) \in \mathscr{C}_k$ can be obtained from $\le k$ chains of $\mathscr{F}$. Indeed, we have to add n to the sets starting at some $2 \le i \le k + 1$; the value $k + 1$ corresponds to the case that n did not occur in any element of the chain in $\mathscr{F}$. Furthermore, $i = 1$ is impossible, since the only initial segment would be $[1, n]$.

Thus

$$|\mathscr{F}| \le (k-1)f(n-1, k-1) + kf(n-1, k) = (k-1)!S(n+1, k). \qquad (1)$$

The uniqueness of the extremal systems can also be proved by induction on n. First, we remark that if $k = n + 1$, every family $\mathscr{F}$ of k-chains must contain the empty set, and maximality implies $\mathscr{F} = \mathscr{F}(\emptyset)$. If $k = 2$ and $|\mathscr{F}| \ge 4$, then $\mathscr{F} \subseteq \mathscr{F}(A)$ for some subset A. Now, $|\mathscr{F}(A)| = 2^{|A|} + 2^{n-|A|} - 2$, which takes its maximum value for $|A| = 0$ and $|A| = n$.

In the case $3 \le k \le n$, we first consider a compressed family $\mathscr{F}$. If $\mathscr{F}$ belongs to Case 1 above, then $\mathscr{F} = \mathscr{F}([1, n])$; otherwise, in Case 2, we must have equality in (1). This implies that $\mathscr{C}_{k-1}$ and $\mathscr{C}_k$ are extremal families in $\mathscr{B}_{n-1}$, and, by the induction hypothesis, they must be the $\mathscr{F}(\emptyset)$ of $(k-1)$-chains and k-chains in $\mathscr{B}_{n-1}$, respectively. So $\mathscr{F}$ must be identical with $\mathscr{F}(\emptyset)$ in $\mathscr{B}_n$.

Finally, we observe that any family whose compressed image is $\mathscr{F}(\emptyset)$ or $\mathscr{F}([1,n])$ is itself one of these families.

We remark that in a preliminary version of the present paper [4], we proved Theorem 1.1 for $n > k \ln k$. Since then, Ahlswede and Cai [1] have also found a proof for Theorem 1.1 (i), but their method does not seem to generalize to the truncated case.

4. Chains in $\mathscr{B}_n^-$

Again, we use induction on n to prove that $f^-(n,k) = k!S(n-1,k)$. The base case $n = 2$ is trivial. Suppose we are done for $n-1$ and let $\mathscr{F}$ be a compressed family of chains in $\mathscr{B}_n^-$. We distinguish two cases:

Case 1: If there exists a chain $\mathscr{L} \in \mathscr{F}$ such that $n - 1 \in L_1$, then $\mathscr{L}$ may contain only one initial segment, namely $[1, n-1]$. Then, since each chain in $\mathscr{F}$ must intersect $\mathscr{L}$ in an initial segment (see Lemma 2.2), all chains contain $[1, n-1]$ and we are done.

Case 2: If each $\mathscr{L} \in \mathscr{F}$ has no $n-1 \notin L_1$, then, in particular, we never have $L_1 \neq \{n-1\}$. Define

$$\mathscr{F}_i = \{\mathscr{L} \in \mathscr{F} : L_{i+1} - L_i = \{n-1\}\}, \qquad (i = 1, 2, ..., k-1)$$
$$\mathscr{F}_k = \{\mathscr{L} \in \mathscr{F} : L_k = \{1, 2, ..., n-2, n\}\},$$
$$\mathscr{F}_0 = \mathscr{F} - \cup_{j=1}^k \mathscr{F}_j.$$

Deleting $n-1$ from each element of each chain of $\mathscr{F}_0$, we obtain a family $\mathscr{F}_0'$ of intersecting k-chains in the truncated Boolean algebra on the underlying set $\{1, 2, ..., n-2, n\}$. By hypothesis, $|\mathscr{F}_0'| \leq f^-(n-1, k)$. Each $(L_1, ..., L_k) \in \mathscr{F}_0'$ can be obtained from $\leq k$ chains of $\mathscr{F}_0$, since $n-1$ could have been inserted starting at $L_2, L_3, ..., L_k$, or could have been an element of $[1, n] \setminus L_k$.

Deleting $n-1$ from every set in every chain in $\mathscr{F}_i$ (for any $i = 1, 2, ..., k-1$), we obtain a family $\mathscr{F}_i'$ of intersecting $(k-1)$-chains in the truncated Boolean algebra on the underlying set $\{1, 2, ..., n-2, n\}$. By hypothesis, $|\mathscr{F}_i| = |\mathscr{F}_i'| \leq f^-(n-1, k-1)$.

Finally, define $\mathscr{F}_k'$ by deleting the largest set $L_k = \{1, 2, ..., n-2, n\}$ from every chain in $\mathscr{F}_k$. Observe that $\mathscr{F}_k'$ is a family of intersecting $(k-1)$-chains in the truncated Boolean algebra on the underlying set $\{1, 2, ..., n-2, n\}$, since the set that we dropped is not an initial segment in the original underlying set. Therefore, by hypothesis, $|\mathscr{F}_k| = |\mathscr{F}_k'| \leq f^-(n-1, k-1)$.

Hence, $|\mathscr{F}| \leq k \cdot k!S(n-2, k) + (k-1)(k-1)!S(n-2, k-1) + (k-1)!S(n-2, k-1) = k!S(n-1, k)$.

This finishes the proof of Theorem 1.1. □

We remark that, analogously to the discussion at the end of Section 3, it can be shown that the only compressed extremal families in $\mathscr{B}_n^-$ are $\mathscr{F}^-([1])$ and $\mathscr{F}^-([1, n-1])$. The extension that the only extremal families are $\mathscr{F}^-(A)$ with $|A| = 1$ or $|A| = n-1$ is still missing.

Acknowledgements

We are indebted to Ulrich Faigle, Zoltán Füredi, and Walter Kern for stimulating conversations on the subject of the paper.

Note added in proof

We have just learned of a research program initiated by Miklós Simonovits and Vera T. Sós on 'structured intersection theorems' [10, 11], which has a fairly large literature. They studied the maximum number of graphs on n vertices such that any two intersect in a prescribed graph, *e.g.* a path or cycle. The following problem fits into their scheme: given a graph G what is the maximum number of pairwise intersecting complete k-subgraphs. In this paper we have studied the comparison graphs of some partially ordered sets.

References

[1] Ahlswede, R. and Cai, N. (1993) Incomparability and intersection properties of Boolean interval lattices and chain posets, preprint.

[2] Deza, M. and Frankl, P. (1983) Erdős-Ko-Rado theorem – 22 years later, *SIAM J. Alg. Disc. Methods* **4**, 419–431.

[3] Erdős, P. L., Faigle, U. and Kern, W. (1992) A group-theoretic setting for some intersecting Sperner families, *Combinatorics, Probability and Computing* **1**, 323–334.

[4] Erdős, P. L., Seress, Á. and Székely, L. A. (1993) *On intersecting k-chains in Boolean algebras*, Preprint, April 1993.

[5] Erdős, P., Ko, C. and Rado, R. (1961) Intersection theorems for systems of finite sets, *Quart. J. Math. Oxford Ser. 2* **12**, 313–318.

[6] Frankl, P. (1987) The shifting technique in extremal set theory, In: Whitehead, C. (ed.) *Surveys in Combinatorics 1987*, Cambridge University Press, 81–110.

[7] Frankl, P. and Füredi, Z. (1980) The Erdős-Ko-Rado theorem for integer sequences, *SIAM J. Alg. Disc. Methods* **1**, 376–381.

[8] Füredi, Z. (1991) Turán type problems, In: Keedwell, A. D. (ed.) *Surveys in Combinatorics 1991*, Cambridge University Press 253–300.

[9] Lovász, L. (1977) *Combinatorial Problems and Exercises*, Akadémiai Kiadó, Budapest and North-Holland, Amsterdam.

[10] Simonovits, M. and Sós, V. T., Intersection theorems for graphs. Problèmes Combinatoires et Theorie des Graphes, Coll. *Internationaux C.N.R.S.* **260** 389–391.

[11] Simonovits, M. and Sós, V. T. (1978) Intersection theorems for graphs II. *Combinatorics, Coll. Math. Soc. J. Bolyai* **18** 1017–1030.

On the Maximum Number of Triangles in Wheel-Free Graphs

ZOLTÁN FÜREDI[†], MICHEL X. GOEMANS[‡]
and DANIEL J. KLEITMAN[§]

Department of Mathematics,
Massachusetts Institute of Technology,
Cambridge, MA 02139

Gallai [1] raised the question of determining $t(n)$, the maximum number of triangles in graphs of n vertices with acyclic neighborhoods. Here we disprove his conjecture ($t(n) \sim n^2/8$) by exhibiting graphs having $n^2/7.5$ triangles. We improve the upper bound [11] of $(n^2 - n)/6$ to $t(n) \leq n^2/7.02 + O(n)$. For regular graphs, we further decrease this bound to $n^2/7.75 + O(n)$.

1. Introduction

Let WFG_n be the class of graphs on n vertices with the property that the neighborhood of any vertex is acyclic. A graph G is given by its vertex set $V(G)$ and edge set $E(G)$. The subgraph *induced* by $X \subset V(G)$ is denoted by $G[X]$. The neighborhood $N(v)$ of vertex v is the set of vertices adjacent to v. Note that $v \notin N(v)$. The *degree* of $v \in V(G)$, denoted by d_v or $d_v(G)$, is the size of the neighborhood: $d_v = |N(v)|$. The maximum (minimum) degree is denoted by Δ (δ), or $\Delta(G)$ ($\delta(G)$, respectively) to avoid misunderstandings. A *matching* $M \subset E(G)$ is a set of pairwise disjoint edges. A *wheel* W_i is obtained from a cycle C_i by adding a new vertex and edges joining it to all the vertices of the cycle; the new edges are called the *spokes* of the wheel ($i \geq 3$, $W_3 = K_4$). Therefore, WFG_n consists of all graphs on n vertices containing no wheel. Let $t(G)$ denote the number of triangles

[†] Research supported in part by the Hungarian National Science Foundation under grant No. 1812.
New address: Dept. Math., Univ. Illinois, Urbana, IL 61801-2917.
E-mail: zoltan@math.uiuc.edu
[‡] Research supported by Air Force contract F49620-92-J-0125 and by DARPA contract N00014-89-J-1988.
E-mail: goemans@math.mit.edu
[§] Research supported by Air Force contract F49620-92-J-0125 and by NSF contract 8606225.
E-mail: djk@math.mit.edu

in G and let $t(n)$ be the maximum of $t(G)$ over WFG_n. Gallai (see [1]) raised the question of determining $t(n)$.

Take a complete bipartite graph $K_{2a,n-2a}$, where a is the closest integer to $n/4$, and add a maximum matching on the side of size $2a$. We obtain a wheel-free graph G_n^1, having $\lfloor n^2/8 \rfloor$ triangles [1]. Gallai and, independently, Zelinka [10] in 1983 conjectured that this is the maximum possible. However, Zhou [11] recently constructed wheel-free graphs having $(n^2 + n)/8$ triangles whenever n is of the form $8q + 7$. He also found an upper bound, $t(n) \leq (n^2 - n)/6$. In this paper we improve both bounds.

Theorem 1. *There exists a wheel-free graph on n vertices with $n^2/7.5 + n/15$ triangles whenever n is a multiple of 15, i.e., $t(n) \geq n^2/7.5 + n/15$.*

This theorem is proved by giving a construction, G_n^2, in Section 2. As $t(n)$ is monotone, we get $t(n) \geq n^2/7.5 - O(n)$ for all n. P. Haxell observed that G_n^2 has the additional property that it is *locally tree-like*, i.e., every neighborhood induces a tree. (More exactly, she improved it so.) Zelinka [10] proved that any locally tree-like graph with n vertices has at least $2n - 3$ edges and posed the question; what is the maximum number of edges of these graphs? As G_n^2 has $n^2/5 + O(n)$ edges, we got a counterexample for a conjecture of Fronček [5], who believed that a locally tree-like graph on n vertices contains at most $\lfloor n(3n + 8)/16 \rfloor$ edges (for $n \geq 8$).

Theorem 2. *Every wheel-free graph on n vertices contains at most $n^2/7.02 + O(n)$ triangles, more exactly, $t(n) \leq n^2/7.02 + 5n$ for all n.*

The main tool of the proof is Proposition 11 (proved in Section 4), which gives $t(G) \leq n^2/8 + o(n^2)$ for several types of graphs. One example is given by the following theorem.

Theorem 3. *Let $G \in \text{WFG}_n$ be a wheel-free graph on n vertices, $n \geq 100$. If $\delta > (2/5)n + 16/5$, then $t(G) \leq n^2/8$.*

Looking at regular wheel-free graphs one can observe that, if n is of the form $4a - 1$, the previously mentioned construction, G_n^1, is regular. Hence, there exist regular graphs in WFG_n having $\lfloor n^2/8 \rfloor$ triangles. But the graph constructed in Section 2 is *not* regular. In fact, the upper bound we prove for regular graphs in Section 7 is lower than the lower bound for general graphs.

Theorem 4. *If G is a regular wheel-free graph, $t(G) \leq n^2/7.75 + O(n)$.*

We conjecture that Gallai's conjecture holds for regular graphs.

Conjecture 5. *If G is a regular wheel-free graph, $t(G) \leq n^2/8$.*

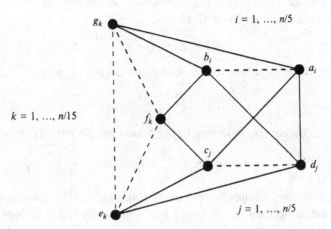

Figure 1 A wheel-free graph having $n^2/7.5 + n/15$ triangles.

2. Wheel-free graphs having more than $n^2/7.5$ triangles

Define the graph G_n^2 on n vertices, where n is a multiple of 15 (see Figure 1) as follows. Its vertex set $V(G_n^2)$ consists of a_i, b_i, c_i, d_i for $i = 1, \ldots, n/5$, and e_k, f_k, g_k for $k = 1, \ldots, n/15$. Its edge set $E(G_n^2)$ consists of

— two matchings of size $n/5$: (a_i, b_i) and (c_i, d_i),
— three matchings of size $n/15$: (e_k, f_k), (f_k, g_k) and (e_k, g_k), and
— all the edges of types: (a_i, c_j), (a_i, d_j), (a_i, g_k), (b_i, d_j), (b_i, f_k), (b_i, g_k), (c_j, e_k), (c_j, f_k) and (d_j, e_k). (Here, again, $1 \le i, j \le n/5$ and $1 \le k \le n/15$.)

It is easy to verify that this graph belongs to WFG_n. For example, the neighborhood of the vertex a_i consists of the matching $\{(c_j, d_j) : 1 \le j \le n/5\}$ as well as a star rooted at b_i with edges $\{(b_i, g_k) : 1 \le k \le n/15\}$ and $\{(b_i, d_j) : 1 \le j \le n/5\}$.

Each triangle in G_n^2 contains an edge from the matchings, and its vertices are in three different classes. An easy calculation shows that $t(G_n^2) = n^2/7.5 + n/15$.

3. Improved upper bounds for $t(n)$

In this section we prove a series of Lemmas that lead to Theorem 6, an upper bound for all n. This theorem was independently proved by P. Haxell [8].

Theorem 6. *Every wheel-free graph on n vertices contains at most $(1/7)n^2 + (9/7)n$ triangles, i.e., $t(n) \le \dfrac{1}{7}n^2 + \dfrac{9}{7}n$.*

Let $G \in \mathrm{WFG}_n$ be an arbitrary wheel-free graph. By definition, $G[N(v)]$, the neighborhood of any vertex v, is acyclic. Hence, the number of edges in $N(v)$ is less than or equal to $d_v - 1$, where $d_v = |N(v)|$. By summing $d_v - 1$ over all vertices of G, we obtain an upper bound for $3t(G)$. On the other hand, the summation of the degrees over all vertices is precisely twice the number of edges of G. Hence,

$$3t(G) \leq 2|E(G)| - n. \tag{1}$$

Since $2|E(G)| \leq n\Delta$, where Δ stands for the maximum degree, (1) gives

$$t(G) \leq \frac{n(\Delta - 1)}{3}. \tag{2}$$

Our next aim is to obtain the following upper bound on the number of edges of G:

$$|E(G)| \leq \frac{n^2}{4} + \frac{n}{4}. \tag{3}$$

Together with (1), this gives Zhou's upper bound, $t(n) \leq (n^2 - n)/6$. The upper bound (3) follows from the following theorem of Erdős and Simonovits [3]: if G does not contain a W_3 or a W_4, then $|E(G)| \leq \lfloor (n^2 + n)/4 \rfloor$, whenever $n > n_0$. They wrote: '... are easy to prove by induction and can be left for the reader'. For completeness we reconstruct their argument in a somewhat simplified form.

A graph is called (k, ℓ)-free if $|E(G[K])| < \ell$ holds for every k-element subset $K \subset V(G)$. Let $f(n; k, \ell)$ be the maximum number of edges of a (k, ℓ)-free graph with n vertices. Turán's classical theorem determines $f(n; k, \binom{k}{2})$, for example, $f(n; 3, 3) = \lfloor n^2/4 \rfloor$. He proposed the problem of determining $f(n; k, \ell)$, but it is still unsolved in a number of cases. Erdős [2] investigated, first, all the cases $k \leq 5$, he also proved that excluding only W_4 implies $|E(G)| \leq \lfloor n^2/4 \rfloor + \lfloor (n + 1)/2 \rfloor$ (for $n > n_0$). For recent accounts, see [6, 7]. A wheel-free graph has neither W_3 nor W_4, so it is $(5, 8)$-free. Thus (3) follows from the following.

Claim 7. *If G is a graph with n vertices such that any 5 vertices span at most 7 edges, $|E(G)| \leq (n^2 + n)/4$ holds for $n \geq 5$.*

Proof. The case $n = 5$ is trivial. Let G be a $(5, 8)$-free graph with n vertices, $n \geq 6$. Considering all the n subgraphs $G \setminus x$, we have $(n - 2)|E(G)| = \sum_{x \in V} |E(G \setminus x)|$, implying

$$f(n; 5, 8) \leq \left\lfloor \frac{n}{n - 2} f(n - 1; 5, 8) \right\rfloor. \tag{4}$$

Let $s_n = \lfloor (n^2 + n)/4 \rfloor$, $n \geq 1$. It is easy to see (by induction, distinguishing four subcases according to the residue of n modulo 4) that $s_n = \lfloor (n/(n - 2))s_{n-1} \rfloor$ holds for all $n \geq 2$. As $f(5; 5, 8) = s_5 = 7$, (4) implies the desired upper bound. $\square$

A *vertex cover* C of the graph H is a subset of vertices with the property that all edges of H are incident to at least one vertex in C.

Lemma 8. *If C is a vertex cover of $G[N(w)]$, where w is a vertex of maximum degree Δ, then*

$$t(G) \leq \frac{1}{2}(\Delta - 1)(n - \Delta + |C|).$$

Proof. Since C is a vertex cover of $G[N(w)]$, the set $S = (V - N(w)) \cup C$ is a vertex cover of G. Hence, all triangles of G contain at least two vertices belonging to S. Summing $(d_v - 1)$ over all vertices v in S, we obtain an upper bound on twice the number of triangles in G:

$$2t(G) \le \sum_{v \in S}(d_v - 1) \le (\Delta - 1)|S| = (\Delta - 1)(n - \Delta + |C|).$$

$\square$

Observe that if $|C| = o(n)$, the function $(1/2)(p-1)(n-p+|C|)$ is maximized at $p = (1/2)n + o(n)$. In this case, we get

$$t(G) \le \frac{n^2}{8} + o(n^2).$$

On the other hand, if the acyclic graph $G[N(w)]$ has no vertex cover of 'small' cardinality, we can use the following two lemmas to isolate two vertices of G that are contained in a small number of triangles. The *total* degree of the subset S (in the graph H) means the sum of degrees: $\sum_{x \in S} d_x(H)$.

Lemma 9. *Let H be an acyclic graph on p vertices with at least one edge, and let b be any positive integer. Then either there exist two adjacent vertices whose total degree is at most $b + 1$, or there exists a vertex cover of H of cardinality at most $(p-2)/b$.*

Proof. Suppose that the size of each cover exceeds $(p-2)/b$. As H is a bipartite graph, König's theorem implies that one can find a matching M of size $m > (p-2)/b$. If we sum up all the degrees of the vertices of $\cup M$, we count all edges of H at most once, except the edges included in $\cup M$. Hence $\sum\{d_u : u \in \cup M\}$ is at most $(p-1) + (2m-1)$. This guarantees the existence of a pair $(a, b) \in M$ such that $d_a + d_b \le (p-2+2m)/m$, which is less than $b + 2$. $\square$

Lemma 10. *Suppose that $G \in \mathrm{WFG}_n$ and $w \in V(G)$. Suppose also that the $d_u(G[N(w)]) + d_v(G[N(w)]) \le s$, where u and v are adjacent vertices in $G[N(w)]$. Then G has at most $n - d_w + s - 2$ triangles containing at least one of $\{u, v\}$.*

Proof. We claim that given any two adjacent vertices u and v, the total number of triangles containing u or v is at most $|N(u) \cup N(v)| - 2$. Indeed, u is contained in at most $|N(u)| - 1$ triangles and v is contained in at most $|N(v)| - 1$ triangles, while the number of triangles containing both u and v is exactly $|N(u) \cap N(v)|$. Hence, the total number of triangles containing u or v is bounded above by $|N(u)| - 1 + |N(v)| - 1 - |N(u) \cap N(v)| = |N(u) \cup N(v)| - 2$.

The assumption in the Lemma implies that $|N(u) \cup N(v)| \le |V(G) \setminus N(w)| + s$. Combining this bound with the previous observation completes the proof. $\square$

Proof of Theorem 6. We prove the bound $(n^2 + 9n)/7$ by induction on the number of vertices n. Let G be a graph in WFG_n with maximum degree Δ, and let $w \in V(G)$ be such that $d_w = \Delta$. We consider three cases.

If $\Delta < (3/7)n + 5$, i.e. $\Delta \leq (3n + 34)/7$, the result follows from the inequality (2).

If there exists a vertex cover of $G[N(w)]$ of cardinality less than or equal to $\Delta/8$, Lemma 8 implies that

$$t(G) \leq \frac{1}{2}(\Delta - 1)\left(n - \Delta + \frac{\Delta}{8}\right) < \frac{\Delta(n - \frac{7}{8}\Delta)}{2} \leq n^2/7.$$

In these two cases, we have not used the inductive hypothesis.

Finally, assume that $\Delta \geq (3/7)n + 5$, and that there is no vertex cover of $G[N(w)]$ of cardinality less than or equal to $\Delta/8$. By Lemma 9, there exist two adjacent vertices u and v in $N(w)$ whose total degree in $G[N(w)]$ is at most 9. Lemma 10 now implies that there are at most $n - \Delta + 7 \leq (4/7)n + 2$ triangles containing u or v. By deleting u and v, we obtain a graph in WFG_{n-2} that contains at most

$$\frac{(n-2)^2}{7} + \frac{9}{7}(n-2)$$

triangles, by the inductive hypothesis. Hence,

$$t(G) \leq \frac{(n-2)^2}{7} + \frac{9}{7}(n-2) + \frac{4n}{7} + 2 = \frac{n^2}{7} + \frac{9}{7}n.$$

$\square$

4. Triangles from a matching

In order to prove the slight additional improvement described in Theorem 2, we first prove a weaker form of Gallai's conjecture. This result is also crucial in improving the upper bound in the case of regular graphs. Let d_{uv} denote $|N(u) \cap N(v)|$, the number of triangles containing the edge uv.

Proposition 11. *Let G be a graph that contains no wheel with 3 or 4 spokes, and let M be a matching in it. Then*

$$\sum_{(u,v) \in M} d_{uv} \leq \frac{n^2}{8}. \tag{5}$$

Proof. For simplicity, throughout this proof, a *triangle* always refers to a triangle containing an edge in M. Thus, the left-hand side of (5) is the number of triangles in G. Let m denote the cardinality of M, let P denote the set of unmatched vertices, and let $p = |P| = n - 2m$.

Observation 1. For two edges (u, v) and (x, y) in M, the induced graph $G[u, v, x, y]$ can contain at most two triangles (Figure 2).

Indeed, more than two triangles would imply that $\{u, v, x, y\}$ induces a K_4, *i.e.*, a wheel with 3 spokes.

Consider first the graph H with a vertex uv for each edge (u, v) of M, and with an edge (uv, xy) whenever $\{u, v, x, y\}$ induces two triangles. Let Q be a maximum matching in H. Let S be the set of vertices in G belonging to edges of M that are saturated by Q, and let R be the remaining vertices in M. Hence (S, R, P) is a partition of $V(G)$. Let q denote the

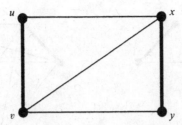

Figure 2 Observation 1 in the proof of Proposition 11

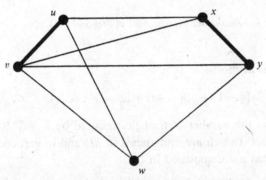

Figure 3 Observation 2 in the proof of Proposition 11

cardinality of Q, and let r denote the number of unmatched vertices in H, i.e., $r = m - 2q$. Clearly, $|S| = 4q$ and $|R| = 2r$. Thus, $p + 2r + 4q = n$.

Observation 2. If $(uv, xy) \in Q$ and w is any vertex of G, then $\{u, v, x, y\}$ and w can connect with one another in at most 1 triangle (Figure 3).

Indeed, if $\{u, v, x\}$ is one of the two triangles induced by $\{u, v, x, y\}$, and w forms two triangles with $\{u, v, x, y\}$, then $\{u, v, x, w\}$ induces a K_4.

Observation 3. If $(uv, xy) \in Q$ and $(a, b) \in M$, then $\{u, v, x, y\}$ and $\{a, b\}$ connect with one another in at most two triangles (Figure 4).

Assume without loss of generality that $(u, x), (v, x)$ and (v, y) are in $E(G)$. If $\{u, v, x, y\}$ and $\{a, b\}$ connect with one another in three or more triangles, we can assume without loss of generality that $\{u, v\}$ and $\{a, b\}$ connect with one another in two triangles. Without loss of generality, we can assume that (u, a) and (v, b) are in $E(G)$. We consider two cases. If $(v, a) \in E(G)$ (see Figure 4.a), then $\{x, y, a, b\}$ is included in $N(v)$, implying that any triangle between $\{x, y\}$ and $\{a, b\}$ would create a K_4. If $(u, b) \in E(G)$ (see Figure 4.b), a triangle of the form $\{a, b, z\}$ with $z \in \{x, y\}$ would create the cycle $u - a - z - v - u$ in the neighborhood of b, while a triangle of the form $\{x, y, c\}$ with $c \in \{a, b\}$ would create the cycle $u - c - y - v - u$ in the neighborhood of x.

From Observations 2 and 3, there are at most $q(n - 4q)$ triangles that connect S to $V - S$. Within S, in addition to the $2q$ triangles corresponding to the edges in Q, there are at most $(1/2)4q(q - 1)$ triangles, by Observation 3. Therefore, the number of triangles

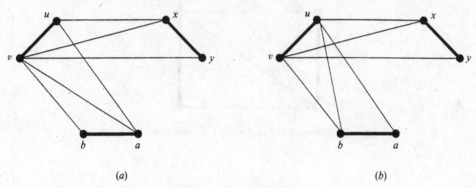

Figure 4   Observation 3 in the proof of Proposition 11

containing vertices in S is at most

$$2q(q-1) + q(n-4q) + 2q = q(n-2q). \tag{6}$$

We now concentrate on the number of triangles induced by $V - S$. Recall that $V - S$ corresponds to vertices in P (which are unmatched in M) and to vertices in R (which are incident to edges in M that are unmatched in Q).

Observation 4. If uv and xy are unmatched vertices in Q, then $\{u,v\}$ and $\{x,y\}$ can connect with one another in at most one triangle by the maximality of Q.

Observation 5. Consider a vertex $w \in P$. We say that two edges (u,v) and (x,y) in M are *independent* if $\{u,v,x,y\}$ does not induce any triangle. Since the neighborhood of w cannot contain any triangle, w can induce triangles only with independent edges in M.

Let s be the cardinality of a largest set of independent edges in $G[R]$. The number of triangles in $G[R]$ is at most $s(r-s) + (1/2)(r-s)(r-s-1) \le (r^2 - s^2)/2$ by Observation 4. Moreover, the number of triangles between P and R is at most ps by Observation 5. Therefore, there are at most $T = ps + (r^2 - s^2)/2$ triangles in $G[V - S]$. We have that $ps + (r^2 - s^2)/2 \le (2r + p)^2/8$ for all real $r \ge s$, $p \ge 0$ (because this is equivalent to $4p(s-r) \le (p-2s)^2$). Hence, the number of triangles in $G[V - S]$ is at most $(n - 4q)^2/8$. Combined with (6), this implies that the number of triangles containing edges in M is at most $q(n-2q) + (n-4q)^2/8 = n^2/8$. $\qquad\square$

5. Graphs with large minimum degree

In this section we prove Theorem 3.

Partition the edge set of G into two classes. We say that an edge (u,v) of $E(G)$ is a *fat* edge if at least $\sqrt{n}$ triangles contain it, i.e., $d_{uv} \ge \sqrt{n}$. An edge that is not fat is said to be *lean*. In the next two lemmas, we show that the set of fat edges is an ideal candidate to play the role of the matching M in Proposition 11.

Lemma 12. *Let G be a graph in* WFG_n *with* $d_u + d_v + d_w \geq n + 3\sqrt{n}$ *for every triangle* uvw. *(For example, $\delta \geq n/3 + \sqrt{n}$.) Then every triangle of G contains a fat edge.*

Proof. Consider a triangle with vertices u, v and w. Observe that $N(u) \cap N(v) \cap N(w) = \emptyset$. Indeed, a vertex adjacent to u, v and w would have a cycle in its neighborhood. Hence,

$$n \geq |N(u) \cup N(v) \cup N(w)| = d_u + d_v + d_w - d_{uv} - d_{uw} - d_{vw}. \tag{7}$$

Since $d_u + d_v + d_w \geq n + 3\sqrt{n}$, at least one of the quantities d_{uv}, d_{uw} and d_{vw} is greater than or equal to $\sqrt{n}$. $\qquad\square$

Lemma 13. *Every fat edge of $G \in \mathrm{WFG}_n$ belongs to a triangle with two lean edges.*

Proof. Let (u,v) be a fat edge. Suppose, on the contrary, that for each $x \in N(u) \cap N(v)$, at least one of the edges (u,x) and (v,x) is fat. This implies, that

$$\sum_{x \in N(u) \cap N(v)} ((d_{ux} - 1) + (d_{vx} - 1)) \geq |N(u) \cap N(v)|(\sqrt{n} - 1) \geq n - d_{uv}. \tag{8}$$

Consider all the triangles of G of the forms uxy and vxy, where $x \in N(u) \cap N(v)$ and $y \notin N(u) \cap N(v)$, $y \neq u, v$. The number of these triangles is exactly the left-hand side of (8). However, all of these triangles have a distinct third vertex outside $(N(u) \cap N(v)) \cup \{u,v\}$, so their number is at most $n - 2 - d_{uv}$, contradicting (8). Indeed, for example, if uxy and $ux'y$ are triangles with $x, x' \in N(v)$, the cycle $xyx'v$ forms a wheel with center u. $\qquad\square$

Proof of Theorem 3. Let G be a graph in WFG_n with $d_u \geq (2/5)n + 17/5$ for each vertex u. For $n \geq 96$, Lemma 12 implies that each triangle contains a fat edge.

We claim that the fat edges form a matching.

Let m be the maximum number of fat edges incident to a vertex of G. We shall prove that $m = 1$. Consider a lean edge (v,w). By the above argument, any triangle containing (v,w) must contain a fat edge. Since there are at most $2m$ fat edges incident to either v or w, we obtain that

$$d_{vw} \leq 2m \tag{9}$$

for any lean edge (v,w).

Let u be a vertex with m fat edges incident to it, say $(u,v_1), (u,v_2), \ldots, (u,v_m)$. Since G is wheel-free, there exist at most $m - 1$ triangles containing two fat edges incident to u. By summing d_{uv_i} over $i = 1, \ldots, m$, we count every triangle containing u at most once, except for those containing two fat edges incident to u. Hence,

$$\sum_{i=1}^{m} d_{uv_i} \leq d_u - 1 + (m - 1). \tag{10}$$

The left-hand side is at least $m\sqrt{n}$, hence we get $m < \sqrt{n} + 1$. For any fat edge (u,v), Lemma 13 implies that there is a triangle uvw with two lean edges. Then (7) and (9) give

$$d_{uv} \geq d_u + d_v + d_w - n - 4m. \tag{11}$$

We derive that

$$\sum_{i=1}^{m} d_{uv_i} \geq md_u + \sum_{i}^{m} d_{v_i} + \sum_{i}^{m} d_{w_i} - nm - 4m^2. \tag{12}$$

Comparing (10) and (12), we get $m - 2 \geq (3m - 1)\delta - nm - 4m^2$. If $2 \leq m < \sqrt{n} + 1$, $n \geq 500$, $\delta > (2n + 16)/5$, which leads to a contradiction. If $100 \leq n < 500$, we can use $d_{vw} < \sqrt{n}$ instead of (9) to get a contradiction in exactly the same way. Therefore m must be equal to 1, i.e., the fat edges form a matching.

Every triangle contains a fat edge, so, by Proposition 11, there are at most $n^2/8$ triangles in G. $\square$

6. Proof of the upper bound

Here we prove Theorem 2 by induction on n. If $n \leq 9126$, it follows from Theorem 6. Let $c = 1/2457 = 1/7 - 1/7.02$.

If $\Delta > ((3/7) + 4c) n$ the proof is similar to the proof of Theorem 6. Either there exists a vertex cover of $G[N(w)]$ of cardinality less than or equal to $\Delta/9$, in which case $t(G) < 9/64n^2 < (1/7 - c)n^2$, or there exist two adjacent vertices u and v in $N(w)$ whose total degree in $G[N(w)]$ is at most 10. In the latter case, Lemma 10 implies that we destroy less than

$$\left(\frac{4}{7} - 4c\right) n + 8$$

triangles by deleting u and v. The claim therefore follows by induction.

If $|E(G)| < (3/2) ((1/7) - c) n^2$, the result follows from (1).

Assume now that

Assumption 1. $\Delta \leq \left(\frac{3}{7} + 4c\right) n$,

Assumption 2. $|E(G)| \geq \frac{3}{2} \left(\frac{1}{7} - c\right) n^2$.

If G satisfies the hypotheses of Theorem 3, we are done. Otherwise, we must have $\delta \leq (2/5)n + 16/5$. Consider the graph G' obtained from G by repeatedly deleting a vertex of minimum degree until

$$\delta(G') > \frac{2}{5}|V(G')| + \frac{16}{5}.$$

Let $v_0, v_1, \ldots, v_{t-1}$ be the sequence of vertices that we delete, and let G_i be the graph obtained from G by deleting $\{v_0, \ldots, v_{i-1}\}$. In particular, $G_0 = G$ and $G_t = G'$. By definition, we have that

$$d_{v_i}(G_i) \leq \frac{2}{5}|V(G_i)| + \frac{16}{5}$$

and

$$d_{v_i}(G) \leq \frac{2}{5}(n - i) + \frac{16}{5} + f_i,$$

where f_i denotes the number of edges in G joining v_i to $\{v_0, \ldots, v_{i-1}\}$. Let $\varepsilon = t/n$. In order to give an upper bound on ε, we consider the number of edges of G. Using Assumptions 1 and 2, we derive

$$3\left(\frac{1}{7} - c\right) \leq \frac{2|E(G)|}{n^2} = \frac{1}{n^2} \sum_{v \in V} d_v(G)$$

$$\leq \left(\frac{3}{7} + 4c\right)(1 - \varepsilon) + \frac{1}{n^2} \sum_{i=0}^{t-1} \left(\frac{2}{5}n - \frac{2}{5}i + \frac{16}{5} + f_i\right)$$

$$= \left(\frac{3}{7} + 4c\right)(1 - \varepsilon) + \frac{2}{5}\varepsilon - \frac{1}{5}\varepsilon^2 + \frac{|E(G[\{v_0, \ldots, v_{t-1}\}])|}{n^2} + \frac{17}{5}\frac{\varepsilon}{n}$$

$$\leq \frac{3}{7} + 4c - \left(\frac{1}{35} + 4c\right)\varepsilon + \frac{1}{20}\varepsilon^2 + \frac{73}{20}\frac{\varepsilon}{n},$$

where we have used (3), *i.e.* that the number of edges of $G[\{v_0, \ldots, v_{t-1}\}]$ is at most $(\varepsilon n)^2/4 + \varepsilon n/4$. For n sufficiently large, say larger than $n_0 = 9126$, and given the value of c, the above inequality can be seen to imply $\varepsilon < 0.12$. Since G' satisfies the hypotheses of Theorem 3, we have $t(G') \leq (n^2/8)(1 - \varepsilon)^2$. Moreover,

$$t(G) - t(G') \leq \sum_{i=0}^{t-1} (d_{v_i}(G_i) - 1) \leq \sum_{i=0}^{t-1} \left(\frac{2}{5}n - \frac{2}{5}i + \frac{11}{5}\right) = \frac{2}{5}\varepsilon n^2 - \frac{1}{5}\varepsilon^2 n^2 + \frac{12}{5}\varepsilon n.$$

Therefore,

$$t(G) \leq \left(\frac{1}{8}(1 - \varepsilon)^2 + \frac{2}{5}\varepsilon - \frac{\varepsilon^2}{5}\right)n^2 + \frac{12}{5}\varepsilon n < \frac{1}{7.02}n^2 + 5n.$$

□

Remark. The result can be improved to $t(n) \leq n^2/7.03 + O(n)$, as shown below. Let $c = 1/1540$. We execute the first two steps of the previous proof, so from now on we may suppose that Assumptions 1 and 2 hold. We delete from G a vertex x if $d_x < (n-i)/3 + \sqrt{n}$. Lemma 13 implies that we obtain a graph where each triangle contains a fat edge. Delete a fat-lean-lean triangle, uvw, if $d_u + d_v + d_w \leq 1.2(n-i) + 12$. We get that for each fat edge, $d_{uv} > 0.2(n - i) + O(1)$. If there exists two adjacent fat edges, (u, x_1) and (u, x_2), for some fat-lean-lean triangles ux_1w_1 and ux_2w_2 we get that $d_u + d_{x_1} + d_{x_2} + d_{w_1} + d_{w_2} \leq 2(n-i) + O(1)$. Delete these 5 points and repeat these steps.

The upper bound for the number of edges implies $\varepsilon \leq .235$. At each of the above steps, by deleting ℓ vertices ($1 \leq \ell \leq 5$) we have destroyed only $\ell(n - i)/3 + O(1)$ triangles at most. We get that $t(G)/n^2 \leq (1/8)(1 - \varepsilon)^2 + (1/3)\varepsilon - (1/6)\varepsilon^2 + o(1)$. □

7. Upper bound for regular graphs

In this last section, we prove Theorem 4, the upper bound for d-regular graphs.

If $d \leq (12/31)n$, equation (2) implies that $t(G) < n^2/7.75$.

If $d > (2/5)n + 16/5$, then Theorem 3 implies that $t(G) \leq n^2/8$, (for $n \geq 100$).

Assume now that $(12/31)n < d \leq (2/5)n + 16/5$. From (12) we can deduce that any vertex of G is incident to at most two fat edges. (The details are left to the reader.) If no vertex is incident to two fat edges, the result follows from Proposition 11. Otherwise, let r be the maximum, over all vertices u, of the number of triangles containing a fat edge that is incident to u, and let w be a vertex attaining this maximum. From the definition of r, we must have $d_{uv} \leq r$ for all edges $(u,v) \in E$. Moreover, since a fat edge is contained in at least $3d - n - 8$ triangles, by (11), and since there exists a vertex incident to two fat edges, r must be at least $6d - 2n - 17$. Let $S = V - \{w\} - N(w)$, and let T_i for $i = 0,\ldots,3$ be the number of triangles having exactly i vertices of S. Clearly, $T_0 \leq d - 1$, since any such triangle must involve w. Moreover, by summing $d_v - 1$ over all vertices $v \in S$, we observe that $3T_3 + 2T_2 + T_1 \leq (d-1)(n-d)$. Finally, we claim that $T_1 \leq r(d-r) + O(n)$. To see this, observe that the number of lean edges contained in $N(w)$ is at least $r - 1$ if w is incident to just one fat edge, or $r - 2$ if w is incident to two fat edges. Thus the number of fat edges contained in $N(w)$ is at most $d - r + 1$. To compute an upper bound on T_1, we sum $d_{uv} - 1$ over all edges contained in $N(w)$ (the -1 term comes from the fact that we do not need to count triangles involving vertex w):

$$T_1 \leq 3(r-2) + (r-1)(d-r+1) = r(d-r) + O(n),$$

since lean edges are contained in at most 4 triangles by (9), while fat edges are contained in at most r triangles by definition of r. Therefore,

$$t(G) \leq T_0 + \frac{1}{2}(T_1 + 2T_2 + 3T_3) + \frac{1}{2}T_1 \leq \frac{1}{2}(d(n-d) + r(d-r)) + O(n).$$

When $r \geq 6d - 2n - 17$ ($\geq d/2$), the right-hand side is maximized for $r = 6d - 2n - O(1)$, giving $t(G) \leq (1/2)(d(n-d) + (6d-2n)(2n-5d)) + O(n)$. Under the constraint $(12/31)n \leq d$, this is in turn maximized for $d = (12/31)n + O(1)$, proving that $t(G) \leq n^2/7.75 + O(n)$. $\square$

8. Wheel-free triple systems

A family of 3-element sets is called *wheel-free* if it contains no k triples isomorphic to $\{\{0,1,2\},\{0,2,3\},\ldots,\{0,i,(i+1)\},\ldots\{0,k,1\}\}$, where $k \geq 3$ is an arbitrary integer. For example, the vertex sets of the triangles in a wheel-free graph form such a system. But the general case is different. Let $\text{ex}(n;W)$ denote the largest cardinality of a wheel-free triple system on an n-element set. V. T. Sós, Erdős and Brown [9] proved that $\lim_{n\to\infty} \text{ex}(n;W)/n^2 = 1/3$.

For further problems of this type, see, for example, [4] and the references therein.

Another interesting question is whether (and how) our results can be extended to $\{W_3, W_4\}$-free graphs, or even more generally for $(5,8)$-free graphs.

9. Acknowledgements

The authors are indebted to P. Haxell for helpful remarks and for improvements to the construction in Section 2. We are also grateful for the referees' conscientious reading.

References

[1] Erdős, P. (1988) Problems and results in combinatorial analysis and graph theory. *Discrete Math.*, **72** 81–92.

[2]   Erdős, P. (1965) Extremal problems in graph theory. In: Fiedler, M. (ed.) *Theory of Graphs and its Appl.*, (Proc. Symp. Smolenice, 1963), Academic Press, New York 29–36.

[3]   Erdős, P. and Simonovits, M. (1966) A limit theorem in graph theory. *Studia Sci. Math. Hungar.* **1** 51–57.

[4]   Frankl, P. and Rödl, V. (1988) Some Ramsey-Turán type results for hypergraphs. *Combinatorica* **8** 323–332.

[5]   Fronček, D. (1990) On one problem of B. Zelinka (manuscript).

[6]   Golberg, A. I. and Gurvich, V. A. (1987) On the maximum number of edges for a graph with n vertices in which every subgraph with k vertices has at most ℓ edges. *Soviet Math. Doklady* **35** 255–260.

[7]   Griggs, J. R., Simonovits, M. and Thomas, G. R. (1993) Maximum size graphs in which every k-subgraph is missing several edges (manuscript).

[8]   Haxell, P. (1993) PhD Thesis, University of Cambridge, Cambridge, England.

[9]   Sós, V. T., Erdős, P. and Brown, W. G. (1973) On the existence of triangulated spheres in 3-graphs, and related problems. *Periodica Math. Hungar.* **3** 221–228.

[10] Zelinka, B. (1983) Locally tree-like graphs. *Čas. pěst. mat.* **108** 230–238.

[11] Zhou, B. (to appear) A counter example to a conjecture of Gallai. *Discrete Math.*

Blocking Sets in $SQS(2v)$

MARIO GIONFRIDDO[†], SALVATORE MILICI[†]
and ZSOLT TUZA[‡]

[†]Dipartimento di Matematica, Città Universitaria, Viale A. Doria 6, 95125 Catania, Italy.
[‡]Computer and Automation Institute, Hungarian Academy of Sciences,
H-1111 Budapest, Kende u. 13–17, Hungary

A Steiner quadruple system $SQS(v)$ of order v is a family $\mathscr{B}$ of 4-element subsets of a v-element set V such that each 3-element subset of V is contained in precisely one $B \in \mathscr{B}$. We prove that if $T \cap B \neq \emptyset$ for all $B \in \mathscr{B}$ (i.e., if T is a *transversal*), then $|T| \geq v/2$, and if T is a transversal of cardinality exactly $v/2$, then $V \setminus T$ is a transversal as well (i.e., T is a *blocking set*). Also, in respect of the so-called 'doubling construction' that produces $SQS(2v)$ from two copies of $SQS(v)$, we give a necessary and sufficient condition for this operation to yield a Steiner quadruple system with blocking sets.

1. Introduction

A *hypergraph* $\mathscr{H}$ is a pair $(V, \mathscr{B})$, where V is a finite v-set (= a set of v elements) and $\mathscr{B} \neq \emptyset$ is a family of nonempty subsets $B \subseteq V$ such that $\bigcup_{B \in \mathscr{B}} B = V$. The integer $v = |V|$ is the *order* of $\mathscr{H}$; the elements of V and $\mathscr{B}$ are called the *vertices* (or *points*) and the *blocks* (or *edges*) of the hypergraph, respectively. If $|B| = r(\mathscr{H}) = r$ for each block $B \in \mathscr{B}$, then $\mathscr{H}$ is called an *r-uniform* hypergraph, or a *uniform hypergraph of rank* $r(\mathscr{H})$.

Given a hypergraph $\mathscr{H} = (V, \mathscr{B})$ and a nonempty subset $W \subseteq V$, $\ll W \gg$ denotes the subhypergraph whose blocks are the blocks of $\mathscr{H}$ contained in W (i.e., subsets B of W such that $B \in \mathscr{B}$). The points of $\ll W \gg$ are those contained in the blocks of $\ll W \gg$. The hypergraph $\ll W \gg$ is called the *subhypergraph of* $\mathscr{H}$ *induced by* W.

Given the complete graph K_v on v vertices (v even), a *1-factor* F_i of K_v is a set of $v/2$ pairwise disjoint edges of K_v. A *factorization* (also called a *1-factorization*) $\mathscr{F}$ of K_v is a set of $v - 1$ 1-factors of K_v such that each edge occurs in precisely one F_i. A factorization

† Research supported by MURST and GNSAGA, CNR.
‡ Research supported in part by the OTKA Research Fund of the Hungarian Academy of Sciences, grant no. 2569, and in part by C.N.R. Italia while the author visited Università di Catania.

$\mathscr{F}$ of the complete *bipartite* graph $K_{v,v}$ of order $2v$ (v arbitrary) is a set of v 1-factors (of v edges each) in $K_{v,v}$ containing no edge twice.

A $t - (v, k, \lambda)$ *design* is a uniform hypergraph $(V, \mathscr{B})$ of rank k such that every t-subset of V is contained in exactly λ blocks of $\mathscr{B}$. If $\lambda = 1$, a $t - (v, k, 1)$ design is also called a *Steiner system*. For $t = 2$ and $k = 3$, a Steiner system is called a *Steiner triple system*, abbreviated by $STS(v)$. For $t = 3$ and $k = 4$, a Steiner system is called a *Steiner quadruple system*, $SQS(v)$ for short. Sometimes a hypergraph will be denoted by $\mathscr{D}$ or by $\mathscr{S}$ (instead of $\mathscr{H}$) if we want to emphasize that it is a design or a Steiner system, respectively.

Concerning the construction of Steiner systems, Hanani [5] proved in 1960 that an $SQS(v)$ exists if and only if $v \equiv 2$ or 4 (mod 6), while it is well known that an $STS(v)$ exists if and only if $v \equiv 1$ or 3 (mod 6).

Given a hypergraph $\mathscr{H} = (V, \mathscr{B})$, a *transversal* of $\mathscr{H}$ is a set $T \subseteq V$ such that $T \cap B \neq \emptyset$ for each $B \in \mathscr{B}$. The *transversal number* $\tau(\mathscr{H})$ of $\mathscr{H}$ is defined as the minimum number of points in a transversal. Moreover, a *blocking set* is a set $T \subseteq V$ such that T and $V \setminus T$ are both transversals. Hence, T is a blocking set of $\mathscr{H} = (V, \mathscr{B})$ if and only if

$$B \cap T \neq \emptyset \quad \text{and} \quad B \setminus T \neq \emptyset \quad \text{for each } B \in \mathscr{B}.$$

Very few facts are known so far about the existence of blocking sets in Steiner systems. Let us recall three important results. In what follows, $\mathscr{P}_k(V)$ will denote the set of all k-subsets of a finite nonempty set V.

Theorem 1.1. (Tallini [8], Berardi and Beutelspacher [1]) *If $\mathscr{D} = (V, \mathscr{B})$ is a $2 - (v, 3, \lambda)$ design, there exist blocking sets in $\mathscr{D}$ only for $v = 4$ and $\mathscr{B} = \mathscr{P}_3(V)$. In this case the blocking sets are all the 2-subsets of V.*

As an important particular case of Theorem 1.1, for Steiner systems we obtain the following corollary.

Corollary 1.1. *There are no blocking sets in Steiner triple systems $STS(v)$.*

Theorem 1.2. (Tallini [7], Berardi and Beutelspacher [1]) *If $\mathscr{D} = (V, \mathscr{B})$ is a $3 - (v, 4, \lambda)$ design and T is a blocking set in $\mathscr{D}$, then either $\mathscr{D}$ has an even order v and $|T| = v/2$, or $v = 5$, λ is even, $\mathscr{B} = \mathscr{P}_4(V)$ and $|T| \in \{2, 3\}$.*

We can see that Theorem 1.2 does not exclude the possibility of the existence of blocking sets in $3 - (v, 4, \lambda)$ designs. On the other hand, a particular case of Theorem 1.2 yields the following corollary.

Corollary 1.2. *If T is a blocking set in a Steiner quadruple system $SQS(v)$, then $|T| = v/2$.*

There are only a very few known examples of blocking sets in Steiner quadruple systems, or in $3 - (v, 4, \lambda)$ designs for $\lambda > 1$. There exist blocking sets in the unique $SQS(8)$ (see [7]), as well as in the unique $SQS(10)$ (see [3]). On the other hand, the four systems $SQS(14)$ have no blocking sets (see [6]).

Regarding the affine Galois spaces $AG(r, 2)$ of dimension r and order 2 (*i.e.*, all lines having two points), we obtain Steiner quadruple systems $SQS(2r)$ when considering the *planes* of $AG(r, 2)$ as blocks of the system. For these particular designs, denoted $AG_2(r, 2)$ (with the subscript referring to the dimension of each block), the following holds.

Theorem 1.3. (Tallini [8]) *If $r \geq 4$, then $AG_2(r, 2)$ contains no blocking set.*

In [4] and [6], the authors study the existence of blocking sets in the systems $SQS(v)$. In particular, in [2], Doyen and Vandensavel give $SQS(v)$ with blocking sets. It seems that there are no other known results for the existence of blocking sets in Steiner quadruple systems.

In this paper we first prove a strong relationship between transversals and blocking sets of a Steiner quadruple system (Theorem 2.1). Then we investigate those systems $SQS(v)$ that can be obtained by the 'doubling construction' (see Section 3 for the definition) and characterize the existence of blocking sets in them, giving a method to determine whether or not $SQS(2v)$ has a blocking set (Section 4). To derive these results, we need to study the properties of factorizations of complete graphs (Section 3).

2. Transversals and blocking sets in quadruple systems

In this section we prove a result on the relationship between transversals and blocking sets of Steiner quadruple systems. Note that the first part of the following theorem also implies Corollary 1.2, since if T is a blocking set, T and its complement $V \setminus T$ are two disjoint transversals of the system.

Theorem 2.1. *Let $\mathscr{S} = (V, \mathscr{B})$ be an $SQS(v)$, $v \geq 8$.*

(i) The transversal number of $\mathscr{S}$ is at least $v/2$.
(ii) A set $T \subset V$, $|T| = v/2$ is a blocking set if and only if T is a transversal.

Proof. For any subset $W \subseteq V$, let $x_i(W)$ denote the number of blocks of $\mathscr{B}$ having exactly i points in common with W :

$$x_i(W) = |\{B \in \mathscr{B} : |B \cap W| = i\}|.$$

Let us prove (ii) first. The 'only if' part is clear by definition. Considering the converse statement, let T be a transversal of $\mathscr{S}$ such that $|T| = v/2$, and let $x_i = x_i(V \setminus T)$. Since T is a transversal, and since any three points of the complement of T are in a unique block, we obtain

$$x_4 = 0, \quad x_3 = \binom{v/2}{3} = \frac{v-4}{6} \binom{v/2}{2}.$$

Furthermore, each pair $x, y \in V$ is contained in exactly $(v-2)/2$ blocks, hence

$$x_2 + 3x_3 = \frac{v-2}{2} \binom{v/2}{2}.$$

Finally, a point $x \in V$ is contained in exactly $(v-1)(v-2)/6$ blocks; therefore,

$$x_1 + 2x_2 + 3x_3 = \frac{(v-1)(v-2)}{6} \frac{v}{2}.$$

Counting the number of blocks that have a nonempty intersection with $V \setminus T$, we obtain

$$
\begin{aligned}
x_1 + x_2 + x_3 + x_4 &= (x_1 + 2x_2 + 3x_3) - (x_2 + 3x_3) + x_3 \\
&= \frac{(v-1)(v-2)}{6} \frac{v}{2} - \frac{v-2}{2}\binom{v/2}{2} + \frac{v-4}{6}\binom{v/2}{2} \\
&= \frac{1}{4}\binom{v}{3},
\end{aligned}
$$

which is equal to the number of blocks of $\mathscr{S}$. Thus, $V \setminus T$ is a transversal, too, and T is a blocking set of $\mathscr{S}$.

In order to prove (i), one can apply a similar computation, which is just a little more complicated than the proof of (ii) above. Now we let $x_i = x_i(T)$. Assuming that T is a transversal, we obtain

$$
\begin{aligned}
x_0 &= 0, \\
x_1 + 2x_2 + 3x_3 + 4x_4 &= \frac{(v-1)(v-2)}{6}|T|, \\
x_2 + 3x_3 + 6x_4 &= \frac{v-2}{2}\binom{|T|}{2}, \\
x_3 + 4x_4 &= \binom{|T|}{3}.
\end{aligned}
$$

Consequently,

$$
\begin{aligned}
\frac{1}{4}\binom{v}{3} &= x_1 + x_2 + x_3 + x_4 \\
&= (x_1 + 2x_2 + 3x_3 + 4x_4) - (x_2 + 3x_3 + 6x_4) + (x_3 + 4x_4) - x_4 \\
&\le \frac{1}{3}\binom{v-1}{2}|T| - \frac{v-2}{2}\binom{|T|}{2} + \binom{|T|}{3}
\end{aligned}
$$

should hold. One can observe that for $v \ge 3 + \sqrt{5}$ the right-hand side is an increasing function of $|T|$ (as its first derivative is $1/2(v/2 - |T|)^2 + 1/24(v^2 - 6v + 4) \ge 0$), and equality holds if $|T| = v/2$ and $x_4 = 0$. Thus, if T is a transversal, it has to contain at least $v/2$ points.   □

3. Factorizations and separation number

In this section we study 1-factorizations of the complete graph K_{2m}. Our motivation to do so is the following well-known operation on Steiner quadruple systems.

Doubling construction. Let $\mathscr{S}' = (V', \mathscr{B}')$ and $\mathscr{S}'' = (V'', \mathscr{B}'')$ be Steiner quadruple systems of the same even order $v = 2m$, $V' \cap V'' = \emptyset$, and let $\mathscr{F}' = \{F_1', F_2', \ldots F_{v-1}'\}$ and $\mathscr{F}'' = \{F_1'', F_2'', \ldots, F_{v-1}''\}$ be arbitrary factorizations of the complete graph of order $2m$

on vertex sets V' and V'', respectively. Define the quadruple system $\mathscr{S} = (V, \mathscr{B})$ with $V = V' \cup V''$ and

$$\mathscr{B} = \mathscr{B}' \cup \mathscr{B}'' \cup \{e' \cup e'' : e' \in F'_i, e'' \in F''_i, 1 \le i \le v - 1\}.$$

One can see that $\mathscr{S}$ is a Steiner quadruple system. For this new system, we use the notation $\mathscr{S} = \mathscr{S}' \times \mathscr{S}''$. Also, if the factorizations $\mathscr{F}', \mathscr{F}''$ are given, we write $\mathscr{S}' = (V', \mathscr{B}', \mathscr{F}')$ and $\mathscr{S}'' = (V'', \mathscr{B}'', \mathscr{F}')$. Blocks of the form $e' \cup e''$ are referred to as *crossing blocks*.

Turning now to the study of factorizations of complete graphs, let us introduce some definitions.

Definition. A factorization $\mathscr{F} = \{F_1, \ldots, F_{2m-1}\}$ of a complete graph K_{2m} is called *decomposable* if there is a partition $X' \cup X''$ of the vertex set, $|X'| = |X''| = m$, such that for each i, $1 \le i \le 2m - 1$, either $|e \cap X'| = 1$ for all $e \in F_i$, or $|e \cap X'| \in \{0, 2\}$ for all $e \in F_i$. In this case we call $\{X', X''\}$ a *decomposition* of $\mathscr{F}$. Note that $m > 1$ must be even here in order to admit $|e \cap X'| \in \{0, 2\}$ for some F_i.

Definition. An equipartition $\{X', X''\}$ *separates* a factor F_i if there is an edge $e \in F_i$ such that $|e \cap X'| \ne 1$. (We sometimes say that the *set* X' separates F_i if the vertex set is understood.)

Lemma 3.1. *Let $\mathscr{F}$ be a factorization of K_{2m}. If an equipartition $\{X', X''\}$ is not a decomposition of $\mathscr{F}$, then it separates at least m factors of $\mathscr{F}$; otherwise it separates precisely $m - 1$ factors of $\mathscr{F}$.*

Proof. Since $|X'| = m$, each factor has at most $m/2$ edges in X'. Suppose that $\{X', X''\}$ separates at most $m - 1$ factors F_i. Each of these factors can contain at most $m/2$ edges in X'. Since we have a factorization of K_{2m}, each of the $m(m - 1)/2$ edges of X' has to occur in some F_i (in precisely one of them), and this implies that each F_i should contain exactly $m/2$ (or 0) edges in X', and in X'' as well. In this case, however, the F_i separated by X' form a factorization of X' and also of X'' (because the other $m/2$ edges of F_i not contained in X' must be contained in X''), so that $\{X', X''\}$ is a decomposition of $\mathscr{F}$ whenever it separates just $m - 1$ factors F_i. $\square$

We note that if $\{X', X''\}$ separates a factor F of the vertex set, there are at least *two* edges $e', e'' \in F$ such that $e' \subset X'$ and $e'' \subset X''$. The reason is that denoting by n' (by n'') the number of edges of F that are in X' (in X''), F contains precisely $m - 2n'$ edges having just one point in X', so that $n' = n''$ holds (and $n' + n'' > 0$ by the definition of separation).

For small values of $2m$, the situation is as follows. There is a unique 1-factorization of K_4 and of K_6. The former is (trivially) decomposable; in fact each equipartition of the vertex set is a decomposition. On the other hand, the factorization of K_6 is indecomposable (since $m = 3$ is odd). We shall see later that the existence or nonexistence of a decomposable factorization depends on the *parity* of m only.

The nonexistence of blocking sets in some class of quadruple systems will be proved by applying the following observation.

Lemma 3.2. _For every $n = 2m$ ($m \geq 3$), K_n has an indecomposable factorization._

Proof. Let $V = \{v_0, v_1, \ldots, v_{n-1}\}$ be the vertex set of K_n. Define a factorization $\mathscr{F} = \{F_1, \ldots F_{n-1}\}$ as follows: $F_i = \{v_0 v_i\} \cup \{v_{i+j} v_{i-j} : 1 \leq j \leq (n/2) - 1\}$ for $i = 1, 2, \ldots, n - 1$, where subscript addition is taken modulo $n - 1$.

We prove that $\mathscr{F}$ is indecomposable. Suppose on the contrary that $\{X', X''\}$ decomposes $\mathscr{F}$. Assume $v_0 \in X''$. Then X' has its $n/2$ vertices in $\{v_1, \ldots, v_{n-1}\}$ so that it contains two _consecutive_ elements v_k and v_{k+1} for some k; let $v_k v_{k+1} \in F_i$. Then F_i contains a factor of X', i.e. $v_{i+j} \in X'$ if and only if $v_{i-j} \in X'$ (and $v_i \notin X'$). Note further that this symmetry property holds in each edge class of the factorization of X'.

Consider now the vertex v_l 'next' to v_{k+1} in X'. By 'next' we mean that $v_l \in X'$ but v_{k+1} and v_l are the only elements of X' in the set $\{v_{k+1}, v_{k+2}, \ldots, v_{l-1}, v_l\}$. The edge $v_k v_l$ is contained in some factor, F_j say. Then $k - j \equiv j - l \pmod{n - 1}$. Since F_j is a factor in X', v_{k+1} is also incident to some edge of F_j. By the symmetry of F_j, the edge should be $v_{k+1} v_{l-1}$, contradicting the assumption that v_l is next to v_{k+1}. Thus, $\mathscr{F}$ is not decomposable. $\square$

Our next objective is to give a more explicit description of quadruple systems, obtained by a doubling construction, that have a blocking set. The structures of these $\mathscr{S}' \times \mathscr{S}''$ are determined by the factorizations $\mathscr{F}'$ and $\mathscr{F}''$ chosen on the vertex sets V' and V'', respectively, and by the permutation that tells which factor F_i' is paired with which F_i''.

Fix a vertex set V of size $2m$ and a set $X' \subset V$ of size m. For a factorization $\mathscr{F}$ of K_{2m}, denote by $s(\mathscr{F})$ the _separation number_ of $\mathscr{F}$, defined as the the number of edge classes F_i separated by $\{X', V \setminus X'\}$. We have seen in Lemma 3.1 that $s(\mathscr{F}) \geq m - 1$ for all $\mathscr{F}$. We note that $s(\mathscr{F}) = m$ can also hold: for instance, we can obtain such an $\mathscr{F}$ by taking two isomorphic factorizations $\mathscr{F}'$ and $\mathscr{F}''$ on m points, say on $\{v_1', \ldots, v_m'\}$ and $\{v_1'', \ldots, v_m''\}$, when m is even, together with the factors $E_j = \{v_i' v_{i+j}'' : 0 \leq i \leq m - 1\}$. So far this factorization has separation number $m - 1$. Taking two isomorphic edge classes F' and F'', $F' \cup F'' \cup E_0$ consists of $m/2$ cycles of length four, and this union can be modified to obtain another factorization of K_{2m}, which then has separation number m.

The following result characterizes the cases when K_{2m} has a decomposable factorization. (For applications to quadruple systems, we shall not need the negative statement for m odd.)

Theorem 3.1. _A complete graph of order $2m$ has a decomposable factorization if and only if m is even._

Proof. The 'only if' part follows from the fact that if $\{X', V \setminus X'\}$ is a decomposition of some factorization $\mathscr{F}$, then some factors $F_i \in \mathscr{F}$ induce factors of X' as well, so $|X'|$ must be even.

To prove existence for m even, let $m = 2k$ and denote the vertices by $v_0, v_1, \ldots, v_{4k-1}$, taking the subscripts modulo $4k$ (i.e., $v_{i+4k} = v_i$ for all i). Define $2k - 1$ factors F_i as follows. For $0 \leq i \leq 2k - 1$ let $F_i = \{v_{i-j} v_{i+1+j} : 0 \leq j \leq 2k - 1\}$. One can see that these F_i cover (precisely once) those edges $v_s v_t$ for which $s - t$ is odd. Thus, we can partition the vertex set into two classes $X' = \{v_{2i+1} : 0 \le i \le 2k - 1\}$ and $X'' = \{v_{2i} : 0 \le i \le 2k - 1\}$, with the property that an edge e is contained in some F_i ($0 \le i \le 2k - 1$) if and only if $|X' \cap e| = 1$. Consequently, if $\{F_0, \ldots, F_{2k-1}\}$ is completed to a factorization $\mathscr{F}$ of K_{2m} in *any* fashion, then $s(\mathscr{F}) = m - 1$ holds and X' decomposes $\mathscr{F}$. One possible way is to take the factors of even subscripts constructed in the proof of Lemma 3.2. □

4. Blocking sets in factorized quadruple systems

Given an even positive integer w, let $\mathscr{F}$ be a factorization of K_w on vertex set $W = \{1, 2, \ldots, w\}$ and let $F_1, F_2, \ldots, F_{w-1}$ be its 1-factors. Let $\mathscr{G}$ be another factorization of K_w on vertex set $W' = \{1', 2', \ldots, w'\}$, where $W \cap W' = \emptyset$, with 1-factors $G_1, G_2, \ldots, G_{w-1}$.

Bearing in mind the doubling construction described at the beginning of Section 3, each permutation α on $\{1, 2, \ldots, w - 1\}$ yields a family $\Gamma = \Gamma(\mathscr{F}, \mathscr{G}, \alpha)$ of 4-subsets of $W \cup W'$ such that

$$\{x, y, x', y'\} \in \Gamma \quad \text{if and only if} \quad \{x, y\} \in F_i \text{ and } \{x', y'\} \in G_{\alpha(i)}$$
$$\text{for some } i \in \{1, \ldots, w - 1\}.$$

The sets in Γ may then provide the collection of crossing blocks, since if $(W, \mathscr{B}_1)$ and $(W', \mathscr{B}_2)$ are two quadruple systems of order w, then the pair $(V, \mathscr{B})$ with $V = W \cup W'$ and $\mathscr{B} = \mathscr{B}_1 \cup \mathscr{B}_2 \cup \Gamma(\mathscr{F}, \mathscr{G}, \alpha)$ is an $SQS(2w)$.

Definition. For $v \equiv 0 \pmod 4$, we say that an $SQS(v)$ $\mathscr{S} = (V, \mathscr{B})$ is a *factorized Steiner quadruple system*, briefly an $FQS(v)$, if $\mathscr{B}$ contains a family $\Gamma(\mathscr{F}, \mathscr{G}, \alpha)$, where α is a permutation on $\{1, 2, \ldots, (v/2) - 1\}$ and $\mathscr{F}, \mathscr{G}$ are two factorizations of $K_{v/2}$ on two disjoint sets whose union is V.

Let $\mathscr{S} = (V, \mathscr{B})$ be an $FQS(v)$ containing a family $\Gamma(\mathscr{F}, \mathscr{G}, \alpha)$, and let

$$F = \bigcup_{\{x,y\} \in F_i} \{x, y\} \quad \text{and} \quad G = \bigcup_{\{x',y'\} \in G_j} \{x', y'\}$$

for some arbitrarily chosen $F_i \in \mathscr{F}$ and $G_j \in \mathscr{G}$.

Proposition 4.1. *The hypergraphs $\ll F \gg$ and $\ll G \gg$ are sub-$SQS(v/2)$'s of $\mathscr{S}$.*

Proof. If x, y, z are three distinct points of F (respectively of G), there exists exactly one block B in $\mathscr{S}$ that contains them. Let $B = \{x, y, z, u\} \in \mathscr{B}$. If $u \in F$, then $B \in \ll F \gg$ as required. Suppose that $u \in G$, and let $i \in \{1, 2, \ldots, v/2 - 1\}$ be such that $\{x, y\} \in F_i$. Then there exists an element $u' \in G$ satisfying $\{u, u'\} \in G_{\alpha(i)}$ and $\{x, y, u, u'\} \in \mathscr{B}$. Since $\mathscr{S}$ is a Steiner system, we obtain $u' = z$, $u' \in F$, a contradiction. □

Proposition 4.2. *An $FQS(v)$ exists if and only if $v \equiv 4$ or $8 \pmod{12}$.*

Proof. By Proposition 1, in every $FQS(v)$ there exist two subsystems $SQS(v/2)$. By the result of Hanani [5], $v/2 \equiv 2$ or $4 \pmod 6$, hence the condition $v \equiv 4$ or $8 \pmod{12}$ is necessary. Conversely, if $v \equiv 4$ or $8 \pmod{12}$, we can consider two quadruple systems $(W, \mathscr{F})$ and $(W', \mathscr{G})$ of order $v/2$, with $W \cap W' = \emptyset$. The doubling construction $SQS(w) \to SQS(2w)$ then yields an $FQS(v)$. $\square$

In what follows, $\mathscr{S}' = (V', \mathscr{B}', \mathscr{F}')$ and $\mathscr{S}'' = (V'', \mathscr{B}'', \mathscr{F}'')$ denote two Steiner quadruple systems of the same order, with $V' \cap V'' = \emptyset$ and with factorizations $\mathscr{F}'$ and $\mathscr{F}''$ on V', respectively on V''. The next assertion follows by definition.

Lemma 4.1. *If T is a blocking set of $\mathscr{S} = \mathscr{S}' \times \mathscr{S}''$, then $T \cap V'$ and $T \cap V''$ are blocking sets in $\mathscr{S}'$ and in $\mathscr{S}''$, respectively.*

Lemma 4.2. *If $\mathscr{S} = \mathscr{S}' \times \mathscr{S}''$ has a blocking set, then at least one of $\mathscr{F}'$ and $\mathscr{F}''$ is decomposable.*

Proof. Let T' (T'') be a blocking set in $\mathscr{S}'$ ($\mathscr{S}''$), $T' = T \cap V'$, $T'' = T \cap V''$ (where T is a blocking set of $\mathscr{S}$). If neither $\{T', V' \setminus T'\}$ decomposes $\mathscr{F}'$ nor $\{T'', V'' \setminus T''\}$ decomposes $\mathscr{F}''$, then, by Lemma 3.1, T' and T'' separate at least $v/2$ factors of $\mathscr{F}'$ and of $\mathscr{F}''$, respectively. Since $|\mathscr{F}'| = |\mathscr{F}''| = v - 1$, there must exist a subscript i such that T' separates F_i', and T'' separates F_i''. Say, $T' \supset e' \in F_i'$ and $T'' \supset e'' \in F_i''$ (such e' and e'' exist – see the comment after Lemma 3.1). Then $e' \cup e''$ is a block in $\mathscr{S}' \times \mathscr{S}''$, but $e' \cup e''$ is a subset of T, contradicting the assumption that T is a blocking set in $\mathscr{S}' \times \mathscr{S}''$. $\square$

Now we are in a position to prove two closely related results. The first presents an infinite family of designs *without* blocking sets, while the second characterizes under precisely what conditions the doubling construction yields a quadruple system *with* blocking sets.

Theorem 4.1. *Let $\mathscr{S}' = (V', \mathscr{B}')$ and $\mathscr{S}'' = (V'', \mathscr{B}'')$ be Steiner quadruple systems of order $v \geq 8$ on disjoint point sets. Then there are factorizations $\mathscr{F}'$ and $\mathscr{F}''$ of V' and of V'', respectively, such that the $SQS(2v)$, $\mathscr{S}' \times \mathscr{S}'' = (V', \mathscr{B}', \mathscr{F}') \times (V'', \mathscr{B}'', \mathscr{F}'')$ has no blocking set.*

Proof. By Lemma 3.2, we can take indecomposable factorizations $\mathscr{F}'$ and $\mathscr{F}''$ on V' and on V''. Then Lemma 4.2 guarantees that $\mathscr{S}' \times \mathscr{S}''$ has no blocking set. $\square$

The main result of this section is the following characterization theorem. Let us recall that the cardinality of a blocking set (if it exists) in a Steiner quadruple system $\mathscr{S}$ is half the order of $\mathscr{S}$.

Theorem 4.2. *Let $\mathscr{S} = (V, \mathscr{B})$ be a factorized quadruple system of order $4v$, such that $\mathscr{S} = \mathscr{S}' \times \mathscr{S}''$, $\mathscr{S}' = (V', \mathscr{B}', \mathscr{F}')$, $\mathscr{S}'' = (V'', \mathscr{B}'', \mathscr{F}'')$. Then a set $T \subset V$ of cardinality $2v$ is a blocking set if and only if the following five conditions are all satisfied:*

(i) The sets $T' = V' \cap T$ and $T'' = V'' \cap T$ are blocking sets in $\mathscr{S}'$ and $\mathscr{S}''$, respectively.
(ii) $\{T', V' \setminus T'\}$ decomposes $\mathscr{F}'$ or $\{T'', V'' \setminus T''\}$ decomposes $\mathscr{F}''$.
(iii) The separation numbers satisfy $s(\mathscr{F}') \leq v$ and $s(\mathscr{F}'') \leq v$.

(iv) If $s(\mathscr{F}') = s(\mathscr{F}'') = v - 1$, there is precisely one subscript i, $1 \leq i \leq 2v - 1$, with the property that F_i' is not separated by T' and F_i'' is not separated by T'', and for every $j \in \{1, \ldots, 2v - 1\} \setminus \{i\}$, F_j' is separated by T' if and only if F_j'' is not separated by T''.

(v) If $s(\mathscr{F}') = v - 1$ and $s(\mathscr{F}'') = v$ (or vice versa), then for each i, $1 \leq i \leq 2v - 1$, F_i' is separated by T' if and only if F_i'' is not separated by T''.

Proof. Sufficiency is easy to see: if the assumptions (i) through (v) are satisfied by $\mathscr{S}'$ and $\mathscr{S}''$, then the blocks of $\mathscr{B}'$ and $\mathscr{B}''$ are partitioned according to (i); and any other block of $\mathscr{S}$ is of the form $e' \cup e''$ where $e' \in F_i'$ and $e'' \in F_i''$ for some i. Since at least one of F_i' and F_i'' is *not* separated (by (iv) or by (v), according to the actual values of $s(\mathscr{F}')$ and $s(\mathscr{F}'')$), $e' \cup e''$ is indeed split into two nonempty parts by T.

The necessity of (i) and (ii) has already been verified in Lemmas 4.1 and 4.2. To prove (iii), (iv) and (v), recall that $\mathscr{F}'$ and $\mathscr{F}''$ contain $2v - 1 - s(\mathscr{F}')$ and $2v - 1 - s(\mathscr{F}'')$ factors not separated by T', respectively, T''. Moreover, the definition of $\mathscr{S}' \times \mathscr{S}''$ implies that in each of the $2v - 1$ pairs $\{F_i', F_i''\}$, at least one of the two factors F_i' and F_i'' is non-separated. Indeed, otherwise there would exist edges $e' \in F_i'$ and $e'' \in F_i''$ with $e' \subset T'$ and $e'' \subset T''$, and hence the block $e' \cup e'' \in \mathscr{S}' \times \mathscr{S}''$ would be a subset of T, contradicting the assumption that T is a blocking set. Consequently, the number s^* of subscripts i ($1 \leq i \leq 2v - 1$) such that neither F_i' nor F_i'' is separated, is equal to $((2v - 1 - s(\mathscr{F}')) + (2v - 1 - s(\mathscr{F}''))) - (2v - 1)$, i.e., $s^* = 2v - 1 - (s(\mathscr{F}') + s(\mathscr{F}''))$. On the other hand, $s(\mathscr{F}') \geq v - 1$ and $s(\mathscr{F}'') \geq v - 1$ hold by Lemma 3.1. Thus, $0 \leq s^* \leq 1$, implying the necessity of (iii), (iv), and (v). □

We have to note that Theorem 4.2 gives a necessary and sufficient condition for a *specified* T of size $2v$ to be a blocking set in a factorized $SQS(4v)$. Some factorizations may have several decompositions, and our characterization theorem says that *at least one* of them should satisfy the requirements (i)–(v) if we wish to obtain a blocking set.

We conclude this paper with a construction that provides an example in which the *FQS* obtained does have a blocking set, but, for some suitably chosen decomposition of the corresponding factorizations, the requirement that precisely one of F_i' and F_i'' is separated is satisfied by *none* of the pairs $\{F_i', F_i''\}$. This structure will be an $FQS(16k)$ by doubling some $FQS(8k)$, where the ordering of the factors in the originally isomorphic factorizations will be permuted. The 'twist' in the ordering will admit an $8k$-element set that separates all factors in the union.

Construction 4.1. Consider a set of $8k$ points, divided into four groups X_1, X_2, X_3, X_4, of cardinality $2k$ each. Let $\mathscr{F}_1$ be a factorization on X_1. Taking a one-to-one mapping between X_1 and X_i ($i = 2, 3, 4$), we find factorizations $\mathscr{F}_i$ on X_i isomorphic to $\mathscr{F}_1$. The isomorphic 1-factors of $\mathscr{F}_1 \cup \mathscr{F}_2$ (of $\mathscr{F}_3 \cup \mathscr{F}_4$) provide 1-factors on $X' = X_1 \cup X_2$ (on $X'' = X_3 \cup X_4$). Now $\mathscr{F}_1 \cup \mathscr{F}_2$ can be extended to a factorization $\mathscr{F}'$ of X' by factorizing the pairs that join X_1 with X_2 (call these pairs '$X_1 - X_2$ edges'; they form a complete bipartite graph on $4k$ vertices). The isomorphisms among the X_i extend $\mathscr{F}_3 \cup \mathscr{F}_4$ to a factorization $\mathscr{F}''$ of X'' such that $\mathscr{F}'' \cong \mathscr{F}'$. Taking the isomorphic mapping between $\mathscr{F}_2$ and $\mathscr{F}_3$ (but fixing $\mathscr{F}_1$ and $\mathscr{F}_4$ for the moment) we obtain $2k$ further factors that cover the $X_1 - X_3$ and $X_2 - X_4$ edges, and the isomorphism between X_1 and X_3 (while fixing X_2

and X_4) covers the remaining edges of $X_1 \cup X_4$ and $X_2 \cup X_3$. Denote this factorization by $\mathscr{F}$. Obviously, both of the partitions $(X_1 \cup X_2, X_3 \cup X_4)$ and $(X_1 \cup X_3, X_2 \cup X_4)$ decompose $\mathscr{F}$ (and so does $(X_1 \cup X_4, X_2 \cup X_3)$, also, but this third partition is not needed for our purpose).

We are going to observe that two disjoint isomorphic copies $\mathscr{F}(1)$ and $\mathscr{F}(2)$ of $\mathscr{F}$ (on $16k$ points in all) can be mixed in such a way that $X_1(1) \cup X_1(2) \cup X_3(1) \cup X_3(2)$ satisfies the requirement described in (iv), but on the other hand the set $X_1(1) \cup X_1(2) \cup X_2(1) \cup X_2(2)$ separates *all* factors, and therefore this latter union can by no means become a blocking set in the corresponding *FQS* obtained by doubling. To ensure this, we have to find a suitable permutation that defines a one-to-one correspondence between the factors of $\mathscr{F}(1)$ and $\mathscr{F}(2)$.

There are $4k$ factors between $X_1(i) \cup X_2(i)$ and $X_3(i) \cup X_4(i)$ $(i = 1, 2)$. A permutation of the desired properties for these factors is provided by an isomorphism $X_3(2) \leftrightarrow X_4(2)$. Indeed, $X_1(2) \cup X_3(2)$ separates precisely those factors whose pairs are not separated by $X_1(1) \cup X_3(1)$, while $X_1(i) \cup X_2(i)$ does not separate any factor of this type. Next, assign the $2k - 1$ factors of $\mathscr{F}_1(1) \cup \mathscr{F}_2(1)$ (of $\mathscr{F}_3(1) \cup \mathscr{F}_4(1)$) to all but one of the factors of $\mathscr{F}'(2) \setminus (\mathscr{F}_1(2) \cup \mathscr{F}_2(2))$ (of $\mathscr{F}''(2) \setminus (\mathscr{F}_3(2) \cup \mathscr{F}_4(2))$), and do the same for the factors of $\mathscr{F}_1(2) \cup \mathscr{F}_2(2)$ and $\mathscr{F}_3(2) \cup \mathscr{F}_4(2)$. There is just one factor left in $\mathscr{F}(1)$ and in $\mathscr{F}(2)$, and they are assigned to each other.

It can be verified that this permutation of the factors satisfies the properties given above.

References

[1] Berardi, L. and Beutelspacher, A. (to appear) *On blocking sets in some block designs.*
[2] Doyen, J. and Vandensavel, M. (1971) Non-isomorphic Steiner quadruple systems. *Bull. Soc. Math. Belg.* **23** 393–410.
[3] Eugeni, F. and Mayer, E. (1988) On blocking sets of index two. *Annals of Discrete Math.* **37** 169–176.
[4] Gionfriddo, M. and Micale, B. (1989) Blocking sets in 3-designs. *J. of Geometry*, **35** 75–86.
[5] Hanani, H. (1960) On quadruple systems, *Canad. J. Math.* **12** 145–157.
[6] Phelps, K. T. and Rosa, A. (1980) 2-chromatic Steiner quadruple systems. *European J. Comb.* **1** 253–258.
[7] Tallini, G. (1983) Blocking sets nei sistemi di Steiner e d-blocking sets in PG(r,q). *Quaderno n. 3 Sem. Geom. Combinatorie Univ. L'Aquila.*
[8] Tallini, G. (1988) On blocking sets in finite projective and affine spaces. *Annals of Discrete Math.* **37** 433–450.

(1,2)-Factorizations of General Eulerian Nearly Regular Graphs

ROLAND HÄGGKVIST and ANDERS JOHANSSON

Department of Mathematics, University of Umeå, S-901 87 Umeå, Sweden

E-mail address: rolandh@biovax.umdc.umu.se, andersj@zeus.cs.umu.se

Every general graph with degrees $2k$ and $2k-2$, $k \geqslant 3$, with zero or at least two vertices of degree $2k-2$ in each component, has a k-edge-colouring such that each monochromatic subgraph has degree 1 or 2 at every vertex.

In particular, if T is a triangle in a 6-regular general graph, there exists a 2-factorization of G such that each factor uses an edge in T if and only if T is non-separating.

1. Introduction

In this paper we will characterize those general graphs with degrees $2k-2$ and $2k$ that can be decomposed into spanning subgraphs with degrees 1 and 2 everywhere. Before we state the result, it is perhaps of some interest to review some related problems and their history.

1.1. Background

One of the starting points of graph theory is a classic investigation by the Danish mathematician Julius Petersen who in 1891 published a paper [7]: 'Die Theorie der regulären graphs', which contains a wealth of material on the problem of factorizing regular graphs into graphs of uniform degree k. An excellent source of information concerning Julius Petersen and problems spawned by his 1891 paper is the conference volume [1].

The motivation for Petersen's work, as given in the first few lines of his article, came from Hilbert's proof of the finiteness of the system of invariants associated with a binary form. Petersen notes that Hilbert's proof employs a theorem by Gordan, which, among other things, implies that for a given n one can construct a finite number of products of the type

$$(x_1 - x_2)^\alpha (x_1 - x_3)^\beta (x_2 - x_3)^\gamma \dots (x_{n-1} - x_n)^\varepsilon,$$

so that all other products of the same type can be built up by multiplying them together, the type in question being the property that the exponents are positive integers, possibly zero, and that the degree in $x_1, x_2, \ldots, x_n$ is constant for each product. Petersen calls a product thus constructed a *ground-factor*, and sets himself the problem of determining the ground-factors for every form of given degree and order. He notes a remarkable difference between forms of even or odd degree: the first have all ground-forms of degrees 1 or 2, whereas in the second case there exist infinitely many examples, the smallest with $n = 10$ and of degree 3, found by Sylvester (writes Petersen), for which this is not true.

Returning to graphs, Petersen shows two theorems, namely, every $2k$-regular graph admits a 2-factorization, and every 3-regular graph with at most one separating edge has a 1-factor (and consequently a 2-factor). Petersen also set himself the task of determining when two edges in a 4-regular graph always belong to the same 2-factor in the 2-factorization, or when they always belong to different 2-factors. This was to some extent motivated by a statement in a letter from Sylvester, who erroneously believed that if a simple 4-regular graph admits a Hamilton decomposition, then every pair of edges can be separated by 2-factors in some 2-factorization. Petersen worked with an auxiliary graph, called the *stretched graph*, and obtained a slightly cumbersome criterion (see Sabidussi [8] for references and a discussion of this particular problem). In this context, note the following characterization, from Sabidussi, determining when two edges in a 4-edge-connected Eulerian graph have the same or different parity in every Euler tour of the graph: if two edges e and f are parity equivalent, $G-e$ and $G-f$ are both nonbipartite, while $G-\{e,f\}$ is bipartite. The similar, easier, problem for diregular digraphs was settled completely by a simple lemma in [2] concerning regular bipartite graphs: two given edges in a k-regular bipartite graph with $k > 2$ can be separated by a proper edge-colouring if and only if they do not form a separating set. When $k = 2$ the condition is obvious: the edges should not be of the same parity in a common component.

The above prompts the question determining conditions ensuring that three given edges e, f and g in a bipartite regular graph belong to different colours in some edge-colouring, or the more general problem of determining when a given partial three edge-colouring can be completed to all of B. Unfortunately this question has not yet been resolved, despite some notable efforts, in particular by Hilton and Rodger [3]. Equivalently, we could ask for criteria ensuring that a diregular digraph admits a 1-difactorization such that three given edges are completely separated by the 1-difactors (a 1-difactor is a spanning set of cycles).

It is therefore of some interest that the corresponding question for general graphs admits a solution, as long as the prescribed edges form a triangle. The general question would be: when do three prescribed edges e, f and g in a regular Eulerian graph G lie in three different 2-factors in some 2-factorization of G? The answer when the three prescribed edges form a triangle is, as shall be seen here, that such a 2-factorization exists if and only if the triangle is nonseparating and the degree is at least 6. In this theorem, loops and multiple edges are allowed. In fact a more general theorem will be proved, which determines when it is possible to find a balanced k-edge-colouring of an Eulerian graph with degrees $2k$ and $2k-2$ everywhere (*i.e.* every colour must appear at least once and at most twice at each vertex). The condition is that if $k = 2$, no component has an odd number of vertices of degree 2, and if $k > 2$, no component contains exactly one vertex of degree $2k-2$.

An auxiliary motivation for a resolution of this particular problem comes from a problem about embedding partial Steiner triple systems with multiplicity λ on r vertices into Steiner triple systems on n vertices. It is well known that this problem is *NP*-complete for $\lambda = 1$, and indeed for odd λ, but it will be seen elsewhere [6], that, as conjectured by Hilton and Rodger in [4], there do exist natural conditions that are necessary and sufficient for such an embedding when λ is even.

1.2. Definitions and the theorem

General graphs (sometimes called pseudographs or multigraphs with loops) are considered, *i.e.*, graphs are allowed to have multiple edges and loops. The *degree* of a vertex v belonging to a graph G, denoted by $d_G(v)$, is the number of edges, with loops counted twice, that contain v. If all vertices have the same degree r, then G is called *r-regular*. All graphs are finite.

An edge-colouring σ of a graph is a mapping $\sigma : E(H) \mapsto \Omega$ of the edges into some set Ω of "colours". It is called a *k-edge-colouring* if $|\sigma(E(G))| \leqslant k$, *i.e.*, when at most k colours are used. In this paper we will, somewhat sloppily, sometimes refer to edge-colourings as colourings; vertex colourings do not appear.

We use the notation $d_\alpha(v) = d_{G_\alpha}(v)$ for the chromatic degree of a vertex v in an (edge-) coloured graph, where G_α is the *monochromatic factor* $G_\alpha = G[\sigma^{-1}(\alpha)]$ (the graph induced by the edges assigned colour α). For bichromatic factors, the notation $G_{\alpha\beta} = G[\sigma^{-1}(\{\alpha, \beta\})]$ is used, and so on.

If $d_\alpha(v) = 2$, for any colour α and any vertex v, the colouring is said to be a *2-factorization*. If

$$d_\alpha(v) = 1 \text{ or } 2, \qquad \forall \alpha \in \Omega \quad \forall v \in V(G),$$

we have a *(1, 2)-factorization*.

A colouring that satisfies

$$||E(G_\alpha)| - |E(G_\beta)|| \leqslant 1$$

for all pairs of colours α and β is said to be *equalized*. Let us call a colouring *vertex-balanced* if the degree-difference is at most 1, *i.e.*,

$$|d_\alpha(v) - d_\beta(v)| \leqslant 1$$

for all pairs α, β of colours and for all vertices v in the graph.

The following theorem states the main result. It was stated as a conjecture in a somewhat different form by Hilton and Rodger in [4] and [5, Conjecture 2]. These authors are mainly interested in the extension-properties of certain *partial Steiner triple systems*, a problem we will not attempt to settle in this paper.

Theorem 1. *Let G be a connected general graph, such that all vertices have degree $2k$ or $2k - 2$, for some $k > 1$. Then G admits an equalized (1, 2)-factorization if and only if the number of vertices of degree $2k - 2$ is either 0 or at least 2, and not an odd number if $k = 2$.*

We note that this immediately implies the following corollary.

Corollary. *Let G be a connected 6-regular general graph and let $T \subset G$ be a triangular* subgraph of G. A 2-factorization such that all three edges of T are coloured differently exists if and only if T is non-separating.

2. Proof of Theorem 1

It is quite evident that the conditions in Theorem 1 are necessary. Each vertex of degree $2k-2$ is the endvertex of exactly two monochromatic paths, and since the other endvertex of one of these paths must have degree $2k-2$, we have, if any, at least 2 vertices of degree $2k-2$. Also, since G is finite, the monochromatic paths make up a collection of cycles. If $k=2$, we have only two colours, and the paths along the cycle alternate in colour, changing colours at the $2k-2$-vertices, which clearly means that the number of these vertices is even.

So, the real issue is then to prove that these conditions are sufficient. We first give some lemmas, and then, ultimately, the proof of Theorem 1.

The problem of finding $(1,2)$-factorizations in Eulerian graphs is closely related to the theory of Eulerian trails and Eulerian orientations. An example of this is the proof of the following lemma, which is needed as a starting point in the proof of Theorem 1.

Lemma 1. *Theorem 1 is true if the number of vertices of degree $2k-2$ is even.*

Proof of Lemma 1. Let $E = x_1 x_2 \ldots x_{m-1} x_m x_1, x_i \in V$ be an Eulerian tour, and give the edges in G the corresponding forward orientation. Assume that $S = \{a_1, \ldots, a_{2k}\}$ is the set of vertices of degree $2k-2$, and assume, without loss of generality, that they occur in E in the given order, *i.e.*, starting at a_1 the first vertex in S, distinct from a_1, that occurs is a_2, and so on.

The edges on the segments $[a_1, a_2], [a_3, a_4], \ldots, [a_{2k-1}, a_{2k}]$ of E are now given the *reverse* orientation; the vertices in S thereby obtain the (oriented) degrees:

$$d^+(a_{2i}) = k, \quad d^-(a_{2i}) = k-2$$
$$d^+(a_{2i+1}) = k-2, \quad d^-(a_{2i+1}) = k.$$

All other vertices still have out- and in-degree k. If an oriented *alternating* cycle C on the vertex-set S, with degrees

$$d_C^+(a_{2i}) = 2, d_C^-(a_{2i+1}) = 2$$

is added to the graph, the result will be a regular general di-graph with in- and out-degree k. That any such has an (oriented) 1-factorization is a well-known fact. This induces a 2-factorization on the underlying undirected graph, and since the added cycle C is alternating, any two consecutive edges on this must get different colours. Consequently, we obtain a $(1,2)$-factorization after deleting C.   □

2.1. Eulerian 2-colourings.
Given a connected *Eulerian* graph $G = (V, E)$, *i.e.*, a graph with vertices of even degree, we may give the graph an *Eulerian 2-edge-colouring* as follows: pick

any Eulerian tour and colour the edges alternately α and β along the tour, starting and ending at a prescribed vertex x. The resulting 2-edge-colouring satisfies

$$d_{\alpha}(v) = d_{\beta}(v)$$

at all vertices, except possibly at x (if $|E(G)|$ is odd), where the difference in the exceptional case will be 2.

Suppose now that we have a graph with at least one (and hence at least two) vertices of odd degree. Since the number of odd vertices is even in any graph, the odd vertices may be paired off as

$$x_1, y_1, x_2, y_2, \ldots, x_t, y_t.$$

By joining each of these paired odd vertices by a subdivided edge $x_i w_i y_i$, for some new vertex w_i, with the possible exception of one pair, x_t, y_t say, which instead is joined by a new edge $x_t y_t$, an Eulerian graph with an *even* number of edges can be constructed. An Eulerian colouring of the new graph clearly induces a vertex-balanced 2-colouring on the original graph. Moreover, this colouring is equalized, since the subdivided edges have one edge of each colour. We state this observation as a lemma.

Lemma 2. *Let $G = (V, E)$ be a connected graph and assume that the number of edges $|E(G)|$ is even or that G has at least one vertex of odd degree. Then G admits an equalized vertex-balanced 2-colouring. If the degrees of G are all 2, 3 and 4, this colouring is also a $(1, 2)$-factorization.*

Note that if $d_G(V) \subseteq \{2, 3, 4\}$, a vertex-balanced 3-colouring is immediately a $(1, 2)$-factorization.

Remark. Note that Lemma 2 implies that

a graph has a $(1, 2)$-factorization if and only if it has an equalied one.

This is easily seen, since if we have a $(1, 2)$-factorization, then by applying Lemma 2 on all non-equalized components of the bichromatic factors $G_{\alpha\beta}$, we eventually end up with an equalized colouring of all such components, and, by a suitable renaming of the colours in each component, we obtain an equalized colouring. Hence, in the following, we may dismiss the discussion of the equalized property altogether.

The following technical lemma (which actually is contained in the preceding proof) is needed in a key step of the proof.

Lemma 3. *Let F be a connected graph with two distinct vertices x and s of odd degree, where the degree of x is at least three. Suppose x and s are in the same component of the graph $F \backslash xy$, for some edge $xy, y \neq x, s$. There is then a vertex-balanced 2-colouring of F such that the monochromatic factor containing the edge xy has degree one more at x than the other factor.*

Proof of Lemma 3. Split the vertex x into two vertices x' and x'' in such a way that x' has degree 1 and is only joined to y by the edge $x'y$. This may or may not split the graph into two components, but the component containing x'' has the odd-degree vertex s and the component containing x' also has odd vertices, one of which is x' itself. Lemma 2 implies that the graph admits a vertex-balanced colouring. Since the degree of x'' is even, the edges incident with x'' are equally divided between each factor. Restoring x by identifying the vertices x' and x'', we have obtained the sought for colouring of the original graph. □

2.1. Proof of Theorem 1

We now turn to the proof of Theorem 1. First a definition: given a coloured graph G, with colouring σ, a *recolouring* of a subgraph H is a colouring σ' of G such that $\sigma'|_{G \setminus H} = \sigma|_{G \setminus H}$, i.e., a new colouring that only differs on the edges in H.

The idea behind the proof of the theorem is that if S has an odd number of vertices, we can at the very least, by Lemma 1, find a colouring that is a $(1,2)$-factorization except for one vertex. This colouring we then transform by a sequence of Eulerian recolourings of bichromatic components, so that we eventually can apply Lemma 2 on the (by that time altered) bichromatic component that misses one colour at a vertex. Implicit in the proof is a polynomial algorithm for the problem of finding a $(1,2)$-factorization in our type of graph.

Proof of Theorem 1. Assuming the theorem to be false, we choose G as a graph that fulfils the conditions of Theorem 1, but fails the conclusion that it admits a $(1,2)$-factorization. Let S be the set of vertices of degree $2k-2$. We may assume that $|S|$ is odd, and hence at least 3, since the case with $|S|$ even is handled by Lemma 1. Consequently, Theorem 1 is true for $k = 2$, and therefore we have at least 3 colours.

Pick an S-path P, i.e., a path between two distinct vertices $a, b \in S$, with all interior vertices in the complement of S.

If we add one loop to any prescribed vertex $z \in S$ not equal to b, we are again in the situation covered by Lemma 1, so the resulting graph has a $(1,2)$-factorization. This induces a k-edge-colouring σ on G that clearly satisfies

$$d_\delta(v) = 1 \text{ or } 2 \tag{A}$$

for all colours δ and all vertices v, *except* at the unique vertex $z = z(\sigma)$ in S, where one colour, which we name α, say, is missing.

We have at least three colours, and all colours have degree 1 or 2 at b, so there is also at least one colour β, such that

$$d_\alpha(b) + d_\beta(b) = 3. \tag{B}$$

Let $H = H(\sigma)$ be the component of $G_{\alpha\beta}$ that contains z. By Lemma 2, we may assume that $|E(H)|$ is odd and that no vertex in H is of odd degree, since any $(1,2)$-factorization of H gives a $(1,2)$-factorization of G. So, by (B), b is not a vertex of H.

We may assume though, that

$$V(H) \text{ intersects } V(P),$$

since this holds if the chosen vertex z is the endpoint a of P. It is therefore possible to assume that the colouring (and also β) is chosen such that

$$\text{the distance between } V(H) \text{ and } b \text{ along } P \text{ is } minimal. \tag{C}$$

Hence there is a vertex x in $V(P) \cap V(H)$ of minimal distance to b along P. Since b is not a vertex of H, x is joined in P to a vertex $y \notin V(H)$ closer to b by an edge $xy \in P$. The colour of xy is called γ and is distinct from α and β (since otherwise y would be nearer to b than x is along P). The typical situation is illustrated in Figure 1.

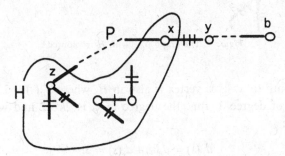

Figure 1. The colours α, β and γ are distinguished by one, two and three crossbars respectively.

We may make some further assumptions on this colouring: we first note that since the component of $G_{\alpha\beta}$ containing b is disjoint from H, we may *interchange* the colours α and β in this component, without violating (A), (B) or (C). This observation, together with (B), makes it legal to assume that

$$d_\beta(b) + d_\gamma(b) \text{ is even.} \tag{1}$$

This means that (B) will still hold after any $(1, 2)$-factorization of the graph $G_{\beta\gamma}$.

If $x = a$, we may also assume that $z = a$ and then interchange the colours β and γ globally, which is legal since (1) implies that (B) then still holds. However, y is now in the same component of $G_{\alpha\beta}$, as $z = a$, which contradicts (C). It is therefore established that x must be an interior vertex of P, and hence that

$$d_G(x) = 2k \text{ and, by (A), } d_\delta(x) = 2 \tag{2}$$

for all colours δ, since P is a S-path.

We now fix this colouring and call it σ, and in the rest of the proof we recolour the graph in three steps. In this process we will only change the colours α, β and γ, and the result will be a colouring σ', that satisfies (B) and (A) with the exception of the colour α at some vertex $z' = z(\sigma')$. But this vertex will be in the same $\alpha\beta$-component as the vertex y, contradicting (C).

The first step is to recolour H in colours α and β, starting at x with colour β. Since we know that $|E(H)|$ is odd, the resulting colouring satisfies (A) for all colours and all vertices, except at x where the discrepancy is 2, which by (2), means that $d_\beta(x) = 3$ and $d_\alpha(x) = 1$.

Now, let F be the component of the current $G_{\beta\gamma}$ that contains the edge xy. The degree of x in F is 5, by the previous recolouring in α and β and by (2), but for any other vertex $v \neq x$ the degree is between 2 and 4, since (A) is satisfied at v. The situation is illustrated in Figure 2.

Note also that, as $d_\beta(x) = 3$ is odd, there is another vertex s of G_β-degree 1 joined by a

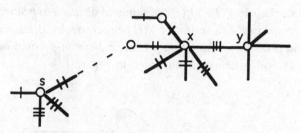

Figure 2. The graph $G_{\alpha\beta\gamma}$ with H recoloured.

monochromatic β-path to x. This vertex is also in H, where $d_H(s) = 2$, since H does not contain any vertex of degree 3. But, the degree in G is $2k-2$ and we have k colours, $d_\gamma(s) = 2$, and hence

$$d_F(s) = d_\beta(s) + d_\gamma(s) = 3.$$

This means that the conditions in Lemma 3 are fulfilled for the edge xy and the vertices x and s, therefore we may recolour F in such a way that the degree $d_\beta(x)$ is still 3. At the same time, the edge xy is coloured β instead of γ. Since the recolouring is also vertex-balanced on F and at all vertices except x the degree is 2, 3 or 4, (A) is now satisfied for all vertices and colours except at x, where $d_\beta(x) = 3$ and $d_\alpha(x) = 1$. The condition (1) ensures moreover that: if b should be a vertex of F, this recolouring has not changed the condition (B).

As the last step of the proof, we consider the component of the current $G_{\alpha\beta}$ that contains the edge xy, and call this graph H'. Note that $d_{H'}(x) = 4$, and for the other vertices it is between 2 and 4, since (A) is valid there. If H' has an even number of edges, or some vertex in H' is of odd degree, we find a $(1, 2)$-factorization of H' by Lemma 2. This contradicts the choice of G, since we have thus obtained a $(1, 2)$-factorization.

Consequently, H' has an odd number of vertices of degree 2, and the rest have degree 4. But, by adding a loop at some vertex z' of degree 2, we can find a recolouring of H' in α and β that satisfies (A), for all vertices except at z', where $d_\alpha(z') = 0$. Moreover, this recolouring has not changed (B), as this vertex cannot be in H', since its $\alpha\beta$-degree is 3.

However, the resulting colouring σ' now contradicts (C) since y is in the same component $H' = H(\sigma')$ as $z' = z(\sigma')$ is. This proves that, contrary to assumption, the graph G must have a $(1, 2)$-factorization. $\square$

References

[1] Andersen, L. D. *et al.* (1992) Special volume to mark the centennial of Julius Petersen's 'Die Theorie der regulären Graphs'. *Discrete Mathematics* **100** 101.

[2] Häggkvist, R. (1976) A solution of the Evans conjecture for Latin squares of large size. In: Proceedings Fifth Hungarian Colloquium, Keszthely 1976, vol. 1. *Combinatorics* **18**.

[3] Hilton, A. J. W. and Rodger, C. A. (1991) Edge-colouring regular bipartite graphs. *Graph theory (Cambridge, 1981)* **56**. North-Holland, Amsterdam–New York, 139–158.

[4] Hilton, A. J. W. and Rodger, C. A. (1990) Edge-Colouring Graphs and Embedding Partial Triple Systems of Even Index. *Cycles and Rays* (NATO ASI Series, eds.), Kluwer, 101–112.

[5]   Hilton, A. J. W. and Rodger, C. A. (1991) The Embedding of Partial Triple Systems when 4 Divides lambda. *Journal of Combinatorial Theory* Series A **56** 109–137.

[6]   Johansson, A. (1993) *A Note on Embedding Partial Triple Systems of Even Index* (in preparation).

[7]   Petersen, J. (1891) Die Theorie der regulären Graphs. *Acta Mathematica* **15** 193–220.

[8]   Sabidussi, G. (1993) *Parity Equivalence in Eulerian Graphs* (preprint).

Oriented Hamilton Cycles in Oriented Graphs

ROLAND HÄGGKVIST[1] and ANDREW THOMASON[2]

[1]Department of Mathematics, University of Umeå, S-901 87 Umeå, Sweden
[2]DPMMS, 16 Mill Lane, Cambridge CB2 1SB, England

We show that, for every $\epsilon > 0$, an oriented graph of order n will contain n-cycles of every orientation provided each vertex has indegree and outdegree at least $(5/12 + \epsilon)n$ and $n > n_0(\epsilon)$ is sufficiently large.

1. Introduction

Dirac's theorem states that every graph G with minimum degree $\delta(G) \geqslant |G|/2$ has a hamilton cycle. The simplest analogue for digraphs is given by the theorem of Ghouila-Houri [3]. Given a digraph G of order n and a vertex $v \in G$, we denote the outdegree of v by $d^+(v)$ and the indegree by $d^-(v)$. We also define $d^o(v)$ to be $\min\{d^+(v), d^-(v)\}$, and $\delta^o(G)$ to be $\min\{d^o(v) : v \in G\}$. Ghouila-Houri's theorem [3] implies that G contains a *directed* hamilton cycle if $\delta^o(G) \geqslant n/2$. Only recently has a constant $c < 1/2$ been established such that every *oriented graph* satisfying $\delta^o(G) > cn$ has a directed hamilton cycle; Häggkvist [5] has shown that $c = (1/2 - 2^{-15})$ will suffice. He also showed that the condition $\delta^o(G) \geqslant n/3$ proposed by Thomassen [9] is inadequate to guarantee a hamilton cycle, and conjectured that $\delta^o(G) \geqslant 3n/8$ is sufficient.

When considering hamilton cycles in digraphs there is no reason to stick to directed cycles only; we might ask for any orientation of an n-cycle. For tournaments G, Thomason [8] has shown that G will contain every oriented cycle (except the directed cycle if G is not strong) regardless of the degrees, provided n is large. For general digraphs, Grant [4] proved that G contains an *antidirected* hamilton cycle if $\delta^o(G) \geqslant 2n/3 + \sqrt{n \log n}$; an antidirected cycle is one in which the edge orientations alternate (of course n has to be even).

We know of no published result in this vein which covers *all* oriented n-cycles. However we recently proved [6] that any digraph with $\delta^o(G) \geqslant n/2 + n^{5/6}$ contains *every* oriented n-cycle, provided n is large enough. Our purpose in this paper is to consider the analogous problem for oriented graphs. We believe that the condition $\delta^o(G) \geqslant (3/8 + \epsilon)n$ will be enough to guarantee all oriented n-cycles in any sufficiently large oriented graph. Here we shall prove that the condition $\delta^o(G) \geqslant (5/12 + \epsilon)n$ is enough.

The proof of our theorem is based on the expansion properties of a graph with large minimum degree. As such, we expect that the machinery developed (by refining the ideas of [6]) could be used to prove similar results for any directed graph having a suitable expansion property. In particular, the methods here give an alternative way to prove the digraph result of [6], though we shall not make this explicit. It is our intention to explore elsewhere a possible extension to a more general context. We hope also to prove the present theorem under the weaker constraint that $\delta^o(G) \geqslant (3/8 + \epsilon)|G|$. The reason for not proving this stronger result here is a purely technical one, which we have not yet had the opportunity to tackle. The problem is pointed out in section 6.

Let ϵ be some constant which remains fixed throughout the paper. Note that the definition of δ^o implies $\epsilon < 1/8$ throughout. Several times we shall claim that a statement is true "provided n is sufficiently large". This will mean that there exists an $n_0 = n_0(\epsilon)$ such that the statement is true provided $n > n_0$. In particular, just how large is "sufficiently large" will depend on ϵ only, and not on any other parameters.

Here is some notation. Given an oriented graph G and a vertex $v \in G$ we denote by $\Gamma^+(v)$ the set of out-neighbours of v and by $\Gamma^-(v)$ the set of in-neighbours. Given an oriented path P, the length of P is denoted $l(P)$; the two paths $A(P;k)$ and $Z(P;k)$ are the paths spanned by the first k edges and the last k edges of P respectively. If two paths P and Q are isomorphic we may write $P \cong Q$ or even $P = Q$. The path PQ is the path of length $l(P) + l(Q)$ formed by identifying the end of P with the beginning of Q. We may also identify the end of Q with the beginning of P to form a cycle, also denoted PQ; whether PQ denotes a path or a cycle will be clear from the context.

2. A first strategy

The proof of the main theorem and the number of supporting lemmas might appear, at first sight, to be a mass of technical details. Indeed, the technical difficulties encountered in implementing our basic strategy are considerable. Nevertheless the essence of the strategy is very straightforward. Consequently, it is worth devoting a paragraph or two to an outline of the idea underlying our construction of a given n-cycle C in an oriented graph G. The reader will thereby be able, later on, to distinguish the wood from the trees. Crucial to the method are two devices for finding collections of paths, namely *pipelines* and *sorters*.

Definition. *A pipeline of width s and length t is an oriented graph whose vertex set comprises $t+1$ subsets $S_0, \ldots, S_t$, each of order s, such that $S_{i-1} \cap S_i = \emptyset$, $1 \leqslant i \leqslant t$. It has the property that for any s oriented paths $P_j = x_{j,0} x_{j,1} \ldots x_{j,t}$ of length t, $1 \leqslant j \leqslant s$, there exist vertex disjoint copies of P_j with $x_{j,i} \in S_i$, $1 \leqslant j \leqslant s$, $0 \leqslant i \leqslant t$. (The set S_0 is called the start of the pipeline and the set S_t the end. Usually the sets S_i will be mutually disjoint, for otherwise the P_j may be realised as trails rather than as paths.)*

We will show that a randomly chosen sequence of subsets $S_i \subseteq V(G)$ almost surely span a pipeline, provided $\delta^o(G)$ is large. Note that a pipeline guarantees the existence of given paths P_j but does not allow us to specify the endvertices within S_0 and S_t. If such a specification were possible, it would be easy to find C in G, at least if n were a multiple of t, as follows. Partition $V(G)$ into t subsets $S_0 = S_t, S_1, \ldots, S_t$ forming a pipeline. Let C consist of paths $P_1 P_2 \ldots P_s$ and let $S_0 = S_t = \{y_1, \ldots, y_s\}$. Then there would be copies of P_j in the pipeline joining y_j to y_{j+1}, $1 \leqslant j \leqslant s$ (where y_{s+1} denotes y_1), and so we would have a copy of C. It is possible that a randomly chosen collection of subsets S_i would allow such specification of endvertices, but we do not investigate this here. Instead we achieve a similar effect by using a sorter.

Definition. *Let $P = x_0 x_1 \ldots x_t$ be an oriented path of length t. A sorter for P of width s, or (s, P)-sorter, is an oriented graph whose vertex set comprises $t + 1$ disjoint subsets $S_i = \{y_{1,i}, \ldots, y_{s,i}\}$, $0 \leqslant i \leqslant t$, such that for any permutation σ of $\{1, \ldots, s\}$ there exist s vertex disjoint copies $P_j = x_{j,0} x_{j,1} \ldots x_{j,t}$ of P, $1 \leqslant j \leqslant s$, with $x_{j,i} \in S_i$, $x_{j,0} = y_{j,0}$ and $x_{j,t} = y_{\sigma(j),t}$, $0 \leqslant i \leqslant t$, $1 \leqslant j \leqslant s$.*

Note that a sorter is stronger than a pipeline in that it allows specification of endvertices, but weaker in that it requires the paths P_j to be isomorphic. However, if we can build both a pipeline and a sorter in G we can still, under suitable circumstances, construct C as follows. Suppose we can write $C = P_1 Q_1 P_2 Q_2 \ldots P_s Q_s$, where $P_1 \cong \ldots \cong P_s$, $l(Q_1) = \ldots = l(Q_s)$ and both $l(P_1)$ and $l(Q_1)$ are not too small. We shall show in section 3 that, if $\delta^o(G) \geqslant (3/8 + \epsilon)|G|$, then there exists a constant $k = k(\epsilon)$ such that we can construct, between any two given vertices, a path of length k with any desired orientation. We call such a path a *handbuilt* path. Write $P_j = A_j P'_j Z_j$, where $l(A_j) = l(Z_j) = k$. The following plausible construction for C is, in fact, feasible. Construct a sorter for P'_1 of width s and construct a pipeline of width s and length $l(Q_1)$. Join the end of the sorter to the start of the pipeline with s handbuilt paths isomorphic to Z_1 and join the end of the pipeline to the start of the sorter with s handbuilt paths isomorphic to A_1. Then C can be found in G.

The reason the above strategy fails in general is that it is not possible to find isomorphic paths $P_1, \ldots, P_s$ equally spaced around C. We therefore have to accept that the lengths of the Q_j may be unequal. To construct paths of differing lengths from a pipeline requires some extra manipulation with handbuilt paths. To obtain the elbow room for such manipulation we shall use two sorters and another pipeline to find $P_1, \ldots, P_s$, while dropping the constraint that these paths be isomorphic. But by now it is time to begin the proof in detail.

3. Expansion properties

In this section we give some elementary lemmas describing how expansion properties yield short oriented paths with prescribed endvertices. We first show that graphs with large minimal degree are expanders.

Lemma 1. *Let G be an oriented graph of order n with $\delta^o(G) \geqslant (3/8 + \epsilon)n$. Let $A \subseteq V(G)$ satisfy $|A| \geqslant 3n/8$, and let $0 \leqslant \eta \leqslant 1/80$. Then*

$$|\{ y \in G ;\ |\Gamma^-(y) \cap A| > \eta n\}| \geqslant (1/4 + 2\epsilon - 11\eta)n + |A|/2.$$

In particular $|\Gamma^+(A)| \geqslant (1/4 + 2\epsilon)n + |A|/2$.

Proof. Let $A^* = \{ y \in G \; ; \; |\Gamma^-(y) \cap A| > \eta n\}$, let $B = A \cap A^*$ and let $C = A \setminus A^*$. Since the indegrees in C are all at most ηn, there are at most $|C|\eta n$ edges in C. But each vertex in G has total degree at least $(3/4 + 2\epsilon)n$, so there are at least $(3n/4 + 2\epsilon n - (n - |C|))|C|/2$ edges in C. Therefore $|C| \leqslant n/4 + 2\eta n - 2\epsilon n$, and so $|B| = |A| - |C| \geqslant |A| - (1/4 + 2\eta - 2\epsilon)n \geqslant n/10$.

Within B there are at most $\binom{|B|}{2}$ edges, and the number of edges from B into C is at most $|C|\eta n$. Thus $e(B, G \setminus A)$, the number of edges from B into $G \setminus A$, is at least $|B|(\delta^o - |B|/2) - |C|\eta n$. Writing D for $A^* \setminus A$ we have

$$|D||B| + (n - |A| - |D|)\eta n \geqslant e(B, G \setminus A) \geqslant |B|(\delta^o - |B|/2) - |C|\eta n,$$

and since $n - |A| - |D| + |C| = n - |B| - |D| \leqslant n \leqslant 10|B|$ we have $|D| + 10\eta n \geqslant \delta^o - |B|/2$. Therefore $|A^*| = |B| + |D| \geqslant \delta^o + |B|/2 - 10\eta n \geqslant (1/4 + 2\epsilon - 11\eta)n + |A|/2$. □

Note that Lemma 1 implies that sets of size less than $(1/2 + 4\epsilon)n$ expand, in the sense that $|\Gamma^+(A)| > |A|$. This fact means the graph has small diameter, as we now show.

Lemma 2. *Let G be an oriented graph of order n with $\delta^o(G) \geqslant (3/8 + \epsilon)n$ and let x and y be vertices of G. Let $4\lceil \log_2(1/\epsilon) \rceil \leqslant k \leqslant \epsilon n/4$, and let P be an oriented path of length k. Then, if n is large enough, there will be a path from x to y isomorphic to P. Moreover there can be found a set of at least $\epsilon n/4k$ disjoint such paths.*

Proof. We show first that a copy of P exists if $k = 2\lceil \log_2(1/\epsilon) \rceil$ and $n > (12/\epsilon)^k$.

Suppose, for the sake of argument, that the first three edges of P go backwards, forwards and backwards. Let $A = \Gamma^-(x)$, let $\eta = \epsilon/12$ and define μ by $|A| = (1/2 + 4\epsilon - 22\eta - \mu)n$. Note that $\mu < 1/2$. Let $A^* = \{ y \in G \; ; \; |\Gamma^-(y) \cap A| > \eta n\}$ and let $A^{**} = \{ y \in G \; ; \; |\Gamma^+(y) \cap A^*| > \eta n\}$. Each vertex of A^* can be reached from x by a forward-backward trail in ηn ways and, by Lemma 1, $|A^*| \geqslant (1/2 + 4\epsilon - 22\eta - \mu/2)n$. Likewise each vertex of A^{**} can be reached from x by $(\eta n)^2$ backward-forward-backward trails and $|A^{**}| \geqslant (1/2 + 4\epsilon - 22\eta - \mu/4)n$. Hence after $\lfloor k/2 \rfloor$ such steps there are at least $(1/2 + 3\epsilon - 22\eta)n \geqslant (1/2 + \epsilon)n$ vertices which can each be reached in $(\eta n)^{\lfloor k/2 \rfloor - 1}$ ways by trails oriented like the first half of P. Likewise there are $(1/2 + \epsilon)n$ vertices which can each be reached in $(\eta n)^{\lceil k/2 \rceil - 1}$ ways by trails oriented like the second half of P. Since $2\epsilon \geqslant \eta$, x can reach y by at least $(\eta n)^{k-1}$ trails oriented like P. At most n^{k-2} of these trails can be self-intersecting and the rest (a positive quantity if $n > (1/\eta)^{k-1}$) must be paths.

To find longer paths we apply induction on k; an x–y path of length k is found by selecting a suitable neighbour z of y and finding an x–z path of length $k - 1$. The lower bound of $4\lceil \log_2(1/\epsilon) \rceil$ claimed in the lemma allows for the reduction from ϵ to $\epsilon/2$ in the expression for δ^o as this process is used up to $\epsilon n/4$ times.

A set of disjoint paths can be found by repeatedly removing from the graph the internal vertices of any path found, and reapplying the above argument to find another path in the remaining graph. Even after $\epsilon n/4k$ applications there still remains a graph G' with $\delta^o(G') \geqslant (3/8 + \epsilon/2)|G'|$, so at least $\epsilon n/4k$ paths can be found. □

Definition. *Let $0 < \lambda < 1$. A pair of disjoint subsets S, T of the vertices of an oriented graph is said to be λ-expanding if, for every subset $A \subseteq S$ with $0 < |A| \leqslant |S|/2$, the inequalities*

$|\Gamma^+(A) \cap T| \geqslant (1 + \lambda)|A|$ and $|\Gamma^-(A) \cap T| \geqslant (1 + \lambda)|A|$ both hold, and for every subset $B \subseteq T$ with $0 < |B| \leqslant |T|/2$, the inequalities $|\Gamma^+(B) \cap S| \geqslant (1 + \lambda)|B|$ and $|\Gamma^-(B) \cap S| \geqslant (1 + \lambda)|B|$ both hold.

The next lemma shows that, if λ-expanding pairs of subsets are available, small oriented paths can be built wherein the choice of the internal vertices is constrained. Paths built by means of this lemma will be nicknamed *handbuilt* paths.

Lemma 3. *Let G be an oriented graph and let $0 < \lambda < 1$. Let $S_0, \ldots, S_k$ be disjoint subsets of $V(G)$, each of order s, such that each pair S_{i-1}, S_i is λ-expanding, $1 \leqslant i \leqslant k$. Let P be some oriented path of length k, let $v_0 \in S_0$ and let $v_k \in S_k$. Then there exists a copy $v_0 \ldots v_k$ of P joining v_0 to v_k in G, such that $v_i \in S_i$, $0 \leqslant i \leqslant k$, provided $k \geqslant \lfloor (4/\lambda) \log_2 s \rfloor$.*

Proof. We show that the path P exists provided $k \geqslant 2t$, where t is the smallest integer such that $(1 + \lambda)^t \geqslant s/2$. This is sufficient to prove the lemma, since $2t = 2\lceil (\log_2 s - 1)/\log_2(1 + \lambda) \rceil \leqslant 2 \log_2 s / \log_2(1 + \lambda) \leqslant (4/\lambda) \log_2 s$.

Consider first the case $k = 2t$. Since the pair S_0, S_1 is λ-expanding there are at least $(1 + \lambda)$ choices for $v_1 \in S_1$. Let $A \subset S_1$ be the set of these choices. Since the pair S_1, S_2 is λ-expanding there are at least $(1 + \lambda)^2 s$ choices for $v_2 \in S_2$ (that is, there are at least $(1 + \lambda)^2$ vertices b in S_2 for which there is a vertex $a \in A$ such that the path $v_0 a b$ is isomorphic to $A(P; 2)$). Continuing in this way we see that there are at least $(1 + \lambda)^t \geqslant s/2$ vertices $w \in S_t$ such that there exists a path $v_0 v_1 \ldots v_t = w$ isomorphic to $A(P; t)$ with $v_i \in S_i$, $0 \leqslant i \leqslant t$. Similarly there are at least $s/2$ choices of w for which there is a copy $w = v_t v_{t+1} \ldots v_{2t}$ of $Z(P; t)$ with $v_i \in S_i$, $t \leqslant i \leqslant 2t$. Hence some choice of w offers a copy both of $A(P; t)$ and of $Z(P; t)$; in other words the path P exists if it is of length $2t$.

The proof is completed by induction on k. For $k > 2t$ select a vertex $v_{k-1} \in S_{k-1}$ so that the edge $v_{k-1} v_k$ has the orientation required; this can be done because the pair S_{k-1}, S_k is λ-expanding. The induction hypothesis ensures the existence of a copy of $A(P; k - 1)$ joining v_0 to v_{k-1}, which extends to the desired copy of P. $\square$

Later we shall see that it is quite easy to find λ-expanding pairs, after which Lemma 3 will prove very useful. In fact the rate of expansion we can achieve in oriented graphs of high minimum degree is much greater than that required by Lemma 3, and handbuilt paths of constant length (that is, $O(1/\epsilon)$ independently of $|S|$) are achievable. However the notion of λ-expansion as defined is one which is customary in the literature and more natural in other contexts.

4. Sorters

A *sorting network* of width s with t stages is an undirected graph consisting of disjoint sets of vertices $S_0, S_1, \ldots, S_t$, each of size s, say $S_i = \{x_{i,1}, \ldots, x_{i,s}\}$. The edges of the i'th *stage* consist of $\lfloor s/2 \rfloor$ disjoint 4-cycles $x_{i-1,j} x_{i,j} x_{i-1,k} x_{i,k}$ for various pairs of indices j and k. The sorter has the property that the input vertices S_0 can be joined to the output vertices S_t in any prescribed order by s vertex disjoint paths of length t. As is well known,

Ajtai, Komlós and Szemerédi [1] have described a sorting network using only $C \log s$ stages. Batcher [2] has described a simple sorting network using $\lceil (\log_2 s)^2 \rceil$ stages, and this is the one we shall use. Any reasonable sorting network would suffice for the present application.

It is clear that if the 4-cycles $x_{i-1,j} x_{i,j} x_{i-1,k} x_{i,k}$ in the sorting network were replaced by disjoint paths $x_{i-1,j} = y_0 y_1 \ldots y_m = x_{i,j}$ and $x_{i-1,k} = z_0 z_1 \ldots z_m = x_{i,k}$, along with two edges $y_{m-1} z_m$ and $y_m z_{m-1}$, then the resultant graph would still function as a sorter. Let $\mathbf{m} = (m_1, \ldots, m_t)$. An $(s, \mathbf{m})$-*sorter* will be a sorting network in which the 4-cycles are replaced in this way, the paths used in the i'th stage all having length m_i. The *length* of the $(s, \mathbf{m})$-sorter will then be $\sum_i m_i$. Suppose now P is an oriented path of the same length as the $(s, \mathbf{m})$-sorter. It is clear that the edges of the $(s, \mathbf{m})$-sorter can be oriented in such a way that the s paths from S_0 to S_t will always be oriented like P. Such an oriented $(s, \mathbf{m})$-sorter is therefore an (s, P)-sorter.

Theorem 4. *Let G be an oriented graph of order n with $\delta^o(G) \geqslant (3/8 + \epsilon)n$. Let s be a natural number with $s \leqslant n^{1/2}$ and let P be an oriented path of length $(3\lceil \log_2(1/\epsilon) \rceil + 1)\lceil (\log_2 s) \rceil^2$. Then G contains an (s, P)-sorter, provided n is large.*

Proof. From the discussion above it can be seen that (s, P)-sorters exist; let us fix our minds on one such and construct a copy of it in G. Suppose the first few stages of the sorter have been constructed, up to say the class S_{i-1}, so that the length of every stage is $k + 1$, where $k = 3\lceil \log_2(1/\epsilon) \rceil$. We construct the next stage as follows. Let P' be the subpath of P, of length $k + 1$, which will need to traverse the gap between S_{i-1} and S_i. Let $Q = A(P'; k)$ and let Q^* be the path of length $2k$ made from two copies of Q by identifying their terminal vertices. If the sorting network, from which the (s, P)-sorter was derived, requires a 4-cycle based on $x_{i-1,j}$ and $x_{i-1,k}$, select $h = \lfloor \epsilon n/8k \rfloor$ vertex disjoint $x_{i-1,j}$–$x_{i-1,k}$ paths each oriented like Q^* and which avoid all vertices used so far in the construction. These h paths can be found by applying Lemma 2 to the graph consisting of G after the removal of the vertices used so far in the sorter. Let H be the set of the h midpoints of these paths.

Assume now that the final edge of P' is a forward edge (the argument is very similar if the edge is a backward edge). Each vertex of H has at least $p = (3/8 + \epsilon/2)n$ neighbours among the set U of vertices not so far used in the sorter or in the h paths linking $x_{i-1,j}$ to $x_{i-1,k}$ via H. Since $u\binom{hp/u}{2} > \binom{h}{2}$, where $u = |U|$, at least two vertices of U, call them $x_{i,j}$ and $x_{i,k}$, will receive edges from the same two vertices of H, called say y and z. From the two copies of Q^* joining $x_{i-1,j}$ to $x_{i-1,k}$ via y and z, select copies of Q joining $x_{i-1,j}$ to y and $x_{i-1,k}$ to z. Hence we have copies of P' joining $x_{i-1,j}$ to $x_{i,j}$ and $x_{i-1,k}$ to $x_{i,k}$, plus the two extra edges needed for the sorter. The vertices of S_i are now created by performing this operation for all pairs j, k which are the bases of 4-cycles in the i'th stage. $\qquad \square$

5. Pipelines

The main tool in our proof is the pipeline, defined earlier. In this section we shall describe how our pipelines are to be constructed. In fact a condition on a sequence of sets $S_0, \ldots, S_t$

which is nearly sufficient to guarantee that a pipeline is formed is that the pairs S_{i-1}, S_i be λ-expanding. The property of λ-expansion can be used in two ways. One is to enable us to form handbuilt paths, as in Lemma 3. The other is to derive matchings.

Definition. *Let S and S' be two subsets of the vertex set of an oriented graph, with $|S| = |S'|$. We say S matches S' if there is a set of $|S|$ independent edges from S to S'.*

Lemma 5. *Let S and S' be disjoint sets of vertices in an oriented graph with $|S| = |S'|$, such that the pair S, S' is λ-expanding. Then S matches S'.*

Proof. The König-Hall theorem tells us that if S does not match S' then there is a set A with $|A| \leqslant |S|/2$, such that either $A \subset S$ and $|\Gamma^+(A)| < |A|$ or $A \subset S'$ and $|\Gamma^-(A)| < |A|$. However, these are both ruled out by the λ-expansion of the pair S, S'. $\qquad\square$

It is clearly necessary that two consecutive sets S_{i-1} and S_i in a pipeline match each other. More generally, when trying to find copies of s paths in a pipeline, we may need to extend partial paths from S_{i-1} to S_i via u forward edges and $s - u$ backward edges. The reader is reminded that the sets S_i we shall eventually use will be chosen at random. It would be ideal if we were able to guarantee that every u-subset $A \subseteq S_{i-1}$ matched some u-subset $B \subseteq S_i$, but that is not the case. We can, however, ensure that almost every u-subset $A \subseteq S_{i-1}$ matches almost every u-subset $B \subseteq S_i$, which will do. Nevertheless, even this is too much to hope for if u is small (say $u = 1$); in practice we will have to extend from S_{i-1} to S_i via handbuilt paths if u is small. These considerations lead to the next definitions, describing the property we actually need to build pipelines.

Definition. *Let $0 < \alpha < 1$ and let r be a natural number. A pair of disjoint w-subsets W, W' of the vertex set of an oriented graph is said to be (r, α)-wed if, for every integer u with $u = 0$, $u = w$ or $r \leqslant u \leqslant w - r$, and for each of at least $(1 - \alpha)\binom{w}{u}$ of the u-subsets $B \subset W'$, there are at least $(1 - \alpha)\binom{w}{u}$ u-subsets $A \subset W$ such that both A matches B and $W' \setminus B$ matches $W \setminus A$.*

Definition. *Let $0 < \lambda, \alpha < 1$ and let r, s be integers. A pair of disjoint s-subsets S, S' of the vertex set of an oriented graph is said to be a (λ, r, α)-matched pair if the following two conditions hold:*

$(16r/\lambda)\log_2 s \leqslant s$ and $(8/\lambda)\log_2 s < 1/\alpha$, and
for every integer $h \leqslant (16r/\lambda)\log_2 s$, for every $(s-h)$-subset $T \subseteq S$ and for every $(s-h)$-subset $T' \subseteq S'$, the pair T, T' is both λ-expanding and (r, α)-wed.

The following theorem shows that matched pairs of sets will form a pipeline.

Theorem 6. *Let $S_0, \ldots, S_t$ be a sequence of $t+1$ subsets forming the vertex set of an oriented graph, each subset having order s. Suppose that each pair S_{i-1}, S_i is a (λ, r, α)-matched pair, $1 \leqslant i \leqslant t$. Then $S_0, \ldots, S_t$ form a pipeline of width s and length t, provided $t < 1/(2\alpha)$.*

Proof. Let $k = \lfloor (4/\lambda) \log_2 s \rfloor$. Observe that $k < 1(2\alpha)$ by the definition of a (λ, r, α)-matched pair. Moreover we may assume that $t \geqslant k$. For if $t < k$ we may extend the sequence of sets S_i by adding extra sets of vertices and edges forming (λ, r, α)-matched pairs, so that the length of the sequence becomes at least k. If now the extended sequence forms a pipeline, so did the original sequence.

Let $P_j = x_{j,0} x_{j,1} \ldots x_{j,t}$, $1 \leqslant j \leqslant s$, be s oriented paths of length t, and let $0 \leqslant m \leqslant t$. We say a labelling of S_m by $x_{j,m}$, $1 \leqslant j \leqslant s$, is *reachable* if there exist vertex disjoint copies of the subpaths $x_{j,0} x_{j,1} \ldots x_{j,m}$ with $x_{j,i} \in S_i$, $0 \leqslant i \leqslant m$. We need to show there is a reachable labelling of S_t. Of the $q!$ labellings of S_m let $(1 - \delta_m) q!$ be reachable. Clearly $\delta_0 = 0$; if we show that $\delta_{m+k} \leqslant \delta_m + 2k\alpha$ (or $\delta_t \leqslant \delta_m + 2(t - m)\alpha$ if $t - m < 2k$), then $\delta_t \leqslant 2t\alpha < 1$, which proves the theorem. We shall show then that $\delta_{m+l} \leqslant \delta_m + 2l\alpha$, provided $k \leqslant l < 2k$.

For $1 \leqslant i \leqslant l$ let F_i be the set of paths P_j with the edge $x_{j,m+i-1} x_{j,m+i}$ oriented forward from $x_{j,m+i-1}$ to $x_{j,m+i}$, and let B_i be the $s - |F_i|$ other paths. If either F_i or B_i is non-empty but small we will have to take special care with those paths. We form the set H of paths needing care by the following simple algorithm. Initially $H = \emptyset$; now repeat the following step. If, for some i, $0 < |F_i \setminus H| < r$ holds, replace H by $H \cup F_i$. Likewise, if $0 < |B_i \setminus H| < r$ holds, replace H by $H \cup B_i$. If no index i has one of these two properties, stop. Observe that, since $s \geqslant 4kr > 2lr$, no index i can be used in more than one step of the first l steps of the algorithm. Therefore the process terminates after at most l steps with $|H| = h$, where $h \leqslant 2kr$. We may suppose that the paths in H are $P_1, \ldots, P_h$.

For each choice $\mu = (y_1, \ldots, y_h)$ of a sequence of h vertices from S_m, let there be $(1 - \beta_\mu)(s - h)!$ reachable labellings of S_m with $x_{j,m} = y_j$, $1 \leqslant j \leqslant h$. Then $\sum_\mu (1 - \beta_\mu)(s - h)! = (1 - \delta_m) s!$, the sum being over all $s(s-1) \ldots (s-h+1)$ choices for μ. It follows that $\beta_\mu \leqslant \delta_m$ for some choice of μ; we make such a choice now and keep it fixed for the remainder of the proof.

Make a choice $v = (z_1, \ldots, z_h)$ of h vertices from S_{m+l}. By applying Lemma 3 h times (making use of the λ-expansion of pairs of subsets of the S_i) we construct handbuilt copies of the subpaths $x_{j,m} \ldots x_{j,m+l}$, $1 \leqslant j \leqslant h$, with $x_{j,m} = y_j$, $x_{j,m+l} = z_j$ and $x_{j,m+i} \in S_{m+i}$, $0 \leqslant i \leqslant l$. Let T_{m+i} be the set of $s - h$ vertices in S_{m+i} not used in these subpaths, $0 \leqslant i \leqslant l$. We will show that $(1 - \delta_m - 2l\alpha)(s - h)!$ of the labellings of T_{m+l} are reachable from T_m, via the sets T_{m+i}, by paths not in H. Therefore $(1 - \delta_m - 2l\alpha)(s - h)!$ of the labellings of S_{m+l} will be reachable in such a way that $x_{j,m+l} = z_j$, $1 \leqslant j \leqslant h$. This holds true for any of the $s(s-1) \ldots (s-h+1)$ choices for v, and summing over these choices we see that at least $\sum_v (1 - \delta_m - 2l\alpha)(s - h)! = (1 - \delta_m - 2l\alpha) s!$ labellings of S_{m+l} are reachable. Thus $\delta_{m+l} \leqslant \delta_m + 2l\alpha$, as claimed.

To complete the proof, therefore, it is enough to show that $(1 - \delta_m - 2l\alpha)(s - h)!$ of the labellings of T_{m+l} are reachable via $T_m, \ldots, T_{m+l-1}$. In fact we shall show that $(1 - \delta_m - 2\alpha)(s - h)!$ labellings of T_{m+1} can be reached via T_m; analogously $(1 - \delta_m - 4\alpha)(s-h)!$ labellings of T_{m+2} can be reached, and so on until the desired outcome is achieved. Let $u = |F_1 \setminus H|$, so $|B_1 \setminus H| = s - h - u$. By the definition of H, either $u = 0$ or $u = s - h$ or $r \leqslant u \leqslant s - h - r$. For each u-subset $B \subseteq T_{m+1}$ there are $u!(s - h - u)!$ labellings (not necessarily reachable) of T_{m+1} so that $x_{j,m+1} \in B$ if $P_j \in F_1 \setminus H$ and $x_{j,m+1} \notin B$ if $P_j \in B_1 \setminus H$; let L_B of these labellings be reachable. Likewise, for each u-subset $A \subseteq T_m$, there are $u!(s-h-u)!$ labellings of T_m so that $x_{j,m} \in A$ if $P_j \in F_1 \setminus H$ and $x_{j,m} \notin A$ if $P_j \in B_1 \setminus H$; let L_A of these be reachable by paths not in H.

Suppose now that for some choice of $A \subseteq T_m$ and $B \subseteq T_{m+1}$ it happens that A matches B and $T_{m+1} \setminus B$ matches $T_m \setminus A$. Fix these two matchings. Each of the L_A reachable labellings of T_m can now be extended to a different reachable labelling of T_{m+1}, so it follows that $L_B \geqslant L_A$. Since (S_m, S_{m+1}) is a (λ, r, α)-matched pair, there are at least $(1-\alpha)\binom{s-h}{u}$ choices for B for which there are at least $(1-\alpha)\binom{s-h}{u}$ choices for A such that $L_B \geqslant L_A$.

The sum of all $\binom{s-h}{u}$ values of L_A is at least $(1-\delta_m)(s-h)!$, by choice of μ. Denote the sum of the $(1-\alpha)\binom{s-h}{u}$ smallest values of L_A by L. Since $L_A \leqslant u!(s-h-u)$ for every A, we have

$$L \geqslant (1-\delta_m)(s-h)! - \alpha \binom{s-h}{u} u!(s-h-u)! = (1-\delta_m-\alpha)(s-h)!.$$

Therefore

$$\sum_B L_B \geqslant \sum_B \max\{L_A : A \text{ matches } B \text{ and } T_{m+1} \setminus B \text{ matches } T_m \setminus A\}$$
$$\geqslant \binom{s-h}{u}^{-1} \sum_B \sum\{L_A : A \text{ matches } B \text{ and } T_{m+1} \setminus B \text{ matches } T_m \setminus A\}$$
$$\geqslant \binom{s-h}{u}^{-1} (1-\alpha)\binom{s-h}{u} L$$
$$\geqslant (1-\alpha)(1-\delta_m-\alpha)(s-h)! \geqslant (1-\delta_m-2\alpha)(s-h)!.$$

Notice that the number of labellings of T_{m+1} which can be reached via T_m is precisely $\sum_B L_B$, which we now see is at least $(1-\delta_m-2\alpha)(s-h)!$, as claimed. This completes the proof of the theorem. $\qquad\square$

6. Robust pipelines

Our aim in this section is to show how to find a pipeline within an oriented graph. In fact, we shall show that a randomly chosen sequence of sets S_i form a pipeline. This cannot happen if $\delta^o(G) < (3/8 - \epsilon)|G|$, since examples of such graphs exist which are not expanders, and in that case a randomly chosen pair of subsets is very unlikely to be λ-expanding. We believe that if $\delta^o(G) > (3/8 + \epsilon)|G|$, in which case the graph is an expander (by Lemma 1), it is likely that a randomly chosen sequence of sets will yield a pipeline. Given such a pipeline, the machinery for proving our main theorem for graphs with this value of δ^o is all in place.

However, our present proof that randomly chosen sets form matched pairs (and so, by Theorem 6, a pipeline) does not make full use of the expansion properties of G. Rather, we achieve the required effect by using only properties of the neighbourhoods of pairs of vertices. In consequence, our proof works only for $\delta^o(G) > (5/12 + \epsilon)|G|$. We hope to have the opportunity to repair this deficiency in the future. In the meantime, we shall need the following definition.

Definition. *Let $\epsilon > 0$ and let G be an oriented graph of order n. A set of vertices $S \subset V(G)$ is said to represent a set $X \subset V(G)$ if $|S \cap X| > (|X|/n - \epsilon/2)|S|$. The set S is said to be*

ϵ-typical if it represents all the following sets:

$$\Gamma^+(x), \quad \Gamma^-(x), \quad \Gamma^+(x) \cup \Gamma^+(y), \quad \Gamma^-(x) \cup \Gamma^-(y),$$

$$Y^+(x,m) = \{\, y \in G \;:\; |\Gamma^+(x) \cup \Gamma^+(y)| > m \,\},$$

$$Y^-(x,m) = \{\, y \in G \;:\; |\Gamma^-(x) \cup \Gamma^-(y)| > m \,\},$$

for all $x, y \in G$ and for all integers $0 \leqslant m \leqslant n$.

To exploit the property of typicality we need a simple estimate for the size of the sets $Y^+(x,m)$ defined above.

Lemma 7. *Let x be a vertex of an oriented graph G and let $m \geqslant d^+(x)$. Then*

$$|Y^+(x,m)| \geqslant 4\delta^o(G) + 2d^+(x) - 4m.$$

Proof. Let $A = \Gamma^+(x) \setminus Y^+(x,m)$. If $A \neq \emptyset$, then some vertex of A sends at least $\delta^o(G) - (d^+(x) - |A|) - (|A| - 1)/2 = \delta^o - d^+ + (|A| + 1)/2$ edges out of $\Gamma^+(x)$. But by the definition of $Y^+(x,m)$, no vertex in A can send more than $m - d^+$ edges out of $\Gamma^+(x)$, so $|A|/2 \leqslant m - \delta^o$. This inequality remains true even if $A = \emptyset$.

Let $B = V(G) \setminus (Y^+(x,m) \cup \Gamma^+(x))$. If $B \neq \emptyset$, then the subgraph induced on B has minimum total degree at least $2\delta^o(G) - (|G| - |B|)$, so some vertex in B sends at least $\delta^o - (|G| - |B|)/2$ edges to vertices within B. This quantity must be at most $m - d^+$, so $|B|/2 \leqslant m - d^+ - \delta^o + |G|/2$. Once again, this inequality holds even if $B = \emptyset$.

Adding the two inequalities which have been derived we obtain that $|A \cup B| \leqslant 4m - 4\delta^o - 2d^+ + |G|$. But $Y^+(x,m) = V(G) \setminus (A \cup B)$ so the proof is complete. $\quad\square$

We can now show that, in a graph with a large value of $\delta^o(G)$, typicality is a guarantee of expansion.

Lemma 8. *Let G be an oriented graph of order n with $\delta^o(G) \geqslant (5/12 + \epsilon)n$. Let S and S' be ϵ-typical subsets with $|S| = |S'|$. Then the pair S, S' is ϵ-expanding.*

Proof. Let $s = |S| = |S'|$ and let $A \subset S$ with $0 < |A| = a \leqslant s/2$. Let $x \in A$ and let $m = \lfloor (a/s + \epsilon)n \rfloor$. We will show $A \cap Y^+(x,m) \neq \emptyset$. This is certainly true if $m < d^+(x)$, since then $Y^+(x,m) = V(G)$. But if $m \geqslant d^+(x)$ we can make use of the estimate given by Lemma 7, and the fact that S represents $Y^+(x,m)$, to obtain that

$$|S \cap Y^+(x,m)| > \big(|Y^+(x,m)|/n - \epsilon/2\big)s \geqslant (6\delta^o - 4m - \epsilon/2)s$$
$$\geqslant (5/2 + 3\epsilon/2)s - 4a \geqslant (1 + 3\epsilon/2)s - a > |S \setminus A|.$$

We may therefore select $y \in A \cap Y^+(x,m)$, the possibility that $y = x$ being permitted. Since $y \in Y^+(x,m)$ we see that $|\Gamma^+(x) \cup \Gamma^+(y)| \geqslant m + 1 > (a/s + \epsilon)n$. But S' is ϵ-typical and so represents $\Gamma^+(x) \cup \Gamma^+(y)$. Hence

$$|\Gamma^+(A)| \geqslant |S' \cap (\Gamma^+(x) \cup \Gamma^+(y))| > ((a/s + \epsilon) - \epsilon/2)s > (a + \epsilon s/2) \geqslant (1 + \epsilon)|A|.$$

Likewise $|\Gamma^-(A)| \geqslant (1+\epsilon)|A|$, and the same two inequalities hold for subsets $A \subset S'$. It follows that the pair S, S' is ϵ-expanding. □

To obtain properties of random subsets we need simple bounds on the tail of the hypergeometric distribution. In fact the usual bounds on the tail of the binomial distribution can be used. The following lemma is proved in [6], following Janson who proves stronger results [7].

Lemma 9. *From an urn containing pn red balls and $(1-p)n$ blue balls, j balls are chosen at random and without replacement. Let X be the number of red balls among the j chosen. Then, for $h \geqslant 0$,*

$$\mathbb{P}\{X \leqslant (p-h)j\} \leqslant \exp(-2jh^2) \quad \text{and} \quad \mathbb{P}\{X \geqslant (p+h)j\} \leqslant \exp(-2jh^2).$$

The next lemma shows that, roughly speaking, a randomly chosen set S will be very typical, as will any small perturbation of it.

Lemma 10. *Let G be an oriented graph of order n and let S be a randomly chosen subset of $V(G)$ of size $s \geqslant (\log n)^9$. Let $r > (50/\epsilon^2)\log n$ be an integer. Then, provided n is sufficiently large, with probability at least $1 - 1/n$ the set S is has the following property:*

Given $y \in S$ and $x \in \{y\} \cup (V(G) \setminus S)$, let $S^* = (S \setminus \{y\}) \cup \{x\}$. Then S^* is ϵ-typical. Moreover, given an h-subset $H \subset S^*$, with $h < (\log n)^3$, and an integer u with $u = 0$, $u = s - h$ or $r \leqslant u \leqslant s - h - r$, then at least $(1 - 1/n)\binom{s-h}{u}$ of the u-subsets of $S^* \setminus H$ are ϵ-typical.

Proof. Let $X \subset V(G)$ and let $J \subset V(G)$ be a randomly chosen subset of size $|J| = j \geqslant r$. By Lemma 9,

$$\mathbb{P}\{|J \cap X| \leqslant (|X|/n - \epsilon/4)j\} \leqslant e^{-\epsilon^2 j/8} < n^{-6}/8.$$

Therefore the probability that J fails to be $(\epsilon/2)$-typical is at most $1/2n^4$, typicality being defined by the representation of at most $4n^2$ subsets. Moreover, if $|J| = j \geqslant (\log n)^5$ and $h < (\log n)^3$ then the probability that J is not $(\epsilon/2)$-typical is at most

$$4n^2 e^{-\epsilon^2 j/8} < s^{-h} e^{-\epsilon^2 j/9 + (\log n)^4} < s^{-h} n^{-4}.$$

Now let S be a random s-subset of $V(G)$ and let $u \geqslant r$. The expected number of u-subsets of S failing to be $(\epsilon/2)$-typical is at most $\frac{1}{2n^3}\binom{s}{u}$, so with probability at least $1 - 1/n^2$ there are at most $\frac{1}{2n}\binom{s}{u}$ such u-subsets. Moreover, if $u \geqslant (\log n)^5$ and H is a specific h-subset of S, the expected number of u subsets of $S \setminus H$ failing to be $(\epsilon/2)$-typical is at most $s^{-h} n^{-4}\binom{s-h}{u}$. We call a set H bad if $S \setminus H$ contains more than $\frac{1}{n}\binom{s-h}{u}$ non-typical u-subsets. Thus the probability of a given H being bad is at most $s^{-h} n^{-3}$. Hence the expected number of bad H within S is at most n^{-3}, so with probability exceeding $1 - 1/n^3$ there are no bad H in S. These calculations were all performed for fixed values of u and h; taking into account all possible values, we conclude that with probability at least $1 - 1/n$ the set S has the following properties: it is itself $(\epsilon/2)$-typical, for each $u \geqslant r$ at least $(1 - 1/2n)\binom{s}{u}$ of the u-subsets of S are $(\epsilon/2)$-typical, and for each $h < (\log n)^3$, each

h-subset $H \subset S$ and each $u > (\log n)^5$ there are at least $(1 - 1/n)\binom{s-h}{u}$ u-subsets of $S \setminus H$ which are $(\epsilon/2)$-typical.

It suffices to show that any set S with the properties just described will also have the property claimed in the lemma. First of all, it is clear that S^* is ϵ-typical because S is $(\epsilon/2)$-typical. Now let y, x and H be as defined in the lemma, and, for a u-subset $A \subset S$, let $A^* = A$ if $y \notin A$ and otherwise let $A^* = (A \setminus \{y\}) \cup \{x\}$. Once again, A^* will be ϵ-typical if A is $(\epsilon/2)$-typical. Therefore, if $u \geqslant (\log n)^5$ and H is any h-subset of S^* there will be at least $(1 - 1/n)\binom{s-h}{u}$ u-subsets of $S^* \setminus H$ which are ϵ-typical. It remains only to verify the same property in the case $r \leqslant u < (\log n)^5$. But in that case $2(h+1)u < s$ and so

$$\binom{s-h}{u}\binom{s}{u}^{-1} \geqslant \left(\frac{s-h-u}{s-u}\right)^u \geqslant 1 - \frac{hu}{s-u} \geqslant \frac{1}{2},$$

and since $(1 - 1/2n)\binom{s}{u}$ u-subsets $A^* \subset S^*$ are ϵ-typical it follows that at least

$$\left(1 - \frac{1}{2n}\right)\binom{s}{u} - \left[\binom{s}{u} - \binom{s-h}{u}\right] = \binom{s-h}{u} - \frac{1}{2n}\binom{s}{u} \geqslant \left(1 - \frac{1}{n}\right)\binom{s-h}{u}$$

sets $A^* \subset S^* \setminus H$ are ϵ-typical. This completes the proof of the lemma.   □

We are now ready to show that a randomly chosen sequence $(S_0, \ldots, S_t)$ of sets forms a pipeline. In fact, when we finally come to work with the pipeline in the proof of the main theorem, we shall need to make a few alterations to the pipeline after we have chosen it but before we make use of it. Naturally, we shall need to know that the alterations we made have not destroyed the pipeline property; this is the motivation behind the next definition.

Definition. *A pipeline* $(S_0, S_1, \ldots, S_t)$ *in an oriented graph G is robust in G if, for any choice of vertices $y_i \in S_i$ and for any choice of distinct vertices $x_i \in \{y_i\} \cup (V(G) \setminus \bigcup_{j=0}^{t} S_j)$, $0 \leqslant i \leqslant t$, the sequence $(S_0^*, \ldots, S_t^*)$ is also a pipeline of width s and length t, where $S_i^* = (S_i \setminus \{y_i\}) \cup \{x_i\}$.*

Theorem 11. *Let G be an oriented graph of order n with $\delta^o(G) \geqslant (5/12 + \epsilon)n$. Then G contains a robust pipeline $(S_0, S_1, \ldots, S_t)$ of width s and length t in which each set S_i is ϵ-typical, provided $s > (\log n)^9$, $3 \leqslant t \leqslant n/s$ and n is sufficiently large.*

Proof. Given a random sequence of $t + 1$ disjoint s-subsets $(S_0, S_1, \ldots, S_t)$, the probability that any set S_i fails to have the property stated in Lemma 10 is at most $1/n$. Thus the expected number of sets failing to have the property is less than one, and so G contains a sequence $(S_0, S_1, \ldots, S_t)$ in which every set has the stated property. We claim that such a sequence forms a robust pipeline. To verify the claim, and so prove the theorem, it suffices by Theorem 6 to show that any pair of sets S_{i-1}^*, S_i^* (defined by the property in Lemma 10) form an $(\epsilon, r, 2/n)$-matched pair, where $r = \lceil (50/\epsilon^2) \log n \rceil$.

Let $S = S_{i-1}$ and $S' = S_i$. Let S^* and S'^* be defined in the obvious way. Note that $(16r/\epsilon) \log_2 s \leqslant s$ and $(8/\epsilon) \log_2 s < n/2$, as needed. Let h be an integer with $h \leqslant (16r/\epsilon) \log_2 s < (\log n)^3$. Let $H \subset S^*$ and $H' \subset S'^*$ be h-subsets. Let $T = S^* \setminus H$ and $T' = S'^* \setminus H'$. Then, for any integer u with $u = 0$, $u = s - h$ or $r \leqslant u \leqslant s - h - r$, there are at least $(1 - 2/n)\binom{s-h}{u}$ u-subsets $A \subset T$ for which both A and $T \setminus A$ are ϵ-typical, and the same is true for $(1 - 2/n)\binom{s-h}{u}$ u-subsets $B \subset T'$. By Lemmas 8 and 5, for such subsets, A matches B and $T' \setminus B$ matches $T \setminus A$. Therefore the pair T, T' is $(r, 2/n)$-wed. Moreover, the case $u = s - h$ shows that the pair T, T' is ϵ-typical and so, by Lemma 8, ϵ-expanding. This completes the proof of the theorem. $\qquad\square$

7. Oriented cycles

We now have all the ingredients needed for the proof of the main theorem.

Theorem 12. *Given $0 < \epsilon < 1$, there exists $n_0 = n_0(\epsilon)$ with the following property. Let G be an oriented graph of order $n > n_0$ with $\delta^o(G) \geqslant (5/12 + \epsilon)n$. Let C be an oriented cycle of length n. Then G contains a copy of C.*

Proof. As indicated in section 2, we shall decompose C into a collection of paths so chosen that they can be found in G using pipelines and sorters joined by handbuilt paths. Let $s = \lfloor \exp\{(\log\log n)(\log n)^{1/4}\} \rfloor > (\log n)^9$, let $k = \lfloor (4/\epsilon)\log_2 s \rfloor + 2$, let $l = (3\lceil\log_2(1/\epsilon)\rceil + 1)\lceil(\log_2 s)^2\rceil$ and let $t = \lfloor n^{1/2} \rfloor$. Some estimates in the proof will hold only if n is large; we will assume without further comment that n is sufficiently large for those estimates to be valid.

First, letting $p = 2l + 2k + t$, split the cycle C into $\lfloor n/p \rfloor$ paths each of length p (except for one whose length is between p and $2p$), and then from these select $\lfloor n/p \rfloor / 2$ non-incident paths of length p. Since $\lfloor n/p \rfloor / 2 > s2^{2l+2k}$, we can choose s of these paths $P'_1, \ldots, P'_s$ such that $l(P'_1) = \ldots = l(P'_s) = p$, $A(P'_1; l + k) \cong \ldots \cong A(P'_s; l + k)$ and $Z(P'_1; l + k) \cong \ldots \cong Z(P'_s; l + k)$. Hence $C = P'_1 Q'_1 P'_2 Q'_2 \ldots P'_s Q'_s$, where $l(Q'_j) \geqslant p$ for each j and $\sum_{j=1}^{s} l(Q'_j) = n - sp$.

In order to find the paths P'_j in G, we write $P'_j = A_j B_j P_j C_j D_j$, $1 \leqslant j \leqslant s$, where $l(A_j) = l(D_j) = l$, $l(B_j) = l(C_j) = k$ and $l(P_j) = t$. Notice that, by the choice of the P'_j, we have $A_1 \cong \ldots \cong A_s$, $B_1 \cong \ldots \cong B_s$, $C_1 \cong \ldots \cong C_s$ and $D_1 \cong \ldots \cong D_s$. We shall later construct two sorters of width s and length l (one for the A_j and one for the D_j), a pipeline of width s and length t (for the P_j) and join them together by handbuilt paths of length k (the C_j and the D_j), thereby realising the paths P'_j.

Unlike the paths P'_j, $1 \leqslant j \leqslant s$, the paths Q'_j may be of unequal length, so requiring more care in their construction. Let $Q'_j = E_j Q_j F_j$ where $l(E_j) = l(F_j) = k$, $1 \leqslant j \leqslant s$, so that our n-cycle can be written

$$C = P'_1 E_1 Q_1 F_1 P'_2 E_2 Q_2 F_2 \ldots P'_s E_s Q_s F_s.$$

We shall find in G a path $Q = Q_1 T_1 Q_2 T_2 \ldots Q_{s-1} T_{s-1} Q_s$, where the paths T_j have length one (they are just edges); the orientations of the T_j are immaterial. We have therefore

$$l(Q) = s - 1 + \sum_{j=1}^{s} l(Q_j) = s - 1 + n - sp - 2sk.$$

Let Q be split into consecutive paths $R'_0, \ldots, R'_s$ where $l(R'_j) = \lfloor l(Q)/s \rfloor$, $1 \leqslant j \leqslant s$, and $l(R'_0) < s$. Finally let $R'_j = H_j R_j$ where $l(H_j) = k$, $1 \leqslant j \leqslant s$, and let $u = l(R_1) = \ldots =$

$l(R_s) = \lfloor l(Q)/s \rfloor - k$; thus $l(R_0) = l(Q) - (u+k)s$. We shall find the cycle C in this way. The short path R'_0 will be handbuilt and the paths $R_1, \ldots, R_s$ will be obtained from a pipeline of length u. The ends of these paths will then be linked by handbuilt paths $H_1, \ldots, H_s$ to form the path Q. By deleting the edges $T_1, \ldots, T_{s-1}$ we obtain $Q_1, \ldots, Q_s$. These latter paths will then be linked to $P'_1, \ldots, P'_s$ via handbuilt paths E_j and F_j, $1 \leqslant j \leqslant s$.

The above describes the manner in which the cycle C can be constructed from various paths, but the order of the operations is critical. The difficulty is that each of the operations (building a pipeline, a sorter or a handbuilt path) requires a large population of vertices from which to choose, and towards the end of the construction of C the available population becomes very small. It is at this point that the notion of robustness will be needed. We are now ready to give the exact method by which the cycle is constructed.

First, using Theorem 4 we may find in G an (s, A_1)-sorter Σ_A. Similarly we may find an (s, D_1)-sorter Σ_D. A copy of the path R_0 may be found in the remaining graph via Lemma 2, since $l(R_0) < s < \epsilon n/4$. Partition the remaining vertices into $t + u + 2$ sets of order s, plus a set X of residual vertices; note that

$$|X| = n - us - l(R_0) - 1 - (t + 2l + 4)s = 5s(k-1).$$

From Theorem 11 it can be seen that the sets of order s can be chosen to form a robust pipeline of length $t + u + 1$, which we break into two robust pipelines, namely Π_R of length u and Π_P of length t. The sets of these pipelines are all ϵ-typical, as provided by Theorem 11.

Find, in Π_R, copies of the paths $R_1, \ldots, R_s$. We are now in need of $5s$ paths of length k, namely, the H_j to form Q and hence $Q_1, \ldots, Q_s$, the E_j to join the end of Σ_D to the start of the Q_j, the F_j to join the ends of the Q_j to the start of Σ_A, the B_j to join the end of Σ_A to the start of Π_P, and the C_j to join the end of Π_P to the start of Σ_D. To construct, say, the path H_1 from the endvertex x of R_0 to the first vertex y of R_1, select $k - 1$ consecutive s-sets from the pipeline Π_P, say $S_{j+1}, \ldots, S_{j+k-1}$. Using the ϵ-typicality of these sets, choose an appropriate neighbour x' of x in S_{j+1} and a neighbour y' of y in S_{j+k-1}. In view of the pairwise ϵ-expansion of these sets proved in Lemma 8, Lemma 3 shows that the rest of H_1 may be found linking x' to y'. Since $t > 5s(k-1)$ all the $5s$ paths required can be handbuilt in this manner without using more than one vertex from any s-set of Π_P. Replace the vertices removed from Π_P with those of X, so forming Π'_P.

The robustness of Π_P implies that Π'_P is itself a pipeline of width s and length t. We can therefore find $P_1, \ldots, P_s$ in Π'_P and extend these via the B_j and C_j to Σ_A and Σ_D. Within Σ_A and Σ_D we can find copies of the A_j and the D_j, so finding copies of $P'_1, \ldots, P'_s$ between the start of Σ_A and the end of Σ_D. Since Σ_A and Σ_D are sorters, we can arrange that the ends of the P'_j link correctly with the E_j and the F_j to form the cycle C as desired. $\square$

References

[1] Ajtai, M., Komlós, J. and Szemerédi, E. (1983) Sorting in $C \log n$ parallel rounds, *Combinatorica* **3** 1–19.

[2] Batcher, K. (1968) Sorting networks and their applications, *AFIPS Spring Joint Conf.* **32** 307–314.

[3] Ghouila-Houri, A. (1960) Une condition suffisante d'existence d'un circuit hamiltonien, *C.R. Acad. Sci. Paris* **251** 495–497.

[4] Grant, D. D. (1980) Antidirected Hamiltonian cycles in digraphs, *Ars Combinatoria* **10** 205–209.

[5] Häggkvist, R. (1993) Hamilton cycles in oriented graphs, *Combinatorics, Probability and Computing* **2** 25–32.

[6] Häggkvist, R. and Thomason, A. Oriented hamilton cycles in digraphs, (to appear in *J. Graph Theory*).

[7] Janson, S. Large deviation inequalities for sums of indicator variables (*preprint*).

[8] Thomason, A.G. (1986) Paths and cycles in tournaments, *Trans. Amer. Math. Soc.* **296** 167–180.

[9] Thomassen, C. (1979) Long cycles in digraphs with constraints on the degrees, in *Surveys in Combinatorics* (B. Bollobás, ed.) *London Math. Soc. Lecture Notes* **38** 211–228. Cambridge University Press.

Minimization Problems for Infinite n-Connected Graphs

R. HALIN

Mathematisches Seminar der Universität Hamburg, Bundesstraße 55, D-20146, Hamburg, Germany

A graph G is called n-minimizable if it can be reduced, by deleting a set of its edges, to a minimally n-connected graph. It is shown that, if n-connected graphs G and H differ only by finitely many vertices and edges, then G is n-minimizable if and only if H is n-minimizable (Theorem 4.12). In the main result, conditions are given that a tree decomposition of an n-connected graph G must satisfy in order to guarantee that the n-minimizability of each of the members of this decomposition implies the n-minimizability of the graph G (Theorem 6.5).

1. Introduction

It is an obvious fact that every finite n-connected graph can be reduced by deleting edges to a minimally n-connected (n-minimal for short) graph. However, the situation changes completely if we consider infinite n-connected graphs for some $n \geqslant 2$. (Throughout this article, n denotes a positive integer $\geqslant 2$.) In the finite case we reach an n-minimal factor simply by deleting edges successively, with the sole restriction that the n-connectedness is preserved at every step. Clearly this method fails in the infinite case. An n-connected graph

Figure 1.

Figure 2.

G is called *n-minimizable* if there is a set of edges L in G such that $G - L$ is *n*-minimal. Each such *n*-minimal factor $G - L$ of G is called an *n-minimization* of G. The infinite ladder of Figure 1 is the classical example of a 2-connected graph that is not 2-minimizable, whereas the locally finite 2-connected graph of Figure 2 has uncountably many pairwise non-isomorphic 2-minimizations. (We get all 2-minimizations by deleting one or two (suitable) edges in each $K_{1,1,2}$.)

While *n*-minimal graphs have been thoroughly studied in the literature, there is, to the author's knowledge, only one major result on *n*-minimizability, namely the following theorem of R. Schmidt [9].

Every n-connected graph that is rayless (i.e. that does not contain a one-sided infinite path) is n-minimizable.

The motivation for the present article arose from the search for some of the deeper reasons why an infinite *n*-connected graph G is, or is not, *n*-minimizable. An edge of G is called redundant if its deletion does not destroy the *n*-connectedness; $R_n(G)$ denotes the set of all redundant edges in G. Clearly, the effect of deleting a redundant edge, or a set of redundant edges, on the 'redundancy character' of the other edges is of basic importance in this context; it is studied in Sections 3, 4 and 5.

In Section 3, sets of edges whose deletion does not destroy the *n*-connectedness are studied from a 'global' point of view. Theorem 3.6 shows that if $R_n(G)$ is infinite, there is an *n*-connected subgraph H of G such that $R_n(G) = R_n(H)$, the order $|H|$ of H equals $|R_n(G)|$, and H is *n*-minimizable if and only if G is *n*-minimizable. From this we see that our minimization problem is reduced to the case that $|G| = |R_n(G)|$; G is then called 'of full redundance'.

The 'local' aspects of the problem are considered in Section 4. It is shown that elementary operations (deleting or adding an edge, deleting or adding a vertex of finite degree) leave *n*-minimizability invariant (provided that we stay in the class of *n*-connected graphs). The key observation for all that follows is Lemma 4.2, which states that deleting a redundant edge of G diminishes $R_n(G)$ only by finitely many edges. (This observation is not as obvious as one might perhaps expect at first glance.) Theorem 4.7 is seminal for the main result in Section 6. Roughly speaking it says that if the *n*-connected graph J is pasted together from two *n*-connected graphs G and H along a common finite subgraph, then $R_n(J)$ differs from $R_n(G) \cup R_n(H)$ only by a finite set of edges.

In Section 5 we try to imitate the reduction procedure of successively deleting redundant edges, which leads to an *n*-minimization in the finite case. We get certain maximal well-ordered sequences of redundant edges, which will be called reducing sequences. The smallest ordinal occurring as the order type of a reducing sequence of an *n*-minimizable graph is defined as its minimization type. All ordinals that may occur as the minimization types of graphs with full redundance are characterized. For *n*-connected graphs that cannot be minimized, a structure theorem is proved (see Theorem 5.6), which in the locally finite case exhibits the presence of an infinite ladder. Section 5 stands by itself, and is not needed elsewhere in this article.

Section 6 investigates *n*-connected graphs G that have a tree-decomposition with *n*-

connected members G_λ. (Roughly speaking, such a representation arises by a well ordered sequence of pasting operations as considered in Theorem 4.7.) In the main result of the present article we show the conditions under which the n-minimizability of each G_λ implies that G is n-minimizable also. These conditions are as follows:

1. The decomposition tree (associated to the decomposition with the members G_λ) is rayless;
2. In each G_λ there are only finitely many vertices with neighbours outside G_λ.

This result may be considered a generalization of Schmidt's theorem, since, by Diestel [9], every rayless n-connected graph has a tree-decomposition with rayless decomposition tree and finite n-connected members.

The simplest case not covered by Schmidt's theorem is that of 2-connected graphs containing a ray but no double ray (two way infinite path). A discussion of this case with respect to 2-minimization is given in Section 7. In Section 8 a few other related minimization problems are considered, and in the final section some open problems are presented.

2. Prerequisites

The graphs considered in this article are undirected and do not contain loops or multiple edges. In general, we adopt the terminology and notation that has become standard in graph theory.

If $G = (V, E)$ is a graph, then $|G|$ denotes the cardinality of V and $\|G\|$ the cardinality of E. An edge joining vertices a, b is denoted ab. If $T \subseteq V$, by $\binom{T}{2}$ we mean the set of all ab with $a \neq b$ in T; $\binom{T}{2}$ is the edge set of the complete graph K_T with vertex set T. For $L \subseteq \binom{V}{2}$, we denote the graphs $(V, E - L)$, $(V, E \cup L)$ by $G - L$ and $G \cup L$, respectively. In the special case $L = \{e\}$, we write $G - e$ and $G \cup e$ instead of $G - \{e\}$, $G \cup \{e\}$. Union and intersection of graphs are formed by joining and intersecting the vertex sets and edge sets of the graphs in question. Δ denotes the symmetric difference for sets. If G, H are graphs, then by $G \Delta H$ we mean the *set* $(V(G) \Delta V(H)) \cup (E(G) \Delta E(H))$.

We consider a subset of $V(G)$ as a subgraph of G with empty edge set. If $H \subseteq G$ is a subgraph of G, then $G - H$ denotes the induced subgraph of G having vertex set $V(G) - V(H)$. For $T \subseteq G$, $G[T]$ denotes the induced subgraph of G having vertex set $V(T)$. A *factor* of G is a subgraph H of G with $V(H) = V(G)$. If L is a set of edges, $V(L)$ denotes the set of end vertices of all these edges.

If $T \subset V$ and $a, b \in V$, we write $a . T . b(G)$ if $a, b \notin T$ and each (a, b)-path in G meets T (or, equivalently, if a, b belong to different components of $G - T$); then T separates a, b in G.

For vertices $a \neq b$ an (a, b)-skein of strength k (k an arbitrary cardinal) is the union of k (a, b)-paths having pairwise nothing but a and b in common. Such a configuration is denoted by $\Theta_k(a, b)$. $\mu_G(a, b)$ denotes the greatest k such that, for given vertices $a \neq b$ of the graph G, there is a $\Theta_k(a, b)$ in G. (This maximum always exists.) For non-adjacent a, b, we have Menger's theorem:

$$\mu_G(a, b) = \min(t \mid \text{there exists a } T \text{ with } a . T . b(G) \text{ and } |T| = t).$$

The connectivity $c(G)$ of G is the minimum of all $\mu_G(a, b)$ with $a \neq b$ (if $|G| \geq 2$); we put $c(K_1) = 0$. G is n-connected if $c(G) \geq n$.

Lemma 2.1. *If $c(G) \geqslant n$ and $L \subseteq E(G)$, then $c(G-L) \geqslant n$ if and only if $\mu_{G-L}(a,b) \geqslant n$ for every $ab \in L$.*

(The if-part is easy to prove as follows. Let $x, y \in V(G)$, $T \subseteq V(G) - \{x, y\}$ with $|T| < n$. Then there is an (x, y)-path P in $G - T$. For each edge $e = ab \in L$ occurring on P, we have an (a, b)-path $P_e \subseteq G - L$ that does not meet T. Replacing every such e by P_e, we find from P a connected subgraph of $G - L$ containing x, y and avoiding T.)

Let $T \subseteq V(G)$ and $a \in V(G) - T$. An (a, T)-*fan* of strength n is the union of paths $P_1, \ldots, P_n$ each starting in a and ending in a vertex of T such that for $i \neq j$ the paths P_i, P_j have only a in common, and, further, each P_i has only its end vertex $\neq a$ with T in common. We denote such a configuration by the symbol $\Psi_n(a, T)$. G with $|G| \geqslant n + 1$ is n-connected if and only if for such a and T a $\Psi_n(a, T)$ always exists in G.

G is n-*minimal* if $c(G) \geqslant n$ and $c(G - e) < n$ for all $e \in E(G)$. The following is well known.

Lemma 2.2. *Let $c(G) \geqslant n$. G is n-minimal if and only if $\mu_G(a, b) = n$ for each $e = ab \in E(G)$. If G is n-minimal and $e = ab \in E(G)$, then $c(G - e) = n - 1$ and for each separating T of $G - e$ with $|T| = n - 1$ we have $a \, . \, T \, . \, b \ (G - e)$.*

Let G be an n-connected graph. An edge or vertex e of G is called n-*redundant* in G if $G - e$ remains n-connected. Let $R_n(G)$ denote the set of n-redundant edges of G. Obviously $ab \in E(G)$ is in $R_n(G)$ if and only if $\mu_G(a, b) > n$, and $H \subseteq G$ implies $R_n(H) \subseteq R_n(G)$. The edges not in $R_n(G)$ are called *necessary*.

G with $c(G) \geqslant n$ is n-*minimizable* if it has an n-minimal factor H. H then must be of the form $G - L$, $L \subseteq R_n(G)$. Clearly, if $R_n(G)$ is finite, then G is n-minimizable.

It is obvious that every n-connected subgraph of an n-minimal graph is n-minimal, too. However, not every n-connected induced subgraph of an n-minimizable graph must be n-minimizable again. For instance, if we add, for each 'step' (horizontal edge) e of the ladder in Figure 1, a new vertex v_e and two edges from v_e to the end vertices of e, we get a 2-minimizable graph G (delete the original 'steps'), whereas the original ladder is not 2-minimizable.

It is easy to show that the union of a non-empty chain of n-minimal graphs (with respect to inclusion) is again n-minimal. Therefore we get by Zorn's Lemma:

Lemma 2.3. *If H is an n-minimal subgraph of the n-connected graph G, then H is contained in an inclusion-maximal n-minimal subgraph of G.*

Of course the statement is no longer true if 'n-minimal' is replaced by 'n-minimizable': every countable 2-connected graph is the union of an ascending chain of finite 2-connected (hence 2-minimizable) graphs.

We use ω to denote the cardinal of the countable sets. For an ordinal σ, $W(\sigma)$ denotes the set of all ordinals $\nu < \sigma$.

Lemma 2.4. *Let* $G_\lambda(\lambda \in W(\sigma))$ *be a well-ordered family of n-connected graphs such that for every* $\lambda > 0$ *we have*

$$\left| G_\lambda \cap \bigcup_{\nu \in W(\lambda)} G_\nu \right| \geq n.$$

Then $\bigcup_{\lambda \in W(\sigma)} G_\lambda$ *is n-connected.*

The proof is routine and left to the reader.

The maximal number of disjoint rays (*i.e.* one-way infinite paths) in G is denoted by $m_1(G)$, see [5]. For a connected G, $m_1(G) = 1$ means that G contains a ray but not a double ray (*i.e.* a two-way infinite path); the structure of these graphs is described in [4].

A profound theory of rayless graphs was developed by R. Schmidt [8], [9]: an ordinal $o(G)$ (called its order) is associated with every rayless graph G such that $o(G) = 0$ if and only if G is finite, and for every rayless graph G with $o(G) > 0$ there exists a finite F in G such that for all components C of $G - F$, $o(C) < o(G)$. This concept allows proofs by transfinite induction on $o(G)$. Also, $G \subseteq H$ always implies $o(G) \leq o(H)$.

3. Sets of redundant edges

In this section some elementary 'global' reductions of infinite n-connected graphs are given. If r and s are cardinals, we write $r \overset{\circ}{=} s$ if either r and s are both finite or r and s are equal infinite cardinals.

Proposition 3.1. *Let* G *be n-connected and* $T \subseteq G$. *Then there is an n-connected subgraph* H *of* G *with the following properties:*

1. $H \supseteq T$;
2. $|H| = |T|$ *if* T *is infinite, and*
   $|H| \leq \omega$ *if* T *is finite;*
3. $R_n(H) = R_n(G) \cap E(H)$.

Proof. If J is any subgraph of G with $|J| \geq 2$, then for a pair $a \neq b$ of $V(H)$ choose a $\Theta_n(a, b) \subseteq G$ and a $\Theta_{n+1}(a, b) \subseteq G$ if $ab \in R_n(G)$; define $\Phi(J)$ as the union of J and all these n-skeins and $(n+1)$-skeins. Then the desired H is obtained in the form $T \cup \Phi(T) \cup \Phi^2(T) \cup \Phi^3(T) \cup \dots$. (It is no restriction to assume $|T| \geq 2$.) □

Corollary 3.2. *If* $c(G) \geq n$, *then the n-minimizability of* G *implies that every* $T \subseteq G$ *with* $|T| \leq \omega$ *is contained in an n-minimizable* $H \subseteq G$ *with* $|H| \leq \omega$.

Problem 3.3. Is the inverse implication of Corollary 3.2 true also?

A positive answer would reduce the problem of n-minimization to the countable case.

Theorem 3.4. *Let* $c(G) \geq n$ *and* L *be an infinite set of edges in* G *such that* $G - L$ *remains n-connected. Let* P *be the set of edges in* $R_n(G)$ *that become necessary in* $G - L$, *i.e.*

$$P = (R_n(G) - L) - R_n(G - L).$$

Then
$$|P| \leqslant |L|.$$

Furthermore, if H is an n-connected subgraph of $G-L$ that contains $V(L)$, then $E(H) \supseteq P$.

Proof. By Proposition 3.1, every such H contains an n-connected subgraph H' with $V(H') \supseteq V(L)$ and $|H'| = |L|$. Assume that there is an edge $e = xy \in P$ that does not belong to H'. Then by $c(H') \geqslant n$ we have $\mu_{G-L-e}(a,b) \geqslant \mu_H(a,b) \geqslant n$ for each $ab \in L$; hence $G - L - e$ would be n-connected by Lemma 2.1, in contradiction to $e \notin R_n(G-L)$. So we have
$$P \subseteq E(H') \subseteq E(H) \quad \text{and} \quad |P| \leqslant |H'| = |L|. \qquad \square$$

Corollary 3.5. *If $c(G) \geqslant n$ and $G-L$ is an n-minimization of G, then*
$$|L| \overset{\circ}{=} |R_n(G)|.$$

Proof. By Theorem 3.4 we have $|R_n(G) - L| \leqslant |L|$, whence the result. $\qquad \square$

Theorem 3.6. *Let G be n-connected with infinite $R_n(G)$. Then there exists an n-connected induced subgraph H of G with $|H| = |R_n(G)|$ such that*
$$R_n(G) = R_n(H),$$
and for each $L \subseteq R_n(G)$ we have that $G-L$ is n-minimal if and only if $H-L$ is n-minimal.

Proof. Let $T = V(R_n(G))$. For each $e = ab \in R_n(G)$ and every finite $F \subseteq R_n(G)$ choose as $H_{e,F}$ a $\Theta_n(a,b) \subseteq G-e$ with $E(\Theta_n(a,b)) \cap R_n(G) = F$ if such an (a,b)-skein exists; if not, let $H_{e,F} = \varnothing$. T and all these $H_{e,F}$ form a graph D with $|D| = |T|$, and by Proposition 3.1, D can be extended to an induced subgraph H of G with $|H| = |T|$ and $c(H) \geqslant n$. By construction we have $E(H) \supseteq R_n(G)$ and (by the choice of the $H_{e,F}$) $R_n(H) = R_n(G)$.

1. Let $L \subseteq R_n(G)$ with $G-L$ n-minimal. We claim that $H-L$ is n-minimal too. To verify $c(H-L) \geqslant n$, by Lemma 2.1, we only have to show $\mu_{H-L}(a,b) \geqslant n$ for every $e = ab \in L$. Now for such an edge, by $c(G-L) \geqslant n$, we find a $\Theta_n(a,b) \subseteq G-L$. Let
$$F = E(\Theta_n(a,b)) \cap R_n(G); \ F \cap L = \varnothing.$$

   By construction of H we have $H_{e,F} \subseteq H$, which is a $\Theta_n(a,b)$ that also shares exactly F with $R_n(G)$, hence avoids L. So we find $\mu_{H-L}(a,b) \geqslant n$.

2. Now assume that $H-L$ is n-minimal for an $L \subseteq R_n(H)$. Since $\mu_{G-L}(a,b) \geqslant \mu_{H-L}(a,b) \geqslant n$ for every $ab \in L$, we conclude (by Lemma 2.1) that $G-L$ is n-connected. Assume that there is an $e = ab \in R_n(G) - L$ such that $\mu_{G-L}(a,b) > n$. Then there is a $\Theta_n(a,b)$ in $G-L-e$. Let F be the set of edges which $\Theta_n(a,b)$ has in common with $R_n(G)$. Then by $H_{F,e} \subseteq H-L-e$ we would get $\mu_{H-L}(a,b) \geqslant n+1$, contradicting the n-minimality of $H-L$. Hence $\mu_{G-L}(x,y) = n$ for each $xy \in E(G-L)$, and $G-L$ must be n-minimal. $\square$

By this theorem, our minimizing problem requires us to consider only such infinite n-

connected graphs with order coinciding with the number of n-redundant edges. We then speak of n-connected graphs *with full redundance*.

Theorem 3.7. *If $c(G) \geqslant n$ and $R_n(G)$ is infinite, then there exists $L \subseteq R_n(G)$ with $|L| = |R_n(G)|$ such that $G - L$ remains n-connected.*

Proof. For each $e = ab \in R_n(G)$ choose a $\Theta_n(a, b) \subseteq G - e$ as H_e. Let $\mathscr{P}$ be the set of all subsets S of $R_n(G)$ such that no $e \in S$ lies in $H_{e'}$ for an $e' \in S - \{e\}$. If $S_i (i \in I)$ is a chain of elements of $\mathscr{P}$, then, clearly, $\bigcup_{i \in I} S_i$ is again in $\mathscr{P}$. By Zorn's Lemma, $\mathscr{P}$ contains a maximal element L.

Then $G - L$ is n-connected. Namely if $e = ab \in L$, then $H_e \subseteq G - L$, hence $\mu_{G-L}(a, b) \geqslant n$, and $c(G - L) \geqslant n$ follows from Lemma 2.1.

Moreover $|L| = |R_n(G)|$. Otherwise $|L| < |R_n(G)| = |R_n(G) - L|$ and $\|\bigcup_{e \in L} H_e\| < |R_n(G)|$; for $e^* \in R_n(G) - \bigcup_{e \in L} E(H_e)$, we would have $L \cup \{e^*\} \in \mathscr{P}$, in contradiction to the maximality of L. $\qquad \square$

4. Elementary operations

In this section we study the effect on the n-minimizability of applying certain elementary operations (deleting and adding vertices or edges).

Let $G = (V, E)$ be an n-connected graph. For distinct $a, b \in V$ and $e \in E$, call e necessary for a, b, if $\mu_{G-e}(a, b) < n$. Let $Q_n(G; a, b)$ be the set of all necessary edges for a, b. Then clearly

Lemma 4.1. *$Q_n(G; a, b)$ is the intersection of the edge-sets of all $\Theta_n(a, b) \subseteq G$, and is hence finite.*

The following observation is the key to what follows.

Lemma 4.2. *For each $e = ab \in R_n(G)$,*

$$R_n(G) - R_n(G - e) \subseteq Q_n(G - e; a, b) \cup \{e\}.$$

Hence, if an n-redundant edge $e = ab$ is omitted, only finitely many edges $\in R_n(G)$ become necessary in $G - e$ (and these must all be necessary for a and b).

Proof. Let $e' = xy \in R_n(G) - R_n(G - e)$ with $e' \neq e$. Since e' is necessary in $G - e$, we have $\mu_{G-e}(x, y) = n$; hence there is a $T \subseteq V$ with $|T| = n - 1$ and $x \cdot T \cdot y \ (G - e - e')$. $(G - e - e') - T$ has exactly two components, say C_x and C_y with $x \in V(C_x)$, $y \in V(C_y)$. e must lead from C_x to C_y; for otherwise $x \cdot T \cdot y \ (G - e')$, contradicting $e' \in R_n(G)$. Hence

$$a \cdot T \cdot b \, (G - e - e'),$$

but $\mu_{G-e}(a, b) > n - 1$. So we conclude $e' \in Q_n(G - e; a, b)$, and the proof is complete. $\qquad \square$

Our proof also yields the following for edges $e \neq e'$.

Lemma 4.3. $e' \in R_n(G) - R_n(G-e) \Rightarrow e \in R_n(G) - R_n(G-e')$.

So the pairs $(e, e') \in R_n(G) \times R_n(G)$ with $e \neq e'$ and $e' \in R_n(G) - R_n(G-e)$ define an irreflexive and symmetric relation on $R_n(G)$, or a graph with vertex set $R_n(G)$. We denote this graph by $\mathscr{R}_n(G)$ and call it the *n-redundance graph* of G. By Lemma 4.2, $\mathscr{R}_n(G)$ is locally finite. If $e \in R_n(G)$, then $\mathscr{R}_n(G-e)$ arises from $\mathscr{R}_n(G)$ by deleting e and all its neighbours, and adding (eventually) finitely many new edges. The local finiteness of $\mathscr{R}_n(G)$ can also be read in the following way.

Lemma 4.4. *For each $e \in R_n(G)$ there are only finitely many $e' = xy \in R_n(G)$ such that e is necessary for x, y in $G-e'$.*

Furthermore, we have

Lemma 4.5. *If $e = ab \in \binom{V}{2} - E$, then in $G \cup e$ there are only finitely many edges that are necessary in G but n-redundant in $G \cup e$.*

(By Lemma 4.2 we have $R_n(G \cup e) - R_n(G) \subseteq Q_n(G; a, b) \cup \{e\}$, which implies the result.)

Lemma 4.6. *If $G = (V, E)$ is n-connected and, for an $e = ab \in \binom{V}{2} - E$, $G \cup e$ is n-minimizable, then G is also n-minimizable.*

Proof. Let $G' \cup e$ be an *n*-minimization of $G \cup e$, where G' is a factor of G. By $c(G) \geq n$, we find a $\Theta_n(a, b) \subseteq G$; and by Lemma 2.1, $G' \cup \Theta_n(a, b)$ is *n*-connected. By Lemma 4.5, $G'' = G' \cup e \cup \Theta_n(a, b)$ has only finitely many *n*-redundant edges and e is one of them. By deleting e and then an appropriate finite sequence of *n*-redundant edges from G'', we find an *n*-minimal graph that is a factor of G. $\qquad\square$

Theorem 4.7. *Let G, H be n-connected graphs with $n \leq |G \cap H| < \infty$. Then $G \cup H$ is n-connected (Lemma 2.4) and we have*

$$R_n(G \cup H) = R_n(G) \cup R_n(H) \cup L$$

with a finite L.

Proof. Let $V(G \cap H) = F$. By Lemma 4.5 we have

$$R_n\left(G \cup \binom{F}{2}\right) = R_n(G) \cup L_1,$$

$$R_n\left(H \cup \binom{F}{2}\right) = R_n(H) \cup L_2$$

with finite L_1, L_2.

We only have to show that if e is in $R_n(G \cup H)$, but not in $R_n(G) \cup R_n(H)$, then e is in $L_1 \cup L_2 \cup \binom{F}{2}$. Assume $e = xy$ not in $\binom{F}{2}$, say $x \in V(G) - F$ (without loss of generality).

There exists a $\Theta_n(x, y) \subseteq (G \cup H) - e$ with the (x, y)-paths $P_1, ..., P_n$. If a P_i leaves G, it has a first and a last vertex with F in common; then we replace the subpath of P_i between these two vertices by the edge $\in \binom{F}{2}$ joining them. In this way we get an (x, y)-skein of strength n in $(G \cup \binom{F}{2})) - e$, and we find that e is n-redundant in $G \cup \binom{F}{2}$. As e is not in $R_n(G)$, it lies in L_1, and our proof is complete. ☐

Corollary 4.8. *Let G, H be n-connected graphs with $n \leqslant |G \cap H| < \infty$. Then $G \cup H$ is n-minimizable if and only if G and H are n-minimizable.*

Proof

1. Let G', H' be n-minimizations of G and H, respectively. Then $G' \cup H'$ is an n-connected factor of $G \cup H$. By Theorem 4.7, $R_n(G' \cup H')$ is finite, hence $G' \cup H'$ (and therefore also $G \cup H$) is n-minimizable.
2. Let $J = G \cup H$ and let J' be an n-minimization of J. Let G', H' denote the subgraphs of J' induced by $V(G)$, $V(H)$ (respectively). With $F = V(G) \cap V(H)$ put

$$G^* = G' \cup \binom{F}{2}, \quad H^* = H' \cup \binom{F}{2}.$$

G^* and H^* are n-connected (this follows by an argument similar to that at the end of the proof of 4.7). By 4.5, $R_n(J' \cup \binom{F}{2})$ is finite. Hence also $R_n(G^*) \subseteq R_n(J' \cup \binom{F}{2})$ is finite, and therefore G^* is n-minimizable. We see that $G \cup \binom{F}{2}$ is n-minimizable, and from 4.6 it follows that G must be n-minimizable. Analogously, we find that H is n-minimizable. ☐

Corollary 4.8 is no longer true if we allow $G \cap H$ to be infinite. This is shown by

Example 4.9. The graphs G and H of Figure 3 are 2-minimal, but it we take their union and identify each i with i', we get a 2-connected graph that is not 2-minimizable.

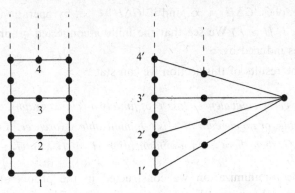

Figure 3.

We define the operation of *adding a vertex of degree k* to a graph G by: add a new vertex x to G, choose a k-element set F of vertices in G, and join x to all the vertices of F by edges. By applying Lemma 4.6 and Corollary 4.8 to G and to the complete graph H with vertex set $F \cup \{x\}$ we get

Corollary 4.10. *If G is n-connected, adding a vertex of finite degree $k \geqslant n$ leads to an n-minimizable graph J if and only if G is n-minimizable.*

If k is allowed to be infinite, J can be n-minimizable whereas G is not. This is shown in the case $n = 2$ by the graph in Figure 4.

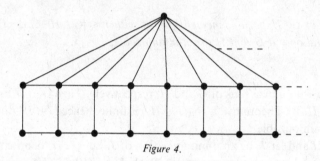

Figure 4.

Two graphs G, H are called *finitely n-related* (written $G \simeq_n H$) if G and H are both n-connected and G can be transformed into H by a finite sequence of the following operations:

1. adding a new edge,
2. adding a vertex of degree n,
3. deleting an n-redundant edge,
4. deleting an n-redundant vertex of degree n.

Clearly $\simeq_n$ is an equivalence relation in the class of n-connected graphs, and we have

Lemma 4.11. $G \simeq_n H$ *if and only if G, H are n-connected and the set $G \Delta H$ is finite.*

(Clearly $G \simeq_n H$ implies $|G\Delta H| < \infty$, and if $|G\Delta H| < \infty$, by operations 1 and 2 we get a graph J with $G \simeq_n J$, $H \simeq_n J$.) We see that the finite n-connected graphs form one of the equivalence-classes induced by $\simeq_n$.

Summarizing the results of this section we can state:

Theorem 4.12. *For every equivalence class $\mathscr{C}$ of finitely n-related graphs, either every element of $\mathscr{C}$ is n-minimizable, or no element of $\mathscr{C}$ is n-minimizable. Moreover, if $G \simeq_n H$ and G' is an n-minimization of G, then there is an n-minimization H' of H with $G' \simeq_n H'$.*

(Notice that the n-minimization we constructed in the proofs of Lemma 4.6 and Corollary 4.10 from a given one always remains finitely n-related to the latter.)

Also we can interpret Corollary 4.8 (and its proof) in the following way: if G, H are n-connected and $n \leqslant |G \cap H| < \infty$, we get all n-minimizations of $G \cup H$ modulo $\simeq_n$ as the unions of n-minimizations of G and H.

From Theorem 4.12 we immediately conclude

Corollary 4.13. *If $c(G) \geqslant n$, either G is n-minimizable or for every n-minimal subgraph H of G we have $|G - H| \geqslant \omega$.*

Corollary 4.14. *Let $e = xy$ be an edge of an n-connected graph G such that x is of finite degree and such that the graph H arises from G by contracting e to a single vertex. Then G is n-minimizable if and only if H is n-minimizable.*

5. Reducing sequences

In this section G is n-connected and of infinite order α (where α is identified with the initial ordinal of cardinality $|G|$), and G is supposed to be of full redundance.

Let σ be an ordinal and W be a well-ordering of $R_n(G)$ with order type σ; then we can consider W as a family $(e_\nu)_{\nu < \sigma}$, where the e_ν are the elements of $R_n(G)$. Such a family $(e_\nu)_{\nu < \sigma}$ is called a *reducing sequence of G.*

We now define a subset $L = L_W$ of $R_n(G)$ by the following inductive selection procedure (which generalizes the natural method of minimizing a finite n-connected graph). Let $e_0 \in L$. Now let $0 < \lambda < \sigma$ and assume that for all e_ν with $\nu < \lambda$ it is decided whether $e_\nu \in L$ or not and let the set of the selected e_ν ($\nu < \lambda$) be denoted by $L(\lambda)$. Then, if $c(G - L(\lambda)) < n$, put $L = L(\lambda)$ and let the procedure be finished. If $c(G - L(\lambda)) \geqslant n$, let $e_\lambda \in L$ if and only if $e_\lambda \in R_n(G - L(\lambda))$.

Lemma 4.5 guarantees that the procedure does not terminate before ω, so L is an infinite set that may be considered as a well-ordered subsequence of W.

Lemma 5.1. *G is n-minimizable if and only if there is a reducing sequence W such that $G - L_W$ remains n-connected.*

Proof. If $G - L$ is an n-minimization of G, choose a well-ordering W of $R_n(G)$ such that L is an initial segment of W; then $L = L_W$.

If W is such that $G - L_W$ is n-connected, then $G - L_W$ must be n-minimal; for if it contained an e_ν, it would have to be selected in the νth selection step. $\qquad\square$

Now we consider the case that G is n-minimizable. We define the smallest ordinal τ for which there is a well-ordering W of $R_n(G)$ of type τ such that $G - L_W$ is n-minimal as the *minimization type* of G (with respect to n-connectedness), and denote it by $\tau_n(G)$.

If $G - L$ is an n-minimization of G, we well-order L according to α and put behind it a well-ordering of $R_n(G) - L$ of type $\leqslant \alpha$; we see that $\tau_n(G) \leqslant \alpha + \alpha$.

Assume that $\tau_n(G) = \alpha + \beta$ with $\beta < \alpha$. Let W be a corresponding reducing sequence $(e_\lambda)_{\lambda < \alpha + \beta}$ and $L = L_W$, so that $G - L$ is n-minimal. Let L' be the subsequence of the e_λ with $\lambda \geqslant \alpha$ and $e_\lambda \in L$. Put L' before e_0, and re-order the set L'' of elements $e_\nu \in R_n(G) - L$ with $\nu \geqslant \alpha$ according to the initial ordinal γ of $|L''|$; clearly $\gamma \leqslant \beta$. The new well-ordering W' obtained in this way is of type $\alpha + \gamma$ with $L_{W'} = L$; therefore $\beta = \gamma$, and β is seen to be an initial ordinal. So we can state:

Lemma 5.2. *If G is n-connected with full redundance and n-minimizable, $\tau_n(G) = \alpha + \beta$, where α is the initial ordinal of cardinality $|G|$ and β is another initial ordinal $\leqslant \alpha$.*

We now show by examples that all these ordinals $\alpha + \beta$ also occur as minimization-types in the case $n = 2$.

We choose two disjoint copies K_α and K'_α of the complete graph of order α with vertices x_ν, $x'_\nu (\nu < \alpha)$, respectively. We subdivide each edge of K_α and K'_α by inserting a new vertex and draw all the edges $e_\nu = x_\nu x'_\nu$. We now add three vertices a, b, c and the edges $x_0 a$, ab, bc, cx'_0; the graph constructed in this way is called X. X is 2-connected and $R_2(X)$ consists of the edges e_ν. A 2-minimization of X is obtained if all the e_ν with the exception of exactly one are deleted. If W is a well-ordering of $R_n(X)$ of type α, $G - L_W$ will not remain 2-connected; but if W is chosen of order type $\alpha + 1$, we get a 2-minimization. Hence $\tau_2(X) = \alpha + 1$.

If $1 \leqslant \beta \leqslant \alpha$, we take β disjoint copies X_λ of X and paste them together along the path with the vertices a, b, c. We get a 2-connected graph Z_β of order α with full redundance; $R_n(Z_\beta) = \bigcup_{\lambda < \beta} R_n(X_\lambda)$. If W is a well-ordering of $R_n(Z_\beta)$ of type $\alpha + \gamma$ with $\gamma < \beta$, there is a λ such that all elements of $R_n(X_\lambda)$ lie in the initial segment of type α; hence in $Z_\beta = L_W$ all these edges are missing. On the other hand, it is easy to find a well-ordering of type $\alpha + \beta$ such that $Z_\beta - L_W$ is 2-minimal. So we have $\tau_2(Z_\beta) = \alpha + \beta$. If Z_0 arises from X by subdivision of the edge e_1, then clearly $\tau_2(Z_0) = \alpha + 0$. So we see:

Theorem 5.3. *The possible minimization-types for 2-connected graphs of infinite order α with full redundance are exactly the ordinals $\alpha + \beta$, where β is an initial ordinal $\leqslant \alpha$.* $\square$

Now we assume that G is not n-minimizable. Let $W = (e_\lambda)_{\lambda < \sigma}$ be a reducing sequence.

Lemma 5.4. *L_W has no greatest element, so its order type is a limit ordinal.*

Proof. Assume e_λ to be the last element of L_W. Then $c(G - L(\lambda)) \geqslant n$ and $e_\lambda \in R_n(G - L(\lambda))$, by construction of L_W. But then $G - L_W$ would be n-connected, hence n-minimal by Lemma 5.1, giving a contradiction. $\square$

So we have:

Lemma 5.5. *$G - F$ remains n-connected for every finite $F \subset L_W$.*

By Lemma 5.1, $c(G - L_W) < n$. Hence there exists a smallest $T \subset V(G)$ with $|T| < n - 1$ (possibly empty) such that $G - L_W - T$ has distinct components $C_i (i \in I)$ with $|I| \geqslant 2$. For every partition of I into two non-empty sets I', I'' there must be infinitely many edges in L_W connecting a vertex of $\bigcup_{i \in I'} C_i$ with a vertex of $\bigcup_{i \in I''} C_i$ (by Lemma 5.5). It is possible to choose this partition such that $H = G[\bigcup_{i \in I'} C_i]$ and $J = G[\bigcup_{i \in I''} C_i]$ are both connected. Let B be the set of edges $\in L_W$ connecting H and J; B is a bond (or edge-cut) in $G - T$. $B \cup T$ forms a 'mixed cut' of G (consisting of vertices and edges), which minimally separates H and J. So we have

Theorem 5.6. *If G with $c(G) \geqslant n$ is no n-minimizable, then there is a mixed cut in G consisting of at most $n-1$ vertices and infinitely many edges such that the deletion of finitely many of these edges never destroys the n-connectivity.*

If G is locally finite, H and J must be infinite and we can find rays U in H and U' in J that are connected (in G) by infinitely many pairwise disjoint paths (this follows by means of [3, Satz 5]) and on each of these paths lies at least one edge $\in L_W$. We see that in this way the infinite ladder must be present in G.

A more thorough study of the components C_i of $G - L_W - T$ should lead to a deeper insight into the structure of non n-minimizable graphs.

6. Tree-decompositions and *n*-minimization

Before we proceed to our main result some further preparations are necessary.

Let $G = (V, E)$ be a graph with $|G| \geqslant n+1$ and $T \subset V$ with $n \leqslant |T| < \infty$. G is called (n, T)-*connected* if $G \cup \binom{T}{2}$ is n-connected.

The following can be seen by easy applications of Menger's theorem and its well-known variations.

Lemma 6.1. *The following statements are equivalent:*

(a) *G is (n, T)-connected;*
(b) *there is a graph H with $V(H) \cap V(G) = T$ such that $G \cup H$ is n-connected;*
(c) *for every $v \in V - T$ there is a $\Psi_n(v, T) \subseteq G$;*
(d) *for every H with $c(H) \geqslant n$ and $V(G) \cap V(H) = T$ we have $G \cup H$ is n-connected.*

A graph G is called (n, T)-*minimal*, if it is (n, T)-connected, but $G - e$ is not (n, T)-connected for every $e \in E - \binom{T}{2}$. G is called (n, T)-*minimizable* if there is an $L \subseteq E$ such that $G - L$ is (n, T)-minimal.

Lemma 6.2. *If $G = (V, E)$ is n-minimal and $T \subseteq V$ with $n \leqslant |T| < \infty$, then there exists a finite $L \subseteq E - \binom{T}{2}$ such that $G - L$ is (n, T)-minimal.*

Proof. By Lemma 4.5, $R_n(G \cup \binom{T}{2})$ is finite. Let L be a maximal subset of $R_n(G \cup \binom{T}{2}) - \binom{T}{2}$ such that $G - L$ remains (n, T)-connected. We claim that $G - L$ is (n, T)-minimal.

Let $e \in E - L - \binom{T}{2}$; $e = xy$ with $x \notin T$. Assume that $G - L - e$ remains (n, T)-connected. Then there is a $\Psi_n(x, T)$ and, if $y \notin T$, also a $\Psi_n(y, T)$ in $G - L - e$. With $H := (G \cup \binom{T}{2}) - L - e$, we then conclude $\mu_H(x, y) \geqslant n$ and, by Lemma 2.1, $c(H) \geqslant n$, contradicting the maximal choice of L. $\qquad\square$

Lemma 6.3. *Let G and G_i ($i \in I$) be n-minimizable graphs. Let T be a finite subset of $V(G)$ such that $V(G) \cap V(G_i) = T_i \subseteq T$ and $|T_i| \geqslant n$ for every $i \in I$ holds. For all $i \neq j$ assume that $V(G_i \cap G_j) \subseteq T$. Then*

$$G \cup \bigcup_{i \in I} G_i$$

is n-minimizable.

Proof. By assumption we have n-minimizations G' of G and G_i'' of G_i ($i \in I$); by Lemma 6.2 each G_i'' has an (n, T_i)-minimization G_i'. Using Lemma 6.1 (c) we see that $H = T \cup \bigcup_{i \in I} G_i'$ is (n, T)-minimal. By Lemma 6.1 (d), $G' \cup H$ is an n-connected factor of $G \cup \bigcup_{i \in I} G_i$. Now we put

$$J = G' \cup H \cup \binom{T}{2}.$$

By Lemma 4.5, $R_n(G' \cup \binom{T}{2})$ is finite. Let $e \in R_n(J)$, $e \notin \binom{T}{2}$. The edge e must lie in $G' \cup \binom{T}{2}$, because otherwise it joins vertices x, y of some G_i' not both in T, and therefore $G_i' - e$ would not be (n, T)-connected, which implies $c(J - e) < n$. Therefore e joins vertices x, y in G', with $x \notin T$ (say): There is a $\Theta_n(x, y)$ in $J - e$ by choice of e, and this can be formed already in $G' \cup \binom{T}{2}$. Hence $e \in R_n(G' \cup \binom{T}{2})$.

We find $R_n(J) \subseteq \binom{T}{2} \cup R_n(G' \cup \binom{T}{2})$, hence it is finite, and therefore J is n-minimizable. By Lemma 4.6 $G' \cup H$ is n-minimizable also, and our proof is complete. □

Example 6.4. The graph H of Figure 5 is 2-connected and $R_2(H) = \{e_1, e_2, \ldots\}$. Obviously H is not 2-minimizable.

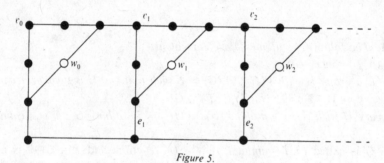

Figure 5.

Let $G = H - \{w_0, w_1, w_2, \ldots\}$ and G_i be the circuit through w_i and v_i ($i = 0, 1, 2, \ldots$). G and the G_i are all 2-minimal. So we see that Lemma 6.3 is no longer true if we drop the hypothesis that the T_i are all contained in a finite subset of $V(G)$.

Let G be a graph, $\sigma > 0$ an ordinal and $\mathcal{F} = (G_\lambda)_{\lambda \in W(\sigma)}$ be a family of induced subgraphs of G. Put

$$G|_\mu = \bigcup_{\lambda < \mu} G_\lambda \quad \text{and} \quad S_\mu = G|_\mu \cap G_\mu$$

for each μ with $0 < \mu < \sigma$. $\mathcal{F}$ is called a *tree-decomposition* of G if the following conditions are satisfied:

1. $G = \bigcup_{\lambda < \sigma} G_\lambda$,
2. for every λ with $0 < \lambda < \sigma$ there is a $\mu < \lambda$ such that $S_\lambda \subseteq G_\mu$.

If λ_- denotes the smallest μ with $S_\lambda \subseteq G_\mu$, then $T(\mathcal{F}) = (W(\sigma), \{\lambda \lambda_- | 0 < \lambda < \sigma\})$ is a tree, called the *decomposition tree* of $\mathcal{F}$. We consider the ordinal 0 as the *root* of $T(\mathcal{F})$. For each subgraph H of $T(\mathcal{F})$ we denote the subgraph $\bigcup_{\nu \in V(H)} G_\nu$ of G by G_H.

$\mathcal{F}$ is of *strength* n if all the 'attachments' S_λ have at least n vertices, and it is called *finitary*, if $|S_\lambda| < \infty$ for all $\lambda < \sigma$. We speak of a *rayless* tree-decomposition if its decomposition-tree is rayless.

We say that $\mathscr{F}$ satisfies the *finite attachment condition* $[FAC]$ if for every $\lambda < \sigma$ there is a finite subset F_λ of $V(G_\lambda)$ such that for every $\mu \neq \lambda$ we have $V(G_\lambda) \cap V(G_\mu) \subseteq F_\lambda$.

It is easily seen that $[FAC]$ is equivalent to each of the following statements:

(a) The union of all attachments S_μ contained in G_λ is finite (for all $\lambda < \sigma$).

(b) In each G_λ there are only finitely many vertices with neighbours in G outside G_λ.

Clearly, if $\mathscr{F}$ satisfies $[FAC]$, it must be finitary. By Lemma 2.4, if $\mathscr{F}$ is of strength n and all the G_λ are n-connected, then G is n-connected. On the other hand, if all the G_λ are n-connected and no G_λ coincides with S_λ, then from $c(G) \geqslant n$ we conclude that $\mathscr{F}$ must be of strength n. For a more thorough treatment of tree-decompositions we refer the reader to [1] and [2].

Now we can prove our main result.

Theorem 6.5. *Let $\mathscr{F} = (G_\lambda)_{\lambda < \sigma}$ be a rayless tree-decomposition of a graph G in which the members G_λ are n-connected. Assume $\mathscr{F}$ to be of strength n and to satisfy $[FAC]$. Then G is n-minimizable if and only if each G_λ is n-minimizable.*

Proof

1. Assume each G_λ to be n-minimizable. We show that G is n-minimizable, too, by induction on $o(T(\mathscr{F}))$, the order of $T(\mathscr{F})$ in the sense of R. Schmidt.

   If $o(T(\mathscr{F})) = 0$, then $\mathscr{F}$ is finite, and the assertion follows by a finite series of applications of Corollary 4.8.

   Let $o(T(\mathscr{F})) > 0$. There exists a finite subgraph B of $T(\mathscr{F})$ such that all the components of $T(\mathscr{F}) - B$ have order smaller than $o(T(\mathscr{F}))$; without loss of generality we may suppose that B is a subtree of $T(\mathscr{F})$ containing the root 0. Let $T_i (i \in I)$ be the components of $T(\mathscr{F}) - B$. Let μ_i be the minimum of the ordinals in $V(T_i)$; the tree-order in T_i with respect to the root μ_i is a restriction of the tree-order in $T(\mathscr{F})$ (with respect to the root 0). For each $i \in I$, $\mathscr{F}_i = (G_\lambda)_{\lambda \in V(T_i)}$ is clearly a tree-decomposition of $H_i = \bigcup_{\lambda \in V(T_i)} G_\lambda$, and we have $T(\mathscr{F}_i) = T_i$; by the induction hypothesis, we may assume that each H_i is n-minimizable. Furthermore $H = \bigcup_{\lambda \in V(B)} G_\lambda$ is n-minimizable by Corollary 4.8. Clearly, $H \cap H_i = S_{\mu_i}$, and by $[FAC]$ there is a finite $D \subseteq V(H)$ such that $V(S_{\mu_i}) \subseteq D$ for all $i \in I$. We see that the conditions of Lemma 6.3 are satisfied; hence $G = H \cup \bigcup_{i \in I} H_i$ is n-minimizable.

2. Assume now that G has an n-minimization G'. We select a member G_λ of $\mathscr{F}$. We use G'_λ to denote the subgraph of G' induced by $V(G_\lambda)$. Choose a finite $F \subseteq V(G_\lambda)$ covering all attachment $\subseteq G_\lambda$ according to $[FAC]$. Then $G'_\lambda \cup \binom{F}{2}$ is n-connected, and $R_n(G' \cup \binom{F}{2})$ is finite by Lemma 4.5. Clearly $R_n(G'_\lambda \cup \binom{F}{2}) \subseteq R_n(G' \cup \binom{F}{2})$. Thus $G'_\lambda \cup \binom{F}{2}$ is n-minimizable, hence $G_\lambda \cup \binom{F}{2}$ and G_λ itself are n-minimizable also. $\quad\square$

Diestel [2] showed that every rayless n-connected graph has a rayless tree-decomposition of strength n with finite n-connected members. By applying Theorem 6.5, we find that each such graph is n-minimizable. So we may consider Theorem 6.5 to be a generalization of Schmidt's theorem.

None of the hypotheses in Theorem 6.5 can be omitted. The infinite ladder has a tree-decomposition with 4-circuits as its members, which satisfies all conditions in 6.5 with the exception of raylessness. The graph in Example 6.4 shows that $[FAC]$ cannot be weakened to the requirement that the given tree-decomposition be finitary.

7. The case $n = 2$ and $m_1(G) = 1$

The question of how possible it is to tackle the minimization problem for simpler cases not covered by Schmidt's Theorem now suggests itself, i.e. the problem for a graph G with a small but positive number $m_1(G)$ of disjoint rays and small connectivity number.

Let $W(G)$ denote the set of vertices x of G such that there exists a ray U and infinitely many paths from x to U having pairwise only x in common; we put

$$|W(G)| = w(G).$$

In [5], §2, the following was shown:

Theorem 7.1. If $1 \leqslant m_1(G) = m < \infty$, then there exists a finite $F \subset G$ and a tree-decomposition $G_i (i < \omega)$ of $G - F$ with attachments $S_i = G|_i \cap G_i$ such that the following conditions are satisfied:

1. $|S_i| = m$ for all $i \in \mathbb{N}$,
2. the S_i are pairwise disjoint,
3. S_i is contained in G_{i-1} and has no vertex in a G_j with $j < i-1$,
4. there is an m-tuple of disjoint paths matching S_i with S_{i+1} in each G_i $(i \geqslant 1)$,
5. the G_i are rayless.

It is easy to see that F can be chosen as $W(G)$.

From Theorem 7.1 we see immediately that for $m_1(G) \geqslant 1$ we have

Corollary 7.2. $c(G) \leqslant m_1(G) + w(G)$.

If for a rayless G we define $w(G)$ as the number of vertices with infinite degree, then Corollary 7.2 is also true in the case $m_1(G) = 0$. (For $c(G) = 1$ it is then König's Lemma, and for $c(G) \geqslant 2$ it follows recursively by deleting an arbitrary vertex of infinite degree.)

As the rayless graphs can be handled by Schmidt's theorem and the structure of a graph G with $m_1(G) < \infty$ is rather transparent by Theorem 7.1, it seems there is hope for the study of the n-minimization problem in these cases.

Let us consider the case $n = 2$ and $m_1(G) = w(G) = 1$; then $W(G)$ consists of a single vertex w and the S_i reduce to vertices s_i that are articulations of the connected graph $G - w$ [4]. We further see that the block-decomposition of $G - w$ is a refinement of the decomposition in Theorem 7.1.

Each G_i must be connected (otherwise w would separate G). G_i may have infinitely many blocks, but the decomposition-tree of its block decomposition must be rayless. We denote the articulation of each endblock B of G_i by a_B. An endblock B of G_i that does not contain s_i or s_{i+1} is called *proper*; from each proper B we can select a vertex v_B in $B - a_B$ adjacent to w.

Theorem 7.3. If infinitely many G_i have proper endblocks, G is 2-minimizable.

Proof. We delete all edges from w to vertices different from the v_B selected above. Clearly the resulting graph G' remains 2-connected and the edges wv_B are all necessary in G'. For a subgraph J of $G' - w$ we denote by $J + w$ the graph J together with w and all edges from w to J. Furthermore we find a sequence of integers $0 \leqslant i_0 < i_1 < i_2 < \ldots$ such that $c((G_0 \cup \ldots \cup G_{i_k}) + w) \geqslant 2$ for all $k \geqslant 0$. Without restriction, we can assume $G = G'$ and $i_k = k$ for all k. In particular, there are at least two edges from w into G_0 and at least one edge from w into $G_i - s_i$ for every $i \geqslant 1$.

Let $H_0 + w$ be a 2-minimization of $G_0 + w$. Then let $H_1 + w$ be a 2-minimization of $(H_0 \cup G_1) + w$, choose a 2-minimization $H_2 + w$ of $(H_1 \cup G_2) + w$, and so on. Let $G_i' = H_{i+1}[V(G_i)]$. Then clearly each G_i' must be connected and each edge xy of G_i' is preserved in all G_{i+k}' with $k \geqslant 2$. Namely, if $xy \in R_2(H_{i+k} + w)$ for a $k \geqslant 2$, then it would be in $R_2(H_{i+1} + w)$, since every x,y-path in $H_{i+k} + w$ either stays in $H_i + w$ or passes through s_{i+1} into G_{i+1}' and contains w; then we can go from s_{i+1} to a v_B in $V(G_{i+1}')$ and from there to w, so avoiding the G_{i+k}' with $k \geqslant 2$.

Thus we have a sequence of connected graphs $G_0', G_1', G_2', \ldots$ such that

$$G^* = (G_0' \cup G_1' \cup G_2' \cup \ldots) + w$$

is a connected factor of G. From this representation we see $c(G^*) \geqslant 2$. By the above consideration for edges xy of G_i', we also see $R_2(G^*) = \emptyset$. □

If a G_i has no proper endblock, it may be possible to create one by deleting appropriately chosen edges. We call a G_i $(i \geqslant 1)$ *feasible* if we can find $L_i \subseteq E(G_i)$ such that

1. $G_i - L_i$ is connected,
2. there is an endblock of $G_i - L_i$ neither containing s_i nor s_{i+1},
3. if B is such an endblock then there is v_B in $B - a_B$ adjacent to w.

Clearly, if we replace each feasible G_i by its $G_i - L_i$, the resulting graph remains 2-connected, and if infinitely many G_i are feasible, we know from Theorem 7.3 that then G is 2-minimizable. On the other hand, if there is an n_0 such that no G_i with $i \geqslant n_0$ is feasible, we consider an arbitrary 2-connected factor H of G. Let $G_i' = H[G_i]$. Clearly there must be edges from w to infinitely many G_i' in H. If $wx \in E(H)$ with x in G_i', $i > n_0$, then $wx \in R_2(H)$, as H has no proper endblock. Therefore H is not 2-minimal. So we can state:

Theorem 7.4. *A graph G with $c(G) \geqslant 2$ and $m_1(G) = w(G) = 1$ is 2-minimizable if and only if in any representation of G according to Theorem 7.1 there occur infinitely many feasible members.*

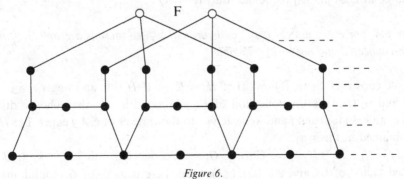

Figure 6.

If we allow $w(G) \geqslant 2$, the condition of Theorem 7.4 in its essence remains necessary for the 2-minimizability, but examples show that it is not sufficient (Figure 6). There is an enormous jump of difficulty if any parameter in Theorem 7.4 is increased.

8. Other kinds of minimization

If P_n is a property that every n-minimal graph must have, one may ask whether an n-connected graph that is not n-minimizable has at least an n-connected factor with P_n.

Every n-minimal graph has a vertex of degree n [6]. An n-connected graph G is called n-degree minimizable if it has a factor G' with $c(G') = n$ containing a vertex of degree n.

Theorem 8.1. *Every $(n+1)$-connected graph is n-degree minimizable.*

(Take an arbitrary vertex v and delete all but n edges incident with v.)

Theorem 8.2. [8, (10)] *If G is n-connected and there is a T in G with $|T| = n$ such that $G - T$ has a finite component, then G is n-degree minimizable.*

Every infinite n-minimal graph G with $n \geqslant 2$ has $|G|$ vertices of degree n [7, Satz 2]. We call an infinite G with $c(G) \geqslant n$ *fully n-degree minimizable* if it has a factor G' with $c(G') \geqslant n$ and $|G'|$ vertices of degree n.

The ladder, for instance, is not 2-minimizable, but fully 2-degree minimizable. If T_k is the k-regular tree, then for integers $k \geqslant 3$, $T_k \times K_2$ is not 2-degree minimizable. In [6] it is proved (see [3] for the notion of free end) that:

Theorem 8.3. *If $c(G) \geqslant n \geqslant 2$ is locally finite with at least one free end, then G is fully n-degree minimizable.*

For a cardinal $k > n$, a graph G with $c(G) \geqslant n$ is called (n, k)-*minimal* if for all $ab \in E(G)$ it is the case that $\mu_G(a, b) < k$. With $k = n+1$, we get the notion 'n-minimal'. G is (n, k)-*minimizable* if it has an (n, k)-minimal factor. Furthermore, we call G [n, k]-*reducible* if G has an n-connected factor G' with an edge $e = ab$ such that $\mu_G(a, b) < k$. Clearly (n, k)-minimizability implies [n, k]-reducibility. One could hope that an n-connected infinite graph G would be at least [$n, |G|$]-reducible. But we have:

Theorem 8.4. *For every integer $n \geqslant 2$ and cardinal $k \geqslant \omega$, there is a graph G_k with $|G_k| = k$ that is n-connected and not [n, k]-reducible.*

Proof. We construct G_k as follows. Let $H_0 = K_{n+1}$. If H_r for an integer $r \geqslant 0$ is already defined and S_i ($i \in I$) is the collection of K_n contained in H_r, then choose distinct new vertices v_{iv} ($v < k$) for each i and join each v_{iv} to the vertices of S_i by edges. Let H_{r+1} be the graph obtained in this way.

Let $G_k = H_0 \cup H_1 \cup H_2 \cup \dots$. Clearly G_k is n-connected and $|G_k| = k$. If J is an n-connected factor of G_k and $e = ab \in E(J)$, then there is a K_n of G_k containing a, b. By construction $G_k - K_n$ and hence also $J - V(K_n)$ contains k components each of which sends edges to a and to b. Hence $\mu_J(a,b) = k$. $\qquad\square$

9. Some open questions

The following question is suggested by Lemma 2.3:

Problem 9.1. What can be said about the existence of an n-minimal subgraph in an n-connected graph? The answer is trivially positive for $n = 2$, but the problem seems to be hard for all $n \geqslant 3$.

In the light of Theorem 8.1, one may ask:

Problem 9.2. Is every $(n+1)$-connected graph n-minimizable? Even the following much more modest question seems to be very hard: Is there an $n \geqslant 3$ such that every n-connected graph is 2-minimizable?

If G is n-minimal, the graph arising from G by adding a new vertex x and joining it to every vertex of G by edges, is $(n+1)$-minimizable. (Leave only those edges from x that lead to a vertex of degree n.) In this connection we put:

Problem 9.3. If G is $(n+1)$-connected and there is a finite F such that $G - F$ is n-minimizable, is G $(n+1)$-minimizable?

Let us call a graph G purely n-connected if it does not contain an $(n+1)$-connected subgraph.

Clearly the n-minimal graphs are purely n-connected.

Problem 9.4. What can be said about the existence of purely n-connected factors or subgraphs in n-connected graphs?

An infinite sequence $F_1, F_2, F_3, \ldots$ of finite subsets of $V(G)$ is called an infinite chain of finite cuts in G if no F_i contains an F_j for $i \neq j$ and every F_i $(i \geqslant 2)$ minimally separates each vertex of $F_{i-1} - F_i$ from each vertex of $F_{i+1} - F_i$.

Problem 9.5. Does every n-connected graph that is not n-minimizable contain an infinite chain of finite cuts? Or at least, can it be reduced by deleting edges to an n-connected graph with such a chain?

As such a chain of cuts leads to the existence of a ray, we would get another approach to Schmidt's theorem and a strengthening of it.

Problem 9.6. Does every infinite $(n+1)$-connected graph contain an infinite set T of vertices (or also of edges) such that $G - T$ remains at least n-connected?

By Diestel's theorem, the answer is certainly 'yes' for rayless graphs.

References

[1]   Diestel, R. (1990) *Graph decompositions: a study in infinite graph theory*. Oxford University Press, Oxford.

[2]   Diestel, R. (to appear) On spanning trees and *k*-connectedness in infinite graphs. *J. Combin. Theory*.

[3]   Halin, R. (1964) Über unendliche Wege in Graphen. *Math. Ann.* **157** 125–137.

[4]   Halin, R. (1965) Charakterisierung der Graphen ohne unendliche Wege. *Arch. Math.* **16** 227–231.

[5]   Halin, R. (1965) Über die Maximalzahl fremder unendlicher Wege in Graphen. *Math. Nachr.* **30** 63–85.

[6]   Halin, R. (1971) Unendliche minimale *n*-fach zusammenhängende Graphen. *Abh. Math. Sem. Univ. Hamburg* **36** 75–88.

[7]   Mader, W. (1972) Über minimal *n*-fach zusammenhängende, unendliche Graphen und ein Extremalproblem. *Arch. Math.* **23** 553–560.

[8]   Schmidt, R. (1982) *Ein Reduktionsverfahren für Weg-endliche Graphen*, PhD-Thesis Hamburg.

[9]   Schmidt, R. (1983) Ein Ordnungsbegriff für Graphen ohne unendliche Wege mit einer Anwendung auf *n*-fach zusammenhängende Graphen. *Arch. Math.* **40** 283–288.

On Universal Threshold Graphs

P. L. HAMMER and A. K. KELMANS[†]

RUTCOR, Rutgers University
New Brunswick, New Jersey 08903

A graph G is *threshold* if there exists a 'weight' function $w : V(G) \to R$ such that the total weight of any stable set of G is less than the total weight of any non-stable set of G. Let $\mathcal{T}_n$ denote the set of threshold graphs with n vertices. A graph is called $\mathcal{T}_n$-*universal* if it contains every threshold graph with n vertices as an induced subgraph. $\mathcal{T}_n$-universal *threshold* graphs are of special interest, since they are precisely those $\mathcal{T}_n$-universal graphs that do not contain any non-threshold induced subgraph.

In this paper we shall study *minimum* $\mathcal{T}_n$-universal (threshold) graphs, *i.e.* $\mathcal{T}_n$-universal (threshold) graphs having the minimum number of vertices.

It is shown that for any $n \geq 3$ there exist minimum $\mathcal{T}_n$-universal graphs, which are themselves threshold, and others which are not.

Two extremal minimum $\mathcal{T}_n$-universal graphs having respectively the minimum and the maximum number of edges are described, it is proved that they are unique, and that they are threshold graphs.

The set of all minimum $\mathcal{T}_n$-universal threshold graphs is then described constructively; it is shown that it forms a lattice isomorphic to the $n - 1$ dimensional Boolean cube, and that the minimum and the maximum elements of this lattice are the two extremal graphs introduced above.

The proofs provide a (polynomial) recursive procedure that determines for any threshold graph G with n vertices and for any minimum $\mathcal{T}_n$-universal threshold graph T, an induced subgraph G' of T isomorphic to G.

1. Introduction

Given a class of graphs $\mathscr{C}$ it is natural to find and to study *extremal $\mathscr{C}$-universal graphs, i.e.* those graphs that *contain* every graph from $\mathscr{C}$ (*e.g.* as subgraphs, as induced subgraphs, as a homeomorphic image, *etc.*), and have some extremal properties (*e.g.*, are of the minimum size).

[†] The authors gratefully acknowledge the partial support of the National Science Foundation under Grants NSF–STC88–09648 and NSF–DMS–8906870, the Air Force Office of Scientific Research under Grants AFOSR–89–0512 and AFOSR–90–0008 to Rutgers University, the Office of Naval Research under Grant N00014-92-J1375 and the DIMACS Center.

The idea of *universal graphs*, conceived in 1964 by R. Rado [16], considered today as fundamental in extremal graph theory, is the subject of numerous investigations (*e.g.* [3, 6, 7, 8, 14, 15]). Beside its mathematical interest, several recent studies (*e.g.* [2, 4]) deal with its applications to various aspects of computer science and engineering.

In this paper we shall concentrate on $\mathscr{C}$-universal graphs, where $\mathscr{C}$ is the set $\mathscr{T}_n$ of all *threshold graphs* with n vertices. The concept of threshold graphs was introduced in [9, 10], and various properties and characterizations of such graphs are known (*e.g.* [5, 9, 10, 11, 13]).

We shall study here $\mathscr{T}_n$–*universal graphs*, *i.e.* graphs that contain as induced subgraphs all threshold graphs with n vertices. The study of *minimum $\mathscr{T}_n$-universal graphs*, *i.e.* $\mathscr{T}_n$-universal graphs having the minimum number of vertices, is one of the main topics of this paper. Minimum $\mathscr{T}_n$-universal graphs may contain non-threshold graphs as induced subgraphs. *Irredundant* minimum $\mathscr{T}_n$-universal graphs could be defined as those minimum $\mathscr{T}_n$-universal graphs that do not contain as induced subgraphs any non-threshold graph. It is easy to notice that they are precisely those minimum $\mathscr{T}_n$-universal graphs that are themselves threshold. The second central topic of this paper consists of the study of minimum $\mathscr{T}_n$-universal *threshold* graphs.

A graph obtained from a $\mathscr{T}_n$-universal graph G by deleting or adding an edge, may or may not be $\mathscr{T}_n$-universal. This leads to the question of finding those minimum $\mathscr{T}_n$-universal graphs that have the minimum or the maximum number of edges. We construct two graphs, M^n and W^n, satisfying, respectively, the above two extremal requirements. These graphs are shown to be themselves threshold, implying that M^n and W^n are also minimum $\mathscr{T}_n$-universal *threshold* graphs having the minimum and the maximum number of edges respectively. We prove that any minimum $\mathscr{T}_n$-universal graph with the minimum (maximum) number of edges is isomorphic to M^n (respectively W^n): in other words M^n and W^n are the unique minimum $\mathscr{T}_n$-universal graphs having, respectively, the minimum and the maximum number of edges.

We give a constructive description of all minimum $\mathscr{T}_n$-universal threshold graphs, show that they form a lattice, the minimum and the maximum elements of which are respectively the extremal graphs M^n and W^n, and show that this lattice is isomorphic to the $n-1$ dimensional Boolean cube. Hence there are 2^{n-1} minimum $\mathscr{T}_n$-universal threshold graphs. We show in particular that a minimum $\mathscr{T}_n$-universal graph has a surprisingly small order equal to $2n-1$ (the lower bound is easy, and so the interesting bit is the upper bound).

A special class of split graphs called *n-stair graphs* turns out to play an essential role in the study of minimum $\mathscr{T}_n$-universal graphs. It is proved that any minimum $\mathscr{T}_n$-universal graph is an *n*-stair graph, and that any *n*-stair graph is a minimum $\mathscr{T}_n$-universal graph if $n \leq 4$, or if the graph has an isolated vertex. It is also shown that a *threshold* graph is a minimum $\mathscr{T}_n$-universal graph if and only if it is an *n*-stair graph. A family of graphs is constructed showing that for any $n \geq 5$ there exists an *n*-stair graph that is not $\mathscr{T}_n$-universal.

It is easy to see that every minimum $\mathscr{T}_n$-universal graph with $n = 1$ or 2 is threshold. We construct a family of graphs that for any $n \geq 3$ provides a minimum $\mathscr{T}_n$-universal graph that is not threshold.

Given an arbitrary threshold graph G with n vertices and an arbitrary minimum $\mathcal{T}_n$-universal threshold graph T, a simple recursive procedure for finding an induced subgraph G' of T isomorphic to G, along with an isomorphism of G and G', can be derived from the description of minimum $\mathcal{T}_n$-universal threshold graphs. For the special case when the minimum $\mathcal{T}_n$-universal graph is the extremal graph M^n, this imbedding can be described directly in terms of the degree sequence of G.

Finally it is shown that the set $\mathcal{T}_n$ of all *threshold graphs* with n vertices contains a proper subset $\mathcal{W}_n$ (of so called *uniform threshold graphs*), which can be viewed as a *kernel* of $\mathcal{T}_n$ because the size of extremal $\mathcal{T}_n$-universal graphs is actually determined by $\mathcal{W}_n$. As a byproduct we get for $\mathcal{W}_n$ the main results established for $\mathcal{T}_n$.

2. Main concepts and notations

We consider undirected graphs without loops or multiple edges [1]. Let $V(G)$ and $E(G)$ denote the set of vertices and edges of G, respectively. Let $\mathcal{G}_n$ denote the set of all graphs with n vertices. Two vertices x and y of G are *adjacent* if $(x, y) \in E(G)$. A subset X of vertices of G is called *stable* if no two vertices of X are adjacent, and *non-stable* otherwise.

Let $N(x)$ denote the set of vertices of G adjacent to x:

$$N(x) = \{z \in V(G) : (x, z) \in E(G)\},$$

and let $d(x) = |N(x)|$ be the *degree* of x. Two vertices x and y of a graph G are called *equivalent* if $N(x) \setminus y = N(y) \setminus x$.

Given $X \subseteq V(G)$, the *subgraph of G induced by X*, denoted by $G(X)$, is the graph whose vertex set is X, and whose edges are those edges of G that have both of their end-vertices in X.

The graph $\overline{G}$ is said to be the *complement of a graph G* if there is a one-to-one mapping $\varphi : V(G) \to V(\overline{G})$ such that (u, v) is an edge of G if and only if $(\varphi(u), \varphi(v))$ is not an edge of $\overline{G}$. A *self-complementary graph* is a graph that is the complement of itself.

Let K_n be the *complete graph* on n vertices, and let its complement $\overline{K}_n$ be the *empty graph* on n vertices. Let P_n and C_n denote the *path* and the *cycle* with n vertices, respectively.

Given two graphs A and B, a graph G is called the *union of A and B*, $G = A \cup B$ if $V(G) = V(A) \cup V(B)$ and $E(G) = E(A) \cup E(B)$.

If A and B are disjoint graphs, their sum $A + B$ will simply denote the graph $A \cup B$, while their product $A \times B$ will denote the graph obtained from $A + B$ by adding all the edges (a, b) with $a \in V(A)$ and $b \in V(B)$. In particular, if B consists of a single vertex b, we write $A + b$ and $A \times b$ instead of $A + B$ and $A \times B$: in these cases b is called an *isolated*, and, respectively, a *universal* vertex.

If $V(K_n) = V(\overline{K}_n) = \{g_1, \ldots, g_n\}$, then $K_n = g_1 \times g_2 \times \ldots \times g_n$ and $\overline{K}_n = g_1 + g_2 + \ldots + g_n$: we shall write simply $K_n = g^n$ and $\overline{K}_n = ng$.

A graph G is called *threshold* if there exists a 'weight' function $w : V(G) \to R$ defined on the set of vertices of G such that $w(X) < w(Y)$ for any stable vertex-set X and any non-stable vertex-set Y of G [9, 10]. Let $\mathcal{T}_n$ denote the set of threshold graphs with n vertices.

A graph is *uniform threshold* if it can be obtained from a complete graph either by deleting the edges of a complete subgraph, or by adding some isolated vertices. Let $\mathcal{W}_n$ denote the set of uniform threshold graphs with n vertices, *i.e.*

$$\mathcal{W}_n = \mathcal{W}_n^0 \cup \mathcal{W}_n^1,$$

where

$$\mathcal{W}_n^0 = \{g^k + sg : k, s \in \{0, 1, \ldots, n\}, k + s = n\},$$

and

$$\mathcal{W}_n^1 = \{g^s(kg) : k, s \in \{0, 1, \ldots, n\}, k + s = n\}.$$

Put $\mathcal{K}_n = \{K_n, \overline{K}_n\}$.

Given a set $\mathcal{C}$ of graphs, a graph G is called $\mathcal{C}$-*universal* if it contains every graph from $\mathcal{C}$ as an induced subgraph.

A $\mathcal{C}$-universal graph is called *minimum* if it has the minimum number of vertices among all finite $\mathcal{C}$-universal graphs (if any). A minimum $\mathcal{C}$-universal graph will be called simply a $\mathcal{C}$-*mug*.

A $\mathcal{C}$-mug is called *minimum* (*maximum*) if it has the minimum (respectively, the maximum) number of edges among all the $\mathcal{C}$-mugs.

Let $\mathcal{U}(\mathcal{C})$, $\mathcal{U}_{min}(\mathcal{C})$, and $\mathcal{U}_{max}(\mathcal{C})$ denote the sets of all $\mathcal{C}$-mugs, minimum $\mathcal{C}$-mugs, and maximum $\mathcal{C}$-mugs, respectively. Let $\mathcal{U}t(\mathcal{C})$, $\mathcal{U}t_{min}(\mathcal{C})$, and $\mathcal{U}t_{max}(\mathcal{C})$ denote the sets of all threshold $\mathcal{C}$-mugs, minimum threshold $\mathcal{C}$-mugs, and maximum threshold $\mathcal{C}$-mugs, respectively.

Let $v(\mathcal{C})$ denote the number of vertices of a $\mathcal{C}$-mug, and let $\underline{e}(\mathcal{C})$ and $\overline{e}(\mathcal{C})$ denote the number of edges of a minimum and of a maximum $\mathcal{C}$-mug, respectively.

Let $v_t(\mathcal{C})$ denote the number of vertices of a threshold $\mathcal{C}$-mug, and let $\underline{e}_t(\mathcal{C})$ and $\overline{e}_t(\mathcal{C})$ denote the number of edges of a minimum and of a maximum threshold $\mathcal{C}$-mug, respectively.

3. Preliminaries

The following characterizations of threshold graphs were given in [9, 10].

Theorem 3.1. *For every finite graph G the following conditions are equivalent.*

(c_1) G *is threshold.*

(c_2) $\overline{G}$ *is threshold.*

(c_3) G *does not contain four vertices* a_1, a_2, b_1, b_2 *such that* $(a_1, a_2), (b_1, b_2)$ *are edges and* $(a_1, b_1), (a_2, b_2)$ *are not (or equivalently, G does not contain* $2K_2, P_4$ *and* C_4 *as induced subgraphs).*

(c_4) *There exists a partition of the set $V(G)$ into two parts X and Y such that no two vertices in X are adjacent, any two vertices in Y are adjacent, there are orderings* $(x_1, x_2, \ldots, x_k)$ *of X and $(y_1, y_2, \ldots, y_l)$ of Y such that*

$$N(x_1) \subseteq N(x_2) \subseteq \cdots \subseteq N(x_k),$$

and

$$N^x(y_1) \supseteq N^x(y_2) \supseteq \cdots \supseteq N^x(y_l),$$

where $N^x(y) = X \cap N(y)$.

Let $\mathcal{T}_n^0$ (respectively, $\mathcal{T}_n^1$) denote the set of all threshold graphs with n vertices that do (respectively, do not) have an isolated vertex; clearly $\mathcal{T}_n = \mathcal{T}_n^0 \cup \mathcal{T}_n^1$ and $\mathcal{T}_n^0 \cap \mathcal{T}_n^1 = \emptyset$.

Corollary 1.

(t1)   *Each disconnected threshold graph G has an isolated vertex g, hence $G = (G \setminus g) + g$.*

(t2)   *Each connected threshold graph G has a universal vertex g, hence $G = (G \setminus g) \times g$.*

(t3)   *If F is a threshold graph and f is an additional vertex, then $F + f$ and $F \times f$ are threshold graphs.*
   Clearly $\mathcal{T}_n^0 = \{G + g : G \in \mathcal{T}_{n-1}\}$, and $\mathcal{T}_n^1 = \{G \times g : G \in \mathcal{T}_{n-1}\}$.

By using Theorem 3.1($c1$),($c3$) and Corollary 1 one can easily prove the following corollary.

Corollary 2.

(a1)   *If $n \leq 3$, then $\mathcal{T}_n = \mathcal{W}_n = \mathcal{G}_n$.*

(a2) $\mathcal{T}_4 = \mathcal{G}_4 \setminus \{2K_2, P_4, C_4\}$
   $= \{4g, 2g + g^2, g + g(2g), g + g^3, g(3g), g(g + 2g), g^2(2g), g^4\}$, *and*
   $\mathcal{W}_4 = \mathcal{T}_4 \setminus \{g + g(2g), g(g + g^2)\} = \{4g, 2g + g^2, g + g^3, g(3g), g^2(2g), g^4\}$,

(a3) $\mathcal{T}_5 = \{5g, 3g + g^2, 2g + g(2g), 2g + g^3, g + g(3g), g + g(g + g^2), g + g^2(2g),$
   $g + g^4, g(4g), g(2g + g^2), g(g + g(2g)), g^2(3g), g^2(g + g^3), g^2(g + g^2), g^3(2g), g^5\}$, *and*
   $\mathcal{W}_5 = \{5g, 3g + g^2, 2g + g^3, g + g^4, g(4g), g^2(3g)g^3(2g), g^5\}$.

4. Universal graphs and stair graphs

In this section we shall describe some properties of $\mathcal{C}$-universal graphs and minimum $\mathcal{C}$-universal graphs for some special classes $\mathcal{C}$. We introduce the concepts of *stair graphs* and *selfstair graphs*, and show that under certain conditions a minimum $\mathcal{C}$-universal graph is a selfstair graph.

Clearly if $\mathcal{C}_1 \subseteq \mathcal{C}_2$, a $\mathcal{C}_2$-universal graph is also a $\mathcal{C}_1$-universal graph. Therefore a $\mathcal{T}_n$-universal graph is also a $\mathcal{W}_n$-universal graph.

Given a set $\mathcal{C}$ of graphs, put $\overline{\mathcal{C}} = \{\overline{G} : G \in \mathcal{C}\}$. Clearly $\mathcal{K}_n = \overline{\mathcal{K}}_n$, and $\mathcal{W}_n^0 = \overline{\mathcal{W}}_n^1$. Therefore $\mathcal{W}_n = \overline{\mathcal{W}}_n$. From Theorem 3.1($c1$),($c2$) we see that $\mathcal{T}_n = \overline{\mathcal{T}}_n$.

Obviously, we have the following propositions.

Proposition 1.   *Let $\mathcal{C}$ be a set of graphs, and $\mathcal{C} = \overline{\mathcal{C}}$. Then G is a $\mathcal{C}$-universal (minimum $\mathcal{C}$-universal) graph if and only if $\overline{G}$ is a $\mathcal{C}$-universal (respectively, minimum $\mathcal{C}$-universal) graph.*

Corollary 3.   *Let $\mathcal{C}$ be a set of graphs, and $\mathcal{C} = \overline{\mathcal{C}}$. G is a minimum $\mathcal{C}$-mug if and only if $\overline{G}$ is a maximum $\mathcal{C}$-mug.*

Lemma 1.   Let $\mathscr{C}$ be a set of graphs containing K_n and $\overline{K}_n$, and let G be a $\mathscr{C}_n$-universal graph, $n \geq 2$. Then

(a1) G has at least $2n - 1$ vertices, and

(a2)   if G has exactly $2n - 1$ vertices, then $V(G)$ consists of three disjoint subsets A_{n-1}, B_{n-1}, and c such that A_{n-1} and B_{n-1} have $n - 1$ vertices, c is a single vertex, no two vertices of A_{n-1} are adjacent, any two vertices of B_{n-1} are adjacent, and the vertex c is connected with no vertex from A_{n-1} and with every vertex from B_{n-1} (so that $G(A_{n-1} \cup c) = \overline{K}_n$ and $G(B_{n-1} \cup c) = K_n$).

Proof. Since K_n and $\overline{K}_n$ are members of $\mathscr{C}_n$, the graph G contains K_n and $\overline{K}_n$ as induced subgraphs. Hence G has two subsets S_n and $\overline{S}_n$ with n vertices such that any two vertices of S_n are adjacent and no two vertices of $\overline{S}_n$ are adjacent. Clearly S_n and $\overline{S}_n$ have at most one vertex in common. Therefore G has at least $2n - 1$ vertices. If G has exactly $2n - 1$ vertices, then obviously S_n and $\overline{S}_n$ have exactly one vertex in common, say c, and $A_{n-1} = \overline{S}_n \setminus c$ and $B_{n-1} = S_n \setminus c$. □

We shall assume from now on that the vertices of the subsets A_{n-1} and B_{n-1} of $V(G)$ described in Lemma 1 are ordered: $A_{n-1} = \{a_1, a_2, \ldots, a_{n-1}\}$ and $B_{n-1} = \{b_1, b_2, \ldots, b_{n-1}\}$ such that $d(a_1) \geq d(a_2) \geq \ldots \geq d(a_{n-1})$ and $d(b_1) \leq d(b_2) \leq \ldots \leq d(b_{n-1})$.

Let Q_n^k be the graph obtained from the complete graph K_{n-k} on the vertex set $Y_{n-k} = \{y_1, y_2, \ldots, y_{n-k}\}$ by adding the set $X_k = \{x_1, x_2, \ldots, x_k\}$ of k new vertices, and the set of edges $\{(x_j, y_i) : i = 1, 2, \ldots, n - k\; ; j = 1, 2, \ldots, k\}$. By using the operations $+$ and $\times$ on graphs, the graph Q_n^k is simply

$$Q_n^k = (y_1 \times \ldots \times y_{n-k}) \times (x_1 + x_2 + \ldots + x_k).$$

(see Fig. 1)

Let R_n^k be the graph obtained from the complete graph K_{n-k} on the vertex set $Y_{n-k} = \{y_1, y_2, \ldots, y_{n-k}\}$ by adding a set $X_k = \{x_1, x_2, \ldots, x_k\}$ of k isolated vertices. In other words

$$R_n^k = (y_1 \times \ldots \times y_{n-k}) + (x_1 + x_2 + \ldots + x_k).$$

Obviously, R_n^k and Q_n^{n-k} are complementary uniform threshold graphs, and

$$\mathscr{W}_n = \{Q_n^{n-k}, R_n^k : k = 0, 1 \ldots, n\}.$$

From Corollary 3 and Lemma 1 we easily obtain the following result.

Proposition 2.   Let $n = 2, 3, \ldots$. Then R_{2n-1}^{n-1} and Q_{2n-1}^{n-1} are the minimum $\mathscr{K}_n$-mug and the maximum $\mathscr{K}_n$-mug, respectively. In particular $v(\mathscr{K}_n) = 2n - 1$.

Lemma 2.   Let G be a $\mathscr{W}_n$-universal graph with $2n - 1$ vertices, $n \geq 2$. Let A_{n-1}, B_{n-1}, and c be the subsets of $V(G)$ described in Lemma 1(a2). Then $d(a_{n-k-1}) \geq k$ for any $k \in \{1, \ldots, n - 2\}$.

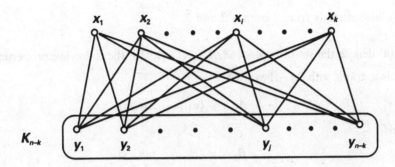

Figure 1 The graph Q_n^k

Proof. Suppose that $d(a_{n-k-1}) < k$ for some $k \in \{1, \dots, n-2\}$. Then in A_{n-1} there are at most $n - k - 2$ vertices of degree at least k. Since G is a $\mathscr{W}_n$-universal graph and $Q_n^{n-k} \in \mathscr{W}_n$, the graph G contains Q_n^{n-k} as an induced subgraph. We can assume that $V(Q_n^{n-k}) \subset V(G)$. Since X_{n-k} is a stable set of Q_n^{n-k} and $B_{n-1} \cup c$ induces a clique in G, clearly X_{n-k} and $B_{n-1} \cup c$ have at most one vertex in common. Therefore X_{n-k} has at least $n - k - 1$ vertices in common with A_{n-1}. Every vertex of X_{n-k} is of degree k in Q_n^{n-k}. Therefore every vertex of X_{n-k} is of degree at least k in G, and so A_{n-1} should contain at least $n - k - 1$ vertices of degree at least k, a contradiction. □

Lemma 3. *Let G be a $\mathscr{W}_n$-universal graph with $2n - 1$ vertices, $n \geq 2$. Let A_{n-1}, B_{n-1}, and c be vertex subsets of G described in Lemma 1(a2). Then for any $k \in \{1, 2, \dots, n-1\}$ there exist at least k vertices in A_{n-1} of degree at most k in G.*

Proof. Every vertex in A_{n-1} is of degree at most $n - 1$. Since A_{n-1} has $n - 1$ vertices, the statement of the lemma holds for $k = n - 1$. Let $k \in \{1, \dots, n-2\}$. Since G is a $\mathscr{W}_n$-universal graph and $R_n^k \in \mathscr{W}_n$, the graph G contains R_n^k as an induced subgraph. We can assume that $V(R_n^k) \subset V(G)$. Consider the k-vertex set X_k of isolated vertices of R_n^k and the $(n-k)$-vertex set Y_{n-k} that induces a clique in R_n^k. Since R_n^k is an induced subgraph of G, the vertex set X_k is stable in G, and Y_{n-k} induces a clique in G. Since $A_{n-1} \cup c$ is stable in G and Y_{n-k} induces a clique in G, clearly Y_{n-k} and $A_{n-1} \cup c$ have at most one vertex in common. Therefore $|Y_{n-k} \cap B_{n-1}| \geq n - k - 1$. Since $k \leq n - 2$, we have $n - k - 1 \geq 1$, so Y_{n-k} and $A_{n-1} \cup c$ have at least one vertex in common. Therefore $X_k \subset A_{n-1}$. We know that every vertex adjacent to a vertex from A_{n-1} in G belongs to B_{n-1}. Since R_n^k is an induced subgraph of G, every vertex a from X_k should be an isolated vertex in $G \setminus (B_{n-1} \setminus Y_{n-k})$. Since $B_{n-1} = n - 1$ and $|Y_{n-k} \cap B_{n-1}| \geq n - k - 1$, it follows that $|B_{n-1} \setminus Y_{n-k}| \leq k$. Therefore every vertex from X_k is of degree at most k in T. Since $X_k \subset A_{n-1}$ and $|X_k| = k$, the statement follows. □

Lemma 4. *Let G be a $\mathscr{W}_n$-universal graph with $2n - 1$ vertices, $n \geq 2$. Let A_{n-1}, B_{n-1}, and c be the subsets of $V(G)$ described in Lemma 1(a2). Then $d(a_{n-k-1})$ is either k or $k + 1$ for any $k \in \{1, \dots, n-2\}$.*

Proof. Follows directly from Lemmas 2 and 3.   □

Let $\mathscr{S}_*[n]$ denote the set of triples (G^n, A_{n-1}, B_{n-1}) with the following properties.

(p1) G^n is a graph with $2n - 1$ vertices,

$$A_{n-1} = \{a_1, a_2, \ldots, a_{n-1}\},$$

and

$$B_{n-1} = \{b_1, b_2, \ldots, b_{n-1}\}$$

are disjoint subsets of vertices of G^n, and

$$d(a_1) \geq d(a_2) \geq \ldots \geq d(a_{n-1}),$$

and

$$d(b_1) \leq d(b_2) \leq \ldots \leq d(b_{n-1})$$

(clearly $V(G^n) \setminus (A_{n-1} \cup B_{n-1})$ consists of a single vertex, say c).

(p2)   No two vertices of A_{n-1} are adjacent, any two vertices of B_{n-1} are adjacent, and the vertex c is adjacent to no vertex from A_{n-1} and to every vertex from B_{n-1} (i.e. $G(A_{n-1} \cup c) = \overline{K}_n$ and $G(B_{n-1} \cup c) = K_n$),

(p3) $d(a_{n-k-1})$ is either k or $k + 1$ for any $k \in \{1, \ldots, n - 2\}$.

Given (G, A_{n-1}, B_{n-1}), let $\overline{G}$ be the complement of G, $V(G) = V(\overline{G})$, $\overline{A}_{n-1} = B_{n-1}$, and $\overline{B}_{n-1} = A_{n-1}$. Let $\mathscr{F}_*[n]$ denote the set of triples (G, A_{n-1}, B_{n-1}) from $\mathscr{S}_*[n]$ such that $(\overline{G}, \overline{A}_{n-1}, \overline{B}_{n-1})$ also belongs to $\mathscr{S}_*[n]$.

Let $\mathscr{S}[n]$ denote the set of graphs G such that (G, A, B) is isomorphic to a triple from $\mathscr{S}_*[n]$ for some subsets A and B of $V(G)$. The graph set $\mathscr{F}[n]$ is defined similarly.

A graph from $\mathscr{S}[n]$ is called an *n-stair graph*, and a graph from $\mathscr{F}[n]$ is called an *n-selfstair graph*.

From Proposition 1, and Lemmas 1 and 4 we have

Proposition 3. *Suppose that G is a $\mathscr{C}$-universal graph with $2n - 1$ vertices, and $\mathscr{W}_n \subseteq \mathscr{C}$. Then $G \in \mathscr{F}[n]$ (i.e. G is an n-selfstair graph).*

5. Stair graphs with given degree sequences

In this section we are going to classify the stair graphs according to their degree sequences. We shall also describe the set of all threshold stair graphs. This description will be used in Section 8 to characterize all threshold minimum $\mathscr{T}_n$-universal graphs.

Given $(G, A_{n-1}, B_{n-1}) \in \mathscr{S}_*^n$, let the non-increasing sequence

$$(d(a_1), d(a_2), \ldots, d(a_{n-1}))$$

of the degrees of vertices from A_{n-1} in G be called the A_{n-1}-*sequence of* (G, A_{n-1}, B_{n-1}).

Let v^{n-1} be the vector $(n - 2, \ldots, 1, 0)$ and z^{n-1} be an arbitrary $\{0, 1\}$-vector of length

$n-1$. Let $\mathscr{S}_\ast(z^{n-1})$ denote the set of triples (G^n, A_{n-1}, B_{n-1}) from $\mathscr{S}_\ast[n]$ with the A_{n-1}-sequence $v^{n-1} + z^{n-1}$, and let $\mathscr{S}(z^{n-1})$ denote the set of the corresponding graphs. Clearly $\mathscr{S}[n] = \bigcup \{\mathscr{S}(z) : z \in \boldsymbol{B}^{n-1}\}$.

From the definition of the set $\mathscr{S}(z^{n-1})$ we have

Proposition 4. *If* $G \in \mathscr{S}(z^{n-1})$, *then*

$$|E(G)| = \binom{n}{2} + |v^{n-1}| + |z^{n-1}|,$$

where $|y|$ *is the sum of coordinates of a vector* y.

Let us denote by $\boldsymbol{0}^n$ and $\boldsymbol{1}^n$ the n-vectors z of length n having all coordinates equal to 0 and 1 respectively.

From the above proposition we have the following corollary.

Corollary 4. *Among all the graphs in* $\mathscr{S}[n]$, *the graphs from* $\mathscr{S}(\boldsymbol{0}^{n-1})$ *and from* $\mathscr{S}(\boldsymbol{1}^{n-1})$ *have the minimum and the maximum number of edges, respectively.*

From Theorem 3.1(c1), (c4) it follows easily that if $(G^n, A_{n-1}, B_{n-1}) \in \mathscr{S}_\ast^n$ and if G^n is a threshold graph, then G^n is uniquely defined (up to an isomorphism) by its A_{n-1}-sequence. Therefore we have

Proposition 5. *The set* $\mathscr{S}_\ast(z^{n-1})$, $z^{n-1} \in \boldsymbol{B}^{n-1}$, *has exactly one triple* (G^n, A_{n-1}, B_{n-1}) *(up to an isomorphism) such that* G^n *is a threshold graph.*

Let us denote this triple by $(T(z^{n-1}), A_{n-1}, B_{n-1})$. By using the operations $+$ and $\times$ on graphs, the graph $T(z^{n-1})$ can be described as follows:

$$T(z^{n-1}) = F^1(z_1^{n-1})F^2(z_2^{n-1})\dots F^i(z_i^{n-1})\dots F^{n-1}(z_{n-1}^{n-1})a_0)_{n-1}\dots)_1,$$

where $a_0 = c$, $z^{n-1} = (z_1^{n-1}, z_2^{n-1}, \dots, z_{n-1}^{n-1})$, and for every $i = 1, 2, \dots, n-1$

$$F^i(0) = (_i\, a_{n-i} + b_{n-i} \times,$$

and

$$F^i(1) = b_{n-i} \times (_i\, a_{n-i} +.$$

Putting $T(\boldsymbol{0}^{n-1}) = M^n$ and $T(\boldsymbol{1}^{n-1}) = W^n$, we notice that

$$M^n = (_1\, a_{n-1} + b_{n-1} \times (_2\, a_{n-2} + b_{n-2} \times (_3\, a_{n-3} + \dots (_i\, a_{n-i} + b_{n-i} \times \dots +$$
$$b_{n-2} \times (_{n-1}\, a_1 + b_1 \times a_0)_{n-1} \dots)_1$$

$$W^n = b_{n-1} \times (_1\, a_{n-1} + b_{n-2} \times (_2\, a_{n-2} + b_{n-3} \times (_3\, a_{n-3} + \dots +$$
$$b_{n-i} \times (_i\, a_{n-i} + \dots b_1 \times (_{n-1}\, a_1 + a_0)_{n-1} \dots)_1$$

(see Figs. 2 and 3).

Two $\{0,1\}$-vectors z^{n-1} and $\bar{z}^{n-1}$ are called *complements* if $\bar{z}^{n-1}$ can be obtained from z^{n-1} by replacing each 0 by 1 and each 1 by 0, i.e. $z^{n-1} + \bar{z}^{n-1} = \boldsymbol{1}^{n-1}$.

Obviously we have the following proposition.

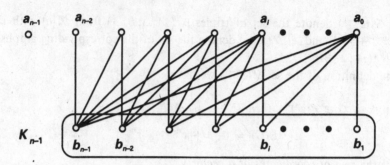

Figure 2   The graph M^n

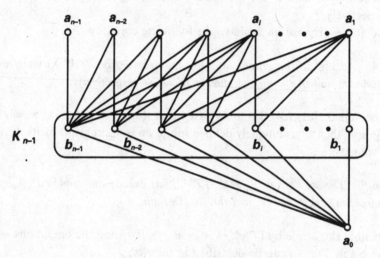

Figure 3   The graph W^n

Proposition 6. *$T(x^{n-1})$ and $T(y^{n-1})$ are complement graphs if and only if x^{n-1} and y^{n-1} are complement $\{0,1\}$-vectors. In particular, M^n and W^n are complement graphs.*

Given a subset $\mathscr{R}_\bullet[n]$ of triples from $\mathscr{S}_\bullet[n]$, put $\mathscr{R}_\bullet(z) = \mathscr{R}_\bullet[n] \cap \mathscr{S}_\bullet(z)$; clearly

$$\mathscr{R}_\bullet[n] = \bigcup \{\mathscr{R}_\bullet(z) : z \in \boldsymbol{B}^{n-1}\}.$$

Let $\mathscr{S}t[n]$ and $\mathscr{F}t[n]$ denote the sets of all threshold graphs in $\mathscr{S}[n]$ and $\mathscr{F}[n]$, respectively. Put $\mathscr{S}t(z^{n-1}) = \mathscr{S}t[n] \cap \mathscr{S}(z^{n-1})$.

From Propositions 5 and 6 we have the following proposition.

Proposition 7.

$$\mathscr{S}t(z^{n-1}) = \{T(z^{n-1})\},$$

and

$$\mathscr{S}t[n] = \mathscr{F}t[n] = \{T(z) : z \in \boldsymbol{B}^{n-1}\}.$$

for every $n = 1, 2, \ldots$.

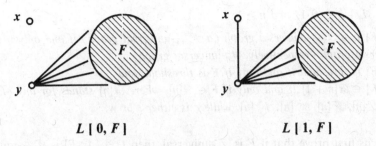

$$L\,[\,0,F\,] \qquad\qquad L\,[\,1,F\,]$$

Figure 4 L-operations on graphs

6. Operations on stair graphs and universal graphs

In this section we shall introduce two operations $L[0,F]$ and $L[1,F]$ on graphs. These operations will play an essential role because several classes of graphs we are interested in (classes of stair graphs, selfstair graphs, universal graphs, etc.) turn out to be closed under these operations. Moreover, we shall see that all threshold stair graphs can be generated by these operations from a small one.

Given a graph F and two distinct vertices x and y not belonging to F, let us put

$$L[0,F] = x + (y \times F),$$

and

$$L[1,F] = y \times (x + F)$$

(see Fig. 4).

Let $\mathcal{U}^t_*[n]$ and $\mathcal{V}^t_*[n]$ denote respectively the set of triples (G, A_{n-1}, B_{n-1}) from $\mathcal{S}_*[n]$ such that G is a $\mathcal{T}_n$-universal graph, and, respectively, a $\mathcal{T}_n$-universal threshold graph.

Similarly let $\mathcal{U}^w_*[n]$ and $\mathcal{V}^w_*[n]$ denote, respectively, the set of triples (G, A_{n-1}, B_{n-1}) from $\mathcal{S}_*[n]$ such that G is a $\mathcal{W}_n$-universal graph, and, respectively, a $\mathcal{W}_n$-universal threshold graph.

Obviously

$$\mathcal{V}^c_*[n] \subseteq \mathcal{U}^c_*[n] \subseteq \mathcal{F}_*[n] \subseteq \mathcal{S}_*[n],$$

where $n \geq 1$ and c is either t or w. Since $\mathcal{W}_n \subset \mathcal{T}_n$, we have

$$\mathcal{V}^t_*[n] \subseteq \mathcal{V}^w_*[n]$$

and

$$\mathcal{U}^t_*[n] \subseteq \mathcal{U}^w_*[n].$$

Given a subset $\mathcal{R}_*[n]$ of triples from $\mathcal{S}_*[n]$, let $\mathcal{R}[n]$ denote the set of graphs G such that (G, A, B) is isomorphic to a triple from $\mathcal{R}_*[n]$ for some subsets A and B of $V(G)$. Put $\mathcal{R}_*(z) = \mathcal{R}_*[n] \cap \mathcal{S}_*(z)$; clearly $\mathcal{R}_*[n] = \bigcup\{\mathcal{R}_*(z) : z \in \boldsymbol{B}^{n-1}\}$.

Similar inclusions hold for the graph sets $\mathcal{S}[n]$, $\mathcal{F}[n]$, $\mathcal{U}^t[n]$, $\mathcal{U}^w[n]$, $\mathcal{V}^t[n]$, and $\mathcal{V}^w[n]$.

Lemma 5. *Let $z \in \{0, 1\}$ and $n \geq 1$.*

(a1) $L[z, F]$ *is a $\mathcal{T}_{n+1}$-universal graph (a $\mathcal{W}_{n+1}$-universal graph) if and only if F is a $\mathcal{T}_n$-universal graph (respectively, $\mathcal{W}_n$-universal graph),*

(a2) $L[z, F]$ *is threshold if and only if F is threshold, and*

(a3) $L[z, F] \in \mathcal{R}[n+1]$ *if and only if $F \in \mathcal{R}[n]$, where $\mathcal{R}[n]$ stands for any of the graph sets $\mathcal{S}[n]$, $\mathcal{F}[n]$, $\mathcal{U}^c[n]$, $\mathcal{V}^c[n]$, while c is either t or w.*

Proof. Let us first prove that if F is $\mathcal{T}_n$-universal, then $L = L[z, F]$ is $\mathcal{T}_{n+1}$-universal. To do this, let G be an arbitrary threshold graph with $(n+1)$ vertices. We should prove that the graph L has an induced subgraph isomorphic to G. Since G is threshold, by Corollary 1 it has a vertex g such that G is either $G' + g$ or $G' \times g$, where $G' = G \setminus g$. Since G' has n vertices and F is $\mathcal{T}_n$-universal, F has an induced subgraph F' isomorphic to G'. By the definition of $L = L[z, F]$, the vertices x and y in the graph L are, respectively, adjacent to no vertex of the subgraph $L \setminus \{x, y\} = F$, and to every vertex of it. Therefore $F' + x$ and $F' \times y$ are induced subgraphs of the graph L. By Corollary 1, G is either $G' + g$ or $G' \times g$. Therefore G is isomorphic to either $F' + x$ or $F' \times y$.

Let us prove now that if F is not $\mathcal{T}_n$-universal, then $L = L[z, F]$ is not $\mathcal{T}_{n+1}$-universal. Since F is not $\mathcal{T}_n$-universal, there exists a graph G with n vertices such that F has no induced subgraph isomorphic to G. Obviously G has at least 2 vertices. Suppose that $L = L[0, F]$, that is, $L = x + (y \times F)$. Consider the graph $H = G \times g$ with $n + 1$ vertices. We shall prove that L has no induced subgraph isomorphic to $H = G \times g$. Assume the contrary, *i.e.* that L has an induced subgraph H' isomorphic to H. Then $H' = G' \times g'$, where g' is a vertex of H' and G' is isomorphic to G. Since the vertex g' of H' is adjacent to every other vertex of H', the graph H' is connected. Since x is an isolated vertex of L, and since H' is an induced subgraph of L, it follows that H' is an induced subgraph of $L \setminus x = y \times F$. Clearly G' is not a subgraph of F, because F has no induced subgraph isomorphic to G. Therefore $y \in V(G')$, that is, $N = H' \setminus y$ is an induced subgraph of F. Obviously $N = g' \setminus Z$, where $Z = G' \setminus y$. Since y is adjacent to every vertex of F, and since $N \subset F$, we have $H' = y \times N = y \times g' \times Z$. Thus $G' = H' \setminus g' = y \times Z$, implying that G' is isomorphic to N. But N is an induced subgraph of F, a contradiction.

Suppose now that $L = L[1, F]$, that is, $L = y \times (x + F)$. Then, by using the same arguments as above, one can prove that if F has no induced subgraph isomorphic to a graph G, then L has no induced subgraph isomorphic to a graph $G + g$.

The statements (*a2*) and (*a3*) follow directly from the corresponding definitions, from Theorem 3.1 and from (*a1*).

The proof of the lemma for $\mathcal{W}_n$ is similar to the above proof for $\mathcal{T}_n$. $\square$

Lemma 6. *Let $z^{n-1} \in B^{n-1}$, and let $z^n = z^{n-1}0$ (i.e. the last coordinate z_n^n of z^n equals 0). Then*

$$\mathcal{F}(z^n) = \{L[0, H] : H \in \mathcal{F}(z^{n-1})\}.$$

Proof. Let $G \in \mathcal{F}(z^n)$ and $z_n^n = 0$. Then $(G, A_{n-1}, B_{n-1}) \in \mathcal{F}_*(z^n)$ for some vertex subsets $A_{n-1} = \{a_1, \ldots, a_{n-1}\}$ and $B_{n-1} = \{b_1, \ldots, b_{n-1}\}$ of G, and $d(a_{n-1}) = 0$ and $d(b_{n-1})$

is either $2n - 3$ or $2n - 2$. Since $d(a_{n-1}) = 0$, we have $d(b_{n-1}) = 2n - 3$. Therefore $G = a_{n-1} + (b_{n-1} \times H) = L[0, H]$, where $H \in \mathscr{F}(z^{n-1})$. $\qquad\square$

From Lemmas 5 and 6 we have the following proposition.

Proposition 8. *Let $z^{n-1} \in \boldsymbol{B}^{n-1}$, and let $z^n = z^{n-1}0$ (i.e. the last coordinate z_n^n of z^n equals 0). Then*

$$\mathscr{R}(z^n) = \{L[0, H] : H \in \mathscr{R}(z^{n-1})\},$$

where $\mathscr{R}(z^n)$ stands for any of the graph sets $\mathscr{V}^c(z^n)$, $\mathscr{U}^c(z^n)$, $\mathscr{F}(z^n)$ while c is either t or w.

Given a graph G and a $\{0,1\}$-vector z^n of length n, let us define the graph $z^n(G)$, $n \geq 1$, recursively:

$$z^n(G) = L[z_n^n, z^{n-1}(G)],$$

where the vector z^{n-1} is obtained from z^n by deleting the last coordinate z_n^n of z^n, and where $z^0(G) = G$.

By using Lemma 5, one can easily prove the following proposition.

Proposition 9.

(a1) $z^n(G)$ *is threshold if and only if G is threshold,*

(a2) $z^n(G)$ *is a $\mathscr{T}_{n+k}$-universal graph if and only if G is a $\mathscr{T}_k$-universal graph, and*

(a3) $z^n(G) \in \mathscr{R}[n + k]$ *if and only if $G \in \mathscr{R}[k]$; here $\mathscr{R}[k]$ stands for any of the graph sets $\mathscr{S}[k]$, $\mathscr{F}[k]$, $\mathscr{U}^c[k]$, $\mathscr{V}^c[k]$, while c is either t or w.*

7. The strict hierarchy of universal, selfstair and stair graphs

In this section we investigate the hierarchy of stair, selfstair, and universal graphs of some type. We will show that this hierarchy is strict for $n \geq 5$. In Section 4 we proved that if a minimum $\mathscr{W}_n$-universal or $\mathscr{T}_n$-universal graph has $2n - 1$ vertices, then it should belong to $\mathscr{F}[n]$ (i.e. it should be an n-selfstair graph). Here we will see that $\mathscr{F}[n]$ does contain minimum $\mathscr{W}_n$-universal graphs and minimum $\mathscr{T}_n$-universal graphs for small n. Later (in Section 8) we will prove that this is true for any n.

By using the descriptions of $\mathscr{T}_n$ for $n \leq 5$ in Corollary 2, we can find the graph sets $\mathscr{F}[n]$, $\mathscr{S}[n]$ and $\mathscr{V}^c[n]$, $\mathscr{U}^c[n]$ for $n \leq 5$, where c is either t or w. This information enables us to establish the following proposition.

Proposition 10. *For any $z^{n-1} \in \boldsymbol{B}^{n-1}$ and $n \in \{1, \ldots, 5\}$*

$$\mathscr{V}^t(z^{n-1}) = \mathscr{V}^w(z^{n-1}) \neq \emptyset,$$

and

$$\mathscr{U}^t(z^{n-1}) = \mathscr{U}^w(z^{n-1}) \neq \emptyset.$$

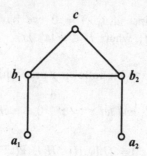

Figure 5 The graph A^3

The validity of analogous results for an arbitrary n will be discussed in Section 8 (see Theorem 8.1).

Put $\mathscr{V}(z^{n-1}) = \mathscr{V}^t(z^{n-1})$, $\mathscr{U}(z^{n-1}) = \mathscr{U}^t(z^{n-1})$, $\mathscr{V}[n] = \mathscr{V}^t[n]$, and $\mathscr{U}[n] = \mathscr{U}^t[n]$.

Proposition 11.

(a1) $\mathscr{V}[2] = \mathscr{U}[2] = \mathscr{F}[2] = \mathscr{S}[2] = \{M^2, W^2\}$, and

(a2) $\mathscr{V}[3] \subset \mathscr{U}[3] = \mathscr{F}[3] = \mathscr{S}[3]$; moreover $\mathscr{U}[3] \setminus \mathscr{V}[3] = \mathscr{U}(10) \setminus \mathscr{V}(10) = \{A^3\}$ (implying that A^3 is the unique minimum $\mathscr{T}_3$-universal graph that is not threshold, see Fig. 5).

Note that according to Proposition 1 A^3 must be self–complementary.

Proposition 12. $\mathscr{V}[4] \subset \mathscr{U}[4] = \mathscr{F}[4] \subset \mathscr{S}[4]$. Moreover, $|\mathscr{U}^4 \setminus \mathscr{V}^4| = 8|$, and $|\mathscr{S}[4] \setminus \mathscr{F}[4]| = 2$.

From Propositions 8 and 12 we have $\mathscr{U}(z^30) = \mathscr{F}(z^30)$. By Lemma 5, $\{L[z, G] : z \in \{0, 1\}, G \in \mathscr{U}[4] \setminus \mathscr{V}[4]\} \subseteq \mathscr{U}[5] \setminus \mathscr{V}[5]$. Since $\mathscr{U}[4] \setminus \mathscr{V}[4] \neq \emptyset$, it follows that $\mathscr{U}[5] \setminus \mathscr{V}[5] \neq \emptyset$.

Proposition 13. $\mathscr{V}[5] \subset \mathscr{U}[5] \subset \mathscr{F}[5] \subset \mathscr{S}[5]$. Moreover, $|\mathscr{U}[5] \setminus \mathscr{V}[5]| = 34$, $|\mathscr{F}[5] \setminus \mathscr{U}[5]| = 7$, and $|\mathscr{S}[5] \setminus \mathscr{F}[5]| = 31$.

From Propositions 9 and 13 we have the following theorem.

Theorem 7.1.

$$\mathscr{V}^s[n] \subset \mathscr{U}^s[n] \subset \mathscr{F}[n] \subset \mathscr{S}[n]$$

for any $n \geq 5$ and $s \in \{t, w\}$. Moreover, $|\mathscr{U}^s[n] \setminus \mathscr{V}^s[n]| \geq c_u 2^{n-5}$, $|\mathscr{F}^n \setminus \mathscr{U}^n| \geq c_f 2^{n-5}$, and $|\mathscr{S}^n \setminus \mathscr{F}^n| \geq c_d 2^{n-5}$, where $c_u \geq 34$, $c_f \geq 7$, and $c_d \geq 31$.

In particular, the number of minimum $\mathscr{T}_n$-universal graphs that are not threshold grows exponentially as a function of n.

8. Characterizations of threshold universal graphs

Proposition 14. *Let* $z^{n-1} \in B^{n-1}$, *and* $n = 2, 3, \dots$. *Then* $T(z^{n-1})$ *is a* $\mathcal{T}_n$-*universal graph.*

Proof. Obviously the graph D of one vertex is $\mathcal{T}_1$-universal, and $T(z^{n-1}) = z^{n-1}(D)$. Therefore the proposition follows from Proposition 9. $\qquad\square$

Clearly $\mathcal{K}_n \subseteq \mathcal{W}_n \subseteq \mathcal{T}_n$. Since $T(z^{n-1})$ has $2n - 1$ vertices, Lemma 1 and Propositions 2 and 14 imply the following proposition.

Proposition 15. *Let* $n = 1, 2, \dots$.
(a1) *If* G *is a* $\mathcal{T}_n$-*mug then* G *is a* $\mathcal{W}_n$-*mug, i.e.* $\mathcal{U}(\mathcal{T}_n) \subseteq \mathcal{U}(\mathcal{W}_n) \subseteq \mathcal{U}(\mathcal{K}_n)$.
(a2) $T(z^{n-1})$ *is a* $\mathcal{T}_n$-*mug, a* $\mathcal{W}_n$-*mug, and a* $\mathcal{K}_n$-*mug.*
(a3) $T(z^{n-1})$ *is a threshold* $\mathcal{T}_n$-*mug, a threshold* $\mathcal{W}_n$-*mug, and a threshold* $\mathcal{K}_n$-*mug.*
(a4) $v(\mathcal{K}_n) = v(\mathcal{T}_n) = v(\mathcal{W}_n) = v_t(\mathcal{T}_n) = v_t(\mathcal{W}_n) = 2n - 1$.

The next proposition follows from Propositions 1, 3, 14, and 15, and gives a necessary condition for a graph to be minimum $\mathcal{T}_n$-universal.

Proposition 16. *A* $\mathcal{T}_n$-*mug and a* $\mathcal{W}_n$-*mug are selfstair graphs.*

We are ready now to give a description of all minimum $\mathcal{T}_n$-universal graphs and all minimum $\mathcal{W}_n$-universal graphs that are threshold (compare with Proposition 10).

Let us recall that $\mathscr{S}t[n]$ and $\mathscr{F}t[n]$ denote the sets of all threshold graphs in $\mathscr{S}[n]$ and in $\mathscr{F}[n]$, respectively, and $\mathcal{U}t(\mathcal{T}_n)$ and $\mathcal{U}t(\mathcal{W}_n)$ denote the sets of all threshold $\mathcal{T}_n$-mugs and threshold $\mathcal{W}_n$-mugs respectively.

Theorem 8.1. *Let* $n = 1, 2, \dots$. *The following conditions are equivalent:*
(c1) G *is a minimum threshold* $\mathcal{T}_n$-*universal graph,*
(c2) G *is a minimum threshold* $\mathcal{W}_n$-*universal graph,*
(c3) G *is a threshold n-stair graph,*
(c4) G *is a threshold n-selfstair graph, and*
(c5) G *is a graph* $T(z^{n-1})$ *for some* $\{0, 1\}$-*vector* z^{n-1} *of length* $n - 1$.

In other words,
$$\mathcal{V}^t = \mathcal{V}^w = \mathcal{U}t(\mathcal{T}_n) = \mathcal{U}t(\mathcal{W}_n) = \mathscr{F}t[n] = \mathscr{S}t[n] = \{T(z^{n-1}) : z^{n-1} \in B^{n-1}\}.$$

Proof. According to Proposition 5, $\mathscr{S}t[n] = \{T(z^{n-1}) : z^{n-1} \in B^{n-1}\}$.
Therefore

$$
\begin{array}{rcll}
\mathscr{F}t[n] & = & \mathscr{S}t[n] & \text{(by Proposition 6),} \\
\mathscr{S}t[n] & \subseteq & \mathcal{U}t(\mathcal{T}_n) & \text{(by Proposition 14),} \\
\mathcal{U}t(\mathcal{T}_n) & \subseteq & \mathcal{U}t(\mathcal{W}_n) & \text{(by Proposition 15), and} \\
\mathcal{U}t(\mathcal{W}_n) & \subseteq & \mathscr{F}t[n] & \text{(by Proposition 16).}
\end{array}
$$

Thus $\mathcal{U}t(\mathcal{T}_n) = \mathcal{U}t(\mathcal{W}_n) = \mathscr{S}t[n]$. $\qquad\square$

From the above theorem we see that the necessary condition of Proposition 16 for a graph to be minimum $\mathcal{T}_n$-universal (minimum $\mathcal{W}_n$-universal) is also sufficient if the graph is required to be threshold.

Corollary 5. *There are exactly 2^{n-1} minimum $\mathcal{T}_n$-universal graphs which are threshold, i.e. $|\mathcal{U}t(\mathcal{T}_n)| = 2^{n-1}$.*

Given two graphs G and H, an injection $\varphi : V(G) \to V(H)$ is called *an induced embedding of G into H* if φ is an isomorphism of G onto the subgraph of H induced by $\varphi(V(G))$.

The proof of Lemma 5 provides a simple recursive procedure for finding an induced embedding of any given threshold graph with n vertices into the minimum $\mathcal{T}_n$-universal threshold graph $T(z^{n-1})$, where z^{n-1} is an arbitrary element of $\boldsymbol{B}^{n-1}$.

9. Extremal universal graphs

In this section we shall describe all minimum $\mathcal{W}_n$-universal graphs and all minimum $\mathcal{T}_n$-universal graphs having the minimum or the maximum number of edges.

Lemma 7. *Let (G, A_{n-1}, B_{n-1}) be a triple from $\mathcal{S}_*(0^{n-1})$ and G be a $\mathcal{W}_n$-universal graph. Then G is a threshold graph, implying that (G, A_{n-1}, B_{n-1}) is isomorphic to (M^n, A_{n-1}, B_{n-1}).*

Proof. Let $(G, A_{n-1}, B_{n-1}) \in \mathcal{S}_*(0^{n-1})$. Then from Theorem 3.1$(c1), (c4)$ it follows that G is threshold if and only if $N(a_{n-1}) \subset N(a_{n-2}) \subset \cdots \subset N(a_1) \subset N(a_0)$, where $a_0 = c$. Let us prove by induction on k that for any $k = 1, \ldots, n-1$

$$N(a_k) \subset N(a_{k-1}) \subset \cdots \subset N(a_0).$$

Clearly $N(a_1) \subseteq N(a_0)$. Suppose that $N(a_{k-1}) \subset \cdots \subset N(a_0)$. Since G is a $\mathcal{W}_n$-universal graph and Q_n^k is a uniform threshold graph with n vertices, we have: G contains Q_n^k as an induced subgraph. We may assume that $V(Q_n^k) \subset V(G)$.

Obviously $Y_{n-k} \subseteq B_{n-1}$ and $X_k \subseteq A_{n-1}$. Since $d(a_j) = n - j$ in G, and $d(x_i) = n - k$ in Q_n^k for any $i = 1, \ldots, k$, we have $X_k = \{a_k, a_{k-1}, \ldots, a_0\}$, and therefore $N(a_k) = Y_{n-k} \subset N(a_{k-1})$. $\qquad\square$

Remark 1. *It is easy to prove that $\mathcal{F}(0^{n-1}) = \{M^n\}$ and $\mathcal{F}(1^{n-1}) = \{W^n\}$. Hence Proposition 16 and the above statement also imply Lemma 7.*

Theorem 9.1. *Let $n = 1, 2, \ldots$. The graphs $M^n = T(0^{n-1})$ and $W^n = T(1^{n-1})$ have the following properties:*

(p1) M^n *and* W^n *are* $\mathcal{T}_n$*-mugs and* $\mathcal{W}_n$*-mugs,*

(p2) M^n *is a minimum* $\mathcal{T}_n$*-mug and a minimum* $\mathcal{W}_n$*-mug,*

(p3) W^n *is a maximum* $\mathcal{T}_n$*-mug and a maximum* $\mathcal{W}_n$*-mug,*

(p4) *If G is a minimum* $\mathcal{T}_n$*-mug or a minimum* $\mathcal{W}_n$*-mug, then G is isomorphic to M^n, i.e. M^n is the unique minimum* $\mathcal{T}_n$*-mug and the unique minimum* $\mathcal{W}_n$*-mug.*

(p5) *If G is a maximum $\mathcal{T}_n$-mug or a maximum $\mathcal{W}_n$-mug, then G is isomorphic to W^n, i.e.*
W^n is the unique maximum $\mathcal{T}_n$-mug and the unique maximum $\mathcal{W}_n$-mug,

(p6) *M^n and W^n are threshold graphs.*

In other words,

$$\mathcal{U}_{min}(\mathcal{T}_n) = \mathcal{U}_{min}(\mathcal{W}_n) = \mathcal{U}t_{min}(\mathcal{T}_n) = \mathcal{U}t_{min}(\mathcal{W}_n) = \{M^n\},$$

and

$$\mathcal{U}_{max}(\mathcal{T}_n) = \mathcal{U}_{max}(\mathcal{W}_n) = \mathcal{U}t_{max}(\mathcal{T}_n) = \mathcal{U}t_{max}(\mathcal{W}_n) = \{W^n\}.$$

Proof. The properties (p1), (p2) and (p5) follow from Theorem 8.1. The property (p3) follows from Corollary 4. By Lemma 7, M^n satisfies (p4). Therefore by Corollary 3 and Proposition 6, W^n satisfies (p4). By Lemma 7, M^n is the unique minimum $\mathcal{W}_n$-mug. Since M^n is a minimum $\mathcal{T}_n$-mug, every minimum $\mathcal{T}_n$-mug has the same number of edges as M^n. Suppose that there exist a minimum $\mathcal{T}_n$-mug G non-isomorphic to M^n. Since every $\mathcal{T}_n$-mug is also a $\mathcal{W}_n$-mug, G is also a minimum $\mathcal{W}_n$-mug. This contradicts the fact that M^n is the unique minimum $\mathcal{W}_n$-mug. Therefore we have proved property (p4). Now (p5) follows from (p4) and Corollary 3. Property (p6) follows easily from Theorem 3.1. $\quad\square$

Corollary 6.

$$\underline{e}(\mathcal{W}_n) = \underline{e}_t(\mathcal{W}_n) = \underline{e}(\mathcal{T}_n) = \underline{e}_t(\mathcal{T}_n) = (n-1)^2$$

and

$$\bar{e}(\mathcal{W}_n) = \bar{e}_t(\mathcal{W}_n) = \bar{e}(\mathcal{T}_n) = \bar{e}_t(\mathcal{T}_n) = n(n-1).$$

References

[1] Bondy, J. A. and Murty, U. S. R. (1976) *Graph Theory with Applications*, Macmillan.

[2] Bhat, S. N. and Leiserson, C. E. (1984) How to assemble tree machines. In: Preparata, F. (ed.) *Advances in Computing Research* 2, JAI Press.

[3] Bhat, S. N., Chung, F. R. K., Leighton, T. and Rosenberg, A. L. (1989) Universal graphs for bounded-degree trees and planar graphs. *SIAM J. Discrete Math.* 2 145–155.

[4] Bhat, S. N., Chung, F. R. K., Leighton, T. and Rosenberg, A. L. (1988). Optimal simulations by butterfly networks. *Proc. 27th Annual ACM Symposium on Theory of Computing*, Chicago 192–204.

[5] Brandstädt, A. (1991) *Special Graph Classes – A Survey*, Section Math, Fredrich Schiller Universität, Jena, Germany.

[6] Goldberg, M. K. and Lifshitz, E. M. (1968) On minimum universal trees. *Mat. Zametki* 4 371–379.

[7] Chung, F. R. K. (1990) Universal graphs and induced-universal graphs. *Journal of Graph Theory* 14 443–454.

[8] Chung, F. R. K., Graham, R. L. and Shaearer, J. (1981) Universal Caterpillars. *J. Combinatorial Theory* B 31 348–355.

[9] Chvátal, V. and Hammer, P. L. (1973) *Set-packing problem and threshold graphs*, University of Waterloo, CORR 73–21.

[10] Chvátal, V. and Hammer, P. L. (1977) Aggregation of inequalities in integer programming. *Annals of Discrete Mathematics* 1 145–162.

[11] Erdős, P., Ordan, E. T. and Zalcstein, Y. (1987) Bounds on the threshold dimension and disjoint threshold coverings. *SIAM J. of Algebra and Discrete Methods* **8** 151–154.

[12] Friedman J. and Pippenger, N. (1987) Expanding graphs contain all small trees. *Combinatorica* **7** 71–76.

[13] Hammer, P. L., Ibaraki, T. and Peled, U. N. (1981) Threshold numbers and threshold completion. *Annals of Discrete Mathematics* **11** 125–145.

[14] Kannan, S., Naos, M. and Rudich, S. (1988) Implicit representation of graphs. *Proceedings of the Twentieth Annual ACM Symposium on Theory of Computing* 334–343.

[15] Moon, J. W. (1965) On minimal n-universal graphs. *Proc. Glasgow Math. Soc.* **7** 32–33.

[16] Rado, R. (1964) Universal graphs and universal functions. *Acta Arith.* **9** 331–340.

Image Partition Regularity of Matrices

NEIL HINDMAN[†] and IMRE LEADER[‡]

†Department of Mathematics, Howard University, Washington, D.C. 20059, U.S.A.

‡Department of Pure Mathematics and Mathematical Statistics, Cambridge University, England

Many of the classical results of Ramsey Theory, including those of Hilbert, Schur, and van der Waerden, are naturally stated as instances of the following problem: given a $u \times v$ matrix A with rational entries, is it true, that whenever the set $\mathbb{N}$ of positive integers is finitely coloured, there must exist some $\vec{x} \in \mathbb{N}^v$ such that all entries of $A\vec{x}$ are the same colour? While the theorems cited are all consequences of Rado's theorem, the general problem had remained open. We provide here several solutions for the alternate problem, which asks that $\vec{x} \in \mathbb{Z}^v$. Based on this, we solve the general problem, giving various equivalent characterizations.

1. Introduction

Consider van der Waerden's Theorem [8]: whenever $\mathbb{N} = \{1, 2, 3, \ldots\}$ is finitely coloured and $\ell \in \mathbb{N}$ is given, there exist a and d in $\mathbb{N}$ such that $a, a+d, a+2d, \ldots, a+\ell d$ are all the same colour (or 'monochrome'). (By a 'finite colouring' we mean, of course, a function defined on $\mathbb{N}$ with finite range.)

Given ℓ, let

$$A = \begin{pmatrix} 1 & 0 \\ 1 & 1 \\ 1 & 2 \\ \vdots & \\ 1 & \ell \end{pmatrix}.$$

Then van der Waerden's theorem asserts that whenever $\mathbb{N}$ is finitely coloured, there is some $\vec{x} = \begin{pmatrix} a \\ d \end{pmatrix}$ in $\mathbb{N}^2$ such that the entries of $A\vec{x}$ are monochrome. In terminology suggested by

† This author gratefully acknowledges support received from the National Science Foundation (USA) via grant DMS90-25025.

Walter Deuber, we are talking about the *image* partition regularity of A, *i.e.* asking that the image of $\bar{x}$ under the map defined by A be monochrome.

By contrast to image partition regularity, the question of which matrices are *kernel* partition regular was completely settled by Rado in 1933 [6]. (Here a $u \times v$ matrix A is kernel partition regular if and only if, whenever $\mathbb{N}$ is finitely coloured, there is some $\bar{x} \in \mathbb{N}^v$ with all entries monochrome such that $A\bar{x} = \bar{0}$. That is, there is a monochrome member of the kernel of the map defined by A.) For anyone not familiar with it, we will present Rado's Theorem later in this introduction.

Now van der Waerden's Theorem can be proved as a consequence of Rado's Theorem as follows: given ℓ, one takes $x_1, x_2, \ldots, x_{\ell+1}$ as the terms of an arithmetic progression and characterizes the fact that they are in an arithmetic progression by the equations $x_2 - x_1 = x_3 - x_2 = x_4 - x_3 = \ldots = x_{\ell+1} - x_\ell$. We can rewrite these as

$$-x_1 + 2x_2 - x_3 = 0$$
$$-x_1 + x_2 + x_3 - x_4 = 0$$
$$-x_1 + x_2 + x_4 - x_5 = 0$$
$$\vdots$$
$$-x_1 + x_2 + x_\ell - x_{\ell+1} = 0$$

so we are asking for the kernel partition regularity of the matrix

$$\begin{pmatrix} -1 & 2 & -1 & 0 & 0 & \ldots & 0 & 0 \\ -1 & 1 & 1 & -1 & 0 & \ldots & 0 & 0 \\ -1 & 1 & 0 & 1 & -1 & \ldots & 0 & 0 \\ \vdots & \vdots & \vdots & \vdots & \vdots & & \vdots & \vdots \\ -1 & 1 & 0 & 0 & 0 & & 1 & -1 \end{pmatrix}.$$

Alternatively, one can rewrite the equations as

$$-x_1 + 2x_2 - x_3 = 0$$
$$-x_2 + 2x_3 - x_4 = 0$$
$$-x_3 + 2x_4 - x_5 = 0$$
$$-x_{\ell-1} + 2x_\ell - x_{\ell+1} = 0$$

in which case we are asking for the kernel partition regularity of the matrix

$$\begin{pmatrix} -1 & 2 & -1 & 0 & \ldots & 0 & 0 & 0 \\ 0 & -1 & 2 & -1 & \ldots & 0 & 0 & 0 \\ 0 & 0 & -1 & 2 & \ldots & 0 & 0 & 0 \\ \vdots & \vdots & \vdots & \vdots & & \vdots & \vdots & \vdots \\ 0 & 0 & 0 & 0 & & -1 & 2 & -1 \end{pmatrix}.$$

But there is a problem here! Rado's Theorem does indeed tell us that both of these matrices are kernel partition regular. But, unfortunately, one monochrome solution has $x_1 = x_2 = \ldots = x_{\ell+1} = 1$; not exactly what we had in mind for our arithmetic progression. (This is not a far-fetched example. The first author made this very error in a talk a few years ago until it was brought to his attention by Deuber.)

A cure in this case can be obtained by strengthening the conclusion of van der Waerden's Theorem to require that the increment d also have the same colour as the terms of the arithmetic progression. With this addition, the original matrix for the image partition regular statement becomes

$$\begin{pmatrix} 0 & 1 \\ 1 & 0 \\ 1 & 1 \\ 1 & 2 \\ \vdots \\ 1 & \ell \end{pmatrix},$$

while one conversion to a kernel partition regular matrix is

$$\begin{pmatrix} 1 & 1 & -1 & 0 & \dots & 0 & 0 \\ 1 & 0 & 1 & -1 & \dots & 0 & 0 \\ 1 & 0 & 0 & 1 & \dots & 0 & 0 \\ \vdots & \vdots & \vdots & \vdots & & \vdots & \vdots \\ 1 & 0 & 0 & 0 & \dots & 1 & -1 \end{pmatrix}.$$

But one can surely imagine potential problems. Conceivably the original statement could have been valid, while the strengthened one was not. For this reason, as well as for the ability to answer a question in the form in which it is naturally stated, we claim our problem is interesting: determine which matrices are image partition regular (in the sense stated earlier).

The theorem of van der Waerden is not the only classical result that is naturally stated in this form. Schur's Theorem [7] says that whenever $\mathbb{N}$ is finitely coloured there exist x_1 and x_2 with x_1, x_2, and $x_1 + x_2$ monochrome. In this case the matrix corresponding to the statement is

$$\begin{pmatrix} 1 & 0 \\ 0 & 1 \\ 1 & 1 \end{pmatrix},$$

i.e. the first three rows of our strengthened version of van der Waerden's Theorem. Even older is the 1892 result of Hilbert [4]: given any $\ell \in \mathbb{N}$, whenever $\mathbb{N}$ is finitely coloured, there exist $a \in \mathbb{N}$ and $x_1, x_2, \dots, x_\ell$ in $\mathbb{N}$ such that all sums of the form $a + \sum_{n \in F} x_n$, where $\varnothing \neq F \subseteq \{1, 2, \dots, \ell\}$, are monochrome. Thus, when $\ell = 3$, this theorem asserts that the matrix

$$\begin{pmatrix} 1 & 1 & 0 & 0 \\ 1 & 0 & 1 & 0 \\ 1 & 1 & 1 & 0 \\ 1 & 0 & 0 & 1 \\ 1 & 1 & 0 & 1 \\ 1 & 0 & 1 & 1 \\ 1 & 1 & 1 & 1 \end{pmatrix}$$

is image partition regular.

Image partition regular matrices have played an important role in Ramsey Theory. In the terminology of Deuber [2] (modified only slightly), say that a matrix A is an (m, p, c) matrix (where m, p, and c are in $\mathbb{N}$) if the rows of A consist of all vectors $\vec{r} \in \mathbb{Z}^m \backslash \{\vec{0}\}$ such that

(1) for each $i \in \{1, 2, ..., m\}$, $|r_i| \leqslant p$, and

(2) if $t = \min\{i : r_i \neq 0\}$, then $r_t = c$.

(Note: two (m, p, c) matrices differ only by the order of their rows.) Deuber showed [2] that any (m, p, c) matrix is image partition regular. He further showed that if B is any kernel partition regular matrix, there exist some m, p and c such that, given an (m, p, c) matrix A and given any $\vec{x} \in \mathbb{N}^m$, one can choose entries for $\vec{y}$ from among the entries of $A\vec{x}$ such that $B\vec{y} = \vec{0}$. Since one can also show that (m, p, c) matrices are image partition regular using Rado's Theorem, one might be led to believe that a matrix is image partition regular if and only if it consists of some of the rows of an (m, p, c) matrix. We shall see, however, that even weakened versions of this hypothesis are false.

As promised earlier, we now present Rado's Theorem. It depends on a notion called the 'columns condition'.

Definition 1.1. Let A be a $u \times v$ matrix with entries from $\mathbb{Q}$, and let $\vec{c}_1, \vec{c}_2, ..., \vec{c}_v$ be the columns of A. Then A satisfies the columns condition if and only if there exist $m \in \mathbb{N}$ and $I_1, I_2, ..., I_m$ such that

(a) $\{I_1, I_2, ..., I_m\}$ partitions $\{1, 2, ..., v\}$,

(b) $\sum_{i \in I_1} \vec{c}_i = \vec{0}$, and

(c) for each $t \in \{2, 3, ..., m\}$ (if any), let $J_t = \bigcup_{j=1}^{t-1} I_j$: there exist $\delta_{t,i} \in \mathbb{Q}$ for each $i \in J_t$ such that $\sum_{i \in I_t} \vec{c}_i = \sum_{J_t} \delta_{t,i} \cdot \vec{c}_i$.

Theorem 1.2. (Rado [6].) *Let A be a $u \times v$ matrix with entries from $\mathbb{Q}$. Then A is kernel partition regular (i.e. whenever $\mathbb{N}$ is finitely coloured, there exists monochrome $\vec{y} \in \mathbb{N}^v$ such that $A\vec{y} = \vec{0}$) if and only if A satisfies the columns condition.*

To describe the results of this paper, we introduce a weaker notion of image partition regularity (so that what we have been calling 'image partition regular' now becomes 'strongly image partition regular').

Definition 1.3. Let A be a $u \times v$ matrix with rational entries.

(a) A is strongly image partition regular if and only if, whenever $\mathbb{N}$ is finitely coloured, there exists $\vec{x} \in \mathbb{N}^v$ such that the entries of $A\vec{x}$ are monochrome.

(b) A is weakly image partition regular if and only if, whenever $\mathbb{N}$ is finitely coloured, there exists $\vec{x} \in \mathbb{Z}^v$ such that the entries of $\bar{A}x$ are monochrome.

Since we have allowed the entries of $\vec{x}$ (in weakly image partition regular matrices) to range over $\mathbb{Z}$, one could reasonably ask what happens if we talk about colourings of $\mathbb{Z}$. First, of course, one would need to be talking about colourings of $\mathbb{Z}\backslash\{0\}$. (Otherwise any matrix would be partition regular, letting $\vec{x} = \vec{0}$.) If then in (b), one replaces colourings of $\mathbb{N}$ with colourings of $\mathbb{Z}\backslash\{0\}$, one arrives at a statement equivalent to (b). Indeed, one implication is trivial. For the other implication, let a colouring of $\mathbb{N}$ be given with say r colours. Colour the negative members of $\mathbb{Z}$ with r new colours, agreeing that a and b get the same colour if and only if $-a$ and $-b$ had the same colour. If $A\vec{x}$ is monochrome, so is $A(-\bar{x})$. (There is a fourth possibility: in (a) one could replace colourings of $\mathbb{N}$ with colourings of $\mathbb{Z}\backslash\{0\}$. This results in a proposition equivalent to the assertion that either A or $-A$ is strongly image partition regular.)

In Section 2 of this paper we present several characterizations of weak image partition regularity. Effective solutions are given by statements (II) and (III) of Theorem 2.2. (As far as we know they are new, although the ideas are not: they are in the spirit of the proofs that Rado's Theorem implies the partition regularity of (m, p, c) matrices.) The solution in either case involves constructing another matrix, and verifying that the new matrix satisfies the columns condition. This is a routine, if lengthy, process. (The problem of determining which matrices satisfy the columns condition is NP complete, because it implies the ability to determine whether a set of numbers has a subset summing to 0.)

In Section 3 we turn our attention to the more difficult problem of characterizing strong partition regularity. We present several analogues to statements in Theorem 2.2 and some new conditions, and prove that they are each equivalent to strong partition regularity.

We conclude the introduction with a remark about vector notation. We take $\mathbb{N}^v$ or $\mathbb{Z}^v$ or $\mathbb{Q}^v$ to consist of column or row vectors as appropriate for the context. Given a row vector $\bar{p} \in \mathbb{Q}^v$ and a $u \times v$ matrix A, we denote by $\binom{A}{\bar{p}}$ the $(u+1) \times v$ matrix whose first u rows are those of A, and whose $(u+1)$st row is $\bar{p}$. The meaning of other similar notation should be obvious. We let $\omega = \mathbb{N} \cup \{0\}$.

2. Weak image partition regularity

We begin by introducing a notion based on Deuber's (m, p, c) matrices.

Definition 2.1. Let A be a $u \times v$ matrix with rational entries. A satisfies the *first entries condition* if and only if each row of A is not $\bar{0}$, and whenever $i, j \in \{1, 2, ..., u\}$ and $t \in \{1, 2, ..., v\}$ and $t = \min\{k : a_{i,k} \neq 0\} = \min\{k : a_{j,k} \neq 0\}$, one has $a_{i,t} = a_{j,t} > 0$.

It is a fact (Theorem 2.11) that if A satisfies the first entries condition, then A is strongly image partition regular. One also easily sees that rearranging the columns of a matrix does not affect its partition regularity. Thus one would be tempted to conjecture that a matrix A is strongly or weakly partition regular if and only if the columns of A could be rearranged so that the resulting matrix satisfied the first entries condition. This is easily seen to be false, however. Consider $A = \begin{pmatrix} 2 & 1 \\ 1 & 2 \end{pmatrix}$. Then neither A nor $\begin{pmatrix} 1 & 2 \\ 2 & 1 \end{pmatrix}$ satisfy the first entries condition, while A is in fact strongly partition regular. (Simply let $x_1 = x_2$.)

We now state the main result of this section. Its proof will be pieced together as we proceed through the section.

Theorem 2.2. *Let A be a $u \times v$ matrix with rational entries. Then the following statements are equivalent*:

(I) *A is weakly image partition regular.*

(II) *Let $\ell = \text{rank}(A)$. Rearrange the rows of A so that the first ℓ rows are linearly* *independent. Let* $\vec{r}_1, \vec{r}_2, \ldots, \vec{r}_u$ *denote the rows of A. For each* $t \in \{\ell+1, \ell+2, \ldots, u\}$ *(if any), let* $\gamma_{t,1}, \gamma_{t,2}, \ldots, \gamma_{t,\ell}$ *be the members of* $\mathbb{Q}$ *determined by* $\vec{r}_t = \sum_{i=1}^{\ell} \gamma_{t,i} \vec{r}_i$. *If* $u > \ell$, *let D be the* $(u-\ell) \times u$ *matrix such that, for* $t \in \{1, 2, \ldots, u-\ell\}$ *and* $i \in \{1, 2, \ldots, u\}$,

$$d_{t,i} = \begin{cases} \gamma_{\ell+t,i} & \text{if } i \leqslant \ell \\ -1 & \text{if } i = \ell+t \\ 0 & \text{otherwise.} \end{cases}$$

Then either $\ell = u$ *or the matrix D satisfies the columns condition.*

(III) *Let* $\vec{c}_1, \vec{c}_2, \ldots, \vec{c}_v$ *be the columns of A. Then there exist* $t_1, t_2, \ldots, t_v$ *in* $\{x \in \mathbb{Q}: x \neq 0\}$ *such that the matrix*

$$\begin{pmatrix} & & & -1 & 0 & \cdots & 0 \\ & & & 0 & -1 & & 0 \\ t_1 \vec{c}_1 & t_2 \vec{c}_2 & \cdots & t_v \vec{c}_v & & \vdots & \vdots & & \vdots \\ & & & 0 & 0 & \cdots & -1 \end{pmatrix}$$

is kernel partition regular.

(IV) *For each* $\vec{p} \in \mathbb{Z}^v \setminus \{\vec{0}\}$, *there exists* $b \in \mathbb{Q} \setminus \{\vec{0}\}$ *such that* $\begin{pmatrix} A \\ b\vec{p} \end{pmatrix}$ *is weakly image partition regular.*

(V) *There exist* $b_1, b_2, \ldots, b_v$ *in* $\mathbb{Q} \setminus \{0\}$ *such that*

$$\begin{pmatrix} & & A & & \\ b_1 & 0 & 0 & \cdots & 0 \\ 0 & b_2 & 0 & \cdots & 0 \\ 0 & 0 & b_3 & \cdots & 0 \\ \vdots & \vdots & \vdots & & \vdots \\ 0 & 0 & 0 & \cdots & b_v \end{pmatrix}$$

is weakly image partition regular.

(VI) *There exist an* $m \leqslant u$ *and a* $u \times m$ *matrix B that satisfies the first entries condition such that for each* $\vec{y} \in \mathbb{Z}^m$ *there exists* $\vec{x} \in \mathbb{Z}^m$ *such that* $A\vec{x} = B\vec{y}$.

Before beginning the proof of Theorem 2.2, a few remarks about the special features of each of the equivalent conditions are in order. As we observed in the introduction, statements (II) and (III) allow us to determine in finite time whether a matrix is weakly image partition regular. The added information conveyed by statement (IV) is clear, but we feel remarkable: a weakly image partition regular matrix can be expanded almost at will. Statement (V) tells us, for example, that given any weakly image partition regular $u \times v$ matrix A, there is a subset P of $\{1, 2, \ldots, v\}$ such that whenever $\mathbb{N}$ is finitely coloured, there is an $\vec{x} \in \mathbb{Z}^v$ such that the entries of $A\vec{x}$ are monochrome and if $i \in P$, $x_i > 0$, and if $i \in \{1, 2, \ldots, v\} \setminus P$, $x_i < 0$. (In particular we may insist that the entries of $\vec{x}$ are not 0.) Finally, statement (VI) tells us that the first entries condition, which one might have hoped was necessary for weak image partition regularity, does provide a characterization.

The argument in the proof of the following lemma is standard. At various stages in subsequent arguments we shall need to consider common multiples in order to make some variables integers. We remark to the interested reader that an alternative approach is to replace $\mathbb{Z}$ by $\mathbb{Q}$ and $\mathbb{N}$ by $\mathbb{Q}^+ = \{x \in \mathbb{Q} : x > 0\}$ and at the end use a compactness argument.

Lemma 2.3. *Let A be a $u \times v$ matrix with rational entries. Then statements* (I) *and* (II) *of Theorem 2.2 are equivalent.*

Proof. Assume (I) holds, assume $\ell < u$, and let D be as defined in statement (II). We show that D is kernel partition regular, so that, by Rado's Theorem (Theorem 1.2), D satisfies the columns condition. Let $\mathbb{N}$ be finitely coloured, and pick $\bar{x} \in \mathbb{Z}^v$ such that the entries of $A\bar{x}$ are monochrome. Let $\bar{w} = A\bar{x}$. We claim that $D\bar{w} = \bar{0}$. To see this, let $t \in \{1, 2, ..., u - \ell\}$ be given. Then

$$\sum_{i=1}^{u} d_{t,i} \cdot w_i = \sum_{i=1}^{u} d_{t,i} \cdot \sum_{j=1}^{v} a_{i,j} \cdot x_j$$

$$= \sum_{i=1}^{\ell} \gamma_{\ell+t,i} \cdot \sum_{j=1}^{v} a_{i,j} \cdot x_j - \sum_{j=1}^{v} a_{\ell+t,j} \cdot x_j$$

$$= \sum_{j=1}^{v} x_j \cdot \left(\sum_{i=1}^{\ell} \gamma_{\ell+t,i} \cdot a_{i,j} - a_{\ell+t,j} \right)$$

$$= \sum_{j=1}^{v} x_j \cdot 0$$

$$= 0.$$

Now assume (II) holds and at first that $\ell = u$. Then we may assume that the first ℓ columns of A are linearly independent. Let A^* consist of the first ℓ columns of A and choose $x_1, x_2, ..., x_\ell$ in $\mathbb{Q}$ such that

$$A^* \bar{x} = \begin{pmatrix} 1 \\ 1 \\ \vdots \\ 1 \end{pmatrix}.$$

Let d be a common multiple of the denominators in $\bar{x}$. For $i \in \{1, 2, ..., \ell\}$, let $y_i = dx_i$, and for $i \in \{\ell + 1, \ell + 2, ..., v\}$ (if any), let $y_i = 0$. Then

$$A\bar{y} = \begin{pmatrix} d \\ d \\ \vdots \\ d \end{pmatrix}.$$

Now assume that $u > \ell$ and that the matrix D of statement (II) satisfies the columns condition. We may assume that the upper left $\ell \times \ell$ corner A^* of A has rank ℓ, by rearranging rows and columns if necessary. Let c be the absolute value of the determinant of A^*. It is immediate (or see Theorem 2.11) that $\mathbb{N}c$ is 'large', that is, whenever $\mathbb{N}c$ is finitely coloured $\mathbb{N}c$ contains monochrome solutions to any kernel partition regular matrix. So let $\mathbb{N}$ be finitely coloured and pick monochrome $x_1, x_2, ..., x_u$ in $\mathbb{N}c$ such that

$$D\bar{x} = \begin{pmatrix} 0 \\ 0 \\ \vdots \\ 0 \end{pmatrix}.$$

For $i \in \{1, 2, ..., u\}$, let $z_i = x_i/c$, and choose $w_1, w_2, ..., w_\ell$ in $\mathbb{Q}$ solving

$$A^* \vec{w} = \begin{pmatrix} z_1 \\ z_2 \\ \vdots \\ z_\ell \end{pmatrix}.$$

For $j \in \{1, 2, ..., \ell\}$ let $y_j = w_j c$, and observe that since $c = |\det A^*|$, each $y_j \in \mathbb{Z}$. For $j \in \{\ell+1, \ell+2, ..., v\}$ (if any), let $y_j = 0$. We show that $A\vec{y} = \vec{x}$, which will complete the proof. If $i \in \{1, 2, ..., \ell\}$, one has immediately that $\sum_{j=1}^{v} a_{i,j} y_j = \sum_{j=1}^{\ell} a_{i,j} w_j c = z_i c = x_i$. Now let $t \in \{\ell+1, \ell+2, ..., u\}$ be given. Then given j we have $a_{t,j} = \sum_{i=1}^{\ell} \gamma_{t,i} \cdot a_{i,j}$. Also, $D\vec{x} = \vec{0}$, so

$$0 = \sum_{i=1}^{u} d_{t-\ell,i} \cdot x_i = \sum_{i=1}^{\ell} \gamma_{t,i} \cdot x_i - x_t$$

so

$$x_t = \sum_{i=1}^{\ell} \gamma_{t,i} \cdot x_i.$$

Thus we have

$$\sum_{j=1}^{v} a_{t,j} \cdot y_j = \sum_{j=1}^{\ell} a_{t,j} \cdot w_j \cdot c$$

$$= \sum_{j=1}^{\ell} \sum_{i=1}^{\ell} \gamma_{t,i} \cdot a_{i,j} \cdot w_j \cdot c$$

$$= \sum_{i=1}^{\ell} \gamma_{t,i} \cdot \sum_{j=1}^{\ell} a_{i,j} \cdot w_j \cdot c$$

$$= \sum_{i=1}^{\ell} \gamma_{t,i} \cdot x_i$$

$$= x_t. \qquad \square$$

We have already observed that statement (II) of Theorem 2.2 is one that is effectively decidable. It is also easy to work with, and as a consequence it will be heavily utilized throughout the rest of the paper, beginning with the next lemma.

Lemma 2.4. *Let A be a $u \times v$ matrix with rational entries that satisfies statement (II) of Theorem 2.2, and let $\vec{p} \in \mathbb{Z}^v \setminus \{0\}$. There exists $b \in \mathbb{Q} \setminus \{0\}$ such that $\begin{pmatrix} A \\ b\vec{p} \end{pmatrix}$ satisfies statement (II) of Theorem 2.2.*

Proof. Let $\ell = \operatorname{rank}(A)$. We may presume the first ℓ rows of A are linearly independent. Let the rows of A be $\vec{r}_1, \vec{r}_2, ..., \vec{r}_u$.

Case 1. ($\ell = u$) If $\vec{p} \notin \operatorname{span}\{\vec{r}_1, \vec{r}_2, ..., \vec{r}_\ell\}$, then $\operatorname{rank}\begin{pmatrix} A \\ \vec{p} \end{pmatrix} = \ell+1 = u+1$, so $\begin{pmatrix} A \\ \vec{p} \end{pmatrix}$ satisfies statement (II) of Theorem 2.2. Thus we assume $\vec{p} \in \operatorname{span}\{\vec{r}_1, \vec{r}_2, ..., \vec{r}_\ell\}$. Let $\alpha_1, \alpha_2, ..., \alpha_\ell \in \mathbb{Q}$ such that $\vec{p} = \sum_{i=1}^{\ell} \alpha_i \vec{r}_i$. Since $\vec{p} \neq \vec{0}$, we may pick $j \in \{1, 2, ..., \ell\}$ such that $\alpha_j \neq 0$ and let

$b = 1/\alpha_j$. Then $b\bar{p} = \sum_{i=1}^{\prime} b\alpha_i \bar{r}_i$, so the matrix D determined by statement II for $\begin{pmatrix} A \\ b\bar{p} \end{pmatrix}$ is

$(b\alpha_1, b\alpha_2, ..., b\alpha_\ell, -1)$. Let $I_1 = \{j, \ell+1\}$, $I_2 = \{1, 2, ..., \ell\}\backslash\{j\}$, $\delta_{2,j} = 0$, and let $\delta_{2,\ell+1} = -\sum_{i \in I_2} b\alpha_i$. Then we have shown that D satisfies the columns condition.

Case 2. ($\ell < u$) Let D be the matrix determined by statement (II) for D. Let $\bar{c}_1, \bar{c}_2, ..., \bar{c}_u$ be the columns of D. Then D satisfies the columns condition, so pick $m \in \{1, 2, ..., u\}$ and $I_1, I_2, ..., I_m$ such that $\{I_1, I_2, ..., I_m\}$ is a partition of $\{1, 2, ..., u\}$ and $\sum_{i \in I_1} \bar{c}_i = \bar{0}$. For $t \in \{2, 3, ..., m\}$, if any, let $J_t = \bigcup_{j=1}^{t-1} I_j$ and pick $\langle \delta_{t,i} \rangle_{i \in J_t}$ in $\mathbb{Q}$ such that $\sum_{i \in I_t} \bar{c}_i = \sum_{i \in J_t} \delta_{t,i} \cdot \bar{c}_i$.

Assume first that $\bar{p} \notin \text{span}\{\bar{r}_1, \bar{r}_2, ..., \bar{r}_\ell\}$. Then let $b = 1$. Then rearrange the rows of $\begin{pmatrix} A \\ \bar{p} \end{pmatrix}$

by adding $\bar{p}$ as $\bar{r}_0$. Let D' be the matrix determined by statement (II) for $\begin{pmatrix} \bar{p} \\ A \end{pmatrix}$. Then D' is

D with a new column $\bar{0}$ added in front as $\bar{c}_0$. Then letting $I_1' = I_1 \cup \{0\}$ and letting $I_t' = I_t$ and $\delta_{t,i}' = \delta_{t,i}$ for $t \in \{2, ..., m\}$ and $i \in J_t$, one sees that D' satisfies the columns condition.

Thus we assume $\bar{p} \in \text{span}\{\bar{r}_1, \bar{r}_2, ..., \bar{r}_\ell\}$, and pick $\alpha_1, \alpha_2, ..., \alpha_\ell$ in $\mathbb{Q}$ such that $\bar{p} = \sum_{i=1}^{\prime} \alpha_i \cdot \bar{r}_i$. For $i \in \{\ell+1, \ell+2, ..., u\}$, let $\alpha_i = 0$. If $\sum_{i \in I_1} \alpha_i \neq 0$, let $b = 1/\sum_{i \in I_1} \alpha_i$ and let $k = 1$. If $\sum_{i \in I_1} \alpha_i = 0$ and there is some $k \in \{2, 3, ..., m\}$ such that $\sum_{i \in I_k} \alpha_i \neq \sum_{i \in J_k} \delta_{k,i} \cdot \alpha_i$, let k be the first such, and let $b = 1/(\sum_{i \in I_k} \alpha_i - \sum_{i \in J_k} \delta_{k,i} \cdot \alpha_i)$. If $\sum_{i \in I_1} \alpha_i = 0$ and for all $t \in \{2, 3, ..., m\}$, $\sum_{i \in I_t} \alpha_i = \sum_{i \in J_t} \delta_{t,i} \cdot \alpha_i$, let $k = m+1$ and let $b = 1$.

Define the matrix D' as follows: for $t \in \{1, 2, ..., u-\ell\}$ and $i \in \{1, 2, ..., u\}$, let $d_{t,i}' = d_{t,i}$ and let $d_{t,u+1}' = 0$; for $i \in \{1, 2, ..., u\}$, let $d_{u-\ell+1,i}' = b\alpha_i$ and let $d_{u-\ell+1,u+1}' = -1$. Then D' is the

matrix determined by statement (II) for $\begin{pmatrix} A \\ b\bar{p} \end{pmatrix}$. Let $\bar{c}_1', \bar{c}_2', ..., \bar{c}_{u+1}'$ be the columns of D'. We

need to show that D' satisfies the columns condition. To do so, we consider the possibilities $k = m+1$ and $k \leqslant m$ separately.

Assume first that $k = m+1$. For $t \in \{1, 2, ..., m\}$ let $I_t' = I_t$ and $J_t' = J_t$, and let $I_{m+1}' = \{u+1\}$. For $t \in \{2, 3, ..., m\}$ and $i \in J_t$, let $\delta_{t,i}' = \delta_{t,i}$. Then $\sum_{i \in I_1'} \bar{c}_i' = 0$, and for $t \in \{2, 3, ..., m\}$, $\sum_{i \in I_t'} \bar{c}_i' = \sum_{i \in J_t'} \delta_{t,i}' \cdot \bar{c}_i'$, so we only need to define $\delta_{m+1,i}'$ for $i \in \{1, 2, ..., u\}$. Since $\bar{p} \neq \bar{0}$, pick $j \in \{1, 2, ..., \ell\}$ such that $\alpha_j \neq 0$, let $\delta_{m+1,j}' = -1/\alpha_j$, and for $i \in \{1, 2, ..., \ell\}\backslash\{j\}$, let $\delta_{m+1,i}' = 0$. For $t \in \{1, 2, ..., u-\ell\}$, let $\delta_{m+1,\ell+t}' = -d_{t,j}'/\alpha_j$. Then $\bar{c}_{u+1}' = \sum_{i=1}^{u} \delta_{m+1,i}' \cdot \bar{c}_i'$, as required.

Now assume $k \leqslant m$. Let $I_k' = I_k \cup \{u+1\}$, and for $t \in \{1, 2, ..., m\}\backslash\{k\}$ let $I_t' = I_t$. For $t \in \{2, 3, ..., m\}$, let $J_t' = \bigcup_{s=1}^{t-1} I_s'$. For $t \in \{2, 3, ..., m\}$ and $i \in J_t$, let $\delta_{t,i}' = \delta_{t,i}$. For $t \in \{k+1, k+2, ..., m\}$, let $\delta_{t,u+1}' = \sum_{i \in J_t} \delta_{t,i}' \cdot b \cdot \alpha_i - \sum_{i \in I_t} b \cdot \alpha_i$. We see then that D' does satisfy the columns condition. $\square$

Lemma 2.5. *Statements* (III) *and* (V) *of Theorem* 2.2 *are equivalent.*

Proof. We show first that statement (III) implies statement (V). Let $t_1, t_2, ..., t_r$ be as in statement (III), and for $j \in \{1, 2, ..., v\}$, let $b_j = 1/t_j$. Let d be a common multiple of the denominators of the t_js. To see that $b_1, b_2, ..., b_r$ are as required by statement (V), let $\mathbb{N}$ be finitely coloured. As we remarked in the proof of Lemma 2.3, $\mathbb{N}d$ is large, so we may choose monochrome $z_1, z_2, ..., z_v, w_1, w_2, ..., w_u$ in $\mathbb{N}d$ such that

$$\begin{pmatrix} & & & -1 & 0 & \cdots & 0 \\ t_1\vec{c} & t_2\vec{c}_2 & \cdots & t_v\vec{c}_v & 0 & -1 & \cdots & 0 \\ & & & \vdots & \vdots & & \\ & & & 0 & 0 & \cdots & -1 \end{pmatrix} \begin{pmatrix} \vec{z}_1 \\ \vec{z}_2 \\ \vdots \\ \vec{z}_v \\ \vec{w}_1 \\ \vec{w}_2 \\ \vdots \\ \vec{w}_u \end{pmatrix} = \vec{0}.$$

Then given any $i \in \{1, 2, ..., u\}$, one has $w_i = \sum_{j=1}^{v} t_j a_{i,j} \vec{z}_j$. For $j \in \{1, 2, ..., v\}$, let $x_j = t_j z_j$ and observe that each $x_j \in \mathbb{Z}$, since $z_j \in \mathbb{N}d$. Then

$$\begin{pmatrix} & & A & & \\ b_1 & 0 & \cdots & 0 \\ 0 & b_2 & \cdots & 0 \\ \vdots & \vdots & & \vdots \\ 0 & 0 & & b_v \end{pmatrix} \begin{pmatrix} x_1 \\ x_2 \\ \vdots \\ x_v \end{pmatrix} = \begin{pmatrix} w_1 \\ w_2 \\ \vdots \\ w_u \\ z_1 \\ z_2 \\ \vdots \\ z_v \end{pmatrix}.$$

The proof that statement (V) implies statement (III) is similar, though somewhat easier, since we do not need to worry about 'large' sets. Given a finite colouring of $\mathbb{N}$, one picks $x_1, x_2, ..., x_v$ in $\mathbb{Z}$ such that if

$$\begin{pmatrix} & & A & & \\ b_1 & 0 & \cdots & 0 \\ 0 & b_2 & & 0 \\ \vdots & \vdots & & \\ 0 & 0 & \cdots & b_v \end{pmatrix} \begin{pmatrix} x_1 \\ x_2 \\ \vdots \\ x_v \end{pmatrix} = \begin{pmatrix} y_1 \\ y_2 \\ \vdots \\ y_{u+v} \end{pmatrix},$$

then $\vec{y}$ is monochrome. Letting $t_j = 1/b_j$ for $j \in \{1, 2, ..., v\}$, one sees that

$$\begin{pmatrix} & & & -1 & 0 & \cdots & 0 \\ t_1\vec{c}_1 & t_2\vec{c}_2 & \cdots & t_v\vec{c}_v & 0 & -1 & \cdots & 0 \\ & & & \vdots & \vdots & & \\ & & & 0 & 0 & & -1 \end{pmatrix} \begin{pmatrix} y_{n+1} \\ y_{n+2} \\ \vdots \\ y_{n+v} \\ y_1 \\ y_2 \\ \vdots \\ y_u \end{pmatrix} = \vec{0}. \qquad \square$$

We can now establish most of Theorem 2.2.

Lemma 2.6. *Statements* (I), (II), (III), (IV) *and* (V) *of Theorem 2.2 are equivalent.*

Proof. By Lemma 2.3, statements (I) and (II) are equivalent. Consequently, Lemma 2.4 tells us that statement (I) implies statement (IV) (which trivially implies statement (I)). Applying

Lemma 2.4 v times in succession to the vectors $(1, 0, ..., 0)$, $(0, 1, ..., 0)$, ..., $(0, 0, ..., 1)$ shows us that statement (I) implies statement (V), which in turn implies statement (I). By Lemma 2.5, statements (III) and (V) are equivalent. $\qquad\square$

We now set out to establish the equivalence of statement (VI). We prove in Lemma 2.7 a statement stronger than needed here, but which will be used in the next section. As a consequence, our proof that statement (II) implies statement (VI) may seem more complicated than it really is.

Lemma 2.7. *Let A be a $u \times v$ matrix with rational entries such that* rank $A = \ell < u$ *and assume that A satisfies statement (II) of Theorem 2.2. Let $I_1, I_2, ..., I_m$ and for $t \in \{2, 3, ..., m\}$ let J_t and $\langle \delta_{t,i} \rangle_{i \in J_t}$ be as given in the columns condition for the matrix D of statement (II). Then there is a $u \times m$ matrix B satisfying the first entries condition such that for each $\bar{y} \in \mathbb{Z}^m$ there exists $\bar{x} \in \mathbb{Z}^v$ such that $A\bar{x} = B\bar{y}$. If for each $t \in \{2, 3, ..., m\}$ and each $i \in J_t \cap \{1, 2, ..., \ell\}$, $\delta_{t,i} < 0$, then for each $i \in \{1, 2, ..., \ell\}$ and each $t \in \{1, 2, ..., m\}$, $b_{i,t} \geqslant 0$, where $b_{i,t}$ is the entry in row i and column t of B.*

Proof. Assume, as in statement (II), that the first ℓ rows of A are linearly independent. Now the matrix

$$\begin{pmatrix} \vec{r}_1 \\ \vec{r}_2 \\ \vdots \\ \vec{r}_\ell \end{pmatrix}$$

has rank ℓ, so we may rearrange the columns of A so that the upper $\ell \times \ell$ corner, A^*, has nonzero determinant.

Let d be a common multiple of the denominators in A, and let $E = Ad$. Then D is also the matrix determined for E by statement (II). Let E^* be the upper left $\ell \times \ell$ corner of E, and let $w = |\det (E^*)|$. Let $\vec{c}_1, \vec{c}_2, ..., \vec{c}_u$ be the columns of D. Now D satisfies the columns condition, so pick $m \in \{1, 2, ..., u\}$ and $I_1, I_2, ..., I_m$ such that $\{I_1, I_2, ..., I_m\}$ is a partition of $\{1, 2, ..., u\}$ and $\sum_{i \in I_1} \vec{c}_i = \vec{0}$. For $t \in \{2, 3, ..., m\}$, if any, let $J_t = \bigcup_{j=1}^{t-1} I_j$ and pick $\langle \delta_{t,i} \rangle_{i \in J_t}$ in $\mathbb{Q}$ such that $\sum_{i \in I_t} \vec{c}_i = \sum_{i \in J_t} \delta_{t,i} \vec{c}_i$. Let $J_1 = \varnothing$ and let n be a common positive multiple of the denominators in $\delta_{t,i}$ for $t \in \{2, 3, ..., m\}$ and $i \in J_t$. Define the $u \times m$ matrix B by

$$b_{i,t} = \begin{cases} -\delta_{t,i} \cdot n \cdot w & \text{if } i \in J_t \\ nw & \text{if } i \in I_t \\ 0 & \text{if } i \notin \bigcup_{j=1}^{t} I_j. \end{cases}$$

Then B satisfies the first entries condition. Further, if $\delta_{t,i} < 0$, then $b_{i,t} > 0$, as claimed.

Now let $\bar{y} \in \mathbb{Z}^m$ be given and let $\bar{z} = B\bar{y}$. Since each $\delta_{t,i} \cdot n \in \mathbb{Z}$, we have that w divides each entry of $\bar{z}$. Let $\bar{v} \in \mathbb{Q}^\ell$ be such that

$$E^* \bar{v} = \begin{pmatrix} z_1/w \\ z_2/w \\ \vdots \\ z_\ell/w \end{pmatrix}$$

and note that (for example by Cramer's rule) each $v_j . w \in \mathbb{Z}$. Define $\bar{x} \in \mathbb{Z}^v$ by

$$x_j = \begin{cases} w . v_j & \text{if } j \in \{1, 2, ..., \ell\} \\ 0 & \text{if } j \in \{\ell+1, \ell+2, ..., v\}. \end{cases}$$

We claim that $E\bar{x} = \bar{z}$ (so $A(d\bar{x}) = E\bar{x} = \bar{z} = B\bar{y}$ as required). For $i \in \{1, 2, ..., \ell\}$, we have

$$\sum_{j=1}^{v} e_{i,j} . x_j = \sum_{j=1}^{\ell} e_{i,j} . w . v_j = w . (z_i/w) = z_i,$$

as required. So let $t \in \{\ell+1, \ell+2, ..., u\}$ be given. Now we have $\bar{r}_t = \sum_{i=1}^{\ell} \gamma_{t,i} . \bar{r}_i$, so for each $j \in \{1, 2, ..., v\}$, $e_{t,j} = \sum_{i=1}^{\ell} \gamma_{t,i} . e_{i,j}$. Thus

$$\sum_{j=1}^{v} e_{t,j} . x_j = \sum_{j=1}^{\ell} \sum_{i=1}^{\ell} \gamma_{t,i} . e_{i,j} . x_j$$

$$= \sum_{i=1}^{\ell} \gamma_{t,i} \sum_{j=1}^{\ell} e_{i,j} . x_j$$

$$= \sum_{i=1}^{\ell} \gamma_{t,i} . z_i,$$

so it suffices to show that $\sum_{i=1}^{\ell} \gamma_{t,i} . z_i = z_t$, that is, we want to show that $\sum_{i=1}^{u} d_{t-\ell,i} . z_i = 0$. Now, given any $s \in \{1, 2, ..., m\}$, we have $\sum_{i \in J_s} \delta_{s,i} . \bar{c}_i = \sum_{i \in I_s} \bar{c}_i$ (where, if $s = 1$, we treat $\sum_{i \in \varnothing} \delta_{s,i} . \bar{c}_i$ as $\bar{0}$). Thus, for each $s \in \{1, 2, ..., m\}$,

$$\sum_{i \in J_s} \delta_{s,i} . d_{t-\ell,i} - \sum_{i \in I_s} d_{t-\ell,i} = 0,$$

so

$$0 = \sum_{s=1}^{m} y_s . (-n) . w . \left(\sum_{i \in J_s} \delta_{s,i} . d_{t-\ell,i} - \sum_{i \in I_s} d_{t-\ell,i} \right)$$

$$= \sum_{s=1}^{m} y_s . \left(\sum_{i \in J_s} d_{t-\ell,i} . \delta_{s,i} . (-n) . w + \sum_{i \in I_s} d_{t-\ell,i} . n . w \right)$$

$$= \sum_{s=1}^{m} y_s . \left(\sum_{i \in J_s} d_{t-\ell,i} . b_{i,s} + \sum_{i \in I_s} d_{t-\ell,i} . b_{i,s} \right)$$

$$= \sum_{s=1}^{m} y_s . \left(\sum_{i=1}^{u} d_{t-\ell,i} . b_{i,s} \right)$$

$$= \sum_{i=1}^{u} d_{t-\ell,i} . \sum_{s=1}^{m} b_{i,s} . y_s$$

$$= \sum_{i=1}^{u} d_{t-\ell,i} . z_i,$$

as required. $\qquad \square$

Lemma 2.8. *Let A be a $u \times v$ matrix with rational entries. Then statement* (II) *of Theorem* 2.2 *implies statement* (VI) *of Theorem* 2.2.

Proof. Let $\ell = \mathrm{rank}\, A$. If $\ell > u$, this follows from Lemma 2.7, so we assume that $\ell = u$. Let $\bar{w} \in \mathbb{Q}^\ell$ be such that $A^* \bar{w} = \bar{1}$, where

$$\bar{1} = \begin{pmatrix} 1 \\ 1 \\ \vdots \\ 1 \end{pmatrix},$$

and let d be a common positive multiple of the denominators in $\bar{w}$. Let

$$B = \begin{pmatrix} d \\ d \\ \vdots \\ d \end{pmatrix}.$$

Then B satisfies the first entries condition. Let $y \in \mathbb{Z}$ be given, and define $\bar{x} \in \mathbb{Z}^v$ by

$$x_i = \begin{cases} ydw_i & \text{if} \quad i \in \{1, 2, \ldots, \ell\} \\ 0 & \text{if} \quad i \in \{\ell + 1, \ell + 2, \ldots, v\} \end{cases}$$

Then

$$A\bar{x} = A^*(yd\bar{w}) = \begin{pmatrix} yd \\ yd \\ \vdots \\ yd \end{pmatrix} = By. \qquad \square$$

The following lemma completes the proof of Theorem 2.2.

Lemma 2.9. *Let A be a $u \times v$ matrix with rational entries. Then statement* (VI) *of Theorem 2.2 implies statement* (I).

Proof. Let $\mathbb{N}$ be finitely coloured. For each $j \in \{1, 2, \ldots, m\}$ pick $d_j \in \mathbb{N}$ such that for any $i \in \{1, 2, \ldots, u\}$, if $j = \min\{t : b_{i,t} \neq 0\}$, then $b_{i,j} = d_j$ (which we can do, since B satisfies the first entries condition). Let c be a common multiple of $d_1, d_2, \ldots, d_m$. Define a new matrix E as follows: for $i \in \{1, 2, \ldots, u\}$ and $j \in \{1, 2, \ldots, m\}$, $e_{i,j} = b_{i,j} \cdot (c/d_j)$. Let $p = \max\{|e_{i,j}| : i \in \{1, 2, \ldots, u\}$ and $j \in \{1, 2, \ldots, m\}\}$. Then E consists of some of the rows of an (m, p, c) matrix, so pick, by Deuber's (m, p, c)-sets theorem [3], $\bar{w} \in \mathbb{N}^m$ such that the entries of $E\bar{w}$ are monochrome. Define $\bar{y} \in \mathbb{N}^m$ by $y_j = w_j \cdot (c/d_j)$ for $j \in \{1, 2, \ldots, m\}$. Then $B\bar{y} = E\bar{w}$. Pick $\bar{x} \in \mathbb{Z}^v$ such that $A\bar{x} = B\bar{y}$. Then the entries of $A\bar{x}$ are monochrome. $\qquad \square$

In the course of proving Lemma 2.9 we showed, using Deuber's (m, p, c)-sets theorem, that any matrix satisfying the first entries condition is strongly image partition regular. We digress now to prove a much stronger assertion in Theorem 2.11, which we believe is interesting in its own right. Our proof is similar to the proof of Rado's Theorem in [3, Theorem 8.22].

We shall have need of a result from [3]. The result refers to the notion of a 'central' subset of $\mathbb{N}$. The definition of central in either of two equivalent forms involves the introduction of considerable terminology, and we will not do this here. For our purposes two facts about central sets are all we need to know. First, if $\mathbb{N}$ is finitely coloured, there is some colour such that the set of n receiving that colour is central. (See [3] or [1].) Second, if C is a central set and $d \in \mathbb{N}$, then $C \cap \mathbb{N}d$ is central. (See [5, Theorem 2.7].)

Theorem 2.10. *Let C be a central subset of $\mathbb{N}$, let $\ell \in \mathbb{N}$, and for $i \in \{1, 2, ..., \ell\}$ let $\langle y_{i,n} \rangle_{n=1}^{\infty}$ be a sequence in $\mathbb{N}$. There exists a sequence $\langle b_n \rangle_{n=1}^{\infty}$ in $\mathbb{N}$ and a sequence $\langle H_n \rangle_{n=1}^{\infty}$ of pairwise disjoint finite nonempty subsets of $\mathbb{N}$ such that, whenever F is a finite nonempty subset of $\mathbb{N}$*

$$\left\{ \sum_{n \in F} b_n, \sum_{n \in F} \left(b_n + \sum_{t \in H_n} y_{1,t} \right), \sum_{n \in F} \left(b_n + \sum_{t \in H_n} y_{2,t} \right), ..., \sum_{n \in F} \left(b_n + \sum_{t \in H_n} y_{\ell,n} \right) \right\} \subseteq C.$$

Proof. [3, Proposition 8.21] or see [1, Theorem 4.12]. □

Theorem 2.11. *Let A be a $u \times v$ matrix with rational entries that satisfies the first entries condition, and let C be a central set in $\mathbb{N}$. There exist sequences $\langle x_{i,n} \rangle_{n=1}^{\infty}$ in $\mathbb{N}$ for $i \in \{1, 2, ..., v\}$ so that, whenever F is a finite nonempty subset of $\mathbb{N}$ and*

$$\vec{x} = \begin{pmatrix} \sum\limits_{n \in F} x_{1,n} \\ \sum\limits_{n \in F} x_{2,n} \\ \vdots \\ \sum\limits_{n \in F} x_{v,n} \end{pmatrix},$$

one has $A\vec{x} \in C^u$. In particular, if $\mathbb{N}$ is finitely coloured, there is a colour-class C as above.

Proof. We proceed by induction on v. Assume first that $v = 1$. Then there is some positive rational d such that $A = (d)$. (We may presume A has no repeated rows.) Write $d = p/q$, where $p, q \in \mathbb{N}$. Then, as we have observed, $C \cap \mathbb{N}p$ is central, so choose, by Theorem 2.7, some sequence $\langle b_n \rangle_{n=1}^{\infty}$ with $\sum_{n \in F} b_n \in C \cap \mathbb{N}p$ whenever F is a finite nonempty subset of $\mathbb{N}$. Let $x_{1,n} = (b_n/p) \cdot q$.

Now let $v \in \mathbb{N}$ and assume the statement is valid for v. Let A be a $u \times (v+1)$ matrix satisfying the positive first entries condition. We may assume we have some $t \in \{1, 2, ..., u-1\}$, and some positive rational d such that if $i \in \{1, 2, ..., t\}$, then $a_{i,1} = 0$, while if $i \in \{t+1, t+2, ..., u\}$, then $a_{i,1} = d$. (Additional rows may be added if need be to ensure that such a t exists.) Let B be the $t \times v$ matrix defined by $b_{i,j} = a_{i,j+1}$. Let a central set C be given, and let, for $i \in \{1, 2, ..., v\}$, $\langle y_{i,n} \rangle_{n=1}^{\infty}$ be a sequence in $\mathbb{N}$ as guaranteed by the induction hypothesis for B and C. For each $i \in \{t+1, t+2, ..., u\}$ and each $n \in \mathbb{N}$, let $z_{i,n} = \sum_{j=2}^{v+1} a_{i,j} \cdot y_{j-1,n}$. Write $d = p/q$, where $p, q \in \mathbb{N}$. Then $C \cap \mathbb{N}p$ is central, so choose, by Theorem 2.10, $\langle b_n \rangle_{n=1}^{\infty}$ and $\langle H_n \rangle_{n=1}^{\infty}$ such that, for each finite nonempty $F \subseteq \mathbb{N}$, one has $\sum_{n \in F} b_n \in C \cap \mathbb{N}p$, and for each $i \in \{t+1, t+2, ..., u\}$,

$$\sum_{n \in F} \left(b_n + \sum_{s \in H_n} z_{i,s} \right) \in C \cap \mathbb{N}p.$$

For each n let $x_{1,n} = (b_n/p) \cdot q$, and for $j \in \{2, 3, ..., v+1\}$, let $x_{j,n} = \sum_{s \in H_n} y_{j-1,s}$.

To see that the sequences $\langle x_{j,n} \rangle_{n=1}^{\infty}$ are as required, let finite nonempty $F \subseteq \mathbb{N}$ be given, and let $i \in \{1, 2, \ldots, u\}$. We need to show that $\sum_{j=1}^{v+1} a_{i,j} \cdot \sum_{n \in F} x_{j,n} \in C$. Assume first that $i \in \{1, 2, \ldots, t\}$. Then

$$\sum_{j=1}^{v+1} a_{i,j} \cdot \sum_{n \in F} x_{j,n} = \sum_{j=2}^{v+1} a_{i,j} \cdot \sum_{n \in F} \sum_{s \in H_n} y_{j-1,s}$$

$$= \sum_{j=1}^{v} b_{i,j} \cdot \sum_{s \in G} y_{j,s},$$

where $G = \bigcup_{n \in F} H_n$. This sum is in C by the induction hypothesis. Now assume $i \in \{t+1, t+2, \ldots, u\}$. Then

$$\sum_{j=1}^{v+1} a_{i,j} \cdot \sum_{n \in F} x_{j,n} = d \cdot \sum_{n \in F} (b_n/d) + \sum_{j=2}^{v+1} a_{i,j} \cdot \sum_{n \in F} \sum_{s \in H_n} y_{j-1,s}$$

$$= \sum_{n \in F} \left(b_n + \sum_{s \in H_n} \sum_{j=2}^{v+1} a_{i,j} \cdot y_{j-1,s} \right)$$

$$= \sum_{n \in F} \left(b_n + \sum_{s \in H_n} z_{i,s} \right) \in C. \qquad \square$$

The reader may wonder why we speak of matrices with rational entries rather than integer entries. Indeed. if A is a matrix with rational entries and d is a positive multiple of the denominators in A, it is easy to see that A is weakly (respectively strongly) image partition regular if and only if dA is weakly (respectively strongly) image partition regular. Certainly, if $(dA)\bar{x}$ is monochrome, then $A(d\bar{x})$ is monochrome, so the sufficiency is immediate. To see the necessity, assume that A is weakly (respectively strongly) partition regular, and let $\varphi : \mathbb{N} \to \{1, 2, \ldots, r\}$ be a finite colouring of $\mathbb{N}$. Define $\tau : \mathbb{N} \to \{1, 2, \ldots, r\}$ by $\tau(n) = \varphi(dn)$, and pick $\bar{x} \in \mathbb{Z}^v$ (respectively $\mathbb{N}^v$) such that $A\bar{x}$ is monochrome with respect to τ. Then $(dA)\bar{x}$ is monochrome with respect to φ.

The reason for choosing to use matrices with rational entries is reflected in statements (IV) and (V) of Theorem 2.2. As we shall see in the final result of this section, even if one starts with an integer matrix, one may not end up with one. The proof illustrates the application of statement (II) of Theorem 2.2.

Theorem 2.12. *Let* $A = \begin{pmatrix} 3 & 1 \\ 2 & 3 \end{pmatrix}$. *Then A is strongly image partition regular, but there do not exist integers b_1 and b_2 such that*

$$\begin{pmatrix} 3 & 1 \\ 2 & 3 \\ b_1 & 0 \\ 0 & b_2 \end{pmatrix}$$

is weakly image partition regular.

Proof. The matrix D given by statement (II) of Theorem 2.2 for the matrix

$$\begin{pmatrix} 3 & 1 \\ 2 & 3 \\ b_1 & 0 \\ 0 & b_2 \end{pmatrix}$$

is

$$D = \begin{pmatrix} \frac{3}{7}b_1 & -\frac{1}{7}b_1 & -1 & 0 \\ -\frac{2}{7}b_2 & \frac{3}{7}b_2 & 0 & -1 \end{pmatrix}.$$

Now if $(b_1, b_2) = (7/2, 7)$, we see that $\vec{c}_1 + \vec{c}_2 + \vec{c}_3 + \vec{c}_4 = \vec{0}$. Consequently by Theorem 2.2,

$$\begin{pmatrix} 3 & 1 \\ 2 & 3 \\ 7/2 & 0 \\ 0 & 7 \end{pmatrix}$$

is weakly image partition regular. Then given any x_1, x_2 with $(7/2)x_1 \in \mathbb{N}$ and $7x_2 \in \mathbb{N}$, one must have $x_1 > 0$ and $x_2 > 0$. Consequently the matrix is strongly partition regular.

Now assume we have (b_1, b_2), making

$$\begin{pmatrix} 3 & 1 \\ 2 & 3 \\ b_1 & 0 \\ 0 & b_2 \end{pmatrix}$$

weakly image partition regular, and observe (since $0 + 0 \notin \mathbb{N}$) that $b_1 \neq 0$ and $b_2 \neq 0$. Now D must satisfy the columns condition (by Theorem 2.2). The only possible choices for I_1 (which do not obviously force $b_1 = 0$ or $b_2 = 0$) are $I_1 = \{1, 3, 4\}$, $I_1 = \{2, 3, 4\}$, and $I_1 = \{1, 2, 3, 4\}$. These choices force (b_1, b_2) to be $(7/3, -7/2)$, $(-7, 7/3)$, and $(7/2, 7)$ respectively. $\square$

3. Strong image partition regularity

In this section we turn our attention to strong image partition regularity. Just as in Section 2, our aim is to give several equivalent characterizations. These are analogues of the conditions in Theorem 2.2. However, the proofs are not just analogues of the proofs in Theorem 2.2, because we are now dealing not only with linear algebra: the ordering on $\mathbb{N}$ is important. In fact, when we come to prove that strong image partition regularity implies various properties, we shall need to construct some explicit colourings of $\mathbb{N}$, rather than relying on the columns property.

Theorem 3.1. *Let A be a $u \times v$ matrix with rational entries. Then the following statements are equivalent.*

(A) *The matrix A is strongly image partition regular.*

(B) *Let $\vec{c}_1, \vec{c}_2, \ldots, \vec{c}_v$ be the columns of A. Then there exist $t_1, t_2, \ldots, t_v$ in $\{x \in \mathbb{Q} : x > 0\}$ such that the matrix*

$$\begin{pmatrix} & & & -1 & 0 & \cdots & 0 \\ & & & 0 & -1 & \cdots & 0 \\ t_1\vec{c}_1 & t_2\vec{c}_2 & \cdots & t_v\vec{c}_v & \vdots & \vdots & \\ & & & 0 & 0 & & -1 \end{pmatrix}$$

is kernel partition regular.

(C) *There exist $b_1, b_2, ..., b_v$ in $\{x \in \mathbb{Q} : x > 0\}$ such that*

$$\begin{pmatrix} & & A & & \\ b_1 & 0 & 0 & \cdots & 0 \\ 0 & b_2 & 0 & \cdots & 0 \\ 0 & 0 & b_3 & \cdots & 0 \\ \vdots & \vdots & \vdots & & \vdots \\ 0 & 0 & 0 & & b_v \end{pmatrix}$$

is weakly image partition regular.

(D) *For each $\vec{p} \in \omega^v \setminus \{\vec{0}\}$ there exists $b \in \mathbb{Q}$ with $b > 0$ such that $\begin{pmatrix} A \\ b\vec{p} \end{pmatrix}$ is strongly image partition regular.*

(E) *There exist $m \in \mathbb{N}$ and a $u \times m$ matrix B that satisfies the first entries condition such that for each $\vec{y} \in \mathbb{N}^m$ there exists $\vec{x} \in \mathbb{N}^v$ such that $A\vec{x} = B\vec{y}$.*

We remark that both statements (B) and (C) of Theorem 3.2 provide us with effective means of determining whether a given matrix is strongly image partition regular. In the case of statement (B), one simply determines whether one can find $t_1, t_2, ..., t_v$ such that the specified matrix satisfies the columns condition. Since statement (C) refers to weak image partition regularity, one may utilize statement (II) of Theorem 2.2 to see if there exist $b_1, b_2, ..., b_v$ making the resulting matrix partition regular.

We now record some trivial implications.

Lemma 3.2. *Statements (D) and (E) of Theorem 3.1 each imply statement (A).*

Proof. The only assertion that is not completely obvious is that (E) implies (A). To see this, one simply applies Theorem 2.11. (If $B\vec{y}$ is monochrome and $A\vec{x} = B\vec{y}$, then $A\vec{x}$ is monochrome.) $\square$

We would not characterize the following as 'trivial', but it does follow quickly from (the hardest part of) Theorem 2.2.

Lemma 3.3. *Let A be a $u \times v$ matrix with rational entries. Then statement (C) of Theorem 3.1 implies statement (D).*

Proof. Applying Theorem 2.2 to the weakly image partition regular matrix

$$\begin{pmatrix} & & A & & \\ b_1 & 0 & 0 & \cdots & 0 \\ 0 & b_2 & 0 & \cdots & 0 \\ 0 & 0 & b_3 & \cdots & 0 \\ \vdots & \vdots & \vdots & & \vdots \\ 0 & 0 & 0 & & b_v \end{pmatrix}$$

we obtain some $d \in \mathbb{Q} \setminus \{0\}$ such that the matrix

$$
\begin{pmatrix}
 & & A & & \\
b_1 & 0 & 0 & \ldots & 0 \\
0 & b_2 & 0 & \ldots & 0 \\
0 & 0 & b_3 & \ldots & 0 \\
\vdots & \vdots & \vdots & & \\
0 & 0 & 0 & \ldots & b_v \\
 & & d\bar{p} & &
\end{pmatrix}
$$

is weakly image partition regular. Since given any i, $b_i > 0$ and $b_i x_i > 0$ implies $x_i > 0$, we see that this latter matrix is strongly image partition regular and hence so is $\begin{pmatrix} A \\ d\bar{p} \end{pmatrix}$. Finally, given any $\vec{x} \in \mathbb{N}^v$, we have $\bar{p} . \vec{x} > 0$, since $\bar{p} \in \omega^v \setminus \{\vec{0}\}$. We know there exists $\vec{x} \in \mathbb{N}^r$ such that $(d\bar{p}) . \vec{x} \in \mathbb{N}$ (by the strong partition regularity) so we conclude that $d > 0$. □

We have one more routine implication.

Lemma 3.4. *Let A be a $u \times v$ matrix with rational entries. Then statement* (B) *of Theorem* 3.1 *implies statement* (C).

Proof. This may be taken verbatim from the first half of the proof of Lemma 2.5, noting that $b_j > 0$ since $t_j > 0$ (and further that each $x_j \in \mathbb{N}$). □

We now set out to show, in Lemma 3.6, that statement (C) of Theorem 3.1 implies statement (E).

Lemma 3.5. *Let A be a $u \times (u+v)$ matrix such that for $i, j \in \{1, 2, ..., u\}$,*

$$
a_{i,v+j} = \begin{cases} 0 & \text{if} \quad i \neq j \\ -1 & \text{if} \quad i = j. \end{cases}
$$

That is

$$
A = \begin{pmatrix}
a_{11} & a_{12} & \ldots & a_{1v} & -1 & 0 & \ldots & 0 \\
a_{21} & a_{22} & \ldots & a_{2v} & 0 & -1 & \ldots & 0 \\
\vdots & \vdots & & \vdots & \vdots & \vdots & & \vdots \\
a_{u1} & a_{u2} & \ldots & a_{uv} & 0 & 0 & \ldots & -1
\end{pmatrix}.
$$

If A satisfies the columns condition and $I_1, I_2, ..., I_m$ and, for $t = \{2, ..., m\}$, J_t and $\langle \delta_{t,i} \rangle_{i \in J_t}$ are as given by the columns condition, then one may assume that for $t \in \{2, ..., m\}$ and $i \in J_t \cap \{1, 2, ..., v\}$, $\delta_{t,i} < 0$.

Proof. For each $t \in \{2, 3, ..., m\}$, let $J_t^* = J_t \cap \{1, 2, ..., v\}$, and for each $t \in \{1, 2, ..., m\}$, let $I_t^* = I_t \cap \{1, 2, ..., v\}$. We proceed by induction on t, producing $\langle \mu_{t,i} \rangle_{i \in J_t}$ such that $\sum_{i \in I_t} \bar{c}_i = \sum_{i \in J_t} \mu_{t,i} . \bar{c}_i$, and for $i \in J_t^*$, $\mu_{t,i} < 0$.

Since the columns $\bar{c}_i$ with $i > v$ have no positive entries, we can assume $I_1^* \neq \varnothing$. Pick $k \in I_1^*$. Then for each $t \in \{2, 3, ..., m\}$, $k \in J_t^*$.

Let $\mu_{2,k} = \min\{-1+\delta_{2,k}-\delta_{2,j}: j \in J_2^*\}$, and for $j \in J_2^* \setminus \{k\}$, let $\mu_{2,j} = \mu_{2,k}+(\delta_{2,j}-\delta_{2,k})$. Then for each $j \in J_i^*, \mu_{2,j} \leqslant -1$. For $j \in J_2 \setminus \{1,2,...,v\}$, let

$$\mu_{2,j} = \sum_{i \in J_2^*} \mu_{2,i} \cdot a_{j-v,i} - \sum_{i \in I_2} a_{j-v,i}.$$

Now we show that

$$\sum_{j \in I_2} \bar{c}_j = \sum_{j \in J_2} \mu_{2,j} \cdot \bar{c}_j.$$

We show this line by line, so let $\ell \in \{1,2,...,u\}$ be given. Assume first that $\ell + v \in J_2$. Then

$$\sum_{j \in J_2} \mu_{2,j} \cdot a_{\ell,j} = \sum_{j \in J_2^*} \mu_{2,j} \cdot a_{\ell,j} - \mu_{\ell,\ell+v}$$

$$= \sum_{j \in J_2^*} \mu_{2,j} \cdot a_{\ell,j} - \left(\sum_{j \in J_2^*} \mu_{2,j} \cdot a_{\ell,j} - \sum_{j \in I_2} a_{\ell,j} \right)$$

$$= \sum_{j \in I_2} a_{\ell,j}.$$

Next assume $\ell + v \notin J_2$. Now $J_2^* = I_1^*$ and $\ell + v \notin I_1$, so $0 = \sum_{j \in I_1} a_{\ell,j} = \sum_{j \in J_2^*} a_{\ell,j}$. Thus $a_{\ell,k} = \sum_{j \in J_2^* \setminus \{k\}} (-a_{\ell,j})$. Then

$$\sum_{j \in J_2} \mu_{2,j} \cdot a_{\ell,j} = \sum_{j \in J_2^*} \mu_{2,j} \cdot a_{\ell,j}$$

$$= \sum_{j \in J_2^* \setminus \{k\}} \mu_{2,j} \cdot a_{\ell,j} + \mu_{2,k} \cdot a_{\ell,k}$$

$$= \sum_{j \in J_2^* \setminus \{k\}} \mu_{2,j} \cdot a_{\ell,j} - \sum_{j \in J_2^* \setminus \{k\}} \mu_{2,k} \cdot a_{\ell,j}$$

$$= \sum_{j \in J_2^* \setminus \{k\}} (\mu_{2,j} - \mu_{2,k}) \cdot a_{\ell,j}$$

$$= \sum_{j \in J_2^* \setminus \{k\}} (\delta_{2,j} - \delta_{2,k}) \cdot a_{\ell,j}$$

$$= \sum_{j \in J_2^* \setminus \{k\}} \delta_{2,j} \cdot a_{\ell,j} - \sum_{j \in J_2^* \setminus \{k\}} \delta_{2,k} \cdot a_{\ell,j}$$

$$= \sum_{j \in J_2^* \setminus \{k\}} \delta_{2,j} \cdot a_{\ell,j} + \delta_{2,k} \cdot a_{\ell,k}$$

$$= \sum_{j \in J_2^*} \delta_{2,k} \cdot a_{\ell,j}$$

$$= \sum_{j \in J_2} \delta_{2,j} \cdot a_{\ell,j}$$

$$= \sum_{j \in I_2} a_{\ell,j}.$$

Now let $t > 2$ and assume the induction has proceeded through $t-1$. For $j \in I_{t-1}^*$ let $\mu_{t-1,j} = -1$ and observe that for all $j \in J_t^*$, $\mu_{t-1,j} < 0$. Also observe that if $\ell \in \{1, 2, \ldots, u\}$ and $\ell + v \notin J_t$, then

$$\sum_{j \in I_{t-1}^*} a_{\ell,j} = \sum_{j \in I_{t-1}} a_{\ell,j} = \sum_{j \in J_{t-1}} \mu_{t-1,j} \cdot a_{\ell,j},$$

so

$$-\mu_{t-1,k} \cdot a_{\ell,k} = \sum_{j \in I_{t-1}^*} (-a_{\ell,j}) + \sum_{j \in J_{t-1}^* \setminus \{k\}} \mu_{t-1,j} \cdot a_{\ell,j} = \sum_{j \in J_t^* \setminus \{k\}} \mu_{t-1,j} \cdot a_{\ell,j}.$$

Consequently $a_{\ell,k} = \sum_{j \in J_t^* \setminus \{k\}} ((-\mu_{t-1,j})/\mu_{t-1,k}) \cdot a_{\ell,j}$.

Now let

$$\mu_{t,k} = \min \{(\mu_{t-1,k}/\mu_{t-1,j}) \cdot (-1 - \delta_{t,j}) + \delta_{t,k} : j \in J_t^*\}.$$

For $j \in J_t^* \setminus \{k\}$, let

$$\mu_{t,j} = (\mu_{t-1,j}/\mu_{t-1,k}) \cdot (\mu_{t,k} - \delta_{t,k}) + \delta_{t,j}.$$

For $j \in J_t \setminus \{1, 2, \ldots, v\}$, let

$$\mu_{t,j} = \sum_{i \in J_t^*} \mu_{t,i} \cdot a_{j-v,i} - \sum_{i \in I_t} a_{j-v,i}.$$

We now let $j \in J_t^*$ and show that $\mu_{t,j} < 0$. Note that $\mu_{t-1,j}/\mu_{t-1,k} > 0$. Now

$$\mu_{t,k} \leqslant (\mu_{t-1,k}/\mu_{t-1,j})(-1 - \delta_{t,j}) + \delta_{t,k},$$

so

$$\mu_{t,j} = (\mu_{t-1,j}/\mu_{t-1,k})(\mu_{t,k} - \delta_{t,k}) + \delta_{t,j} \leqslant -1.$$

Now let $\ell \in \{1, 2, \ldots, u\}$, and assume first that $\ell + v \in J_t$. Then

$$\sum_{j \in J_t} \mu_{t,j} \cdot a_{\ell,j} = \sum_{j \in J_t^*} \mu_{t,j} \cdot a_{\ell,j} - \mu_{t,v+\ell}$$

$$= \sum_{j \in J_t^*} \mu_{t,j} \cdot a_\ell - \left(\sum_{j \in J_t^*} \mu_{t,j} \cdot a_{\ell,j} - \sum_{j \in I_t} a_{\ell,j} \right)$$

$$= \sum_{j \in I_t} a_{\ell,j}.$$

Finally assume $\ell + v \notin J_t$. Then

$$\sum_{j \in J_t} \mu_{t,j} \cdot a_{\ell,j} = \sum_{j \in J_t^*} \mu_{t,j} \cdot a_{\ell,j}$$

$$= \sum_{j \in J_t^* \setminus \{k\}} \mu_{t,j} \cdot a_{\ell,j} + \mu_{t,k} \cdot a_{\ell,k}$$

$$= \sum_{j \in J_t^* \setminus \{k\}} \mu_{t,j} \cdot a_{\ell,j} + \mu_{t,k} \sum_{j \in J_t^* \setminus \{k\}} ((-\mu_{t-1,j})/\mu_{t-1,k}) \cdot a_{\ell,j}$$

$$= \sum_{j \in J_t^* \setminus \{k\}} (\mu_{t,j} - \mu_{t,k} \cdot \mu_{t-1,j}/\mu_{t-1,k}) \cdot a_{\ell,j}$$

$$= \sum_{j \in J_t^* \setminus \{k\}} ((\mu_{t-1,j}/\mu_{t-1,k})(\mu_{t,k} - \delta_{t,k}) + \delta_{t,j} - \mu_{t,k} \cdot \mu_{t-1,j}/\mu_{t-1,k}) \cdot a_{\ell,j}$$

$$= \sum_{j \in J_t^* \setminus \{k\}} (\delta_{t,i} - \delta_{t,k} \cdot \mu_{t-1,j}/\mu_{t-1,k}) \cdot a_{/,j}$$

$$= \sum_{j \in J_t^* \setminus \{k\}} \delta_{t,j} \cdot a_{/,j} + \delta_{t,k} \cdot a_{/,k}$$

$$= \sum_{j + J_t^*} \delta_{t,j} \cdot a_{/,j}$$

$$= \sum_{j \in J_t} \delta_{t,j} \cdot a_{/,j}$$

$$= \sum_{j \in I_t} a_{/,j}. \qquad \qquad \Box$$

Lemma 3.6. *Let A be a $u \times v$ matrix with rational entries. Then statement* (C) *of Theorem* 3.1 *implies statement* (E).

Proof. Pick $b_1, b_2, \ldots, b_v$ in $\{x \in \mathbb{Q} : x > 0\}$ so that the matrix

$$A^* = \begin{pmatrix} b_1 & 0 & 0 & \ldots & 0 \\ 0 & b_2 & 0 & \ldots & 0 \\ 0 & 0 & b_3 & \ldots & 0 \\ \vdots & \vdots & \vdots & & \vdots \\ 0 & 0 & 0 & \ldots & b_v \\ & & A & & \end{pmatrix}$$

is weakly partition regular. Note that rank $A^* = v$. Then A^* satisfies statement (II) of Theorem 2.2, so the matrix D given by statement (II) satisfies the columns condition. By Lemma 3.5, we may assume that for $t \in \{2, 3, \ldots, m\}$ and $j \in J_t \cap \{1, 2, \ldots, v\}$, $\delta_{t,j} < 0$. Consequently, by Lemma 2.7, we may pick a $(u + v) \times m$ matrix B^* satisfying the first entries condition, so that for every $\bar{y} \in \mathbb{Z}^m$ there exists $\bar{x} \in \mathbb{Z}^v$ such that $A^* \bar{x} = B \bar{y}$, and such that $b_{i,t} \geqslant 0$ whenever $i \in \{1, 2, \ldots, v\}$ and $t \in \{1, 2, \ldots, m\}$.

Let B consist of the bottom u rows of B^*. Then if $A^* \bar{x} = B^* \bar{y}$, we also have that $A \bar{x} = B \bar{y}$. It thus suffices to show that if $y \in \mathbb{N}^m$, $\bar{x} \in \mathbb{Z}^v$, and $A^* \bar{x} = B^* y$, then all of the entries of $\bar{x}$ are positive. To this end, let $\bar{y} \in \mathbb{N}^m$ and $i \in \{1, 2, \ldots, v\}$ be given. Then the ith entry of $A^* \bar{x}$ is $b_i \cdot x_i$, while the ith entry of $B^* \bar{y}$ is $\sum_{t=1}^{m} b_{i,t} \cdot y_t$. Since each $b_{i,t} \geqslant 0$ and at least one is positive, one then has $b_i \cdot x_i > 0$, so $x_i > 0$. $\qquad \Box$

We have now established the following pattern of implications:

$$(B) \Longrightarrow (C) \Longrightarrow (D)$$
$$\Downarrow \qquad \qquad \Downarrow$$
$$(E) \Longrightarrow (A).$$

To complete the proof of Theorem 3.1, we now set out to show that statement (A) of Theorem 3.1 implies statement (B).

Definition 3.7. Let $\bar{c}_1, \bar{c}_2, ..., \bar{c}_v$ be in $\mathbb{R}^u$ and let $I \subseteq \{1, 2, ..., v\}$. The *I-restricted span* of $(\bar{c}_1, \bar{c}_2, ..., \bar{c}_v)$ is $\{\sum_{i=1}^{v} \alpha_i \bar{c}_i : \text{each } \alpha_i \in \mathbb{R} \text{ and if } i \in I, \text{ then } \alpha_i \geq 0\}$.

We shall need two very easy facts about linear spans, which we present below. We give proofs for the sake of completeness.

Lemma 3.8. *Let $\bar{c}_1, \bar{c}_2, ..., \bar{c}_v$ be in $\mathbb{Q}^u$ and let $I \subseteq \{1, 2, ..., v\}$. Let S be the I-restricted span of $(\bar{c}_1, \bar{c}_2, ..., \bar{c}_v)$.*
(a) S is closed in $\mathbb{R}^u$.
(b) If $\bar{y} \in S \cap \mathbb{Q}^u$, then there exist $\delta_1, \delta_2, ..., \delta_v$ in $\mathbb{Q}$ with $\delta_i \geq 0$ whenever $i \in I$, such that $\bar{y} = \sum_{i=1}^{v} \delta_i \bar{c}_i$.

Proof.
(a) We proceed by induction on $|I|$ (for all v). If $I = \varnothing$, this is simply the assertion that any vector subspace of $\mathbb{R}^u$ is closed. So we assume $I \neq \varnothing$ and assume, without loss of generality, that $1 \in I$. Let T be the $(I \setminus \{1\})$-restricted span of $(\bar{c}_2, \bar{c}_2, ..., \bar{c}_v)$. By the induction hypothesis, T is closed.

To see that S is closed, let $\bar{b} \in \bar{S}$, the closure of S. We show $\bar{b} \in S$. For each $n \in \mathbb{N}$, pick $\langle \alpha_i(n) \rangle_{i=1}^{v}$ such that $\alpha_i(n) \geq 0$ when $i \in I$ and $\|\bar{b} - \sum_{i=1}^{v} \alpha_i(n) \bar{c}_i\| < 1/n$.

Case 1. ($\{\alpha_1(n): n \in \mathbb{N}\}$ is bounded) Pick δ a limit point of the sequence $\langle \alpha_1(n) \rangle_{n=1}^{\infty}$, and note that $\delta \geq 0$. Then $\bar{b} - \delta \bar{c}_1 \in T$. (Given $\epsilon > 0$, pick $n > 2/\epsilon$ such that $|\alpha_1(n) - \delta| < \epsilon/(2 \|\bar{c}_1\|)$). Then

$$\left\| \bar{b} - \delta \bar{c}_1 - \sum_{i=2}^{v} \alpha_i(n) \bar{c}_i \right\| \leq \left\| \bar{b} - \sum_{i=1}^{v} \alpha_i(n) \bar{c}_i \right\| + \left\| \alpha_1(n) \bar{c}_1 - \delta \bar{c}_1 \right\| < \epsilon.)$$

Then $\bar{b} \in \delta \bar{c}_1 + T \subseteq S$, and we are done.

Case 2. ($\{\alpha_1(n): n \in \mathbb{N}\}$ is unbounded) We claim then that $-\bar{c}_1 \in T$. To see this, let $\epsilon > 0$ be given and pick n such that $\alpha_1(n) > (1 + \|\bar{b}\|)/\epsilon$. For $i \in \{2, 3, ..., v\}$, let $\delta_i = \alpha_i(n)/\alpha_1(n)$, and note that for $i \in I \setminus \{1\}$, $\delta_i \geq 0$. Then

$$\left\| -\bar{c}_1 - \sum_{i=2}^{v} \delta_i \bar{c}_i \right\| \leq \left\| \bar{b}/\alpha_1(n) - \bar{c}_1 - \sum_{i=2}^{v} (\alpha_i(n)/\alpha_1(n)) \bar{c}_i \right\| + \left\| \bar{b}/\alpha_1(n) \right\|$$
$$< 1/(n\alpha_1(n)) + \|\bar{b}\|/\alpha_1(n)$$
$$< (1 + \|b\|)/\alpha_1(n)$$
$$< \epsilon.$$

Since T is closed, it follows that $\bar{c}_1 \in T$. Thus $\bar{c}_1$ and $-\bar{c}_1$ are in S, from which it follows immediately that S is in fact the $(I \setminus \{1\})$-restricted span of $(\bar{c}_1, \bar{c}_2, ..., \bar{c}_v)$. Thus S is closed by induction.

(b) Again we proceed by induction on $|I|$. The case $I = \varnothing$ is immediate, being merely the assertion that a rational vector in the linear span of some other rational vectors is actually in their rational linear span (which is true because we are solving linear equations with rational coefficients).

So assume $I \neq \varnothing$. Let $X = \{\bar{x} \in \mathbb{R}^v : \sum_{i=1}^{v} x_i \bar{c}_i = \bar{y}\}$. Thus X is an affine subspace of $\mathbb{R}^v$, and we are told there is some $\bar{x} \in X$ with $x_i > 0$ for all $i \in I$. Also (by the case $I = \varnothing$), there is some $\vec{z} \in X$ with $z_i \in \mathbb{Q}$ for all i. If $z_i \geqslant 0$ for all $i \in I$, then we are done, so suppose that $z_i < 0$ for some $i \in I$. Choose $t \in [0,1]$ maximal such that the vector $\vec{w} = (1-t)\vec{x} + t\vec{z}$ satisfies $w_i \geqslant 0$ for all $i \in I$ we have $w_i = 0$. Say $1 \in I$ and $w_1 = 0$. Then $\vec{y}$ is in the $(I\backslash\{1\})$-restricted span of $(\vec{c}_1, \vec{c}_3, ..., \vec{c}_v)$, and we are done, by induction. $\qquad\square$

To prove that statement (A) implies statement (B), we shall need a special class of colourings, which we introduce now.

Definition 3.9. Let $p \in \mathbb{N}\backslash\{1\}$. The *start base* p colouring is the function $\sigma_p : \mathbb{N} \to \{1, 2, ..., p-1\} \times \{0, 1, ..., p-1\} \times \{0, 1\}$ defined as follows: given $y \in \mathbb{N}$, write $y = \sum_{t=0}^{n} a_t p^t$, where each $a_t \in \{0, 1, ..., p-1\}$ and $a_n \neq 0$; if $n > 0$, $\sigma_p(y) = (a_n, a_{n-1}, i)$, where $n \equiv i \bmod 2$; if $n = 0$, $\sigma_p(y) = (a_0, 0, 0)$.

For example, given $p > 8$, if $x = 8320100$, $y = 503011$, $z = 834$, and $w = 834012$ (all written in base p, of course), then $\sigma_p(x) = \sigma_p(z) = (8, 3, 0)$, $\sigma_p(w) = (8, 3, 1)$, and $\sigma_p(y) = (5, 0, 0)$.

Lemma 3.10. *Let A be a $u \times v$ matrix with rational entries. Then statement* (A) *of Theorem 3.1 implies statement* (B).

Proof. As usual, let $\vec{c}_1, \vec{c}_2, ..., \vec{c}_v$ denote the columns of A, and let $\vec{d}_1, \vec{d}_2, ..., \vec{d}_u$ denote the columns of the $u \times u$ identity matrix. Let B be the matrix

$$(t_1\vec{c}_1 \quad t_2\vec{c}_2 \quad \cdots \quad t_v\vec{c}_v \quad -\vec{d}_1 \quad -\vec{d}_2 \quad \cdots \quad -\vec{d}_u),$$

where $t_1, t_2, ..., t_v$ are as yet unspecified positive rationals. Denote the columns of B by $\vec{b}_1, \vec{b}_2, ..., \vec{b}_{u+v}$. Then

$$\vec{b}_i = \begin{cases} t_i\vec{c}_i & \text{if } i \leqslant v \\ -\vec{d}_{i-v} & \text{if } i > v. \end{cases}$$

Given any $p \in \mathbb{N}\backslash\{1\}$ and any $x \in \mathbb{N}$, let $\gamma(p, x) = \max\{n : p^n \leqslant x\}$. Now let $p \in \mathbb{N}\backslash\{1\}$ be given. We obtain $m = m(p)$ and an ordered partition $(D_1(p), D_2(p), ..., D_m(p))$ of $\{1, 2, ..., u\}$ as follows. Pick $\vec{x} \in \mathbb{N}^v$ such that $A\vec{x} = \vec{y}$ is monochrome with respect to the start base p colouring. Now divide up $\{1, 2, ..., u\}$ according to which of the y_is start furthest to the left in their base p representation. That is, we get $D_1(p), D_2(p), ..., D_m(p)$ so that

(1) if $k \in \{1, 2, ..., m\}$ and $i, j \in D_k(p)$, then $\gamma(p, y_i) = \gamma(p, y_j)$, and

(2) if $k \in \{2, 3, ..., m\}$ and $i \in D_k(p)$ and $j \in D_{k-1}(p)$, then $\gamma(p, y_j) > \gamma(p, y_i)$. (We also observe that since $\sigma_p(y_j) = \sigma_p(y_i)$, we have $\gamma(p, y_j) \equiv \gamma(p, y_i) \bmod 2$, and hence $\gamma(p, y_j) \geqslant \gamma(p, y_i) + 2$.)

There are only finitely many ordered partitions of $\{1, 2, ..., u\}$. Therefore we may pick an infinite subset P of $\mathbb{N}\backslash\{1\}$, $m \in \mathbb{N}$, and an ordered partition $(D_1, D_2, ..., D_m)$ of $\{1, 2, ..., u\}$ so that for all $p \in P$, $m(p) = m$ and $(D_1(p), D_2(p), ..., D_m(p)) = (D_1, D_2, ..., D_m)$. We shall utilize $(D_1, D_2, ..., D_m)$ to find a partition of $\{1, 2, ..., u+v\}$, as required for the columns condition.

We proceed by induction. First we shall find $E_1 \subseteq \{1, 2, ..., v\}$, specify $t_i \in \mathbb{Q}^+ = \{t \in \mathbb{Q} : t > 0\}$ for $i \in E_1$, let $I_1 = E_1 \cup (v + D_1)$, and show that $\sum_{i \in I_1} \vec{b}_i = \vec{0}$. That is, we will show that $\sum_{i \in E_1} t_i \vec{c}_i + \sum_{i \in D_1} (-\vec{d}_i) = \vec{0}$. In order to do this, we show that $\sum_{i \in D_1} \vec{d}_i$ is in the $\{1, 2, ..., v\}$-restricted span of $(\vec{c}_1, \vec{c}_2, ..., \vec{c}_v)$. (For then, by Lemma 3.8(b), one has

$\sum_{i \in D_1} \bar{d}_i = \sum_{i=1}^{v} \alpha_i \bar{c}_i$, where each $\alpha_i \in \mathbb{Q}$ and each $\alpha_i \geqslant 0$. Let $E_1 = \{i \in \{1, 2, ..., v\} : \alpha_i > 0\}$, and for $i \in E_1$, let $t_i = \alpha_i$.)

Let S be the $\{1, 2, ..., v\}$-restricted span of $(\bar{c}_1, \bar{c}_2, ..., \bar{c}_v)$. In order to show that $\sum_{i \in D_1} \bar{d}_i$ is in S, it suffices, by Lemma 3.8(a), to show that $\sum_{i \in D_1} \bar{d}_i$ is in the closure of S. To this end, let $\epsilon > 0$ be given, and pick $p \in P$ with $p > u/\epsilon$. Pick the $\bar{x} \in \mathbb{N}^v$ and $\bar{y} \in \mathbb{N}^u$ that we used to get $(D_1(p), D_2(p), ..., D_m(p))$. That is $A\bar{x} = \bar{y}$ and $\bar{y}$ is monochrome with respect to the start base p colouring, and $(D_1, D_2, ..., D_m)$ is the ordered partition of $\{1, 2, ..., u\}$ induced by the starting positions of the y_is. Pick γ so that for all $i \in D_1$, $\gamma(p, y_i) = \gamma$. Pick $(a, b, c) \in \{1, 2, ..., p-1\} \times \{0, 1, ..., p-1\} \times \{0, 1\}$ such that $\sigma_p(y_i) = (a, b, c)$ for all $i \in \{1, 2, ..., u\}$. Let $\ell = a + b/p$ and observe that $1 \leqslant \ell < p$. For $i \in D_1$, $y_i = a \cdot p^\gamma + b \cdot p^{\gamma-1} + z_i \cdot p^{\gamma-2}$, where $0 \leqslant z_i < p$, and hence $y_i/p^\gamma = \ell + z_i/p^2$; let $\lambda_i = z_i/p^2$ and note that $0 \leqslant \lambda_i < 1/p$. For $i \in \bigcup_{j=2}^{m} D_j$, we have $\gamma(p, y_i) \leqslant \gamma - 2$; let $\lambda_i = y_i/p^\gamma$ and note that $0 \leqslant \lambda_i \leqslant 1/p$.

Now, $A\bar{x} = \bar{y}$ so

$$\sum_{i=1}^{v} x_i \bar{c}_i = \bar{y} = \sum_{i=1}^{u} y_i \bar{d}_i = \sum_{i \in D_1} y_i \bar{d}_i + \sum_{j=2}^{m} \sum_{i \in D_j} y_i \bar{d}_i.$$

Thus

$$\sum_{i=1}^{v} (x_i/p^\gamma) \bar{c}_i = \sum_{i \in D_1} \ell \bar{d}_i + \sum_{i=1}^{u} \lambda_i \bar{d}_i$$

and consequently

$$\left\| \sum_{i \in D_1} \bar{d}_i - \sum_{i=1}^{v} (x_i/(\ell p^\gamma)) \bar{c}_i \right\| = \left\| \sum_{i=1}^{u} (\lambda_i/\ell) \bar{d}_i \right\| \leqslant \sum_{i=1}^{u} |\lambda_i/\ell| < u/p < \epsilon.$$

Since $\sum_{i=1}^{v} (x_i/(\ell p^\gamma)) \bar{c}_i$ is in S, we have that $\sum_{i \in D_1} \bar{d}_i$ is in the closure of S, as required.

Now let $k \in \{2, 3, ..., m\}$, and assume we have chosen $E_1, E_2, ..., E_{k-1} \subseteq \{1, 2, ..., v\}$, and $t_i \in \mathbb{Q}^+$ for $i \in \bigcup_{j=1}^{k-1} E_j$, and $I_j = E_j \cup (v + D_j)$, as required for the columns condition. Let $L_k = \bigcup_{j=1}^{k-1} E_j$ and let $M_k = \bigcup_{j=1}^{k-1} D_j$, and enumerate M_k in order as $(q(1), q(2), ..., q(s))$. We claim that it suffices to show that $\sum_{i \in D_k} \bar{d}_i$ is in the $\{1, 2, ..., v\}$-restricted span of $(\bar{c}_1, \bar{c}_2, ..., \bar{c}_v, \bar{d}_{q(1)}, \bar{d}_{q(2)}, ..., \bar{d}_{q(s)})$, which we will again denote by S. Indeed, assume we have done this and pick by Lemma 3.8(b), $\alpha_1, \alpha_2, ..., \alpha_v$ in $\mathbb{Q}$ with each $\alpha_i \geqslant 0$, and $\delta_{q(1)}, \delta_{q(2)}, ..., \delta_{q(s)}$ in $\mathbb{Q}$ such that

$$\sum_{i \in D_k} \bar{d}_i = \sum_{i=1}^{v} \alpha_i \bar{c}_i + \sum_{i=1}^{s} \delta_{q(i)} \bar{d}_{q(i)}.$$

Let $E_k = \{i \in \{1, 2, ..., v\} \setminus L_k : \alpha_i > 0\}$ and for $i \in E_k$, let $t_i = \alpha_i$. Let $I_k = E_k \cup (v + D_k)$. Then

$$\sum_{i \in I_k} \bar{b}_i = \sum_{i \in E_k} \alpha_i \bar{c}_i + \sum_{i \in D_k} -\bar{d}_i$$

$$= \sum_{i \in L_k} -\alpha_i \bar{c}_i + \sum_{i=1}^{s} -\delta_{q(i)} \bar{d}_{q(i)}$$

$$= \sum_{j=1}^{k-1} \sum_{i \in I_j} \beta_i \bar{b}_i,$$

where $\beta_i = -\alpha_i/t_i$ if $i \in L_k$ and $\beta_i = \delta_i$ if $i \in M_k$.

To see that $\sum_{i \in D_k} \bar{d}_i$ is in S, it suffices, by Lemma 3.8(a), to show it is in the closure of S, so let $\epsilon > 0$ be given and pick $p \in P$ with $p > u/\epsilon$. Pick the $\bar{x} \in \mathbb{N}^v$ and $\bar{y} \in \mathbb{N}^u$ that we used to get $(D_1(p), D_2(p), \ldots, D_m(p))$. Pick γ so that for all $i \in D_k$, $\gamma(p, y_i) = \gamma$. Pick $(a, b, c) \in \{1, 2, \ldots, p-1\} \times \{0, 1, \ldots, p-1\} \times \{0, 1\}$ such that $\sigma_p(y_i) = (a, b, c)$ for all $i \in \{1, 2, \ldots, u\}$. Let $\ell = a + b/p$. For $i \in D_k$, $y_i = a \cdot p^\gamma + b \cdot p^{\gamma-1} + z_i \cdot p^{\gamma-2}$, where $0 \le z_i < p$, and hence $y_i/p^\gamma = \ell + z_i/p^2$; let $\lambda_i = z_i/p^2$ and note that $0 \le \lambda_i < 1/p$. For $i \in \bigcup_{j=k+1}^m D_j$, we have $\gamma(p, y_i) \le \gamma - 2$; let $\lambda_i = y_i/p^\gamma$ and note that $0 \le \lambda_i < 1/p$. (Of course we have no control on the size of y_i/p^γ for $i \in M_k$.)

Now $A\bar{x} = \bar{y}$ so

$$\sum_{i=1}^v x_i \bar{c}_i = \sum_{i=1}^u y_i \bar{d}_i$$

$$= \sum_{i \in M_k} y_i \bar{d}_i + \sum_{i \in D_k} y_i \bar{d}_i + \sum_{j=k+1}^m \sum_{i \in D_j} y_i \bar{d}_i.$$

Thus

$$\sum_{i=1}^v (x_i/p^\gamma) \bar{c}_i = \sum_{i \in M_k} (y_i/p^\gamma) \bar{d}_i + \sum_{i \in D_k} \ell d_i + \sum_{i=k}^m \sum_{i \in D_j} \lambda_i \bar{d}_i.$$

Consequently,

$$\left\| \sum_{i \in D_k} d_i - \left(\sum_{i=1}^v (x_i/(\ell p^\gamma)) \bar{c}_i - \sum_{i \in M_k} (y_i/(\ell p^\gamma)) d_i \right) \right\| = \left\| \sum_{j=k}^m \sum_{i \in D_j} (\lambda_i/\ell) \bar{d}_i \right\|$$

$$\le \sum_{j=k}^m \sum_{i \in D_j} |\lambda_i/\ell|$$

$$< u/p$$

$$< \epsilon,$$

as required.

Having chosen $I_1, I_2, \ldots, I_m$, if $\{1, 2, \ldots, u+v\} = \bigcup_{j=1}^m I_j$, we are done. Otherwise, let $I_{m+1} = \{1, 2, \ldots, u+v\} \setminus \bigcup_{j=1}^m I_j$. Now, $\{1, 2, \ldots, u\} = \bigcup_{j=1}^m D_j$, so $\{\bar{d}_1, \bar{d}_2, \ldots, \bar{d}_u\} \subseteq \{\bar{b}_i : i \in \bigcup_{j=1}^m I_j\}$ and we can write $\sum_{i \in I_{m+1}} \bar{b}_i$ as a linear combination of $\{\bar{b}_i : i \in \bigcup_{j=1}^m I_j\}$. $\square$

Observe that we have in fact shown that a matrix is strongly partition regular if and only if it is partition regular with respect to the start base p colourings for all (or even for infinitely many) p in $\mathbb{N} \setminus \{1\}$.

Let us close with an illustration of the use of the effectively computable conditions, namely conditions (II) and (III) of Theorem 2.2 and condition (B) of Theorem 3.1. We utilize them to determine whether or not

$$\begin{pmatrix} 1 & -1 \\ 3 & 2 \\ 4 & 6 \end{pmatrix}$$

is strongly image partition regular (and at the same time whether or not it is weakly image partition regular). In the process, we will be developing anecdotal evidence that the solution method based on statement (II) of Theorem 2.2 (in conjunction with statement (V) of Theorem 2.2 and statement (C) of Theorem 3.1) is the more efficient of the two methods.

To utilize statement (II) of Theorem 2.2, we let b_1 and b_2 be as yet undetermined non-zero rationals such that

$$\begin{pmatrix} 1 & -1 \\ 3 & 2 \\ 4 & 6 \\ b_1 & 0 \\ 0 & b_2 \end{pmatrix}$$

is weakly partition regular. Then

$$\vec{r}_3 = -2\vec{r}_1 + 2\vec{r}_2$$
$$\vec{r}_4 = (2b_1/5)\,\vec{r}_1 + (b_1/5)\,\vec{r}_2,$$
$$\vec{r}_5 = (-3b_2/5)\,\vec{r}_1 + (b_2/5)\,\vec{r}_2.$$

Thus the matrix D of statement (II) of Theorem 2.2 is

$$\begin{pmatrix} -2 & 2 & -1 & 0 & 0 \\ 2b_1/5 & b_1/5 & 0 & -1 & 0 \\ -3b_2/5 & b_2/5 & 0 & 0 & -1 \end{pmatrix}.$$

We consider possibilities for I_1. Clearly $I_1 \nsubseteq \{3, 4, 5\}$. Thus, looking at row 1 we see we must have I_1 as some one of $\{1, 2\}$, $\{1, 2, 4\}$, $\{1, 2, 5\}$, or $\{1, 2, 4, 5\}$. Any of the first three alternatives requires either that $b_1 = 0$ or $b_2 = 0$. Consequently, one has I_1 must be $\{1, 2, 4, 5\}$. This can happen if and only if $b_1 = 5/3$ and $b_2 = -5/2$. Thus, in one stroke we see that

$$\begin{pmatrix} 1 & -1 \\ 3 & 2 \\ 4 & 6 \end{pmatrix}$$

is weakly but not strongly image partition regular.

For comparison, let us examine the same matrix using statements (III) of Theorem 2.2 and (B) of Theorem 3.1 instead. To this end we let t_1 and t_2 be as yet undetermined non-zero rationals such that

$$\begin{pmatrix} t_1 & -t_2 & -1 & 0 & 0 \\ 3t_1 & 2t_2 & 0 & -1 & 0 \\ 4t_1 & 6t_2 & 0 & 0 & -1 \end{pmatrix}$$

is kernel partition regular. As before, one quickly sees $\{1, 2\} \subseteq I_1$. Further, if $3 \notin I_1$, one has $t_1 = t_2$, so $3t_1 + 2t_2 = 5t_1$ and $4t_1 + 6t_2 = 10t_1$ – conditions that are incompatible with any possibilities for I_1. This leaves the possibilities for I_1 as $\{1, 2, 3\}$, $\{1, 2, 3, 4\}$, $\{1, 2, 3, 5\}$, or $\{1, 2, 3, 4, 5\}$. Each choice leads to an inconsistent set of equations except $I = \{1, 2, 3, 4\}$, which forces $t_1 = 3/5$ and $t_2 = -2/5$. Thus again we see, with somewhat more effort, that the system is weakly but not strongly image partition regular.

Acknowledgements

The authors would like to thank Walter Deuber, Ronald Graham, and Hanno Lefmann for some helpful correspondence.

References

[1] Bergelson, V., Deuber, W. and Hindman, N. (1990) Nonmetrizable topological dynamics and Ramsey Theory. *Trans. Amer. Math. Soc.* **320** 293–320.

[2] Deuber, W. (1973) Partitionen und lineare Gleichungssysteme. *Math. Zeit.* **133** 109–123.

[3] Furstenberg, H. (1981) *Recurrence in ergodic theory and combinatorial number theory*, Princeton Univ. Press, Princeton.

[4] Hilbert, D. (1982) Über die Irreducibilität ganzer Rationaler Funktionen mit ganzzahligen Koeffizienten. *J. Reine Angew Math.* **110** 104–129.

[5] Hindman, N. and Woan, W. (to appear) Central sets in semigroups and partition regularity of systems of linear equations. *Mathematika*.

[6] Rado, R. (1933) Studien zur Kombinatorik. *Math. Zeit.* **36** 424–480.

[7] Schur, I. (1916) Über die Kongruenz $x^m + y^m = z^m \pmod p$. *Jahresbericht der Deutschen Math. – Verein.* **25** 114–117.

[8] van der Waerden, B. (1927) Beweis einer Baudetschen Vermutung. *Nieuw Arch. Wisk.* **15** 212–216.

Extremal Graph Problems for Graphs with a Color-Critical Vertex

CHRISTOPH HUNDACK, HANS JÜRGEN PRÖMEL
and ANGELIKA STEGER

Institut für Diskrete Mathematik, Universität Bonn, Nassestr. 2, 53113 Bonn, Germany

In this paper we consider the following problem, given a graph H, what is the structure of a typical, i.e. random, H-free graph? We completely solve this problem for all graphs H containing a critical vertex. While this result subsumes a sequence of known results, its short proof is self contained.

1. Introduction

What does a typical triangle-free graph look like? This question was answered by Erdős, Kleitman and Rothschild [3] proving that almost every triangle-free graph is bipartite, i.e., is two-colorable.

From the point of view of extremal graph theory, this result resembles an old result of Mantel [6] stating that the complete bipartite graph is the extremal, i.e., edge-maximum, triangle-free graph. Mantel's solution was a kind of forerunner of extremal graph theory. Its starting point is usually considered to be Turán's celebrated generalization [14] of Mantel's result, characterizing the extremal graphs $T_l(n)$ on n vertices which do not contain a complete graph K_{l+1} on $l+1$ vertices as a subgraph. Turán's result stimulated a variety of deep results in graph theory, the reader is referred to [1] and [12], two excellent sources on these problems. For our purposes we will just mention two strengthenings of Turán's theorem.

Let H be a graph of chromatic number $l+1$. By $\mathscr{F}orb_n(H)$ we denote the class of all graphs on n vertices that do not contain H as a weak subgraph, i.e., the class of all H-free graphs. Basic problems in extremal graph theory are the following: if a graph H is given, what is the maximum number of edges a graph in $\mathscr{F}orb_n(H)$ can have, and, provided G is such an extremal graph, what can be said about the structure of G? If $H = K_{l+1}$, Turán's theorem gives the answer to both questions.

Research supported by DFG-project Pr 296/2-1.

An edge e of H is called *critical* if the deletion of e from H results in a graph with chromatic number l. H is called *edge-critical* if H contains a critical edge. The first strengthening of Turán's theorem we are going to mention is Simonovits's inverse extremal theorem [11] which gives a complete characterization of those graphs H sharing the property that $T_n(l)$ is the extremal graph in $\mathcal{F}orb_n(H)$.

Theorem 1.1. [11] *Let $l > 1$ and let H be a graph of chromatic number $l + 1$. Then the Turán graph $T_n(l)$ is an extremal graph in $\mathcal{F}orb_n(H)$ if and only if H is edge-critical.*

If $G_1, \ldots, G_l$ is a family of graphs, we denote by $X_{i=1}^{l} G_i$ the product of the graphs G_i, that is, the graph obtained by taking the disjoint union of $G_1, \ldots, G_l$ and joining every two vertices belonging to different G_i's. Let $K_{l+1}(r_0, \ldots, r_l)$ denote the complete $(l + 1)$-partite graph with r_i vertices in the ith class. In particular, $K_{l+1} = K_{l+1}(1, \ldots, 1)$. A vertex v of H is called *critical* if the deletion of v from H results in a graph with chromatic number less than H. If H contains a critical vertex, H is called *vertex-critical*. Obviously, if H contains a critical edge, it is vertex-critical. Moreover, observe that each $K_{l+1}(1, r_1, \ldots, r_l)$ is vertex-critical.

The second extension of Turán's theorem we mention, also due to Simonovits [10], and later generalized by Erdős and Simonovits [4] gives a characterization of the extremal graphs in classes of graphs that are defined by forbidding certain vertex-critical graphs, viz. $K_{l+1}(1, r_1, \ldots, r_l)$-graphs.

Theorem 1.2. [10] *Let $l > 1, 1 \leq r_1 \leq \ldots \leq r_l$ and n be sufficiently large. If G is an extremal graph in $\mathcal{F}orb_n(K_{l+1}(1, r_1, \ldots, r_l))$, then G admits a representation as $\overset{l}{\underset{i=1}{X}} G_i$, where*

(1) $|V(G_i)| = n/l + o(n)$,   for each $i = 1, \ldots, l$, and
(2) G_i is an extremal graph in $\mathcal{F}orb_n(K_2(1, r_1))$ for each $i = 1, \ldots, l$.

The general random graph question we are considering in this paper is: given a graph H, what is the number of graphs in $\mathcal{F}orb_n(H)$ and, moreover, what does a typical H-free graph look like? In answer to a question of Paul Erdős, a counterpart to Simonovits's [11] inverse extremal theorem was proved in [8]. For convenience, we denote by $\mathcal{C}_n(l)$ the class of l-colorable graphs on n vertices.

Theorem 1.3. [8] *Let $l > 1$ and let H be a graph of chromatic number $l + 1$. Then*

$$|\mathcal{F}orb_n(H)| = (1 + o(1)) \cdot |\mathcal{C}_n(l)|$$

if and only if H is edge-critical.

In particular, choosing $H = K_3$, gives as a corollary the result of Erdős, Kleitman and Rothschild that almost every triangle-free graph is two-colorable.

Let H be a vertex-critical graph and let v be a critical vertex in H. We say that v has *criticallity* d, if there exist d edges $e_1, \ldots, e_d$ incident to v such that $H \setminus \{e_1, \ldots, e_d\}$

has chromatic number l, and d is minimal with respect to this property. Furthermore we denote by $\mathscr{P}_n(l,d)$ the class of graphs on n vertices that are subgraphs of some product of l graphs, each of which has maximal degree at most d, *i.e.*,

$$\mathscr{P}_n(l,d) := \left\{ G \mid G \subseteq \bigtimes_{i=1}^{l} G_i, \text{ where } \sum_{i=1}^{l} |V(G_i)| = n, \; \Delta(G_i) \leq d \text{ for all } i = 1, \ldots, l \right\}.$$

The main result of this paper is the following theorem.

Theorem 1.4. *If H has chromatic number $l+1$ and contains a color-critical vertex v_0 with criticallity d, then*

$$|\mathscr{F}orb_n(H)| = (1 + O(2^{-cn})) \cdot |\mathscr{F}orb_n(H) \cap \mathscr{P}_n(l, d-1)|,$$

for some constant $c > 0$.

Note that this result on random graphs resembles Theorem 1.2 on extremal graphs in a similiar way to that in which the Erdős, Kleitman, Rothschild theorem resembles Mantel's result. For general vertex-critical graphs H, different from a $K_{l+1}(1, r_1, \ldots, r_l)$, Theorem 1.4, together with some additional counting, also implies results on the structure of the extremal graph in $\mathscr{F}orb_n(H)$ that are slightly stronger than the ones following from the Asymptotic Structure Theorem of Erdős and Simonovits (*cf.* [1]).

2. The Proof

We prove Theorem 1.4 by partitioning the class $\mathscr{F}orb_n(H)$ into a finite number of subclasses and showing that all but one of these subclasses are asymptotically negligible. This then proves that almost all members $\mathscr{F}orb_n(H)$ have the properties of this one remaining class. In addition, the rate of convergence can be expressed in terms of two functions that bound the growth rate of the negligible classes. For a proof of the following theorem, compare [13] or [7], where proofs of similar theorems can be found. For the sake of completeness, we also include a sketch of the proof.

Theorem 2.1. [7], [13] *Let k be a positive integer, let $\mathscr{E}(n) \subseteq \mathscr{F}(n)$ and $\mathscr{Z}_i(n)$, $i = 1, \ldots, k$, be $k+2$ families of graphs on n vertices, and let $\alpha, \beta > 0$ be reals such that for all sufficiently large n we have*

$$(1) \qquad \log \frac{|\mathscr{E}(n)|}{|\mathscr{E}(n+1)|} \leq -\alpha \cdot n + \beta\sqrt{n}, \quad and$$

$$(2) \qquad \mathscr{F}(n) \subseteq \mathscr{E}(n) \cup \bigcup_{i=1}^{k} \mathscr{Z}_i(n).$$

Moreover, let $z_i = z_i(n)$ be positive integral functions such that $z_i = o(n)$, and let $\epsilon_i > 0$ be constants such that for all sufficiently large n we have

$$(3) \qquad \log \frac{|\mathscr{Z}_i(n)|}{|\mathscr{F}(n - z_i)|} \leq (\alpha - \epsilon_i) \cdot z_i n \quad \text{for every } i = 1, \ldots, k.$$

Then there exist $c \geq 1$ and $\gamma > 1$ such that the following inequality holds for every $n \in \mathbb{N}$:

$$|\mathscr{F}(n)| \leq \left(1 + c \cdot \gamma^{-n}\right) \cdot |\mathscr{E}(n)|.$$

Proof (Sketch). Let $\epsilon = \min_{1 \leq i \leq k} \epsilon_i$ and $\gamma = 2^{\epsilon/4}$, and choose n_0 sufficiently large such that (1) – (3) of the Theorem and all inequalities below are fulfilled for all $n \geq n_0$. Finally, choose $c \geq 1$ sufficiently large such that the claim of the Theorem holds for all $n \leq n_0$. We proceed by induction on n. Using (2), we conclude that it suffices to show that

$$\frac{|\mathscr{L}_i(n)|}{|\mathscr{E}(n)|} \leq \frac{c}{k} \cdot \gamma^{-n} \qquad \text{for every } i = 1, \ldots, k.$$

Using (3), the induction hypothesis, and (1), we obtain

$$\frac{|\mathscr{L}_i(n)|}{|\mathscr{E}(n)|} \quad \leq \quad \frac{|\mathscr{L}_i(n)|}{|\mathscr{F}(n - z_i)|} \cdot \frac{|\mathscr{F}(n - z_i)|}{|\mathscr{E}(n - z_i)|} \cdot \prod_{j=1}^{z_i} \frac{|\mathscr{E}(n - j)|}{|\mathscr{E}(n + 1 - j)|}$$

$$\leq \quad 2^{(\alpha - \epsilon) \cdot z_i n} \cdot \left(1 + c\gamma^{-(n - z_i)}\right) \cdot 2^{\sum_{j=1}^{z_i} (-\alpha \cdot (n - j) + \beta \sqrt{n - j})}.$$

From this the claim follows by straightforward calculations.   □

In the application of Theorem 2.1 we will have $\mathscr{E}(n) = \mathscr{F}orb_n(H) \cap \mathscr{P}_n(l, d - 1)$ and $\mathscr{F}(n) = \mathscr{F}orb_n(H)$. We first determine the constants α and β such that (1) of Theorem 2.1 is satisfied.

Lemma 2.1. *There exists a constant $D > 0$ such that for all sufficiently large n:*

$$\log \frac{|\mathscr{F}orb_n(H) \cap \mathscr{P}_n(l, d - 1)|}{|\mathscr{F}orb_{n+1}(H) \cap \mathscr{P}_{n+1}(l, d - 1)|} \leq -\frac{l - 1}{l} \cdot n + D\sqrt{n}.$$

Proof. Let $\mathscr{E}(n) = \mathscr{F}orb_n(H) \cap \mathscr{P}_n(l, d - 1)$, let $\mathscr{C}ol_n(l)$ denote the set of all l-colorable graphs on n vertices, and let $\mathscr{X}_n(d)$ denote the set of all graphs of order n with maximal degree at most d. Then, obviously,

$$|\mathscr{C}ol_n(l)| \leq |\mathscr{E}(n)| \leq |\mathscr{C}ol_n(l)| \cdot |\mathscr{X}_n(d - 1)|.$$

It is well-known, *cf. e.g.* [9], that the number of l-colorable graphs satisfies

$$|\mathscr{C}ol_n(l)| = \Theta\left(2^{\frac{l-1}{2l} \cdot n^2 + n \cdot \log l - \frac{l-1}{2} \cdot \log n}\right).$$

Furthermore, we easily obtain the following crude bound for $\mathscr{X}_n(d)$:

$$|\mathscr{X}_n(d)| \leq 2^{dn \log n}.$$

We need some more notation. Let $\mathscr{Q}$ be a property that an l-colorable graph may have. (For example, the existence of a color-critical vertex is such a property.) Let

$$\mathscr{C}_n(\mathscr{Q}) = \{G \in \mathscr{C}ol_n(l) \mid G \text{ has property } \mathscr{Q}\}$$

and

$$\mathscr{E}_n(\mathscr{Q}) = \{G \in \mathscr{F}orb_n(H) \mid G = G_1 \cup G_2, G_1 \in \mathscr{X}_n(d - 1), G_2 \in \mathscr{C}ol_n(l), G_2 \text{ has property } \mathscr{Q}\}.$$

Obviously, for all such properties $\mathscr{Q}$:

$$|\mathscr{E}_n(\mathscr{Q})| \leq |\mathscr{C}_n(\mathscr{Q})| \cdot |\mathscr{X}_n(d-1)| \leq |\mathscr{C}_n(\mathscr{Q})| \cdot 2^{dn\log n}.$$

That is, if

$$|\mathscr{C}_n(\mathscr{Q})| = o(2^{-dn\log n}) \cdot |\mathscr{C}ol_n(l)|, \tag{1}$$

then $|\mathscr{E}_n(\mathscr{Q})| = o(|\mathscr{E}(n)|)$.

We will consider the following two properties:

$\mathscr{Q}_1 = $ there exists a coloring with color classes $P_1, \ldots, P_l$ such that $\min_{1 \leq i \leq l} |P_i| \leq \frac{n}{2l}$, and

$\mathscr{Q}_2 = $ for all colorings with color classes $P_1, \ldots, P_l$ such that $|P_1| = \min_{1 \leq i \leq l} |P_i| \geq \frac{n}{2l}$ there exist sets A_{ij}, $1 \leq i \leq \sqrt{n}$, $2 \leq j \leq l$ such that

— $A_{ij} \subseteq P_j$, $|A_{ij}| = |V(H)|$,

— $A_{i_1 j} \cap A_{i_2 j} = \emptyset$ for all $i_1 \neq i_2$, and

— $|\{v \in P_1 \mid \bigcup_{2 \leq j \leq l} A_{ij} \subseteq \Gamma(v)\}| \leq |V(H)|$ for all $1 \leq i \leq \sqrt{n}$.

One easily checks that $\mathscr{Q}_1$ as well as $\mathscr{Q}_2$ satisfy inequality (1). (Overestimate the number of graphs contained in $\mathscr{C}_n(\mathscr{Q}_i)$ by first choosing appropriate color classes (less than l^n ways), the sets A_{ij} (less than $\binom{n}{|V(H)|}^{(l-1)\sqrt{n}}$ ways), and then choosing the edges between the color classes.) More precisely,

$$|\mathscr{C}_n(\mathscr{Q}_1)| \leq 2^{n(n-n/2l)/2l+\binom{l-1}{2}\left(\frac{n-n/2l}{l-1}\right)^2} = o(2^{-dn\log n}) \cdot |\mathscr{C}ol_n(l)|,$$

and (here we let $h = |V(H)|$)

$$|\mathscr{C}_n(\mathscr{Q}_2)| \leq l^n \cdot \binom{n}{h}^{(l-1)\sqrt{n}} \cdot 2^{\binom{l-1}{2}\left(\frac{n-|P_1|}{l-1}\right)^2+|P_1|\cdot(n-|P_1|-(l-1)h\sqrt{n})} \cdot$$

$$\left[\sum_{i=0}^{h} \binom{|P_1|}{i}\right]^{\sqrt{n}} \cdot (2^{(l-1)h}-1)^{|P_1|\sqrt{n}}$$

$$\leq 2^{\binom{l}{2}(n/l)^2+n\log l-(1/2^{l\cdot h}\cdot l)n\sqrt{n}}$$

$$= o(2^{-dn\log n}) \cdot |\mathscr{C}ol_n(l)|.$$

The cardinality of $\mathscr{E}(n+1)$ can now be bounded as follows. Fix an arbitrary vertex $v_0 \in V_{n+1}$. We claim that there are at least $2^{\frac{l-1}{l}(n-l|V(H)|\sqrt{n})}$ possibilities to connect v_0 to any graph $G \in \mathscr{E}_n(\bar{\mathscr{Q}}_1 \wedge \bar{\mathscr{Q}}_2)$ induced on $V_{n+1} \setminus v_0$. Indeed, by the definition of $\mathscr{Q}_1$ and $\mathscr{Q}_2$ every graph $G \in \mathscr{E}_n(\bar{\mathscr{Q}}_1 \wedge \bar{\mathscr{Q}}_2)$ may be written as $G = G_1 \cup G_2$ with $G_2 \in \mathscr{C}_n(\bar{\mathscr{Q}}_1 \wedge \bar{\mathscr{Q}}_2)$. Let $|P_1| \leq \cdots \leq |P_l|$ be the color classes of an appropriate coloring of G. Then $|P_1| \geq n/2l$, and for $i = 2, \ldots, l$ there exist sets $S_i \subseteq P_i$ of size $|S_i| = \sqrt{n} \cdot |V(H)|$ such that any tuple $(X_2, \ldots, X_l)$ with $X_i \subseteq P_i \setminus S_i$ and $|X_i| = |V(H)|$ has the property that $\bigcup_{2 \leq i \leq l} X_i$ has at least $|V(H)|$ many common neighbors in P_1. In particular, this implies that v_0 may be connected to $\bigcup_{i=2}^{l}(P_i \setminus S_i)$ in an arbitrary way without generating a copy of the forbidden subgraph H. Therefore,

$$\mathscr{E}(n+1) \geq |\mathscr{E}_n(\bar{\mathscr{Q}}_1 \wedge \bar{\mathscr{Q}}_2)| \cdot 2^{\frac{l-1}{l}(n-l|V(H)|\sqrt{n})}$$

$$\geq (|\mathscr{E}(n)| - |\mathscr{E}_n(\mathscr{Q}_1)| - |\mathscr{E}_n(\mathscr{Q}_2)|) \cdot 2^{\frac{l-1}{l}(n-l|V(H)|\sqrt{n})}$$

$$\geq (1 - o(1)) \cdot |\mathscr{E}(n)| \cdot 2^{\frac{l-1}{l}(n-l|V(H)|\sqrt{n})},$$

from which the desired inequality follows easily.   □

For the rest of this section, fix integers $l \geq 2$ and $1 \leq d \leq k$ and let H be an arbitrary but fixed $(l+1)$-partite graph containing a color-critical vertex of criticality d. The main part of the proof is devoted to defining appropriate sets $\mathscr{L}_i(n)$.

We proceed with some definitions. Throughout the rest of this section we fix $\epsilon_l := 1/(2^{3d+8} \cdot l^3)$ and an $\epsilon_0 > 0$ such that

$$-(\epsilon_0 \log \epsilon_0 + (1 - \epsilon_0) \log(1 - \epsilon_0)) = \frac{1}{5}\epsilon_l.$$

Observe that this implies $n^2 \cdot \binom{n}{\epsilon_0 n} < 2^{\frac{1}{4}\epsilon_l n}$ for n sufficiently large. All logarithms in this paper are to base 2. The kth iteration of $\log n$ is denoted by $\log^{(k)} n$, i.e. $\log^{(k)} n = \log(\log^{(k-1)} n)$ and $\log^{(1)} n = \log n$.

For $0 \leq k < l$, let f_k denote the integral function

$$f_k(n) = \lceil \log^{(k^2-k+2)} n \rceil.$$

A p_k-*set* is a subset of the vertex set V_n of size $kf_k(n)$ with a partition into k subsets $P_i = \{v_{(i-1)f_k(n)+1}, \ldots, v_{if_k(n)}\}$, $i = 1, \ldots, k$, of size $f_k(n)$ such that the following conditions are satisfied:

(1) $P_i \cap P_j = \emptyset$ for every $1 \leq i < j \leq k$,
(2) $|P_i| = f_k(n)$ for every $1 \leq i \leq k$ and
(3) every pair of sets P_i and P_j is completely connected, that is $\{x, y\} \in E$ for all $x \in P_i$ and $y \in P_j$, and all $1 \leq i < j \leq k$.

For a p_k-set $P = \bigcup_{i=1}^{k} P_i$, we denote by $T_k(P) = T_k(P_1, \ldots, P_k)$ the set

$$T_k(P) = \left\{ v \in V_n \setminus \bigcup_{i=1}^{k} P_i \;\middle|\; |\Gamma(v) \cap P_i| \geq \epsilon_0 |P_i| \text{ for every } i = 1, \ldots, k \right\}.$$

Observe that, by definition, a p_0-set is empty and $T_0(\emptyset) = V_n$. For simplicity, a p_{l-1}-set Q is also called a q-*set*. The corresponding set $T_{l-1}(Q)$ is also denoted by $R(Q)$ and called the r-*set* of Q. In addition, we let q denote the size of a q-set, i.e. $q = (l-1)f_{l-1}$. A q-*set of a vertex* v is a q-set contained in the neighborhood $\Gamma(v)$ of v.

Lemma 2.2. *Let $1 \leq k \leq l - 1$ be an integer and let $\mathscr{A}_k(n)$ denote the set of all graphs $G \in \mathscr{F}orb_n(H)$ that contain a p_k-set P such that the set*

$$|T_k(P)| \leq \left(\frac{l-k}{l} - \epsilon_l \right) n.$$

Then

$$\log \frac{|\mathscr{A}_k(n)|}{|\mathscr{F}orb_{n-kf_k}(H)|} \leq \left(\frac{l-1}{l} - \frac{1}{2k}\epsilon_l \right) kf_k n$$

for all sufficiently large n.

Proof. Construct all graphs in $\mathscr{A}_k(n)$ as follows. First choose the p_k-set $P = \bigcup_{i=1}^{k} P_i$ and an H-free graph on $V_n \setminus P$ (at most $n^{kf_k} \cdot |\mathscr{F}orb_{n-kf_k}(H)|$ ways). Then choose edges inside P (less than $2^{\frac{1}{2}(kf_k)^2}$ ways), choose the set $T_k(P)$ (less than 2^n ways) and connect P to

$T_k(P)$ (at most $2^{kf_k|T_k(P)|}$ ways). Finally, connect P to $V_n \setminus (P \cup T_k(P))$. Do this by first choosing for each $v \in V_n \setminus (P \cup T_k(P))$ an index i, $1 \leq i \leq k$, so that v has less than $\epsilon_0 f_k$ neighbors in P_i. In this way the total number of ways to connect P to $V_n \setminus (P \cup T_k(P))$ can be bounded, for sufficiently large n, by

$$\left(k \cdot \sum_{j=0}^{\epsilon_0 f_k} \binom{f_k}{j} \cdot 2^{(k-1)f_k} \right)^{n - |T_k(P)|} \leq 2^{(\frac{1}{4}\epsilon_l f_k + (k-1)f_k) \cdot (n - |T_k(P)|)}.$$

Together this gives

$$
\begin{aligned}
\log \frac{|\mathcal{A}_k(n)|}{|\mathcal{F}orb_{n-kf_k}(H)|} &\leq kf_k \log n + \frac{1}{2}(kf_k)^2 + n + kf_k|T_k(P)| \\
&\quad + \left(\frac{1}{4}\epsilon_l f_k + (k-1)f_k \right) \cdot (n - |T_k(P)|) \\
&\leq \left(k - 1 + \frac{l-k}{l} - \epsilon_l + \frac{1}{2}\epsilon_l \right) \cdot f_k n \\
&\leq \left(\frac{l-1}{l} - \frac{1}{2k}\epsilon_l \right) \cdot kf_k n
\end{aligned}
$$

for n sufficiently large. $\qquad\square$

Lemma 2.3. *Let $0 \leq k \leq l - 2$ be an integer and let $\mathcal{B}_k(n)$ denote the set of all graphs $G \in \mathcal{F}orb_n(H)$ that contain a vertex v and a p_k-set P contained in $\Gamma(v)$ such that*

$$|T_k(P)| \geq \left(\frac{2}{l} - \epsilon_l \right) n \quad \text{and} \quad |\Gamma(v) \cap T_k(P)| \leq f_k(n).$$

Then

$$\log \frac{|\mathcal{B}_k(n)|}{|\mathcal{F}orb_{n-1}(H)|} \leq \left(\frac{l-1}{l} - \frac{1}{2l} \right) n$$

for all sufficiently large n.

Proof. Construct all graphs in $\mathcal{B}_k(n)$ as follows. First choose the vertex v and an H-free graph on $V_n \setminus v$ (at most $n \cdot |\mathcal{F}orb_{n-1}(H)|$ ways). Then choose the p_k-set P (less than n^{kf_k} ways) — observe that this implicitly defines the set $T_k(P)$. Next connect v with $T_k(P)$ (less than $\sum_{i=0}^{f_k} \binom{n}{i} \leq 2^{f_k \log n}$ ways) and with $V_n \setminus (P \cup T_k(P))$ (less than $2^{n - |T_k(P)|}$ ways).

Together this gives

$$
\begin{aligned}
\log \frac{|\mathcal{B}_k(n)|}{|\mathcal{F}orb_{n-1}(H)|} &\leq \log n + kf_k \log n + f_k \log n + n - |T_k(P)| \\
&\leq \left(1 - \frac{2}{l} + 2\epsilon_l \right) n \\
&\leq \left(\frac{l-1}{l} - \frac{1}{2l} \right) n
\end{aligned}
$$

for n sufficiently large. $\qquad\square$

Lemma 2.4. *For n sufficiently large, every graph $G = (V_n, E)$ in $\mathscr{F}orb_n(H) \backslash$ $\left(\bigcup_{k=1}^{l-2} \mathscr{A}_k(n) \cup \bigcup_{k=0}^{l-2} \mathscr{B}_k(n) \right)$ has the following property: the neighbourhood $\Gamma(v)$ of every vertex $v \in V_n$ contains a q-set Q.*

The proof of this lemma is analogous to the proof of a similar result in [5]. In particular we need the following lemma from [5], which generalizes a result of [2].

Lemma 2.5. *[5] Let $0 < \epsilon < 1$ and $k \in \mathbb{N}$ be given. Then there exists $N_0 = N_0(\epsilon, k)$ such that the following is true. Let G be a graph with vertex set $A_0 \cup \cdots \cup A_k$, where the A_i are pairwise disjoint sets each of size $N \geq N_0$, and suppose that $|\Gamma(v) \cap A_i| \geq \epsilon N$ for all $v \in A_0$ and all $1 \leq i \leq k$. Then there exist sets $A'_i \subseteq A_i$, $0 \leq i \leq k$, such that $|A'_i| = \lceil \log^{(2k)} N \rceil$ for each $i = 0, \ldots, k$ and such that $\bigcup_{i=1}^{k} A'_i \subseteq \Gamma(v)$ for all $v \in A'_0$.* $\qquad\square$

Proof of Lemma 2.4.   Let v be an arbitrary vertex of G. By definition of the set $\mathscr{B}_0(n)$ we observe that $\Gamma(v)$ contains a set P_{11} of size $f_1(n)$. We conclude the proof by induction on k. Suppose for some k, $1 \leq k \leq l-2$, there exist sets $P_{k1}, \ldots, P_{kk}$ satisfying the following two conditions:

 (i) the P_{ki}, $1 \leq i \leq k$ are pairwise disjoint, and

(ii) $P_{k1} \cup \cdots \cup P_{kk}$ form a p_k-set contained in $\Gamma(v)$.

As G is neither contained in $\mathscr{A}_k(n)$ nor in $\mathscr{B}_k(n)$, we know that

$$|\Gamma(v) \cap T_k(P_{k1} \cup \cdots \cup P_{kk})| > f_k(n).$$

Let $\tilde{P}$ be an arbitrary subset of $\Gamma(v) \cap T_k(P_{k1} \cup \cdots \cup P_{kk})$ of size $f_k(n)$. Apply Lemma 2.5 to $A_0 = \tilde{P}$ and $A_i = P_{ki}$, $1 \leq i \leq k$ to obtain sets $P_{k+1,1}, \ldots P_{k+1,k+1}$ satisfying conditions (i) and (ii) for $k + 1$. Inductively, we thus obtain sets $P_{l-1,1}, \ldots, P_{l-1,l-1}$ that form a q-set contained in $\Gamma(v)$. $\qquad\square$

Observe also that another immediate consequence of Lemma 2.5 is that a vertex may not have many neighbours in the r-set of a q-set contained in its neighbourhood.

Corollary 2.1. *For all sufficiently large n, every graph $G = (V_n, E)$ in $\mathscr{F}orb_n(H)$ has the following property: if $v \in V_n$ is a vertex with q-set Q,*

$$|\Gamma(v) \cap R(Q)| \leq q.$$ $\qquad\square$

Lemma 2.6. *Let $\mathscr{C}(n)$ denote the set of all graphs $G \in \mathscr{F}orb_n(H)$ that contain a vertex v with q-set Q such that*

$$|R(Q)| \geq \left(\frac{1}{l} + \epsilon_l \right) n.$$

Then

$$\log \frac{|\mathscr{C}(n)|}{|\mathscr{F}orb_{n-1}(H)|} \leq \left(\frac{l-1}{l} - \frac{1}{2}\epsilon_l \right) n$$

for all sufficiently large n.

Proof.   Construct all graphs in $\mathscr{C}(n)$ as follows. First choose the vertex v and an H-free graph on $V_n \setminus v$ (at most $n \cdot |\mathscr{F}orb_{n-1}(H)|$ ways). Then choose the q-set Q in $V_n \setminus v$ (less than n^q ways) and connect v to $V_n \setminus R(Q)$ (less than $2^{n-|R(Q)|}$ ways). Finally, connect v to $R(Q)$. Observe that by Corollary 2.1 there are at most

$$\sum_{i=0}^{q} \binom{n}{i} \leq 2^{q \log n}$$

ways to do this.

Together this gives

$$\log \frac{|\mathscr{C}(n)|}{|\mathscr{F}orb_{n-1}(H)|} \leq \log n + q \log n + n - |R(Q)| + q \log n$$

$$\leq \left(\frac{l-1}{l} - \frac{1}{2} \epsilon_l \right) n$$

for n sufficiently large. □

Corollary 2.2. *For n sufficiently large, every graph $G = (V_n, E)$ in $\mathscr{F}orb_n(H) \setminus (\mathscr{A}_{l-1}(n) \cup \mathscr{C}(n))$ has the following property: if v is a vertex with q-set Q,*

$$\left(\frac{1}{l} - \epsilon_l \right) n \leq |R(Q)| \leq \left(\frac{1}{l} + \epsilon_l \right) n.$$

□

Lemma 2.7. *Let $0 \leq k \leq l - 2$ and $m \in \{2, d+1\}$ be integers, and let $\mathscr{D}_{k,m}(n)$ denote the set of all graphs $G \in \mathscr{F}orb_n(H)$ that contain distinct vertices $v_1, \ldots, v_m$ with q-sets $Q_1, \ldots, Q_m$, respectively, and a p_k-set P contained in $\bigcap_{i=1}^{m} \Gamma(v_i)$ such that the following properties are satisfied simultaneously:*

$$|T_k(P)| \geq \left(\frac{2}{l} - \epsilon_l \right) n, \qquad \left| \bigcap_{i=1}^{m} \Gamma(v_i) \cap T_k(P) \right| \leq f_k(n),$$

$$\left| R(Q_1) \cap \bigcap_{i=2}^{m} R(Q_i) \right| \geq 2^{3d+6} \epsilon_l n, \qquad \left| \bigcap_{i=2}^{m} R(Q_i) \right| \geq \left(\frac{1}{l} - 2^{d+4} \epsilon_l \right) n,$$

and

$$\left(\frac{1}{l} - \epsilon_l \right) n \leq |R(Q_i)| \leq \left(\frac{1}{l} + \epsilon_l \right) n \quad for \quad i = 1, \ldots, m.$$

Then

$$\log \frac{|\mathscr{D}_{k,m}(n)|}{|\mathscr{F}orb_{n-m}(H)|} \leq \left(\frac{l-1}{l} - \frac{\epsilon_l}{m} \right) mn$$

for all sufficiently large n.

Proof. Similarly, as in Lemma 2.3, we construct all graphs in $\mathscr{D}_{k,m}(n)$ as follows. We first choose the vertices $v_1, \ldots, v_m$, an H-free graph on $V_n \setminus \{v_1, \ldots, v_m\}$, choose the q-sets $Q_1, \ldots, Q_m$, and the p_k-set P (less than $n^m \cdot |\mathscr{F}orb_{n-m}(H)| \cdot n^{mq} \cdot n^{kf_k}$ ways). Then we choose the edges among the v_i (less than 2^{m^2} ways), connect v_1 to $R_1 := R(Q_1)$ and the v_i, $i \geq 2$, to $\tilde{R} := \bigcap_{i=2}^{m} R(Q_i)$ (less than $2^{mq \log n}$ ways, cf. Corollary 2.1), connect v_i, $i \geq 2$, to $R_1 \setminus \tilde{R}$

and v_1 to $\tilde{R} \setminus R_1$ (less than $2^{(m-1)(|R_1| - |\tilde{R} \cap R_1|)} \cdot 2^{|\tilde{R}| - |\tilde{R} \cap R_1|}$ ways), and connect v_i, $i \geq 1$, to $V_n \setminus (\tilde{R} \cup R_1 \cup T_k(P))$ (less than $2^{m(n - |T_k(P) \setminus (\tilde{R} \cup R_1)| - |\tilde{R}| - |R_1| + |\tilde{R} \cap R_1|)}$ ways). Finally, we connect the v_i, $i \geq 1$, to $T_k(P) \setminus (\tilde{R} \cup R_1)$. This can be done by first choosing the at most $f_k(n)$ many vertices of $T_k(P) \setminus (\tilde{R} \cup R_1)$ connected to all the v_i and then connecting the remaining vertices to at most $m - 1$ of the v_i. That is, there are at most

$$\sum_{j=0}^{f_k} \binom{n}{j} \cdot (2^m - 1)^{|T_k(P) \setminus (\tilde{R} \cup R_1)|} \leq 2^{f_k \log n + (m - 2^{-m-1})|T_k(P) \setminus (\tilde{R} \cup R_1)|}$$

ways to connect the v_i to $T_k(P) \setminus (\tilde{R} \cup R_1)$.

Together this gives

$$\begin{aligned}
\log \frac{|\mathscr{D}_{k,m}(n)|}{|\mathscr{F}orb_{n-m}(H)|} &\leq m \log n + mq \log n + k f_k \log n + m^2 + mq \log n + (m-1)|R_1| \\
&\quad + |\tilde{R}| + m \cdot (n - |T_k(P) \setminus (\tilde{R} \cup R_1)| - |\tilde{R}| - |R_1|) \\
&\quad + f_k \log n + \left(m - \frac{1}{2^{m+1}}\right)|T_k(P) \setminus (\tilde{R} \cup R_1)| \\
&\leq mn - \frac{1}{2^{m+1}}|T_k(P) \setminus (\tilde{R} \cup R_1)| - (m-1)|\tilde{R}| - |R_1| + \frac{1}{2^{m+1}}\epsilon_l n \\
&\leq \frac{l-1}{l} \cdot mn - \underbrace{\left(\frac{2^{3d+6} - 3}{2^{m+1}}\epsilon_l - d 2^{d+4}\epsilon_l - \epsilon_l - \frac{1}{2^{m+1}}\epsilon_l\right)}_{\geq \epsilon_l} \cdot n
\end{aligned}$$

for n sufficiently large. □

The following corollaries can be proved in exactly the same way as Lemma 2.4 if one uses the $\mathscr{D}_{k,m}(n)$-sets instead of the $\mathscr{B}_k(n)$-sets.

Corollary 2.3. *For n sufficiently large, every graph $G = (V_n, E)$ in $\mathscr{F}orb_n(H) \setminus \left(\bigcup_{k=1}^{l-1} \mathscr{A}_k(n) \cup \bigcup_{k=0}^{l-2} \mathscr{B}_k(n) \cup \mathscr{C}(n) \cup \bigcup_{k=0}^{l-2} \mathscr{D}_{k,2}(n)\right)$ has the following property: for every two vertices $v_1, v_2 \in V_n$ that have q-sets Q_1 and Q_2, such that $|R(Q_1) \cap R(Q_2)| \geq 2^{3d+6}\epsilon_l n$, there exists a q-set Q contained in $\Gamma(v_1) \cap \Gamma(v_2)$.*

Corollary 2.4. *For n sufficiently large, every graph $G = (V_n, E)$ in $\mathscr{F}orb_n(H) \setminus \left(\bigcup_{k=1}^{l-1} \mathscr{A}_k(n) \cup \bigcup_{k=0}^{l-2} \mathscr{B}_k(n) \cup \mathscr{C}(n) \cup \bigcup_{k=0}^{l-2} \mathscr{D}_{k,d+1}(n)\right)$ has the following property: for every $d+1$ vertices $v_1, \ldots, v_{d+1}$ that have q-sets $Q_1, \ldots, Q_{d+1}$, respectively, such that $|\bigcap_{i=1}^{d+1} R(Q_i)| \geq (\frac{1}{l} - 2^{d+4}\epsilon_l)n$, there exists a q-set Q contained in $\bigcap_{i=1}^{d+1} \Gamma(v_i)$.*

Lemma 2.8. *Let $\mathscr{E}(n)$ denote the set of all graphs $G \in \mathscr{F}orb_n(H)$ that contain a vertex v with two q-sets Q_1 and Q_2 such that the following two properties are satisfied simultaneously:*

$$|R(Q_1)| \geq \left(\frac{1}{l} - \epsilon_l\right)n \quad and \quad |R(Q_2) \setminus R(Q_1)| \geq 4\epsilon_l n.$$

Then

$$\log \frac{|\mathscr{E}(n)|}{|\mathscr{F}orb_{n-1}(H)|} \leq \left(\frac{l-1}{l} - 24\epsilon_l\right)n$$

for all sufficiently large n.

Proof. Construct all graphs in $\mathscr{E}(n)$ as follows. First choose the vertex v and an H-free graph on $V_n \setminus v$ (at most $n \cdot |\mathscr{F}orb_{n-1}(H)|$ ways). Then choose the q-sets Q_1 and Q_2 (less than n^{2q} ways) and connect v with $V_n \setminus (R(Q_1) \cup R(Q_2))$ (less than $2^{n-|R(Q_1)|-|R(Q_2)\setminus R(Q_1)|}$ ways). Finally connect v to $R(Q_1) \cup R(Q_2)$. Observe that by Corollary 2.1 there are at most $2^{2q \log n}$ ways to do this.

Together this gives

$$\log \frac{|\mathscr{E}(n)|}{|\mathscr{F}orb_{n-1}(H)|} \;\leq\; \log n + 2q \log n + n - |R(Q_1)| - |R(Q_2)| \setminus R(Q_1)| + 2q \log n$$

$$\leq \left(\frac{l-1}{l} - 2\epsilon_l \right) n$$

for n sufficiently large. $\qquad\qquad\square$

Corollary 2.5. *For n sufficiently large, every graph $G = (V_n, E)$ in $\mathscr{F}orb_n(H) \setminus$*
$\left(\bigcup_{k=1}^{l-1} \mathscr{A}_k(n) \cup \bigcup_{k=0}^{l-2} \mathscr{B}_k(n) \cup \mathscr{C}(n) \cup \bigcup_{k=0}^{l-2} \mathscr{D}_{k,2}(n) \cup \mathscr{E}(n) \right)$ has the following property: if v_1, v_2
are two vertices with q-sets Q_1 and Q_2, resp., then the corresponding r-sets either 'coincide'
or are 'disjoint', that is, either

$$|R(Q_1) \cap R(Q_2)| \geq \left(\frac{1}{l} - 9\epsilon_l \right) n \qquad or \qquad |R(Q_1) \cap R(Q_2)| \leq 2^{3d+6}\epsilon_l n.$$

Proof. Let v_1 and v_2 be two vertices with q-sets Q_1 and Q_2, respectively. Assume that $|R(Q_1) \cap R(Q_2)| \geq 2^{3d+6}\epsilon_l n$. According to Corollary 2.3, there exists a q-set Q_{12}, which is a q-set of v_1 as well as of v_2. As G is not in $\mathscr{E}(n)$, this implies that

$$|R(Q_1) \setminus R(Q_{12})| \leq 4\epsilon_l n \qquad \text{and} \qquad |R(Q_{12}) \setminus R(Q_2)| \leq 4\epsilon_l n.$$

Therefore,

$$\begin{aligned} |R(Q_1) \cap R(Q_2)| \;&\geq\; |R(Q_1) \cap R(Q_{12}) \cap R(Q_2)| \\ &\geq\; |R(Q_1)| - |R(Q_1) \setminus R(Q_{12})| - |R(Q_{12}) \setminus R(Q_2))| \\ &\geq\; \left(\frac{1}{l} - 9\epsilon_l \right) n. \qquad\qquad\square \end{aligned}$$

Corollary 2.6. *For n sufficiently large,*

$$\mathscr{F}orb_n(H) \setminus \left(\bigcup_{k=1}^{l-1} \mathscr{A}_k(n) \cup \bigcup_{k=0}^{l-2} \mathscr{B}_k(n) \cup \mathscr{C}(n) \cup \bigcup_{k=0}^{l-2} (\mathscr{D}_{k,2}(n) \cup \mathscr{D}_{k,d+1}(n)) \cup \mathscr{E}(n) \right)$$

$$\subseteq \mathscr{F}orb_n(H) \cap \mathscr{P}_n(l, d-1).$$

Proof. Let $G = (V_n, E)$ be a graph contained in the set on the left-hand side and assume n is large enough for all of the above lemmas and corollaries hold. We now construct a partition $V_n = \bigcup_i X_i$ and show that it has the desired properties.

By Lemma 2.4, the neighbourhood of every vertex $v \in V_n$ contains at least one q-set. For every vertex v, fix one such q-set, let us call it Q_v, and denote the corresponding r-set by $R_v = R(Q_v)$. Choose a maximum number of vertices $v_1, \ldots, v_s \in V_n$ such that

$$|R_{v_i} \cap R_{v_j}| \leq 2^{3d+6} \epsilon_l n \qquad \text{for all pairs } 1 \leq i < j \leq s,$$

and let

$$X_i := \left\{ x \in V_n \,\middle|\, |R_x \cap R_{v_i}| \geq \frac{2}{3} \frac{n}{l} \right\}.$$

Claim 1. The sets X_i partition the set V_n.

Choose $x \in V_n \setminus \{v_1, \ldots, v_s\}$ arbitrarily. By the maximality of the set $\{v_1, \ldots, v_s\}$ there has to exist at least one v_i such that $|R_x \cap R_{v_i}| > 2^{3d+6} \epsilon_l n$. By Corollary 2.5 this implies $|R_x \cap R_{v_i}| \geq \frac{2}{3} \frac{n}{l}$, that is $x \in X_i$. Assume now there also exists an $1 \leq j \leq s$, $j \neq i$, such that $x \in X_j$. Then

$$\frac{4}{3} \frac{n}{l} \leq |R_x \cap R_{v_i}| + |R_x \cap R_{v_j}| \leq |R_x| + |R_{v_i} \cap R_{v_j}| \leq \left(\frac{1}{l} + \epsilon_l + 2^{3d+6} \epsilon_l \right) n,$$

which is a contradiction.

Claim 2. $G[X_i]$ has maximal degree at most $d - 1$, for all $i = 1, \ldots, s$.

Assume there exist $1 \leq i_0 \leq s$ and vertices $x_0, x_1, \ldots, x_d \in X_{i_0}$ such that x_0 is connected to all vertices $x_1, \ldots, x_d$. Then the definition of X_{i_0} and Corollaries 2.2 and 2.5 imply

$$|R_{x_i} \cap R_{x_j}| \geq \left(\frac{1}{l} - 9\epsilon_l \right) n \qquad \text{for all } 0 \leq i < j \leq d.$$

Therefore

$$|R_{x_i} \setminus R_{x_j}| \leq 10\epsilon_l n \qquad \text{for all } 0 \leq i < j \leq d$$

and

$$\left| \bigcap_{i=0}^{d} R_{x_i} \right| \geq |R_{x_0}| - \sum_{i=1}^{d} |R_{x_0} \setminus R_{x_i}| \geq \left(\frac{1}{l} - \epsilon_l - 10d\epsilon_l \right) n \geq \left(\frac{1}{l} - 2^{d+4} \epsilon_l \right) n.$$

This means that by Corollary 2.4 we know that there exists a q-set $\tilde{Q} = \bigcup_{i=1}^{l-1} \tilde{Q}_i$ contained in $\bigcap_{i=0}^{d} \Gamma(x_i)$. Recall now that the definition of the r-set together with Lemma 2.5 implies that there exist sets $H_1 \subseteq R(\tilde{Q})$ and $H_{i+1} \subseteq \tilde{Q}_i$, $i = 1, \ldots, l-1$ so that $\{x_0\}, H_1 \cup \{x_1, \ldots, x_d\}, H_2, \ldots, H_l$ form the parts of the forbidden subgraph H, which is a contradiction.

Claim 3. $s = l$.

By Claim 2, at most $d - 1$ vertices of Q_{v_i} can belong to X_1. Similarly, we observe that the $l - 1$ parts of Q_{v_i} must belong to different sets X_i. This shows $s \geq l$. As, on the other hand,

$$n \geq \left| \bigcup_{i=1}^{s} R_{v_i} \right| \geq \sum_{i=1}^{s} |R_{v_i}| - \sum_{i<j} |R_{v_i} \cap R_{v_j}| \geq s \left(\frac{1}{l} - \epsilon_l \right) n - \binom{s}{2} 2^{3d+6} \epsilon_l n,$$

we immediately observe that also $s \leq l$, that is $s = l$. □

References

[1] Bollobás, B. (1978) *Extremal Graph Theory*, Academic Press, New York, London.

[2] Bollobás, B. and Erdős, P. (1973) On the structure of edge graphs. *Bull. London Math. Soc.* **5** 317–321.

[3] Erdős, P., Kleitman, D.J. and Rothschild, B.L. (1976) Asymptotic enumeration of K_n-free graphs. In: International Colloquium on Combinatorial Theory. *Atti dei Convegni Lincei* **17** (2) Rome 19–27.

[4] Erdős, P. and Simonovits, M. (1971) An extremal graph problem. *Acta Mathematica Academiae Scientiarum Hungaricae* **22** 275–282.

[5] Kolaitis, Ph. G., Prömel, H.J. and Rothschild, B.L. (1987) K_{l+1}-free graphs: asymptotic structure and a $0 - 1$ law. *Trans. Amer. Math. Soc.* **303** 637–671.

[6] Mantel, W. (1907) Problem 28, soln. by H. Gouwentak, W. Mantel, J. Teixeira de Mattes, F. Schuh and W.A. Wythoff. *Wiskundige Opgaven* **10** 60–61.

[7] Prömel, H.J. and Steger, A. (1992) Coloring clique-free graphs in linear expected time. *Random Structures and Algorithms* **3** 375–402.

[8] Prömel, H.J. and Steger, A. (1992) The asymptotic number of graphs not containing a fixed color-critical subgraph. *Combinatorica* **12** 463–473.

[9] Prömel, H.J. and Steger, A. (1993) *Random l-colorable graphs*, Forschungsinstitut für Diskrete Mathematik, Universität Bonn.

[10] Simonovits M. (1966) A method for solving extremal problems in graph theory, stability problems. In: *Theory of Graphs*, Proc. Coll. held at Tihany, Hungary.

[11] Simonovits, M. (1974) Extremal graph problems with symmetrical extremal graphs. Additional chromatic conditions. *Discrete Math.* **7** 349–376.

[12] Simonovits, M. (1983) Extremal graph theory. In: Beineke, L.W. and Wilson R.J. (eds) *Selected Topics in Graph Theory 2*, Academic Press, London, 161–200.

[13] Steger, A. (1990) *Die Kleitman-Rothschild Methode*, Dissertation Universität Bonn.

[14] Turán, P. (1941) Egy gráfelméleti szélsőértékfeladatról. *Mat. Fiz. Lapok* **48** 436–452.

A Note on $\omega_1 \to \omega_1$ Functions

PÉTER KOMJÁTH†

Dept. Comp. Sci. Eötvös University, Budapest, Múzeum krt 6–8, 1088, Hungary
e-mail: kope@cs.elte.hu

A well known and widely investigated statement in model theory is Chang's conjecture. It asserts that whenever M is some structure on ω_2 of countable length then there is an elementary substructure N of cardinal $\aleph_1$ such that $N \cap \omega_1$ is a countable ordinal. See [1]. This statement fails under the axiom of constructibility, e.g., a Kurepa-tree gives a counter-example. Namely, Chang's conjecture implies that there are no $\aleph_2$ functions $\omega_1 \to \omega$ such that any two differ on a co-countable set. It is even true that there are no $\aleph_2$ functions $\omega_1 \to \omega$ such that any two differ on a closed unbounded set as the following argument shows. Assume that $\{f_\alpha : \alpha < \omega_2\}$ is such a family. Let the model M contain the following functions F and G. $F(\alpha, \xi) = f_\alpha(\xi)$, and for $\alpha, \beta < \omega_2$, $\alpha \neq \beta$, $\xi < \omega_1$ $G(\alpha, \beta, \xi)$ is the ξ'th element of a closed unbounded set on which f_α, f_β differ. If N is an uncountable elementary substructure with $\delta = N \cap \omega_1 < \omega_1$ then for $\alpha \neq \beta$ in N f_α, f_β differ on the closed unbounded subset $\{G(\alpha, \beta, \xi) : \xi < \delta\}$ of δ so necessarily $f_\alpha(\delta) \neq f_\beta(\delta)$, that is, the $\aleph_1$ functions $\{f_\alpha : \alpha \in N\}$ get different values at δ which is impossible.

The consistency of Chang's conjecture was first established (from the consistency of the existence of an ω_1-Erdős cardinal) by J. Silver (unpublished, but see e.g., in [2]). See also in [5], pp. 395–400.

In the seventies the investigation of the generalized continuum hypothesis led to the research of $\omega_1 \to \omega_1$ functions under eventual and club domination (see [3]). The "first" ω_2 of them can easily be constructed, in fact they are quite determined (see the Statement). If the axiom of constructibility holds there is a function that eventually dominates all of them and our above argument shows that Chang's conjecture implies that no $\omega_1 \to \omega_1$ function can dominate all of them on a closed unbounded set. Here we show that it is consistent that there is a function dominating in the weak sense, but there is no one which dominates in the strong sense. I suspect that this result is known but have not been able to trace an explicit reference.

† Research partially supported by Hungarian National Science Research Fund OTKA No. 1908

Definition 1. *If* $f, g : \omega_1 \rightarrow \omega_1$ *then* $f <^* g$, $f \leqslant^* g$ *denote that* $\{\xi : f(\xi) < g(\xi)\}$, *resp.* $\{\xi : f(\xi) \leqslant g(\xi)\}$ *is co-countable.* $f <^+ g$, $f \leqslant^+ g$ *denote that these hold for a closed unbounded set.*

Assume that for every limit $\alpha < \omega_2$ a cofinal sequence $\{\alpha_\tau \nearrow \alpha : \tau < \mathrm{cf}(\alpha)\}$ is fixed.

Definition 2. *For* $\alpha < \omega_2$, $h_\alpha : \omega_1 \rightarrow \omega_1$ *is the following function:* $h_0(\xi) = 0$. $h_{\alpha+1}(\xi) = h_\alpha(\xi) + 1$. *If* $\mathrm{cf}(\alpha) = \omega$, *and* $\alpha_n \nearrow \alpha$, *then* $h_\alpha(\xi) = \sup\{h_{\alpha_n}(\xi) : n < \omega\}$. *If* $\mathrm{cf}(\alpha) = \omega_1$ *and* $\alpha_\tau \nearrow \alpha$, *put* $h_\alpha(\xi) = \sup\{h_{\alpha_\tau}(\xi) : \tau < \xi\}$.

Notice that these functions $\{h_\alpha : \alpha < \omega_2\}$ depend on the particular choice of the convergent sequences $\alpha_\tau \nearrow \alpha : \tau < \mathrm{cf}(\alpha)$.

Statement

(a) For $\alpha < \beta < \omega_2$, $h_\alpha <^* h_\beta$ holds.
(b) If $\{f_\alpha : \alpha < \omega_2\}$ is a $<^+$-increasing sequence of functions, then $h_\alpha \leqslant^+ f_\alpha$ holds for every $\alpha < \omega_2$.

Proof. (a) By induction on β. (b) By induction on α. The only non-trivial case is when $\mathrm{cf}(\alpha) = \omega_1$. Assume that $\alpha = \sup\{\alpha_\tau : \tau < \omega_1\}$. $h_{\alpha_\tau}, f_{\alpha_\tau} < f_\alpha$ on some club C_τ. On a club set of $\xi < \omega_1$, $h_\alpha(\xi) = \sup\{h_{\alpha_\tau}(\xi) : \tau < \xi\}$, and also $\xi \in C_\tau$ holds $(\tau < \xi)$. But then, $h_{\alpha_\tau}(\xi) \leqslant f_{\alpha_\tau}(\xi) \leqslant f_\alpha(\xi)$ $(\tau < \xi)$, so $h_\alpha(\xi) \leqslant f_\alpha(\xi)$. $\qquad\square$

Theorem. *It is consistent relative to the existence of an* ω_1-*Erdős cardinal that there is a function* $f : \omega_1 \rightarrow \omega_1$ *such that* $h_\alpha <^+ f$ *for every* $\alpha < \omega_2$, *but there is no function* $g : \omega_1 \rightarrow \omega_1$ *with* $h_\alpha <^* g$ *for every* $\alpha < \omega_2$.

We in fact prove that if CH holds and $\{h_\alpha : \alpha < \omega_2\}$ is an arbitrary family of $\omega_1 \rightarrow \omega_1$ functions that is not $<^*$ bounded by an $\omega_1 \rightarrow \omega_1$ function then there is a countably closed, $\aleph_2$-c.c. notion of forcing which adds a function dominating each h_α on a closed unbounded set but no such function in the extension can dominate each h_α on a co-countable set.

The result is connected to the paper of Kanai [4] where it is claimed that if there is a $<^+$-chain of $\omega_1 \rightarrow \omega_1$ functions of length $\omega_2 + 1$ then there is even a $<^*$-chain of length $\omega_2 + 1$. The proof in [4], however, seems not to be complete (ξ_σ in the proof of Claim 2 really depends on β as well).

Proof. Let V be a model of GCH in which the second statement of the Theorem holds, e.g., if Chang's Conjecture is true.

Let P be the following poset. $p = (\alpha, f, S, C) \in P$, if $\alpha < \omega_1$, $f : \alpha \rightarrow \omega_1$, $S \in [\omega_2]^{\aleph_0}$, C is a function on S, $C(\xi)$ is a closed subset of α for $\xi \in S$, and, if $\beta \in C(\xi)$, then $f(\beta) > h_\xi(\beta)$. $(\alpha', f', S', C') \leqslant (\alpha, f, S, C)$ iff $\alpha' \geqslant \alpha$, $f' \supseteq f$, $S' \supseteq S$, and for $\beta \in S$, $C'(\beta) \cap \alpha = C(\beta)$ holds.

Claim 1. $(P, \leqslant)$ *is* ω_1-*closed.*

Proof. If $p_0 \geqslant p_1 \geqslant \cdots$, $p_n = (\alpha_n, f_n, S_n, C_n)$ then $p = (\alpha, f, S, C)$ extends all p_n where $\alpha = \sup\{\alpha_n : n < \omega\}$, $f = \bigcup\{f_n : n < \omega\}$, $S = \bigcup\{S_n : n < \omega\}$ and, if $\xi \in S_n$, then $C(\xi) = \bigcup\{C_i(\xi) : i \geqslant n\}$. $\square$

Claim 2. *If $\gamma < \omega_1$, then the set $D_\gamma = \{(\alpha, f, S, C) : \alpha \geqslant \gamma\}$ is dense in P.*

Proof. By induction on γ. Using Claim 1, it suffices to show that $p = (\alpha, f, S, C)$ has an extension of the form $(\alpha + 1, f', S, C')$. Take $f'(\alpha)$ bigger than every $h_\xi(\alpha)$ ($\xi \in S$), and $C'(\xi) = C(\xi) \cup \{\alpha\}$. $\square$

Claim 3. *If $\xi < \omega_2$ then $\{(\alpha, f, S, C) \in P : \xi \in S\}$ is dense in $(P, \leqslant)$.*

Proof. If $(\alpha, f, S, C) \in P$, $\xi \notin S$ then $(\alpha, f, S \cup \{\xi\}, C')$ extends it where $C'(\xi) = \emptyset$ and $C'(\zeta) = C(\zeta)$ for $\zeta \in S$. $\square$

Claim 4. *In $(P, \leqslant)$ among any $\aleph_2$ conditions some $\aleph_2$ are pairwise compatible, so $(P, \leqslant)$ has the ω_2-chain condition.*

Proof. By CH and the Δ-system lemma, among $\aleph_2$ conditions there are $\aleph_2$ such that any two are of the form $(\alpha, f, S \cup S', C')$ and $(\alpha, f, S \cup S'', C'')$ with $C'|S = C''|S$. Then $(\alpha, f, S \cup S' \cup S'', C' \cup C'')$ is a common extension. $\square$

Claim 5. *If $G \subseteq P$ is generic, $\xi < \omega_2$, then the set $E_\xi = \bigcup\{C(\xi) : (\alpha, f, S, C) \in G, \xi \in S\}$ is a closed unbounded subset of ω_1.*

Proof. If $p = (\alpha, f, S, C)$ forces that some $\tau < \omega_1$ is a limit point of E_ξ and $\alpha > \tau$ then by the definition of P, $\tau \in C(\xi) = E_\xi \cap \alpha$ and so E_ξ is closed. That E_ξ is unbounded can be proved by an argument as in Claim 2. $\square$

Claim 6. *If $G \subseteq P$ is generic, then $F = \bigcup\{f : (\alpha, f, S, C) \in G\}$ is an $\omega_1 \to \omega_1$ function such that $h_\xi <^+ F$ for every $\alpha < \omega_2$.*

Proof. By Claims 3 and 5. $\square$

Claim 7. *In $V[G]$, there is no $g : \omega_1 \to \omega_1$ function such that $h_\xi <^* g$ for every $(\xi < \omega_2)$.*

Proof. Assume that $1 \Vdash g : \omega_1 \to \omega_1$ is such a function. For every $\xi < \omega_2$ there is a $p_\xi \in P$ which determines the point from which $g(\alpha) > h_\xi(\alpha)$ holds. For a set $X \subseteq \omega_2$, $|X| = \aleph_2$, this point is the same, call it β. We may as well assume that $p_\xi = (\alpha, f, S \cup S_\xi, C_\xi)$ where the sets $\{S, S_\xi : \xi \in X\}$ are disjoint and $C_\xi|S = C$ are the same.

Assume first that for every $\gamma > \beta$, there are $g'(\gamma) < \omega_1$, $Z(\gamma) \in [X]^{\aleph_1}$ such that if $\xi \in X - Z(\gamma)$, then $h_\xi(\gamma) < g'(\gamma)$. Then the function $g' : \omega_1 \to \omega_1$ dominates every h_ξ ($\xi \in X - \bigcup\{Z(\gamma) : \gamma < \omega_1\}$) on a co-bounded subset, so, it dominates every h_ξ ($\xi < \omega_2$), a contradiction to the assumption on V. There is, therefore, a $\gamma > \beta$ such that no such $g'(\gamma)$

exists. Select $p' = (\alpha', f', S', C') \leqslant (\alpha, f, S, C)$ such that $p' \Vdash g(\gamma) = \delta$ and then choose a $\tau \in X$ such that $S_\tau \cap S' = S$, and $h_\tau(\gamma) > \delta$. Put $r = (\alpha', f', S' \cup S_\tau, C'')$ where $C''(v) = C_\tau(v)$ if $v \in S_\tau$, $C''(v) = C'(v)$ for $v \in S'$. r is clearly a condition, as $C_\tau(v)$ has no new element beyond α. r extends p_τ and p', so it forces that $h_\tau(\gamma) > g(\gamma) = \delta$ and also that $h_\tau(\gamma) \leqslant g(\gamma)$ (as $\gamma > \beta$), a contradiction. $\square$

Acknowledgment. The author is grateful to the referee for a very careful job.

References

[1] Chang, C. C. and Keisler, J. (1973) *Model Theory*, North-Holland.

[2] Devlin, K. (1975) A note on a problem of Erdős and Hajnal, *Discrete Math.* **11** 9–22.

[3] Galvin, F. and Hajnal, A. (1975) Inequalities for cardinal powers, *Annals of Mathematics* **101** 491–498.

[4] Kanai, Y. (1991) On a variant of Chang's conjecture, *Zeitschr. für math. Logic und Grundlagen d. Math.* **37** 289–292.

[5] Shelah, S. *Proper Forcing*, Lecture Notes 940, Springer-Verlag.

Topological Cliques in Graphs

JÁNOS KOMLÓS and ENDRE SZEMERÉDI

Rutgers University and Hungarian Academy of Sciences

Let $f(t)$ be the largest integer such that every graph with average degree t has a topological clique with $f(t)$ vertices. It is widely believed that $f(t) > c\sqrt{t}$. Here we prove the weaker estimate $f(t) > c\sqrt{t}/(\log t)^{\eta}$.

1. Introduction

A *subdivision* of a graph G is obtained by replacing some of the edges by independent (vertex disjoint) paths. We say that the graph H is a *topological subgraph* of the graph G (and write $H \prec G$) if there is a subgraph H' of G that is isomorphic to a subdivision of H.

If $K_r \prec G$, where K_r is the complete graph on r vertices, we say that G has a topological r-clique. We also define the *topological clique number*

$$tcl(G) = \max\{r : K_r \prec G\},$$

and write $f(t)$ for the largest integer such that every graph with average degree at least t has a topological clique with $f(t)$ vertices:

$$f(t) = \min\{tcl(G) : t(G) \geq t\},$$

where $t(G)$ is the average degree of the graph G.

It is an easy exercise to show that $tcl(G) \approx c\sqrt{n}$ for *most* graphs G on n vertices. (This is the Erdős–Fajtlowicz theorem [4], see also the papers of Bollobás and Catlin [3], and also [1].)

The following lower bound was conjectured by Mader [8], and Erdős and Hajna [5].

Conjecture 1. $f(t) \geq ct^{1/2}$.

The weaker bound $f(t) > c(\log t)^{1/2}$ was proved by Mader [9].

Theorem 1.1. *For any $\eta > 7$ there is a positive c_η such that*

$$f(t) > c_\eta t^{\frac{1}{2}} (\log t)^{-\eta} \quad for \quad t \geq 2.$$

1.1. Sketch of the proof

We use the following greedy-type algorithm. We pick r points far from each other, and pairwise connect them with the shortest possible paths that avoid a large neighbourhood of the other $r - 2$ points.

To guarantee that these paths do not use up too many vertices, we need to control the diameter. We do this by controlling degrees: we start with a subgraph G' of minimum degree ct and maximum degree at most t^2 (Section 2.1).

To avoid the main obstruction – small bottlenecks in the graph – we select a further subgraph G'' with the following expanding property: any m vertices in G'' have at least $m/(\log m)^\kappa$ neighbours (Section 2.3).

In fact, we show that if the number of vertices in G'' is sufficiently large (in terms of t) then G'' even has a topological clique of order ct. The bottleneck of the problem is when this number of vertices is small (*e.g.* a power of t) but not as small as $O(t)$.

We shall also point out in Section 2.2 a very simple proof of Theorem 1.1 for graphs with positive density. The proof is entirely self-contained; moreover, the general proof of Theorem 1.1 does not make use of any material from Section 2.2.

1.2. Definitions and Notation

All our graphs will be simple and have no isolated vertices. The vertex-set and the edge-set of the graph G will be denoted by $V(G)$ and $E(G)$, and we let $n(G) = |V(G)|$ and $e(G) = |E(G)|$. By an *n-graph*, we mean a graph of order n (that is, a graph with n vertices). We write (A, B, E) for the bipartite graph with bipartition (A, B) and edge set $E \subset A \times B$.

The set of neighbours of $v \in V$ including v itself will be denoted by $N(v)$. Hence $|N(v)| = deg(v) + 1$, where $deg(v)$ is the degree of v. More generally, for $A \subset V(G)$ we write $N(A) = \bigcup_{v \in A} N(v)$. For $v \in V, U, U' \subset V, U \cap U' = \emptyset$, we write $deg(v, U)$ for the number of edges from v to U, and $e(U, U')$ for the number of edges between U and U'. For non-empty A and B, we write

$$d(A, B) = \frac{e(A, B)}{|A||B|}$$

for the *density* of the graph between A and B.

The graph G restricted to the vertex-set S is denoted $G|_S$, and $G - S$ is shorthand for $G|_{V(G)-S}$, the graph obtained by deleting the vertex set S. We also use the abbreviation $e(U) = e(G|_U)$. An important definition is that of the *boundary* of the vertex-set S, which is denoted by ∂S:

$$\partial S = N(S) - S = \{v \in V - S : \{s, v\} \in E(G) \text{ for some } s \in S\}.$$

The notation $H \subset G$ means that H is a subgraph of G, and $H \prec G$ means that H is a topological subgraph of G. When a graph H is subdivided into a graph H', the original vertices of H will be called the *principal vertices* in H'.

The average degree, minimum degree and maximum degree of the graph G will be denoted by $t(G)$, $\delta(G)$ and $\Delta(G)$, respectively. A graph G is called *t-maximal* if $t(G) = \max_{H \subset G} t(H)$ (which implies $\delta(G) \geq \frac{1}{2} t(G)$), and it is *T-maximal* if $t(G) = \max_{H < G} t(H)$.

Finally, $c_0, c_1, c_2, \ldots$ will stand for positive absolute constants. The notation $f(x) \uparrow$ ($f(x) \downarrow$) will signify that the function $f(x)$ is monotone increasing (decreasing).

2. Preliminaries

2.1. Almost regular subgraphs Given a constant c, an *almost regular* graph is one with $\Delta(G) \leq c\delta(G)$. We would like to be able to select, for every graph G, an almost regular subgraph with only a constant loss in average degree. Unfortunately, this is not always possible. (It is not too hard to construct counterexamples.) The following lemma suffices for our present purpose.

Lemma 2.1. *Every graph G has a topological subgraph H such that*

$$t(H) \geq \frac{1}{6} t(G) \quad and \quad \frac{1}{2} t(H) \leq \delta(H) \leq \Delta(H) \leq 72\, t^2(H). \tag{1}$$

In fact, we have the following more detailed picture.

Lemma 2.2. *Let G be a T-maximal graph. Then either G has a **subgraph** H for which (1) holds, or $tcl(G) \geq (1/4)t(G)$.*

Remarks. The bound $72\,t^2(H)$ in (1) can be replaced by $c_\beta t^\beta(H)$ for any $\beta > 1$. Also, with some loss in the average degree, we can get an almost regular topological subgraph. The following lemma may help improve Theorem 1.1 for not too sparse graphs, but we will neither use it nor prove it in this paper.

Lemma 2.3. *Every graph G with average degree $t \geq 2$ has an almost regular topological subgraph H with*

$$t(H) > c_1 t(G)/\log t(G) \quad and \quad \frac{1}{2} t(H) \leq \delta(H) \leq \Delta(H) \leq c_2 t(H).$$

Proofs of Lemmas 2.1 and 2.2. These are trivial if $t(G) < 12$ (if G is a forest, pick an edge, otherwise pick any cycle as H). Thus, we will assume that $t(G) \geq 12$, in which case Lemma 2.2 implies Lemma 2.1.

Now, write $t = t(G)$, and set $S = \{v \in V : deg(v) \leq 2t^2\}$, $L = V - S$.
Since $|L| < n/(2t) \leq n/24$ (by Markov's inequality), and all degrees are at least $t/2$, there are two possibilities:

(A) $\#\{s \in S : deg(s, S) \geq t/4\} \geq \frac{2}{3} n$

(B) $S' = \{s \in S : deg(s, L) \geq t/4\}$ is of size at least $\frac{1}{4} n$

In case (A), we choose $H = G|_S$. We have $t(H) \geq (2/3)(t/4) = t/6$ and $\Delta(H) \leq 2t^2 \leq 72t^2(H)$, as needed.

In case (B), we construct a topological subgraph of G with principal points in L as follows. We pick vertices of S', one-by-one, and for each $s' \in S'$ we find two vertices $x, y \in L$ adjacent to s'. The path $xs'y$ (of length 2) will represent an edge between x and y. We do the selections in an arbitrary fashion; the only thing to keep in mind is not to use a vertex $s' \in S'$ twice, or a pair (x, y), $x, y \in L$ twice (no multiple edges). We only stop when there are no more vertices $s' \in S'$ to use.

Since the resulting topological subgraph is of average degree at most t (by the T-maximality of G), we could pick at most $t|L|/2 < n/4$ such vertices $s' \in S'$. But $|S'| > n/4$. Hence there is at least one $s' \in S'$ that is unusable. Why couldn't we use it? Because any two of its $t/4$ (or more) neighbours have already been connected via previously chosen vertices from S'. In other words, there is a topological clique of size at least $t/4$ in G.

$\square$

2.2. Dense graphs While the content of this section is not necessary for proving Theorem 1.1, it may illuminate the general proof, and it is also interesting in its own right.

It is a standard exercise in graph theory classes to show that $tcl(G) = \Theta(\sqrt{n})$ for *most* n-graphs G. It is not mentioned, however, that in fact $tcl(G) > c\sqrt{n}$ for **all** dense n-graphs.

Theorem 2.1. *For every $c > 0$ there is a $c' > 0$ such that, for all n,*

$$tcl(G) > c'\sqrt{n} \quad \text{for all } n\text{-graphs with at least } cn^2 \text{ edges.}$$

A trivial proof is through the use of the Regularity Lemma [10, 11]. Indeed, every dense n-graph has a subgraph that is a regular pair with minimum degree $c'n$, and such graphs are easily seen to contain topological cliques of size $c''\sqrt{n}$. (Use the greedy algorithm and the fact that such a regular pair has a diameter at most 4.) Here are more details. Let $0 < \varepsilon < 1/2$.

Definition 1. The bipartite graph (A, B, E) is called an ε-**regular** pair if

$$X \subset A, \ Y \subset B, \ |X| > \varepsilon|A|, \ |Y| > \varepsilon|B|$$

imply

$$|d(X, Y) - d(A, B)| < \varepsilon.$$

The Regularity Lemma guarantees, among other things, that for any $c, \varepsilon > 0$ there is a $c'' > 0$ such that every n-graph with more than cn^2 edges contains as a subgraph an ε-regular pair $H = (A, B, E)$, $|A| = |B| > c''n$, with density greater than c.

Let $\varepsilon < \delta < 1/2$ be positive numbers.

Definition 2. The bipartite graph $G = (A, B, E)$, $|A| = |B| = n$, is called an (ε, δ)-**expander**

if $\delta(G) \geq \delta n$, and

$$e(X, Y) > 0 \quad \text{for all} \quad X \subset A, \ Y \subset B, \ |X|, |Y| > \varepsilon n.$$

It easily follows from the above definitions that any ε-regular pair (A, B, E), $|A| = |B| = n$, of density $d > 2\varepsilon$ has an (ε', δ')-expander subgraph (A', B', E'), where

$$|A'| = |B'| \geq (1 - \varepsilon)n, \ \varepsilon' = \varepsilon/(1 - \varepsilon), \ \delta' = d - 2\varepsilon.$$

Now the proof of Theorem 2.1 goes as follows. Let $\varepsilon = c/5$. Apply the Regularity Lemma to find an ε-regular subgraph H of G of order $c''n$ with density $d > c = 5\varepsilon$. Then select a $(2\varepsilon, 3\varepsilon)$-expander subgraph H' of H of order at least $c''n/2$. Apply the following lemma to conclude the proof.

Lemma 2.4. *Let* $H = (A, B, E)$, $|A| = |B| = n$, *be an* (ε, δ)-*expander. Then* $tcl(H) > ((\delta - \varepsilon)n)^{1/2}$.

Proof of the lemma. Pick r arbitrary vertices, where $r = \lfloor ((\delta - \varepsilon)n)^{1/2} \rfloor$. Connect them pairwise with vertex disjoint shortest paths using the greedy algorithm, going through the pairs one-by-one. To show that the deletion of previous paths does not increase the diameter, use the following trivial observation: let $H = (A, B, E)$, $|A| = |B| = n$, be an (ε, δ)-expander. Then H has a diameter at most 4, and this remains true even after deleting arbitrary $(\delta - \varepsilon)n$ arbitrary vertices.

This proves the lemma, and hence Theorem 2.1. $\qquad\qquad\qquad\qquad\qquad$ □

How efficient is this method? The Regularity Lemma says much more than we have used (see [11]), and thus it is not surprising that the constant c' obtained in this way is quite small; for $\varepsilon = c/2$ (say), $1/c'$ is a tower function with about $1/c$ levels, that is, the $1/c$ times iterated logarithm of $1/c'$ is about 1.

Since we only use a small part of the Regularity Lemma, it is more prudent, and fits our present purpose better, to find a more direct approach. Using graph functionals of the form $\psi(G) = \psi_1(\delta(G))\psi_2(n(G))$, we can directly select expanding subgraphs of G, and these subgraphs are much bigger than those guaranteed by the Regularity Lemma (see [6]).

We will generalize this method in the next section.

2.3. ε-expanders We will generalize the expander method used for the case of dense graphs by extending the definition to expanding by a varying degree.

Note that $\underline{\varepsilon} = \varepsilon(1), \varepsilon(2), \dots$ will always denote a non-negative sequence.

Definition 3. A graph $G = (V, E)$ is an $\underline{\varepsilon}$-expander if

$$\frac{|\partial X|}{|X|} \geq \varepsilon(|X|) \quad \text{for all subsets} \quad X \subset V, \ |X| \leq (1/2)|V|. \tag{2}$$

Theorem 2.2. *Let*

$$\varepsilon(k) \downarrow, \ k\varepsilon(k) \uparrow \quad and \quad \sum_{k \geq 1} \varepsilon(k)/k \leq 1/6. \tag{3}$$

Then every graph G has an ε-expander subgraph H with

$$t(H) > \frac{1}{2}t(G) \quad and \quad \delta(H) \geq \frac{1}{2}t(H).$$ (4)

If $\varepsilon(k) \downarrow$, $k\varepsilon(k) \uparrow$ are only assumed for $k \geq k_0$, then some subgraph H (satisfying (4)) will still be expanding in the sense that (2) holds for $3k_0/2 \leq |X| \leq 2n(H)/3$.

Let us define the graph functional

$$\psi(G) = t(G)(1 + \varphi(n(G))),$$

where $\varphi(k)$ is a monotone decreasing positive sequence.

Definition 4. *G is called ψ-maximal if $\psi(G) = \max_{H \subset G} \psi(H)$.*

Note that ψ-maximality implies t-maximality.

Lemma 2.5. *Let G be a ψ-maximal graph on n vertices. Then, for all $Y \subset V$,*

$$\frac{|\partial Y|}{|X|} \geq \frac{\varphi(|X|) - \varphi(n)}{1 + \varphi(|X|)} \geq \frac{\varphi(|X|) - \varphi(n)}{1 + \varphi(1)},$$

where $X = Y \cup \partial Y$.

Proof. By the t-maximality of G, $t(G|_{\overline{Y}}) \leq t(G)$, which implies

$$\begin{aligned} 2e(G) &= 2e(Y) + 2e(Y, \overline{Y}) + 2e(\overline{Y}) \\ &\leq 2e(Y) + 2e(Y, \overline{Y}) + t(G)|\overline{Y}| \leq 2e(X) + t(G)|\overline{Y}|, \end{aligned}$$

whence

$$t(G)|Y| \leq 2e(X) = t(G|_X)|X| \leq t(G)|X|(1 + \varphi(n))/(1 + \varphi(|X|)),$$

and the rest is highschool algebra.   □

Lemma 2.6. *Let ε satisfy (3), and let*

$$\varphi(k) = 6 \sum_{i > 2k/3} \varepsilon(i)/i, \quad \psi(G) = t(G)(1 + \varphi(n(G))).$$

Then every ψ-maximal graph is an ε-expander.

If $\varepsilon(k) \downarrow$, $k\varepsilon(k) \uparrow$ are only assumed for $k \geq k_0$, every ψ-maximal graph G is still expanding in the weaker sense that (2) holds for $3k_0/2 \leq |X| \leq 2n(G)/3$.

Proof. It is easy to see that we have $\varphi(1) \leq 1$, and

$$(1 + \varphi(1))^{-1}[\varphi(k) - \varphi(3k/2)] \geq 3 \sum_{2k/3 < i \leq k} \varepsilon(i)/i \geq \varepsilon(k)$$ (5)

for $k \geq 3k_0/2$. Now let $Y \subset V$, $|Y| \leq |V|/2$, be arbitrary, and let $X = Y \cup \partial Y$. In the case $|X| \leq 2n/3$, Lemma 2.5 and (5) imply

$$|\partial Y| \geq \varepsilon(|X|)|X| \geq \varepsilon(|Y|)|Y|.$$

If $|X| > 2n/3$ then $|\partial Y| > n/6 \geq |Y|/3 > \varepsilon(|Y|)|Y|$. □

Proof of Theorem 2.2. We simply choose a ψ-maximal subgraph H of G. By the previous lemma, H is an ε-expander. We still have to estimate $t(H)$ and $\delta(H)$. We have

$$\psi(H) = t(H)(1 + \varphi(n(H))) \geq \psi(G) > t(G),$$

whence $t(H) > t(G)/(1 + \varphi(1)) \geq t(G)/2$. Also, a ψ-maximal graph is obviously t-maximal, too. Hence $\delta(H) \geq \frac{1}{2} t(H)$. □

Fix a constant $\kappa > 1$. From now on, we will always use the same sequence

$$\varepsilon(i) = \begin{cases} 0 & \text{if } i < t/4, \\ c_3 \log^{-\kappa}(8i/t) & \text{if } i \geq t/4, \end{cases}$$

and $\qquad \varphi(k) = 6 \sum_{i > 2k/3} \varepsilon(i)/i, \quad \psi(G) = t(G)(1 + \phi(n(G))). \qquad (6)$

3. Proof of Theorem 1.1

According to Lemma 2.1 and Theorem 2.2, we can restrict our attention to graphs G satisfying $t(G) = t \geq t_0$ and $t/2 \leq \delta(G) \leq \Delta(G) \leq 72t^2$. Moreover, we may assume G is ψ-maximal, and so is ε-expanding with φ and ε defined in (??).

At this point we introduce some more notation. The *distance* $\rho(u,v)$ between vertices u, v is the length of the shortest path between them. We also define *balls* and *spheres* around a vertex v, in the natural way, as

$$B(v,r) = \{u \in V : \rho(u,v) \leq r\}, \quad S(v,r) = \{u \in V : \rho(u,v) = r\}.$$

Note that $\partial B(v,r) \subset S(v,r+1)$. The *bottleneck for a vertex* v is defined as $\min\{|S(v,r+1)| : r \geq 0, |B(v,r)| \leq (1/2)n(G)\}$, and the *bottleneck of* G is then the minimum of those of its vertices.

Here are some of the properties of G that we shall use. First, by the choice of ψ, the ψ-maximality of G implies that

$$\text{if} \quad |B(v,r)| \leq \tfrac{1}{2} n(G) \quad \text{then} \quad |S(v,r+1)| \geq \varepsilon(|B(v,r)|)|B(v,r)|.$$

Given this expansion inequality for balls, it can then be shown (say by induction on r) that

$$\min\{\exp\{c_4 r^{1/(1+\kappa)}\}, \tfrac{1}{2} n\} < |B(v,r)| < (9t)^{2r} \quad \text{for all } v, r.$$

It now follows fairly straightforwardly that the diameter of G is less than $c_5(\log n)^{1+\kappa}$, and also that G has a bottleneck greater than $c_6 t$.

It is important that the properties of G just mentioned are robust, in the sense that the removal of a few vertices from G will not affect the properties (apart from slightly changed values of the constants). In particular, the properties remain if at most $(1/2)c_6 t$ vertices are removed from G (the bottleneck will still be at least $(1/2)c_6 t$). The properties will also remain if a number of vertices are deleted from G in such a way that the number removed from any sphere $S(v,r)$ is very small in comparison to the size of the sphere itself.

In the next two subsections, we will take $\kappa > 1$ fixed, and use an arbitrary parameter $\alpha > \kappa^2 + 3\kappa + 3 > 7$.

3.1. *Not too sparse graphs*

Here we prove Theorem 1.1 for n-graphs with $\log n < (\log t)^{\alpha}$. In fact, in this case we simply use the greedy algorithm.

Choose arbitrary vertices v_1, v_2 to start with, and connect them with a shortest path. Select any vertex v_3 outside this path. In general, choose an arbitrary non-used vertex, and connect it, one by one, with shortest paths to the old vertices, each time making sure to use only vertices that have not been used before. As remarked earlier, provided we use fewer than $(1/2)c_6 t$ in the total construction, we can always find, between any two vertices, a path of length at most $c_7(\log n)^{1+\kappa}$. For a topological clique of order r, we need less than r^2 such paths, so the greedy algorithm provides a topological clique of order r with $r > c_8\sqrt{t}(\log n)^{-(1+\kappa)/2} > c_8\sqrt{t}(\log t)^{-\eta}$, where $\eta = \alpha(1+\kappa)/2 > 7$.   □

3.2. *Sparse graphs*

Theorem 3.1. *Let G be an ε-expander graph on n vertices, with ε defined in (2.3), with average degree $t \geq t_0$ such that $\log n \geq (\log t)^{\alpha}$. Then $tcl(G) \geq c_9 t$.*

Miki Simonovits noted that what we really prove here is that any graph G as in Theorem 3.1 is highly connected in the following sense:

Let k satisfy $\log k \leq c(\log t)^{\beta}$, for some $\beta < (\alpha-2-\kappa)/(1+\kappa)^2$. Then, given any k vertices in G pairwise far from each other, and given any graph H on k vertices and maximum degree $\Delta \leq c_{10}t$, there is a topological H in G with the given k vertices as principal vertices.

We will only use this for the very small value $k = ct$ (that is, $\beta = 1$).

Here the algorithm is somewhat different. Let

$$R = 0.1 \log n / \log t \quad \text{and} \quad r = c_{11}(\log k + \log\log n)^{1+\kappa}.$$

We start by selecting $k = c_{12}t$ vertices $v_1, ..., v_k$ such that any two are at a distance at least $2R$. These will be our principal points. Then we connect these points pairwise as follows. We surround each of our k vertices with balls of radius r, and whenever we are to connect v_i by v_j, we first connect the spheres $S(v_i, R)$ and $S(v_j, R)$ in such a way that the path goes entirely outside all the balls $B(v_\ell, r)$, $\ell \neq i, j$. We choose a shortest such connection. Then we connect the contact points on these two R-spheres to their respective centres with shortest paths. We continue the algorithm in the graph obtained by deleting all interior vertices of the path constructed.

(At this point, we mentally rebuild the spheres, in that we recompute diameter, bottlenecks, *etc.*, in the new graph obtained by deleting the vertices used. But this only concerns the analysis of the algorithm, not the algorithm itself.)

Lemma 3.1. *The conditions of Theorem 3.1 imply that G has k vertices such that any two are at a distance at least $2R$. (Here k is as defined in the remark after the theorem.)*

Furthermore, for any positive integer $w \leq n/4$, and for any set $S \subset V$, $|S| < w\varepsilon(w)$, the union of all components of $G - S$ of size at most $n/2$ is of size less than w.

In particular, if the pairwise disjoint sets $T_1, T_2, S \subset V$ satisfy $|T_1|, |T_2| \geq w$ and $|S| < w\varepsilon(w)$, then S cannot separate T_1 and T_2.

Proof. First, we can find k such vertices, since the size of a $2R$-ball is at most $(9t)^{4R}$, and $k(9t)^{4R} < n$. The second claim is also true. Indeed, otherwise the union of some of the smaller components would form a set $W \subset V$, $w \leq |W| \leq 2w \leq n/2$ such that $\partial W \subset S$, whence $|\partial W| \leq |S| < w\varepsilon(w) \leq |W|\varepsilon(|W|)$ (since $x\varepsilon(x) \uparrow$), so contradicting expansion. The final claim follows directly, for there can be at most one component of size greater than $n/2$. $\square$

Proof of Theorem 3.1. The crucial observation is that, for any v_i and any $\rho \leq r$, we deleted at most $\Delta \leq ct$ vertices from any one sphere $S(v_i, \rho)$, which is less than the bottleneck in the graph. Also, we deleted at most $\binom{k}{2}d$ vertices from the whole graph altogether, where d is an upper bound on the maximum distance between two of our principal points v_i at the end of the construction after we made all the deletions.

Now, as we said earlier, if the number of deleted vertices turns out to be much less than the number of vertices on any sphere $S(v_i, \rho)$, $\rho > r$, then the lower bounds $|B(v_i, \rho)| > c \exp\{cr^{1/(1+\kappa)}\}$ still hold for all i, so we can take $d = c(\log n)^{1+\kappa}$ for some constant c.

We use Lemma 3.1 with T_1, T_2 being R-spheres around two distant points, and S being the union of r-spheres around all the other $k - 2$ points, and the interior points of all paths constructed before.

Each T_i has at least $\exp\{cR^{1/(1+\kappa)}\}$ points, while S has size less than $k(9t)^{2r} + \binom{k}{2}d$. It is easy to check that the conditions of the lemma hold.

The lengths of the connections are at most the diameters in the current graph, thus it is easy to check that on each sphere $S(v_i, \rho)$, $\rho > r$, there are more points than the total lengths of all the paths combined.

It remains to choose the parameters such that $k \geq ct$. This holds if $\alpha > 2 + \kappa + (1+\kappa)^2$. $\square$

Addendum. Noga Alon has remarked that an unpublished theorem of Robertson and Seymour about graph minors together with a theorem of Mader and a bound of Thomason and Kostochka imply Theorem 1.1 with a better constant ($\eta = 1/4$). The approach seems to be completely different from ours.

References

[1] Ajtai, M., Komlós, J. and Szemerédi, E. (1979) Topological complete subgraphs in random graphs. *Studia Sci. Math. Hung.* **14**, 293–297.

[2] Bollobás, B. (1978) *Extremal Graph Theory*, Academic Press, London.

[3] Bollobás, B. and Catlin, P. (1981) Topological cliques of random graphs. *J. Combinatorial Theory* **30B**, 224–227.

[4] Erdős, P. and Fajtlowicz, S. (1981) On the conjecture of Hajós. *Combinatorica* **1**, 141–143.

[5] Erdős, P. and Hajnal, A. (1969) On complete topological subgraphs of certain graphs. *Annales Univ. Sci. Budapest* **7**, 193–199.

[6] Komlós, J. and Sós, V. (manuscript) Regular subgraphs of graphs.

[7] Lovász, L. (1979) *Combinatorial Problems and Exercises*, Akadémiai Kiadó, Budapest.

[8] Mader, W. (1967) Homomorphieeigenschaften und mittlere Kantendichte von Graphen. *Math. Annalen* **174**, 265–268.

[9] Mader, W. (1972) Hinreichende Bedingungen für die Existenz von Teilgraphen die zu einem vollständigen Graphen homöomorph sind. *Math. Nachr.* **53**, 145–150.

[10] Szemerédi, E. (1976) Regular partitions of graphs. *Colloques Internationaux C.N.R.S. Nº 260 - Problèmes Combinatoires et Théorie des Graphes*, Orsay, 399–401.

[11] Szemerédi, E. (1975) On a set containing no *k* elements in arithmetic progression. *Acta Arithmetica XXVII*, 199–245.

Local-Global Phenomena in Graphs

NATHAN LINIAL

Institute of Computer Science, Hebrew University, Jerusalem, Israel

This is a survey of a number of recent papers dealing with graphs from a geometric perspective. The main theme of these studies is the relationship between graph properties that are local in nature, and global graph parameters. Connections with the theory of distributed computing are pointed out and many open problems are presented.

1. Introduction

How well can global properties of a graph be inferred from observations that are purely local? This general question gives rise to numerous interesting problems that we want to discuss here. Such a *local-global* approach is often taken in geometry, where it has a long and successful history, but a systematic study of graphs from this perspective has not begun until recently. Nevertheless, a number of older results in graph theory do fit very nicely into this framework, as we later point out. Most of the specific problems fall in two categories. In the first, local structural information on the graph is collected and then used to derive certain consequences for the graph as a whole. The other class of problems concerns *consistency* of local data. Namely, one asks to characterize those sets of local data that may come from some graphs.

As the reader will soon see, the local-global paradigm leads to many questions in which graphs are viewed as geometric objects, a point of view that we believe can greatly benefit graph theory. Besides the geometric connection, ties also exist with the theory of combinatorial algorithms. We suggest a specific test case for the heuristic notion that polynomial-time algorithms are capable of examining only local phenomena. In distributed computing, locality of computation is an already recognized and studied notion, and some connections with this discipline are pointed out as well.

2. Packing and covering with spheres and local-global averaging

Let $W \subseteq V(G)$ be a set of vertices in a graph G. If the vertices in W form a majority in every ball of radius between 1 and r in G, does this imply that W has a large cardinality?

As an illustration, consider the following example with $r = 1$. In this graph, W is a clique of $\sqrt{n}$ vertices. Each vertex in W has a set of $\sqrt{n} - 1$ neighbors not in W, each of which has degree 1. It is a routine matter to check that this graph satisfies the assumption for $r = 1$. It is also not hard to modify the construction for any fixed $\alpha < 1$ so that W occupies a fraction $\geq \alpha$ of any 1-neighborhood, while $|W| = O(\sqrt{n})$ (here α was $1/2$).

Let us introduce some notation. The ball[†] of radius k centered at x, denoted $B_k(x)$ consists of all vertices y whose distance from x does not exceed k, and its cardinality $|B_k(x)|$ is denoted $\beta_k(x)$. Our question is how small $|W|$ may be in terms of r and n, the order of G.

If we represent W by its characteristic function, we are led to consider a more general problem. Namely, let f be a nonnegative function defined on the vertices of an n-vertex graph G. Suppose that we have a lower bound on the average of f on every ball in G of radius between 1 and r. What can we conclude for the overall average of f?

This subject has been recently taken up by Linial, Peleg, Rabinovich and Saks [23] who show the following.

Theorem 2.1. (Local Averages) *Let f be a nonnegative function defined on the vertices of an n-vertex graph G. Suppose that the average of f over every ball of radius $r \geq t \geq 1$ in G is at least μ. Then, the average of f over all of V is at least $\mu \cdot n^{-O(1/\log r)}$. The bound is tight.*

Consequently, if we let r be n^c for some positive constant c, local averages do reflect the true global behavior of f. Examples are given in [23] showing that smaller r's will not do. It is natural to ask at this point what happens if we only know a lower bound for the average of f over balls of radius r (and not for every $r \geq t \geq 1$). Examples are given showing that only very weak conclusions can be drawn about the overall average of f, however big r may be. Namely, it may be that the average of f is only $O(n^{-1/3})$. It is also worthwhile noting that the conclusions of the theorems remain unchanged even if we make the assumption only for balls whose radius $r \geq t \geq 1$ is *a power of 2*.

The result for local averages is proved as a consequence of tight theorems about sphere packing and about covering by spheres in general graphs. Either 0–1 or fractional packing and covering results will do for this purpose.

Theorem 2.2. (Covering by Spheres) *For integers $n > r$, the vertices of an n-vertex graph can be covered by a collection of balls with radii in the range $[1, \ldots, r]$, that cover no vertex more than $n^{O(1/\log r)}$ times. The bound is tight.*

Theorem 2.3. (Sphere Packing) *In any n-vertex graph, there is a collection of disjoint balls whose radii are in the range $[1, \ldots, r]$, which together cover at least $n^{1-O(1/\log r)}$ vertices. The bound is tight.*

[†] The words ball and sphere are used interchangeably here.

It would be very interesting to understand how various properties of a graph affect the efficiency of sphere packing and of covering by spheres. Also, it is not hard to extend these results to general finite metric spaces. We still do not know, for example, what happens if the metric space is embedded in a d-dimensional Euclidean space or other low-dimensional normed space. These questions lead us to our next subject.

2.1. Connections with the theory of maximal functions

There is an appealing connection between this class of problems and the theory of maximal functions in analysis (*e.g.* [32]). This observation came up in discussions with Metanya Ben-Artzi.

Briefly, the connection is this: again let $B_r(x)$ denote the ball of radius r centered at $x \in \mathbf{R}^d$, the d-dimensional Euclidean space. Let f be a real function on $\mathbf{R}^d$, and let $a_r(x)$ be the average of f over $B_r(x)$. Define $f^*(x)$ as the supremum of $a_r(x)$ over all $r > 0$. The function f^* is called the *maximal function* of f. Numerous results have been derived over the years concerning maximal functions. Informally speaking, among the most basic findings is that 'f^* is not much larger than f'.

Our proof for Theorem 2.1 shows a significant similarity with the methods used in analysis to compare the p-norms of f^* and f. Specifically, the most traditional proof technique involves some geometric covering arguments (Vitali's Lemma), and a similar argument underlies some of our proofs as well. In analysis, such arguments lead to results of the form

$$\|f^*\|_p < C_{d,p} \|f\|_p$$

where $C_{d,p}$ grows exponentially with the dimension d. This bad dependency is unavoidable in this method, since the bounds in Vitali's lemma do grow this way. More modern results concerning maximal functions (*e.g.* [33]) manage to bypass this difficulty. It is conceivable that these methods may help settle our questions on low-dimensional finite metric spaces. It would also be interesting to see if similar ideas can be developed for other classes of graphs.

3. Locality in distributed systems

The theory of distributed computing concerns a set of processors connected through a communication network. The network is depicted as a graph in whose vertices computers or processors reside. Communication takes place as messages are exchanged between neighboring vertices. The processors' goal is to perform some computational task together. Let us restrict our attention to deterministic and synchronized networks – the simplest among this class of computational models. In such an environment it is easy to see that in t time units a processor can only learn about the situation at processors that are within distance at most t from itself in the graph. This observation gives rise to numerous questions of the local-global type. In studying such questions, some care has to be given to symmetry breaking. If processors 'have no identity' and cannot be told apart by other processors, then almost nothing interesting can be done. We do not elaborate on this, but rather say that the common practice in this area is to assume that processors are equipped with individual (distinct) ID-numbers, and so symmetry causes no problems.

3.1. Low-diameter decompositions of graphs

Perhaps the most fundamental difficulty in distributed processing, as compared with more traditional computational models, is the absence of central control. It is very difficult to have many processors perform in concert when there is no conductor around. Indeed, much research effort in distributed processing concerns efficient and reliable methods for electing a leader. We will not pursue this fascinating subject, and only point out some of the shortcomings of this approach. It creates a communication bottleneck around the elected leader. It is also very sensitive to failures, or latency of the leader and its neighbors. Moreover, if the graph underlying the communication network has a large diameter, this method is also very wasteful in terms of communication.

In view of the difficulties involved with such a 'central government' the next thing to try is a set of cooperative 'local authorities'. Namely, in the previous section we were covering vertices by balls; now we consider *decomposing* the vertices, subject to a certain upper bound on the diameter of each part. Let us introduce some notation: if Π is a decomposition of the vertices of graph G into subsets, $V(G) = \bigcup S_i$. The *diameter* of this decomposition is defined as the maximum over all $diam(S_i)$.

Remark 3.1. In defining the diameter, we may consider the graphs induced by the parts, and compute distances within these graphs. Alternatively, we may consider distances as inherited from the whole graph. Our statements, slightly modified, hold for either definition.

The graph *induced* by Π has one vertex per S_i, with vertices i, j adjacent iff there is a vertex in S_i and one in S_j that are adjacent in G. The goal is to find partitions Π with small diameter and favorable properties for the induced graph. Linial and Saks [25] show (see also [6, 7]):

Theorem 3.2. *An n-vertex graph has a decomposition of diameter r, where the induced graph has chromatic number $\leq \chi$, if both*

$$\chi = \Omega\left(\frac{\log n}{\log r}\right) \ and \ r = \Omega\left(\frac{\log n}{\log \chi}\right)$$

hold. Examples exist showing these bounds are tight. A randomized distributed algorithm of $\log^{O(1)} n$ run time is provided to obtain such decompositions.

We briefly discuss some extreme examples for Theorem 3.2. It is easily seen that there are two interesting ranges to this theorem:

$$r \geq \frac{\log n}{\log \log n} \geq \chi.$$

In this range, the tradeoff between r and χ is given by:

$$\chi = \Omega \left(\frac{\log n}{\log r} \right).$$

The known extreme examples in this range are graphs corresponding to triangulations of Euclidean spaces. For example, the graph whose vertices are all lattice points in $\log n / \log r$ dimensions, with adjacency between $\mathbf{x}, \mathbf{y}$ iff $\|\mathbf{x} - \mathbf{y}\|_\infty = 1$.

$$\chi \geq \frac{\log n}{\log \log n} \geq r,$$

where the condition is

$$r = \Omega \left(\frac{\log n}{\log \chi} \right).$$

Here *trees* and *expander graphs* provide extreme examples.

Remark 3.3. Notice that radius $\log n$ along with $\chi = O(\log n)$ are possible. Consequently, if every ball or radius $\log n$ in G is k-colorable, $\chi(G) = O(k \log n)$. So, up to a logarithmic factor, the coloring number can be inferred from radius $\log n$ views of G.

More on coloring from the local-global perspective will be said later.

So far we have considered only the chromatic number of the graph induced by a decomposition. Other properties of this graph are of interest as well. Let us point out the analogy between these questions and notions from dimension theory in topology [19]. The following question is inspired by the notion of covering dimension of metric spaces. Let $\Pi : V(G) = \bigcup S_i$ be, again, a decomposition of the vertices of graph G. For a vertex x, let $\gamma(x)$ be the number of S_i in which x has a neighbor. $\Delta(\Pi)$ is defined as $max_x(\gamma(x))$.

Problem 3.4. What is the least $D = D(r, n)$, such that any n-vertex graph has a decomposition Π of diameter $\leq r$ with $\Delta(\Pi) \leq D$?

Possibly, the tradeoff between D, r and n is the same as the one for χ, r and n in Theorem 3.2.

3.2. Applications of low-diameter decompositions

Low diameter graph decompositions have found numerous applications in distributed computing. We briefly sketch some of these. We begin with the Maximal Independent Set (MIS) problem. (We mean inclusion-maximal. This problem is not to be confused with the search for an independent set of largest cardinality, which is NP-complete.) There is, of course, a most simple sequential algorithm, which at each step adds a new vertex to the MIS and eliminates all its neighbors from the graph. While such a naive sequential algorithm solves the problem in optimal time, finding efficient *parallel* algorithms for this question is not nearly as obvious. An efficient parallel algorithm was first found by Karp and Wigderson [20] with numerous improvements and ramifications by others (*e.g.* [1, 27]). In fact, Luby's algorithm [27] works also in the distributed model, but it does use randomization, however. One of the tantalizing questions that remain in this area is:

Problem 3.5.   Is there a deterministic, distributed, polylog-time algorithm to find a maximal independent set (MIS) in a graph?

It has been observed in [4] that low-diameter decompositions with a low-chromatic decomposition graph may help provide such an algorithm. Assuming such (an already colored) decomposition is available, we construct, in parallel, an MIS within each part colored 1. Since each part has diameter at most r, an MIS for it can be constructed by an elected leader, where both election and construction take time $O(r)$. Also, there are no edges between different parts of color 1, so these activities in different parts can be performed in parallel without affecting each other. Vertices selected so far for the MIS eliminate their neighbors (also in other parts), and we move on to parts colored 2 etc. Using the terminology of Theorem 3.2, a time bound of $O(r\chi)$ can be achieved, which by proper choice of parameters may be made $O(\log^2 n)$. The difficulty is, of course, that this argument assumes a partition to be already available. Currently, however, only *randomized* distributed algorithms are known that find such decompositions in polylogarithmic time (Theorem 3.2 and [3]). Problem 3.5 thus remains open.

Another problem for which low-diameter decompositions help is *distributed job scheduling* [5]. In this problem, processors try to efficiently share their workloads. Initially, each processor is assigned a number of (unit-cost) jobs to perform. In each step, a processor can perform one of its assigned jobs, as well as send some of its assigned jobs to neighbors. It can also communicate messages to its neighbors. A processor knows only its own history and the contents of the messages it receives. An algorithm is sought where the completion time is early as possible. Moreover, the following strong ('competitive') criterion is applied: the time for completion should compare favorably with the best that can be achieved by an optimal central controller having a complete view of the situation at all times (and not just local views at a certain processor). Using low-diameter decompositions, [5] manages to guarantee a completion time that is only $O(\log n)$ longer than can be attained by a knowledgable central controller. This result is shown to be almost optimal for certain families of graphs in [2]. For further applications see [6, 7].

4. Distributed coloring and related problems

The systematic study of locality in distributed processing was begun in [22]. Our first result concerned the time required to 3-color an n-cycle of processors. A clever algorithm by Cole and Vishkin [11] does this in time $O(\log^* n)$. (Recall that $\log^* n$ is the number of times one has to iterate the log function to come down from n to 1).

The first result in [22] says that this algorithm is optimal.

Theorem 4.1.   *A distributed algorithm that properly colors the n-cycle with only 3 colors requires time $\Omega(\log^* n)$. The bound is tight.*

It was later shown by Naor [30] that the same statement also holds for *randomized* algorithms.

Assuming n is even, how long does it take to 2-color an n-cycle? A huge difference shows up, in comparison with the time complexity of 3-coloring.

Theorem 4.2. *A distributed algorithm that properly colors the n-cycle (n even) with only 2 colors requires time $\Omega(n)$. The bound is tight.*

This example captures a big difference in locality of 2 and 3-coloring of cycles. Of course, a 2-coloring requires perfect *global* coordination, which results in an excessive time complexity.

Other results from [22] are:

Theorem 4.3. *Let T be the d-regular tree of radius r. Any algorithm that properly colors T and runs for time $< 2r/3$ requires at least $\Omega(\sqrt{d})$ colors.*

This result can probably be improved to $\Omega(d/\log d)$. An intriguing open question in this area is:

Problem 4.4. Consider distributed algorithms that properly color n-vertex graphs and take time $\log^{O(1)} n$. What can be said about the least number of colors required by such an algorithm? Specifically, is it possible that $\Delta + 1$ colors suffice, where Δ is the largest vertex degree of the graph?

This question is closely related to Problem 3.5. Some partial results have been provided in [22].

Theorem 4.5. *An $O(\log^* n)$-time algorithm exists to color any n-vertex graph G with $O(\Delta^2)$ colors, where $\Delta = \Delta(G)$ is the largest vertex degree in G.*

See also [34] for some recent progress in this area. Naor and Stockmeyer [31] have recently investigated the limits of what can be computed with a *constant* diameter of locality.

Another related problem is that of finding *happy partitions*. A partition of the vertex set of a graph $V = A \cup B$ is called happy if every $x \in A$ has most of its neighbors in B and vice versa. That such partitions always exist is easy to show, and a sequential algorithm to construct such partitions is easy to find. The distributed time complexity of this is still unknown: Linial and Saks conjecture (unpublished) as follows.

Conjecture 4.6. *There are n-vertex graphs where a distributed algorithm to find a happy partition requires time $\Omega(\sqrt{n})$.*

5. Coloring

The chromatic number of a graph is a good example of a global parameter where the behavior of small induced subgraphs seems to be a weak indicator of global properties.

(But notice Remark 3.3.) Up to this point 'local' has always been taken in the sense of distance. It is also interesting to examine assumptions about the behavior of (cardinality) small sets of vertices. In this section we consider sets that are small in either diameter or cardinality. One easy consequence of Theorem 3.2 is the following.

Theorem 5.1. *If the subgraph spanned by every k vertices in G is 2-colorable, $\chi(G) = O(n^{O(1/k)})$. This bound is tight. Moreover, it is possible to find a proper coloring with this number of colors in polynomial time.*

Dealing with 2-colorabilty is usually much easier than with any larger coloring number. Is there, perhaps, a similar result for graphs that are, say, locally 3-colorable? To simplify our notation, we will only consider 3-colorability, leaving out the obvious extension to more colors.

Problem 5.2. Let $\chi(n, k)$ be the largest chromatic number of an n-vertex graph G if the subgraph spanned by every k vertices in G is 3-colorable. Determine the behavior of $\chi(n, k)$.

It is not hard to see that

$$n^{1/2+o(1)} \geq \chi(n, k) \geq n^{7/20+o(1)},$$

where the $o(1)$ terms tend to zero as k grows. The upper bound follows, *e.g.*, from Wigderson's argument [35] mentioned below. The lower bound combines an argument from [24] with a lower bound due to Gallai [15] on the least number of edges in minimally non-3-colorable graphs. Note the difference compared with locally 2-colorable graphs (Theorem 5.1).

Besides the interest in the problem *per se*, it is related to approximating chromatic numbers in polynomial time. That it is NP-hard to determine the chromatic number has been known for a long time [16]. How well this quantity may be approximated is still unknown, although considerable progress has been made. An early positive result on approximating chromatic numbers is a polynomial-time algorithm by Wigderson [35], which colors any n-vertex 3-colorable graph with $O(\sqrt{n})$-colors. Here is the argument: as long as you can find a vertex x of degree $\geq \sqrt{n}$, allot two fresh colors for the neighbors of x and discard them (they are two-colorable, since $\chi(G) = 3$). When the remaining graph has all degrees $< \sqrt{n}$, it can be $\sqrt{n}$-colored by a greedy algorithm. Altogether, only $O(\sqrt{n})$ colors are utilized.

Observe that the algorithm actually applies not only to 3-colorable graphs, but, in fact, to every graph in which the neighborhood of every vertex is 2-colorable. Now, bounds on Ramsey numbers naturally fit into the local-global framework. For example, the fact that

$$R(3, k) = k^{2-o(1)}$$

(see [17] for the sharpest known bounds) answers the following question: given that the neighbors of any vertex in G form an anti-clique, what is the best lower bound on the largest anti-clique in G (answer: $n^{1/2-o(1)}$). In particular, triangle-free graphs exist of chromatic number $\Omega(n^{1/2-o(1)})$. But in a triangle-free graph, the neighborhood of every vertex is, in fact an independent set, so under the more general assumptions, Wigderson's algorithm is in fact optimal.

These arguments were further improved by A. Blum [9], who showed how to color a 3-colorable graph with $n^{3/8}$ colors in polynomial time. Interestingly, Blum's algorithm (which is much more involved than that of [35]) also exploits only local (neighborhoods of radius 2) properties of 3-colorable graphs. This leads us to ask some questions to capture the heuristic claim that *polynomial-time graph-coloring algorithms can only check local properties*. We first observe that the answer to Problem 5.2 yields a completely trivial algorithm to tell 3-chromatic graphs from those not colorable in n^c colors. We expect this algorithm to be better than Blum's in this respect.

Conjecture 5.3. *Let G be an n-vertex graph in which every induced subgraph of order k is 3-colorable. Then, $\chi(G) < n^{\gamma+o(1)}$ for some $3/8 > \gamma \geq 7/20$, and where the $o(1)$ term tends to zero with $k \to \infty$.*

An exhaustive algorithm running in time $O(n^k)$ can obviously test this condition. It is an interesting possibility that this procedure may be transformed into an algorithm that actually provides a $n^{\gamma+o(1)}$ coloring.

If this conjecture fails, it may be possible to save it by adding an assumption such as that the neighborhood of every vertex is 2-colorable, a condition that is again polynomial-time verifiable.

A more daring conjecture is:

Conjecture 5.4. *If $P \neq NP$, then no polynomial time algorithm can color every 3-chromatic n-vertex graph with fewer than n^θ colors for some $\theta > 0$.*

There have been many new and exciting results on the difficulty of approximating NP-hard problems. The first step in establishing such a result for coloring has been taken by Lund and Yanakakis [28], who establish a separation between coloring numbers n^{c_1} and n^{c_2} for some fixed $1 > c_1 > c_2 > 0$. A simpler proof has been provided recently by Khanna, Linial and Safra [21], who also show that it is NP-hard to 5-color 3-colorable graphs. All this is, obviously, still a far cry from Conjecture 5.4, but some progress in this direction is likely to occur in the foreseeable future.

6. Sizes of neighborhoods

Perhaps the most obvious 'local' information about a graph is the degree sequence, classically characterized by Erdős and Gallai [12]. Briefly, $d_1 \geq \ldots d_n \geq 0$ is such a sequence iff (i) $\sum d_i$ is even and (ii) for all $1 \leq k \leq n$ it is the case that $\sum_1^k d_j \leq k(k-1) + \sum_{j>k} \min\{k, d_j\}$. The necessity of these conditions is easy to establish and the thrust of the theorem is that they are also sufficient. Pursuing our local-global approach we ask: what else can be said about the possible *rate of growth* of (balls in) a graph?

Recall that $\beta_k(x)$ is the number of vertices y whose distance from x does not exceed

k. In a connected n-vertex graph, one obtains n integer sequences, one for each vertex, $1 = \beta_0(x) \leq \beta_1(x) \ldots \leq \beta_n(x) = n$. Following Erdős–Gallai's result, it is appealing to ask:

Problem 6.1. Characterize those sets of n integer sequences of the type

$$1 = \beta_0(x) \leq \beta_1(x) \ldots \leq \beta_n(x) = n$$

that are obtained from connected graphs.

This question, in full generality, is presently too difficult, and at this time one should settle for less. Here are some illustrative special cases of this problem:

— Is it possible to characterize sets of n pairs $\beta_1(x) \leq \beta_2(x)$ that come from graphs? One possible approach would be to get sufficient information on squares of graphs and then resort to Erdős–Gallai. Note, however, that Motwani and Sudan [29] have shown that it is NP-complete to decide whether a given graph is a square.

— In the context of the previous question, it is not hard to derive some necessary conditions, e.g., that

$$\sum_x (\beta_2(x) - 1) \leq \sum (\beta_1(x) - 1)^2,$$

with equality iff $girth(G) \geq 5$. This inequality suggests that there might exist some comparison theorems between norms of the various vectors $\bar{\beta}_i = (\beta_i(x)|x \in V)$.

— Obviously, for any fixed x, the sequence $1 = \beta_0(x) \leq \beta_1(x) \ldots \leq \beta_n(x) = n$ is unrestricted. It may be possible to characterize pairs of such sequences, one for vertex x and one for y. Such an analysis could start by considering for any i, j the number of vertices z that are at distance i from x and j from y.

— For which parameters is it possible that all x satisfy $\beta_1(x) = d + 1$, while for every $i \leq c \log n$ it is the case that $\beta_{i+1}(x) \geq (1 + \delta)\beta_i(x)$? This question is clearly related to the existence of constant-degree expanders. Methods developed in that area may prove helpful in studying growth rates of graphs in general.

— What is the largest girth of a d-regular n-vertex graph? Specifically, is it

$$(2 - o(1))\log n / \log(d - 1)?$$

This is also an instance of the general problem. We conjecture the answer to be negative. The best current lower bound [26, 8] gives $4/3$ instead of the 2.

A problem related to the last item in this list concerns the ratio between girth and diameter. Consider the distance between two vertices that are antipodal in a shortest cycle. This consideration shows that $2 \cdot diameter(G) \geq girth(G)$. Equality holds for even cycles, but what if all degrees are ≥ 3? Examples are known with $girth(G) = 2 \cdot diameter(G)$, where the numbers are small, e.g., the points-lines graph of a projective plane. We are not aware of similar constructions with large girth, so we ask:

Problem 6.2. Consider graphs G with all degrees ≥ 3. What are possible values for the pair $(girth(G), diameter(G))$? In particular, can their ratio be kept as close to 2 as we wish?

See [14] for a related classical work.

Together with S. Hoori [18], we have recently obtained some results concerning the existence of a 'center of mass' in both graphs and sets in Euclidean spaces. Namely, we are looking for a vertex x where we can establish a tight lower bound on the numbers $\beta_k(x)$ $(k = 1, 2, \ldots)$.

7. Cliques

A number of people have investigated how well the clique number of a graph can be inferred from local behavior. The earliest work we are aware of is by Erdős and Rogers [13]. Recent work on the subject can be found in [10, 24] and the references therein. The main question studied here can be stated in terms of computing, or estimating, the quantities related to the following arrow relation. Say that a graph G has property (p, q) if every set of p vertices in G contains a q-clique. We say that $(p, q) \rightarrow (f, n)$, if every G of order $\geq p$ having the (p, q)-property must satisfy (f, n) as well. The question then is, for given p, q, n estimate the least $f = f(p, q, n)$ for which $(p, q) \rightarrow (f, n)$. The exact determination of f includes, as a special case, the exact evaluation of Ramsey numbers, so it is more realistic to ask for estimates, or to settle for special cases. We use both the arrow notation and the function f to describe the results.

Bollobas and Hind [10] concentrate on the case of large p and small q, n. Among their results are:

Theorem 7.1. *For* $s > r \geq 3$,

$$(n, r) \rightarrow (cn^{s-r+1}, s)$$

for some constant $1 > c > 0$. *Also, for* $r \geq 3$ *and* n *large enough*

$$(n, r) \nrightarrow (n^{1+r/(r^2-2)-\epsilon}, r + 1).$$

Linial and Rabinovich [24] consider fixed p, q and n tending to infinity. Their main result breaks down into three cases, roughly according to whether p/q is smaller than, equal to or bigger than 2.

Theorem 7.2.

— For $p \leq 2q - 2$ and all n,

$$f(p, q, n) = n + p - q.$$

— For $p = 2q - 1$ and all $n \geq p$,

$$n^{1+2/(q-3)+o(1)} \geq f(p, q, n) \geq n^{1+1/(8q-5)}.$$

— For all $n \geq p \geq q$,

$$(p, q) \rightarrow (R(r, n) + p - 1, n),$$

where $r = \lceil \frac{p}{q-1} \rceil$, $R(r, n)$ is the Ramsey number and c is an absolute constant. On the other hand,

$$(p, q) \nrightarrow (n^{(T-1)/(p-2)+o(1)}, n),$$

where T is a Turan number: the least number of edges in a p-vertex graph without a q-anticlique.

All o(1) terms are for fixed p, q and growing n.

References

[1]   Alon, N., Babai, L. and Itai, A. (1986) A fast and simple randomized algorithm for the maximal independent set problem. *J. of Algorithms* **7** 567–583.

[2]   Alon, N., Kalai, G., Ricklin, M. and Stockmeyer, L. (1992) Lower bounds on the competitive ratio for mobile user tracking and distributed job scheduling. *FOCS* **33** 334–343.

[3]   Awerbuch, B., Berger, B., Cowen, L. and Peleg, D. (1992) Fast distributed network decomposition. *PODC* **11** 169–178.

[4]   Awerbuch, B., Goldberg, A. V., Luby, M. and Plotkin, S. A. (1989) Network decomposition and locality in distributed computation. *FOCS* **30** 364–369.

[5]   Awerbuch, B., Kutten, S. and Peleg, D. (1992) Competitive distributed job scheduling. *STOC* **24** 571–580.

[6]   Awerbuch, B. and Peleg, D. (1990) Sparse partitions. *FOCS* **31** 503–513.

[7]   Awerbuch, B. and Peleg, D. (1990) Network synchronization with polylogarithmic overhead. *FOCS* **31** 514–522.

[8]   Biggs, N. L. and Boshier, A. G. (1990) Note on the girth of Ramanujan graphs. *J. Combin. Th.* ser. B **49** 190–194.

[9]   Blum, A. (1990) Some tools for approximate 3-coloring. *FOCS* **31** 554–562.

[10] Bollobás, B. and Hind, H. R. (1991) Graphs without large triangle free subgraphs. *Discrete Math.* **87** 119–131.

[11] Cole, R. and Vishkin, U. (1986) Deterministic coin tossing and accelerating cascades: micro and macro techniques for designing parallel algorithms. *STOC* **18** 206–219.

[12] Erdös, P. and Gallai, T. (1960) Graphen mit Punkten vorgeschriebenen Grades. *Mat. Lapok* **11** 264–274.

[13] Erdös, P. and Rogers, C. A. (1962) The construction of certain graphs. *Canad. J. Math.* **14** 702–707.

[14] Feit, W. and Higman, G. (1964) The nonexistence of certain generalized polygons. *J. Algebra* **1** 114–131.

[15] Gallai, T. (1963) Kritische Graphen I. *Publ. Math. Inst. Hungar. Acad. Sci.* **8** 165–192.

[16] Garey, M. R. and Johnson, D. S. (1979) *Computers and Intractability: A Guide to NP-completeness*, W. H. Freeman.

[17] Graham, R. L., Rothschild, B. L. and Spencer, J. (1980) *Ramsey Theory*, Wiley, New York.

[18] Hoori, S. and Linial, N. (March 1993) Work in progress.

[19] Hurewicz, W. and Wallman, H. (1948) *Dimension Theory*, Princeton University Press.

[20] Karp, R. and Wigderson, A. (1985) A fast parallel algorithm for the maximal independent set problem. *J. ACM* **32** 762–773.

[21] Khanna, S., Linial, N. and Safra, S. (to appear) On the hardness of approximating the chromatic number. *ISTCS*.

[22] Linial, N. (1992) Locality in distributed graph algorithms. *SIAM J. Comp.* **21** 193–201. (Preliminary version: Linial, N, (1987) Distributive graph algorithms – global solutions from local data. *FOCS* 331–335.

[23] Linial, N., Peleg, D., Rabinovich, Yu. and Saks, M. (to appear) Sphere packing and local majorities in graphs. *ISTCS*.

[24] Linial, N. and Rabinovich, Yu. (in press) Local and global clique numbers. *J. Combin. Th.* ser. B.

[25] Linial, N. and Saks, M. (to appear) Low diameter graph decompositions. *Combinatorica*. (Preliminary version (1991) published in *SODA* **2** 320–330.)

[26] Lubotsky, A., Phillips, R. and Sarnak, P. (1988) Ramanujan graphs. *Combinatorica* 8 261–278.

[27] Luby, M. (1986) A simple parallel algorithm for the maximal independent set problem. *SIAM J. Comp.* **15** 1036–1053.

[28] Lund, C. and Yanakakis, M. (1992) On the Hardness of Approximating Minimization Problems, manuscript.

[29] Motwani, R. and Sudan, M. (1991) Computing roots of graphs is hard, manuscript, Stanford.

[30] Naor, M. (1991) A lower bound on probabilistic algorithms for distributive ring coloring. *SIAM J. Disc. Math.* **4** 409–412.

[31] Naor, M. and Stockmeyer, L. (1993, to appear) What can be Computed Locally? *STOC* **25**.

[32] Stein, E. M. (1970) Topics in Harmonic Analysis. *Annals of Math. Study* **63**, Princeton University Press.

[33] Stein, E. M. (1985) Three variations on the theme of maximal functions. In: Peral, I. and Rubio de Francia, J.-L. (eds.) *Recent Progress in Fourier Analysis*, North-Holland Mathematics Studies **111**, North-Holland, 229–244.

[34] Szegedy, M. and Vishwanatan, S. (1993, to appear) Locality-based graph coloring. *STOC* **25**.

[35] Wigderson, A. (1983) Improving the performance for approximate graph coloring. *J. ACM* **30** 325–329.

On Random Generation of the Symmetric Group

TOMASZ ŁUCZAK[†*] and LÁSZLÓ PYBER[‡**]

† Mathematical Institute of the Polish Academy of Sciences, Poznań, Poland
‡ Mathematical Institute of the Hungarian Academy of Sciences, Budapest, Hungary

We prove that the probability $i(n, k)$ that a random permutation of an n element set has an invariant subset of precisely k elements decreases as a power of k, for $k \leqslant n/2$. Using this fact, we prove that the fraction of elements of S_n belong to transitive subgroups other than S_n or A_n tends to 0 when $n \to \infty$, as conjectured by Cameron. Finally, we show that for every $\epsilon > 0$ there exists a constant C such that C elements of the symmetric group S_n, chosen randomly and independently, generate *invariably* S_n with probability at least $1 - \epsilon$. This confirms a conjecture of McKay.

1. Introduction

Let π_n be a permutation picked randomly from S_n in such a way that each element of S_n is equally likely to be chosen as π_n, and let $i(n, k)$ denote the probability that some set k elements remains invariant under π_n. We show that there exists an absolute constant A such that, $i(n, k) \leqslant A k^{-0.01}$, whenever $k \leqslant n/2$. This fact has been known to have important implications for the statistical theory of the symmetric group. In particular, we confirm a conjecture of Cameron [3], showing that only a small part of the symmetric group S_n can be covered by non-trivial transitive subgroups.

Theorem 1. *Let t_n denote the number of elements of the symmetric group S_n that belong to transitive subgroups different from S_n and A_n. Then*

$$\lim_{n \to \infty} t_n/n! = 0.$$

This theorem has various applications. A classical result of Dixon [4] states that a random pair of permutations generates either A_n or S_n with large probability. As it is very

* On leave from Adam Mickiewicz University, Poznań, Poland. Research partially supported by KBN grant 2 1087 91 01.
** Research partially supported by the Hungarian National Foundation for Scientific Research, Grant No 4267.

easy to see that such a random pair with probability $1 - o(1)$ generates a transitive group, Theorem 1 can be viewed as a natural extension of Dixon's result.

Another consequence of Theorem 1 is mentioned by Cameron. He observed that it would immediately imply his result from [3] that for almost all Latin squares L the group generated by the rows of L is the symmetric or alternating group.

Dixon [5] noticed another application of non-trivial upper bounds for $i(n, k)$. Let us say that a group G is generated *invariably* by the elements $x_1, x_2, ..., x_m$, if the elements $y_1, y_2, ..., y_m$ generate G whenever y_i is conjugate to x_i for $i = 1, 2, ..., m$ (this definition is motivated by problems emerging in computational Galois theory (see (5))). Dixon [5] showed the existence of a constant b such that, with probability $1 - o(1)$, $b\sqrt{\ln n}$ randomly chosen permutations generate S_n invariably. Furthermore, he noted that a good upper bound for $i(n, k)$ would imply that $O(\ln \ln n)$ random permutations suffice. Using our previous result, we can prove an even stronger statement, confirming a conjecture of McKay (see [5]).

Theorem 2. *For every $\epsilon > 0$ there exists a constant $C = C(\epsilon)$ such that C permutations, chosen from S_n uniformly and independently, generate invariably S_n with probability larger than $1 - \epsilon$.*

Finally, let us mention that our argument does *not* require the classification of finite simple groups.

2. Properties of a random permutation π_n

In this part of the note we study the cyclic structure of π_n, i.e. the asymptotic behaviour of the random variables $L = L(n)$ and $C_i = C_i(n)$, $i = 1, 2, ..., L$, where $C_1 \geqslant C_2 \geqslant ... \geqslant C_L$ denote the lengths of the cycles in the decomposition of the permutation π_n.

Claim 1. *The following statements hold for a random permutation π_n with probability at least $1 - o(n^{-0.05})$.*

(i) $|L - \ln n| \leqslant 0.11 \ln n$,
(ii) *if $i \neq j$, then C_i and C_j have no common divisors larger than $n^{0.9}$.*

Proof. The asymptotic behaviour of L is well studied (see [8] and also [9] and [10]) and (i) follows immediately from known estimates. To verify (ii) note that the probability that π_n has cycles of lengths k_1 and k_2 is bounded from above by

$$\binom{n}{k_1}\binom{n-k_1}{k_2}\frac{(k_1-1)!\,(k_2-1)!\,(n-k_1-k_2)!}{n!} = \frac{1}{k_1 k_2}.$$

Thus, the probability that π_n has two cycles of lengths k_1, k_2 such that $n^{0.9} \leqslant k_1 \leqslant k_2$, and for some $d \geqslant n^{0.9}$ both $d | k_1$ and $d | k_2$, is smaller than

$$\sum_{d=\lceil n^{0.9} \rceil}^{n} \sum_{i_1=1}^{\lfloor n^{0.1} \rfloor} \sum_{i_2=1}^{\lfloor n^{0.1} \rfloor} \frac{1}{di_1}\frac{1}{di_2} \leqslant nn^{0.2}\,(\lceil n^{0.9} \rceil)^{-2} = o(n^{-0.05}). \qquad \square$$

Note that a random permutation π_n can be generated in the following way. First choose $i_1 = \pi_n(1)$ uniformly at random from all elements of $[n]$. Then, for $r \geqslant 2$, pick at random $i_{r+1} = \pi_n(i_r)$ uniformly from all elements of $[n] - \{i_1, i_2, ..., i_r\}$ until $i_{r_1} = 1$ for some r_1. Having constructed the cycle $C^{(1)}$ of π_n containing 1, in the same way one may find the cycle $C^{(2)}$ containing the smallest element of $[n] - \{i_1, i_2, ..., i_{r_1}\}$ and so on. Notice that for any i, $1 \leqslant i \leqslant L$, $C^{(i)}, C^{(i+1)}, ..., C^{(L)}$ can be viewed as cycles of a random permutation generated on the set $\bigcup_{j=i}^{L} C^{(j)} = [n] \setminus \bigcup_{j=1}^{i-1} C^{(j)}$, i.e. before generating $C^{(i)}$ each permutation of the set $[n] \setminus \bigcup_{j=1}^{i-1} C^{(j)}$ is equally likely to appear.

Let $C^{(i)}$ denote the length of $C^{(i)}$ for $i = 1, 2, ..., L$. Then, the probability that $C^{(1)} = r_1$ is given by

$$\binom{n-1}{r_1-1} \frac{(r_1-1)!\,(n-r_1)!}{n!} = \frac{1}{n},$$

and, consequently, the probability that $C^{(i)} = r_i$ is the same for every possible r_i and equal to

$$\frac{1}{n - \sum_{j=1}^{i-1} C^{(j)}} = \frac{1}{\sum_{j=1}^{L} C^{(j)}}. \tag{1}$$

Claim 2. *Let* $m_0 = \lfloor \sqrt{\ln n} \rfloor$. *Then, the probability that* $C^{(1)}$, $C^{(2)}$, ..., $C^{(m_0)}$ *have a common divisor tends to* 0 *as* $n \to \infty$. *Furthermore, with probability tending to* 1 *as* $n \to \infty$, *we have* $C^{(i)} > n^{0.99}$ *for every* $i = 1, 2, ..., m_0$.

Proof. From (1) the probability that some given d divides $C^{(i)}$, is less than

$$\left(1 \Big/ \sum_{j=1}^{L} C^{(j)}\right) \left\lfloor \frac{\sum_{j=i}^{L} C^{(j)}}{d} \right\rfloor \leqslant 1/d.$$

so the probability that $d | C_i$ for some $2 \leqslant d \leqslant n$ and all $i = 1, 2, ..., m_0$ is less than

$$\sum_{d=2}^{\lfloor n \rfloor} d^{-m_0} = (1 + o(1)) \, 2^{-m_0} = o(1).$$

By Claim 1 (i), with probability tending to 1 as $n \to \infty$, we have $C^{(1)} \geqslant 0.5n/\ln n$. Suppose that $C^{(i)} \leqslant n^{0.99}$ for some $1 \leqslant i \leqslant m_0$. This implies the existence of j, $1 \leqslant j \leqslant i-1$, for which $C^{(j)}/C^{(j+1)} \geqslant \exp(0.009 \sqrt{\ln n})$. Thus, we have either

$$C^{(j+1)} \Big/ \left(\sum_{l=j+1}^{L} C^{(l)}\right) \leqslant \exp(-0.004 \sqrt{\ln n}),$$

or

$$\sum_{l=j}^{L} C^{(l)}/C^{(j)} \leqslant 1 + \exp(-0.004 \sqrt{\ln n}),$$

which, in turn, implies

$$C^{(j)} \Big/ \left(\sum_{l=j}^{L} C^{(l)}\right) \geqslant 1 - \exp(-0.004 \sqrt{\ln n}).$$

But the probability that for some j any of the above inequalities holds is, by (1), less than

$$2m_0 \exp(-0{\cdot}004 \sqrt{\ln n}) = 2\lfloor \sqrt{\ln n}\rfloor \exp(-0{\cdot}004 \sqrt{\ln n}) = o(1). \qquad \square$$

Let us define the random variable $M(n, k)$ as the number of cycles one can create before the number of 'unused' vertices drops down under k, i.e.

$$M = M(n, k) = \min\left\{i: \sum_{j=i+1}^{L} C^{(j)} \leqslant k\right\}.$$

Our next result states that, with large probability, $\sum_{j=M(n,k)+1}^{L} C^{(j)}$ is not too small.

Claim 3. *For every k and n, $1 \leqslant k^{1{\cdot}011} \leqslant n$, the probability that $\sum_{j=M(n,k^{1{\cdot}011})+1}^{L} C^{(j)} \leqslant 2k$ is smaller than $2k^{-0{\cdot}01}$.*

Proof. Set $R = M(n, k^{1{\cdot}011})$. Then the formula for the total probability, together with (1), gives

$$\Pr\left(\sum_{j=r+1}^{L} C^{(j)} \leqslant 2k\right) = \sum_{i > k^{1{\cdot}011}} \Pr\left(\sum_{j=r+1}^{L} C^{(j)} \leqslant 2k \,\bigg|\, \sum_{j=r}^{L} C^{(j)} = i\right) \Pr\left(\sum_{j=r}^{L} C^{(j)} = i\right)$$

$$\leqslant \max\left\{\Pr\left(\sum_{j=r+1}^{L} C^{(j)} \leqslant 2k \,\bigg|\, \sum_{j=r}^{L} C^{(j)} = i\right) : i > k^{1{\cdot}011}\right\}$$

$$\leqslant \max\{2k/i : i > k^{1{\cdot}011}\} \leqslant 2k^{-0{\cdot}01}. \qquad \square$$

We conclude this section with an old result of Erdős and Turán [6].

Claim 4. *If $\omega(n) \to \infty$, the probability that the largest prime that divides $\prod_{i=1}^{L} C_i$ is smaller than $n \exp(-\omega(n) \sqrt{\ln n})$ tends to 0 as $n \to \infty$.*

3. An upper bound for $i(n, k)$

This part of the note will be entirely devoted to the proof of the following result.

Lemma. *There exists an absolute constant A such that $i(n, k) \leqslant Ak^{-0{\cdot}01}$ for all $1 \leqslant k \leqslant n/2$.*

Proof. Note first that one only has to show that the assertion holds for $k, n > N$, where N is sufficiently large. We split the proof into two cases.

Case 1. $n^{0{\cdot}99} \leqslant k \leqslant n/2$.

The proof of this part is based on a rather simple idea – by Claim 1 (i), most permutations have less than $1{\cdot}11 \ln n$ cycles, so, one may choose a subset of indices $i_1, i_2, \ldots, i_t$ in at most $2^{1{\cdot}11 \ln n} \leqslant n^{0{\cdot}77}$ ways. Thus, the probability that the sum $\sum_{j=1}^{t} C_{i_j}$ attains a particular value k should decrease as some power of n.

In the proof we shall need some more notation. Let $\mathscr{A}$ denote the family of all permutations of the set $[n]$ that contain less than $1{\cdot}11 \ln n$ cycles. Furthermore, for

$n^{0.99} \leqslant k \leqslant n/2$ and $t = 0, 1, ..., \lfloor n^{0.8} \rfloor$, define $\mathscr{D}(k, t)$ as the family of all permutations such that for some $i_1, i_2, ..., i_l$ we have

$$\sum_{j=1}^{l} C_{i_j} = k + t.$$

We will show that for n large enough,

$$|\mathscr{D}(k, 0) \cap \mathscr{A}|/|\mathscr{A}| \leqslant n^{-0.02}. \tag{2}$$

Suppose that (2) does not hold, *i.e.*

$$|\mathscr{D}(k, 0) \cap \mathscr{A}|/|\mathscr{A}| > n^{-0.02}. \tag{$\bar{2}$}$$

Fix t and $\sigma \in \mathscr{A} \cap \mathscr{D}(k, 0)$. Choose $i_1, i_2, ..., i_l$ in such a way that $\sum_{j=1}^{l} C_{i_j} = n - k$, and $C_{i_1} \geqslant C_{i_j}$ for $2 \leqslant j \leqslant l$. Let C_j denote the length of the longest of the remaining cycles. Note, that because $\sigma \in \mathscr{A}$, both C_{i_1} and C_j are larger than $(k - t)/(1.11 \ln n) > n^{0.9}$. Below we denote the cycles of indices i_1 and j by C_{i_1} and C_j respectively.

Now take t consecutive elements from C_{i_1} and add them to the cycle C_j, *i.e.* choose $m_0 \in C_{i_1}$ and $m'_0 \in C_j$, and define a new permutation $\tau = \tau(m_0, m'_0)$ setting:

(i) $\tau^i(m'_0) = \sigma^i(m_0)$ for $i = 1, 2, ..., t$,
(ii) $\tau^i(m'_0) = \sigma^{i-t}(m_0^*)$ for $i = t+1, t+2, ..., t+C_j$,
(iii) $\tau^i(m_0) = \sigma^{i+t}(m_0)$ for $i = 1, 2, ..., C_{i_1} - t$,
(iv) $\tau(m) = \sigma(m)$ for $m \notin C_{i_1} \cup C_j$.

In this way, for any $\sigma \in \mathscr{A} \cap \mathscr{D}(k, 0)$, one can obtain $C_{i_1} C_j$ different permutations from $\mathscr{A} \cap \mathscr{D}(k, t)$. Clearly, each such permutation τ is a result of the modification of at most $(1.11 \ln n)^2 (C_{i_1} - t)(C_j + t)$ permutations σ from $\mathscr{A} \cap \mathscr{D}(k, 0)$. Hence

$$|\mathscr{A} \cap \mathscr{D}(k, t)| \geqslant \frac{C_{i_1} C_j}{(C_{i_1} - t)(C_j + t)(1.11 \ln n)^2} |\mathscr{A} \cap \mathscr{D}(k, 0)|.$$

Note that as $C_{i_1}, C_j \geqslant n^{0.99}$ and $t \leqslant n^{0.8}$, for sufficiently large n we have

$$\frac{C_{i_1} C_j}{(C_{i_1} - t)(C_j + t)(1.11)^2} \geqslant 0.5,$$

so, from ($\bar{2}$), we get

$$\sum_{t=0}^{\lfloor n^{0.8} \rfloor} \frac{|\mathscr{A} \cap \mathscr{D}(k, t)|}{|\mathscr{A}|} \geqslant 0.5 \frac{n^{0.8}}{(\ln n)^2} \frac{|\mathscr{A} \cap \mathscr{D}(k, 0)|}{|\mathscr{A}|} \geqslant 0.5 n^{0.78}.$$

On the other hand, every single permutation $\sigma \in \mathscr{A}$ can contribute at most $2^{1.11 \ln n + 1}$ to the sum $\sum_t |\mathscr{A} \cap \mathscr{D}(k, t)|$. Hence

$$\sum_{t=0}^{\lfloor n^{0.8} \rfloor} \frac{|\mathscr{A} \cap \mathscr{D}(k, t)|}{|\mathscr{A}|} \leqslant 2^{1.11 \ln n + 1} \leqslant n^{0.77}.$$

Thus, ($\bar{2}$) leads to a contradiction and (2) holds.

To complete the proof of the first case it is enough to note that by Claim 1 (i)

$$\frac{|\mathscr{D}(k,0)|}{n!} \leqslant \frac{|\mathscr{D}(k,0) \cap \mathscr{A}|}{|\mathscr{A}|} + \frac{n!-|\mathscr{A}|}{|\mathscr{A}|} \leqslant n^{-0.02} + \frac{o(n^{-0.05})}{1-o(n^{-0.05})} \leqslant 2n^{-0.02}.$$

Case 2. $1 \leqslant k \leqslant n^{0.99}$.

As we have already mentioned, $C^{(i_0+1)}, C^{(i_0+2)}, ..., C^{(L)}$ might be viewed as cycles of a random permutation on $m = \sum_{i=i_0+1}^{L} C^{(i)}$ vertices, where, by Claim 3, with probability at least $1 - 2k^{-0.01}$, we have $2k < m \leqslant k^{1.011}$. Thus,

$$|i(n,k) - i(m,k)| \leqslant 2k^{-0.01},$$

and, since $m^{0.99} < k < m/2$, the assertion follows from the part of the Lemma we have already shown.   □

4. Proofs of Theorems 1 and 2

Proof of Theorem 1. Let us first look at primitive subgroups of S_n. Babai [1] (see also [12]) showed that the minimal degree of a primitive subgroup not containing A_n is at least $(\sqrt{n}-1)/2$. On the other hand, a well known result of Bovey [2] states that the probability that the minimal degree of a subgroup generated by a random permutation π_n is greater than n^{α} decreases with n as $n^{-\alpha+o(1)}$. Thus, the probability that π_n belongs to a non-trivial primitive subgroup is $n^{-0.5+o(1)}$, and, consequently, primitive subgroups contain not more than $n^{-0.5+o(1)}$ elements of S_n combined.

Now consider permutations with proper blocks, *i.e.* those permutations σ for which one can find a proper divisor r of n and a partition of the set $[n]$ into blocks $A_1, A_2, ..., A_r$ such that

(i) $|A_i| = n/r$ for $i = 1, 2, ..., r$,
(ii) for every $i = 1, 2, ..., r$ there exists an index j such that $\sigma(A_i) = A_j$.

Note that each cycle of σ has the same number of elements in each block it intersects.

We shall prove that, with large probability, for a random permutation π_n such a block system does not exist. We rule out all possible candidates for the number of blocks r in three steps.

Case 1. $2 \leqslant r \leqslant r_0 = \exp(\ln \ln n \sqrt{\ln n})$.

By Claim 2, for every r we can find a cycle C of π_n whose length is not divisible by r. Let B denote the union of all blocks that intersect C. Since r does not divide the length of C, B must be a proper invariant subset of $[n]$ of sn/r elements, for some $1 \leqslant s < r$. By the Lemma, the probability that such a subset exists is bounded from above by

$$\sum_{r=2}^{r_0} \sum_{s=1}^{r} i(n, sn/r) \leqslant A r_0^2 (n/r_0)^{-0.01} = n^{-0.009}.$$

Case 2. $r_0 = \exp(\ln \ln n \sqrt{\ln n}) \leqslant r \leqslant n \exp(-\ln \ln n \sqrt{\ln n})$.

Since each cycle of a permutation shares the same number of elements with each block it intersects, Claim 4 implies that the probability that a random permutation has the above block structure tends to 0 as $n \to \infty$.

Case 3. $n\exp(-\ln\ln n\sqrt{\ln n}) \leqslant r \leqslant n$.

Let C be a cycle longer than $n^{0.99}$ whose length is not divisible by the block size $s = n/r$ (the existence of such a cycle is guaranteed by Claim 2), and let B be the union of blocks that intersect C. Since s does not divide the length of C, C must be a proper subset of B. Thus, π_n must contain another cycle C' that intersects each block of B. Hence, both C and C' should have length divisible by the number of blocks contained in B, which is greater than

$$|B|/s \geqslant |C|/s \geqslant n^{0.99}/s > n^{0.9},$$

contradicting Claim 1 (ii). □

Proof of Theorem 2. Let $x_1, x_2, ..., x_C$ denote permutations chosen randomly and independently from S_n. If $x_1, x_2, ..., x_C$ do not generate invariably S_n, then there exist permutations $y_1, y_2, ..., y_C$, with y_i conjugate to x_i, such that one of the following holds:

(i) there exists a proper subset of $[n]$ invariant under y_i for all $i = 1, 2, ..., C$;
(ii) $y_1, y_2, ..., y_C$ generate a non-trivial transitive subgroup of S_n;
(iii) $y_1, y_2, ..., y_C$ generate A_n.

However, the probability that (i) holds is bounded from above by $\sum_{k=1}^{\lfloor n/2\rfloor}(i(n,k))^C$, and, from the Lemma and the fact that $i(n,k) \leqslant 2/3$ (see [5]), can be made arbitrarily small by choosing C large enough. Furthermore, by Theorem 1, the probability that y_1 is contained in some non-trivial transitive subgroup of S_n tends to 0 as $n \to \infty$, which rules out the second case. Finally, the probability that all permutations $y_1, y_2, ..., y_C$ are even is less than 2^{-C} and tends to 0 with C. Thus, the assertion of Theorems 2 holds for $C = C(\epsilon)$ such that $\sum_{k=1}^{n-1}(i(n,k))^{C_0} < \epsilon/3$ and $2^{-C_0} < \epsilon/3$ whenever n is large enough to make the fraction of elements of S_n belonging to non-trivial transitive groups smaller than $\epsilon/3$. □

5. Final remarks and comments

As a matter of fact, using our argument it is possible to show that the fraction of elements of S_n that belong to non-trivial transitive subgroups decreases with n as $n^{-\alpha}$, for some absolute constant $\alpha > 0$. (One only has to estimate more carefully how fast the probabilities in Claims 2 and 4 tend to 0. Since the proof is somewhat lengthy and not particularly interesting (though not very difficult), we decided to present Theorem 1 in a slightly weaker form). Nevertheless, we cannot prove that, for every $\epsilon > 0$, one can take $\alpha = 0.5 - \epsilon$, as was recently conjectured by Cameron. Indeed, if n is even, the number of possible splits of $(1 + o(1))\ln n$ cycles of a random permutation into two groups is at least $2^{(1+o(1))\ln n}$, so, using our argument, we cannot approximate the probability that the set can be divided into two blocks of equal size by anything better than $2^{(1+o(1))\ln n}/n = n^{-(1+o(1)\alpha)}$ for $\alpha = 1 - \ln 2 = 0.30685...$. Thus, a proof of the stronger version of Cameron's conjecture would require either much more detailed analysis or a new method.

Some analogues of Dixon's theorem were recently obtained by Kantor and Lubotzky [7] for finite simple classical groups. This prompts us to ask whether analogues of our result also hold for families of groups of Lie type. In [12], Stong proved a number of results for the statistical theory of the group $GL(n, p^\alpha)$, which are somewhat similar to that obtained by Erdős and Turán for the symmetric group. Thus, the first question to decide seems to be the following:

Problem. *Suppose that p is a fixed prime and n tends to infinity. Is it true that almost all elements of GL(n, p) do not belong to an irreducible subgroup not containing SL(n, p)?*

Acknowledgements

We would like to thank Boris Pittel and Valentin F. Kolchin who kindly provided us with suitable references to Claim 1 (i), and Peter J. Cameron for stimulating discussions.

References

[1]   Babai, L. (1981) On the order of uniprimitive permutation groups. *Ann. of Math* **113**, 553–568.
[2]   Bovey, J. D. (1980) The probability that some power of a permutation has small degree. *Bull. London Math. Soc.* **12** 47–51.
[3]   Cameron, P. J. (1992) Almost all quasigroups have rank 2. *Discrete Math.* **106/107**, 111–115.
[4]   Dixon, J. D. (1969) The probability of generating the symmetric group. *Math. Z.* **110**, 199–205.
[5]   Dixon, J. D. (1992) Random sets which invariably generate the symmetric group. *Discrete Math.* **105**, 25–39.
[6]   Erdős, P. and Turán P. (1967) On some problems of a statistical group-theory. II. *Acta Math. Acad. Sci. Hung.* **18**, 151–163.
[7]   Kantor, W. M. and Lubotzky, A. (1990) The probability of generating a finite classical group. *Geometriae Dedicata* **36**, 67–87.
[8]   Moser, L. and Wyman M. (1958) Asymptotic development of the Stirling numbers of the first kind. *J. London Math. Soc.* **33**, 133–146.
[9]   Pavlov, Yu. L. (1988) On random mappings with constraints on the number of cycles. *Proc. Steklov Inst. Math.* **177**, 131–142. (Translated from *Tr. Mat. Inst. Steklova* **177** (1986), 122–132).
[10] Pittel, B. (1984) On growing binary trees. *J. Math. Anal. Appl.* **103**, 461–480.
[11] Pyber, L. (1991/1992) Asymptotic results for permutation groups. In: Kantor, W. M. and Finkelstein, L. (eds.) *Groups and Computations*, DIMACS Ser. Discrete Math. (And *Theoretical Comp. Sci.* (1993) **11** 197–219.)
[12] Pyber, L. (to appear) The minimal degree of primitive permutation groups. *Handbook of Combinatorics*.
[13] Stong, R. (1988) Some asymptotic results on finite vector spaces. *Adv. Appl. Math.* **9**, 167–199.

On Vertex-Edge-Critically n-Connected Graphs

W. MADER

Institut für Mathematik, Universität Hanover, 30167 Hanover, Weifengarten 1, Germany

All digraphs are determined that have the property that when any vertex and any edge
that are not adjacent are deleted, the connectivity number decreases by two.

1. Introduction and notation

Whereas the characterization of all graphs having the property that the deletion of any
two edges decreases the connectivity number by two is rather easy, and well known [6] (see
Section 2), the characterization of all graphs with the analogous property for the deletion
of two vertices instead of two edges seems to be hopeless. So the following idea suggests
itself. A graph or digraph G is called *vertex-edge-critically n-connected* (abbreviated to
n-ve-critical), if the deletion of any vertex v and any edge e not incident to v decreases the
connectivity number n of G by two (and such v and e exist). If we do not want to specify
the connectivity number, we write *vertex-edge-critical* or *ve-critical*. When I determined
the minimum number of 1-factors of a $(2k)$-connected graph containing a 1-factor, the
ve-critical graphs played an important role and all ve-critical undirected graphs were
characterized there [2]. It was shown in [2] that every ve-critical undirected graph is
obtained in the following way. For an integer $m \geq 1$, take vertex-disjoint circuits of length
$m + 2$ and vertex-disjoint copies of $\overline{K}_m$ (the complementary graph of the complete graph
K_m on m vertices) and take all edges between these vertex-disjoint graphs. We will give
an easier proof of this characterization in Section 3 by using the characterization of all
minimally n-connected graphs with exactly $n + 1$ vertices of degree n, given in [3]. The
main result of the paper is the characterization of all ve-critical digraphs in Sections 4
and 5: *every vertex-edge-critical digraph arises from a vertex-edge-critical undirected graph
by replacing every edge with a pair of oppositely directed edges.*

First we will put together our notation and definitions. A (directed) multigraph $G =
(V, E)$ consisting of the vertex set $V(G) = V$ and the edge set $E(G) = E$ may have

multiple edges, but no loops. A multidigraph is a directed multigraph. The set of edges between the vertices x and y (in the directed case, from x to y) in G is denoted by $[x, y]_G$, and, for $X, Y \subseteq V(G)$, $[X, Y]_G := \bigcup_{x \in X, y \in Y} [x, y]_G$. Distinct edges from $[x, y]_G$ are distinguished by an upper index, for instance $[x, y]^i$. If G is directed and $e \in [x, y]_G$, then x is the *tail* and y is the *head* of e. The set of edges in G with tail in x (head in x) is denoted by $E^+(x; G)$ $(E^-(x; G))$. A *graph* has no multiple edges and is undirected and a *directed graph* or *digraph* has no multiple edges of the same direction. For emphasis, we sometimes say undirected (multi-)graph for (multi-)graph. In a graph or digraph we write $[x, y]$ for the edge from x to y. An edge $[x, y]$ of a digraph D is *symmetric*, if $[y, x] \in E(D)$ also, and *asymmetric*, otherwise. If every edge of a digraph D is symmetric, we call D *symmetric*. In a drawing of a digraph, a pair of symmetric edges is displayed as a line without an arrow-head. Edges $e \in [x, y]_G$ and $e' \in [x', y']_G$ of a directed multigraph G are *consecutive* if $y = x'$ or $y' = x$ holds. For a multigraph G, the directed multigraph $\overleftrightarrow{G}$ arises from G by replacing every edge of G with a pair of oppositely directed edges. For a directed or undirected multigraph G and a positive integer n, G^n is constructed from G by replacing every edge of G with n edges. The *dual* of a digraph D arises from D by reversing the direction of every edge of D. The vertex number and the edge number of G are denoted by $|G|$ and $\|G\|$, respectively. For a vertex set A, we define $A \cap G := A \cap V(G)$, and $x \in G$ means $x \in V(G)$. For $A \subseteq V(G)$, the submultigraph of G spanned by A is $G(A) := G - (V(G) - A)$. For undirected G and $x \in G$, we use $d(x; G)$ to denote the degree of x in G, and $N(x; G)$ is the set of neighbours of x in G. For directed G and $x \in G$, we use $d^+(x; G)$ $(d^-(x; G))$ to denote the outdegree (indegree) of x in G, and $N^+(x; G)$ $(N^-(x; G))$ is the set of outneighbours (inneighbours) of x in G. For a digraph D and $x \in D$, we define $N_s(x; D) := N^+(x; D) \cap N^-(x; D)$, $N_a^\epsilon(x; D) := N^\epsilon(x; D) - N_s(x; D)$ and $d_a^\epsilon(x; D) := |N_a^\epsilon(x; D)|$ for $\epsilon \in \{+, -\}$, and $\triangle_a(D) := \max\{d_a^\epsilon(x; D) : x \in D \text{ and } \epsilon \in \{+, -\}\}$. A directed multigraph D is called *n-regular* if $d^+(x; D) = d^-(x; D) = n$ for every $x \in D$. If there is no doubt which graph is meant, we drop it in the above notation. $\mathbb{N}$ denotes the set of positive integers, n is always from $\mathbb{N}$, and $\mathbb{N}_m := \{n \in \mathbb{N} : n \leq m\}$ for $m \geq 0$.

A *path* and a *circuit* in G pass through every vertex of G at most once. If G is directed, they are *continuously directed*. For $x, y \in G$, an *x, y-path* P is a path from x to y, and for $u, v \in P$ such that u is before v on P in the directed case, $P[u, v]$ is the u, v-path contained in P and $P[u, v) := P[u, v] - \{v\} =: P[u, v] - v$. We consider paths and circuits as subgraphs, but write them as a sequence of their vertices in the order passed through (for multidigraphs, in the direction of the path or the circuit). We say that the paths $P_1, \ldots, P_n$ in G cover G if $\bigcup_{i \in \mathbb{N}_n} V(P_i) = V(G)$ holds. In a directed or undirected multigraph G, x, y-paths P, Q are *openly disjoint* if they are distinct and $V(P) \cap V(Q) = \{x, y\}$ holds. The maximum number of pairwise openly disjoint x, y-paths in G is denoted by $\kappa(x, y; G)$. The *connectivity number* $\kappa(G)$ of G is defined by $\kappa(G) := \min_{x \neq y} \kappa(x, y; G)$ for $|G| \geq 2$ and $\kappa(G) := |G| - 1$ for $|G| \leq 1$. In an analogous manner, the *edge-connectivity number* $\lambda(G)$ is defined by edge-disjoint paths. A directed or undirected multigraph G is *k-minimally n-(edge-)connected*, for $k \in \mathbb{N}$, iff $\|G\| \geq k$ and, for all $E' \subseteq E(G)$ with $|E'| \leq k$, we have $\kappa(G - E') = n - |E'|$ $(\lambda(G - E') = n - |E'|)$. For 1-minimally n-connected we say *minimally n-connected*. Let us make precise the definition of ve-criticality: a (di-)graph G is *vertex-edge-critically n-connected* iff $\kappa(G) = n \geq 2$, and for every $v \in V(G)$ and $e \in E(G-v)$, $\kappa(G-v-e) = n-2$ holds.

A sequence $v_1, [v_1, \overline{v_1}], \overline{v_1}, [v_2, \overline{v_1}], v_2, [v_2, \overline{v_2}], \ldots, [v_n, \overline{v_n}], \overline{v_n}, [v_1, \overline{v_n}]$ of vertices $v_1, \overline{v_1}, \ldots, \overline{v_n}$ and distinct edges of a digraph D is called an *alternating cycle* in D. Normally, we omit the edges in the notation and write $v_1^+, \overline{v_1}, v_2^+, \ldots, \overline{v_n}$ for an alternating cycle, where the upper index $+$ at v_i means that the edges (cyclically) on either side of v_i have their tails in v_i. Sometimes we consider an alternating cycle as a subdigraph of D. If $G_1, \ldots, G_n$ are graphs (digraphs) with $V(G_i) \cap V(G_j) = \emptyset$ for $i \neq j$, the graph (digraph) $\sum_{i=1}^n G_i = G_1 + G_2 + \cdots + G_n$ is defined as

$$\left(\bigcup_{i=1}^n V(G_i), \bigcup_{i=1}^n E(G_i) \cup \bigcup_{i=1}^n \left\{ [x, y] : x \in G_i \text{ and } y \in \bigcup_{j \in \mathbb{N}_n - \{i\}} V(G_j) \right\} \right).$$

If all G_i are isomorphic, we write $nG_1 := \sum_{i=1}^n G_i$. If $G_1, \ldots, G_n$ are not vertex-disjoint, we define $\sum_{i=1}^n G_i$ by vertex-disjoint copies of $G_1, \ldots, G_n$. If $G = H_1 + H_2$ and $E(H_1) = \emptyset$, we also write $G = V(H_1) + H_2$. For an integer $m \geq 3$, C_m denotes an undirected circuit of length m. For integers $m \geq 3$, $k \geq 0$, $l \geq 0$, the multidigraph $D = C_m^{k,l}$ is defined by $V(D) := \mathbb{N}_m$ and $|[i, i+1]_D| = k$, $|[i+1, i]_D| = l$ for i modulo m.

2. 2-minimally n(-edge)-connected graphs

B. Maurer and P. Slater determined in [6] all 2-minimally n-connected graphs and all 2-minimally n-edge-connected multigraphs. We give a simpler proof of the latter result, and show that, in the proof of the former, it is not necessary to use the fact from [1] that every minimally n-connected graph has at least $n + 1$ vertices of degree n.

Let $G = (V, E)$ be a 2-minimally n-connected multigraph. Consider any $x \in V$. There are an $e \in E$ incident to x, say, $e \in [x, y]_G$ and a system of n openly disjoint x, y-paths $P_1, \ldots, P_n$. From $\kappa(G - e) < n$, it easily follows that $\kappa(x, y; G) = n$ by Menger's Theorem. Hence $e \in \bigcup_{i=1}^n E(P_i)$. For every $e' \in E - \{e\}$, $\kappa(G - \{e, e'\}) = n - 2$ holds and implies $\kappa(x, y; G - \{e, e'\}) = n - 2$ by Menger's Theorem. Hence $e' \in \bigcup_{i=1}^n E(P_i)$ and thus $E = \bigcup_{i=1}^n E(P_i)$ and $d(x) = n$ follow. Hence G is finite and n-regular. If $|G| \geq 3$, there is a $z \in V - \{x, y\}$, and z is on exactly one of the paths $P_1, \ldots, P_n$, since $E = \bigcup_{i=1}^n E(P_i)$ holds and $P_1, \ldots, P_n$ are openly disjoint. Hence $E = \bigcup_{i=1}^n E(P_i)$ implies $d(z) = 2$, and G is 2-regular. So we have somewhat generalized a result from Maurer and Slater.

Theorem 1. [6] *The only 2-minimally n-connected multigraphs are K_2^n and, for $n = 2$, the circuits C_m.*

Let $G = (V, E)$ now be a 2-minimally n-edge-connected multigraph, and choose $x \in V$ and $e \in [x, y]_G$ as above. Now there are edge-disjoint x, y-paths $P_1, \ldots, P_n$. As above, $E = \bigcup_{i=1}^n E(P_i)$ and $d(x) = n$ follow. Again, G is finite and n-regular. Let us assume $|G| \geq 3$ and consider $z \in V - \{x, y\}$. Then every edge incident to z belongs to exactly one of $P_1, \ldots, P_n$. Hence n is even and exactly $n/2$ of $P_1, \ldots, P_n$ pass through z. Suppose $N(x) = \{y, y_1, \ldots, y_k\}$. Then $|G| \geq 3$ and $E = \bigcup_{i=1}^n E(P_i)$ imply $\{y_1, \ldots, y_k\} \neq \emptyset$. We define a directed multigraph D on the vertex set $\{y_1, \ldots, y_k\}$. Every path P_j of length at least 2 generates the following edges of D (and there are no further edges in D):

if z is the first vertex of P_j after x, then $z \in \{y_1, \ldots, y_k\}$ and we add the edges $[z, u]^j$ for all $u \in (P_j - z) \cap \{y_1, \ldots, y_k\}$. We prove that D has no circuit. Suppose there is a circuit in D, and this may have the edges $[z_1, u_1]^{j_1}, \ldots, [z_m, u_m]^{j_m}$ in this cyclic order (hence $u_i = z_{i+1}$). By definition of D, $j_1, \ldots, j_m$ are distinct, since $z_1, \ldots, z_m$ are. If we replace P_{j_i} with the x, y-path $P'_{j_i} := P_{j_i}[x, z_i] \cup P_{j_{i-1}}[u_{i-1}, y]$ for $i = 1, \ldots, m$ (i modulo m), we get from $P_1, \ldots, P_n$ a system $P'_1, \ldots, P'_n$ of edge-disjoint x, y-paths in G. But $\bigcup_{i=1}^n E(P'_i) \subsetneqq \bigcup_{i=1}^n E(P_i)$ holds, contradicting the remarks above. Hence D is acyclic and there is a $z \in V(D)$ with $d^-(z; D) = 0$. This means that all the $n/2$ paths P_i containing z have $E(P_i) \cap [x, z]_G \neq \emptyset$. But this implies $|[x, z]_G| = n/2$. Considering an $e' \in [x, z]_G$ instead of e, we get, in the same way, a vertex $z' \neq z$ with $|[x, z']_G| = n/2$. Since G is n-regular and finite, we get $G \cong C_{|G|}^{n/2}$. The following theorem summarizes what we have proved.

Theorem 2. [6] *The only 2-minimally n-edge-connected multigraphs are K_2^n and, for even $n \geq 4$, also $C_m^{n/2}$.*

Obviously, the deletion of two consecutive edges of a digraph cannot decrease the connectivity number or edge-connectivity number by two (*cf.* [6]). So it is natural to consider only the deletion of non-consecutive edges in a digraph. Let us call a multidigraph D weakly 2-minimally n-connected (weakly 2-minimally n-edge-connected), if $\kappa(D) = n \geq 2$ ($\lambda(D) = n \geq 2$), but for all non-consecutive $e_1 \neq e_2$ from $E(D)$, we have $\kappa(D - \{e_1, e_2\}) = n - 2$ ($\lambda(D - \{e_1, e_2\}) = n - 2$).

Let $D = (V, E)$ be a weakly 2-minimally n-connected multigraph. Choosing $[x, y]_D \neq \emptyset$ and openly disjoint x, y-paths $P_1, \ldots, P_n$, we conclude, as above, $E - (E^-(x) \cup E^+(y)) \subseteq \bigcup_{i=1}^n E(P_i)$. Hence D is n-regular and finite. Let us assume $|D| \geq 3$. Then D has no multiple edges, since D is n-regular and $\kappa(D) = n \geq 2$. Since only one edge of $\bigcup_{i=1}^n E(P_i)$ has its head in $z \in V - \{x, y\}$, we get $n = 2$ and $[y, z] \in E$ for all $z \in V - \{x, y\}$. Now $D \cong \overleftrightarrow{K_3}$ follows easily.

Theorem 1d. *The only weakly 2-minimally n-connected multidigraphs are*

$$\overleftrightarrow{K_2^n} \text{ and } \overleftrightarrow{K_3} \text{ for } n = 2.$$

Let us now consider a weakly 2-minimally n-edge-connected multidigraph $D = (V, E)$. If $[x, y]_D \neq \emptyset$ and $P_1, \ldots, P_n$ are edge-disjoint x, y-paths, $E - (E^-(x) \cup E^+(y)) \subseteq \bigcup_{i=1}^n E(P_i)$ follows as above. Hence D is n-regular and finite. Put $m := \max_{x, y \in V} |[x, y]_D|$ and choose $x, y \in V$ such that $m = |[x, y]_D|$ holds. If $N^+(x) = \{y\}$ holds, then $m = n$ and $E - (E^-(x) \cup E^+(x)) \subseteq [x, y]_D$. But this implies $D \cong \overleftrightarrow{K_2^n}$ or $D \cong C_3^{n,0}$. So we assume $|N^+(x)| \geq 2$. Let $P_1, \ldots, P_n$ be edge-disjoint x, y-paths. As in the proof of Theorem 2, we find a $z \in N^+(x) - \{y\}$ such that $z \in P_i$ implies $[x, z]_D \cap E(P_i) \neq \emptyset$. Set $k := |\{i \in \mathbb{N}_n : z \in P_i\}|$. Since $z \in N^+(x)$ and $d^+(x) = n$, we have $k = |[x, z]_D| \geq 1$. Since $E - (E^-(x) \cup E^+(y)) \subseteq \bigcup_{i=1}^n E(P_i)$ and $d^-(z) = n$, we conclude $|[y, z]_D| = n - k$. Since $m + k \leq d^+(x) = n$ and $n - k \leq m$ by choice of m, it follows that $n - k = m$. Since $E - (E^-(x) \cup E^+(z))$ is contained in the n-edge-disjoint x, z-paths of $D(\{x, y, z\})$, we conclude $E^-(y) - [x, y]_D = [z, y]_D$ and $E^+(y) - [y, z]_D = [y, x]_D$, hence $|[z, y]_D| = k = |[y, x]_D|$. Furthermore, $|D| = 3$ follows, since D is n-regular and $k \geq 1$. So $D \cong C_3^{k, n-k}$, and we have proved the following theorem.

Theorem 2d. *The only weakly 2-minimally n-edge-connected multidigraphs are $\overleftrightarrow{K}_2^n$ and $C_3^{k,n-k}$ for $k \in \mathbb{N}_n$.*

3. Vertex-edge-critical undirected graphs

First we will deduce some common properties of undirected and directed ve-critical graphs. Subsequently, we will determine all ve-critical undirected graphs.

Let $G = (V, E)$ be an n-ve-critical graph or digraph. Consider an edge $e = [x, y] \in E$ and openly disjoint x, y-paths $P_1, \ldots, P_n$ in G. For $v \in V - \{x, y\}$, $\kappa(G - v - e) = n - 2$ by assumption, hence $\kappa(x, y; G - v - e) = n - 2$ follows easily from Menger's Theorem. But this implies $v \in \bigcup_{i=1}^n V(P_i)$, hence $V = \bigcup_{i=1}^n V(P_i)$ and G is finite. On the other hand, if for every edge $[x, y]$ of a graph or digraph G with $\kappa(G) = n \geq 2$, every system of n openly disjoint x, y-paths covers G, obviously G is n-ve-critical. We state this equivalence formally.

Lemma 1. *A graph or digraph G with $\kappa(G) = n \geq 2$ is n-ve-critical, iff for every edge $[x, y]$ of G, every system of openly disjoint x, y-paths $P_1, \ldots, P_n$ covers G.*

From this, the following property is easily deduced.

Lemma 2. *Every n-ve-critical graph or digraph is finite and n-regular.*

Proof. It remains to show that an n-ve-critical G is n-regular. Consider an edge $[x, y]$ of G and openly disjoint x, y-paths $P_1, \ldots, P_n$ in G. Suppose there is an edge $[x, z]$ in G that is not on any P_i. Since $P_1, \ldots, P_n$ cover G by Lemma 1, there is a P_i containing z, say, $z \in P_n$. Then the openly disjoint x, y-paths $P_1, \ldots, P_{n-1}, P_n'$, where $P_n' := x, P_n[z, y]$, do not cover G, contradicting Lemma 1. Hence $d(x) = n$ or $d^+(x) = n$, respectively, and G is n-regular. $\square$

If we delete a vertex v from an n-ve-critical graph or digraph G, then $G - v$ is minimally $(n - 1)$-connected. So one can apply the results on minimally n-connected graphs and digraphs. By Lemma 2, $G - v$ has exactly n vertices of degree $n - 1$ or n vertices of outdegree $n - 1$ and n vertices of indegree $n - 1$, respectively. On the other hand, it is well known [1] that a minimally $(n - 1)$-connected graph has at least n vertices of degree $n - 1$, and in [3] even a characterization of all minimally $(n - 1)$-connected graphs containing exactly n vertices of degree $n - 1$ was obtained. This permits a straightforward proof of the characterization theorem on n-ve-critical undirected graphs, which was first proved in [2] using the fact known from [1] that every circuit in a minimally n-connected graph contains a vertex of degree n. First, we state the above mentioned result for minimally n-connected graphs.

Theorem A. [3] *For $n \geq 2$, all minimally n-connected graphs containing exactly $n+1$ vertices of degree n are obtained in the following way.*

(a) *For an integer $m \in \mathbb{N}_n \cup \{0\}$, let H be an $(n - m)$-regular, $(n - m)$-connected graph on $n + 1$ vertices. Then $\overline{K}_m + H$ is minimally n-connected, containing exactly $n + 1$ vertices of degree n.*

(b) *For an integer m with $4 \le m \le n$, let H be an $(n-m)$-regular, $(n-m)$-connected graph on $n-1$ vertices, and let P be a path with $|P| = m$. Then $P + H$ is minimally n-connected, containing exactly $n + 1$ vertices of degree n.*

For characterizing all ve-critical graphs, we need a further lemma.

Lemma 3. *If $G + H$ is a non-complete, ve-critical, undirected or directed graph, H is ve-critical or $\|H\| = 0$.*

Proof. Set $n := \kappa(G + H)$ and $m := n - |G|$. Using Lemma 2, we see that H is m-regular and m-connected. We assume $\|H\| > 0$. Then $\kappa(H) = m > 0$ holds. Suppose $m = 1$ and consider an edge $[x, y] \in E(H) \ne \emptyset$. There are n openly disjoint x, y-paths in $G + H(\{x, y\})$. This implies $|G + H| = n + 1$ by Lemma 1, hence $G + H$ is complete. This contradiction shows $\kappa(H) \ge 2$. Since for $e \in E(H)$, every separating vertex set S of $(G + H) - e$ with $|S| = n - 1$ must contain $V(G)$, it is easy to see that H is ve-critical, since $G + H$ is. $\square$

Without difficulty, we now get the following result.

Theorem 3. [2] *The vertex-edge-critical graphs are exactly the graphs $G_{m,k,l} := k\overline{K}_m + lC_{m+2}$, where $m \ge 1, k, l$ are non-negative integers such that $\kappa(G_{m,k,l}) \ge 2$ holds.*

Proof. Suppose G is a ve-critical graph of the form $\sum_{i=1}^{k} \overline{K}_{m_i} + \sum_{j=1}^{l} C_{n_j}$ with $m_i \in \mathbb{N}$. Since G is regular by Lemma 2, we get immediately that $m_1 = m_2 = \cdots = m_k$ and $n_1 = n_2 \cdots = n_l$ and $n_1 = m_1 + 2$, if $k > 0$ and $l > 0$. This implies $G \cong G_{m_1,k,l}$. On the other hand, it is easy to check that the graphs $G_{m,k,l}$ with $\kappa(G_{m,k,l}) \ge 2$ are ve-critical. So it remains to show that every ve-critical graph has the form $\sum \overline{K}_{m_i} + \sum C_{n_j}$. We will prove this by induction on the connectivity number.

Let G be an $(n + 1)$-ve-critical graph. If $n = 1$, then G is a circuit by Lemma 2. So suppose $n \ge 2$ and choose $v \in V(G)$. Then $G - v$ is minimally n-connected and has exactly $n + 1$ vertices of degree n, namely $N(v; G)$. So $G - v$ has the structure described in Theorem A. If $G - v = \overline{K}_m + H$, as in case (a) of Theorem A, then $G = \overline{K}_{m+1} + H$, where $V(\overline{K}_{m+1}) = V(\overline{K}_m) \cup \{v\}$. If G is complete or $\|H\| = 0$, then G has the form wanted. Otherwise, Lemma 3 implies that H is ve-critical, and hence, by the induction hypothesis, H has the form $\sum \overline{K}_{m_i} + \sum C_{n_j}$, and hence G does as well. If $G - v = P + H$, as in case (b) of Theorem A, then $G = C + H$, where C is a circuit containing v with the property $C - v = P$. By an application of Lemma 3 and the induction hypothesis as in case (a), the proof is complete. $\square$

4. Vertex-edge-critical directed graphs: introduction and preliminaries.

Of course, every n-ve-critical graph G provides an n-ve-critical digraph $\overleftrightarrow{G}$. The aim of this paper is to show that we get every ve-critical digraph in this way, *i.e.*, that every ve-critical digraph is symmetric. With regard to Theorem 3, we will then have proved the following theorem.

Theorem 4. *The vertex-edge-critical digraphs are exactly the digraphs $\overleftrightarrow{G}_{m,k,l}$ with $\kappa(G_{m,k,l}) \ge 2$.*

The proof of this theorem cannot be based on an analogue of Theorem A: only recently [5], I have shown that every minimally n-connected digraph has at least $n + 1$ vertices of outdegree n (that there are at least n such vertices was known before from [4]), but at the moment there is no hope to characterize all minimally n-connected digraphs containing exactly $n + 1$ vertices of outdegree n and $n + 1$ vertices of indegree n. However, there is an analogue in the directed case to the fact that a minimally n-connected graph does not contain a circuit consisting only of vertices of degree exceeding n, which we will state now.

Let $D = (V, E)$ be a minimally n-connected digraph. Let D_0 be the subdigraph of D given by $V(D_0) := \{v \in V : d^+(v; D) > n \text{ or } d^-(v; D) > n\}$ and $E(D_0) := \{[x, y] \in E : d^+(x; D) > n \text{ and } d^-(y; D) > n\}$. It was proved in [4] that D_0 has no alternating cycle.

Theorem B. [4] *For every minimally n-connected digraph D, D_0 does not contain an alternating cycle.*

To every digraph D, we let correspond a bipartite undirected graph $\overline{D}$ as follows: take vertices $x' \neq x''$ for every $x \in V(D)$ so that $\{x', x''\} \cap \{y', y''\} = \emptyset$ holds for $x \neq y$, and define $\overline{D}$ by $V(\overline{D}) := \bigcup_{x \in D} \{x', x''\}$ and $E(\overline{D}) := \{[x', y''] : [x, y] \in E(D)\}$. The following equivalence is easily seen and was shown in [4].

Lemma C. [4] *A digraph D does not have an alternating cycle iff $\overline{D}$ is a forest.*

In the following, $D = (V, E)$ always denotes an n-ve-critical digraph containing an asymmetric edge that has a minimum number of vertices. Our aim is to show that such a digraph cannot exist. By Lemma 2, D is finite and n-regular and $|D| \geq n + 2$ holds. Since the dual digraph of a ve-critical digraph is ve-critical again, for every result on D, there is a dual one, which we will use, but, in general, not state explicitly. For $x \in V$, $H := D - x$ is minimally $(n - 1)$-connected. So H_0 and $\overline{H}_0$ are defined, and we set $D_x := H_0$ and $F_x := \overline{H}_0 - (\{y'' : y \in N_a^+(x)\} \cup \{y' : y \in N_a^-(x)\})$. Defining $R(x) := V - (N^+(x) \cup N^-(x) \cup \{x\})$, we observe $D_x = (V - (N_s(x) \cup \{x\}), [N_a^+(x) \cup R(x), R(x) \cup N_a^-(x)]_D)$, since D is n-regular. Furthermore, F_x has the partition $F_x' := \{y' : y \in N_a^+(x) \cup R(x)\}$, $F_x'' := \{y'' : y \in R(x) \cup N_a^-(x)\}$ into independent vertex sets. Since D is n-regular by Lemma 2, we have $d_a^+(x) = d_a^-(x) =: d_a(x)$, and hence $|F_x'| = |F_x''|$. By Theorem B and Lemma C, F_x is a forest. Theorem B implies the following important properties of D.

Lemma 4.

(a) If $a_1^+, \overline{a}_1, a_2^+, \ldots, \overline{a}_k$ is an alternating cycle of D, then for every $x \in V - \{a_1, \ldots, \overline{a}_k\}$, there is an $i \in \mathbb{N}_k$ such that $[a_i, x] \in E$ or $[x, \overline{a}_i] \in E$ holds.

(b) If $z \notin N^+(x) \cup N^+(y)$ for distinct $x, y, z \in V$, then $|N^+(z) \cap N^+(x) \cap N^+(y)| \geq |N^+(x) \cap N^+(y)| - 1$.

Proof. (a) If $[a_i, x] \notin E$ and $[x, \overline{a}_i] \notin E$ holds for all $i \in \mathbb{N}_k$, then $a_1^+, \overline{a}_1, \ldots, \overline{a}_k$ is an alternating cycle in D_x, contradicting Theorem B.

(b) For $u \neq v$ in $N^+(x) \cap N^+(y)$, x^+, u, y^+, v is an alternating cycle in D. If $z \notin N^+(x) \cup N^+(y) \cup \{x, y\}$, we get $[z, u] \in E$ or $[z, v] \in E$ by (a). This implies $|N^+(z) \cap N^+(x) \cap N^+(y)| \geq |N^+(x) \cap N^+(y)| - 1$. $\qquad\square$

Lemma 5. *For all vertices $x \neq y$ of D, the following statements are true:*

(a) *if $[x, y] \in E$, then $|N^+(x) \cap N^-(y)| \leq n - 2$;*

(b) *if $[x, y] \in E$, then $|N^+(x) \cap N^+(y)| \leq n - 2$;*

(c) *$N^+(x) \neq N^+(y)$.*

Proof. (a) Suppose $[x, y] \in E$ and $|N^+(x) \cap N^-(y)| \geq n - 1$. Then there are n openly disjoint x, y-paths in $D(N^+(x) \cup \{x\})$. These paths cover D by Lemma 1, which implies the contradiction $|D| = n + 1$.

(b) Suppose $[x, y] \in E$ and $|N^+(x) \cap N^+(y)| \geq n - 1$. Let z be the element of $N^+(y) - N^+(x)$. Suppose $z \neq x$, and consider a system of n openly disjoint y, z-paths in D. Obviously, these paths cannot contain x. So Lemma 1 implies $z = x$. But then $S := N^+(x) - \{y\} = N^+(y) - \{x\}$ with $|S| = n - 1$ is separating, since $|D| \geq n + 2$ holds. This contradiction proves (b).

(c) We suppose there are vertices $x \neq y$ in D with $N^+(x) = N^+(y) =: N$. For $z \in V - (N \cup \{x, y\})$, we get $|N^+(z) \cap N| \geq n - 1$ by Lemma 4 (b). Hence (b) implies $[z, x] \notin E$ and $[z, y] \notin E$ and, therefore, $N^-(x) = N = N^-(y)$ holds. Suppose there is an edge $[z, \bar{z}] \in E(D - (N \cup \{x, y\}))$ and consider n openly disjoint $z, \bar{z}$-paths in D. Since $|N^+(z) \cap N| \geq n - 1$ and $N^+(x) = N^+(y) = N$, these paths cannot contain both the vertices x and y. So Lemma 1 implies $\|D - (N \cup \{x, y\})\| = 0$. In particular, for $z \in V - (N \cup \{x, y\})$, we get $N^+(z) = N$ and so also $N^-(z) = N$. Altogether, we have shown $D = (V - N) + D(N)$. Hence $\|D(N)\| = 0$ holds or $D(N)$ is ve-critical by Lemma 3. If $\|D(N)\| = 0$ holds, D is symmetric, contrary to our assumption. So $D(N)$ is ve-critical. But then $D(N)$ is symmetric by choice of D as a minimal counterexample, hence D is symmetric as well. This contradiction proves (c).   □

Lemma 5 (a) and (b) mean that for every $x \in V$, the maximum outdegree and maximum indegree of $D(N^+(x))$ are at most $n - 2$, and, dually, the same holds for $D(N^-(x))$. We now deduce some properties of F_x from Lemma 5.

Lemma 6.

(a) *For every $v \in F_x$, $d(v; F_x) \geq 1$ and for every $v \in F_x - \bigcup_{y \in R(x)} \{y', y''\}$, $d(v; F_x) \geq 2$ holds.*

(b) *For $F = F_x'$ and $F = F_x''$, $|\{v \in F : d(v; F_x) = 1\}| = c + \sum_{\substack{r \in F \\ d(r; F_x) \geq 3}} (d(v; F_x) - 2)$ holds, where*

c denotes the number of components of F_x.

Proof. (a) By duality, it suffices to consider a $v \in F_x'$. Then there is a $z \in N_a^+(x) \cup R(x)$ such that $z' = v$ holds. Since $[z, x] \notin E$, we have $d(z'; F_x) = n - |N^+(z; D) \cap N^+(x; D)|$. So Lemma 5 (c) implies $d(z'; F_x) \geq 1$. Assume $z \in N_a^+(x)$. Then $d(z'; F_x) \geq 2$ by Lemma 5 (b).

(b) Since F_x', F_x'' is a bipartition of the forest F_x into independent sets F_x' and F_x'' of the same cardinality, we get

$$\sum_{v \in F_x'} d(v; F_x) = \|F_x\| = |F_x| - c = 2|F_x'| - c = \left(\sum_{v \in F_x'} 2\right) - c.$$

This proves assertion (b), since there are no isolated vertices in F_x by (a) and the case $F = F''_x$ is dual. $\qquad\Box$

Lemma 6 (b) provides at least one vertex $z \in N_a^+(x) \cup R(x) \neq \emptyset$ with $d(z'; F_x) = 1$, and by Lemma 6 (a), every such vertex is in $R(x)$. So there is a vertex $z \in R(x)$ with $|N^+(z) \cap N^+(x)| = n - 1$, and we define $R^+(x) := \{z \in R(x) : |N^+(z) \cap N^+(x)| = n - 1\}$. $R^-(x)$ is defined dually as $\{z \in R(x) : |N^-(z) \cap N^-(x)| = n - 1\}$. We emphasize once again that $R^+(x) \neq \emptyset$ and $R^-(x) \neq \emptyset$.

We need a series of preliminary lemmas. Herein, x_0 always denotes any vertex of D.

Lemma 7.

(a) If $x \in N_a^+(x_0)$ such that $R^-(x_0) \not\subseteq N^+(x)$ holds, then $|N^-(x) \cap N^-(x_0)| = n - 2$ and $|N^-(x) \cap (N_a^+(x_0) \cup R(x_0))| = 1$.

(b) For all $x \in N_a^+(x_0)$ but at most one, $|N^-(x) \cap (N_a^+(x_0) \cup R(x_0))| = 1$ holds.

Proof. (a) Suppose there is $z \in R^-(x_0) - N^+(x)$. Since $[x, x_0] \notin E$, we can apply the dual of Lemma 4 (b) for x_0, z, x, and get $|N^-(x) \cap N^-(x_0) \cap N^-(z)| \geq |N^-(x_0) \cap N^-(z)| - 1 = n - 2$, by definition of $R^-(x_0)$. This implies $|N^-(x) \cap N^-(x_0)| = n - 2$ by the dual of Lemma 5(b), so $|N^-(x) \cap (N_a^+(x_0) \cup R(x_0))| = n - |N^-(x) \cap N^-(x_0)| - 1 = 1$ follows.

Since for every $z \in R^-(x_0)$, the definition of $R^-(x_0)$ implies that $|N^-(z) \cap N_a^+(x_0)| \leq 1$ holds, (b) follows from (a), since $R^-(x_0) \neq \emptyset$. $\qquad\Box$

Lemma 8.

(a) If $v \in R^+(x_0)$, $x \in N^+(x_0) - N^+(v)$, and $y \in N^-(x) \cap (N_a^-(x_0) \cup R(x_0))$, then $[v, y] \in E$ or $y \in R^+(x_0)$ holds.

(b) If $x \in N^+(x_0)$ such that $|R^+(x_0) - N^-(x)| \geq 2$ holds, then $N^-(x) \cap N_a^-(x_0) = \emptyset$.

Proof. (a) Suppose $[v, y] \notin E$. Since $[x_0, y] \notin E$ also, we can apply Lemma 4 (b) to x_0, v, y and get $|N^+(y) \cap N^+(x_0) \cap N^+(v)| \geq |N^+(x_0) \cap N^+(v)| - 1 = n - 2$. This implies $|N^+(y) \cap N^+(x_0)| \geq n - 1$, since $x \in N^+(y) - N^+(v)$ holds. Hence, by Lemma 5 (b), $y \notin N^-(x_0)$ holds, so $y \in R(x_0)$ and even $y \in R^+(x_0)$ by Lemma 5 (c).

(b) Suppose there are $v_1 \neq v_2$ in $R^+(x_0) - N^-(x)$ for an $x \in N^+(x_0)$, and there is a $y \in N^-(x) \cap N_a^-(x_0)$. Then (a) implies $[v_i, y] \in E$ for $i = 1, 2$. Since $N^+(v_i) \cap N^+(x_0) = N^+(x_0) - \{x\}$ for $i = 1, 2$, we get $N^+(v_1) = N^+(v_2)$, which contradicts Lemma 5 (c). $\qquad\Box$

For every $[x, y] \in E$, we have $|N^+(x) \cap N^-(y)| \leq n - 2$ by Lemma 5 (a). Let us assume equality holds. Then $(D - [x, y]) - (N^+(x) \cap N^-(y))$ has exactly one x, y-path, since every such path does contain $V - (N^+(x) \cap N^-(y))$ by Lemma 1. This path has length at least 3.

Lemma 9. Let $[x, y] \in E$ be such that $S := N^+(x) \cap N^-(y)$ has exactly $n - 2$ vertices, and let $P : x, x_1, \ldots, x_k, x_{k+1}$ be the x, y-path in $(D - [x, y]) - S$. Then the following statements are true.

(a) $N^+(x_1) = \{x, x_2\} \cup S$ and $N^-(x_k) = \{x_{k-1}, y\} \cup S$;

(b) $S \subseteq N^-(x) \to [y, x] \in E$.

Proof. Since, by Lemma 1, $N^+(x_1) \cap \{x_3, \ldots, x_{k+1}\} = \emptyset$ holds, we must have $N^+(x_1) \subseteq V - \{x_3, \ldots, x_{k+1}\}$, hence $N^+(x_1) = \{x, x_2\} \cup S$. The other claim in (a) follows by duality. Now suppose $S \subseteq N^-(x)$. Then there are $n-1$ openly disjoint x_1, x-paths in $D(\{x_1, x\} \cup S)$ by (a). There is a $z \in N^-(x) \cap \{x_2, \ldots, x_{k+1}\}$, and by Lemma 1, $P[x_1, z], x$ does contain $\{x_2, \ldots, x_{k+1}\}$, which implies $z = x_{k+1} = y$, and hence (b). $\qquad\square$

If we assume that $[x, y] \in E$ is asymmetric, and that $\triangle_a(D) = 1$ holds, then $S := N^+(x) \cap N^-(y)$ is a subset of $N^-(x)$, and Lemma 9 (b) implies $|S| \le n - 3$. It is possible to improve this result.

Lemma 10. *Let $[x, y] \in E$ be asymmetric and assume $\triangle_a(D) = 1$. Then*

(a) $|N^+(x) \cap N^-(y)| \le n - 4$, *and*

(b) $|N^+(x) \cap N^+(y)| \le n - 3$ *hold.*

Proof. (a) We suppose $S := N^+(x) \cap N^-(y)$ has at least $n - 3$ elements. Since $\triangle_a(D) = 1$ holds, $[x, y]$ is the only asymmetric edge with tail in x. Hence, there is exactly one asymmetric edge in D with head in x, say $[y', x]$. In particular, we see $S \subseteq N^-(x)$ and, dually, $S \subseteq N^+(y)$. Then Lemma 9 (b) implies $|S| = n - 3$, since $[x, y]$ is asymmetric. Hence, there are two openly disjoint x, y-paths $P_i : x, x_1^i, \ldots, x_{k_i+1}^i$ $(i = 1, 2)$ in $(D - [x, y]) - S$. Furthermore, $k_i \ge 2$ holds, and P_1, P_2 cover $D - S$ by Lemma 1, in particular, $y' \in P_1[x_2^1, y] \cup P_2[x_2^2, y]$. First, we prove a few properties.

(1) $S \cup \{x_1^{i+1}\} \not\subseteq N^+(x_1^i)$ for $i = 1, 2 \pmod 2$.

Suppose $S \cup \{x_1^2\} \subseteq N^+(x_1^1)$. Then $S \cup \{x_1^2\} \subseteq N^+(x_1^1) \cap N^-(x)$ holds. Applying Lemma 9 to $[x_1^1, x] \in E$, we get from the second equality in Lemma 9 (a) the contradiction that $[y', x]$ is symmetric.

(2) For $i = 1, 2 \pmod 2$, $[x_1^i, x_2^{i+1}] \in E$ and $|N^+(x_1^i) \cap (S \cup \{x_1^{i+1}\})| = n - 3$ hold.

By (1), there is at least one edge from x_1^i to $P_i[x_3^i, y] \cup P_{i+1}[x_2^{i+1}, y]$ for $i = 1, 2$. By Lemma 1, this can be only the edge $[x_1^i, x_2^{i+1}]$, since $k_i \ge 2$. Hence (2) follows.

Dually, we get $[x_{k_i-1}^i, x_{k_{i+1}+1}^{i+1}] \in E$ for $i = 1, 2$ (see Figure 1).

(3) $S \subseteq N^+(x_1^i)$ for $i = 1, 2$.

Suppose, for instance, $S \not\subseteq N^+(x_1^1)$, say $s \in S - N^+(x_1^1)$. Then $S' := (S - \{s\}) \cup \{x_1^2\} \subseteq N^+(x_1^1)$ holds by (2). Set $D' := (D - [x_1^1, x]) - S'$. If $y' \notin \{x_{k_1}^1, x_{k_2}^2\}$, we can easily find openly disjoint x_1^1, x-paths Q_1 with $y' \in Q_1$ and Q_2 with $[y, s] \in E(Q_2)$ in D' such that $Q_1 \cap \{x_{k_1}^1, x_{k_2}^2\} = \emptyset$ and $|Q_2 \cap \{x_{k_1}^1, x_{k_2}^2\}| = 1$, contradicting Lemma 1. Hence, $y' \in \{x_{k_1}^1, x_{k_2}^2\}$ holds. Suppose there is a $z \neq y'$ in $N^-(s) \cap (P_1[x_2^1, y) \cup P_2[x_2^2, y))$. If $\{y', z\} \not\subseteq V(P_1)$ and $\{y', z\} \not\subseteq V(P_2)$, using $[x_1^1, x_1^2] \in E$, we get, obviously, openly disjoint x_1^1, x-paths Q_1, Q_2 in D' with $y \notin V(Q_1) \cup V(Q_2)$, contradicting Lemma 1. So $\{y', z\} \subseteq V(P_1)$ or $\{y', z\} \subseteq V(P_2)$ holds. Then we find, again, two openly disjoint x_1^1, x-paths in $D' - y$, namely in the former case (then $x_{k_1}^1 = y'$) the paths $x_1^1, P_2[x_2^2, x_{k_2-1}^2], y', x$ and $P_1[x_1^1, z], s, x$, and in the latter case (then $x_{k_2}^2 = y'$) the paths $x_1^1, P_2[x_2^2, z], s, x$ and $P_1[x_1^1, x_{k_1-1}^1], y', x$. (Note that $k_2 \ge 3$ in the former case, since in this case $k_1 \ge 3$, hence $x_2^1 \neq x_{k_1}^1$ holds, but there is only the edge $[x_1^2, x_2^1]$ from x_1^2 to $P_1[x_2^1, y]$ by (2)). This contradiction with Lemma 1 shows

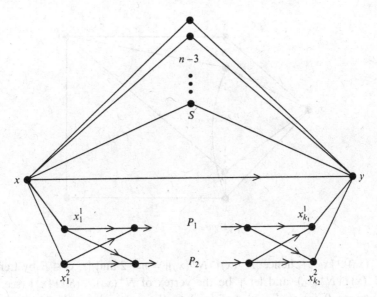

Figure 1

$N^-(s) = (S - \{s\}) \cup \{x, y, y', x_1^2\}$. Now it is easy to find in D n openly disjoint x, s-paths not containing $\{x_{k_1}^1, x_{k_2}^2\} - \{y'\}$. This contradiction to Lemma 1 proves (3).

We may assume $y' \in P_2$, hence $y' \in P_2[x_2^2, y)$. If $[x_2^2, x_1^1] \in E$ holds, by using (3) and (2), it is easy to find n openly disjoint x_1^2, x-paths in $D - y$, contradicting Lemma 1. So $[x_1^1, x_2^1] \in E$ is asymmetric. Hence $[x_1^2, x_1^1] \in E$ is symmetric, since $\triangle_a(D) = 1$, that is $[x_2^1, x_1^2] \in E$ holds. By using (3) and (2) again, we now easily find n openly disjoint x_1^1, x-paths in $D - y$. This contradiction to Lemma 1 proves (a).

(b) If $|N^+(x) \cap N^+(y)| \geq n-2$ holds, then $|N^+(x) \cap N^-(y)| \geq n-3$ follows, since $\triangle_a(D) = 1$, thus at most one of the edges $[y, z]$ for $z \in N^+(x) \cap N^+(y)$ is asymmetric. Hence (b) follows from (a). $\qquad\square$

Lemma 11. $|D| \geq n + 4$ *holds.*

Proof. If $\triangle_a(D) \geq 2$, we choose an x_0 with $d_a(x_0) \geq 2$ and get $|N^+(x_0) \cup N^-(x_0)| \geq n+2$, hence $|D| \geq n+4$, since $R(x_0) \neq \emptyset$. If $\triangle_a(D) = 1$, we choose an asymmetric edge $[x, y] \in E$ and get $|N^+(x) \cup N^+(y)| \geq n+3$ by Lemma 10 (b), hence $|D| \geq n+4$. $\qquad\square$

Lemma 12. *If $d_a(x_0) > 0$, then*

(a) $\bigcup_{v \in R^+(x_0)} N^+(v) \supseteq N^+(x_0)$ *and*
(b) $|R^+(x_0)| \geq 2$ *hold.*

Proof. Since $|N^+(v) \cap N^+(x_0)| = n-1$ for $v \in R^+(x_0)$, it suffices to prove (a). Suppose there is an $x \in N^+(x_0) - \bigcup_{v \in R^+(x_0)} N^+(v)$. There is a $v \in R^+(x_0)$, and $N^+(v) - N^+(x_0)$ has exactly one element, say, z. Consider any $y \in V - (N^+(x_0) \cup \{x_0, v, z\})$ and suppose $[y, x] \in E$. Then Lemma 8 (a) implies $y \in R^+(x_0)$, contradicting the choice of x. This contradiction shows

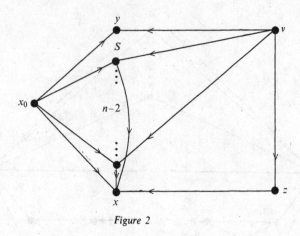

Figure 2

$N^-(x) \subseteq N^+(x_0) \cup \{x_0, z\}$, hence $|N^-(x) \cap N^+(x_0)| = n-2$ and $[z, x] \in E$ by Lemma 5 (a). Set $S := N^-(x) \cap N^+(x_0)$, and let y be the vertex of $N^+(x_0) - (S \cup \{x\})$ (see Figure 2). We apply Lemma 9 (a) to $[x_0, x] \in E$ and to the only x_0, x-path $P : x_0, y, \ldots, z, x$ in $(D - [x_0, x]) - S$, and get $S \subseteq N^-(z)$ and $[y, x_0] \in E, [x, z] \in E$. Since $[v, z] \in E$ and $S \subseteq N^+(v) \cap N^-(z)$, the path v, y, x_0, x, z must cover $D - S$ by Lemma 1. But this implies $|D| = n + 3$, contradicting Lemma 11. $\qquad\square$

5. Proof of Theorem 4

As in section 3, let $D = (V, E)$ be a minimal counterexample to Theorem 4, and $n := \kappa(D)$. We will show that D cannot have an asymmetric edge. First we prove that D must contain many symmetric edges.

(1) $\triangle_a(D) \leq 2$.

Suppose there is an $x_0 \in V$ with $d_a(x_0) \geq 3$. Since $|R^+(x_0)| \geq 2$ by Lemma 12 (b), Lemma 7 (b) implies $d_a(x_0) = 3$ and $|R^+(x_0)| = 2$, since $|N^+(v) \cap N^+(u) \cap N_a^+(x_0)| \geq d_a(x_0) - 2$ for $u \neq v$ in $R^+(x_0)$. Set $N_a^+ = \{x_1, x_2, x_3\}$ and choose $v \in R^-(x_0) \neq \emptyset$. Then $|N^-(v) \cap N_a^+(x_0)| \leq 1$, say, $[x_i, v] \notin E$ for $i = 1, 2$. Hence, Lemma 7 (a) implies $|N^-(x_i) \cap N^-(x_0)| = n-2$ and thus $|N^-(x_i) \cap R^+(x_0)| = 1$ by Lemma 7 (b) (or 12 (a)) for $i = 1, 2$. Therefore, we have $d^-(x_i; D(N_a^+(x_0))) = 0$ for $i = 1, 2$. Furthermore, $|N^-(x_i) \cap R^+(x_0)| = 1$ for $i = 1, 2$ implies $R^+(x_0) \subseteq N^-(x_3)$. Since $[x_3, x_1] \notin E$, and $[x_3, x_0] \notin E$, we get from the dual of Lemma 4 (b) that $|N^-(x_3) \cap N^-(x_0) \cap N^-(x_1)| \geq |N^-(x_0) \cap N^-(x_1)| - 1 = n-3$, so $N^-(x_3) \subseteq N^-(x_0) \cup \{x_0\} \cup R^+(x_0)$. Together, we have seen $||D(N_a^+(x_0))|| = 0$ and so $d(x_i'; F_{x_0}) \geq 3$ for $i \in \mathbb{N}_3$. But this implies $|R^+(x_0)| \geq 4$ by Lemma 6 (b) and (a). This contradiction proves (1). $\qquad\square$

In the next step we show the following.

(2) $\triangle_a(D) = 1$.

Suppose $\triangle_a(D) \geq 2$, hence $\triangle_a(D) = 2$ by (1). Let $x_0 \in V$ with $d_a(x_0) = 2$, say,

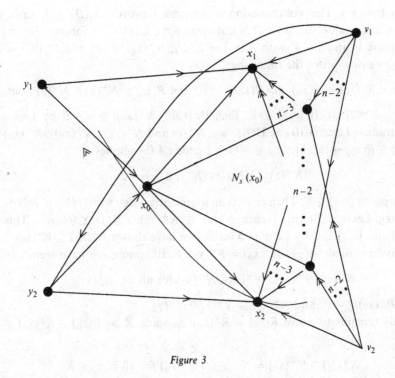

Figure 3

$N_a^+(x_0) = \{x, x_2\}$ and $N_a^-(x_0) = \{y_1, y_2\}$. By Lemma 12 (b), $|R^+(x_0)| \geq 2$ holds. Consider any $v \in R^-(x_0) \neq \emptyset$. Since $|N^-(v) \cap N_a^+(x_0)| \leq 1$, say, $[x_2, v] \notin E$. Then $|N^-(x_2) \cap N^-(x_0)| \geq n - 2$ by Lemma 4 (b). Since $|N^-(x_2) \cap N_s(x_0)| \geq n - 2$ is not possible by Lemma 9 (b), $N^-(x_2) \cap N_a^-(x_0) \neq \emptyset$ follows, say, $[y_2, x_2] \in E$. Since $|N^-(x_2) \cap (N^-(x_0) \cup \{x_0\})| \geq n - 1$ and $y_2 \in N^-(x_2)$, Lemma 8 (b) implies $|R^+(x_0)| = 2$, say, $R^+(x_0) = \{v_1, v_2\}$ and $|N^-(x_2) \cap R^+(x_0)| = 1$, say, $v_2 \in N^-(x_2)$. Hence $[v_1, x_2] \notin E$, so $N^+(v_1) \cap N^+(x_0) = N_s(x_0) \cup \{x_1\}$, and since $y_2 \in N^-(x_2)$, we get $[v_1, y_2] \in E$ from Lemma 8 (a). Hence $[v_1, y_1] \notin E$, and Lemma 8 (a) implies $[y_1, x_2] \notin E$, hence $|N^-(x_2) \cap N_s(x_0)| = n - 3$. Furthermore, $[v_1, y_1] \notin E$ implies $|N^+(y_1) \cap N^+(v_1) \cap N^+(x_0)| \geq n - 2$ by Lemma 4 (b). Since $N_s(x_0) \subseteq N^+(y_1)$ is not possible by the dual of Lemma 9 (b), the last inequality implies $[y_1, x_1] \in E$. (So far, we have got the edges without or with one arrow-head in Figure 3.) Since $[x_1, x_2] \notin E$ and $|N^-(x_2) \cap N^-(x_0)| = n - 2$, we get $|N^-(x_1) \cap N^-(x_2) \cap N^-(x_0)| \geq n - 3$ from Lemma 4 (b). Since $\{x_0, v_1, y_1\} \subseteq N^-(x_1)$, but $\{x_0, v_1, y_1\} \cap N^-(x_2) \cap N^-(x_0) = \emptyset$, we conclude $[v_2, x_1] \notin E$, hence $N_s(x_0) \subseteq N^+(v_2)$. So Lemma 8 (a) implies, as above, that $[v_2, y_1] \in E$ and $[y_2, x_1] \notin E$, hence $|N^-(x_1) \cap N_s(x_0)| = n - 3$.

Since $|R^-(x_0)| = 2$ holds by duality, there are only two $z \in F_{x_0}''$ with $d(z; F_{x_0}) = 1$ by Lemma 6 (a), and so Lemma 6 (b) implies $d(y_i''; F_{x_0}) \leq 2$ for $i = 1$ or $i = 2$. Suppose $d(y_1''; F_{x_0}) \leq 2$. This means $|N^-(y_1) \cap N^-(x_0)| \geq n - 2$. Now we will point out n openly disjoint v_1, x_1-paths that do not cover D. If z denotes the vertex of $N_s(x_0) - N^-(x_1)$, then $\kappa(v_1, x_1; D(\{v_1, x_1\} \cup (N_s(x_0) - \{z\}))) = n - 2$ and $\kappa(v_1, x_1; D(\{v_1, x_1, z, y_2, x_0, y_1\}) - [v_1, x_1])) = 2$ hold, since $N^-(y_1) \cap \{z, y_2\} \neq \emptyset$. So we get n openly disjoint v_1, x_1-paths that do not contain x_2 (and v_2). This contradiction to Lemma 1 proves $\triangle_a(D) \leq 1$, since the case $d(y_2''; F_{x_0}) \leq 2$ is analogous. Since D is not symmetric, $\triangle_a(D) = 1$ follows.

By (2), there is an $x_0 \in V$ with $d_a(x_0) = 1$, say, $N_a^+(x_0) = \{x\}$ and $N_a^-(x_0) = \{y\}$. For such an x_0, we now prove the following.

(3) $R^+(x_0) = R^-(x_0) = R(x_0)$, $\|D(R(x_0))\| = 0$, and $R(x_0) \subseteq N^+(x) \cap N^-(y)$ hold.

Choose $v \in R^+(x_0)$. If $[v, y] \notin E$, then $|N^+(y) \cap N^+(x_0)| \geq n - 2$ by Lemma 4 (b), which contradicts Lemma 10 (b). Hence $y \in N^+(v)$ and $N^+(v) \subseteq N^+(x_0) \cup N^-(x_0)$ follows. Consider $z \in R(x_0) - \{v\}$. Since $z \notin N^+(v)$, Lemma 4 (b) implies

$$|N^+(z) \cap N^+(x_0) \cap N^+(v)| \geq n - 2. \qquad (\alpha)$$

Let us suppose $[z, v] \in E$. Then $[z, v]$ is an asymmetric edge with $|N^+(z) \cap N^+(v)| \geq n - 2$, contradicting Lemma 10 (b). Hence $N^-(v) \subseteq N^+(x_0) \cup N^-(x_0)$ follows. This implies $v \in R^-(x_0)$ and $[x, v] \in E$ by Lemma 5 (c). So we have shown $R^+(x_0) \subseteq R^-(x_0)$. Since the other inclusion is dual, we get $R^+(x_0) = R^-(x_0)$. Furthermore, we have seen

$$N^+(v) \cup N^-(v) \subseteq N^+(x_0) \cup N^-(x_0) \text{ for all } v \in R^+(x_0), \qquad (\beta)$$

hence $\|D(R^+(x_0))\| = 0$, and $R^+(x_0) \subseteq N^+(x) \cap N^-(y)$.

So it only remains to prove $R(x_0) = R^+(x_0)$. Suppose $\overline{R} := R(x_0) - R^+(x_0) \neq \emptyset$. First we show

$$N^+(z_1) \cap N^+(x_0) = N^+(z_2) \cap N^+(x_0) \text{ for all } z_1, z_2 \in \overline{R} \qquad (\gamma)$$

We can choose $v_1 \neq v_2$ from $R^+(x_0)$, since $|R^+(x_0)| \geq 2$ by Lemma 12 (b). Since $y \in N^+(v_1) \cap N^+(v_2)$, Lemma 5 (c) implies $N^+(v_1) \cap N^+(x_0) \neq N^+(v_2) \cap N^+(x_0)$. Using (α) for v_1 and v_2, we conclude $N^+(z) \cap N^+(x_0) = N^+(v_1) \cap N^+(v_2) \cap N^+(x_0)$ for every $z \in \overline{R}$, since $|N^+(z) \cap N^+(x_0)| \leq n - 2$. This implies (γ).

Let us now consider $\overline{D} := D_{x_0} - R^+(x_0)$ and $\overline{F} := F_{x_0} - \bigcup_{v \in R^+(x_0)} \{v', v''\}$. Since $|N^+(z) \cap N^+(x_0)| \leq n - 2$ and $|N^-(z) \cap N^-(x_0)| \leq n - 2$ for $z \in \overline{R} = R(x_0) - R^-(x_0)$, using (β), we see $d(z; \overline{F}) \geq 2$ for all $z \in V(\overline{F}) - \{x', y''\}$. Since $\overline{F}$ is a forest with $|\overline{F}| \geq 4$, it must be an x', y''-path. There are $z_1, z_2 \in \overline{R}$, such that $[x', z_1''] \in E(\overline{F})$ and $[z_2', y''] \in E(\overline{F})$ hold. Using (α) and (γ), Lemma 10 (b) implies, that $D_{x_0}(\overline{R})$ is symmetric. This implies $z_1 = z_2$, since $d^+(z; \overline{D}) = d^-(z; \overline{D})(= 2)$ holds for every $z \in \overline{R}$. Then the undirected graph G with the property $\overleftrightarrow{G} = D_{x_0}(\overline{R})$ has one vertex of degree 1, namely z_1, and all the other vertices of degree 2. This contradiction shows $\overline{R} = \emptyset$, and (3) is proved.

By Lemma 12 (b) and (3), $r := |R(x_0)| = |R^+(x_0)| \geq 2$ holds. Define $S := N^-(y) \cap N^+(x_0)$ and $T := N^+(x_0) - S$. Since $R(x_0) \subseteq N^-(y)$ by (3), $N^-(y) = R(x_0) \cup S$, hence $|S| = n - r$ and $|T| = r$ hold. Since $d_a(y) = 1$, and thus $N^+(y) - N^-(y) = \{x_0\}$, we conclude $R(y) = T$. Since $\|D(T)\| = 0$ by (3), using Lemma 5 (c), we immediately have the following.

(4) For every $t \in T$, $|N^+(t) \cap (S \cup R(x_0))| \geq n - 1$ and for all $t_1 \neq t_2$ from T, $S \cup R(x_0) \subseteq N^+(t_1) \cup N^+(t_2)$ holds.

Now we will complete the proof of Theorem 4 by constructing for an edge $[u, v]$ in D n openly disjoint u, v-paths, that do not cover D. This contradicts Lemma 1.

First, we assume there is a $z_0 \in R(x_0)$ with $N^+(z_0) \not\supseteq S$. Then $[z_0, y] \in E$ by (3). Since $z_0 \in R^+(x_0)$ by (3), there is only one vertex in $S - N^+(z_0)$, say s_0, and $T \subseteq N^+(z_0)$ holds. Then $\kappa(z_0, y; D(\{z_0, y\} \cup (S - \{s_0\}))) = n - r$ holds. By (4), it is easy to find r disjoint edges in $[T, (R(x_0) - \{z_0\}) \cup \{s_0\}]_D$, since $r \geq 2$ holds. This implies $\kappa(z_0, y; D(\{y, s_0\} \cup R(x_0) \cup T) - [z_0, y]) = r$. Together, this gives n openly disjoint z_0, y-paths not containing x_0. Since the asymmetric edge $[y, x_0]$ was arbitrary, this contradiction establishes our next claim.

(5) For every asymmetric edge $[u, v] \in E$ and every $t \in R(v)$, $N^-(u) \cap N^+(v) \subseteq N^+(t)$ holds.

For every $z \in R(x_0)$, therefore, $S \subseteq N^+(z)$ and thus $\kappa(z, y; D(\{z, y\} \cup S)) = n - r + 1$ holds. Choose $z_0 \in R(x_0)$ and define $T' := N^+(z_0) \cap T$. Then $|T'| = r - 1$ holds. If $r \geq 3$, we get, as above, $r - 1$ disjoint edges in $[T', R(x_0) - \{z_0\}]_D$, hence $\kappa(z_0, y; D(\{y\} \cup R(x_0) \cup T') - [z_0, y]) = r - 1$, so there are n openly disjoint z_0, y-paths not containing x_0. This contradiction shows $r = 2$.

So we have $|S| = n - 2$, and Lemma 10 (b) shows that $N^+(y) \supseteq S$ is impossible. Hence there is an asymmetric edge $[s_0, y] \in [S, \{y\}]_D$. Since $d_a(y) = 1$, all the edges $[R(x_0), \{y\}]_D$ are symmetric, hence $R(x_0) \subseteq N^+(y)$ holds by (3). So we have $R(x_0) \subseteq N^-(s_0) \cap N^+(y)$, and (5) implies $R(x_0) \subseteq N^+(t)$ for every $t \in R(y) = T$. So we see $\kappa(z_0, y; D - x_0) = n$, since $\kappa(z_0, y; D(\{y\} \cup R(x_0) \cup T')) = 2$. This contradiction to Lemma 1 completes the proof of Theorem 4.

References

[1] Mader, W. (1972) Ecken vom Grad n in minimalen n-fach zusammenhängenden Graphen. *Archiv der Mathematik* **23**, 219–224.

[2] Mader, W. (1973) 1-Faktoren von Graphen. *Math. Ann.* **201**, 269–282.

[3] Mader, W. (1979) Zur Struktur minimal n-fach zusammenhängender Graphen. *Abh. Math. Sem. Universität Hamburg* **49**, 49–69.

[4] Mader, W. (1985) Minimal n-fach zusammenhängende Digraphen. *J. Combinatorial Theory* (B) **38**, 102–117.

[5] Mader, W. (in preparation) On vertices of outdegree n in n-minimal digraphs.

[6] Maurer, St. B. and Slater, P. J. (1978) On k-minimally n-edge-connected graphs. *Discrete Mathematics* **24**, 185–195.

On a Conjecture of Erdős and Čudakov

A. R. D. MATHIAS

Peterhouse, Cambridge

Let f be an arbitrary function from the set of positive integers to the set $\{-1,+1\}$, C a negative integer and D a positive one. We write $f \ll D$ if for all positive m and d, $\sum_{k=1}^{m} f(kd) < D$, and $C \ll f$ if for all positive m and d, $C < \sum_{k=1}^{m} f(kd)$. In this terminology, a question popularised by Erdős runs:

Are there f, C, D with $C \ll f \ll D$?

The purpose of this note is to prove that for $D = 2$, no such f and C exist:

Proposition. *Let* $f: \{1, 2, 3, \ldots\} \rightarrow \{-1, +1\}$ *be such that for all* m, d, $\sum_{k=1}^{m} f(kd) \leqslant 1$. *Then for all* $C < 0$, *there are* m, d *with*

$$\sum_{k=1}^{m} f(kd) < C.$$

Proof. For each $x \geqslant 1$, $f(x) + f(2x) = -2, 0$ or 2, but the last possibility is excluded by the condition on f, so $f(x) + f(2x) \leqslant 0$. Thus $\phi(n) = \sum_{x=1}^{n} (f(x) + f(2x))$ is a weakly decreasing function of n. If for some n, $\phi(n) < 2C$, then either $\sum_{x=1}^{n} f(x) < C$ or $\sum_{x=1}^{n} f(2x) < C$. We may therefore suppose that for all n, $\phi(n) \geqslant 2C$, and hence that $\phi(n)$ is eventually constant. Thus there is a d such that for all $x \geqslant d$, $f(x) + f(2x) = 0$, and because f takes the value -1 infinitely often, we may, without loss of generality, assume that $f(d) = -1$. We now consider only multiples of d, and since the original condition on f, together with the fact that for all k, $f(2kd) = -f(kd)$, imply that for each $m \geqslant 1$ and each multiple d' of d, $-1 \leqslant \sum_{k=1}^{m} f(kd') \leqslant +1$, we find that the values of $f(kd)$ and $f(2kd)$ for $k = 1, \ldots, 12$ must be as in the following table; but then $\sum_{k=1}^{4} f(3kd) = +2$, a contradiction.

k	1	2	3	4	5	6	7	8	9	10	11	12
$f(kd)$	−	+	+	−	+	−	−	+	+	−	−	+
$f(2kd)$	+	−	−	+	−	+	+	−	−	+	+	−

Remark 1. By compactness the problem may be stated in terms of finite sequences, as is done in [2] Problème 49, a negative answer to the above question being equivalent to the assertion that for all $D > 0$ there is an $N > 0$ such that for all $f : \{1, 2, \ldots, N\} \rightarrow \{-1, 1\}$

there are $m \geqslant 1$, $d \geqslant 1$ with $md \leqslant N$ and $\left| \sum_{k=1}^{m} f(kd) \right| \geqslant D$. In [2] reference is made to the paper [1] which studies related questions.

Remark 2. Repeated attempts to improve the proposition to the case $D = 3$ have failed. Indeed the following is still open:

$$\text{Is there an } f \text{ with } -3 \ll f \ll +3 \text{ ?}$$

References

[1] Čudakov, N.G. (1956) Theory of the characters of number semigroups, *Journal Ind. Math. Soc.* **20** 11–15.

[2] Erdős, P. (1963) *Quelques Problèmes de la Théorie des Nombres*. Monographies de l'Enseignement mathématique, No 6, Génève.

A Random Recolouring Method for Graphs and Hypergraphs

COLIN McDIARMID

Department of Statistics, University of Oxford

We consider a simple randomised algorithm that seeks a weak 2-colouring of a hypergraph H; that is, it tries to 2-colour the points of H so that no edge is monochromatic. If H has a particular well-behaved form of such a colouring, then the method is successful within expected number of iterations $O(n^3)$ when H has n points. In particular, when applied to a graph G with n nodes and chromatic number 3, the method yields a 2-colouring of the vertices such that no triangle is monochromatic in expected time $O(n^4)$.

A *hypergraph* H on a set of *points* V is simply a collection of subsets E of V, the *edges* of H. A *d-graph* is a hypergraph in which each edge has size d. A *weak 2-colouring* of a hypergraph is a partition of the points into two 'colour' sets A and B such that each edge E meets both A and B. Deciding if a 3-graph has a weak 2-colouring is NP-complete [6, 4].

The following simple randomised recolouring method attempts to find a weak 2-colouring of a hypergraph H. It is assumed that we have a subroutine SEEK; which on input of a 2-colouring of the points outputs a monochromatic edge if there is one, and otherwise reports that there are none.

RECOLOUR

    start with an arbitrary 2-colouring of the points

    while SEEK returns a monochromatic edge E

        pick a random point in E and change its colour.

If H has a weak 2-colouring, the method will ultimately find one with probability 1: indeed if H has n points and maximum edge size d, at any stage the probability of success within the next $n/2$ steps is at least $d^{-n/2}$. What is of interest is the expected number of iterations needed. A *fair partial 2-colouring* of a hypergraph is a pair of disjoint 'colour' sets A and B of points such that for each edge E we have

$$|E \cap A| = |E \cap B| > 0.$$

Theorem 1.

Let H be a hypergraph with maximal edge size d, and suppose that H has a fair partial 2-colouring that colours m points. Then RECOLOUR returns a weak 2-colouring with expected number of iterations at most $\frac{1}{8}dm^2$.

Proof. By assumption there is a (fixed) pair of disjoint subsets A_0, B_0 of points such that $|A_0| + |B_0| = m$ and $|E \cap A_0| = |E \cap B_0| > 0$ for each edge E of H. Suppose that at some stage we have a 2-colouring f of H, that is a partition of the points into two sets A and B. Define the *agreement number* $N(f)$ of f to be $|A \cap A_0| + |B \cap B_0|$.

Suppose that SEEK returns a monochromatic edge E. Note that we must have $0 < N(f) < m$. Let $\hat{f}$ be the random colouring obtained from f by changing the colour of a point picked uniformly from E, as in the algorithm RECOLOUR. Then, if we let $k = |E \cap A_0| = |E \cap B_0|$,

$$N(\hat{f}) - N(f) = \begin{cases} -1 & \text{with probability } k/|E| \\ +1 & \text{with probability } k/|E| \\ 0 & \text{with probability } 1 - 2k/|E|. \end{cases}$$

Denote the colourings produced by the algorithm by $f_0, f_1, \ldots$. Then we see that $N(f_0), N(f_1), \ldots$ is a symmetric random walk with 'holding', up to a stopping time. Further, this stopping time is no later than when such a walk would be absorbed at 0 or m. But the expected number of unit steps to absorption here is at most $m^2/4$, and each expected holding time is at most $d/2$, so the expected number of iterations is at most $\frac{1}{8}dm^2$ (see for example [3]). $\qquad\square$

Comments

1   The algorithm RECOLOUR is similar in spirit to the randomised method proposed by Petford and Welsh [7] for seeking a proper 3-colouring of a graph (see also [1, 2, 8]).

2   A *strong k-colouring* of a hypergraph is a colouring of the points with k colours such that in each edge the points all receive distinct colours. Let $d \geq 2$, and let H be a d-graph with n points and with a strong d-colouring. Then H has a fair partial 2-colouring with at most $2n/d$ points coloured, and so RECOLOUR will yield a weak 2-colouring in expected number of iterations at most $n^2/2d$. We consider the particular cases $d = 2$ and $d = 3$ below.

3   Let G be a connected bipartite graph with n nodes. It is of course easy to find a proper 2-colouring of G. However, suppose that we do apply RECOLOUR to G to obtain a proper 2-colouring, starting from a random 2-colouring. Then the expected number of iterations is *exactly* $\frac{1}{4}n(n-1)$.

4   Let G be a graph with n nodes that has a proper 3-colouring. Then we may use

RECOLOUR to obtain a 2-colouring of the nodes such that no triangle is monochromatic, in expected number of iterations at most $n^2/6$ and thus in expected time $O(n^4)$. Of course, if we could find a proper 4-colouring of G, we could amalgamate colours to obtain a 2-colouring without monochromatic triangles. However, it has recently been shown [5] that it is NP-hard to find a proper 4-colouring in a graph with chromatic number 3.

5 We could allow SEEK to be an oracle or adversary, as long as future coin tosses cannot be seen.

6 (added in proof) For a deterministic approach to problems as considered above, see McDiarmid, C. (1993). *On 2-colouring a 3-colourable graph to avoid monochromatic triangles* (manuscript).

Acknowledgements

I would like to thank Noga Alon and Nathan Linial for helpful discussions.

References

[1] Donnelly, P. and Welsh, D. J. A. (1983) Finite particle systems and infection models. *Math. Proc. Cambridge Philosophical Society* **94** 167–182.

[2] Donnelly, P. and Welsh, D. J. A. (1984) The antivoter problem: random 2-colourings of graphs. In: Bollobás, B. (ed.) *Graph Theory and Combinatorics*, Proc. Conference in honour of Paul Erdős, Cambridge, 1983, Academic Press, 133–144.

[3] Feller, W. (1968) *An Introduction to Probability Theory and its Applications*, Volume 1, 3rd edition, Wiley, New York.

[4] Garey, M. R. and Johnson, D. S. (1979) *Computers and Intractability*, W.H. Freeman & Co, San Francisco.

[5] Khanna, S., Linial, N. and Safra, S. (1993) *On the hardness of approximating the chromatic number*. (Manuscript.)

[6] Lovász, L. (1973) Coverings and colorings of hypergraphs. *Proc. 4th S.E. Conference on Combinatorics, Graph Theory and Computing*, Utilitas Mathematica Publishing, Winnipeg, 3–12.

[7] Petford, A.D. and Welsh, D. J. A. (1989) A randomised 3-colouring algorithm. *Discrete Mathematics* **74** 253–261.

[8] J. Žerovnik (1987) A randomised heuristical algorithm for graph colouring. *Proc. 8th Yugoslav Seminar on Graph Theory*, Novi Sad.

Obstructions for the Disk and the Cylinder Embedding Extension Problems

BOJAN MOHAR[†]

Department of Mathematics, University of Ljubljana, Jadranska 19, 61111 Ljubljana, Slovenia
email: bojan.mohar@uni-lj.si

Let S be a closed surface with boundary ∂S and let G be a graph. Let $K \subseteq G$ be a subgraph embedded in S such that $\partial S \subseteq K$. An *embedding extension* of K to G is an embedding of G in S that coincides on K with the given embedding of K. Minimal obstructions for the existence of embedding extensions are classified in cases when S is the disk or the cylinder. Linear time algorithms are presented that either find an embedding extension, or return an obstruction to the existence of extensions. These results are to be used as the corner stones in the design of linear time algorithms for the embeddability of graphs in an arbitrary surface and for solving more general embedding extension problems.

1. Introduction

Let K be a subgraph of G. A *K-component* or a *K-bridge* in G is a subgraph of G that is either an edge $e \in E(G) \backslash E(K)$ (together with its endpoints) that has both endpoints in K, or it is a connected component of $G - V(K)$ together with all edges (and their endpoints) between this component and K. Each edge of a K-component R having an endpoint in K is a *foot* of R. The vertices of $R \cap K$ are the *vertices of attachment* of R. A vertex of K of degree in K different from 2 is a *main vertex* of K. For convenience, if a connected component of K is a cycle, we choose an arbitrary vertex of it and declare it to be a main vertex of K as well. A *branch* of K is any path in K whose endpoints are main vertices and such that no internal vertex on this path is a main vertex. If a K-component is attached at a single branch of K, it is said to be *local*. The number of branches of K is called the *branch size* of K.

Let $K \subseteq G$, and suppose that we are given an embedding of K into a (closed) surface Σ. The *embedding extension problem* asks whether it is possible to extend the given embedding of K to an embedding of G, and any such embedding is said to be an *embedding extension*

† Supported in part by the Ministry of Science and Technology of Slovenia, Research Project P1–0210–101–92.

of K to G. Let Σ be the (closed) disk or the cylinder. Let K be embedded in Σ such that $\partial \Sigma \subseteq K$. An *obstruction* for embedding extensions is a subgraph Ω of $G - E(K)$ such that the embedding of K cannot be extended to $K \cup \Omega$. The obstruction is *small* if $K \cup \Omega$ has bounded branch size. If Ω is small, it is easy to verify that no embedding extension to $K \cup \Omega$ exists, and hence that Ω is a good verifier that there are no embedding extensions of K to G as well. In this paper, minimal obstructions for embedding extension problems in the disk and the cylinder are classified for several 'canonical' choices of K. Although much work has been done on 'embedding obstructions', our results seem to be new, apart from the case of the disk (*cf.* [20]; see also Section 3) or the case when $K = \emptyset$ and Σ is a closed surface [19]. It is interesting that minimal obstructions are not always small. They can be arbitrarily large but their structure is easily described. We also present linear time algorithms to either find an embedding extension, or return a (minimal) obstruction to the existence of extensions.

The basic results of this paper (Theorems 3.1, 4.3, 5.3, and 6.2) are to be used as the basic building stones in the design of linear time algorithms for embedding graphs in general surfaces [10, 12, 16, 11]. Moreover, we are able to solve even more general embedding extension problems in linear time.

Robertson and Seymour (*cf.* [19] and the graph minors papers preceding it) proved a Kuratowski theorem for general surfaces. In our further project [17], results of this paper are used to obtain a reasonably short proof of Robertson and Seymour's result. It is worth mentioning that all our results are direct and constructive, in the tradition of Archdeacon and Huneke [1]. (Recently, Seymour [23] also obtained a constructive proof by using graph minors and tree-width techniques.)

Embeddings in orientable surfaces can be described combinatorially [6] by specifying a *rotation system*: for each vertex v of the graph G we have the cyclic permutation π_v of its neighbors, representing their circular order around v on the surface. In order to make a clear presentation of our algorithm, we have decided to use this description only implicitly. Whenever we say that we have an embedding, we mean such a combinatorial description. Whenever used, it is easy to see how one can combine the embeddings of some parts of the graph described this way into the embedding of larger species.

In discussing the time complexity of our algorithms, we assume a random-access machine (RAM) model with unit cost for basic operations. This model was introduced by Cook and Reckhow [4]. More precisely, our model is the *unit-cost* RAM where operations on integers, whose values are $O(n)$, need only constant time (n is the order of the given graph).

2. Basic definitions

Let G and K be graphs (both subgraphs of some graph H). Then we denote by $G - K$ the graph obtained from G by deleting all vertices of $G \cap K$ and all their incident edges. If $F \subseteq E(G)$, then $G - F$ denotes the graph obtained from G by deleting all edges in F. If K and L are subgraphs of G, then we say that a path P in G *joins* K and L if P is internally disjoint from $K \cup L$ and one of its ends is in K and the other end is in L. Moreover, if an end of P is in both K and L, then P is a trivial path.

A *block* or *2-connected component* of a graph G is either an isolated vertex, a loop, a bond of G, or a maximal 2-connected subgraph of G. One can also define the concept of 3-connected components. A graph G is said to be *k-separable* if it can be written as a union $G = H \cup K$ of (non-empty) edge-disjoint graphs H and K that have exactly k vertices in common, and each of them contains at least k edges. Such a pair $\{H, K\}$ is called a *k-separation* of G. A graph is *nonseparable* if it has no 0- or 1-separations. Let G be a nonseparable graph and let $\{H, K\}$ be a 2-separation of G. Let x, y denote the vertices of $V(H) \cap V(K)$. The 2-separation is *elementary* if either $H - \{x, y\}$ or $K - \{x, y\}$ is non-empty and connected, and either H or K is nonseparable. It turns out [26] that nonseparable graphs without elementary 2-separations are either 3-connected graphs, cycles C_n ($n \geq 3$), $p \geq 1$ parallel edges, K_1, or a loop. Assume now that the 2-separation $\{H, K\}$ of G is elementary. Denote by H' and K' the graphs obtained from H and K, respectively, by adding to each of them a new edge between the vertices of $H \cap K$. The added edges are called *virtual edges*. It is easy to verify that H' and K' are both nonseparable, and we may repeat the process on their elementary 2-separations (if there are any) until no further elementary 2-separations are possible. As mentioned above, the obtained graphs are either 3-connected, cycles, edges in parallel, or rather small. Each of the graphs obtained this way is called a *3-connected component* of G. It was shown by MacLane [14] (*cf.* also [26]) that the set of 3-connected components of the graph is uniquely determined, although different choices of the 2-separations may have been used during the process of constructing them. Every 3-connected component consists of several edges of G and several virtual edges. It is obvious by construction that each edge of G belongs to exactly one 3-connected component, and each virtual edge has a corresponding virtual edge in some other 3-connected component. The 3-connected components of G may be viewed as subgraphs of G, where each virtual edge corresponds to a path in G. These subgraphs are positioned in G in a tree-like way [26]. We also speak of *3-connected components* when the graph is separable. In that case we define them to be the 3-connected components of the blocks of the graph.

A linear time algorithm for obtaining the 3-connected components of a graph was devised by Hopcroft and Tarjan [7].

There are very efficient (linear time) algorithms that for a given graph determine whether the graph is planar or not. The first such algorithm was obtained by Hopcroft and Tarjan [8] back in 1974. There are several other linear time planarity algorithms (Booth and Lueker [2], de Fraysseix and Rosenstiehl [5], Williamson [27, 28]). Extensions of the original algorithms also produce an embedding (rotation system) whenever the given graph is found to be planar [3], or find a small obstruction – a subgraph homeomorphic to K_5 or $K_{3,3}$ – if the graph is non-planar [27, 28] (see also [13]). The subgraph homeomorphic to K_5 or $K_{3,3}$ is called a *Kuratowski subgraph* of G.

Lemma 2.1. *There is a linear time algorithm that, given a graph G, either exhibits an embedding of G in the plane, or finds a Kuratowski subgraph of G.*

We will refer to the algorithm mentioned in Lemma 2.1 as *testing for planarity*. This

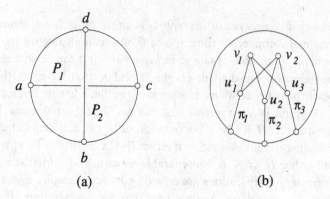

Figure 1   Disjoint crossing paths and a tripod

procedure not only checks the planarity of the given graph but also takes care of exhibiting an embedding, or finding a Kuratowski subgraph.

Let C be a cycle of a graph G. Two C-components B_1 and B_2 *overlap* if either B_1 and B_2 have three vertices of attachment in common, or there are four distinct vertices a, b, c, d that appear in this order on C and such that a and c are vertices of attachment of B_1, and b, d are vertices of attachment of B_2. In the latter case, B_1 and B_2 contain disjoint paths P_1 and P_2 whose ends a, c and b, d, respectively, interlace on C. Such paths will be referred to as *disjoint crossing paths* – see Figure 1(a). We will need another type of subgraph of G that is attached to C. A *tripod* is a subgraph T of G that consists of two main vertices v_1, v_2 of degree 3, whose branches join them with the same triple of vertices u_1, u_2, u_3, together with three vertex disjoint paths π_1, π_2, π_3 joining u_1, u_2, and u_3 with C. Moreover, T intersects C only at the ends of π_1, π_2, and π_3 – see Figure 1(b). One or more of the paths π_i are allowed to be trivial, in which case $u_i \in C$. If all three paths π_1, π_2, and π_3 are trivial (just vertices), then the tripod is said to be *degenerate*. We use the same name for attachments of the tripod in the case when the corresponding path is trivial.

3. The disk

Let D be the closed unit disk in the euclidean plane. Given a graph G and a cycle C in G, we would like to find an embedding of G in D so that C is embedded on ∂D. Of course, this is a case of the embedding extension problem for which an easy answer is at hand. First, we construct the *auxiliary graph* $\tilde{G} = \text{Aux}(G, C)$, which is obtained from G by adding a new vertex v (called the *auxiliary vertex*) and joining it to all vertices on C. It is easy to see that an embedding extension of C on ∂D to G exists if and only if the auxiliary graph $\tilde{G}$ is planar. Its plane embedding also determines an embedding extension. In the case of non-planar $\tilde{G}$, a Kuratowski subgraph $\tilde{K}$ of $\tilde{G}$ determines the subgraph $K = \tilde{K} - v$ of G that is an obstruction for the embedding extension in the disk. Although $K \cup C$ can have arbitrarily large branch size, it can easily be modified to an obstruction Ω for which $\Omega \cup C$ has bounded branch size. Our answer seems to solve the question reasonably well. However, there is a better solution. Namely, it is known that when G

is 3-connected, a pair (G, C) for which there is no disk embedding extension necessarily contains either a pair of disjoint crossing paths or a tripod. This simple but useful result was 'in the air' for quite some time. It seems to have appeared for the first time in a paper by Jung [9] in a slightly weaker version. It also appeared in a paper by Seymour [21] (with the complete proof in [22]), Shiloach [24] and Thomassen [25], all in relation to the non-existence of two disjoint paths between specified vertices. This result recently appeared in a more explicit form in Robertson and Seymour's work on graph minors [20]. In this section we will prove a slightly more specific result by also taking care of the case when G is not 3-connected. Moreover, we will show how to obtain such an obstruction in linear time.

Theorem 3.1. *Let G, C, D be as above. Let $\tilde{G} = \mathrm{Aux}(G, C)$. There is a linear time algorithm that either finds an embedding of G in D with C on ∂D, or returns a small obstruction Ω. In the latter case, Ω is one of the following types of subgraphs of $G - E(C)$:*

(a) *a pair of disjoint crossing paths,*
(b) *a tripod, or*
(c) *a Kuratowski subgraph contained in a 3-connected component of $\tilde{G}$ distinct from the 3-connected component of $\tilde{G}$ containing C.*

Before giving the proof of Theorem 3.1 we state a lemma whose easy proof is left to the reader.

Lemma 3.2. *Let H be a graph with a cycle C and let e be an edge of H that is not a chord of C. If the edge-contracted graph H/e contains a tripod or a pair of disjoint crossing paths with respect to C (or C/e if $e \in E(C)$), then H also contains a tripod or a pair of disjoint crossing paths.*

Proof of Theorem 3.1. By testing $\tilde{G}$ for planarity, we can check if $\tilde{G}$ is planar. If yes, we also get a required disk embedding of G.

Suppose now that $\tilde{G}$ is non-planar. Determine the 3-connected components of $\tilde{G}$, for example, by using the linear time algorithm of Hopcroft and Tarjan [7]. Note that C (with some of its edges having possibly become virtual) and the auxiliary vertex are in the same 3-connected component. Denote this 3-connected component by R. If R is planar, then the obstruction to the planarity of $\tilde{Q}$ lies in one of the other 3-connected components. We get (c) in one of the planarity tests. From now on we may thus assume that R is non-planar. Let us show how to get disjoint crossing paths or a tripod.

Let K be a Kuratowski subgraph of R found by a planarity test on R. Denote by H the graph $(K \cup C) - w$, where w is the auxiliary vertex of $\tilde{G}$. Note that the branch size of H is not necessarily small. We will first try to find disjoint crossing paths or a tripod in H. Consider the C-components in H. First of all we check if two of them overlap. In order to perform efficient checking, we split the bridges into two classes: the bridges that contain main vertices of K (possibly as their vertices of attachment) are *main bridges* of C in H, and the remaining bridges are called *chords*, since they are just paths joining two distinct vertices of C. There are at most 20 main bridges. To be more efficient in the following, we can temporarily replace every bridge by a single vertex joined to all of its vertices of attachment on C.

Step 1: Are there two main bridges that overlap?
If yes, we either have a (degenerate) tripod or disjoint crossing paths. If not, proceed with the next step. Since only the main bridges have to be considered, this step can be carried out in constant time.

Step 2: Is there a main bridge that overlaps with a chord?
If yes, we have disjoint crossing paths. Otherwise continue with Step 3. This question can easily be answered in linear time. Observe that the number of candidates for one of the disjoint crossing paths in the main bridges is small.

Step 3: Are there two overlapping chords?
Note that no chord contains a main vertex. Thus, at most two chords are attached at the same vertex. A simple way to find overlapping chords is to start building a stack by traversing C once around, starting at an arbitrary vertex of C. If there are two chords at the same point on C, we first consider chords that have already been met during the traversal. If both chords are not new, we give priority to the one that is on the top of the stack (if none is on the top, their order is not important). Then we process the new chords. If both are new, we give priority to the one whose other attachment is further away in the direction of the traversal. Every new chord met during the traversal is put to the top of the stack. Meeting the chord for the second time, we check if the top element in the stack is the same chord. If yes, the chord is removed from the stack and the traversal is continued. If not, then we have another chord at the top. It is easy to see that these two chords overlap and they give rise to disjoint crossing paths.

We may now assume that no two distinct bridges of C in H overlap. This means that the obstruction is in one of the main bridges. Such a bridge B can be discovered in $O(1)$ time since the number of main bridges is small, and each of them has small branch size. We will also assume that B is minimal in the sense that for every branch e of B, the graph $(B - e) \cup C$ is planar (if not, we can remove e and repeat the above procedure in order to get (a), (b), or a new bridge B with a smaller number of branches). Note that $B \cup C$ is non-planar and has small branch size. Therefore B can be used as a legitimate obstruction in some applications. However, our goal is to show more: we want a tripod or disjoint crossing paths.

Since $B \cup C$ has constant branch size, it is easy to find a tripod or disjoint crossing paths in B whenever B contains one of them. Assume from now on that this is not the case. We will prove that under this assumption, B has at most two vertices of attachment on C. Let K' be a Kuratowski subgraph of $B \cup C$. By the minimality property of B, K' contains the whole of B plus, possibly, some parts on C. If two main vertices of K' lie on C, they are either non-adjacent in K', or connected by a branch that is contained in C. Therefore it is easy to see that at most three main vertices of K' lie on C (the case of four vertices of $K_{3,3}$ forming a cycle on C is the only possibility, but they give rise to disjoint crossing paths). Similarly, we can exclude three main vertices of K' being on C. (In the case analysis for the last claim, an application of Lemma 3.2, using the 'contraction' argument as also used below, makes the number of cases much smaller.)

Now, if B has a vertex of attachment on C that is not a main vertex of K', we may contract the corresponding branch e of $B \cup C$ and obtain the non-planar graph $(B/e) \cup C = (B \cup C)/e$ with more main vertices of K'/e ($\approx K'$) on C. Inductively, we have a tripod or disjoint crossing paths. Also, by Lemma 3.2, $B \cup C$ contains a tripod or disjoint crossing paths.

Suppose now that B has $t \leq 2$ vertices of attachment on C, and recall that we know how to get in linear time a tripod or disjoint crossing paths in the case of three or more vertices of attachment. Let $\overline{B}$ be the C-component in R that contains B as a subgraph. Since R is 3-connected, there are disjoint paths $e_1, \ldots, e_{3-t}$ in $\overline{B}$, starting at $C - B$ and terminating in $B - C$, whose only vertices in B are their endpoints. Such paths can be found in linear time by applying, for example, the appropriate modification of the augmented paths method used to test connectivity of graphs [18, Chapter 9]. The connectivity test should be applied on the graph $\overline{B} \cup C$ with the t attachments of B removed. Since $t \leq 2$, the graph $H = \overline{B} \cup C \cup e_1 \cup \cdots \cup e_{3-t}$ contains a copy of K' that does not contain the endpoints of $e_1, \ldots, e_{3-t}$ on C (this fact is really needed only when $t = 2$). Therefore, the graph $H' = H/(e_1 \cup \cdots \cup e_{3-t})$ also contains a copy of K'. Note that the only 3-separations in Kuratowski subgraphs intersect at the three vertices of the same color class in $K_{3,3}$. Therefore H' is equal to C plus a single bridge (plus, possibly, a branch between two vertices on C that can be replaced by a segment on C), except when the three vertices of $K' \subseteq H'$ that lie on C are the three vertices of the same color class of $K_{3,3}$. In the latter case, we clearly have a tripod in H'. In the other cases, we can apply the results from above, since H' is of appropriate form and has three attachments on C, we can find a tripod in it. By Lemma 3.2, we have a tripod or disjoint crossing paths in H. $\qquad\square$

4. The cylinder

In this section we will consider the embedding extension problems in the cylinder. Let C_1 and C_2 be disjoint cycles in the graph G, and for an integer $k \geq 0$, let $P_1, P_2, \ldots, P_k$ be vertex disjoint paths in G joining C_1 and C_2 (with no interior points on $C_1 \cup C_2$). Suppose, moreover, that the endpoints of the paths P_i appear on both cycles C_1, C_2 in the same (cyclic) order. The embedding extension problem in the cylinder with respect to the subgraph $K = C_1 \cup C_2 \cup P_1 \cup \cdots \cup P_k$, where K is embedded in such a way that C_1 and C_2 cover the boundary, will be referred to as the *k-prism embedding extension problem*. Note that when $k \leq 2$, we have two essentially different problems depending on the embeddings of $C_1 \cup C_2$ on the boundary of the cylinder.

In testing for the k-prism embedding extension of K to G, we make use of the *auxiliary graph* $\tilde{G}$, which is obtained from G by adding two new *auxiliary vertices* v_1 and v_2, and for $i = 1, 2$, joining v_i to all vertices of C_i. If $k \geq 3$, an embedding extension of K to G exists if and only if $\tilde{G}$ is planar, and a planar embedding of $\tilde{G}$ determines a cylinder extension. Something similar holds also when $k \leq 2$. More details will be provided later. Note that in the cylinder case, the auxiliary graph contains two auxiliary vertices, while the auxiliary graph for disk embeddability has just one. Although we are using the same name and

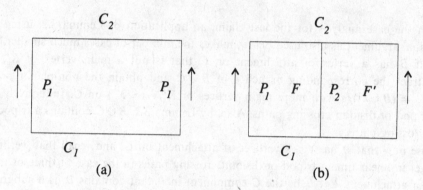

Figure 2   The 1- and 2-prism embedding extension problem

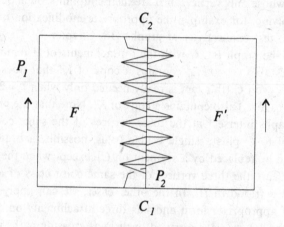

Figure 3   A large obstruction using local bridges

notation, there will be no confusion, since it will always be clear from the context which case is applied.

If there are local bridges attached to one or more of the paths P_i, we may get arbitrarily long chains of successively overlapping local bridges on P_i (see Figure 3). There are examples where, after eliminating any of the branches, there exists an embedding extension. So we can have arbitrarily large minimal obstructions. On the other hand, in applications using the obstructions, certain connectivity conditions on the involved graphs can be achieved. In that case, the local bridges can be eliminated efficiently (in linear time: see [15] and [10] for more details). Since we are usually allowed to change the paths P_i during the pre-processing time, it makes sense to assume that there are no local bridges attached to any of the paths $P_1, \ldots, P_k$.

Obstructions for the k-prism embedding extension problem with $k \geq 3$ are easy to find – they are not much more complicated than the closed disk obstructions classified in theorem 3.1. Besides the disjoint crossing paths and the tripods, we get another type of obstruction. A *dipod* (with respect to the cycle C) is a subgraph H of G consisting of distinct vertices $a, b, c, d \in V(C)$ that appear on C in that order, distinct vertices v, u where $v \notin V(C)$, and $u \notin V(C)$ unless $u = b$, and branches va, vc, vu, ub, and ud (Figure 4).

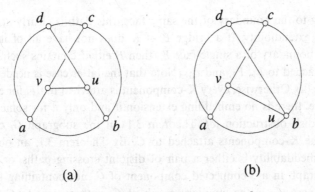

(a)   (b)

Figure 4 A dipod

The branches are internally disjoint from C. If $u = b$, the branch ub vanishes (see Figure 4(b)). We also define a *triad* (with respect to a subgraph K of G) as a subgraph of G consisting of a vertex $x \notin V(K)$ and three paths joining x with K that are pairwise disjoint except at their common end x.

For $K = C_1 \cup C_2 \cup P_1 \cup \cdots \cup P_k$ embedded in the cylinder with C_1 and C_2 on its boundary, let $F_1, \ldots, F_k$ be the faces of K. We suppose that for $i = 1, 2, \ldots, k$, ∂F_i contains P_i and P_{i+1} (index modulo k).

Theorem 4.1. *Let $K \subseteq G$ be the subgraph of G for the k-prism embedding extension problem, where $k \geq 3$. Suppose that no K-component of G is attached to just one of the paths P_i of K, $1 \leq i \leq k$. Then there is no embedding extension of K to G if and only if $G - E(K)$ contains a subgraph Ω of one of the following types:*

(a) *A path joining two vertices of K that do not lie on the boundary of a common face of K, or (with $k = 3$) a triad attached to P_1, P_2, and to P_3.*

(b) *A tripod attached to the boundary of one of the faces F_i. Not all three attachments of the tripod lie on just one of the paths P_i, P_{i+1} on ∂F_i.*

(c) *A pair of disjoint crossing paths with respect to the boundary of one of the faces F_i. None of the two paths is attached to just one of the paths P_i, P_{i+1} on ∂F_i.*

(d) *A dipod with respect to the boundary cycle of some F_i. In this case, the vertices a, c, and d from the definition of the dipods all lie on one of the paths P_i, or P_{i+1}, while $b \in \partial F_i$ does not lie on the same path.*

(e) *A Kuratowski subgraph contained in a 3-connected component L of the auxiliary graph $\tilde{G}$ of G, where L is such that it does not contain auxiliary vertices of $\tilde{G}$.*

There is a linear time algorithm that either finds an embedding extension of K to G, or returns an obstruction Ω which fits one of the above cases.

Proof. We can find embedding extensions, if they exist at all, by testing the auxiliary graph $\tilde{G}$ for planarity. Suppose now that embedding extensions do not exist. Our goal is to show how to find the required obstruction Ω.

Since $k \geq 3$ and there are no local bridges at the paths P_j, every K-component is embeddable in at most one of the faces F_i. If one of the bridges contains a path whose ends do not belong to the boundary of the same face, this path is clearly an obstruction for the embedding extendibility. If a bridge B of K does not have all of its vertices of attachment on the boundary of a single face F_i, then B either contains such a path, or it contains a triad attached to P_1, P_2, and P_3. (Note that the latter case is needed only when $k = 3$.) So, we have (a). Otherwise, every K-component is attached to ∂F_i for exactly one i, $1 \le i \le k$. Therefore, there is no embedding extension if and only if for some i, $1 \le i \le k$, we have a closed disk obstruction (*cf.* Theorem 3.1) in the subgraph G_i consisting of $C = \partial F_i$ and all the K-components attached to C. By Theorem 3.1, an obstruction to the (G_i, C) disk embeddability is either a pair of disjoint crossing paths, or a tripod, or a Kuratowski subgraph in a 3-connected component of $\tilde{G}_i$ not containing the auxiliary vertex. In the last case, $\tilde{G}_i$ is the auxiliary graph of G_i with respect to C for the disk embedding extension problem. Since there are no local bridges attached to the paths P_i and P_{i+1}, the 3-connected components of $\tilde{G}_i$ not containing the auxiliary vertex are also 3-connected components of $\tilde{G}$. Consequently, the Kuratowski subgraph obstruction in G_i gives (e).

Suppose now that in G_i we have a tripod T. If T is not local on P_i and not local on P_{i+1}, we have (b). Otherwise, assume all three attachments of T are on P_i. Denote by v_1, v_2, u_1, u_2, u_3, π_1, π_2, π_3 the elements of T as they are shown on Figure 1, and suppose that π_2 is attached at P_i between π_1 and π_3. Construct a path P, internally disjoint from C, that connects $C - P_i$ with an interior vertex x of T. The existence of P is guaranteed since the bridges containing T are not local on P_i. If x is on π_s for some $s \in \{1, 2, 3\}$, we can replace the segment of π_s from x to P_i by P and get a tripod satisfying (b). If x is an interior vertex of the branch $u_2 v_1$, then $T \cup P$ contains a dipod satisfying (d). By the symmetries of T, the only essentially different remaining case is when x is on the branch $u_1 v_1$, where $x \ne u_1$ but possibly $x = v_1$. Let Q_1 be the path $P x v_1 u_2 \pi_2$ and let Q_2 be the path $\pi_1 u_1 v_2 u_3 \pi_3$ in $T \cup P$. If Q_1 and Q_2 are in the same K-component of G, we can find a path P' from Q_1 to Q_2 that is disjoint from C, and $Q_1 \cup Q_2 \cup P'$ is a dipod satisfying (d). On the other hand, if Q_1 and Q_2 are in different bridges B_1, B_2 of K, respectively, let P' be a path from the interior of Q_2 to C that is disjoint from P_i. Such a path exists, again, because B_2 is not local on P_i. Now, $Q_1 \cup Q_2 \cup P'$ contains disjoint crossing paths satisfying (c), unless the endpoints of P and P' on C coincide. But in this case, $Q_1 \cup Q_2 \cup P'$ is a degenerate dipod with the attachment b (see Figure 4(b)) corresponding to the common point of P and P'.

It remains to consider the case of disjoint crossing paths, say Q_1 and Q_2, obtained as an obstruction in G_i. If both Q_1 and Q_2 are attached locally to P_i, we change one of them so that it has an attachment on $C - P_i$. For this purpose, the same method as above can be applied. If just one of the paths (possibly after the previous change) is local on P_i, the same procedure can be applied, as above, with the paths Q_1 and Q_2 in the case of the tripods. We either get a dipod or disjoint crossing paths satisfying (d) or (c), respectively.

It is easy to perform the above construction in linear time. To find disk obstructions, we use Theorem 3.1, and to find paths P, P', *etc.*, we can use standard graph search. □

Once we know how the case $k \ge 3$ works, we can also solve the 0-prism embedding extension problem. If C_1, C_2 are cycles of G embedded on the boundary of the cylinder, an orientation of the cylinder yields consistent orientations of C_1 and C_2. If $P_1, \ldots, P_k$ are disjoint (C_1, C_2)-paths, they are said to be *attached consistently* if their ends on C_1 follow each other in the inverse cyclic order to that on C_2, i.e., the embedding of $C_1 \cup C_2$ can be extended to $C_1 \cup C_2 \cup P_1 \cup \cdots \cup P_k$. Note that for $k \le 2$, the paths are always attached consistently.

Before stating our next result on obstructions, let us formulate a lemma that will be used in its proof.

Lemma 4.2. *Let G be a 3-connected graph, C a cycle of G, and B a C-component in G. Let $G(B, C)$ be the graph obtained from $B \cup C$ by adding a new vertex adjacent to all vertices on C. Then $G(B, C)$ is 3-connected.*

Proof. The graph is clearly 2-connected. It is also easy to see that it has no 2-separations. $\square$

Theorem 4.3. *Let C_1 and C_2 be disjoint cycles of a graph G that are embedded on the boundary of the cylinder. There is no embedding extension to G if and only if $G - E(C_1) - E(C_2)$ contains a subgraph Ω of one of the following types:*

(a) *Three disjoint paths from C_1 to C_2 that are not attached consistently on C_1 and C_2.*

(b) *Disjoint paths P_1, P_2, P_3, where P_1, P_2 join C_1 with C_2, both endpoints of P_3 are on C_1 (respectively, on C_2) and the endpoints of P_3 interlace with the endpoints of P_1 and P_2 on C_1 (respectively, on C_2).*

(c) *A tripod or a pair of disjoint crossing paths with respect to C_1 (respectively, C_2). If G is not 3-connected, this obstruction may have a vertex, two vertices, or a segment of one of its branches contained in C_2 (respectively, in C_1).*

(d) *A path P from C_1 to C_2 together with a tripod T with respect to $C_1 \cup C_2$ disjoint from P that has two attachments on C_1 and one on C_2, or vice versa.*

(e) *Disjoint paths P_1, P_2, P_3 from C_1 to C_2 attached consistently on C_1 and C_2, together with a triad attached to P_1, P_2, and to P_3.*

(f) *A Kuratowski subgraph contained in a 3-connected component L of the auxiliary graph $\tilde{G}$ of G, where L is such that it does not contain auxiliary vertices of $\tilde{G}$.*

Moreover, there is a linear time algorithm that either finds an embedding extension of $C_1 \cup C_2$ to G, or returns an obstruction Ω for the embedding extendibility. In the latter case, Ω fits one of the above cases (a)–(f).

Proof. First of all we try to find three disjoint (C_1, C_2)-paths in G. If such paths exist, let $k = 3$, and let P_1, P_2, P_3 be the paths. Otherwise, let $k \le 2$ be the maximal number of disjoint paths from C_1 to C_2. All these can be obtained in linear time by standard connectivity algorithms using flow techniques [18, Chapter 9].

Let us first consider the case when $k = 3$. If P_1, P_2, P_3 are not attached consistently at C_1 and C_2, then $\Omega = P_1 \cup P_2 \cup P_3$ is a small obstruction satisfying (a). Otherwise, we first reduce the problem to the 3-connected case. Without loss of generality, we can remove the

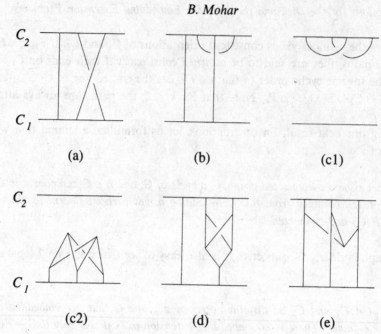

(a)   (b)   (c1)

(c2)   (d)   (e)

Figure 5   0-prism obstructions

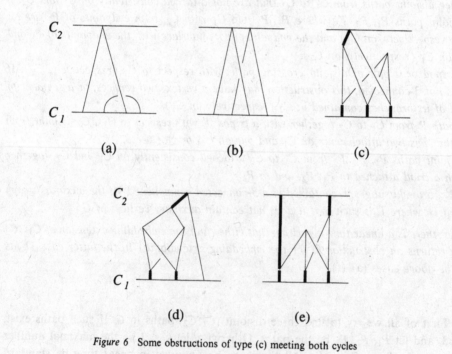

(a)   (b)   (c)

(d)   (e)

Figure 6   Some obstructions of type (c) meeting both cycles

3-connected components of the auxiliary graph that do not contain the auxiliary vertices (or we get (f)). So, we assume from now on that $\tilde{G}$ is 3-connected. Next we try to replace the paths P_1, P_2, and P_3 by disjoint paths joining the same pairs of endpoints so that no local bridge of $K' = C_1 \cup C_2 \cup P_1 \cup P_2 \cup P_3$ will be attached to some P_i only. It can be shown that this is always possible, since $\tilde{G}$ is 3-connected, but it is not entirely obvious how to perform it in linear time. For $i = 1, 2, 3$, let G_i be the graph consisting of P_i together with all its local bridges and with an additional edge joining the ends of P_i. If G_i is planar, an algorithm from [15] replaces P_i with a new path P_i', joining the same endvertices, that is internally disjoint from $K' - P_i$, and such that no local bridge of $(K' - P_i) \cup P_i'$ is attached to P_i'. So, we either achieve our goal, or get one of G_i, say G_1, to be non-planar. Let us first deal with the latter possibility. Let C be the cycle composed of the paths P_2 and P_3 together with the segments on C_1 from P_3 to P_1 and from P_1 to P_2, and the segments on C_2 from P_2 to P_1 and from P_1 to P_3. Denote by B the $(C_1 \cup C_2 \cup P_2 \cup P_3)$-component in G that contains P_1. If B contains a vertex of $(C_1 \cup C_2) - C$, then a path in B from that vertex to an end of P_1, together with P_2 and P_3 determines three non-consistently attached paths from C_1 to C_2, and so we have case (a). Therefore, we may assume that B is attached to C only. Let $H = B \cup C$. It is clear that H is 2-connected, and, by Lemma 4.2, its auxiliary graph $\tilde{H}$ with respect to C is 3-connected. Moreover, $\tilde{H}$ is non-planar, since G_1 is contained in H (with the edge joining the ends of P_1 replaced by a path in C). By Theorem 3.1, we can find in H a tripod T or disjoint crossing paths Q_1 and Q_2 with respect to C. Let us first consider the case when we have disjoint paths Q_1, Q_2. For $j = 1$, 2, denote by e_j the foot of P_1 on C_j, and let C_j° be the open segment of C_j obtained from $C_j \cap C$ by removing its endpoints. If $Q_1 \cup Q_2$ is not attached to C_1°, take a path P in $B - C$ from e_1 to $Q_1 \cup Q_2$. Such a path clearly exists, since Q_1, Q_2 are both contained in the same bridge. Using this path, we can change Q_1 or Q_2 to get disjoint crossing paths that are attached to C_1°. We repeat the same procedure at C_2°. Up to symmetries, there are three possible outcomes:

(i) Q_1 joins C_1° and C_2°:

If Q_2 is attached on P_2 and P_3, take a path P in $B - C$ joining Q_1 and Q_2. Now, the paths Q_1, P_2, P_3 and the triad $Q_2 \cup P$ satisfy (e). Otherwise, it is easy to see that we get a subgraph of type (a), or (b) contained in $Q_1 \cup Q_2 \cup P_2 \cup P_3$.

(ii) Q_1 is attached to C_1° and Q_2 is attached to C_2°:

Excluding the above possibility (i), we may assume that the other attachment of Q_1 is on $P_2 - C_1$. Then $Q_1 \cup Q_2 \cup P_2$ contains disjoint crossing paths between C_1 and C_2. Together with P_3 they determine a subgraph of type (a).

(iii) Q_1 is attached to C_2° and both endpoints of Q_2 are on P_2:

Let P be a path in $B - C$ joining Q_1 and Q_2. Then $Q_1 \cup Q_2 \cup P \cup P_2$ is a tripod on $C_1 \cup C_2$, and, together with P_3, we have (d).

Suppose now that T is a tripod with respect to C that is contained in B. If T is not attached on C_1°, let P be a path in $B - C$ from e_1 to T. Then $T \cup P$ either contains a pair of disjoint crossing paths (which we have already covered above), or a tripod T' with an attachment on C_1°. If T' is not attached to P_3, then $T' \cup P_2$ contains a tripod T'' with respect to $C_1 \cup C_2$ that is either attached to C_1 only (case (c)), or is attached to C_1 and

C_2 (in this case $T'' \cup P_3$ satisfies (d)). Similarly, if T' is attached only to $C - P_2$. We are left with the case when T' is attached to C_1° and to P_2 and P_3. In this case we construct a path in $B - C$ from e_2 to T'. It gives rise to disjoint crossing paths, or to a tripod that are disjoint from P_2 or P_3, and both of these cases have already been covered above.

From now on we may assume that we have P_1, P_2, P_3 without local bridges. Let $K' = C_1 \cup C_2 \cup P_1 \cup P_2 \cup P_3$. By Theorem 4.1, we either extend the embedding of K' to G, or find an obstruction. The first outcome is fine, while in the second case we get one of the obstructions (a)–(d) of Theorem 4.1. Obstruction (a) of Theorem 4.1, together with P_1, P_2, P_3, necessarily contains one of our cases (a), (b), or (e). Case (b) of Theorem 4.1, together with P_1, P_2, P_3, implies our cases (c) or (d). The possibility (c) of Theorem 4.1 yields either (a), (b), or (c). Finally, a dipod D of type (d) in Theorem 4.1 attached three times on P_1, say, gives rise to a tripod with respect to $C_1 \cup C_2$ contained in $D \cup P_1$ (plus a segment on P_2 (or P_3) if D is attached to P_2 (respectively, on P_3)). This tripod is disjoint from one of the paths, and fits our case (d).

Finally, we have reached the cases $k = 0, 1, 2$. The first two ($k = 0, 1$) are easy. We are faced with two disk embedding extension problems, and to solve each of them, we apply Theorem 3.1. The resulting obstruction fits (c). If a cutvertex v of G separating C_1 and C_2 is on C_2 (assuming that the block containing C_1 is non-planar), the obstruction may contain v. (The possible cases are shown in Figure 6(a), (c), and (d).) Note that this is the first time that disjoint crossing paths or a tripod with respect to C_1 have a vertex on C_2.

Suppose now that $k = 2$. Let Q be the block of G containing C_1 and C_2. If the embedding of $C_1 \cup C_2$ extends to Q, we test the other blocks of G for planarity. We either get an embedding extension to G, or one of the blocks is non-planar. In the latter case, we have (f). So we may assume from now on that G is 2-connected and that there is no embedding extension. Since $k = 2$, Menger's theorem guarantees that C_1 and C_2 are in distinct 3-connected components of the auxiliary graph $\tilde{G}$. If all 3-connected components of $\tilde{G}$ are planar, $\tilde{G}$ is also planar. (This is easily seen by constructing a plane embedding of a graph by using plane embeddings of graphs forming its 2-separation.) Unfortunately, it may happen that the plane embedding of $\tilde{G}$ obtained in this way will not determine an embedding extension, since C_1 and C_2 may not be oriented consistently. In this case, let $Q_1 \supset C_1$ and $Q_2 \supset C_2$ be the graphs used in merging at the time when C_1 and C_2 merge in the same part, and let e be the corresponding virtual edge. Fixing the embedding of Q_1, there are two possibilities for the embedding of Q_2 that differ from each other only by the choice of orientation. One of them gives the consistent orientation of C_1 and C_2.

We may assume now that one of the 3-connected components of $\tilde{G}$ is non-planar. If the two 3-connected components containing C_1 and C_2, respectively, are planar, we get (f). Suppose now that the 3-connected component $Q_1 \supset C_1$ of $\tilde{G}$ is non-planar. This is equivalent to the property that Q_1 minus the auxiliary vertex has no embedding into the disk having C_1 on its boundary. By Theorem 3.1, we know how to handle this case. Since Q_1 is 3-connected, we get disjoint crossing paths or a tripod in it. This almost always gives rise to a subgraph of G satisfying (c). The only trouble may arise if our obstruction in Q_1 contains the virtual edge e having its pair in the 3-connected component Q_2 of $\tilde{G}$ that contains C_2. In this case, the replacement of e by a path P in $Q_2 - e$ should be done carefully so that $P \cap C_2$ is either empty, a vertex, or a segment on C_2. Since this is easy to achieve, we are through with our case analysis. It is worth remarking that some of the possibilities when $P \cap C_2$ is non-empty lead to cases (b) or (d). Some of the really new cases are shown in Figure 6, where each of the bold segments can be contracted to a point. The cases shown in Figure 6 include all possibilities that arise when the obstruction in Q_1 is either a pair of disjoint crossing paths, or a tripod whose intersection with C_2 is a vertex, or a segment.

Finally, we remark that all the steps of the algorithm that follows the above proof are easy to implement in linear time. $\qquad\square$

It is worth remarking that all cases of Theorem 4.3 are indeed obstructions for the 0-prism embedding extension problem, and that they are minimal (except in some cases of (c) when the intersection with the other cycle is non-empty) in the sense that if any of the branches is removed from such an obstruction, there exists an embedding extension. Note that the branch size of minimal 0-prism obstructions is at most 12. The obstructions (a)–(e) (without showing their 'degenerate' versions) are presented in Figure 5.

5. The 2-prism embedding extension problem

It may happen that minimal obstructions for the k-prism embedding extension problems are arbitrarily large. However, under the additional assumption that there are no local bridges attached to the paths P_i $(1 \leq i \leq k)$, large minimal obstructions are unavoidable only for the k-prism embedding extension problem with $k = 1$ or 2. An example of such an obstruction is shown in Figure 7. Since the general case of large minimal obstructions look like our example in Figure 7, we use the name millipede. More precisely, a *millipede* M for the 2-prism embedding extension problem is a subgraph of $G - E(K)$ that can be expressed as $M = B_1^\circ \cup B_2^\circ \cup \cdots \cup B_m^\circ$ $(m \geq 2)$, where $B_1^\circ, \ldots, B_m^\circ$ are subgraphs of distinct K-bridges $B_1, B_2, \ldots, B_m$ (respectively) and satisfy the following conditions.

(1) Each of B_1° and B_m° is embeddable in exactly one of the faces of K. If m is even, then B_1° and B_m° are embeddable in the same face of K. If m is odd, then B_1° and B_m° are embeddable in distinct faces of K.
(2) For $2 \leq i \leq m - 1$, B_i° is embeddable in both faces of K.
(3) For each $i = 1, 2, \ldots, m - 1$, B_i° and B_{i+1}° cannot be embedded simultaneously in the same face of K.
(4) No other pair B_i°, B_j° $(1 \leq i < i + 2 \leq j \leq m)$ interferes with each other, *i.e.*, for any embedding of B_i°, there is an embedding of B_j° in the same face of K unless such an embedding is not possible by (1) (when $i = 1$ or $j = m$).
(5) B_i° $(1 \leq i \leq m)$ are minimal in the sense that the removal of any branch from B_i° destroys either (1), or (3).

It is easy to see that a millipede is a minimal obstruction for embedding extendibility. It follows from the minimality property (5) that each B_i° $(1 \leq i \leq m)$ contains at most 6 feet (at most a triple for overlapping with B_{i-1}° and possibly another triple for overlapping with B_{i+1}°) and has at most 11 branches. (We will see that it suffices to consider only millipedes in which every B_i° contains at most 4 feet.) Let us remark that the millipedes constructed

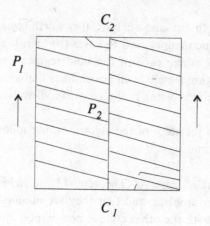

Figure 7 A millipede

by our succeeding theorems will satisfy an even stronger 'minimality' condition: Properties
(1), (2) and (4) will hold not only for the subgraphs B_i° but also for their 'master-bridges' B_i.

Given the 2-prism embedding extension problem with C_1, C_2, P_1, P_2, F, F', as in
Figure 2(b), and $K = C_1 \cup C_2 \cup P_1 \cup P_2$, we define the *overlap graph* $O(G, K)$ of K-bridges
in G as follows. Its vertices are the K-bridges, and two of them are adjacent in $O(G, K)$ if
they overlap in one of the faces, *i.e.*, they can be embedded in the same face, say F, but
their union cannot be embedded in F. The *extended overlap graph* $AO(G, K)$ is obtained
from the overlap graph by adding two new vertices, w and w', that are adjacent to each
other. Moreover, w is adjacent to all bridges of K that are not embeddable in F, and w'
is adjacent to all bridges that are not embeddable in F'.

Lemma 5.1. *The embedding of K for the 2-prism embedding extension problem has an
extension to G if and only if the extended overlap graph $AO(G, K)$ is bipartite.*

Proof. If a K-bridge B cannot be embedded in any of the faces F, F', then B together
with w and w' determines a triangle, and the extended overlap graph is not bipartite.
Therefore we may assume from now on that every bridge can be embedded in at least
one of the faces, F, or F'.

Suppose now that we have an embedding extension. Color the bridges that are embedded in F using color 1, and color the bridges in F' using color 2. Moreover, let w be
colored 1, and let w' be colored 2. It is easy to see that this determines a 2-coloring of
$AO(G, K)$, so the extended overlap graph is bipartite.

Conversely, if the extended overlap graph is bipartite, choose one of its 2-colorings
having w and w' colored 1 and 2, respectively. Consider the bridges colored 1. Each of
them can be embedded in F, since otherwise it would be adjacent in $AO(G, K)$ to w that
also has color 1. Moreover, all these bridges can be embedded in F simultaneously, since
no two of them overlap. Similarly, the bridges colored 2 have an embedding in F', and
we get a required embedding extension.   □

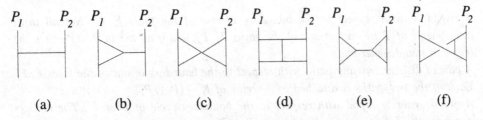

(a)	(b)	(c)	(d)	(e)	(f)

Figure 8 H-graphs

The above lemma provides a simple answer for the 2-prism embedding extension problem. It also yields an algorithm that is linear in the number of edges of $AO(G,K)$. Having a 2-coloring, we easily get an embedding extension. Otherwise, an odd cycle in $AO(G,K)$ determines an obstruction. Unfortunately, the usual 2-coloring algorithm can be of quadratic complexity in terms of the size of G, since the number of edges of $AO(G,K)$ may be quadratic in terms of the number of bridges, and this number can be linear in terms of $|E(G)|$. Therefore, we have to solve the biparticity problem of $AO(G,K)$ with some additional care in order to fulfil our linearity goal. One possible approach is explained in more detail in [12].

In the following results, we will use some special subgraphs of K-bridges. Let B be a K-bridge in G. For each branch e of K that B is attached to, let e_1 and e_2 be feet of B attached as close as possible on e to one and the other end of e (including the possibility of being attached to the end). Furthermore, let these feet be chosen in such a way that their total number is as small as possible, *i.e.*, if there is just one attachment on e, we select $e_1 = e_2$, and similarly when different branches of K share an attachment of B. Let $H = H(B)$ be a minimal subtree of B that contains all chosen feet. The obtained graph H is said to be an H-*graph* of B. Suppose now that B is attached only to $P_1 \cup P_2$. Then H contains at most 4 feet. If there are three or just two distinct feet in H, then H is unique up to homeomorphisms. But in the case of four distinct edges, there are four homeomorphically distinct cases for H (see Figure 8). Let us remark that the last case of Figure 8 is excluded if B can be embedded in F, since it contains disjoint crossing paths. Note that H-graphs can be constructed in linear time by standard graph search algorithms. The following simple result justifies the introduction of H-graphs.

Lemma 5.2. *Let G be a graph and K be a subgraph that is 2-cell embedded in some surface. Let B and B' be K-bridges in G that can be embedded in the same face F of K. If ∂F is a cycle in G and neither B nor B' is a local bridge, then B and B' overlap in F if and only if their H-graphs overlap in F.*

Theorem 5.3. *Let $K \subseteq G$ be the subgraph of G for the 2-prism embedding extension problem, and let F, F' be the faces of K. Suppose that no K-component of G is attached just to one of the paths P_1, P_2 of K. Then there is no embedding extension of K to G if and only if $G - E(K)$ contains a subgraph of one of the following types:*

(a) *A path that is internally disjoint from K and connects a vertex of $\partial F - (P_1 \cup P_2)$ with a vertex of $\partial F' - (P_1 \cup P_2)$.*

(b) A tripod T with respect to the boundary of one of the faces F, F'. Not all three attachments of T lie on just one of the paths P_1, P_2, and if all are in $P_1 \cup P_2$, then the tripod is non-degenerate.

(c) A pair of disjoint crossing paths with respect to the boundary of one of the faces F, F'. Each of the two paths is attached to a vertex of $K - (P_1 \cup P_2)$.

(d) A non-degenerate dipod with respect to the boundary cycle of F or F'. The vertices a, c from the definition of a dipod lie on $P_1 \cup P_2$, and not all attachments of the dipod lie on just one of the paths P_1 or P_2.

(e) Internally disjoint triads T_1, T_2 attached to the same triple of vertices on $P_1 \cup P_2$, together with a path joining the main vertices of T_1 and T_2. Not all three attachments of $T_1 \cup T_2$ are on just one of the paths P_1, or P_2.

(f) Subgraphs H_1, H_2, H_3 that are pairwise overlapping in F or in F'. They are minimal pairwise overlapping subgraphs of H-graphs of distinct K-bridges. H_2 and H_3 are attached to $P_1 \cup P_2$ only.

(g) A millipede.

(h) A Kuratowski subgraph contained in a 3-connected component L of the auxiliary graph $\tilde{G}$ of G, where L is such that it does not contain auxiliary vertices of $\tilde{G}$.

Moreover, there is a linear time algorithm that either finds an embedding extension of K to G, or returns an obstruction Ω of one of the above types.

Proof. We may check the embedding extendibility by testing the planarity of the auxiliary graph $\tilde{G}$ (cf. the $k = 2$ case in the proof of Theorem 4.3 for details). Moreover, if there is no embedding extension, we can reduce the problem to the case where G is 2-connected (or we get (h)). Note that C_1 and C_2 are in the same block of G. If C_1 and C_2 are in the same 3-connected component of $\tilde{G}$, we can also reduce the problem to the case where $\tilde{G}$ is 3-connected (or we get (h)). If this is not the case, let $\tilde{G}'$ be the graph obtained from $\tilde{G}$ by adding the edge joining the auxiliary vertices. By considering the 3-connected components of $\tilde{G}'$, we can also in this case either get (h) (since the 3-connected components of $\tilde{G}'$ are also 3-connected components of $\tilde{G}$ except for the one containing C_1 and C_2), or reduce the problem to the case when $\tilde{G}'$ is 3-connected. The latter case will be assumed henceforth.

Consider the K-components of G. Suppose that B is one of them, and that the embedding of K cannot be extended to $K \cup B$. We either have (a), or B is attached to the boundary of one of the faces of K, say to ∂F. In the latter case, let $L = \partial F \cup B$. Since $\tilde{G}'$ is 3-connected, the auxiliary graph $\tilde{L}$ of L (with respect to ∂F) is 3-connected by Lemma 4.2. Clearly, $\tilde{L}$ is non-planar, and by Theorem 3.1, B contains a tripod or a pair of disjoint crossing paths with respect to ∂F. Since B is not local on P_1 or P_2, we can use the same strategy as in the proof of Theorem 4.1 to change the obtained obstacle so that not all of its attachments are on just one of P_1 or P_2. Usually we will get case (b), or (c), but there are also two exceptions. The first possibility is when we get a degenerate tripod with all attachments on $P_1 \cup P_2$. Add a path P in B between the two triads in the tripod. If P joins the two main vertices of the tripod, then we have case (e). In all other cases, the union of P and the tripod contains a non-degenerate tripod, *i.e.*, a subgraph satisfying (b). The other case is when we have disjoint crossing paths that do not satisfy (c). One of the paths is then attached only to $P_1 \cup P_2$. Hence, by adding a path in B that joins the two paths, we get a non-degenerate dipod satisfying (d).

From now on we may assume that every K-component is embeddable either in F, or in F' (or both). A linear time algorithm of [12] shows how to solve this problem. That algorithm finds an induced odd cycle Γ in the extended overlap graph $AO(G, K)$. There are 4 cases to be distinguished.

(i) *Both vertices w and w' lie on Γ:*

In this case, the edge ww' is on Γ, since Γ is an induced cycle. Let $B_1, \ldots, B_m$ be the K-bridges corresponding to the sequence of vertices of Γ from w to w' (but not including these two). By our assumptions, $m > 1$. By the definition of $AO(G, K)$, B_1 cannot be embedded in F, and B_m cannot be embedded in F'. Note that m is odd, so the conclusion here fits condition (1) of the definition of a millipede. Next we describe how to get the subgraph B_i° of B_i, for $i = 1, 2, \ldots, m$. Since the cycle Γ of $AO(G, K)$ is induced, no bridge B_i, $2 \le i \le m - 1$, is adjacent to w, or w'. This means that B_i itself is embeddable in F and in F'. Therefore, arbitrary subgraphs B_i° of B_i also satisfy (2). The bridges B_i and B_{i+1} ($i < m$) cannot be simultaneously embedded in the same face, and at least one of them is embeddable in both faces of K. By Lemma 5.2, their H-subgraphs (which are easy to find) overlap in the same way as the bridges themselves. By taking such obstructions for all bridges B_i, we get small subgraphs of $B_1, \ldots, B_m$ satisfying (1) and (3). Since these subgraphs are small, we can check whether each of them satisfies the minimality requirement (5), and remove the superfluous branches whenever necessary. Finally, (4) is satisfied automatically, since Γ is an induced cycle of $AO(G, K)$. Therefore we have a millipede.

(ii) $w \in V(\Gamma)$:

We get a millipede in the same way as in Case (i), except that m is even.

(iii) $w' \in V(\Gamma)$:

Same as Case (ii).

(iv) $w, w' \notin V(\Gamma)$: We will show that in this case the length m of Γ is rather small. Let $B_1, \ldots, B_m$ be the K-bridges corresponding to the successive vertices on Γ. Suppose first that $m = 3$. If two of the bridges, say B_1, B_2, are adjacent in $AO(G, K)$ to w (or w'), we replace Γ by the triangle B_1, B_2, w (respectively, B_1, B_2, w'), and by (ii) (respectively, (iii)) we get a millipede. If one of them is adjacent to w, another to w', we get a millipede of length 3, as was the case in (i). (This works even though the corresponding cycle in Γ obtained by replacing the edge B_1B_2 by the path $B_1ww'B_2$ is not induced.) We may therefore assume that B_2 and B_3 are embeddable in F and F' and that B_1 is embeddable at least in F. For $i = 1, 2, 3$, let H_i be an H-graph of B_i. By Lemma 5.2, the H-graphs overlap as much as the original bridges. We therefore have (f).

Suppose now that $m \ge 5$. As above, the cases when two of the bridges B_i are adjacent to w or w' (possibly one to w, another to w') can be reduced to the previously treated cases. We may thus assume that at most one of the bridges is not embeddable in both F and in F'. If there is a bridge that cannot be embedded in one of the faces, we will assume that this is B_1, and that this bridge can be embedded in F. Let us

Figure 9 One-sided millipedes

write $B_i \prec B_j$ if $B_i \cup B_j$ can be embedded in F and B_i is embedded closer to C_1 than B_j. Since none of the bridges is local on P_1 or P_2, the relation $\prec$ is well defined. The relation $\prec$ is transitive, and since $\tilde{G}'$ is 3-connected, it is also asymmetric. Therefore it has minimal elements. We may assume that B_1 is a minimal element for this relation. (If B_1 is only embeddable in F, we thus assume that it is attached to $C_1 - (P_1 \cup P_2)$, and then B_1 is clearly minimal by the definition of the relation $\prec$.) We claim that for $i = 1, 2, \ldots, m - 2$, $B_i \prec B_{i+2}$. Suppose that this is not true. Let i be the smallest index for which $B_{i+2} \prec B_i$. Since B_j and B_{j+2} do not overlap, they are $\prec$-comparable for every j and thus such an index i exists. By our choice of B_1, we have $i > 1$. Since B_{i+2} is attached closer to C_1 than B_i, and B_i overlaps with B_{i-1}, B_{i+2} has an attachment on P_1 or P_2 that is closer to C_1 than one of the attachments of B_{i-1} on the same path. Similarly, since B_{i+2} overlaps with B_{i+1} and $B_{i+1} \succ B_{i-1}$, the bridge B_{i+2} has an attachment that is further away from C_1 than an attachment of B_{i-1} on the same path P_1, or P_2. This implies that B_{i+2} overlaps with B_{i-1}. But this is not possible since $m \geq 5$. This proves the claim. In particular, we know that $B_{m-2} \prec B_m$. Since B_1 is $\prec$-minimal and $\prec$-comparable with B_j if $j \neq 1, 2, m$, we have $B_1 \prec B_{m-2}$. By transitivity we have $B_1 \prec B_m$. This contradicts the fact that B_1 and B_m overlap. The proof is thus complete.   □

   In the last part of the above proof, we have learned even more than needed. A straightforward extension gives the following result. Let us call a millipede *two-sided* if it is attached to $C_1 - (P_1 \cup P_2)$ and to $C_2 - (P_1 \cup P_2)$. Otherwise it is *one-sided*. Some one-sided millipedes are shown in Figure 9.

Proposition 5.4. *If M is a one-sided millipede, the number of K-bridges it includes is at most 4. In particular, M is a small obstruction.*

6. The 1-prism embedding extension problem

It remains to determine minimal obstructions for the 1-prism embedding extension problem. Let us first extend a few definitions used in previous sections for the purpose of this section. If F is a face of an embedded graph $K \subseteq G$, and P, P' are paths in G with endpoints on ∂F but otherwise disjoint from K, they are said to be *disjoint crossing paths* with respect to F if they cannot be simultaneously embedded in F. The essentially different cases of disjoint crossing paths with respect to the face F of a 1-prism embedding extension problem are shown in Figure 10 (up to symmetries). The only case where we

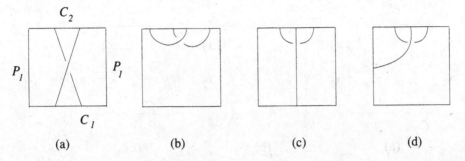

Figure 10 Disjoint crossing paths with respect to F

have one of the paths attached to P_1 is (d), which also includes the possibility of the attachment at $C_1 \cap P_1$ or $C_2 \cap P_1$. *Tripods* with respect to the face F are another kind of obstruction for the 1-prism embedding extension problem. They are defined in the same way as in the case when the face is bounded by a cycle, with the additional requirement that it should be an obstruction. It turns out that tripods with respect to F can be divided into four classes as follows.

1. Attached twice to $C_1 - P_1$ or twice to $C_2 - P_1$ with the third attachment anywhere else on ∂F and with no restrictions on non-degeneracy.
2. Attached to $C_1 - P_1$, to $C_2 - P_1$, and to P_1. The attachment on P_1 is non-degenerate.
3. Attached once to $C_1 - P_1$ (or to $C_2 - P_1$) and twice to P_1. The attachment on P_1 that is closer to C_1 (respectively, closer to C_2) is non-degenerate.
4. Attached only to P_1. The middle attachment on P_1 is non-degenerate.

The *1-millipedes* are another type of obstruction for the 1-prism embedding extension problem. These obstructions are of the same type as the millipedes are for the 2-prism embedding extension problem, and they can be arbitrarily large, though minimal, as well. A 1-millipede is a subgraph consisting of a path P_2 joining C_1 and C_2 and disjoint from P_1, together with a millipede for the 2-prism problem with respect to $K \cup P_2$. Moreover, the following additional requirement is imposed on 1-millipedes.

(6) For $j = 3, 4, \ldots, m-2$, denote by l_j^- and r_j^- the vertices of attachment of $B_{i-1}^\circ \cup B_{i-2}^\circ$ on K closest to C_1 and C_2, respectively. Similarly, let l_i^+ and r_i^+ be the extreme vertices of attachment of $B_{i+1}^\circ \cup B_{i+2}^\circ$. Then r_i^+ is strictly closer to C_2 than r_i^-, and l_i^- is strictly closer to C_1 than l_i^+.

Note that (6) is void if $m < 5$. It should also be pointed out that we assume that $l_3^- \in C_1 - P_1$ (an attachment of B_1°), and this is considered as being strictly closer to C_1 than any vertex on P_1. Similarly on the other side, where $r_{m-2}^+ \in C_2 - P_1$.

Yet another type of obstruction will be needed. Let $x_1, x_2 \in V(P_1)$ and suppose that x_2 is closer on P_1 to C_1 than x_1 is. A subgraph Ω of $G - E(K)$ is a *left side obstruction* with respect to x_1 and x_2 if it satisfies:

(i) Ω contains a path joining $C_1 - P_1$ with x_1 (respectively, a path joining $C_2 - P_1$ with x_2).
(ii) No attachment of Ω to K is closer to C_2 than x_1 (respectively, closer to C_1 than x_2), and no attachment of Ω is on $C_2 - P_1$ (respectively, $C_1 - P_1$).

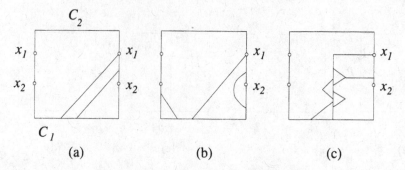

Figure 11   The minimal left side obstructions

(iii) Ω can be embedded in F in such a way that all feet of Ω attached to P_1 (strictly) between x_1 and x_2 are touching P_1 at the right side of F. (The left and the right are well defined with respect to Figure 2.)

(iv) Ω cannot be embedded in F in such a way that all feet of Ω attached to P_1 (strictly) between x_1 and x_2 are touching P_1 at the left side of F.

We define the *right side obstructions* similarly. Their attachments on P_1 between x_1 and x_2 can be embedded on the left side of F, but cannot be embedded on the right side.

Examples of left side obstructions are given in Figure 11. We will prove in Theorem 6.3 that Figure 11 contains all minimal left side obstructions attached to $C_1 - P_1$, where case (c) of Figure 11 represents arbitrary two-sided millipedes for the following 2-prism embedding extension problem. Add the edge $x_1 x_2$ (x_1 and x_2 are formerly non-adjacent) and embed it across the face F so that it is attached to x_1 on the left and to x_2 on the right. Add also a path P_2 from $C_1 - P_1$ to x_1. Let P_1' be the segment of P_1 from C_1 till x_2 and let C_2' be the cycle consisting of the segment $x_2 x_1$ on P_1 together with the new edge. Then we consider the 2-prism problem with respect to $K' = C_1 \cup C_2' \cup P_1' \cup P_2$ embedded in the cylinder as described above. Note that the first bridge in a two-sided millipede for this problem will be attached between P_1 and P_2 on C_1, while the last one will be attached to the segment of P_1 between x_2 and x_1. It is easy to see that such a millipede is a left side obstruction. Note that one-sided millipedes do not give rise to left side obstructions.

Having a left side obstruction Ω_1 attached to $C_1 - P_1$ and a left side obstruction Ω_2 attached to $C_2 - P_1$ (with respect to the same pair x_1, x_2), which do not intersect out of P_1, their union $\Omega_1 \cup \Omega_2$ cannot be embedded in F. This way we get a rich family of 1-prism embedding obstructions.

Before stating the main result of this section, we will prove a lemma about 3-connected subgraphs that will be needed in the proof. Let us recall that a graph H is *nodally 3-connected* if the graph obtained from H by replacing each branch with an edge between the corresponding main vertices is 3-connected.

Lemma 6.1. *Let G be a graph with disjoint nodally 3-connected subgraphs K and L. Let π_1, π_2, π_3 be disjoint paths in G joining three main vertices of K with a triple of main vertices of L. Let $J = K \cup L \cup \pi_1 \cup \pi_2 \cup \pi_3$. If for every branch e of $K \cup L$, no two consecutive (on*

e) connected components of $e \cap (\pi_1 \cup \pi_2 \cup \pi_3)$ belong to the same path π_i, then J is nodally 3-connected.

Proof. J is clearly 2-connected. Suppose now that there is a 2-separation $J = J_1 \cup J_2$, $J_1 \cap J_2 = \{x, y\}$, where x and y are main vertices of J and each of J_1, J_2 contains two or more branches of J. One of the paths π_i is disjoint from x, y and is thus totally contained in J_2, say. Since K, L are nodally 3-connected, J_2 contains all main vertices of $K \cup L$. By our assumption, J has no parallel branches. Thus J_1 contains a main vertex z of J. Clearly, z is obtained as the intersection of one of the paths, say π_1, with a branch e in K, say. Since J_2 contains all main vertices of $K \cup L$, x, y both lie on e and both lie on π_1. Follow π_1 from z in a direction out of the branch e. The first intersection with $K \cup L$ must be on e. Otherwise we could reach a main vertex of $K \cup L$ different from x, y. By our assumption on J, there is a main vertex w between the two intersections of π_1 with e such that $w \notin \pi_1$. That vertex is neither x nor y and belongs to J_1. By repeating the above arguments with w, we see that x, y also belong to another path, π_2, or π_3. This is a contradiction. $\qquad\square$

Our next result describes minimal obstructions for 1-prism embedding extension problems. In cases (f)–(h) of Theorem 6.2, obstructions (and, in particular, 1-millipedes) are defined with respect to the following 2-prism embedding extension problem. Suppose that in $G - P_1$, there is no path from $C_1 - P_1$ to $C_2 - P_1$. Let P_2 be a path from $C_1 - P_1$ to a vertex x on P_1 such that x is as close as possible to C_2. Let y be the neighbor of x on P_1 that is closer to C_1 than x. Then let P_1' be the segment of P_1 from C_1 to y, and let C_2' be a cycle yx_1xx_2y, where x_1 and x_2 are new vertices. We consider the 2-prism problem for the subgraph $K' = C_1 \cup C_2' \cup P_1' \cup P_2$ of the graph G' obtained from K' by adding all K-bridges in G with an attachment on $C_1 - P_1$. (In particular, no attachment on P_1 of these K-bridges is closer to C_2 than x.) In case (g) (and (h)) an additional edge xz is added into G'. The vertex $z \in V(P_1')$ has the property that in G there is a path internally disjoint from P_1 joining $C_2 - P_1$ with z. The 2-prism problems for (f)–(h) in the case when P_2 joins $C_2 - P_1$ with P_1 are defined similarly. It should also be mentioned that the millipedes appearing in (h) are defined with respect to the above 2-prism embedding extension problems.

Theorem 6.2. *Let G and $K = C_1 \cup C_2 \cup P_1 \subseteq G$ be graphs for a 1-prism embedding extension problem. Suppose that no K-component of G is attached just to P_1. Then there is no embedding extension to G if and only if $G - E(K)$ contains a subgraph Ω of one of the following types:*

(a) *Disjoint crossing paths with respect to the face of K.*

(b) *A tripod with respect to the face of K. If $G - P_1$ contains a path joining C_1 and C_2, at least one attachment of the tripod is not on P_1.*

(c) *A path P_2 joining C_1 and C_2 and disjoint from P_1 together with a dipod attached three times to P_1 and once to $(C_1 \cup C_2) - (P_1 \cup P_2)$. If the dipod is degenerate, its degenerate attachment is not on P_1.*

(d) *A path P_2 from C_1 to C_2 disjoint from P_1 together with a 2-prism embedding extension obstruction of type (b)–(f) of Theorem 5.3 with respect to $K \cup P_2$. (It may happen that such an obstruction is not minimal. In this case, a part of P_2 can be removed.)*

(e) *A 1-millipede.*

(f) *Same as (d) but with P_2 joining $C_1 - P_1$ (or $C_2 - P_1$) with a vertex $x \in P_1$. No other attachment of the obstruction is closer to C_2 (respectively, to C_1) than x.*

(g) *Same as (f) where P_2 joins $C_1 - P_1$ with $x \in P_1$ (respectively, $C_2 - P_1$ with $x \in P_1$) where one of the branches of the obstruction joins x with another vertex $z \in P_1$. In Ω, this branch is replaced by a branch joining $C_2 - P_1$ (respectively, $C_1 - P_1$) with z.*

(h) *A one-sided 1-millipede attached to C_1 (or C_2) and to a segment of P_1. The path P_2 of the 1-millipede joins $C_1 - P_1$ (respectively, $C_2 - P_1$) with the attachment x on P_1 closest to C_2 (respectively, C_1). If the 1-millipede contains a branch joining x with another vertex z on P_1, this branch is possibly replaced in Ω by a branch joining $C_2 - P_1$ (respectively, $C_1 - P_1$) with z.*

(i) *Union $\Omega = \Omega_1 \cup \Omega_2$, where $\Omega_1 \cap \Omega_2 \subseteq P_1$. For $i = 1, 2$, Ω_i contains a path π_i joining $C_i - P_1$ with P_1. The end x_2 of π_2 on P_1 is closer to C_1 than the end x_1 of π_1. Moreover, Ω_1 and Ω_2 are both left side, or both right side obstructions with respect to x_1 and x_2.*

(j) *A Kuratowski subgraph contained in a 3-connected component of the auxiliary graph $\tilde{G}$. The 3-connected component does not contain auxiliary vertices.*

Cases (f), (g), (h), and (i) appear only when in $G - P_1$ there is no path from C_1 to C_2. Moreover, there is a linear time algorithm that either finds an embedding extension of K to G, or returns an obstruction Ω for the embedding extendibility. In the latter case, Ω fits one of the above cases (a)–(j).

Proof. First of all, we can check the embedding extendibility by applying Theorem 4.3. Suppose now that there is no embedding extension of K to G. Consider the block(s) of G containing C_1 and C_2. If there is an embedding extension of K to this (these) block(s), there is also an embedding extension to G, unless we have (j). If C_1 and C_2 are in different blocks of G, there is no embedding extension if and only if one of them, say the one containing C_1, has no embedding extension in the disk with C_1 on the boundary. We leave this case to the end of the proof, since we will reduce it to the 2-connected case. Suppose now that C_1 and C_2 are in the same block of G. To simplify notation, we will assume that this block is G itself, *i.e.*, G is 2-connected. Then there is no embedding extension if and only if one of the 3-connected components of the cylinder auxiliary graph $\tilde{G}$ is non-planar. (See the details in the proof of Theorem 4.3, case $k = 2$.) If the 3-connected component(s) of $\tilde{G}$ containing the auxiliary vertices is (are) planar, then we have (j). Otherwise, let $\overline{G}$ be the graph obtained from $\tilde{G}$ by adding an edge between the auxiliary vertices. Since G is 2-connected, the cycles C_1 and C_2, and the auxiliary vertices will be in the same 3-connected component of $\overline{G}$. The other 3-connected components of this graph are also 3-connected components of $\tilde{G}$, and can thus be eliminated.

We assume from now on that $\overline{G}$ is 3-connected. First of all, we try to find two disjoint paths in $G - P_1$ joining C_1 and C_2. The search for such paths can be performed in linear time by standard flow techniques, which were also used in our previous results. Suppose first that we have found such paths P_2 and P_3. If P_1, P_2, P_3 are not attached consistently on C_1 and C_2, then we have (a). Suppose now that this is not the case. Then we try to change P_2 and P_3 in such a way that there are no local bridges of $K \cup P_2 \cup P_3$ attached only to P_2 or only to P_3. To achieve this goal, we use the same technique as in the proof of Theorem 4.3. Let $K' = K \cup P_2 \cup P_3$. For $i = 2, 3$, let G_i be the graph consisting of P_i together with all its local K'-bridges and with an additional edge joining the ends of P_i. If G_i is planar, an algorithm from [15] replaces P_i with a new path that has no local bridges attached to it. So, we either achieve our goal, or get one of G_i, say G_3, to be non-planar. Let C be the cycle composed of the paths P_1 and P_2 together with the segments on C_1 from P_1 to P_3 and from P_3 to P_2, and the segments on C_2 from P_1 to P_3 and from P_3 to P_2. Denote by B the $(K \cup P_2)$-component in G that contains P_3. If B contains a vertex of $(C_1 \cup C_2) - C$, then a path in B from that vertex to an end of P_3 together with P_2 determines an obstruction of type (a). Therefore, we may assume that B is attached only to C. Since G_3 is non-planar, $C \cup B$ is non-planar. By Lemma 4.2, the auxiliary graph of $C \cup B$ with respect to C is 3-connected. By Theorem 3.1, we can find in B a pair of disjoint crossing paths or a tripod with respect to C. We can change the obtained obstruction as in the proof of Theorem 5.3 to get a 2-prism obstruction in the graph $H = K \cup P_2 \cup B$ with respect to its subgraph $K \cup P_2$. It follows from the proof of Theorem 5.3 that the only obstructions that appear in this case are types (b)–(e) of Theorem 5.3. Thus we have our case (d).

Suppose now that P_2 and P_3 do not have local K'-bridges. By Theorem 4.1, the obstructions to the non-extendibility of the embedding of K' to G are rather simple. Since $\overline{G}$ is 3-connected and in G there are three disjoint paths from C_1 to C_2, the auxiliary graph $\tilde{G}$ of G is 3-connected also. Therefore, we need only consider cases (a)–(d) of Theorem 4.1. Case (a) of Theorem 4.1 gives our case (a) or (d) (the latter one being case (d) of Theorem 5.3). Case (b) yields case (b) of Theorem 5.3, thus our case (d). (Here we need to be careful in selecting two paths among P_1, P_2, P_3 for which we get the 2-prism obstruction. We need to take P_1. The other path is P_3 if the tripod obstruction of case (b) is attached to P_2. Otherwise, we take P_2.) Case (c) of Theorem 4.1 yields case (c) of Theorem 5.3, thus our case (d). Finally, in case (d) of Theorem 4.1, we either have cases (b) or (d) of Theorem 5.3 (thus our case (d)), or we have a degenerate dipod attached three times to P_1 and disjoint from P_2, say. In the latter case, we get our possibility (c). (Note that (c) is contained in (d) if the dipod is non-degenerate.)

Next we suppose that there are no two paths P_2, P_3 as asked for above. Suppose that there is a path P_2 disjoint from P_1 joining C_1 and C_2. Let B be the K-bridge containing P_2. We will show that B either contains a subgraph of type (a) or (b), or P_2 can be changed so that the local bridges on P_2 will disappear. In the latter case, we will be able to get an obstruction of type (a), (d), or (e).

For $i = 1, 2$, let S_i be the segment on C_i between the 'leftmost' and the 'rightmost' attachment of B to $C_i - P_1$. Let $B' = (B - P_1) \cup S_1 \cup S_2$. For $i = 1, 2$, if S_i is not just a vertex, let x_i be the end of P_2 on S_i. Otherwise, we add to B' a pendant edge attached to the vertex S_i, and we let x_i be the new vertex on this edge. Let $Q_0, Q_1, \ldots, Q_t$ be the shortest sequence of blocks of B' satisfying the following conditions:

(i) $x_1 \in V(Q_0)$, $x_2 \in V(Q_t)$, and

(ii) for $i = 1, 2, \ldots, t$, Q_{i-1} and Q_i intersect at a cutvertex w_i of B'.

By the minimality requirement for t, the blocks Q_i and the cutvertices w_i are all distinct and uniquely determined. We also define $w_0 = x_1$ and $w_{t+1} = x_2$. Let us mention that the sequence w_0, Q_0, w_1, Q_1, $\ldots$, w_t, Q_t, w_{t+1} can be determined in linear time by standard biconnectivity algorithms. We also note that w_1, w_2, $\ldots$, w_t are vertices of P_2. By our assumption, B' contains no two disjoint paths from C_1 to C_2. Thus, by Menger's Theorem, we have $t \geq 1$.

Suppose that one of the blocks Q_i ($1 \leq i < t$) cannot be embedded in the plane with w_i and w_{i+1} on the boundary of the outer face. Let K_i be its subgraph obtained from the Kuratowski subgraph of $Q_i + w_i w_{i+1}$ by deleting its edge $w_i w_{i+1}$ if necessary. Next, find in G three disjoint paths π_1, π_2, π_3 from K to the main vertices of K_i (in linear time by using flow techniques). This is possible since $\overline{G}$ is 3-connected. We will show next that we can change the paths π_j in such a way that one of them will be attached to the end of P_2 on C_1 and one of them to the end of P_2 on C_2. Since Q_i is 2-connected, there are two disjoint paths in Q_i from $\{w_i, w_{i+1}\}$ to main vertices of K_i. Let π_1' and π_2' be obtained from these paths by adding a segment of P_2 from w_i to C_1 and from w_{i+1} to C_2, respectively. If π_1' is disjoint from π_1, π_2, π_3, it can replace π_1. Otherwise, suppose that π_1' first meets π_1 (in the direction from C_1). Do the same with π_2': if it does not intersect any of the paths, it can replace π_2. If its first intersection from C_2 with $\pi_1 \cup \pi_2 \cup \pi_3$ is on π_2 (or similarly π_3), we replace π_1 with the segment of π_1' up to its intersection with π_1 followed by the remaining segment of π_1, and we replace π_2 with a path consisting of the initial segment of π_2' and the terminal segment of π_2. Our goal for the paths to attach to C_1 and to C_2 is then satisfied. The remaining possibility is when π_2' first intersects π_1 as well as π_1' does. In this case, let y_1, y_2, $\ldots$ be the consecutive intersections of π_1' with $\pi_1 \cup \pi_2 \cup \pi_3$. Similarly, let z_1, z_2, $\ldots$ be the intersections of π_2' with the union of the paths. By our assumption, y_1, $z_1 \in V(\pi_1)$. Suppose that p, q are the largest indices such that all y_1, $\ldots$, y_p and z_1, $\ldots$, z_q belong to π_1. Let y be the vertex among y_1, $\ldots$, y_p, z_1, $\ldots$, z_q that is closest to the end of π_1 in K'. Suppose that $y \in \pi_1'$. Now replace π_1 by the segment of π_1' till y and the segment of π_1 from y to its end at a main vertex of K_i. Now, π_2' either does not intersect the paths π_j at all, or intersects first a path distinct from π_1, and we can apply the above procedure to fulfil our task.

Now we have three disjoint paths π_1, π_2, π_3 joining K with three of the main vertices of K_i, where one of the paths starts at $C_1 \cap P_2$ and another starts at $C_2 \cap P_2$. Note that these two paths pass through w_i and w_{i+1}, respectively. Let H be the graph obtained from K_i by adding the three paths and the cycle C obtained as follows. Let e_1 and e_2 be edges of C_1 and C_2, respectively, that are adjacent to P_1 (both at the same side of P_1 with respect to the given embedding of K in the cylinder). Then let C be the cycle obtained from $K - e_1 - e_2$ by adding a new edge between the two vertices of degree one. We can change π_1, π_2, π_3 so that the graph $H = C \cup K_i \cup \pi_1 \cup \pi_2 \cup \pi_3$ has no parallel branches. By Lemma 6.1, the auxiliary graph of H with respect to the cycle C is nodally 3-connected. By Theorem 3.1, H contains a tripod T (since there are only three attachments on C). By construction of H, the tripod T is also a tripod in G with respect to the face of K, since it is attached twice to $K - P_1$, and if the third attachment lies on P_1, it is non-degenerate. Thus we have our case (b).

Consider now Q_0 and suppose that it is non-trivial, *i.e.*, S_1 is not just a vertex. Let p and q be the endpoints of the segment S_1, and let Q_0' be the graph obtained from Q_0 by adding the edges pw_1 and qw_1 to it. If Q_0' has no embedding in the plane such that the cycle $S_1 + pw_1 + w_1q$ bounds the outer face, we can get an obstruction Ω_0 by using Theorem 3.1. If Ω_0 is attached to the cycle at w_1, we add to it the segment on P_2 from w_1 to C_2. Now Ω_0 is either a pair of disjoint crossing paths with respect to the face of K (case (a)) or a tripod attached to $K - P_1$ (case (b)).

We perform the similar procedure with Q_t. Not having obtained an obstruction, we know that the graph $Q = Q_0 \cup Q_1 \cup \cdots \cup Q_t$ can be embedded in the face F of K. Since $P_2 \subseteq Q$, every $(K \cup P_2)$-bridge in G that is locally attached to P_2 is totally contained in Q. Now, since Q can be embedded in F, the algorithm of [15] enables us to remove the local bridges at P_2 in linear time. Note also that after a possible change of P_2 by another path in Q with the same endpoints, P_2 still passes through $w_1, \ldots, w_t$, and that for every $(K \cup P_2)$-bridge B, there is some i, $0 \le i \le t$, such that B is attached to P_2 only between w_i and w_{i+1}.

Now we can apply Theorem 5.3 for the subgraph $K' = K \cup P_2$ of G. We note that P_2 plus a 2-prism embedding obstruction of Theorem 5.3 is not necessarily a minimal obstruction for our embedding extension problem. Clearly, the deletion of any branches not in P_2 is not possible, since we have a minimal obstruction for the embedding extension of K'. However, P_2, or a part of it, may be superfluous in the obstruction and may then be omitted. Note that case (a) of Theorem 5.3 gives our case (a), and cases (b)–(f) give (d). Case (h) of Theorem 5.3 can be excluded because of our initial connectivity reductions. In the remaining case (g) of millipedes, we claim that we really get a 1-millipede. We need to show that the corresponding millipede for K' satisfies (6). In order to achieve this property, we change P_2 before applying Theorem 5.3, as explained in the next paragraph.

For $i = 0, 1, \ldots, t$, let Q_i' be K' together with all K'-bridges attached to the segment w_iw_{i+1} of P_2, except for those K'-bridges whose only attachment on P_2 is one of w_i, w_{i+1}. If the embedding of K' cannot be extended to the obtained subgraph Q_i' of G, we get an obstruction of type (a), (d), or a millipede. Having a 1-sided millipede, its length m is at most 4 (Proposition 5.4), and thus it is clear that it satisfies the required property (6). Two-sided millipedes are excluded, since in Q_i', bridges of K' are either not attached to $C_1 - P_1$ (if $i \ne 0$), or are not attached to $C_2 - P_1$ (if $i \ne t$), since $t \ge 1$. Thus we may assume that we have an extension of the embedding of K' to Q_i'. Suppose first that $i \ne 0, t$. Consider the induced embedding of $Q_i \subseteq Q_i'$ and change the segment w_iw_{i+1} of P_2 to be the leftmost path in Q_i from w_i to w_{i+1}. After this change of P_2, there is just one bridge R_i attached to the right side of the segment (with respect to the embedding of Q_i') that has an attachment on P_1, since Q_i is 2-connected. Unfortunately, some local bridges attached at the right side of the segment may arise. In such a case, replace the obtained segment of P_2 by the rightmost path through such local bridges. It is easy to see that, because of the 3-connectivity, local bridges disappear after this change. On the right side of the new segment, the same bridge R_i remains as the only bridge on the right of it, while on the left, we can get more than one bridge. No two of the left bridges overlap.

Moreover, every left bridge overlaps with R_i since R_i is attached to w_i and to w_{i+1}. We perform similar changes for $i = 0, t$ (and possibly change the ends of P_2 on C_1 and C_2).

Suppose that, after the above change of P_2, we get a millipede M when applying Theorem 5.3. If M is one sided, it satisfies (6) since $m \leq 4$. If M is two sided, it can contain at most three bridges from Q'_i. If it contains two or three such bridges, R_i is among them. Suppose now that for some j, $3 \leq j \leq m - 4$, $r_j^+ \leq r_j^-$ (*i.e.*, (6) is violated). Notation $\leq$ means being closer to C_1 than to C_2 on P_1. Then $r_j^+ = r_j^-$. By (4), $r = r_j^+$ is the only attachment of $B_{j+1}^\circ \cup B_{j+2}^\circ$ on P_1. Since $B_j^\circ \leq B_{j+2}^\circ$, B_{j+1}° overlaps with B_j° and with B_{j+2}° on P_2. This implies that B_j°, B_{j+1}°, B_{j+2}° are the three bridges of some Q'_i and that $B_{j+1}^\circ = R_i$. Since B_{j+2}° is attached only to $P_1 \cup P_2$, we have $j + 2 < m$. Thus, there is a bridge B_{j+3}° overlapping with B_{j+2}° and not overlapping with B_{j+1}°. But clearly, this is not possible. We have a contradiction, so $r_j^- < r_j^+$. The proof that $l_j^- < l_j^+$ is similar. Hence, (6) holds.

We have covered the case when, in $G - P_1$, there is a path from C_1 to C_2. Suppose now that this is not the case. Then the K-bridges can be partitioned into classes $\mathcal{B}_1$, $\mathcal{B}_2$ such that $\mathcal{B}_i$ ($i = 1, 2$) contains exactly those bridges that are attached to $C_i - P_1$. Let x_1 be the vertex of P_1 as close to C_2 as possible such that there is a bridge in $\mathcal{B}_1$ that is attached to x_1. (If none of the bridges in $\mathcal{B}_1$ is attached to P_1, we let x_1 be the end of P_1 on C_1.) Define x_2 similarly for the bridges in $\mathcal{B}_2$. For $i = 1, 2$, we let G_i be the graph consisting of C_i, the segment of P_1 from C_i to x_i, and the bridges in $\mathcal{B}_i$. We will use the same notation later when providing details for the case when C_1 and C_2 are in distinct blocks of G. Clearly, this is not the case if and only if on P_1, x_1 is strictly closer to C_2 than x_2.

Let π_1 be a path in G_1 joining C_1 with x_1 such that $\pi_1 \cap P_1 = \{x_1\}$. Define π_2 similarly in G_2. Suppose that we have an embedding extension of K to G. If π_1 is attached to x_1 at the left side of F (with the obvious meaning of the 'left' with respect to Figure 2), then π_2 is attached at the right side. Then all the attachments of G_1 to P_1 at x_1 and between x_1 and x_2 are also on the left. Thus we say that G_1 has the *left side embedding* and, similarly, G_2 has the *right side embedding* with respect to x_1 and x_2. There are two possibilities for the non-existence of embedding extensions: either G_1 (or G_2) admits neither the left nor the right side embedding, or each of G_1 and G_2 does not admit the left side embedding (respectively, the right side embedding), but each of them admits the right (left) side embedding.

Suppose first that G_1 admits neither side embeddings. What are the possible obstructions? Define the graph G'_1 as follows. Let $y \in V(P_1)$ be the neighbor of x_1 that is closer to C_1 than x_1. Replace the edge yx_1 of G_1 by a pair of paths of length two between these two vertices and denote by C'_2 the obtained cycle of length 4. In addition to this, add the edge x_1x_2. It is easy to see that G_1 has neither side embeddings if and only if the obtained graph G'_1 has no embedding extending the 1-prism embedding of $K' = C_1 \cup C'_2 \cup (P_1 \cap G'_1)$ with C_1 and C'_2 on the boundary. (Note: the two embeddings of K' are really equivalent to each other.) This type of 1-prism embedding extension problem has been covered above – the case when there is a path π_1 disjoint from P_1 joining the two cycles. Since C'_2 is joined to C_1 only through y and x_1, we have just one such path. It is easy to see that G'_1 is 2-connected and that the auxiliary graph G'_1 with the auxiliary vertices joined to each other is 3-connected (since there are no local bridges on P_1). As shown above, this gives rise to obstructions of types (a), (b), (d), or (e) for the extension problem of G_1'. The obtained obstruction possibly contains the new edge x_1x_2, which is not present in G. By replacing this edge by π_2, we get an obstruction Ω for our original 1-prism embedding extension problem. If Ω is of type (a) (with respect to G_1'), it is easy to see that $\pi_2 \not\subseteq \Omega$, and that Ω fits our case (a) as well. If Ω is of type (b), then, again, π_2 is not a part of Ω. Thus Ω fits case (b). Note that in this case Ω can be a tripod attached only to P_1. In the case of type (d), the path appearing in (d) is π_1 (which may have been changed when constructing the obstruction). We have our case (f) or (g) with $P_2 = \pi_1$. In case (e) in G_1', note that Ω cannot be attached to $C_2' - (P_1 \cup P_2)$. Thus the corresponding 1-millipede is one-sided, and we have (h).

Up to symmetries, the only remaining case is when neither G_1 nor G_2 admits the left side embedding. We may as well assume that G_1 and G_2 admit the right side embeddings. Then we are looking for minimal left side obstructions. It is clear that we get case (i).

It remains to consider the case when C_1 and C_2 are in distinct blocks of G. Let G_1 and G_2 be the corresponding blocks. We suppose that G_1 has no embedding in the plane with C_1 bounding a face. An obstruction for this will obstruct the original embedding extension problem. Let us remark that using Theorem 3.1 is not a straightforward success since an obstruction obtained by using that result could intersect P_1 too many times. However, the application of Theorem 3.1 is possible if $G_1 \cap P_1$ is just the end of P_1 on C_1. Then the obstruction is either our case (a), or (b) (or (j)). Thus we assume that this is not the case.

Note that $G_1 \cap P_1$ is a segment of P_1 from C_1 to x_1. Let y be the neighbor of x_1 that is closer to P_1 than x_1. By the assumption made above, y is well-defined. Define G_1' to be the graph obtained from G_1 by replacing the edge yx_1 with two paths of length two between x_1 and y. Denote by C_2' the obtained cycle of length four consisting of these two paths. Clearly, G_1' has an embedding in the plane with C_1 and C_2' bounding faces if and only if G_1 has an embedding with C_1 bounding a face. By our assumption, this is not the case. Thus G_1' has no 1-prism embedding extension with respect to $K' = C_1 \cup C_2' \cup (P_1 \cap G_1)$. Possible obstructions have been classified above since in this problem we have a path disjoint from the first one. We get obstructions of types (a), (b), (f), or (h). $\qquad\square$

Case (i) of the previous theorem is well-described if we know what the minimal left side obstructions are and how we get them in linear time. Their discovery was covered in the theorem. The proof of the last theorem was also detailed enough to yield a simple classification of these obstructions.

Theorem 6.3. *Let Ω be a minimal left side obstruction with respect to x_1 and x_2. Then Ω is one of the graphs shown in Figure 11, where case (c) represents an arbitrary two-sided millipede for the 2-prism embedding extension problem described before Lemma 6.1.*

Proof. We will use all the notation and assumptions introduced in the preceding proof up to the point where we encountered the case (i). We suppose that there is a left side obstruction in G_1.

Let P_1' be the segment of P_1 from C_1 to x_2, and let C_2' be the segment of P_1 from x_2 to x_1 together with an additional edge x_1x_2. (If x_2 is just preceding x_1 on P_1, then we add a path of length two in order not to get parallel edges.) Define G_1' to be the graph obtained from $C_1 \cup P_1' \cup C_2'$ by adding all bridges from $\mathscr{B}_1$. Then G_1 has no left side embedding if and only if G_1' has no embedding in the cylinder with C_1 and C_2' on the boundary (with C_1 at the bottom and C_2' with its new edge on the right side). Thus we are looking for 1-prism embedding extension obstructions in G_1' with respect to $K' = C_1 \cup C_2' \cup P_1'$. Since in $\mathscr{B}_1$ there is a bridge joining $C_1 - P_1'$ with $x_1 \in C_2' - P_1'$, we get the case with two or three disjoint paths from C_1 to C_2' (counting also the path P_1'). If there are three disjoint paths, they obstruct the right side embeddings of G_1. By applying Theorem 6.2, we get an obstruction Ω_1 of type (a), (b), (d), or (e) with respect to G_1', since these are the possible cases that arise when in addition to P_1', there is only one path. In cases (d) and (e), we may suppose that the corresponding path P_2 is π_1 (which has possibly been changed during the procedure of constructing the obstruction, but its end x_1 has remained unchanged). We know, moreover, that Ω_1 has right side embeddings, since G_1 has such an embedding in F. Let us now consider particular cases for the resulting left side obstruction.

If Ω_1 is of type (a) (in G_1'), we claim that the disjoint crossing paths Q_1, Q_2 can be changed in such a way that one of them is attached to x_1. If this is not already the case in Ω_1, add the path π_1. Note that each of Q_1, Q_2 is attached to $C_1 - P_1$ and to the segment of P_1 between x_2 and x_1, since otherwise they also obstruct the right side embeddings. If π_1 intersects Q_1 or Q_2 in an internal vertex, we can change Q_1 or Q_2 so that one of them is attached to x_1. Otherwise, we can either replace one of the paths by π_1, or $\Omega_1 \cup \pi_1$ obstructs the right side embeddings. Hence the claim. Consequently, we have a left side obstruction represented in Figure 11(a).

If Ω_1 is of type (b), it is also the right side obstruction. If it is of type (d), corresponding to one of the cases (b), (d), (e), or (f) of Theorem 5.3, it is a right side obstruction as well. Type (d), case (c) is a right side obstruction except in two cases. One of them is represented in Figure 11(b), while the other contains case (a) of Figure 11 after removing the middle part of π_1. Finally, if Ω_1 is a millipede, it does not obstruct the right side embeddings if and only if it is two-sided. So, the last type of minimal left side obstruction is as claimed. $\square$

7. Conclusion

There is an additional property that the millipedes may be assumed to have. This property is essential for our further applications of the results of this paper and will be stated in our last results.

An *extended millipede* (or an *extended 1-millipede*) is defined by the same conditions (1)–(4) (and (6)) as the millipede, but (5) is replaced by the requirement that B_i° ($1 \le i \le m$) is an H-graph of a K-bridge in G. This ensures, in particular, that B_i° ($2 \le i \le m - 1$) are attached to P_1 and to P_2, but we lose the minimality property of millipedes as obstructions.

For $\Omega \subseteq G - E(K)$, we define $b(\Omega)$ to be the number of branches of Ω where all vertices of attachment of Ω (including those of degree 2 in Ω) are considered to be main vertices of Ω.

Theorem 7.1. *Let $K \subseteq G$ be a subgraph for a 2-prism embedding extension problem. There is a linear time algorithm that either finds an embedding extension of K to G, or returns an obstruction $\Omega \subseteq G - E(K)$ for such extensions. In the latter case, the obstruction Ω satisfies one of the following conditions:*

(a) *Ω is small, $b(\Omega) \leq 20$, $K \cup \Omega$ has at most four K-bridges, and at most 8 vertices of Ω are on P_2.*

(b) *$\Omega = B_1^{\circ} \cup B_2^{\circ} \cup \cdots \cup B_m^{\circ}$ is an extended millipede of length $m \geq 5$. Then $b(\Omega) \leq 5m$ and at most $2m$ vertices of Ω are on P_2. Moreover, if D ($\supseteq B_2^{\circ} \cup \cdots \cup B_{m-1}^{\circ}$) is the union of all K-bridges in G that are attached only to $P_1 \cup P_2$, there is an embedding extension of K to $K \cup D$.*

Proof. Consider first the 2-prism problem for $K \cup D$. By applying Theorem 5.3, we either get an embedding extension to $K \cup D$ or a small obstruction, since millipedes have attachments out of $P_1 \cup P_2$. In the latter case we have (a). Otherwise, we apply Theorem 5.3 again, this time for the original 2-prism problem. What we have gained, is that in the case of millipedes, we can guarantee the property stated in (b). The stated bounds follow from Theorem 5.3. Note that the case of millipedes having $m < 5$ is hidden in our case (a). □

Theorem 7.2. *Let $K = C_1 \cup C_2 \cup P_1 \subseteq G$ be a subgraph for a 1-prism embedding extension problem. Suppose, moreover, that there is a path in G disjoint from P_1 that joins C_1 and C_2. There is a linear time algorithm that either finds an embedding extension of K to G, or returns an obstruction $\Omega \subseteq G - E(K)$ for such extensions. In the latter case, the obstruction Ω satisfies one of the following conditions.*

(a) *Ω is small, and $b(\Omega) \leq 29$.*

(b) *$\Omega = P_2 \cup B_1^{\circ} \cup B_2^{\circ} \cup \cdots \cup B_m^{\circ}$ is an extended 1-millipede of length $m \geq 5$. Then $b(\Omega) \leq 7m$. Moreover, if D ($\supseteq B_2^{\circ} \cup \cdots \cup B_{m-1}^{\circ}$) is the union of all $(K \cup P_2)$-bridges in G that are attached only to $P_1 \cup P_2$, there is an embedding extension of K to $K \cup P_2 \cup D$.*

Proof. Apply the algorithm described in the proof of Theorem 6.2 except that instead of using Theorem 5.3 within that proof, we use Theorem 7.1 instead. The stated bounds on the branch size also follow from Theorem 7.1. □

Acknowledgement

We are greatly indebted to the referee for a very detailed checking of the manuscript and for his numerous comments that improved the presentation.

References

[1] Archdeacon, D. and Huneke, P. (1989) A Kuratowski theorem for non-orientable surfaces. *J. Combin. Theory*, Ser. B **46** 173–231.

[2] Booth, K. S. and Lueker, G. S. (1976) Testing for the consecutive ones property, interval graphs, and graph planarity using PQ-tree algorithms. *J. Comput. System Sci.* **13** 335–379.

[3] Chiba, N., Nishizeki, T., Abe, S. and Ozawa, T. (1985) A linear algorithm for embedding planar graphs using PQ-trees. *J. Comput. System Sci.* **30** 54–76.

[4] Cook, S. A. and Reckhow, R. A. (1973) Time bounded random access machines. *J. Comput. System Sci.* **7** 354–375.

[5] de Fraysseix, H. and Rosenstiehl, P. (1982) A depth-first search characterization of planarity. *Ann. Discrete Math.* **13** 75–80.

[6] Gross, J. L. and Tucker, T. W. (1987) *Topological Graph Theory*, Wiley-Interscience.

[7] Hopcroft, J. E. and Tarjan, R. E. (1973) Dividing a graph into triconnected components. *SIAM J. Comput.* **2** 135–158.

[8] Hopcroft, J. E. and Tarjan, R. E. (1974) Efficient planarity testing. *J. Assoc. Comput. Mach.* **21** 549–568.

[9] Jung, H. A. (1970) Eine Verallgemeinerung des *n*-fachen Zusammenhangs für Graphen. *Math. Ann.* **187** 95–103.

[10] Juvan, M., Marinček, J. and Mohar, B. (submitted) *Embedding graphs in the torus in linear time.*

[11] Juvan, M., Marinček, J. and Mohar, B. (in preparation) *Efficient algorithm for embedding graphs in arbitrary surfaces.*

[12] Juvan, M. and Mohar, B. (submitted) *A linear time algorithm for the 2-restricted embedding extension problem.*

[13] Karabeg, A. (1990) Classification and detection of obstructions to planarity. *Linear and Multilinear Algebra* **26** 15–38.

[14] MacLane, S. (1937) A structural characterization of planar combinatorial graphs. *Duke Math. J.* **3** 460–472.

[15] Mohar, B. (1993) Projective planarity in linear time. *J. Algorithms* **15** 482–502.

[16] Mohar, B. (submitted) *Universal obstructions for embedding extension problems.*

[17] Mohar, B. (in preparation) *A Kuratowski theorem for general surfaces.*

[18] Papadimitriou, C. H. and Steiglitz, K. (1982) *Combinatorial Optimization: Algorithms and Complexity*, Prentice-Hall.

[19] Robertson, N. and Seymour, P. D. (1990) Graph minors. VIII. A Kuratowski theorem for general surfaces. *J. Combin. Theory*, Ser. B **48** 255–288.

[20] Robertson, N. and Seymour, P. D. (1990) Graph minors. IX. Disjoint crossed paths. *J. Combin. Theory*, Ser. B **49** 40–77.

[21] Seymour, P. D. (1980) Disjoint paths in graphs. *Discrete Math.* **29** 293–309.

[22] Seymour, P. D. (1986) Adjacency in binary matroids. *European J. Combin.* **7** 171–176.

[23] Seymour, P. D. (submitted) *A bound on the excluded minors for a surface.*

[24] Shiloach, Y. (1980) A polynomial solution to the undirected two paths problem. *J. Assoc. Comput. Mach.* **27** 445–456.

[25] Thomassen, C. (1980) 2-linked graphs. *European J. Combin.* **1** 371–378.

[26] Tutte, W. T. (1966) *Connectivity in Graphs*, Univ. Toronto Press, Toronto, Ontario; Oxford Univ. Press, London.

[27] Williamson, S. G. (1980) Embedding graphs in the plane – algorithmic aspects. *Ann. Discrete Math.* **6** 349–384.

[28] Williamson, S. G. (1984) Depth-first search and Kuratowski subgraphs. *J. Assoc. Comput. Mach.* **31** 681–693.

A Ramsey-Type Theorem in the Plane

JAROSLAV NEŠETŘIL[†] and PAVEL VALTR[‡]

[†‡]Department of Applied Mathematics, Charles University,
Malostranské nám. 25, 118 00 Praha 1, Czech Republic
[‡]Graduiertenkolleg 'Algorithmische Diskrete Mathematik',
Fachbereich Mathematik, Freie Universität Berlin,
Takustrasse 9, 14194 Berlin, Germany

We show that, for any finite set P of points in the plane and for any integer $k \geq 2$, there is a finite set $R = R(P,k)$ with the following property: for any k-colouring of R there is a monochromatic set $\tilde{P}, \tilde{P} \subseteq R$, such that $\tilde{P}$ is combinatorially equivalent to the set P, and the convex hull of $\tilde{P}$ contains no point of $R \setminus \tilde{P}$. We also consider related questions for colourings of p-element subsets of R ($p > 1$), and show that these analogues have negative solutions.

1. Introduction and statement of results

In this paper we investigate geometrical Ramsey-type results that are related to the celebrated Erdős–Szekeres Theorem.

Theorem 1. (Erdős–Szekeres Theorem) *For every positive integer n there exists a positive integer $ES(n)$ such that any set X of $ES(n)$ points in general position in the plane (i.e., no three lie on a line) contains vertices of a convex n-gon.*

The Erdős–Szekeres Theorem is one of the original gems of Ramsey theory. By combining it with Ramsey's theorem [17] (see also [5] or [14]) itself we get the following corollary.

Corollary 2. *For every choice of positive integers p, k, n, there exists a positive integer $ES(p, k, n)$ with the following property: for any set X of at least $ES(p, k, n)$ points in general*

‡ The author was supported by 'Deutsche Forschungsgemeinschaft', grant We 1265/2-1.

position in the plane, and for any partition $\binom{X}{p} = C_1 \cup \ldots \cup C_k$, *there exists a set* $Y \subseteq X, |Y| = n$, *such that all p-subsets of* Y *belong to one class* C_{i_0} *of the partition and* Y *is the set of vertices of a convex n-gon.*

To verify the corollary, one simply puts $ES(p, k, n) = r(p, k, ES(n))$, where $r(p, k, N)$ is the usual Ramsey number for p-tuples and k colours.

We are interested in a generalization of Corollary 2 to sets Y of a given configuration. Somewhat surprisingly, this generalization can be made for $p = 1$ (i.e., for partition of points), while in general ($p > 1$) a similar statement is not true.

Here is the key concept of this paper: two finite planar point sets P and Q are called *combinatorially equivalent* if there exists a bijection $i : P \to Q$ such that $p \in \operatorname{conv}(P')$ if and only if $i(p) \in \operatorname{conv}(i(P'))$ for any $p \in P$ and $P' \subseteq P$. Here $\operatorname{conv}(X)$ is the convex hull of the set X and $i(X)$ denotes the set $\{i(x) : x \in X\}$.

A finite planar point set X is said to be *convex independent* if $\operatorname{conv}(X') \neq \operatorname{conv}(X)$ for every proper subset X' of X (or, equivalently, if points of X are vertices of a convex polygon). Otherwise X is said to be *convex dependent*. It is easy to see that two sets are combinatorially equivalent if and only if there is a bijection between them that preserves both convex dependent and convex independent sets.

Combinatorially equivalent sets and similar concepts have already been studied explicitly in several papers (see [3] or [4] for a survey). Implicitly they play a very important role, mainly in discrete and computational geometry. See [10] also for a survey on more general structures. In Section 2 we prove the following Ramsey-type theorem.

Theorem 3. *For any finite set P of points in the plane, and for any integer $k \geq 2$, there is a finite set $R = R(P, k)$ of points in the plane such that for any partition of R into k colour classes there is a subset $\tilde{P}$ of R with the following three properties:*
(i) $\tilde{P}$ is monochromatic (i.e., it is a subset of one of the colour classes),
(ii) $\tilde{P}$ is combinatorially equivalent to P,
(iii) $\operatorname{conv}(\tilde{P}) \cap R = \tilde{P}$.

Moreover, as we shall see from the proof, the set $\tilde{P}$ in Theorem 3 may be required to be an affine transform of P.

A subset X of a finite planar point set P is called a *hole in P*, or simply a *P-hole*, or an *empty polygon*, if X is convex independent and $\operatorname{conv}(X) \cap P = X$. Horton [8] (see also [1] or [19]) proved that, for $n \geq 7$, the Erdős–Szekeres Theorem cannot be strengthened to guarantee the existence of an *n*-hole. The existence of a 5-hole in any set of 10 points in general position in the plane was shown by Harborth [7], while the case $n = 6$ is still open.

Now Theorem 3 can be rephrased as follows.

Theorem 4. *For any finite set P of points in the plane and for any integer $k \geq 2$, there is a finite set $R = R(P, k)$ of points in the plane such that for any partition $R = C_1 \cup \ldots \cup C_k$ there is an injection $i : P \to R$ with the following three properties:*
(i) $i(P) \subseteq C_{i_0}$ for some $i_0 \in \{1, \ldots, k\}$,

(ii) i preserves convex independent and convex dependent sets,
(iii) i maps P-holes to R-holes.

Motivated by Corollary 2, one can also consider a higher-order Ramsey-type theorem (for $p > 1$). If we are partitioning the set $\binom{R}{2}$ into k colour classes, a result analogous to Theorem 3 cannot be valid in general: the pairs of any finite set R can be coloured as follows. On every line with at least two points of R we colour pairs of consecutive points alternately by two colours while all the other pairs are coloured arbitrarily. In this way we avoid a monochromatic triple of collinear points of R containing no other point of R in the convex hull. It follows that if we are partitioning the set $\binom{R}{2}$, a result analogous to Theorem 3 cannot be valid for planar point sets P that are not in general position. More generally, for $p \geq 2$, if we are partitioning the set $\binom{R}{p}$, a result analogous to Theorem 3 cannot be valid for planar point sets P with $p + 1$ points on a line.

However, the situation is more difficult if we restrict our attention to point sets P in general position. It turns out that Theorem 3 has no higher-order analogues even in this case. In fact, this remains true even if we drop the hole-preserving condition:

Theorem 5. *For every $p \geq 2$, there exists a finite planar point set $P(p)$ in general position with the following property: there exists a partition $\binom{\mathbb{R}^2}{p} = C_1 \cup C_2$ of all p-element subsets of the plane into two colour classes such that no monochromatic subset of $\mathbb{R}^2$ is combinatorially equivalent to $P(p)$.*

Thus Theorem 3 fails to have a higher-order analogue in general. However, such an analogue holds in some particular cases. Apart from Corollary 2 (which yields such an analogue for any finite convex independent set P), we have the following result, which deals with the configuration Q containing the three vertices and an inner point of a triangle.

Theorem 6. *Let Q be a convex dependent set of four points in general position in the plane. Then, for every integer $k \geq 1$, there exists a finite planar point set $R = R(k)$ such that for every partition of pairs of R into k classes, one of the classes contains all 6 pairs of points of some 4-point set combinatorially equivalent to Q.*

The paper is divided into sections as follows: Section 2 contains the proof of Theorem 3; Section 3 contains the proofs of Theorems 5 and 6; and Section 4 contains concluding remarks.

2. Proof of the main result

We begin with a short outline of the proof of Theorem 3. Given a planar point set, we find an equivalent set $\tilde{P}$ whose points are placed inside a small neighborhood of a line. Then we construct a set R containing many subsets combinatorially equivalent to $\tilde{P}$ and apply the Hales–Jewett Theorem to show that at least one of these subsets is monochromatic.

Fix a set P in the plane, and let $n = |P|$. Without loss of generality (or by a slight rotation of P), we may assume that all x-coordinates of the set P are different. Let the points of P, ordered according to their x-coordinate, be $p(1), \ldots, p(n)$.

Let $M > 0$ be any number such that all points of the set P lie inside the M-neighborhood of the x-axis (i.e., the y-coordinates of the points in P lie in the interval $(-M, M)$). For any $\varepsilon \geq 0$, let P_ε be the set obtained from the set P by replacing each point (x, y) by the point $(x, \frac{\varepsilon}{M} y)$. If $\varepsilon \neq 0$, the set P_ε is equivalent to P and is placed inside the ε-neighborhood of the x-axis. Otherwise (if $\varepsilon = 0$) the points of P_0 are collinear. The points of P_ε, listed according to increasing x-coordinate, will be denoted by $p_\varepsilon(1), \ldots, p_\varepsilon(n)$.

Let $\alpha \in [0, \pi)$ and let r_α be the rotation of the plane by the angle α around the origin. Thus $r_\alpha(x)$ denotes the point that we get by rotating x by r_α. Put $p_{\varepsilon,\alpha}(i) = r_\alpha(p_\varepsilon(i))$ (for $\varepsilon \geq 0, i = 1, \ldots, n$) and $P_{\varepsilon,\alpha} = \{p_{\varepsilon,\alpha}(1), \ldots, p_{\varepsilon,\alpha}(n)\}$ (for $\varepsilon \geq 0$). Thus we could also write $P_{\varepsilon,\alpha} = r_\alpha(P_\varepsilon)$.

To invoke the Hales–Jewett Theorem we need some notation. Put $A = \{1, \ldots, n\}$ (and think of A as an alphabet). Given a positive integer N, we consider the set A^N of all mappings $\{1, \ldots, N\} \to A$. One can also think of A^N as the N-dimensional cube over A. A (combinatorial) line L in A^N is defined as an n-element subset of A^N satisfying the following condition: there exists a proper subset $\omega \subset \{1, \ldots, N\}$ and a mapping $f_0 : \omega \to A$ such that the line L is formed by all mappings $f : \{1, \ldots, N\} \to A$ that satisfy $f(i) = f_0(i)$ for $i \in \omega$ and $f(i) = f(j)$ whenever $i, j \notin \omega$. More explicitly, $L = \{x(1), \ldots, x(n)\}$, where (for $j = 1, \ldots, n$)

$$x(j) = (x_1(j), \ldots, x_N(j))$$

and

$$x_i(j) = \begin{cases} f_0(i) & \text{for } i \in \omega \\ j & \text{for } i \notin \omega. \end{cases}$$

Now we can formulate the Hales–Jewett Theorem [6] (see also [5] or [14]).

Theorem 7. (Hales–Jewett Theorem) *For any two positive integers $n, k \geq 2$, there exists an integer $N = N(n, k)$ such that for any k-colouring of the points of the cube A^N, $A = \{1, \ldots, n\}$, there exists a monochromatic line.*

Denote by $\mathscr{L}(A^N)$ the set of all (combinatorial) lines in A^N. For two points $x = (x_1, x_2), y = (y_1, y_2)$ in the plane, we define $x + y = (x_1 + y_1, x_2 + y_2)$.

We shall embed the cube A^N into the plane using appropriately chosen rotations determined by 'independent' angles. For simplicity, for $\varepsilon \geq 0$, $\alpha = (\alpha_i)_1^N$, and $x = (x_i)_1^N \in A^N$, put $q_{\varepsilon,\alpha}(x) = \sum_{i=1}^N p_{\varepsilon,\alpha_i}(x_i)$ and $q_\alpha(x) = q_{0,\alpha}(x)$. Now let $L \in \mathscr{L}(A^N)$ be a combinatorial line such that the points $q_\alpha(x)$ ($x \in L$) are distinct. It is easy to check that the points $q_\alpha(x)$ ($x \in L$) all lie on a straight line on the plane. Let us denote this line by $q_\alpha(L)$. We say that $\alpha = (\alpha_1, \ldots, \alpha_N)$ above is P-independent if the points $q_\alpha(x)$ ($x \in A^N$) are distinct, and $q_\alpha(x) \notin q_\alpha(L)$ for all lines $L \in \mathscr{L}(A^N)$ and all $x \in A^N \setminus L$.

Lemma 8. *For any finite planar point set P with different x-coordinates, and for any positive integer N, there is a P-independent N-tuple $\alpha = (\alpha_1, \ldots, \alpha_N)$.*

Proof. Let $\alpha = (\alpha_1, \ldots, \alpha_N)$ be an arbitrary tuple of N angles. We shall show that one can get a P-independent N-tuple by a small change of the angles α_i. Suppose two points $q_\alpha(x_1), q_\alpha(x_2)$ $(x_1, x_2 \in A^N, x_1 \neq x_2)$ coincide. Let x_1 and x_2 differ in the i-th coordinate. For $\varepsilon > 0$, set $\overline{\alpha} = (\alpha_1, \ldots, \alpha_{i-1}, \alpha_i + \varepsilon, \alpha_{i+1}, \ldots, \alpha_N)$. The points $q_{\overline{\alpha}}(x_1)$ and $q_{\overline{\alpha}}(x_2)$ are different, and, if $\varepsilon > 0$ is sufficiently small, any two points $q_{\overline{\alpha}}(x), q_{\overline{\alpha}}(x'), x, x' \in A^N$, are different whenever the points $q_\alpha(x)$ and $q_\alpha(x')$ are different. Thus, if α is replaced by $\overline{\alpha}$ (with $\varepsilon > 0$ sufficiently small), the number of coincidences between the points $q_\alpha(x)$ $(x \in A^N)$ drops. Repeating this procedure, we obtain a tuple α such that the points $q_\alpha(x)$ $(x \in A^N)$ are distinct.

Suppose now that, for $L_0 \in \mathscr{L}(A^N)$ and for $x_0 \in A^N \setminus L_0$, the point $q_\alpha(x_0)$ lies on the line $q_\alpha(L_0)$. Let $i \in \{1, \ldots, N\}$ be such that the n points of L_0 have distinct i-th coordinates (i.e., their i-th coordinates are 1 through n). For $\varepsilon > 0$, set $\overline{\alpha} = (\alpha_1, \ldots, \alpha_{i-1}, \alpha_i + \varepsilon, \alpha_{i+1}, \ldots, \alpha_N)$ as above.

Suppose $q_\alpha(L_0)$ determines the angle β with the positive x-axis. Picking small enough $\varepsilon > 0$, we may assume that $(\alpha_i + (\alpha_i + \varepsilon))/2 \neq \beta - \pi/2$ and $(\alpha_i + (\alpha_i + \varepsilon))/2 \neq \beta + \pi/2$. Then, $q_\alpha(L_0)$ and $q_{\overline{\alpha}}(L_0)$ are not parallel. Let $\hat{x}$ be the point of L_0 whose i-th coordinate equals the i-th coordinate of x_0. Then, the points $q_{\overline{\alpha}}(\hat{x})$ and $q_{\overline{\alpha}}(x_0)$ lie on a line that is parallel to $q_\alpha(L_0)$ and, hence, not parallel to $q_{\overline{\alpha}}(L_0)$. Note that the point $q_{\overline{\alpha}}(\hat{x})$ lies on the line $q_{\overline{\alpha}}(L_0)$ by definition. It follows that the point $q_{\overline{\alpha}}(x_0)$ does not lie on the line $q_{\overline{\alpha}}(L_0)$ (since otherwise $q_{\overline{\alpha}}(x_0) = q_{\overline{\alpha}}(\hat{x})$ and, consequently, $q_\alpha(x_0) = q_\alpha(\hat{x})$). If $\varepsilon > 0$ is sufficiently small, the points $q_{\overline{\alpha}}(x)$ $(x \in A^N)$ are distinct and, for any $L \in \mathscr{L}(A^N)$ and $x \in A^N \setminus L$, the point $q_{\overline{\alpha}}(x)$ does not lie on the line $q_{\overline{\alpha}}(L)$ whenever the point $q_\alpha(x)$ does not lie on the line $q_\alpha(L)$. Thus, if α is replaced by $\overline{\alpha}$ (with $\varepsilon > 0$ sufficiently small), the number of line–point incidences drops.

Repeating the procedure described in the above two paragraphs, we finally obtain a P-independent tuple α. □

Now we are in a position to conclude the proof of Theorem 3.

Proof of Theorem 3. Let P be a set of n points in the plane with distinct x-coordinates. Put $A = \{1, \ldots, n\}$. Let $N = N(n, k)$ be a number guaranteed by the Hales–Jewett Theorem. Let α be a P-independent N-tuple. Thus for any line $L \in \mathscr{L}(A^N)$, and any $x \in A^N \setminus L$, we have $q_\alpha(x) \notin q_\alpha(L)$. By an obvious continuity argument, there is a sufficiently small $\varepsilon > 0$ such that, for any line $L \in \mathscr{L}(A^N)$, the convex hull of the points $q_{\varepsilon,\alpha}(x)$ $(x \in L)$ contains no point $q_{\varepsilon,\alpha}(x)$ with $x \in A^N \setminus L$. For any line $L \in \mathscr{L}(A^N)$, the set $\{q_{\varepsilon,\alpha}(x) : x \in L\}$ is combinatorially equivalent to the set P. Now we are ready to show that the set $R = \{q_{\varepsilon,\alpha}(x) : x \in A^N\}$ has all the required properties.

Let $R = C_1 \cup \ldots \cup C_k$ be a k-colouring of R. It induces a k-colouring of the set A^N. According to the Hales–Jewett Theorem, there exists a monochromatic line $L_0 \in \mathscr{L}(A^N)$. The set $\tilde{P} = q_{\varepsilon,\alpha}(L_0) = \{q_{\varepsilon,\alpha}(x) : x \in L_0\}$ is a monochromatic subset of R, it is combinatorially equivalent to P, and its convex hull contains no point of $R \setminus \tilde{P}$. □

Note that the set $\tilde{P}$ in the above proof is actually an affine transform of the set P.

3. Proofs of related results

In this section we prove Theorems 5 and 6. The proof of Theorem 5 follows from the following geometric result, which we proved in [15].

Theorem 9. *([15]) For any integer $k > 0$ and for any $k+1$ positive real numbers $\varepsilon, r_1, r_2, \ldots, r_k > 0$, there exists a finite planar point set P in general position such that any set combinatorially equivalent to P determines $k + 1$ distances d_i $(i = 0, 1, \ldots, k)$ such that $|r_i - d_i/d_0| < \varepsilon$ for any $i = 1, 2, \ldots, k$. Moreover, one may require the distances d_i $(i = 0, 1, \ldots, k)$ to be determined by pairwise disjoint pairs of points.*

Proof of Theorem 5. First we prove Theorem 5 for $p = 2$. Let P be a set satisfying Theorem 9 for $k = 2$, $\varepsilon = 0.01$, $r_1 = 1.9$, and $r_2 = 2.5$. We find a 2-colouring of all pairs of points of the plane such that no set combinatorially equivalent to P has all pairs coloured by the same colour. Any pair $(x, y) \in \binom{\mathbb{R}^2}{2}$ of points of the plane with Euclidean distance $|xy| > 0$ will be coloured blue if $\lfloor \log_2 |xy| \rfloor$ is an even integer. Otherwise (x, y) will be coloured red. In other words, a pair of points is coloured blue if and only if the distance between the two points belongs to some interval $[2^t, 2^{t+1})$, where t is an even integer.

Now let P' be a set combinatorially equivalent to P and let $d_i, i = 0, 1, 2$, be the three distances in P' ensured by Theorem 9. Thus $|d_1/d_0 - 1.9| < 0.01$. If the two pairs of points determining d_0 and d_1 are coloured by the same colour, the numbers d_0 and d_1 belong to the same interval $[2^t, 2^{t+1}), t \in \mathbb{Z}$, and, consequently, d_2 belongs to the next interval $[2^{t+1}, 2^{t+2})$. It follows that all the three pairs determining the distances d_0, d_1, d_2 cannot be coloured by the same colour. Theorem 5 for $p = 2$ follows.

Now let $p > 2$. Fix an arbitrary linear order $<$ of the points of the plane, and colour every p-tuple of points of the plane by the colour in which the pair of the two smallest (in the order $<$) points of the p-tuple was coloured above. A short argument shows that Theorem 5 holds for this 2-colouring and for the set $P(p) = P$ obtained from Theorem 9 for $k = 3p - 4$, $\varepsilon = 0.01$, $r_1 = \ldots = r_{p-2} = 1$, $r_{p-1} = \ldots = r_{2p-3} = 1.9$, and $r_{2p-2} = \ldots = r_{3p-4} = 2.5$. $\square$

Thus, an analogue of Theorem 3 fails to be true for partitions of p-tuples in a very strong sense. However, some particular cases are valid. One such example is provided by Theorem 6. We could denote the statement of Theorem 6 by $R \to (Q)^2_k$. Despite the simplicity of the configuration Q, our proof is quite involved, and, in particular, it makes use of Theorem 3. We shall only sketch the proof here as we intend to return to this topic elsewhere.

Proof of Theorem 6 (Sketch). First we prove the following lemma.

Lemma 10. *For every given point sets P_1, P_2 in general position, and for every $k \geq 1$, there are point sets R_1, R_2 with the following two properties:*

(i) $R_1 \cup R_2$ is in general position,

(ii) for every partition $C_1 \cup \ldots \cup C_k$ of all pairs $(x_1, x_2), x_1 \in R_1, x_2 \in R_2$, there exist two sets $P_1' \subseteq R_1, P_2' \subseteq R_2$, P_i' combinatorially equivalent to P_i for $i = 1, 2$, such that all pairs $(x_1, x_2), x_1 \in R_1, x_2 \in R_2$, are monochromatic.

Proof. We apply Theorem 3. Put $R_1 \to (P_1)_k^1, R_2 \to (P_2)_K^1$, where $K = k^{|R_1|}$. By a standard Ramsey theory argument we get the statement.   □

Clearly Lemma 10 may be generalized (from bipartite to multipartite graphs with more sets $P_1, P_2, P_3, \ldots$). We shall use this for r-partite graphs, where $r = r_k(3)$ is the classical Ramsey number for a monochromatic triangle in any k-colouring of the edges of the complete graph.

Somewhat surprisingly, we shall prove Theorem 6 by induction on k. For $k = 1$ Theorem 6 trivially holds. Let us assume that we have already found a planar point set S such that $S \to (Q)_{k-1}^2$. Now let $R_1, \ldots, R_r$ be r planar point sets such that for any partition of all pairs $(x, y), x \in R_i, y \in R_j, i \neq j$, there are r sets $S_1, \ldots, S_r, S_1 \subseteq R_1, \ldots, S_r \subseteq R_r$, equivalent to S such that for every choice of indices $i, j, i \neq j$, all pairs between S_i and S_j are coloured by the same colour $c(i, j)$. Now we can suppose that the set R_i is placed in a very small neighborhood of the vertical line $L_i = \{(i, y) : y \in \mathbb{R}\}$ with all its y-coordinates distinct. Assume that the y-coordinates of all the points in R_i are in the interval $(i^2, i^2 + 1)$. According to Ramsey's Theorem $(r = r_k(3))$, there are three indices $i, j, l, i < j < l$, such that all pairs x, y between S_i, S_j, S_l are coloured by the same colour c (in the above notation $c = c(i, j) = c(i, l) = c(j, l)$). If no pair of distinct points of S_j is coloured by the colour c, we can use the inductive assumption $S_j \to (Q)_{k-1}^2$ to get a copy of Q with all pairs of points coloured by the same colour. Thus we may assume that there exists a pair (x_j, x_j') of points of S_j coloured by the colour c. Choose $x_i \in S_i, x_l \in S_l$ arbitrarily. By our construction, both x_j and x_j' lie below the line $x_i x_l$ and, if R_i is in a small enough neighborhood of L_i, the line $x_j x_j'$ separates the points x_i and x_l. Thus $\{x_i, x_j, x_j', x_l\}$ is a homogeneous set equivalent to Q.   □

4. Concluding remarks

1. Theorem 3 may be generalized to any fixed finite dimension. Since Theorem 9 holds in a higher dimension (see [15]), Theorem 5 may also be generalized to a higher dimension.

2. Another way of rephrasing Theorem 3 (for sets in general position) is by means of geometric graphs (which were studied in various contexts, *e.g.* in [11], [13], [9], [16], [12]). A *geometric graph* is defined as a pair (V, E), where V is a set of points in general position in the plane and E is a subset of the set of all line segments connecting points of V. Two geometric graphs $(V, E), (V', E')$ are said to be *isomorphic* if there exists a bijection $f : V \to V'$ satisfying the following two conditions:

(i) $v_1 v_2 \in E$ if and only if $f(v_1)f(v_2) \in E'$,
(ii) two line segments $v_1 v_2, v_3 v_4 \in E$ cross if and only if the corresponding line segments $f(v_1)f(v_2), f(v_3)f(v_4) \in E'$ cross.

The following theorem may be proved by the method used in the proof of Theorem 3.

Theorem 11. *For every geometric graph (V, E) there exists a geometric graph (W, F) such that, for every partition $W = C_1 \cup \ldots \cup C_k$, there exists a set $V' \subseteq W$ with the following three properties:*

(i)   *The subgraph of (W, F) induced by V' is isomorphic to (V, E) as a geometric graph,*
(ii) $V' \subseteq C_{i_0}$ *for some i_0,*
(iii) *the convex hull of V' contains no point of $W \setminus V'$.*

3. The above proof of Theorem 6 does not guarantee that $\mathrm{conv}(Q') \cap R = Q'$. We do not know whether Theorem 6 with the extra condition $\mathrm{conv}(Q') \cap R = Q'$ holds. In general we have the following question: does there exist a planar point set $\overline{Q}$ for which an analogue of Theorem 6 holds but such an analogue does not hold if we further require that the corresponding set $\overline{Q}'$ should satisfy $\mathrm{conv}(\overline{Q}') \cap R = \overline{Q}'$?

4. The minimal size of the set R in Theorem 3 is bounded by a primitive recursive function (by Shelah's proof of the Hales–Jewett Theorem, [18]). However, the best lower bound we have is only quadratic (in $|P|$). The quadratic lower bound holds even if we delete condition (iii) in Theorem 3.

5. If we delete condition (iii) in Theorem 3, then, for $k = 2$ and for any set P in general position, it is possible to find a set R of size $O(n^2)$ satisfying Theorem 3 (without (iii)), where $n = |P|$. On the other hand, for any positive integer n, there is a set P of size n in general position for which the size of any set R satisfying Theorem 3 (without (iii)) is at least $\Omega(n^2/\log n)$.

Acknowledgment

We would like to thank the referee for his very careful work.

References

[1] Bárány, I. and Füredi, Z. (1987) Empty simplices in Euclidean space. *Canadian Math. Bull.* **30** 436–445.
[2] Erdős, P. and Szekeres, G. (1935) A combinatorial problem in geometry. *Compositio Math.* **2** 463–470.
[3] Goodman, J. E. and Pollack, R. (1991) The complexity of point configurations. *Discrete Appl. Math.* **31** 167–180.
[4] Goodman, J. E. and Pollack, R. (1993) Allowable sequences and order types in discrete and computational geometry, in: Pach, J. (ed.), *New trends in discrete and computational geometry*, Springer-Verlag.
[5] Graham, R. L., Rothschild, B., and Spencer, J., (1980) *Ramsey Theory*, J. Wiley & Sons, New York.
[6] Hales, A. W., and Jewett, R. I., (1963) Regularity of Positional Games, *Trans. Am. Math. Soc.* **106** 222-229.

[7]   Harborth, H., (1978) Konvexe Fünfecke in ebenen Punktmengen. *Elem. Math.* **33** 116–118.

[8]   Horton, J. D., (1983) Sets with no empty convex 7-gons. *Canadian Math. Bull.* **26** 482–484.

[9]   Korte B., and Lovász, L., (1985) Posets, matroids, and greedoids. in: Lovász, L. and Recski, A. (eds.), *Matroid theory*, North-Holland, 239–265.

[10] Korte, B. Lovász, L. and Schrader, R. (1991) *Greedoids*, Springer-Verlag, Berlin.

[11] Lovász, L. (1979) Topological and algebraic methods in graph theory. In: Bondy, A. and Murty, U. S. R. (eds.), *Graph Theory and Related Topics*, Academic Press 1–14.

[12] Moser, W. and Pach, J. (1993) Recent developments in combinatorial geometry. In: Pach, J. (ed.) *New trends in discrete and computational geometry*, Springer-Verlag.

[13] Nešetřil, J. Poljak, S. and Turzík, D. (1981) Amalgamation of matroids and its applications, *J. Comb. Th. B* **31** 9–22.

[14] Nešetřil, J. and Rödl, V. (eds.) (1990) *Mathematics of Ramsey Theory*, Springer-Verlag (1990).

[15] Nešetřil, J. and Valtr, P. (preprint) Order types containing approximately an affine transformation of the grid $k \times k$.

[16] Pach, J. (1991) Notes on Geometric Graph Theory. *DIMACS Series in Discrete Mathematics and Theoretical Computer Science* **6** 273–285.

[17] Ramsey, F. P. (1930) On a problem of formal logic. *Proc. Lond. Math. Soc.*, II. Ser. **30** 264–286.

[18] Shelah, S. (1988) Primitive recursive bounds for van der Waerden numbers. *Journal AMS* **1** 683–697.

[19] Valtr, P. (1992) Convex independent sets and 7-holes in restricted planar point sets. *Discrete Comput. Geom.* **7** 135–152.

The Enumeration of Self-Avoiding Walks and Domains on a Lattice

H. N. V. TEMPERLEY

Emeritus Professor of Applied Mathematics, Swansea, Wales, U.K.
Thorney House, Thorney, Langport, Somerset, U.K.

1. Introduction

The domain problem is of interest in magnetism and the walks problem in polymer science and molecular biology. Almost the only analytic information is due to Edwards – see for example Edwards (1) and many earlier papers. His method is approximate, but is reliable enough to give correct critical exponents.

2. The Cluster Method

We use an approach based on Mayer's classical work on the imperfect gas, namely we begin by counting *all* chain configurations and then remove those configurations in which one or more points coincide. Our chains consist of links lying along the lines of a plane square or simple cubic lattice, the ends of the links being on lattice points. A cluster consists of two or more links connected up in succession with two or more end-points coinciding. Figure 1 shows some of the smallest clusters.

The full lines correspond to links on the chain, the dotted lines join points that are constrained to be coincident. Cluster (a) corresponds to two links lying together on a line of the lattice, cluster (b) to three links lying together on a line. Cluster (c) may lie on a square of lines on the lattice or on two adjacent lines, or on one line. Cluster (d) can only lie along one line of the lattice. To form the cluster series we weight each cluster with

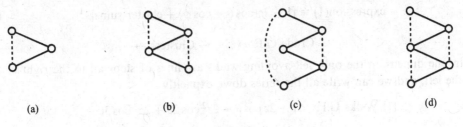

(a)   (b)   (c)   (d)

Figure 1

z to the power of the number of links, multiplied by the number of embeddings of the cluster in the lattice. We also introduce a factor -1 for each dotted line, corresponding to the fact that such configurations are eventually to be subtructed out of the "walks" generating function. It turns out that the only types of cluster that we need are what Mayer called "irreducible" but most graph theorists would call them multiply connected. (A graph containing an articulation point can be factored into two smaller graphs and we do not need such a graph in our cluster generating function.

In Temperley (4) (1988) it is formally proved that the generating function for non-self intersecting walks with end-to-end displacement specified by selector variables is the same as the reciprocal of the generating function for chains of clusters and single links.

For example, for the plane square lattice, if z is the variable whose power counts the number of steps, and the powers of $e^{i\theta}$ and $e^{i\phi}$ count horizontal and vertical steps, the first few terms of the generating function for self-avoiding walks on this lattice are easily seen to be

$$1+z(2\cos\theta+2\cos\phi)+z^2[2\cos 2\theta+2\cos 2\phi+4\cos(\theta+\phi)+4\cos(\theta-\phi)]$$
$$+z^3[2\cos z\theta+2\cos 3\phi+8\cos(2\theta+\phi)+8\cos(2\theta-\phi)+8\cos(\theta+2\phi)+8\cos(\theta-2\phi)]$$
$$+z^4[2\cos 4\theta+2\cos 4\phi+\ldots$$

$$(1)$$

whereas the crude generating function for walks with self-intersections allowed is simply

$$[1-z(2\cos\theta+2\cos\phi)]^{-1}. \qquad (2)$$

The term in z^2 in (1) differs from $z^2(2\cos\theta+2\cos\phi)^2$ by the term $4z^2$ which corresponds to four two-step walks involving "immediate reversals" the only two-step walks that intersect themselves. These immediate reversals are enumerated by the cluster sum (a) in Figure 1. The generating function for walks of two steps without immediate reversals may be written $z^2(2\cos\theta+2\cos\phi)^2-4z^2$ simply subtracting the immediate reversals. It is also the coefficient of z^2 in the modified generating function $[1-2(2\cos\theta+2\cos\phi)+4z^2]^{-1}$. However if we want self-avoiding walks involving more steps, we have to introduce further connecting terms, the first of which is z^3 times the cluster sum in Figure 1(b), and there are further terms in z^4 involving clusters of four steps etc. In Temperley (4) (1988) it is formally proved that a generating function such as (1) is obtained from the crude generating function (2) by subtracting out clusters involving self-intersections and that all walks involving self-intersections can be expressed as products of single links and irreducible clusters.

Explicitly we obtain the cluster generating function simply by taking the reciprocal of the walks generating function, e.g.,

$$\text{expression(1)} \equiv [1-2z(\cos\theta+\cos\phi)+\text{cluster sums}]^{-1} \qquad (3)$$

$$\leftarrow \text{Crude G.F.} \rightarrow \qquad \leftarrow \text{Clusters} \rightarrow$$

In one dimension the only self-avoiding walks are lines of steps all to the right or all to the left, and we can write all the series down explicitly.

[1] Walks G.F. $= 1 + 2z\cos\theta + 2z^2\cos 2\theta + 2z^3\cos 3\theta + \ldots$

$$\equiv \left[1 - 2z\cos\theta + \frac{2z^2}{1-z^2} - \frac{2z^3\cos\theta}{1-z^2}\right]^{-1}$$

Table 1   Terms up to z^{15} for the plane square lattice

	z^2	z^4	z^6	z^8	z^{10}	z^{12}	z^{14}
$C_{0,0}(z)$	$-4z^2$	$-12z^4$	$-60z^6$	$-332z^8$	$-1948z^{10}$	$-11708z^{12}$	$-71789z^{14}$
$C_{1,1}(z)$			$-8z^6$	$-64z^8$	$-424z^{10}$	$-2608z^{12}$	$-16184z^{14}$
$C_{2,0}(z)$				$-4z^8$	$-36z^{10}$	$-256z^{12}$	$-1268z^{14}$
$C_{2,2}(z)$							$-32z^{14}$

	z	z^3	z^5	z^7	z^9	z^{11}	z^{13}	z^{15}
$C_{1,0}(z)$	$2z$	$+2z^3$	$+14z^5$	$+78z^7$	$+482z^9$	$+2926z^{11}$	$+18006z^{13}$	
$C_{2,1}(z)$					$4z^9$	$+20z^{11}$	$+120z^{13}$	
$C_{3,0}(z)$								$16z^{15}$
$C_{3,2}(z)$								$4z^{15}$

Table 2   Terms up to z^{14} for the simple cubic lattice

	z^2	z^4	z^6	z^8	z^{10}	z^{12}	z^{14}
$C_{0,0,0}(z)$	$-6z^2$	$-30z^4$	$-366z^6$	$-5022z^8$	$-76062z^{10}$	$-1230462z^{12}$	$-20787102z^{14}$
$C_{1,1,0}(z)$			$-8z^6$	$-96z^8$	$-1032z^{10}$	$-9840z^{12}$	$-69512z^{14}$
$C_{2,0,0}(z)$				$-8z^8$	$-232z^{10}$	$-3888z^{12}$	$-58176z^{14}$
$C_{2,2,0}(z)$							$-128z^{14}$
$C_{3,1,0}(z)$							$-8z^{14}$

	z	z^3	z^5	z^7	z^9	z^{11}	z^{13}
$C_{1,0,0}(z)$	$2z$	$+2z^3$	$+26z^5$	$+394z^7$	$+5778z^9$	$+90714z^{11}$	$+1490378z^{13}$
$C_{1,1,1}(z)$					$-48z^9$	$-578z^{11}$	$+5424z^{13}$
$C_{1,1,1}(z)$					$-14z^9$	$+60z^{11}$	$+1904z^{13}$

where the crude generating function is $(1 - 2z\cos\theta)^{-1}$ and the effect of the cluster terms is to remove walks with self-intersections. In (4) the term $2z^2$ corresponds to the cluster 1(a) in Figure 1, the term $-2z^3\cos\theta$ to the cluster 1(b) in Figure 1, the term $2z^4$ to the sum of all cluster sums involving four links and so on.

Since generating functions for self-avoiding walks on the plane square lattice, weighted according to the end distances of the walks, were available in King's College, London, having been originally obtained by Watson for up to 15 steps, a programme was written by G. Evans of Swansea Computing Centre to take the reciprocal of the walks generating function, thus obtaining the cluster series. The results for the plane square lattice for walks of up to 16 steps are tabulated in Temperley (5) (1989). Guttman meanwhile verified Watson's original work on the plane square lattice and extended the series out to 25 steps. He also obtained the generating function for walks on the simple cubic lattice for up to 15 steps. The results of taking the reciprocal of the extended generating function for the plane square lattice and of the newly available generating function for the simple cubic lattice are reported in Temperley (6) (1991).

The results (up to z^{14}) for the plane square lattice and simple cubic lattice are reproduced in Table 1. Here $C_{0,0}(z)$ gives the generating function corresponding to clusters corresponding to zero end to end spacing, $C_{1,0}(z)$ to clusters corresponding to end to end spacing of one lattice distance, coefficient $(\cos\theta + \cos\phi)$, $C_{1,1}(z)$ to clusters corresponding to one horizontal and one vertical lattice distance, coefficient $(\cos\theta\cos\phi)$ etc. For the simple cubic lattice (Table 2), we use an obvious extension of the notation for the plane square lattice, for example $C_{1,0,0}(z)$ is the coefficient of $(\cos\theta + \cos\phi + \cos\psi)$ etc.

3. Discussion of the results

Two things stand out on looking at these data. First, the coefficients of the corresponding powers of z in the cluster series are very much smaller than the corresponding terms in the original "walks" generating function. (The sum of the coefficients of z^N in the original generating functions is of the order of 3^N for the plane square lattice and 5^N for the simple cubic lattice.) Second, the corresponding coefficients in the cluster series decrease very rapidly as the end to end distance in lattice spacings increases. It is clear from these results that we have obtained a very rapidly converging set of successive approximations to the generating function series. In fact in the plane square lattice we can replace the generating function by

$$[1 - 2z(\cos \theta + \cos \phi) - C_{0,0}(z) - C_{1,0}(z)]^{-1}$$

and still have the generating function reproduced up to the term in z^6, with only a small correction to this and higher terms arising from the $C_{1,1}(z)$ and later series. Similarly for the simple cubic lattice it will be a good approximation to work with just the cluster series $C_{0,0,0}(z)$ and $C_{1,0,0}(z)$ and neglect the rest. Recall that the values of z that we are interested in for series analysis are small fractions, $[2.638\ldots]^{-1}$ for the plane square lattice and $[4.638\ldots]^{-1}$ for the simple cubic lattice. (The numbers 2.638 and 4.638 are the expected numbers of ways of adding another step to a long walk without causing a self-intersection. These are slightly less than the numbers 3 and 5 which would correspond just to avoiding "immediate reversals".)

In statistical mechanics we are mainly interested in obtaining the limiting number of walks as $(2.638\ldots)^N \phi(N)$ where $\phi(N)$ is a slowly varying function of N, usually a power or logarithm. Sometimes we are interested in quantities like the mean square end-to-end displacement, obtained from the second derivative of the generating function with respect to θ or ϕ.

As in problems in the theory of numbers, our first approximation just determines the number 2.638... and gives a preliminary estimate of $\phi(N)$. Higher approximations, introducing $C_{1,1}(z)$ and later series should give better estimates of the function of $\phi(N)$. This approach to the generating function seems to give a great deal more insight than direct analysis of the generating function. As the result of very complicated analysis, Guttman concluded that the generating function for the plane square lattice had a confluent singularity of logarithmic type. Looking at our series $C_{0,0}(z)$ and $C_{1,0}(z)$, Guttman concluded that they both diverge logarithmically at the critical value $z = (2.638\ldots)^{-1}$, which seems to confirm Guttman's conclusion.

4. Conclusion

A mathematical physicist would prefer an exact result for the generating function, such as those obtained by Onsager for the two-dimensional ferromagnetic, and by Lieb and Baxter for the six and eight vertex models and the hard hexagon model. However, such exact results are still scarce and we usually have to be content with series analysis or renormalization type approximations.

Only a few theory of numbers type problems have exact treatments, for example the Hardy-Ramanujan-Rademacher recipes for calculating exactly the number of partitions of N. Our work suggests that our self-avoiding walk problem is amenable to successive improvements of the asymptotic results as in other theory of numbers and lattice-point type problems.

5. Acknowledgement

The references acknowledge help by A.J. Guttman and others.

References

[1] Edwards, S.F. (1986) *Theory of Polymer Dynamics* (Doi, M. and Edwards, S. F., eds.), Oxford University Press

[2] Guttmann, A. J. (1987) On the critical behaviour of self-avoiding walks, *J. Phys. A: Math. Gen.* **20** 1839–1854.

[3] Temperley, H. N. V. (1957) On the statistical mechanics of non-crossing chains: part 1, *Trans. Faraday Soc.* **53** 1065–1073.

[4] Temperley, H. N. V. (1988) New results on the enumeration of non-intersecting random walks, *Discrete Appl. Maths.* **19** 367–379.

[5] Temperley, H. N. V. (1989) On the statistical mechanics of non-crossing chains: part 2 *J. Phys. A: Math. Gen.* **22** L843–L847.

[6] Temperley, H. N. V. (1991) On the statistical mechanics of non-crossing chains: part 3 *J. Phys. A: Math. Gen.* **24** L609–L613.

An Extension of Foster's Network Theorem

PRASAD TETALI

AT & T Bell Labs, Murray Hill, NJ 07974.
Email: prasad@research.att.com

Consider an electrical network on n nodes with resistors r_{ij} between nodes i and j. Let R_{ij} denote the *effective resistance* between the nodes. Then Foster's Theorem [5] asserts that

$$\sum_{i \sim j} \frac{R_{ij}}{r_{ij}} = n - 1,$$

where $i \sim j$ denotes i and j are connected by a finite r_{ij}. In [10] this theorem is proved by making use of random walks. The classical connection between electrical networks and reversible random walks implies a corresponding statement for reversible Markov chains. In this paper we prove an elementary identity for ergodic Markov chains, and show that this yields Foster's theorem when the chain is time-reversible.

We also prove a generalization of a *resistive inverse* identity. This identity was known for resistive networks, but we prove a more general identity for ergodic Markov chains. We show that time-reversibility, once again, yields the known identity. Among other results, this identity also yields an alternative characterization of reversibility of Markov chains (see Remarks 1 and 2 below). This characterization, when interpreted in terms of electrical currents, implies the *reciprocity theorem* in single-source resistive networks, thus allowing us to establish the equivalence of *reversibility* in Markov chains and *reciprocity* in electrical networks.

1. Foster's Theorem

Let $P = (P_{ij})$ denote the n by n transition probability matrix of an ergodic Markov chain with stationary distribution π, and let us assume that $P_{ii} = 0$ for all i. Furthermore, let $H = (H_{ij})$ denote the expected first-passage matrix (also of size $n \times n$) of the above chain. Thus H_{ij} denotes the expected time it takes to reach state j from state i. We call the H_{ij} the *hitting times*. Here then is our result, which will easily imply Foster's Theorem.

Theorem 1. *With the notation above,*

$$\sum_{i,j} \pi_j P_{ji} H_{ij} = n - 1.$$

We give two elementary proofs of this identity. Having found the first, the author realized that the shorter second proof is hidden in a proof of [2, Theorem 1], unbeknown to the authors of [2].

Proof 1. Let N_k^{ij} denote the expected number of visits to k in a random walk from i to j. Then, [8, equation (34) (page 221)] implies that, for $k \neq j$,

$$\sum_i P_{ji} N_k^{ij} = \frac{\pi_k}{\pi_j}.$$

That is

$$\sum_i \pi_j P_{ji} N_k^{ij} = \pi_k.$$

Summing both sides over k ($\neq j$),

$$\sum_{k \neq j} \sum_i \pi_j P_{ji} N_k^{ij} = 1 - \pi_j,$$

which implies

$$\sum_i \pi_j P_{ji} \sum_{k \neq j} N_k^{ij} = 1 - \pi_j,$$

so

$$\sum_i \pi_j P_{ji} H_{ij} = 1 - \pi_j.$$

Finally, summing over j, we obtain

$$\sum_{i,j} \pi_j P_{ji} H_{ij} = \sum_j (1 - \pi_j) = n - 1. \qquad \square$$

Proof 2. Note simply that

$$\sum_{i,j} \pi_j P_{ji} H_{ij} = \sum_j \pi_j \left(\sum_i P_{ji} H_{ij} \right) = \sum_j \pi_j [H_{jj} - 1].$$

Since $H_{jj} = 1/\pi_j$, this implies that

$$\sum_{i,j} \pi_j P_{ji} H_{ij} = \sum_j \pi_j [1/\pi_j - 1] = n - 1. \qquad \square$$

It is well known that a Markov chain is time-reversible if, and only if, it can be represented as a random walk on an undirected weighted graph. Moreover, if the weights are interpreted to be electrical conductors (inverses of *resistors*), there is a pleasant correspondence between the electrical properties of such resistor networks and reversible Markov chains ([4, 1, 10]).

More precisely, given an undirected graph with weight $c_{ij} = c_{ji}$ on the edge ij, define a random walk with transition probability matrix $P = (P_{ij})$, where $P_{ij} = c_{ij}/\sum_j c_{ij}$. If ij is not an edge then $P_{ij} = 0$; in particular, $P_{ii} = 0$ for all i. This walk has the stationary measure $\pi_i = \sum_j c_{ij}/C$, where $C = \sum_{i,j} c_{ij}$. Reversibility follows from the fact that $\pi_i P_{ij} = \pi_j P_{ji} = c_{ij}/C$. Using the classical interpretation of c_{ij} as the conductance $1/r_{ij}$, Chandra *et al.* [1] showed that

$$H_{ij} + H_{ji} = CR_{ij},\tag{1}$$

where R_{ij} is the *effective resistance* between i and j.

Given all this, it is easy to deduce Foster's Theorem from Theorem 1. Indeed, as P is reversible and (1) holds, we have

$$
\begin{aligned}
\sum_{i,j} \pi_j P_{ji} H_{ij} &= \sum_{i<j} [\pi_i P_{ij} H_{ji} + \pi_j P_{ji} H_{ij}] \\
&= \sum_{i<j} \frac{c_{ij}}{C} [H_{ji} + H_{ij}] \\
&= \sum_{i<j} \frac{c_{ij}}{C} [C \cdot R_{ij}] \\
&= \sum_{i<j} \frac{R_{ij}}{r_{ij}}.
\end{aligned}
$$

2. Reciprocity and reversibility

The following resistive inverse identity is well known in electrical network theory. Given conductances c_{ij} and an all-pairs effective resistance matrix $R = \{R_{ij}\}$, with $R_{ii} = 0$ for all i, define two $(n-1) \times (n-1)$ matrices $\bar{c} = (\bar{c}_{ij})$ and $\bar{R} = (\bar{R}_{ij})$ by the formulae

$$\bar{c}_{ii} = \sum_{\substack{j=1 \\ j \neq i}}^{n} c_{ij}, \quad \bar{c}_{ij} = -c_{ij}, \quad 1 \leq i, j \leq n-1,$$

$$\bar{R}_{ij} = [R_{in} + R_{nj} - R_{ij}]/2, \quad 1 \leq i, j \leq n-1.$$

Then $\bar{c}$ is the inverse of $\bar{R}$:

$$\bar{c}\bar{R} = \bar{R}\bar{c} = I_{n-1},$$

where I_{n-1} is the identity matrix of order $n-1$. This identity can be generalized as follows.

Let $P = (P_{ij})$ be a probability transition matrix of an ergodic Markov chain on n states, with $P_{ii} = 0$ for all i. Define an $(n-1) \times (n-1)$ matrix $\bar{P} = (\bar{P}_{ij})$ by setting, for $1 \leq i, j \leq n-1$,

$$\bar{P}_{ii} = \pi_i \left(= \sum_{\substack{j=1 \\ j \neq i}}^{n} \pi_i P_{ij} \right) \text{ and } \bar{P}_{ij} = -\pi_i P_{ij}.$$

Furthermore, for $1 \leq j, k \leq n-1$ and $j \neq k$, let $\bar{H}_{jj} = H_{jn} + H_{nj}$, and $\bar{H}_{jk} = H_{jn} + H_{nk} - H_{jk}$.

Theorem 2. *With the notation above,*

$$\bar{P}\bar{H} = \bar{H}\bar{P} = I_{n-1}.$$

Proof. The basic identity we use is the triangle inequality for hitting times. [9, Proposition 9-58] asserts that

$$H_{xz} + H_{zy} - H_{xy} = \frac{N_y^{xz}}{\pi_y}, \tag{2}$$

$$H_{xz} + H_{zx} = \frac{N_x^{xz}}{\pi_x}. \tag{3}$$

Recall that N_y^{xz} denotes the expected number of visits to y in a random walk from x to z. From (2) and (3) we have, for all j and k,

$$\bar{H}_{jk} = \frac{N_k^{jn}}{\pi_k}.$$

Now consider the summation implicit in the statement of the theorem:

$$\sum_{j=1}^{n-1} \bar{P}_{ij} \bar{H}_{jk} = \bar{P}_{ii} \bar{H}_{ik} + \sum_{\substack{j=1 \\ j \neq i}}^{n-1} \bar{P}_{ij} \bar{H}_{jk}$$

$$= \pi_i \frac{N_k^{in}}{\pi_k} - \sum_{\substack{j=1 \\ j \neq i}}^{n-1} \pi_i P_{ij} \frac{N_k^{jn}}{\pi_k}$$

$$= \frac{\pi_i}{\pi_k} \left[N_k^{in} - \sum_{\substack{j=1 \\ j \neq i}}^{n-1} P_{ij} N_k^{jn} \right].$$

By taking means conditional on the first outcome, we see that the last expression is equal to

$$\frac{\pi_i}{\pi_k} [\delta_{ik}] = \delta_{ik}.$$

An analogous argument shows that $\sum_{j=1}^{n-1} \bar{P}_{ji} \bar{H}_{kj} = \delta_{ik}$, completing the proof. $\square$

Remark 1. For reversible chains, we have $\bar{H}_{jk} = \bar{H}_{kj}$. This is because, for all i and j,

$$H_{jn} + H_{nk} + H_{kj} = H_{jk} + H_{kn} + H_{nj}, \tag{4}$$

A proof of this can be found in [3], alternatively, (4) can be verified directly by using the formula for the hitting times in terms of either resistances (see [10]) or the *fundamental matrix* (see [8]). Thus the proof of Theorem 2 becomes simpler for the reversible case; in particular, we do not need to use equations (2) and (3). Note that the resistive inverse identity follows by using the analogs mentioned in the previous section: essentially, $\pi_i P_{ij} = c_{ij}/C$ and $H_{ij} + H_{ji} = CR_{ij}$, for all i, j.

Remark 2. Another interesting consequence of Theorem 2 is that the property in (4) is not only necessary but also sufficient to imply reversibility. Indeed, (4) implies that $\bar{H}$ is symmetric, which, in turn, implies that $\bar{P}$ is symmetric, i.e., $\pi_i P_{ij} = \pi_j P_{ji}$ for all i, j.

We now show that identity (4) has an interesting electrical interpretation. First, recall the following *reciprocity theorem* from electrical networks (see [7] for a proof).

Theorem 3. *The voltage V across any branch of a network, due to a single current source I anywhere else in the network, is equal to the voltage across the branch at which the source was originally located if the source is placed at the branch across which the voltage V was originally measured.*

Using the techniques from [4], it was shown in [10] that, for any network of unit resistors,

(a) the induced voltage V_{zy} with a unit current flowing into x and out of y is equal to $N_z^{xy}/d(z)$, where $d(z)$ is the degree of z, and

(b) the reciprocity theorem is equivalent to the fact that $N_z^{xy}/d(z) = N_x^{zy}/d(x)$ for all x, z and $y (\neq x, z)$.

Essentially the same proof can be used to show that the reciprocity theorem in *general* resistive networks (*i.e.*, not necessarily with unit resistors) is equivalent to the statement that

$$N_z^{xy}/\pi(z) = N_x^{zy}/\pi(x)$$

holds for all x, z, and $y (\neq x, z)$. Using relation (2) above, it is easy to see that this is the same as identity (4). In view of Remarks 1 and 2 above, we have thus established the following assertion.

Corollary 4. *Reversibility in ergodic Markov chains is equivalent to reciprocity in electrical networks.*

Corollary 5. *Given P and π, the hitting times (H_{ij}) can be computed with a single matrix inversion, and conversely, given the hitting times, P and π can be computed with a single matrix inversion.*

Proof. In view of Theorem 2, we only need to show:

(a) how to compute H from $\bar{H}$, and

(b) to compute P from $\bar{P}$.

(a) For $1 \leq i, j \leq n - 1$, we have

$$H_{in} = \sum_k N_k^{in} = \sum_k \pi_k \bar{H}_{ik},$$

$$H_{ni} = \bar{H}_{ii} - H_{in},$$

and

$$H_{ij} = H_{in} + H_{nj} - \bar{H}_{ij}.$$

Thus we can first compute H_{in} and H_{ni}, for all $i < n$, and then compute H_{ij} for $1 \le i, j \le n-1$.

(b)   We need to compute π_n and P_{ni}, since the rest of the information is available in $\bar{P}$. Since π is stochastic and $\pi P = \pi$, we have

$$\pi_n = 1 - \sum_{i<n} \pi_i = 1 - \sum_{i<n} \bar{P}_{ii}$$

and

$$\pi_n P_{ni} = \pi_i - \sum_{j \ne i,n} \pi_j P_{ji} = \sum_{j \ne n} \bar{P}_{ji}. \qquad \square$$

Remark 3. We defined $\bar{P}$ and $\bar{H}$ by treating n as a special state of the chain. Clearly, we could have chosen any other state j and carried out a similar analysis.

Remark 4. In the reversible case, we can interpret $\bar{H}$ as $\bar{R}$, and using part(a) of the proof of Corollary 5, we can write a formula for the hitting times in terms of effective resistances. This gives an alternative proof of the main result in [10]:

$$H_{ij} = \frac{1}{2} \sum_k c(k) \cdot [R_{ij} + R_{jk} - R_{ik}],$$

where $c(k)$ is the sum $\sum_w c_{kw}$ of the conductances at node k.

Remark 5. [8, Theorem 4.4.12] gives an alternative way of computing the chain, given all-pairs hitting times. However, the method outlined above seems simpler, since the solution can be written in essentially one equation – Theorem 2.

Finally we comment that these identities are useful in designing randomized *on-line algorithms* (essentially extending several results of [2]), and we refer to [11] for this work. The author thought of Theorem 2 while trying to extend the results of [2].

Note added in proof

Thanks to David Aldous, Theorem 1 has further been generalized as follows (see [11]). For any n-state ergodic Markov chain, we have $\sum_{i,j} \pi_i P_{ij} H_{ij} \le n-1$, with equality only under reversibility of the chain.

References

[1]   Chandra, A. K., Raghavan, P., Ruzzo, W. L., Smolensky, R. and Tiwari, P. (1989) The electrical resistance of a graph captures its commute and cover times. *Proc. of the 21st Annual ACM Symp. on Theory of Computing* 574–586.

[2]   Coppersmith, D., Doyle, P., Raghavan, P. and Snir, M. (to appear) Random Walks on Weighted Graphs, and Applications to On-line Algorithms. *Jour. of the ACM.*

[3]   Coppersmith, D., Tetali, P. and Winkler, P. (1993) Collisions among random walks on a graph. *SIAM J. on Discrete Math.* **6** 363–374.

[4]   Doyle, P. G. and Snell, J. L. (1984) *Random Walks and Electric Networks,* The Mathematical Association of America.

[5] Foster, R. M. (1949) The Average impedance of an electrical network. *Contributions to Applied Mechanics* (Reissner Anniversary Volume), Edwards Bros., Ann Arbor, Mich. 333–340.

[6] Foster, R. M. (1961) An extension of a network theorem. *IRE Trans. Circuit Theory* **8** 75–76.

[7] Hayt Jr., W. H. and Kemmerly, J. E. (1978) *Engineering Circuit Analysis*, McGraw-Hill, 3rd ed.

[8] Kemeny, J. G. and Snell, J. L. (1983) *Finite Markov Chains*, Springer-Verlag.

[9] Kemeny, J. G., Snell, J. L. and Knapp, A. W. (1976) *Denumerable Markov Chains*, Springer-Verlag.

[10] Tetali, P. (1991) Random Walks and the effective resistance of networks. *J. Theoretical Probability* **4** 101–109.

[11] Tetali, P. (1994) Design of on-line algorithms using hitting times. *Proceedings of the 5th annual ACM-SIAM Symp. on Discrete Algorithms*, Virginia 402–411.

Randomised Approximation in the Tutte Plane

D. J. A. WELSH

Mathematical Institute and Merton College, University of Oxford

It is shown that unless NP collapses to random polynomial time RP, there can be no fully polynomial randomised approximation scheme for the antiferromagnetic version of the Q-state Potts model.

1. Introduction

Exact evaluation of the Tutte polynomial is a computational problem that contains the problems of computing the partition function of the Potts model in statistical physics, determining the weight enumerator of a linear code, the Jones polynomial of an alternating knot and many other well-known combinatorial problems. Unfortunately, even for very restricted classes it is known to be computationally infeasible, that is #P-hard, except at a few very special points or along a few special curves. For details see [1].

Nevertheless, the problem does not go away, and in the same way as a randomised polynomial time (RP)-algorithm is a very attractive and practical solution in the case of decision problems such as primality testing, it would be extremely interesting and practically important if a fully polynomial time randomised approximation scheme (fpras) could be shown to exist for most of the Tutte plane. Here we consider this problem.

2. Statement of results

For a graph G with edge set E the Tutte polynomial can be defined as a 2-variable polynomial,

$$(2.1) \qquad T(G; x, y) = \sum_{A \subseteq E} (x - 1)^{r(E) - r(A)} (y - 1)^{|A| - r(A)}.$$

The *rank function* r is defined by $r(A) = |V(G)| - k(A)$, where $k(A)$ denotes the number of connected components of the subgraph having A as its edge set.

Here we shall be concentrating on just a few of the specialisations of T along particular curves of the (x, y)-plane.

First we define the *Q-state Potts model* on a general graph G. A *state* σ of the model consists of a mapping of the vertex set V into the set $\{1, 2, ..., Q\}$ and the *energy* or *Hamiltonian* of that state is

$$H(\sigma) = \sum K(1 - \delta(\sigma(i), \sigma(j))$$

where the sum is over all (i, j) that are joined by an edge, and δ is the usual delta function. The *partition function Z* is then given by

$$Z = Z_Q(G) = \sum_\sigma \exp(-H(\sigma)),$$

where the sum is over all possible states σ. It is not difficult to show (see [6]), that,

$$Z = Q(e^K - 1)^{|V|-1} e^{-K|E|} T\left(G; \frac{e^K + Q - 1}{e^K - 1}, e^K\right).$$

Its importance is that in the stochastic version of the Potts model, the Gibbs measure μ representing the probability that the system finds itself in the state σ is given by

$$\mu(\sigma) = Z^{-1} \exp(-H(\sigma)).$$

Now consider the effect of the parameter K. The 'high probability states' are those for which $H(\sigma)$ is low. But

$$H(\sigma) = \sum_{(ij) \in E} K \, |\{\sigma(i) \neq \sigma(j)\}|,$$

and thus if K is positive, it favours σ in which neighbouring spins are the same – that is the *attractive* or *ferromagnetic* case. Conversely, if K is negative, the favoured states are those in which neighbourhood spins are different, in other words we have the *repulsive* or *antiferromagnetic* case. For more details of this, see [6].

We can summarise the above in the following statements.

(2.2) For positive integer Q, along the hyperbola

$$H_Q = \{(x, y) : (x - 1)(y - 1) = Q\},$$

T equals (up to an easily computed constant) the partition function Z of the Q-state Potts model. We denote this by

$$\text{`}T \sim Z_Q \text{ along } H_Q.\text{'}$$

Note also that H_Q has 2 branches, the *positive branch*

$$H_Q^+ = \{(x, y) : (x - 1)(y - 1) = Q, \quad x > 1, \quad y > 1\}$$

corresponds to the *ferromagnetic* Potts model. The *negative branch* $H_Q^- = H_Q \backslash H_Q^+$. Many other specialisations of T are given in [6]. However, here we emphasize just two.

(2.3) Along the line $y = 0$, T specialises to the chromatic polynomial $P(G; \lambda)$. In particular,

$$P(G; k) \sim T(G; 1 - k, 0).$$

Now let Γ denote a finite Abelian group and let ω be any orientation of the edges of the graph G. A *nowhere zero* Γ*-flow* on G is a map $\phi : E(G) \rightarrow \Gamma \backslash \{0\}$ such that Kirchhoff's laws (under the group operation) hold at each vertex. It is a remarkable fact that the number of such flows depends only on G and the order of the group Γ.

The simplest proof of this is to show that if $|\Gamma|$ denotes the order of Γ, then the number of nowhere zero Γ-flows on G is given by $(-1)^{|E|-k(G)+1} T(G; 0, 1 - |\Gamma|)$, where $k(G)$ is the number of connected components of G. It follows that the number of such flows in a group of order λ is a polynomial in λ, which we call the *flow polynomial* of G, and denote by $F(G; \lambda)$. In other words:

(2.4) Along the line $x = 0$, T specialises to the flow polynomial, $F(G; \lambda)$, which for positive integer λ counts the number of nowhere zero λ-flows on G. In particular,

$$F(G; k) \sim T(G; 0, 1 - k).$$

When G is a planar graph there is a well-known exact correspondence between k-colourings of G and k-flows on any plane dual G^* of G. This correspondence does not work for nonplanar G, since G^* does not exist. However, it is just an example of a general correspondence between the Tutte polynomial of a matroid M (defined exactly as in (2.1), with rank interpreted as matroid rank) and its dual matroid M^*. This correspondence is that

$$T(M; x, y) = T(M^*; y, x).$$

Thus, from (2.3) and (2.4) we get the duality between chromatic and flow polynomials.

A *fully polynomial randomised approximation scheme* (fpras) for estimating $T(G; x, y)$ is a randomised algorithm that takes as input a graph G, a pair of rationals (a, b), and $\epsilon > 0$, and produces as output a random variable Y such that

$$Pr\left((1 - \epsilon)T(G; a, b) \leq Y \leq (1 + \epsilon)T(G; a, b)\right) \geq \frac{3}{4},$$

and, moreover, does so within time that is bounded by a polynomial function of the size of input and ϵ^{-1}.

The existence of such a scheme is the exact analogue for counting problems of an RP algorithm for a decision problem, hence it is a very positive statement.

In an important paper, Jerrum and Sinclair [2], have shown that there exists a fpras for the ferromagnetic Ising problem. This corresponds to the $Q = 2$ Potts model and thus, their result can be restated in the terminology of this paper as follows.

(2.5) There exists a fpras for estimating T along the positive branch of the hyperbola H_2.

However, it seems to be difficult to extend the argument to prove a similar result for the Q-state Potts model with $Q > 2$, and this remains one of the outstanding open problems in this area.

A second result of Jerrum and Sinclair is the following.

(2.6) There is no fpras for estimating the antiferromagnetic Ising partition function unless $NP = RP$.

Since it is regarded as highly unlikely that $NP = RP$, this can be taken as evidence of the intractability of the antiferromagnetic problem.

Examination of (2.6) in the context of its Tutte plane representation shows that it can be restated as follows.

(2.7) Unless $NP = RP$, there is no fpras for estimating T along the curve

$$\{(x, y) : (x - 1)(y - 1) = 2, \quad 0 < y < 1\}.$$

The following is an extension of this result.

Theorem. *On the assumption that $NP \neq RP$, the following statements are true.*

(2.8) *Even in the planar case, there is no fully polynomial randomised approximation scheme for T along the negative branch of the hyperbola H_3.*

(2.9) *For $Q = 2, 4, 5, \ldots$, there is no fully polynomial randomised approximation scheme for T along the curves*

$$H_Q^- \cap \{x < 0\}.$$

It is worth emphasising that the above statements do not rule out the possibility of there being a fpras at *specific points* along the negative hyperbolae. For example;

(2.10) T can be evaluated exactly at $(-1, 0)$ and $(0, -1)$, which both lie on H_2^-.

(2.11) There is no inherent obstacle to there being a fpras for estimating the number of 4-colourings of a planar graph.

I do not believe such a scheme exists, but cannot see how to prove it. It certainly is not ruled out by any of our results. I therefore pose the following specific question.

(2.12) **Problem.** Is there a fully polynomial randomised approximation scheme for counting the number of k-colourings of a planar graph for any fixed $k \geq 4$?

I conjecture that the answer to (2.12) is negative.

Similarly, since, by Seymour's theorem [4], every bridgeless graph has a nowhere zero 6-flow, there is no obvious obstacle to the existence of a fpras for estimating the number of k-flows for $k \geq 6$. Thus a natural question, which is in the same spirit as (2.12), is the following.

(2.13) Show that there does not exist a fpras for estimating T at $(0, -5)$. More generally, show that there is no fpras for estimating the number of k-flows for $k \geq 6$.

Again, although, because of (2.9), a large section of the relevant hyperbola has no fpras, there is nothing to stop such a scheme existing at isolated points.

Another point of special interest is $(0, -2)$. Mihail and Winkler [3] have shown, among other things, that there exists a fpras for counting the number of ice configurations in a 4-regular graph.

An *ice configuration* is an assignment of directions to the edges of G such that at each vertex, there are exactly the same number of edges pointing in as pointing out. Counting ice-configurations is a longstanding problem in statistical physics.

In the special case where G is 4-regular it is not difficult to verify that ice configurations correspond to nowhere zero flows in the group $\mathbb{Z}_3$, and thus:

(2.14) For 4-regular graphs counting nowhere-zero 3-flows has a fpras.

In other words, because of (2.4):

(2.15) There is a fpras for computing T at $(0, -2)$ for 4-regular graphs.

If we dualise this (and assume planarity) it gives:

(2.16) For planar graphs in which every face is a quadrangle there is a fpras for counting 3-colourings.

However, the degree restriction is quite severe. For a start, it demands 2-colourability and suggests a question that may be easier to settle.

(2.17) Is there a fpras for counting Q-colourings in bipartite planar graphs?

Again, exact counting is hard (see Vertigan and Welsh (1992)), and I suspect the answer to this too is negative.

3. Proof of the Theorem

We say that a set X of edges of G is a Q-*cut* if there is a partition of the vertex set V of G into Q, possibly empty, subsets $V_1, ..., V_Q$, such that each member of X joins a pair of vertices belonging to different members of the partition. We claim that the following is true.

Lemma. *For $Q \geq 2$, determining the maximum size of a Q-cut in a graph is NP-hard.*

When $Q = 2$, this is just the problem *MAX CUT*, which is well known to be NP-hard. For $Q \geq 3$, the result may also be well known, but I cannot find it in the literature, so supply the following proof.

Proof of the Lemma. Let $\mathscr{A}$ be a polynomial time algorithm for max Q-cut, $Q \geq 3$. The following transformation will give a polynomial time algorithm for max $(Q-1)$-cut. Given a graph G, on, say, n vertices, form a new graph G' by adding a set U of m new vertices and joining each of them to each vertex of G. For m at least $\binom{n}{2}$, it is easy to prove that there exists a Q-cut of maximum size in G' that has U as one of the sets in the Q-partition. Thus it gives a polynomial time algorithm for finding the maximum size of a $(Q-1)$-cut in G. Since max 2-cut is NP-hard, the result follows by induction on Q. □

Returning now to the proof of the main theorem, let $R_Q^1 = H_Q^- \cap \{x < 0, y > 0\}$. Suppose that there exists a fpras for T along R_Q^1. For $Q = 2$, we know from [2, result (2.7)] that this is impossible, unless $NP = RP$. For integer Q, strictly greater than 2, exactly the same argument as used in [2] for the $Q = 2$ case will work, the only difference being that *MAX Q-CUT* takes the role that *MAX-CUT (= MAX 2-CUT)* took in their argument. In other words, the existence of a fpras in the region R_Q^1 would mean a random polynomial time decision procedure for max Q-cut, and thus would imply $NP = RP$.

Now let $R_Q^2 = H_Q^- \cap \{x < 0, y < 0\}$. I show that if there exists a fpras $\mathscr{A}$ for T in this region, it will give a fpras for T in the region R_Q^1. But as we have seen, this will imply $NP = RP$. To prove the assertion, we use the tensor product developed in [1]. Let N be any graph and let e be a special edge of N. The *tensor-product* $G \oplus N$ is formed by glueing a copy of N to each edge f of G by identifying the edges e and f, and then deleting both. The following relationship holds between $T(G)$ and $T(G \oplus N)$;

$$(3.1) \qquad T(G \oplus N; x, y) = cT(G; X, Y),$$

where c, X and Y are easily computable functions of Q, x, y and N. Explicitly

$$(3.2) \qquad X = \frac{(Q-1)T'}{(x-1)T'' - T'},$$

$$(3.3) \qquad Y = \frac{(Q-1)T''}{(y-1)T' - T''},$$

where T', T'' are respectively given by $T' = T(N_e'; x, y)$ and $T'' = T(N_e''; x, y)$, and $N_e' (N_e'')$ denote the deletion (contraction) of e from N, and

$$c = (xT'' - T)^{r(E)}(yT'' - T)^{|E|-r(E)}(Q-1)^{-|E|}$$

where $T = T' + T''$.

In the special case where N is the graph on two vertices and three parallel edges, this gives the *2-thickening* of G and (3.2) and (3.3), after some simplification, reduce to

$$(3.4) \qquad X = 1 + \frac{Q}{y^2 - 1}$$

$$(3.5) \qquad Y = y^2.$$

Thus, let $(a, b) \in R_Q^1$, so that $0 < b < 1$. Form the 2-thickening of G to get G' and apply the postulated fpras $\mathscr{A}$ to G' at the point $(a', b') \in R_Q^2$, where b' is the negative square root of b. Using (3.1) with $(a, b) = (X, Y)$ and $(a', b') = (x, y)$ gives the fpras in the region R_Q^1. But this cannot exist unless $NP = RP$, by our previous argument.

All of the above holds for general positive integer Q. Now we turn to the particular case $Q = 3$, which is where we can say more.

Let $R_Q^3 = H_Q^- \cap \{0 < x < 1\}$. Suppose that $\mathscr{A}$ is a polynomial time algorithm that gives a fpras for T on R_Q^3. When G is planar we may use $\mathscr{A}$ to obtain a fpras for T on R_Q^1 by the simple expedient of forming the dual graph G^* and using the duality relation $T(G; x, y) = T(G^*; y, x)$.

But when $Q = 3$, we notice that the whole of the previous argument goes through just for the class of planar graphs. This is because deciding *MAX 3-CUT* is NP-hard in the planar case. It contains the problem of deciding whether a planar graph is 3-colourable, and this is a known NP-hard problem. Hence for $Q = 3$, the duality argument will work and shows that the whole of the negative hyperbola H_3^- has no fpras, unless $NP = RP$. □

It is frustrating that even for general graphs, I cannot see how to prove non-approximability along the whole negative hyperbola, except when $Q = 3$. It would be highly surprising if it were not true. However, the technique used to prove the $Q = 3$ case will not work. Nor will the obvious idea of trying to combine k-thickenings and k-stretches as in [1]. Thus, if there does exist a proof using 'tensor transformations' it will have to involve transformations that are very much more complicated. For example, the smallest tensor transformation that cannot be expressed in terms of a composition of thickenings and stretches is to take $N = K_4$. This leads to the transformation $(x, y) \mapsto (X, Y)$, where

(3.6) $$X = \frac{(Q-1)(x^3 + 2x^2 + 2xy + y^2 + x + y)}{y^3(x-1) + 2x^2(y-1) - 3y(x+1) + 2xy^2 - 2(x+y)}$$

and Y is obtained by interchanging x and y.

Apart from the questions already raised, we should also emphasise that the above says nothing about the inability to estimate T along H_Q^- for fractional Q. It would be very surprising if such a result were not true, but again it is likely to need a new technique, since a tensor transformation will only shift a point along the hyperbola H_Q containing it.

Acknowledgement

I am very grateful for very helpful discussions with Martin Grötschel, Mark Jerrum, David Johnson and Paul Seymour, and acknowledge also the support of Esprit Working Group 'RAND'.

References

[1] Jaeger, F., Vertigan, D. L. and Welsh, D. J. A. (1990) On the computational complexity of the Jones and Tutte polynomials. *Math. Proc. Camb. Phil. Soc.* **108**, 35–53.

[2] Jerrum, M. and Sinclair, A. (1990) Polynomial-time approximation algorithms for the Ising model (Extended Abstract). *Proc. 17th ICALP*, Springer-Verlag, 462–475.

[3] Mihail, M. and Winkler, P. (1991) On the number of Eulerian orientations of a graph. *Bellcore Technical Memorandum*, TM-ARH-018829.

[4] Seymour, P. D. (1981) Nowhere-zero 6-flows. *J. Combinatorial Theory* (B) **30**, 130–135.

[5] Vertigan, D. L. and Welsh, D. J. A. (1992) The computational complexity of the Tutte plane: the bipartite case. *Combinatorics, Probability and Computing* **1**, 181–187.

[6] Welsh, D. J. A. (1993) Complexity: Knots Colourings and Counting. *London Mathematical Society Lecture Note Series* **186**, Cambridge University Press.

On Crossing Numbers, and Some Unsolved Problems

HERBERT S. WILF[†]

University of Pennsylvania, Philadelphia, PA 19104-6395

This paper is about two rather disjoint subjects. First I want to discuss a new result of Ed Scheinerman and myself [4], in which we found a connection between the rectilinear crossing number problem for graphs and an old question in geometric probability that was asked by J.J. Sylvester [5]. Second I want to propose a number of unsolved problems in combinatorics and graph theory, some of which I have previously aired, but which have not appeared in print before.

1. On probability and crossing numbers

Here is the rectilinear crossing number problem. For a given graph G, suppose we draw G in the plane with straight line edges. Among all ways of doing so, we define $\bar{\kappa}(G)$, the rectilinear crossing number of G, to be the smallest number of crossings of edges that can be achieved.

For the complete graph K_n the values $\bar{\kappa}(K_n)$ are known for $1 \leqslant n \leqslant 9$, and they are $0, 0, 0, 1, 3, 9, 19, 36$, respectively. For $n = 10$ the number is 61 or 62.

It is also well known that

$$c_1 = \lim_{n \to \infty} \frac{\bar{\kappa}(K_n)}{\binom{n}{4}}$$

exists, and that $0 < c_1 < \infty$, though its exact value is unknown.

Here is Sylvester's question about geometric probability. Let K be a convex set in the plane. Choose four points independently uniformly at random (iuar) in K. Then with probability 1 there are two possibilities: either the convex hull of the four points is a quadrilateral or it is a triangle. Let $q(K)$ denote the probability that the convex hull is a quadrilateral. Sylvester asked for the minimum and maximum values of $q(K)$ over all convex sets K in the plane.

That question was answered some time ago [1]. The maximum is

$$1 - \frac{35}{12\pi^2} = 0.704\ldots,$$

† Supported in part by the United States Office of Naval Research

which is achieved on an ellipse, and the minimum is 2/3, attained when K is a triangle.

Generalize Sylvester's question as follows. Let K be an open set in the plane, of finite measure (but not necessarily convex, or even connected). Define $q(K)$ as before. Now determine the *inf* and the *sup* of $q(K)$ over all such sets K.

The supremum is certainly 1, for we can take K to be a very thin annulus, in which case four points selected iuar in K will almost surely span a quadrilateral.

The infimum is another matter entirely, so define

$$c_2 = \inf_K q(K),$$

over all open plane sets K of finite measure.

Theorem [4]. $c_1 = c_2$

Proof. Indeed, let us show first that $c_1 \leqslant c_2$. Fix some region K, choose n points iuar in K, and join each pair of them by a straight line segment. We call this the *sample drawing*.

The rectilinear crossing number of K_n cannot exceed the *average* number of crossings observed, over all such sample drawings. The latter average is the expectation of a sum of $\binom{n}{4}$ random variables, one for each 4-subset of vertices in the chosen drawing of K_n. The random variable that is attached to a 4-subset is 1 if they span a quadrilateral and 0 otherwise, since each crossing that occurs belongs to one and only one such quadrilateral. These random variables all have expectation $q(K)$, hence $\bar{\kappa}(K_n) \leqslant \binom{n}{4} q(K)$. If we divide by $\binom{n}{4}$ and take the limit as $n \to \infty$, we find that $c_1 \leqslant q(K)$. If we now take the infimum over all K, the claimed upper bound follows.

Next we claim that $c_1 \geqslant c_2$. For this, fix an optimum drawing of K_n, i.e., a drawing of K_n with straight line segment edges, in which the smallest number of edge crossings occur. At each vertex of this drawing place an open disk of radius ϵ, where ϵ is small enough so that for all choices of n vertices, one from each disk, the resulting drawing will still be optimum.

Let K be the union of these n open disks, and consider the following question. If we choose four distinct disks of K, and choose iuar a point from each of them, what is the probability q that the resulting quadrilateral is convex?

We answer this question in two ways, and then compare the answers. First, q is the number of convex quadrilaterals in our optimum drawing, divided by $\binom{n}{4}$. Since the former are in 1-1 correspondence with edge crossings, there are exactly $\bar{\kappa}(K_n)$ of them. Thus $q = \bar{\kappa}(K_n)/\binom{n}{4}$.

Second, $q(K)$ is the probability that four points chosen iuar in K will form a convex quadrilateral. But four points so chosen will lie in four distinct disks of K with probability $1 - O(1/n)$. Thus

$$q = q(K) + O(1/n) \geqslant c_2 + O(1/n).$$

If we compare these two answers to the question we obtain

$$q = \frac{\bar{\kappa}(K_n)}{\binom{n}{4}} \geqslant c_2 + O(1/n),$$

and if we let $n \to \infty$, the result follows. $\qquad\square$

As a corollary we get an estimate for the rectilinear crossing number of any graph.

Corollary. *Let G be a graph for which exactly M pairs of edges span four distinct vertices. Then the rectilinear crossing number of G (and therefore its crossing number also) is at most $c_1 M/3$.*

To get numerical estimates of the universal constant c_1 we proceed as follows. For a lower bound we can use $\bar{\kappa}(K_n)/\binom{n}{4}$ for any particular value of n for which a lower bound on the crossing number is known. Since it is known [(David Singer, personal communication)] that $\bar{\kappa}(K_{10}) \geq 61$, we have $c_1 \geq 61/210 = 0.290...$

In the other direction, we need a clever drawing of K_n, i.e., one with few crossings, for a sequence of $n \to \infty$. The best drawing that is known is due to David Singer, and since it is not available in the open literature, we give here a summary of his idea, which is quite simple to describe.

Theorem [David Singer]. *We have $\lim \bar{\kappa}(K_n)/\binom{n}{4} \leq 5/13$.*

Proof. We give a recursive description of drawings of K_n when n is a power of 3. When $n = 3$, draw a triangle. Inductively, suppose we have described the drawing D_n of K_n when $n = 3^q$. Then take the drawing D_n and put it in the upper half plane, with the y-coordinates of its vertices all distinct. Next, carry out a mapping of the plane to itself by means of $(x, y) \to (\epsilon x, y)$, which will map the drawing D_n into another rectilinear drawing with the same number of crossings, all of which is contained in a thin vertical strip about the y-axis. Make three copies of this thin drawing, and place one of them on each of the rays $\theta = 0, 2\pi/3, 4\pi/3$, and connect all pairs of points with straight line segments. The parameter ϵ is small enough so that if two vertices are chosen in the same copy of the drawing, then the infinite line through them separates the other two copies of the drawing.

In this drawing we count the convex quadrilaterals, i.e., the crossings, and we are thereby led to the recurrent inequality satisfied by $f(r) = \bar{\kappa}(K_{3^r})$, which is

$$f(r + 1) \leq 3f(r) + r^2(r - 1)(5r - 7)/4.$$

This implies that if n is a power of 3 then K_n can be drawn with straight line edges and no more than

$$\frac{5n^4 - 39n^3 + 91n^2 - 57n}{312}$$

crossings, which proves Singer's theorem. □

These best known estimates combine to give

$$0.290... = \frac{61}{210} \leq c_1 = c_2 \leq \frac{5}{13} = 0.3846...$$

2. Some unsolved problems

2.1. Graceful permutations

A permutation σ of n letters is *graceful* if

$$\{|\sigma(i) - \sigma(i + 1)| : i = 1, 2, \ldots, n - 1\} = \{1, 2, \ldots, n - 1\},$$

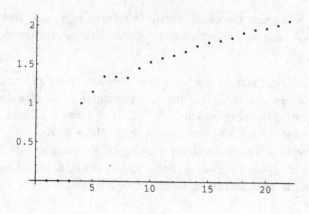

Figure 1

which is to say that no unsigned consecutive difference occurs twice. Such a permutation is, of course, a graceful labelling of a path, in the sense of the well known "all trees are graceful" problem. The permutation

$$\begin{pmatrix} 1 & 2 & 3 & 4 & 5 & 6 \\ 6 & 1 & 5 & 2 & 4 & 3 \end{pmatrix}$$

is graceful.

The problem is to count these permutations, a question that was first raised in [6]. If $f(n)$ is the number of them on n letters, then the following is a table of $f(n)/4$ for $n = 4, 5, 6, \ldots, 23$, the values having been computed by Dennis Deturck:

$$1, 2, 6, 8, 10, 30, 74, 162, 332, 800, 2478, 6398, 13980, 35798, 127674,$$
$$362824, 874336, 2612956, 9642676, 29728748$$

Deturck has also shown that $\liminf f(n)^{1/n} \geq 2^{\frac{1}{2}}$. Does the limit exist, as a finite, nonzero number? A plot of $f(n)^{1/n}$, for $n \leq 22$, is shown in Figure 1.

2.2. Balanced binomial coefficients

Let p be a prime. We will say that an integer n is p-balanced if, among the nonzero values of the binomial coefficients $\{\binom{n}{k}\}_{k=0}^{n}$, taken modulo p, there are equal numbers of quadratic residues and quadratic nonresidues modulo p.

Next, define the set T_p as the set of integers n, $0 \leq n \leq p-1$, that are p-balanced.

In [3] it was shown that if T_p is empty, then no integer n is p-balanced, and that if T_p is nonempty, then n is p-balanced iff its p-ary expansion contains at least one digit $d \in T_p$.

The question is: for which primes p is the set T_p empty? What is known (A. M. Odlyzko (p.c.)) is that among all primes less than 1,000,000, the only empty T_p are those with $p = 2, 3, 11$. Are there any others?

2.3. Graphical partitions

If G is a graph of m edges, then the degrees of the vertices of G constitute a partition of the even integer $2m$. Conversely, among the partitions of the integer $2m$, some are graphical and some are not. Let $p_g(2m)$ be the probability that a partition of $2m$ is graphical.

Table 1 Some values of $p_g(n)$

n	$p_g(n)$	n	$p_g(n)$
2	0.5	22	0.386228
4	0.4	24	0.385397
6	0.454545	32	0.380046
8	0.409091	34	0.37831
10	0.404762	36	0.378205
12	0.402597	38	0.376821
14	0.4	50	0.372646
16	0.38961	100	0.3586 (*est.*)
18	0.392208	200	0.3354 (*est.*)
20	0.389155		

The questions are:
(a) Does $\lim_{m \to \infty} p_g(2m)$ exist?
(b) Is it 0?

I asked these questions first in 1980, and two of my graduate students at that time worked on them, but they seem extremely difficult. Numerically, the values of $p_g(2m)$ decrease slowly from 0.5 at $2m = 2$, down to 0.3726, at $2m = 50$. Beyond that I have done some random sampling to estimate the values 0.3586, from 8000 trials at $2m = 100$, and 0.3354, from 1100 random samples with $2m = 200$.

Erdős and Richmond [2] have shown that

$$\liminf_{n \to \infty} \sqrt{2n} p_g(2n) \geq \frac{\pi}{\sqrt{6}}.$$

It is clear that $\limsup_n p_g(2n) \leq 0.5$, since not both a partition and its conjugate can be graphical. In [2] the upper bound is reduced to 0.4258.

2.4. Triangular matrices

Let A be an $m \times n$ matrix of 0's and 1's. Consider the computational problem: do there exist permutations P of the rows of A, and Q, of the columns of A such that after carrying out these permutations, A is triangular?

The question we ask concerns the complexity of the problem. Is this problem NP-complete? Or, does there exist a polynomial time algorithm for doing it? The question is related to job scheduling with precedence constraints, a well known problem in theoretical computer science, but it falls, in difficulty, between a known easy case and a known hard case of the general problem.

2.5. Patterns of permutations

Let $k \leq n$ be given positive integers and let τ and σ be given permutations, of $1, \ldots, k$, and of $1, \ldots, n$, respectively. We say that σ *contains the pattern* τ if there are integers $1 \leq i_1 < i_2 < \cdots < i_k \leq n$ such that for all $1 \leq \mu < \nu \leq k$ we have $\sigma(i_\mu) < \sigma(i_\nu)$ if and only if $\tau(\mu) < \tau(\nu)$.

For example, a permutation contains the pattern (123) iff it has an ascending subsequence (i.e., a not necessarily consecutive ascending triple of values) of length 3.

Given a pattern τ of k letters, we let $f(n, \tau)$ denote the maximum number of occurrences of the pattern τ that can be packed into a permutation of n letters. If the limit

$$\lim_{n \to \infty} \frac{f(n, \tau)}{\binom{n}{k}} \overset{\text{def}}{=} f(\tau)$$

exists, then we call it the packing density of the permutation τ.

Does every permutation τ have a packing density? That is, does the limit always exist? If it does exist, what is it, expressed in terms of accessible properties of τ?

It is easy to see that the lower packing density is always strictly positive. Indeed, for a fixed pattern τ of k letters, we have

$$\liminf_{n \to \infty} \frac{f(n, \tau)}{\binom{n}{k}} \geqslant \frac{k!}{k^{k+1}(k^{k-1} - 1)} > 0.$$

To see why, suppose first that $n = k^m$, and construct a permutation σ as follows. Divide the letters $1, \ldots, n$ into k intervals of consecutive letters. Arrange these intervals in the pattern τ. That is, if the largest value of τ is first, for instance, then put the interval that has the largest letters first, etc. With no further effort this already guarantees us that the pattern τ will appear in σ at least $(n/k)^k = k^{k(m-1)}$ times.

But we can do more. *Within each interval* we can arrange the letters to have the maximum number of τ's that are possible for permutations with n/k letters.

Example. For the pattern $\tau = (132)$ we can construct a permutation of 9 letters with many occurrences of (132) as follows:

$$\sigma = (132\,798\,465)$$

In general, this construction shows that

$$f(k^m, \tau) \geqslant k^{k(m-1)} + kf(k^{m-1}, \tau),$$

which leads to the result stated. The construction does not, however, always give the best result. The packing density of the pattern (132) has been determined by W. Stromquist (p.c.), and it is $2\sqrt{3} - 3 = 0.4641 \ldots$.

References

[1] Blaschke, W. (1917) *Leipziger Berichte* **69** 436–453.
[2] Erdős, P. and Richmond, L. B. (1989) On graphical partitions, *Combinatorics and optimization*, Research Report CORR 89-42, University of Waterloo, 13pp.
[3] Garfield, R. and Wilf, H. S. (1992) The distribution of the binomial coefficients modulo p, *J. Number Theory* **41**, 1–5.
[4] Scheinerman, E. R. and Wilf, H. S. The rectilinear crossing number of a complete graph and Sylvester's "Four Point Problem" of geometric probability, *Amer. Math. Monthly*, to appear.
[5] Sylvester, J. J. (1908) *Mathematical Papers*, Vol. II (1854-1873), Cambridge University Press, Cambridge.
[6] Wilf, H. S. and Yoshimura, N. (1986) Ranking rooted trees, and a graceful application, in *Perspectives in Computing*, Proc. Japan–US Joint Seminar June 4–6, 1986, in Kyoto, Academic Press, Boston.